Safety in the Process Industries

Related titles from Butterworth–Heinemann

Safety in the Process Industries

Ralph King BSc, CEng, FIChemE, FInstP

Butterworth–Heinemann

London Boston Singapore Sydney Toronto Wellington

First published 1990

© **Butterworth–Heinemann Ltd, 1990**

British Library Cataloguing in Publication Data
King, Ralph
  Safety in the process industries
  1. Industrial safety
  1. Title
  658.382
  ISBN 0-7506-1019-0

Library of Congress Cataloging in Publication Data
King, Ralph W. (Ralph William), 1918–
  Safety in the process industries/Ralph King.
      p.   cm.
  Includes bibliographical references (p.
  Includes index.
  ISBN 0-7506-1019-0:
  1. Chemical engineering – Safety measures.   I. Title.
TP149.K56   1990
660'.2804—dc20   90-1924

Photoset by Genesis Typesetting, Laser Quay, Rochester, Kent
Printed and bound in Great Britain by Courier International Ltd, Tiptree, Essex

Foreword

I am delighted to have been asked by Ralph King to write the foreword to his excellent book. Although I do not necessarily share all his expressed views I do wholeheartedly support the substance contained therein, which is not only readable but is packed full of vital information and learning experience. If it receives the attention it deserves it will, in my opinion, help to make the process industry both a healthier and safer place to work in and a better neighbour.

As a practising safety officer in the process industry I welcome this book and wish it had been available when I began my safety career; it will join a select few in my bookcase.

Bill Sampson
The Dow Chemical Company, King's Lynn, Norfolk

Preface

It may seem presumptuous for one writer to attempt to cover all the hazards of the process industries, with their many different technologies. Yet when considering the causes of past accidents, most appear to be well within the understanding of anyone with a broad technical background. The single writer has at least one advantage over a panel of authors in that he or she can present the subject as a logical whole and in a consistent style.

The book is written in a sincere attempt to help all those involved in the management, development, planning, design, construction, operation, inspection and maintenance of process plant, as well as safety professionals. It is hoped that it will also be read by insurers, lawyers, MPs, local councillors, civil servants, journalists and producers of TV programmes concerned with process hazards and disasters.

We humans have developed a love–hatred relationship with our process industries. We depend on them for cheap, standard and usually stable (sometimes too stable!) bulk products which supply much of the material needs of our bourgeoning population. Yet many fear and curse the process industries because of their potential for death, destruction and pollution caused by the escape of flammable, toxic or otherwise harmful chemicals and intermediates used in them. Escapes may be sudden and massive, causing such disasters as Flixborough, Seveso, Bhopal and the poisoning of the Rhine, or they may be small but persistent, leading to disease and premature death. The dangers are liable to increase when the industries and their technologies cross national frontiers and especially when they are set up in Third World countries which lack the resources, trained personnel and infrastructure needed to control them. Here situations which are seldom found in more industrialised countries, except in war-time, are common.

The hazard potential in many process industries is like a time-bomb. This potential is inescapable, although it can sometimes be reduced by minimising the quantities of harmful substances present or using less dangerous ones. The main hazards with which we have to deal are the several different ways in which harmful substances can escape. That is what this book is about.

Safety in the process industries depends first, on ensuring high integrity in the plant as designed and built; second, in maintaining that integrity throughout its working life in the face of wear and corrosion; third, in operating it skilfully and safely to avoid conditions which increase the likelihood of releasing harmful substances; and fourth, in thoroughly checking all modifications for unsuspected hazards which they may introduce.

Past disasters have engendered much of today's concern over safety within our process industries and the monitoring of their hazards. The numbers and total (inflation adjusted) value of very large losses worldwide in the hydrocarbon processing and chemical industries more than quadrupled between the first and last of three consecutive ten-year periods ending in 1987, according to Marsh and McLennan's eleventh 30-year survey (summarised in Appendix E). However, much has been and is being done. Safety and safety training are high on the priorities of professional groups on both sides of the Atlantic and in international organisations such as the International Labour Office. Legislation in European countries arising from EEC directives (e.g. the UK CIMAH and COSHH Regulations) is having a marked effect in raising standards. Penalties for infringements have become tougher. The number of conferences held annually on some aspect of health and safety in the process industries has increased several times during the last 15 years. Many excellent training films and other material now available are referred to here in Appendix M.

In many ways it was the indelible memory of the Flixborough disaster of 1974 and of my involvement in the subsequent investigations that spurred me to write this book. Yet memory, like a *jinnee*, is a provoking companion, constantly trying to throw its owner off-balance. I was fortunate in having Don Goodsell, formerly Butterworth's commissioning editor, as a guide, philosopher and friend. Having commissioned me to write the book, Don often had to wrestle with the *jinnee* and replace it in its bottle before I could complete the manuscript on an even keel. While writing it, many fresh hazards have come to light and aspects of which I was at first only dimly aware were thrust upon me. As a result, the book has grown considerably beyond its intended size.

Besides Don, the generous financial help of the Colt Foundation was vital. I am also indebted to many people and organisations for providing source materials and allowing me to make full use of them, and for reading parts of the manuscript, correcting errors and making suggestions. Only the following to whom I owe a special debt are listed here, although they are in no way responsible for my own opinions expressed in the book or for its shortcomings:

- Bill Sampson, safety officer of Dow Chemical Company Limited, King's Lynn, and The Dow Chemical Company itself;
- Professor Trevor Kletz, formerly safety advisor to ICI's petrochemical division;
- Glynne Evans, senior engineering inspector and pressure vessel expert in the UK Health and Safety Executive, Bootle, and many others in HSE;
- Dr Barry Turner, author of the book *Man-made Disasters* and head of the Sociology Department of Exeter University;

- Laurie Flynn, producer of the *World in Action* television documentary on the Bhopal disaster, and Granada Television who made it;
- Professor Frank Lees, author of *Loss Prevention in the Process Industries* and Professor of Plant Engineering in the Department of Chemical Engineering, Loughborough University;
- David Brown and other former and present officials of the International Labour Office, Geneva;
- Mr Robertson, secretary of the Industrial Safety Protective Equipment Manufacturer's Association, Mr Simpson, managing director of Bellas Simpson Ltd and other members of the Association;
- Dr Doran, manager of the Explosion Hazards Team of ICI Chemicals and Polymers, Northwich;
- Mr G. C. Wilkinson, formerly chief inspector of aircraft accidents in the Department of Transport;
- Arthur Robertson and Joyce Hainsworth, directors of Furmanite Engineering Limited of Kendal, Cumbria;
- David Gee and Steve Rabson, former and present national health and safety officers of the General Municipal, Boilermakers and Allied Trades Union, Claygate, Surrey;
- Mr J. Clifton, head of the Major Hazards Group of the Safety and Reliability Directorate of the UKAEA, Warrington;
- Peter Syrett, mechanical engineer of Howard Humphreys and Sons, Consulting Engineers, Leatherhead;
- Ken Palmer, consultant to Sedgwick International Ltd, insurance brokers, London;
- David Lewis, consultant and explosion expert, Liverpool;
- Antony Cuming, safety specialist of Atkins Research and Development, Epsom;
- The Institution of Chemical Engineers and Brian Hancock, head of its health and safety group, Rugby;
- The Chemical Industries Association and the Society of Chemical Industry, London;
- Past and present editors of *Process Engineering* and *The Engineer*, London.

I must apologise to and thank many other helpers and organisations whose names are not included in this list. Sources of illustrations and quotations reproduced here (with permission) are acknowledged elsewhere in the text.

Errata

Page 62 In Figure 4.1, the reactors should be labelled 1, 2, 3, *4* (R2524), *5* (R2525), 6.

Page 69 In line 3, 'R4' should read 'R5'.

Contents

Part III   Hazard control in design and maintenance

Chapter 14   Reliability and risk analysis   336

Chapter 15   Active protective systems and instrumentation   381

Chapter 16   Designing for safety   429

Chapter 17   Maintenance and inspection   478

Appendices

Introduction

Most human activities carry special risks. Steel erectors and roof workers are most at risk of falling while machine operatives are more at risk of lacerations. The risk profiles of particular industries change with time as certain hazards (such as boiler explosions) are conquered and new ones (such as gamma-rays) appear.

The main hazards of the process industries arise from the escape of process materials which may be inherently dangerous (e.g. flammable or toxic) and/or present at high pressure and high or low temperatures. Large and sudden escapes may cause explosions, toxic clouds and pollution whose effects extend far beyond the works perimeter. Such major accidents include the explosion of liquefied petroleum gas in Mexico City in 1984 which resulted in 650 deaths and several thousand injuries, followed two weeks later by the release of toxic methyl isocyanate gas in Bhopal, India, which caused over 2000 deaths and over 200 000 injuries. These have rightly attracted world attention. Small and persistent escapes may lead to chronic ill-health and environmental pollution. Their insidious effects which have taken longer to arouse the public have contributed to the present prominence of 'green' issues.

Hazards differ widely between processes. Their magnitude depends mainly on the process materials and their quantity. The probability of an accident depends more on the process conditions and their complexity.

To prevent repetition of past disasters, correct diagnosis and exposure of the relevant hazards is essential. The lessons then need to be incorporated in the training of managers and staff who may be faced with these hazards, in company rules, in codes of practice and sometimes in legislation. Diagnosis is often difficult and controversial and one seldom knows whether it is quite correct or complete. It first requires all known and possibly relevant facts to be disclosed, related and assessed. To do this a broad scientific and technical background is more important than a legal one. As Sir Geoffrey De Havilland and P.B.Walker pointed out after the early Comet disasters[1], most accidents in the technical sphere are caused by combinations of (relatively straightforward) hazards. Unfortunately, the legal and political connotations of many accident inquiries may put investigators under pressure to give undue attention to explanations which would exonerate parties whom they represent. Yet even the most objective investigation may succeed only in identifying several possible causes, each of which must be treated to minimise the probability of future failures.

The process industries

I cannot better the definition issued by the journal *Process Engineering*[2]:

> The process industries are . . . involved in changing by chemical, physical or other means raw materials into intermediate or end products. They include gas, oil, metals, minerals, chemicals, pharmaceuticals, fibres, textiles, food, drinks, leather, paper, rubbers and plastics. In addition the important service areas of energy, water, plant contracting and construction are included.

From this we can visualise the process industries as an intermediate stage in the transformation of raw materials of every kind – animal, vegetable or mineral – into materials and finished goods. The process industries convert these diversified raw materials into standardised bulk products. Some are sold direct to the customer (e.g. motor fuel), some merely packaged before sale (e.g. milk and lubricating oil), and some (e.g. wood-pulp and polyethylene) supplied to factories which make finished products.

Clear dividing lines cannot always be drawn between process industries and those which precede or follow them. Those preceding include mineral dressing at mines, water treatment (e.g. for injection into oil wells to assist recovery) and milk pasteurisation and cereal treatment at the farm. Those following include thermal and mechanical forming and cooking processes such as the casting and cold drawing of metals, the spinning and weaving of fibres, the moulding of plastics and the baking of bread. Many typical hazards of the process industries discussed here are found again in the industries which precede or follow them. This book is also addressed to those working in them.

The process industries account for about a tenth of the working populations of many industrialised countries. Those so-classified in the UK are listed in Appendix A with the numbers of employees. Of these about 30% work in oil, gas and chemicals.

Chemical hazards

Today there is widespread concern over the hazards of chemicals, not only to those who work with them but also to the environment and the general public. However well-designed a plant may be, it is very difficult to entirely prevent some dangerous materials from escaping. The longevity and concentration in nature of chemicals such as chlorinated biphenyls and chlorofluorocarbons, whose hazards only became apparent after they had been in production for many years, has heightened this awareness. One major problem in dealing with hazardous chemicals is that there are so many of them. There are now about *five million* chemicals listed in *American Chemical Abstracts*, and over 100 000 compounds in NIOSH's *Registry of Toxic Effects of Chemical Substances*[3].

Apart from the general problems of manufacture, special ones arise in bulk transport by road, rail and water, and when pregnant women are employed in manufacture and packing[4]. Much new legislation has followed this public concern. If its results are often disappointing, this is largely because of the wide gaps in understanding and experience between the legislators and those most at risk[5].

Many readers will surely be familiar with the *Handbook of Reactive Chemical Hazards* by Bretherick[6], which covers some 7000 chemicals, and *Dangerous Properties of Industrial Materials* edited by Sax[7] which refers to more than 19 000 such materials. A useful classification of hazardous chemicals for quick reference is that published by the National Fire Protection Association of America[8] (Appendix B). This provides a numerical rating of 0 to 4 for three regular hazards of every chemical – health, flammability and reactivity.

All those using, handling or making chemicals should have full information, which the supplier should provide, about their properties and possible hazards. Material safety data sheets (MSDS) giving this information should be brought to the special attention of persons and departments in need of it (e.g. fire, safety, medical, operations, maintenance, cleaning and transport). An EC directive and MSDS form, whose draft headings (June 1990) are given in Appendix C, is expected to be issued in 1991. Compliance with the directive will be judged on whether the user has sufficient information to work safely rather than on the provision of lists of specific data. An MSDS form issued by OSHA[9] for use in ship repairing, shipbuilding and shipbreaking is also shown in Appendix C. MSDSs for a wide range of chemicals are available from the on-line data base OHS.MS produced by Occupational Health Services Inc. [M.12].

In considering chemical hazards, we must think not only of chemicals in their restrictive sense but of all materials which may display hazards designated as chemical. They can include soils, minerals, metals, mineral waters, gases, food, drink, fuels, building materials, pharmaceuticals, photographic materials, textiles, fertilisers, pesticides, herbicides and lubricants. Each is composed of one or more of the 92-plus chemical elements and may well have hazards resulting from its composition.

Safety and technical competence

Safety in the process industries cannot be treated as a separate subject like design, production or maintenance, but is inextricably interwoven into these and other activities. It depends on both the technical competence and safety awareness of all staff and employees.

At least one company tries to solve this problem by assigning its key production and maintenance personnel for periods, usually of several months, during the early part of their careers, to work in the safety or loss-prevention department under a permanent safety manager. This gives them a new outlook and philosophy on safety which they do not easily lose when they return to face the myriad pressures of production. Furthermore, they are aware that their superiors in the management structure share their experience and outlook. It is hardly surprising that this company has an exceptional safety record.

Of the various specialists involved, the process engineer occupies a central position. While not always recognised in terms of his authority, he should by education and experience be able to appreciate, on the one hand, the chemistry of the process and the materials processed, and on the other, the factors involved in the mechanical design and construction, and in the materials of construction used. He should be thoroughly familiar with hazards inherent in the process and should be aware of those arising from the detailed engineering and other areas outside his direct concern.

About this book

Although many specialised books and papers have been written about specific facets of hazard control in the process industries, only a few have attempted to cover the whole field. One (published in the UK) is Lees's *Loss Prevention in the Process Industries*[10] which was written mainly as a reference book for students. It has a very comprehensive bibliography and gives quite a detailed mathematical treatment of reliability theory, gas dispersion and some protective systems. While it does not define the process industries, it is clearly slanted to the oil, gas, petrochemical and heavy chemical industries. Another is *Safety and Accident Prevention in Chemical Operations*[11], with chapters by 28 specialists, edited by Fawcett and Wood and published in the USA. I have drawn extensively on both books and refer to them frequently. Like them, this book does not attempt to cover the special hazards of nuclear energy, biochemical engineering or offshore oil and gas production.

Having spent about a third of my working life in several different countries, I have tried to write from an international viewpoint. I have also tried to adopt a multi-disciplinary approach while using a minimum of mathematics. To avoid repetition, each subject is treated as far as possible in a single appropriate place, with extensive cross-references to other chapters, sections and subsections. Here square brackets [ ] are used for '(see) chapter, section, appendix, etc.'.

The 23 chapters of this book fall loosely into four parts.

Part I, 'Setting the stage', includes the first five chapters. These deal with history (mainly recent), the legal background and five major accidents, their causes and lessons.

Part II, 'Hazards – chemical, mechanical and physical', comprises the next eight chapters. Five of these deal with the toxic, reactive, explosive, flammable and corrosive hazards of process materials.

Part III, 'Hazard control in design and maintenance', consists of the next five chapters. These include discussions of modern ideas about reliability, active and passive protection, control instruments and permit-to-work systems.

Part IV, 'Management, production and related topics', comprises the last five chapters and includes training, personal protection and hazards which arise in the transfer of modern technologies.

The book has several appendices, one of which, Appendix M, lists sources of help and information, particularly for safety training. A glossary of abbreviations used is given at the end of the book.

References

1. De Havilland, G. and Walker, P. B., 'The Comet failure' in *Engineering Progress Through Trouble*, edited by Whyte, R. R., The Institution of Mechanical Engineers, London (1975)
2. Sales brochure, *Process Engineering – The Market Leader*, Morgan-Grampian (Process Press Ltd), London (September 1984)
3. *NIOSH Registry of Toxic Effects of Chemical Substances* (revised annually), National Institute for Occupational Safety and Health, Rockville, Md. 20857
4. Brown, M. L., *Occupational Health Nursing – Principles and Practices*, Springer, New York (1980)
5. Ashford, N. A., *Crisis in the Workplace – Occupational Disease and Injury*, MIT Press, Cambridge, Mass. (1966)
6. Bretherick, L., *Bretherick's Handbook of Reactive Chemical Hazards*, 4th edn, Butterworths, London (1990)
7. Sax, N. I., *Dangerous Properties of Industrial Materials*, 6th edn, Van Nostrand, New York (1984)
8. National Fire Protection Association, *Standard 704 M, Identification systems for fire hazards of materials*, NFPA, Boston, Mass. (1975)
9. National Research Council, *Evaluation of the Hazards of Bulk Water Transportation of Industrial Chemicals – A Tentative Guide*, National Academy of Sciences, Washington DC (January 1974)
10. Lees, F. P., *Loss Prevention in the Process Industries*, Butterworths, London (1980)
11. Austin, G. T., 'Hazards of commercial chemical operations' and 'Hazards of commercial chemical reactions', in *Safety and Accident Prevention in Chemical Operations*, 2nd edn, edited by Fawcett, H. H. and Wood, W. S., Wiley-Interscience, New York (1982)

Chapter 1

From past to present

History has several lessons for us about the hazards of the modern process industries. One is the toxicity of many useful metals and other substances won from deposits in the earth's crust. Although the dangers of extracting and using lead have been known from the earliest times, these seem later to have been forgotten. A second lesson, then, is that past lessons are sometimes forgotten after a lapse of a few years, when history has an unfortunate habit of repeating itself.

A third lesson is that there is often a time-lag between the initial manufacture of hazardous substances and general appreciation of the dangers. Today we are all aware of the hazards of asbestos, CFC refrigerants and aerosols, chlorinated hydrocarbon insecticides, benzene and the bulk storage of ammonium nitrate. Only recently all were considered to be safe and needing no special precautions. Similar time-lags occurred a few decades ago before the hazards of yellow phosphorus (matches) and benzidine (dyestuffs) were appreciated.

A fourth lesson is that it is usually the lowest and least articulate strata in society who bear the brunt of industry's hazards.

A fifth lesson is that the capacities of process plants and the magnitude of major losses involving them have increased continuously *and are still increasing*. Related to this and to the high capital:worker ratio of these plants is the high ratio of capital loss:human fatalities in most major fires and explosions. This does not, however, apply to poisoning and pollution incidents which spread well beyond the works boundary (cf. Seveso and Bhopal).

The recent world record of large losses in the process industries, especially oil and chemicals, is truly alarming and gives us no grounds for complacency. The numbers of losses in excess of $10 million (adjusted to 1988 values), and their total value, have increased greatly in each successive decade since 1958, particularly in the second decade as the following figures show[1]:

Period	Number of losses over $10 million at 1986 values	Total ($10 million)
1958–1967	13	442
1968–1977	33	1438
1978–1987	58	2086

Having persisted for so long, it would be very surprising if these themes did not continue into the future. If history teaches us nothing else, it should warn us to be sceptical of claims that the use of some new material in an industrial process is entirely safe. Usually only time will tell.

This chapter falls into seven largely unrelated sections each forming a brief historical sketch. (The reader can skip any of these without losing the thread of the argument.)

1.1 Origins of process hazards

Several typical hazards of the process industries have a very long history. This is because a number of the 92-plus chemical elements (particularly metals and semi-metals) of which all matter is compounded are poisonous, and are naturally concentrated here and there on and below the earth's surface.

As human prowess developed, the mastery of fire, and through it the invention of smelting to obtain bronzes and other metals, released fumes which affected the health of the craftsmen. Another early 'industrial' hazard was the making of flint tools, where abundant archaeological evidence of silicosis has been found.

As human occupations became more specialised, it was clear that some were more dangerous and less healthy than others. Thus Hephaistos, the Greek god of fire and patron of smiths and craftsmen, was lame and of unkempt appearance, while Vulcan, the Roman god of metal workers and fire, was also ugly and misshapen. It is now thought that the lameness of the smith-gods was the result of arsenic poisoning, since many of the ores from which copper and bronze articles were made contained arsenic, which improved the hardness of the resulting articles.

From the earliest times, there has been a strong prejudice among the articulate elite against such craftsmen. Socrates was reported to have passed the following judgement:

> What are called the mechanical arts, carry a social stigma and are rightly dishonoured in our cities. For these arts damage the bodies of those who work at them or who have charge of them, by compelling the workers to a sedentary indoor life, and in some cases spending the whole day by the fire. This physical degeneration results in degeneration of the soul as well.

The social cleavage illustrated by such attitudes to industrial hazards has persisted through human history. Despite the recent elimination of many of these hazards, our social and economic structure and the mental habits that go with it are slow to adjust to the possibilities of a golden age, free from occupational hazards and excessive working hours, in which all can enjoy our common heritage of knowledge, invention and accumulated technical progress. It is tragic, as the British miners' strike of 1984/1985 showed, that working people still feel compelled to fight for the right to continued employment in an occupation notorious for accidents and disease.

In more recent times, some occupational diseases were so common as to have acquired well-known names, such as those quoted by Hunter[2]:

> Brassfounders' ague, copper fever, foundry fever, iron puddlers' cataract, mule-spinners' cancer, nickel refiners' itch, silo-workers' asthma, weavers' deafness and zinc oxide chills.

1.2 Toxic hazards of ancient metals[2]

Several of these hazards have persisted to the present day, although their forms have changed. To say that any metal is poisonous is an over-simplification. Metals usually occur combined in nature and few are found in their free state. While several inorganic compounds of a metal display the same characteristic toxic features, the degree of toxicity of such compounds depends on their solubility in water and body fluids, as well as on the ionic and complex state of the metal. Insoluble elements and compounds are seldom toxic in themselves. The first lead ore worked at Broken Hill in Australia was the relatively soluble cerussite, $PbCO_3$, the dust of which caused much disease among the miners. Fortunately this was soon worked out, and the ore subsequently mined was galena, PbS, which is very insoluble, and has caused few cases of lead poisoning.

Besides the toxic inorganic compounds whose effects are typical of the metal present, there are many man-made organo-metallic compounds, which have different and more acute toxic effects. An example is the volatile tetraethyllead, used as a petrol additive, which produces cerebral symptoms. Nickel carbonyl, used in the purification of nickel, is another example. Even metals which exhibit no marked toxicity in their inorganic compounds, such as tin, can form highly toxic organo-metallic ones (tetramethyltin).

The hazards of two metals used since antiquity, lead and mercury, are next considered. They are discussed again [23] in the context of technology transfer to developing countries.

1.2.1 Lead

The symptoms of inorganic lead poisoning – constipation, colic, pallor and ocular disturbances – were recognised by Roman and earlier physicians. The symptoms of poisoning by organo-lead compounds include insomnia, hallucinations and mania. Lead ores have been smelted since early Egyptian times. Being soft, dense, easily worked and fairly resistant to corrosion, lead was long the favourite metal for water pipes, roof covering and small shot. With the invention of printing, it became the principal metal used for casting type. White lead (a basic carbonate) and red lead (an oxide) were long used as paint pigments and ingredients of glass and pottery glazes. The smelting of lead ores (Figure 1.1) and the manufacture and use of lead compounds increased greatly during the Industrial Revolution, together with an increase in death and injury among workers exposed to them.

Figure 1.1 Sixteenth century furnaces for smelting lead ore

The health hazards to lead workers featured in Victorian factory legislation and it was eventually recognised by Sir Thomas Legge that:

Practically all industrial lead poisoning is due to the inhalation of dust and fume; and if you stop their inhalation, you will stop the poisoning.

Although the conditions in established lead processes improved considerably after this, newer large-scale uses of lead, first, the manufacture of lead-acid car batteries, and second, the manufacture and use of volatile organo-lead compounds for incorporation into petrol to improve its performance, brought further hazards. Several multiple fatalities occurred during the cleaning of large tanks which had contained leaded petrol. The worst happened at Abadan refinery during World War II while I was working there. There were then about 200 cases of lead poisoning with 40 deaths among Indian and Iranian workers.

As the hazards of lead are better appreciated today, its use has declined. One special hazard in its production is that many lead ores also contain arsenic, which is even more toxic [23.3.2].

1.2.2 Mercury

Mercury, the only liquid metal, is highly toxic and has an appreciable vapour pressure at room temperatures. Symptoms of poisoning from mercury vapour are salivation and tenderness of the gums, followed in chronic cases by a tremor. Another symptom is *erethism*, a condition in which the victim becomes both timid and quarrelsome (Figure 1.2), easily upset and embarrassed, and neglects his or her work and family. Merely to be in an unventilated room where mercury is present and exposed to the atmosphere can, in time, lead to mercury poisoning.

Figure 1.2 The Mad Hatter, drawn by Tenniel

Mercury occurs as its sulphide in the ore cinnabar, which has been mined in Spain since at least 415 BC, and mercury poisoning has long been prevalent among workers employed in such mines and reduction plants.

Mercury has long been used as such in the manufacture of thermometers and barometers, and more recently in the electrical industry for contact breakers, rectifiers and direct current meters. New compounds of mercury have been invented and commercialised, including mercury fulminate, used in detonators, and organo-mercury compounds used as antiseptics, seed disinfectants, fungicides and weedkillers. Mercury is used as cathode and solvent for metallic sodium in the Castner–Kelner process for the production of chlorine and caustic soda. This process has now been largely replaced by others which are free of the mercury hazard. In most of its industrial applications there are well-authenticated cases of poisoning by exposure to mercury vapour, or dusts containing its compounds. Mercury poisoning has been notorious in the felt-hat industry for centuries, where mercuric nitrate was used to treat rabbit and other furs to aid felting.

An infamous case of mass poisoning from organo-mercury compounds occurred among the fishermen and their families living along the shores of Minamata Bay in the south of Japan in the 1950s. This was ultimately traced to the discharge of spent mercury-containing catalyst into the bay from a nearby chemical factory which made vinyl chloride monomer. Organo-mercury compounds settled in the silt of the bay and were ingested by fish which were caught, sold and eaten by the local inhabitants and their cats. By July 1961 there had been 81 victims, of whom 35 died. The symptoms included numbness in the extremities, slurred speech, unsteady gait, deafness and disturbed vision. The mud of the bay remained loaded with mercury compounds for many years afterwards.

1.3 Changing attitudes to health and safety in chemical education

The last fifty years have shown great changes in attitudes to chemical safety in schools and colleges. This is clear from my own education in the 1930s. The first hazardous chemical to which I was exposed was mercury. This was in our school chemistry laboratory-cum-classroom (1932–1936). Our chemistry master, a middle-aged bachelor, had studied under Rutherford and had a penchant for research. For this he needed copious supplies of mercury, which he purified in his spare time in the school laboratory. Although I did not recognise his symptoms at the time, in retrospect the tremor of his hands, his high-pitched nervous twittering laugh, general shyness and odd mannerisms were typical of *erethism*. Globules of mercury, which was used for many juvenile pranks, were scattered on the laboratory benches and floor. Perhaps the fact that the laboratory was underheated saved me from serious mercurial poisoning.

At college (1936–1940), I was exposed to blue asbestos, from which we made mats for filter crucibles used in inorganic analysis, hydrogen sulphide, benzene, which was used as a common laboratory reagent and solvent, and again mercury, of which I used several kilograms for a research project. My most serious exposure was probably to a complex mixture of polynuclear aromatic compounds containing sulphur, which I was asked to prepare for a professor during a long vacation by bubbling acetylene through molten sulphur. The professor contracted cancer from his researches and died in middle age a few years later.

Chemical research was then held up to students as a vocation, demanding sacrifice of time and, where necessary, of health, in order to advance the frontiers of knowledge in the service of mankind. Madame Curie was quoted as a noble and inspiring example, whose work somehow justified the cancer which finally killed her. Other scientists such as J. B. S. Haldane and Dr C. H. Barlow who carried out dangerous and often painful experiments on their own bodies were also regarded as heroes.

It was only later that I realised that such sacrifices can only be justified if they improve the health of others. Often the reverse has been the case. A survey by Li et al.[15] of causes of death in members of the American Chemical Society between 1943 and 1967 showed that deaths from cancer of the pancreas and malignant lymphomas were significantly higher than among the general population. Scientists have tended to regard working conditions which they readily tolerate as quite good enough for their laboratory assistants. This is but a short step to expecting the same acceptance from industrial workers and the general public.

In spite of many exposures to harmful chemicals throughout my training and subsequent career, I am fortunate to be alive and in excellent health and still a keen squash player in my seventies. As most of my former colleagues are dead, I must be the exception which proves the rule!

The situation in schools today is very different to that in the 1930s. Safety policy has been greatly tightened over the past 20 years and most heads of science take their safety responsibilities very seriously indeed. A chemistry teacher was recently prosecuted and fined £500 for failing to take adequate safety precautions. The Association for Science Education has a Laboratory Safeguards Committee and its journal *Education in Science* carries regular updates on potential hazards. Local education authorities publish safety guidelines and individual schools are often required to have such guides to suit their particular situations. As examples of the changing situation, traditional asbestos bench mats were phased out during the 70s, and the safe handling of chemicals and manipulation of apparatus is one of the features of pupils' practical chemistry work which is assessed for the new GCSE examination.

1.4 Insurance losses in the US chemical industry

This section is based on a survey of 1028 accidents in the US chemical industry over the three-year period 1978–1980, given by Norstrom of Industrial Risk Insurers in Fawcett and Woods's book[3]. These resulted in insurance losses of $152 million, exclusive of deductibles and self-retention by the insured. Only one catastrophic loss of over $20 million was included, and no vapour cloud explosions. The period appears to have been one in which chemical plant losses were relatively light, considering that the loss from a single incident, the Flixborough disaster in the UK in 1974 [4], was reported at about $100 million. The survey did not include losses by those major international companies which carry their own insurance.

The following general conclusions were drawn from the survey:

1. The most frequent and severe losses in the chemical industry are caused by fire and explosion.
2. Explosion causes more severe losses than fire.
3. The main causes of explosion losses are accidental and uncontrolled chemical reactions.
4. Most explosion losses occur in enclosed process buildings and involve batch reactions.
5. Rupture of vessels, pipes and equipment contribute greatly to the magnitude of fire and explosion losses.
6. Most fire losses result from the release of flammable gases and liquids.
7. Lack of sprinklers and water spray was a major contributory factor in 38% of fire losses.

While only 12.7% of the individual losses exceeded $100000, these together accounted for nearly 95% of the total monetary value of all claims. The following details therefore apply mainly to losses greater than $100000, which are referred to subsequently as 'large losses'.

Figure 1.3 shows the percentages of the numbers and total value of large losses grouped by cause (explosion, fire, windstorm and all other). Over 50% of the losses and over 70% of their total value were caused by explosions, with fire accounting for 33% of the losses and 20% of their total value. Windstorm and other causes accounted, however, for a higher proportion of minor losses.

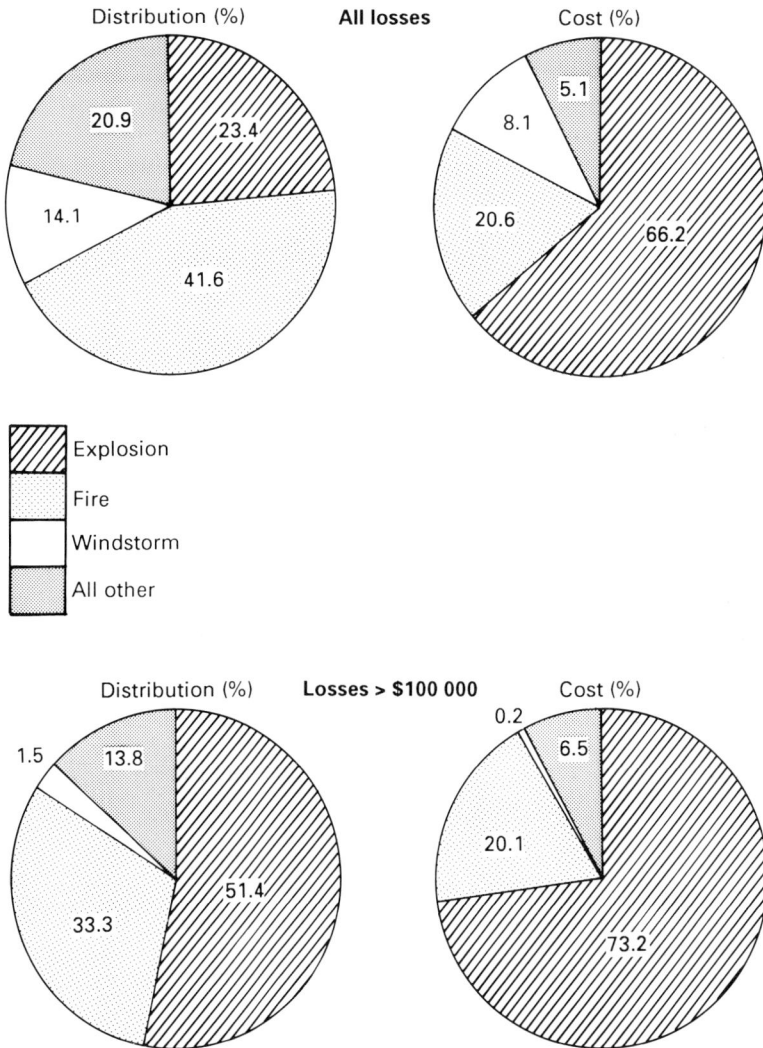

Figure 1.3 Peril analysis of US chemical losses, 1978–1980 (data from reference 3)

Table 1.1 Occupancy analysis of US chemical losses 1978–1980

Occupancy class	Frequency (%)	Percent of total $ loss
Extra-heavy hazard. Organic peroxide and explosive manufacture, nitrations and other very hazardous processes	4.9	4.2
Petrochemicals. Processes using hydrocarbon feedstock, mainly olefin plants	8.9	10.9
Heavy hazard. Polymerisation, solvent extraction, sulphonation, hydrogenation and processes involving flammable and combustibles	32.1	59.6
Light hazard. Mainly inorganics	22.9	9.6
Paint, dyestuffs, inks	11.4	3.3
Soaps and vegetable oils	6.2	0.9
Pharmaceuticals and fine chemicals	13.6	2.5

Table 1.1 gives the 'occupancy' analysis of losses, i.e. according to the branch of the industry where they occurred.

The preponderance of losses in the 'heavy hazard' class which in the UK are for the most part regarded as petrochemicals should be noted. The small dollar losses of the last two categories may have been partly exaggerated by the high 'self-retention' of risks which is common in these industries.

Figure 1.4 shows the percentage frequencies of large fire and explosion losses by location.

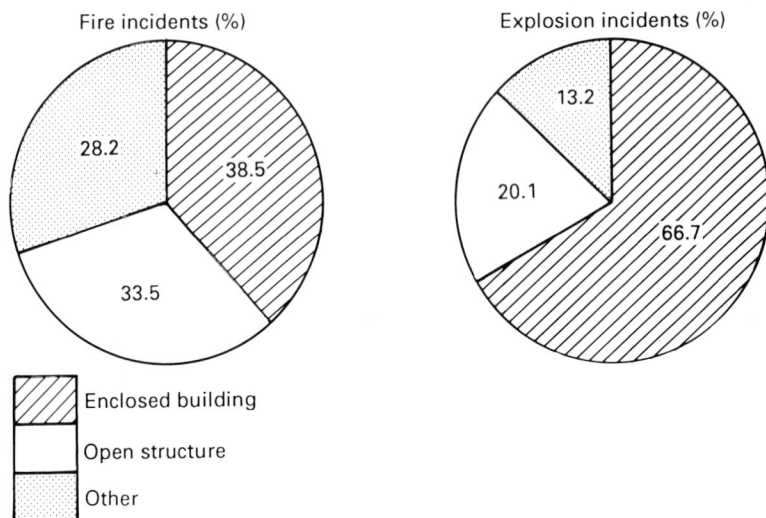

Figure 1.4 Location analysis of large fire and explosion losses (data from reference 3)

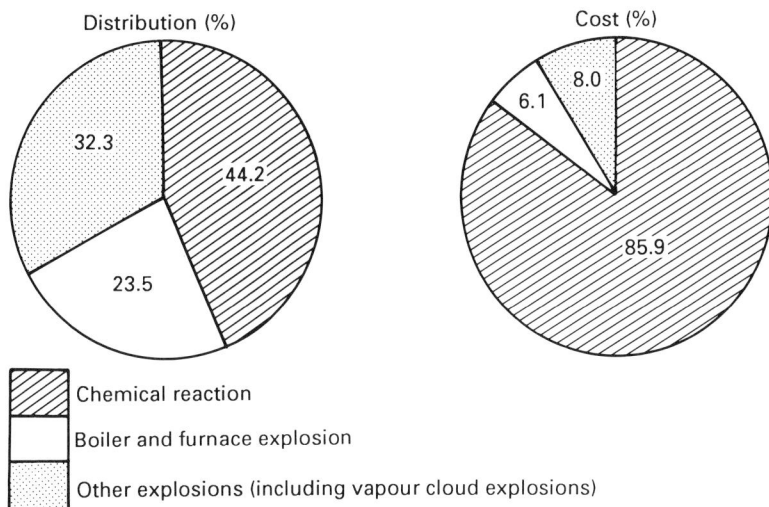

Figure 1.5 Analysis of large explosion losses by cause (data from reference 3)

1.4.1 Explosion losses

Because of the importance of explosions as a cause of loss, these are further analysed. Figure 1.5 gives a percentage breakdown of the numbers and total value of large explosion losses by cause (chemical reaction, boiler and furnace explosions, and other causes). Table 1.2 gives an analysis of those large losses caused by chemical reactions.

Table 1.3 analyses explosion losses by the type of process where they occurred.

Table 1.2 Analysis of chemical reaction losses

Cause	Frequency (%)
Accidental reaction[a]	33.3
Uncontrolled reaction[b]	40.0
Decomposition of unstable materials	13.3
Other causes	13.4

[a] Due to accidental contact of material(s).
[b] Intended reactions which become uncontrollable.

Table 1.3 Analysis of explosion losses by type of process

Type of process	Frequency (%)
Batch reaction	60.0
Continuous reaction	13.6
Recovery unit	6.6
Evaporation unit	6.6
Other	13.2

1.4.2 Large fire losses

These are analysed by cause in Table 1.4.

Table 1.4 Analysis of large fire losses by cause

Cause	Frequency (%)
Release or overflow of flammable liquids/gases	28.5
Overheating	24.3
Failure of pipe or fitting	14.3
Static electricity or spark	9.5
Electrical or mechanical breakdown	9.4
All other	14.0

1.5 Recent UK experience

Here we consider, first the general safety record of the 'mainstream' process industries in the UK (listed with their employment statistics in Appendix A), and second, some special features of accidents in the chemical sector. Both are based on information published by HSE[4,5].

1.5.1 Safety record for 'mainstream' process industries

Incidence rates for fatal plus major injuries in the various sectors of these industries are given in Table 1.5 for the years 1981–1983/1984, based on the 1968 standard industrial classification (SIC) which was still in use by HSE.

The industries listed in Table 1.5 are in two groups:

1. The 'chemical sector' as recognised by the Factory Inspectorate[5]. Employment figures (1984) for this sector are included in Table 1.5.
2. Other 'mainstream' process industries listed in Appendix A which are not included in the chemical sector. These include most of the food, drink and tobacco processing industries, most of the metal-producing industries, and paper and board. Most involve chemical reactions of some kind.

The fatal and major accident incidence rates of most industries listed in Table 1.5 are higher than the average (87.8) for all UK manufacturing industries in those years. The safest, with average incidence rates below 50 per 100 000 employees, were toilet preparations and tobacco, which involve little or no chemical reactions and only moderate temperatures. The more dangerous, with average incidence rates above 150, were lubricating oils and greases, dyestuffs and pigments, vegetable and animal oils and fats, coke ovens and manufactured fuel, iron and steel general, other base metals and paper and board.

The hazards of coke ovens, etc. and the metallurgical industries are well known and include those of high-temperature operations, dust, solids handling, and gassing from carbon monoxide.

Table 1.5 Fatal and major injury incidence rates for UK process industries

SIC Order	Description MLH[b]	Employment in 1984 ($\times 10^3$)	Fatal and major injury incidence rates			

Chemical sector[c]

SIC Order	Description MLH[b]	Employment in 1984 ($\times 10^3$)	1981	1982	1983	1984
IV	262 Mineral oil refining	18.2	125.0	189.6	117.6	82.4
IV	263 Lubricating oils and greases	4.9	158.7	180.3	137.0	204.1
V	271 General chemicals	111.5	95.2	120.6	138.8	131.8
V	272 Pharmaceuticals	73.8	70.8	69.2	43.5	58.3
V	273 Toilet preparations	20.6	28.8	20.1	39.0	34.0
V	274 Paint	25.0	58.8	51.5	67.5	64.0
V	275 Soaps and detergents	16.7	67.5	93.2	89.8	119.8
V	276 Synthetic resins, plastics materials and synthetic rubber	42.6	75.9	96.4	135.2	105.6
V	277 Dyestuffs and pigments	8.7	150.8	181.8	269.7	264.4
V	278 Fertilizers	7.8	74.8	50.5	125.0	128.2
V	279 Other chemical industries[d]	53.3	54.9	47.9	75.5	80.7
	Total Order V chemicals	360.0				
XIII	411 Man-made fibres	15.1	49.9	63.1	118.4	92.7
	Total chemical sector A	398.2				

Other mainstream process industries

SIC Order	Description MLH[b]		1981	1982	1983
III	211 Grain milling		93.4	100.6	123.6
III	212 Bread and flour confectionery		74.5	78.6	81.9
III	213 Biscuits		29.9	54.7	29.6
III	214 Bacon curing, meat and fish products		98.0	104.7	116.8
III	216 Sugar		104.2	146.1	163.0
III	217 Cocoa, chocolate and sugar confectionery		48.2	45.3	63.2
III	218 Fruit and vegetable products		65.5	74.7	97.5
III	219 Animal and poultry foods		128.1	149.6	157.0
III	221 Vegetable and animal oils and fats		164.2	133.3	213.3
III	229 Food industries n.e.s.		90.4	86.2	84.5
III	232 Soft drinks		56.6	116.1	107.9
III	239 Other drink industries		74.4	65.5	15.2
III	240 Tobacco		54.0	49.6	44.6
IV	261 Coke ovens and manufactured fuel		228.1	280.0	267.9
VI	311 Iron and steel general		243.2	218.8	211.2
VI	321 Aluminium and aluminium alloys		100.0	114.0	132.3
VI	322 Copper, brass and other copper alloys		88.8	111.1	170.9
VI	323 Other base metals		159.0	158.4	200.0
XVIII	481 Paper and board		142.0	167.4	206.2

[a] Per 100 000
[b] Minimum list headings
[c] As listed in Table 13 of Report by HM Chief Factory Inspector, 1985[5]
[d] Comprising polishes, adhesives, etc., explosives and fireworks, pesticides, etc., printing ink, surgical bandages, etc., photographic chemical materials

1.5.2 Special features of accidents in the chemical sector

An analysis of the more serious accidents in this sector made by the Chemical National Industry Group (NIG) of the HSE in 1983[5] is summarised in Table 1.6. This showed that 65% of these accidents were process related.

Table 1.6 Main types of incidents in the chemical industry in 1983

Incident type	Number	%
Release of chemicals (including 97 toxic, 51 corrosive, 45 flammable, 23 hot and six other materials)	222	34.5
Machinery incidents	77	12.0
Process-related fires and explosions	66	10.3
Falls from a height	56	8.7
Falls at same level and striking against objects	39	6.1
Pressure system and other equipment failures (where main risk was not from chemicals released)	24	3.7
Hit by falling objects	23	3.6
Failure or overturning of lifting equipment	22	3.4
Struck or trapped by vehicle	15	2.3
Affected by chemicals during work (e.g. by decanting, charging, etc. without significant escape or spill)	13	2.0
Run-away exothermic reactions (with no major release of chemicals)	9	1.4
Manual handling and strains	8	1.2
Non-process-related fires and explosions	7	1.1
Confined space incidents – people overcome	5	0.8
Electric short circuits	5	0.8
Not elsewhere classified	52	8.1
Total	643	100.0

This study also brought out the following significant points:

- Of the 222 releases of chemicals which occurred, 127 (57%) affected people directly or indirectly; 92 (41%) occurred during normal process operation with no immediate direct involvement of workpeople; 64 (30%) happened during repair, maintenance or cleaning operations.
- Sixty-two of the 643 incidents (9.6%) involved personnel other than those employed by the factory occupier, i.e. contractors, visitors, etc.
- Forty-nine cases of acute ill health were recorded in the year, mostly associated with the release of chemicals or confined-space incidents.
- Thirty pipework failures occurred in addition to 37 incidents involving flexible hoses or insecure temporary joints.
- Seventeen of the 643 incidents involved clear failures of permit-to-work procedures.
- Seventeen incidents involved tanker vehicles or tank containers.

1.5.3 Further features of accidents in the chemical sector

The findings of the above survey were confirmed by a more extensive three-year survey reported by Robinson of the Chemical NIG[6], which revealed several additional points including the following:

1. *Releases*. 55% of chemical releases affected personnel; 15% of chemical releases involved failure of flexible hoses or insecure joints between flexible hoses and fixed connections; 35% of releases of flammable materials ignited.
2. *Substances involved in incidents*. One or more chemicals were involved in about 50% of the incidents; 184 different substances were involved, of which 108 were only involved once.The numbers of incidents involving the same substance more than ten times are given in Table 1.7.

Table 1.7 Number of times the same chemical was involved 1983–1985

Times	Chemical	Times	Chemical
52	Unspecified hydrocarbons	16	Hydrogen chloride
45	Sulphuric acid	14	Hydrochloric acid
38	Hot water or steam	13	Petrol
35	Chlorine	12	Phosgene
32	Caustic soda or potash	12	Xylene/toluene
28	Explosives or pyrotechnics	11	Nitrous fumes
28	Ammonia	10	Hydrogen sulphide
20	Vinyl chloride		

3. *Maintenance-related incidents*. 30% of reported incidents were maintenance based. Of these, 65% involved injury, over 50% involved release of harmful substances and over 25% involved maintenance of pipes, pumps and valves. In 75% of cases investigated, management failed to take all reasonable precautions, particularly over permit-to-work systems and the provision of adequate protective equipment. Mineral oil refining and the dyestuffs and pigments industries had considerably higher incidence rates than other chemical industries.
4. *Accidents involving permits-to-work*. 8% occurred where no permit existed; 55% involved inadequate permits; 37% occurred when permits were not followed by employees.
5. *Injuries from machines*. Machinery accidents, although less frequent than in other industries, were noticeably concentrated in the pharmaceutical and plastics sectors.

1.6 Vapour cloud explosions (VCEs) and other major world losses in the hydrocarbon-chemical industries

The incidents discussed here are summarised in Appendix D, based on a paper[7] by Davenport of Industrial Risk Insurers, and in Appendix E, based on annual reviews of large property losses in the hydrocarbon-chemical industries published by Marsh and McLennan[1, 1a]. Most of the incidents are of the types discussed in section 10.5.

Appendix D[7] lists 25 VCEs, each causing a property loss in excess of $10 million, between 1950 and 1983, during which a total of 69 VCEs were reported, i.e. about two per year. Each of the 25 large VCEs caused an average of 4 fatalities and an average property loss of $40 million. From this it is clear that the economic incentives to reduce the incidence of VCEs are at least as strong as the need to save lives. Methods of recognising installations where VCEs are possible and of evaluating the maximum probable property damage if one occurs are described in section 12.4.

Appendix E lists 97 major world property losses (including VCEs, and each in excess of $10 million at 1988 values) in the oil, natural gas and chemical industries from 1963 to 1988, excluding those in communist countries. The loss figures relate only to property damage, debris removal and cleanup. Claims for business interruption, employee injuries and liability are excluded. While most of these losses involved fires and explosions, some other incidents involving collapse, pressure rupture, implosion, flooding and windstorms are included. Offshore accidents involving gas/oil production and ships at sea are excluded and no information is given about about human casualties. Bhopal [5.4], in which 2500 people died, is not included since the property damage in that incident was comparatively light.

1.6.1 Analysis of large losses

The twelfth edition[1a] of Marsh and McLennan's annual review analyses the 150 largest losses since 1959. The number and total magnitude of the losses during each consecutive five-year period are shown in Figure 1.6. The progressive increases are due partly to the availability of more complete loss data in recent years but more to the dramatic increase in the size of process units. Thus the capacity of single-train ethylene plants has risen from 20 000 t/a to 700 000 t/a. More congested plant layouts resulting from efforts to minimise energy needs, piping and instrumentation have also increased the magnitude of individual losses.

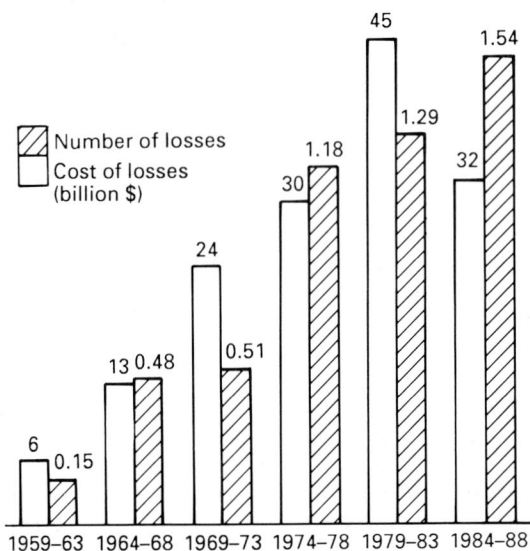

Figure 1.6 Number and total magnitude of large losses, 1959–1988 (data from reference 1a)

The distribution of loss and average value of losses over $10 million at 1988 values for various types of complex are shown in Figure 1.7. Most of such losses occurred in oil refineries while the highest average losses occurred in natural gas processing plants.

The primary causes of loss, broken down into seven headings, are shown in Figure 1.8. Mechanical failure of equipment was the most frequent of these causes. Many of these failures resulted from metal corrosion, erosion, embrittlement and/or fatigue. Most of these could have been avoided by proper inspection and maintenance. The next most frequent cause was stated to be operational errors made on the spur of the moment. Most of these could have been avoided by providing more thorough written operating procedures and guidance and more careful selection and training of operators themselves.

The relative frequency of involvement of eleven different types of equipment in origin of loss is shown in Figure 1.9. Piping systems, which include hose, tubing, flanges, gauges, strainers and expansion joints, were the most frequent origin of loss. The low frequency of losses originating at pumps and compressors was unexpected.

An analysis of whether the loss occurred while the installation was in normal operation or not showed that 24 per cent of the losses occurred during start-up, shutdown, during maintenance or while the plant was idle. In some cases operators had become aware of trouble (leaks etc.) while the plant was running and were in the process of shutting it down when the loss occurred.

Table 1.8 shows the frequency and average cost of four types of large losses.

The most devastating losses involved the delayed ignition of vapour clouds of accidentally released materials. Vapour clouds often cover a large area before igniting, thus causing widespread fires and blast damage. They also cause flying missiles which cause fires and damage remote from the point of vapour release. The miscellaneous type included a wind-storm, rupture of a steam pipe and a tank collapse.

Figure 1.10 shows the frequency of different types of loss for different types of complex. Here there is a marked contrast between oil refineries and chemical plants: 75% of losses in chemical plants were initiated by explosions and 8% by fires, compared with 13% by explosions and 52% by fires in refineries.

Attempts to classify losses by source of ignition proved of limited value since in most cases the source of ignition remained unknown.

Most of the installations where these losses occurred can truly be called 'self-destructing'. By this I mean that unless they are maintained in a high state of integrity, even minor failures can quickly escalate into catastrophes.

When a plant is operating unprofitably in a period of depressed markets and high oil prices, like a redundant ship it can barely be sold for its scrap value. There is a strong temptation to skimp on maintenance and run it on a shoestring, until times improve or it goes up in smoke. The conscious decision to close it down may be more difficult to take both politically and economically. Hard times are also dangerous times. New safety legislation and increased insurance premiums may have little effect on this situation.

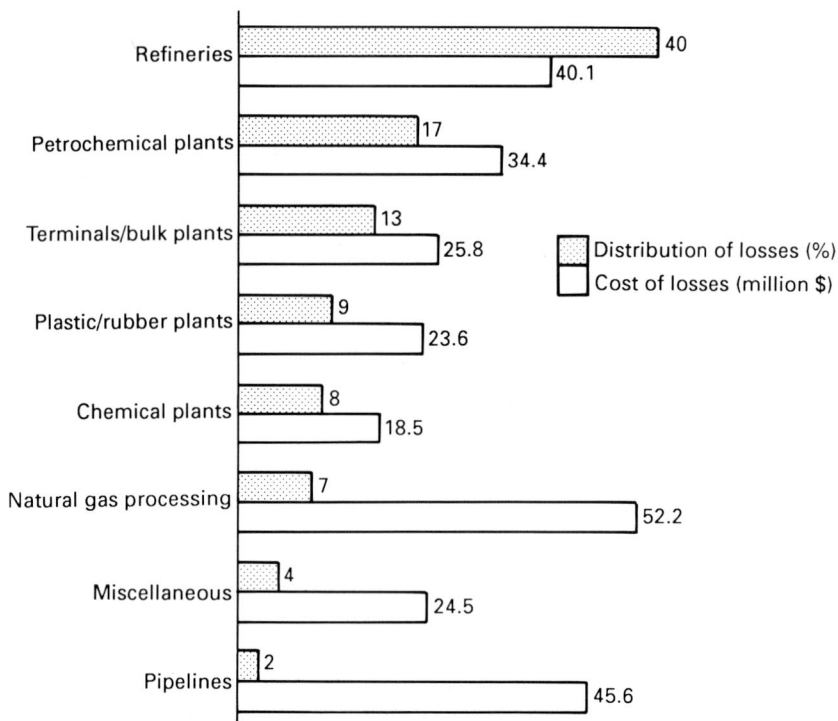

Figure 1.7 Distribution and average cost of large losses by type of complex (data from reference 1a)

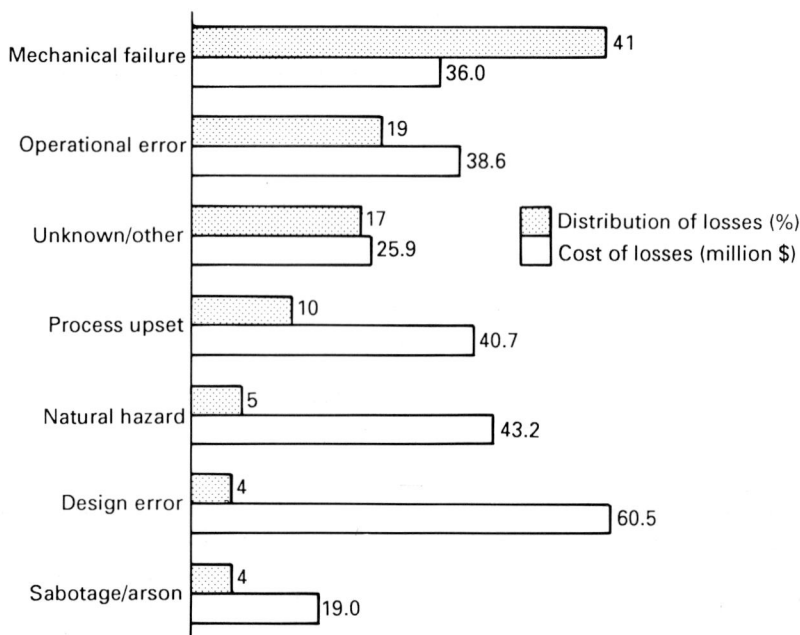

Figure 1.8 Primary causes of large losses (data from reference 1a)

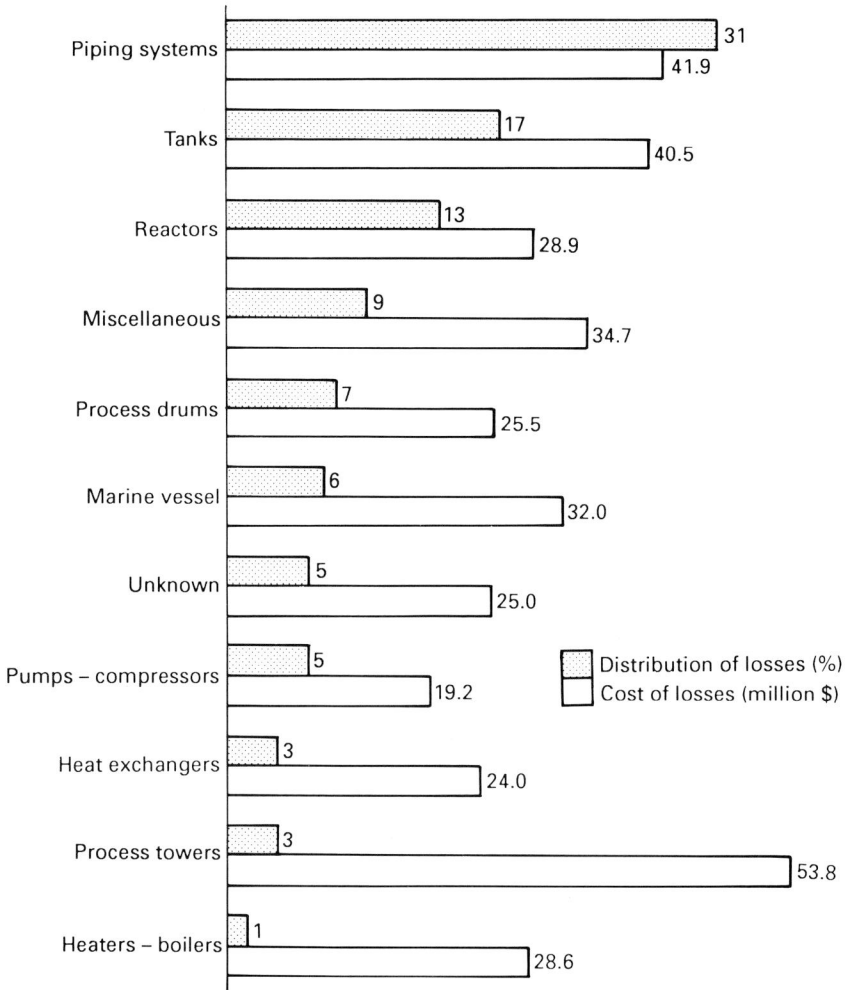

Figure 1.9 Types of equipment involved in large losses (data from reference 1a)

Table 1.8 Frequency and average cost of four types of large loss

Type of loss	Per cent of losses	Average loss at 1988 values ($ million)
Fire	38	31
Explosions		
Vapour cloud	36	46
Other	24	25
Miscellaneous	2	16

Cause of loss

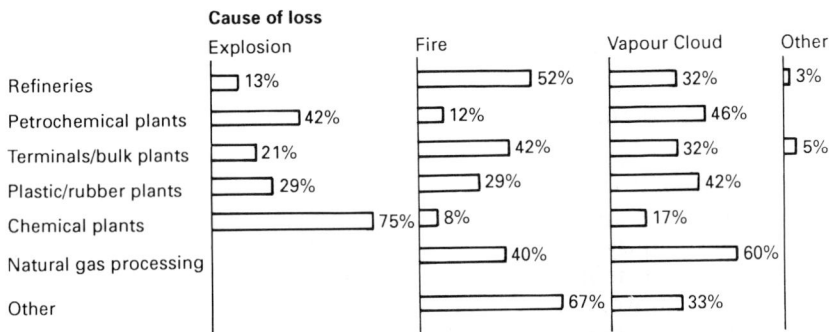

	Explosion	Fire	Vapour Cloud	Other
Refineries	13%	52%	32%	3%
Petrochemical plants	42%	12%	46%	
Terminals/bulk plants	21%	42%	32%	5%
Plastic/rubber plants	29%	29%	42%	
Chemical plants	75%	8%	17%	
Natural gas processing		40%	60%	
Other		67%	33%	

Figure 1.10 Analysis of large losses by type of complex (data from reference 1a)

Such questions lie outside the scope of this book, however, and can probably only be tackled on a global scale. Like most problems of our time, their burden falls most heavily on developing countries which desperately need capital and modern technology to feed, clothe, house and provide work for their growing populations [23].

1.7 Health in the process industries

Information about occupational ill-health of workers in the process industries is extremely patchy. Although numerous studies have been made by epidemiologists and medical specialists of ill-health caused by exposure to particular substances in particular factories and industries, the number of substances involved is so large and the conditions of exposure are so varied that it is difficult to obtain a clear overall impression of the problem. The latest edition of Hunter's classic book[2] probably still gives as good a general picture as any.

There are no occupational exposure limits for the great majority of pharmaceutical drugs and their intermediates, although discussions between HSE and the Association of the British Pharmaceutical Industry (ABPI) began in 1983 about standards for the industry.

Both the recent COSHH (Control of Substances Hazardous to Health)[8] [2.5.6] and RIDDOR (Reporting of Injuries, Diseases and Dangerous Occurrences)[9] [2.5.5] regulations should eventually improve the information available.

At the same time, the exposure of workers outside the process industries to solvents, resins, agro-chemicals and other chemicals appears to be at least as great as that of those engaged in their manufacture, where control measures can be more effectively applied. Those particularly at risk include painters, farm workers, tyre factory workers and furniture makers.

The National Joint Advisory Committee (NJAC), a joint trade union/employer body, cooperates with the Chemical Industry Safety and Health Council (CISHEC) which was set up by the Chemical Industries Association (CIA). There is, however, no tripartite body which involves the HSE through the NIG for the chemical industry[5], as there is for most other industries.

1.7.1 Official information on occupational ill-health

The introduction to HSC's earlier Consultative Document on COSHH[10] contains the following information on occupational ill-health caused or exacerbated by occupational exposure to toxic substances:

1. In 1979/1980, the (then-named) Department of Health and Social Security (DHSS) awarded (in respect of diseases attributable to occupational exposure to substances other than lead and asbestos) over 5700 new injury benefits for short spells of incapacity, 850 new longer-term benefits, and 710 death benefits. Over the same period absence from work stemming from the new benefit cases of this type amounted to approximately 187 000 working days[11]. These figures relate only to clearly established cases of these diseases for which benefit is payable and do not describe the full extent of the problem.
2. There are no statistics to show what proportion of chronic bronchitis, asthma and emphysema is occupationally linked. In 1980, there were over 20 000 deaths[12] and over 30 million lost working days from these causes.
3. Cancers (except mesothelioma and bladder cancer) are not covered by the DHSS statistics either. The proportion of cancer deaths which are occupationally related is a source of some controversy, but a recent review[13] suggested that 2–8% of cancer deaths each year could be prevented if all occupational hazards were removed. Other sources have estimated that up to 30% of disease is occupationally related[14]. There were 130 000 cancer deaths in Britain in 1980[11].

References

1. Garrison, W. G., *100 Large Losses – A thirty-year review of property damage losses in the hydrocarbon-chemical industries*, 11th edn, Marsh and McLennan Protection Consultants, 222 South Riverside Plaza, Chicago, Illinois 60606 (1988)
1a. Garrison, W. G., *Large Property Damage Losses in the Hydrocarbon-Chemical Industries – A Thirty-Year Review*, 12th edn, Marsh and McLennan, Chicago (1989)
2. Hunter, D., *The Diseases of Occupations*, 6th edn, Hodder and Stoughton, London (1978)*
3. Norstrom II, G. P., 'An insurer's perspective of the chemical industry' in *Safety and Accident Prevention in Chemical Operations*, edited by Fawcett, H. H. and Wood, W. S., 2nd edn, Wiley-Interscience, New York (1982)
4. HSE, *Manufacturing and Service Industries*, 1982, 1983 and 1984, HMSO, London
5. *Report by HM Chief Inspector of Factories 1985*, HMSO, London
6. Robinson, B. J., 'A three year survey of accidents and dangerous occurrences in the UK chemical industry', paper for *World Conference on Chemical Accidents*, Rome, July (1987)
7. Davenport, J. A., 'A study of vapour cloud incidents – an update', paper in *4th International Symposium on Loss Prevention and Safety Promotion in the Process industries, Harrogate*, The Institution of Chemical Engineers, Rugby (1983)
8. S.I. 1988, No. 1657, *The Control of Substances Hazardous to Health (COSHH) Regulations 1988*, HMSO, London
9. S.I. 1985, No. 2023, *The Reporting of Injuries, Diseases and Dangerous Occurrences Regulations (RIDDOR)*, HMSO, London

* Sections 1.1 and 1.2 are based on this reference and other historical references given in it.

10. HSC Consultative Document, *The Control of Substances Hazardous to Health: Draft Regulations and Draft Approved Codes of Practice*, HMSO, London (1984)
11. Government Statistical Services, *DHSS Social Security Statistics 1981*, HMSO, London
12. *1980 Mortality Statistics: Cause – OPCS Series DH2 No. 7*, HMSO, London
13. Doll, R. and Peto, R., *The Causes of Cancer*, Oxford University Press, Oxford (1982)
14. Disher, Kleinman and Foster, *Pilot Study Developments of an Occupational Disease Surveillance Method*, University of Washington (for US Department of Health, Education and Welfare (1975)
15. Li, F. P., Fraumeni, J. F., Mantel, N. and Miller, R. W., 'Cancer mortality among chemists', *Journal of the National Cancer Institute,* **43**, 1159 (1969)

Chapter 2

Laws, codes and standards

Not only laws, but entire legal systems change, as one crosses a national boundary (even between England and Scotland). North American law, like English law, is largely Common Law, based on precedent rather than codes. Most law in Western Europe, Scandinavia and Latin America is based on the Romano-Germanic system and heavily codified. Legal inquiries in these countries follow an inquisitorial procedure, rather than the adversarial one common in the UK. Different outlooks and legal systems complicate health and safety problems when modern industrial technologies cross national boundaries [23.4].

Here, after a brief coverage of international trends [2.1], I discuss the background to the UK position, with an outline of selected UK legislation affecting health and safety (HS) in the process industries [2.2–2.5]. I next discuss the role of lawyers in major accident inquiries in the UK [2.6], and finally the importance of standards to health and safety [2.7–2.9].

2.1 Present international trends

Since 1945 there has been a growing trend for more unified HS legislation, both globally and within regions with common political and economic systems. On the international level, the International Labour Office (ILO) and the World Health Organisation (WHO) strive for unified world regulations and standards to protect workers, by voluntary conventions duly ratified by the governments of their member states. Some of their achievements are discussed in Chapter 23.

2.1.1 The European Community (EC)

After the creation of the European Community (EC) – formerly known as the European Economic Community (EEC) – with its Parliament, Commission and Council[1], the Council issued directives to member countries on common legislation on a range of issues, including HS. Proposals for such directives are drafted by the Commission and submitted to the Council, who refer them for advice to a working party of civil service experts. In the light of this the proposals may be amended by the Commission before being approved and issued by the Council.

Governments of EC countries are required to implement any such directive by making and applying appropriate regulations in their homelands, unless their existing legislation already complies with the directive's requirements. The Council may later issue amendments to previous directives. All corresponding national legislation then has to be amended accordingly.

The Commission also monitors the progress and effectiveness of the resulting national legislation. Three important sets of HS regulations in the UK have come about in this way:

- The Control of Industrial Major Accident Hazards (CIMAH) Regulations 1984. (The original EC directive has now been amended twice, with the national regulations following suit.)
- The Classification, Packaging and Labelling of Dangerous Substances (CPL) Regulations 1984.
- The Control of Substances Hazardous to Health (COSHH) Regulations 1988.

Equivalent regulations are in force throughout the EC. They aim to provide the same protection to workers in all EC countries and to prevent manufacturers in one EC country with low HS standards gaining an advantage over those in another with higher standards. Unfortunately, lower standards and less effective legislation, which sometimes attract foreign capital, still prevail in many Third World countries [23].

2.1.2 The US

Prior to 1970, American legislation for industrial health and safety was mainly a state responsibility and varied widely. Although state legislation continues to lay special responsibility on industry, the Occupational Safety and Health Act of 1970 transferred most responsibility for regulation to the Federal government and led to the first uniform HS standards throughout the nation. The Act has a special administration (OSHA) (within the US Department of Labour) with its own research and development agency, the National Institute for Occupational Safety and Health (NIOSH). It promulgates standards which have the force of law. Many of these are specifications with precise descriptions of what must be done to comply with the standard, although some are performance codes stating only the required result with no indication how this is to be achieved. The first type makes it easier for the inspector to determine whether or not the process complies with the law, but makes the introduction of any innovation in the process more difficult[2].

Inspection plays a large part of OSHA's activities. Four main areas of concern of its inspectors are:

- The risk management plans of the enterprises inspected and their effectiveness. Inspectors review loss prevention, communications, design, hazard evaluation, reactive materials, inspection, maintenance, modifications and training.
- Hazard identification and assessment. This includes the properties of materials and reactions with high energy potential and extreme operating conditions.

- Process design and controls. For these there are special and very detailed checklists which cover normal operation, emergency relief and special safety protection.
- Emergency response procedures involving the local community and services as well as the manufacturing enterprise.

Besides the OSHA, the Environmental Protection Agency (EPA) plays a major role in protecting the public and the environment from emissions of hazardous materials. Its Emergency Planning and Community Right-To-Know Act of 1986 is the first attempt at federal level to prepare manufacturing enterprises and local communities for emergency response to accidental releases of hazardous chemicals. It has four main provisions:

- Emergency planning,
- Emergency notification,
- Community right-to-know reporting requirements,
- Toxic chemical release emissions inventory.

Whilst differing in details, this Act has many parallels with our own CIMAH regulations [2.4.2].

Important new state legislation includes the New Jersey Toxic Catastrophe Prevention Act of 1986 and the California Hazardous Material Planning Programme of 1987. These place similar but generally stricter requirements on manufacturing enterprises than the federal legislation. Pressure groups such as GASP (Group Action to Stop Poison) play an important role in the US in influencing public opinion and both local and federal legislation. Partly to head off stringent regulations which might result in loss of jobs, industry groups and others have initiated voluntary programmes such as CAER (community awareness and emergency response), which was promoted by the Chemical Manufacturers' Association, and the Center for Chemical Process Safety (CCPS), which was established by the American Institute of Chemical Engineers.

2.1.3 Other countries

Throughout the world Sweden is generally regarded of having the best industrial HS record, legislation and standards. Standards in some former communist bloc countries are, on paper higher, although their effectiveness is open to question [7.5.2].

2.2 The UK background

While UK factory legislation, including that on health and safety, dates from the beginning of the last century, most of the early legislation is mainly of historical interest and has limited relevance today. The prevailing hopelessness and ignorance about the health and safety problems of the process industries at the beginning of this century is illustrated by Alan Bennett's play *The Insurance Man*. Lees[3] gives a useful summary of UK safety legislation relevant to the process industries up to 1979. Comprehensive and frequently updated coverage of UK legislation on HS is given in Sweet and Maxwell's encyclopedia[4].

As the terms 'practicable' and 'reasonably practicable' are found repeatedly in British legislation, the following interpretation[5] is quoted:

1. A duty qualified by 'so far as is practicable' has to be complied with regardless of cost or difficulty, so far as the means of complying with the duty are 'possible within the light of current knowledge and invention';
2. A duty qualified by 'so far as is reasonably practicable' has to be complied with until the cost of additional control measures becomes grossly disproportionate to the further reduction in the risks which the duty is designed to eliminate or control.

For much of our present HS legislation, we do not need to go further back than the Factories Act 1961. Some earlier legislation which is still in force is discussed briefly in 2.2.3. Most previous piecemeal legislation to protect the health of workers against harmful substances has now been replaced by the COSHH regulations[5] [2.5.6].

2.2.1 The Factories Act 1961

The Factories Act 1961, the last of many earlier ones, deals with health, safety, welfare, enforcement and other matters and is supplemented by a number of detailed regulations.

The Factory Inspectorate, first set up to enforce an earlier Factories Act of 1833, had grown by 1976 to about 700 general inspectors, who were responsible for over 200 000 industrial premises. Specialist branches or units within the Inspectorate were formed to inspect or advise on health, engineering, electrical installations and dangerous trades. These included the Chemical Branch and the Industrial Hygiene Division. The Employment Medical Advisory Service (EMAS) was set up to improve health coverage.

2.2.2 The Robens Report of 1972[6]

The British approach to industrial health and safety changed in the early 1970s following the report of Lord Robens's committee on safety at work (excluding transport). Hitherto most industrial health and safety legislation, which was very fragmented, had dealt in detail with the processes and practices which it tried to control. Nine sets of legislation had been administered by five government departments with seven separate inspectorates. Legislation had neither kept up with changes in technology and industrial practices nor with the very large increase in scale of many process plants.

Three difficulties were:

- The shortage of legal draughtsmen, lawyers and judges who understood the many different technologies involved;
- The frequent updating of legislation needed when technologies change rapidly;
- The delays involved in achieving a consensus of different interests before legislation could be agreed and passed (by which time it was often out of date).

The report recommended sweeping changes in HS law and its administration and suggested that detailed regulations were not appropriate to modern technology. It emphasised the importance of management attitudes and industrial organisation and recommended that more reliance be placed on self-regulation by industry itself. It stressed the need for a more unified authority for regulation and inspection, and recommended that:

- More voluntary standards and codes of practice should be developed by industry itself;
- Greater recognition of their safety responsibilities was needed by industrial enterprises. This included statements of safety policy and identification of their own hazards;
- More disclosure on hazards was needed, particularly to workers who may be exposed to them;
- The interests of the public needed greater recognition;
- Explosive, flammable and toxic substances needed special provisions.

2.2.3 Some pre-HSWA 1974 legislation and later legislation arising from it

Virtually all earlier regulations on the protection of employees from harmful substances have now been replaced by COSHH [2.5.6]. Others still in force which deal with explosives and pollution are discussed here.

The Explosives Acts of 1875 and 1923
These Acts and the regulations associated with them relate to the manufacture, handling, storage and transport of explosives [9.9]. They contain detailed technical provisions for the construction and spacing of buildings used for manufacture and storage of explosives. These led to generous plant spacing in this industry which contrasted with the cramped spacing prevalent in much of the chemical industry.

The Alkali etc. Works Regulation Act 1906 and other anti-pollution measures
This, the last in a series of Alkali Acts, is mainly concerned with industrial air pollution. It has been followed by a series of other anti-pollution legislation which includes

- The Public Health Act 1936;
- The Clean Air Acts of 1956 and 1968;
- The Rivers (Prevention of Pollution) Act 1970;
- The Control of Pollution Act 1974.

The last of these extends to the control of noise which affects the general public.
    Subsequent anti-pollution legislation includes:

- The Health and Safety (Emissions into the Atmosphere) Regulations 1983 (S.I. 1983, No. 934) and their 1989 amendment (S.I. 1989, No. 319),
- The Water Act 1989.

The many man-made problems of the natural environment, including ozone, sulphur dioxide and oxides of nitrogen, are now the concern of HM Pollution Inspectorate in the Department of the Environment.

The Public Health Act 1936

This enables local authorities to serve abatement orders on persons, owners and occupiers of premises where statutory nuisances exist and offensive trades are carried out which affect the neighbouring inhabitants. Before such nuisances and trades are allowed, the consent of the local authority is required. The emission of excessive smoke is one of the nuisances covered by the act.

The Fire Services Acts of 1947 and 1959 and the Fire Precautions Act 1971

Fire regulations which apply to industry are very fragmented. The main legislation is contained in the Factories Act 1961. The Chemical Works Regulations 1922 and the Petroleum (Consolidation) Act 1928 also apply to fire hazards.

The Fire Certificates (Special Premises) Regulations 1976 (S.I. 1976, No. 2003)

These regulations require persons in control of premises containing more than specified quantities of certain flammable and otherwise hazardous substances to apply to HSE for a fire certificate. The substances include any highly flammable liquid, LPG, phosgene, ethylene oxide, carbon disulphide, acrylonitrile, hydrogen cyanide, ethylene, propylene and sources of ionising radiation. The application must show that the following provisions are adequate and are maintained in constant readiness for an outbreak of fire:

- Means of escape;
- Means of firefighting;
- Means of raising an alarm.

On receiving an application, HSE is required to inspect the premises to satisfy itself that the provisions are adequate before issuing a fire certificate. Work may not be done in any 'Special Premises' unless a fire certificate has been issued or after one has been withdrawn. These regulations run parallel with and reinforce the NIHHS and CIMAH regulations discussed in 2.4.

2.3 The Health and Safety at Work etc. Act 1974 (HSWA)

This Act came into force in 1975 as the direct outcome of the Robens Report. It created the Health and Safety Commission (HSC) which is responsible for policy, and its executive arm, the Health and Safety Executive (HSE). The Act is an enabling measure which provides a framework for a new, comprehensive and integrated system of health and safety legislation, in the form of regulations and 'Approved Codes of Practice', to replace previous legislation. The Act has the following headings: scope, regulations, codes of practice, general duties, enforcement and the HSC.

The Act is intended to protect all persons at work, and the public so far as they are affected by work activities. It contains no detailed regulations, but provides powers to allow them to be made by the appropriate minister on the basis of proposals formulated by the HSC, after consulting interested parties. They are to be supplemented by Approved Codes of Practice which have a special legal status. The Act imposes general duties on employers, employees, the self-employed and manufacturers and suppliers. While the HSE is the enforcing authority for factories, local authorities are responsible for enforcement in offices and service industries, particularly shops, hotels and restaurants.

Factory Inspectors have four options for enforcement under the Act – seizure, improvement notice, prohibition notice and prosecution.

The HSWA brought together a number of inspectorates and functions which previously operated under different Acts and authorities, e.g. the Alkali, Mines and Quarries and Nuclear Inspectorates, a Hazardous Substances Branch and a Safety and General Group. In 1987 the responsibility for pollution control was transferred from the HSE to the Department of the Environment. Advisory committees have been set up in special areas such as major hazards and asbestos. The HSE (which has special access to several government research and consultancy establishments) issues Guidance Notes on a number of topics.

2.4 Legislation on the Control of Major Hazards

The increase in scale of oil and chemical plants over the past 40 years has increased the possibilities of large but infrequent accidents which affect people, property and the environment outside as well as inside the factory fence, e.g. the Flixborough disaster of 1974 [4]. This led to the appointment of HSC's Advisory Committee on Major Hazards. It published three reports[7], which in turn led to the 'Notification of Installations Handling Hazardous Substances Regulations 1982' (NIHHS)[8]. These and subsequent legislation which is discussed below have been reviewed by Morgan[9].

2.4.1 The Notification of Installations Handling Hazardous Substances (NIHHS) Regulations 1982 (S.I. 1982, No.1357)

These require that all fixed installations, including factories, warehouses, transit depots, ports, pipelines, and moored ships and vehicles used for storage, where more than specified quantities of 35 named substances or classes of substances are present, be notified to the HSE. (The regulations do not apply to transport.)

The specified quantities of the various substances so far as possible represent equivalent hazards, although this is more difficult for toxic materials than for flammable or explosive ones. LPG and sodium chlorate are both listed at 25 t, some ammonium nitrate fertilisers at 500 t, and chlorine at 25 t. The regulations are intended:

- To give the HSE a clearer picture of the major hazard sites in the UK, and help it to define its inspection priorities;
- To remind employers and employees of the hazards of their activities.

The information to be notified is limited, but includes a description of the activities carried out and the maximum quantity of each hazardous substance for which notification is made.

2.4.2 The Control of Industrial Major Accident Hazards (CIMAH) Regulations 1984 (S.I. 1984, No. 1902)

The CIMAH regulations of 1984 'implement in full an EEC Directive the scope of which is determined by definitions of a major accident, industrial activity and dangerous substance'[8]. Similar regulations are in force in all EC countries. The directive was inspired by EC concern at the possible repetition of a disaster like Seveso [5.3] which caused long-term environmental damage, just as our NIHHS regulations were inspired by the lessons of the Flixborough disaster.

The main object of the regulations is to keep the managements of process industries in which major accidents could possibly occur 'on their toes'. The regulations do not apply to installations under the control of the Ministry of Defence (or international and allied organisations as provided by appropriate Acts). Nor do they apply to 'a factory, magazine or store licensed under the Explosives Act 1875'.

The regulations embrace most of the chemical and petrochemical industries which use substances with highly toxic, reactive, explosive or flammable properties. Substances listed in the regulations are discussed in Chapters 7–10.

The regulations apply in the first place to the 'processing' and 'isolated storage' of a number of 'dangerous substances', most of which are individual organic compounds. Their requirements for 'processing' are more stringent than for 'isolated storage'.

The term 'processing' includes a number of stated chemical and physical processes (Schedule 4), such as polymerisation, nitration, distillation and mixing. Under 'processing' the regulations apply to extensive lists of substances, for each of which there is a threshold quantity which must be present before the regulations apply. These threshold quantities range from as little as 1 kg for dioxin and other 'super-toxic' chemicals to 1000 t for sulphur dioxide and 50 000 t for petrol.

The regulations apply at two levels. At the general or lower level, regulation 4 requires manufacturers to provide a 'demonstration of safe operation'. This should provide evidence at any time to the HSE that hazards to both man and the environment have been identified and properly guarded against.

At the higher or specific level at sites involving larger inventories and more dangerous substances, regulations 7 to 12 apply. Measures called for here include:

- The manufacturer to notify the HSE initially, and later send them a written report or 'safety case'[9] (regulation 7);

- The manufacturer to prepare an on-site emergency plan (regulation 10);
- The local county (or equivalent) authority to prepare an off-site emergency plan (regulation 11);
- The manufacturer to provide information for those near the site who might be affected by a major accident (regulation 12); this is to be passed on to them by a route agreed with the local district council.

Schedule 3 is a list of substances arranged in five groups, to which regulations 7 to 12 apply. The first two groups are toxic substances. Both are discussed in 7.8. The third group are highly reactive substances which are discussed in 8.8. Group 4 are explosive substances which are discussed in 9.10. Group 5 are flammable substances. These are defined in Schedule 1 and are discussed in 10.3.

About 200 large installations in the UK are subject to these requirements. Most of these installations are also notifiable under NIHHS. The advent of CIMAH has necessitated changes in the laws on planning controls. Guidance on the regulations and on the preparation of emergency plans is available from the HSE. Amendments to CIMAH (and to other EC-inspired regulations) follow from further EC directives.

As with most regulations with detailed schedules of substances and processes, it is probably inevitable that some which should be included slip through the net. It is, however, a little disturbing to find that in the case of the chemical plant explosion discussed in 5.2, neither the chemical involved (3,5-dinitro-2-toluamide) nor the process in which it was used (drying) are included in the relevant schedules. There may be a risk that managements find their hands so full complying with the letter of the regulations, that they miss other serious hazards not covered by them.

2.5 Other relevant legislation and its problems

2.5.1 Pressurised systems

One serious gap in the detailed provisions of the Factories Act 1961 concerns the inspection of pressurised systems other than steam boilers and compressed air receivers. The importance of this can hardly be exaggerated, since so many process hazards today involve the accidental release of flammable and toxic gases (e.g. propane, ammonia, chlorine) from pressurised systems.

The Flixborough disaster of 1974, as we show in Chapter 4, might have been prevented had there been effective legislation requiring the monitoring of the design of pressurised systems by a competent design inspector backed by inspection and testing of each pressurised system as a whole. The design fault would then surely have come to light before the plant was started up. (This may surprise readers whose main knowledge of that disaster is based on the official report[10].)

The first Boiler Explosions Act of 1882 resulted from numerous deaths caused by boiler explosions, following the introduction of steam engines for factory power. Many boilers were located in or adjacent to congested factory workrooms. It was not, however, until the 1901 Factories and Workshop Act that the periodic inspection of boilers was required by law.

This requirement was refined and amended by subsequent Factory Acts, the last being that of 1961. This covers steam boilers, steam receivers, air receivers and (incidentally) gas holders. The Act deals with standards of construction and maintenance, protective features and periodic inspections.

The risks resulting from the failure of some pressurised process units which contain flammable, toxic and corrosive fluids are far greater today than from the largest (non-nuclear) boilers. Yet there is no specific legislation calling for the inspection of such pressure vessels, let alone pressurised systems. Apart from company and insurance standards [17.3.2], reliance depends only on a (non-mandatory) British Standard[11] for the design and construction of pressure vessels.

In 1977 the HSC published proposals for new legislation for pressurised systems as a consultative document[12], which contained the following passage:

> The Commission believe there is a pressing need for a new homogeneous framework of legislation which takes full account of the problems of modern pressurised systems. Such a framework should be flexible enough to deal with specific requirements of design, construction and use both now and in the future.

These proposals, while applying to virtually all systems under pressure including transportable vessels, require the user to assess the hazards involved, to prepare detailed schemes of examination, testing and record keeping, and to appoint competent persons to carry these out. They are based on the philosophy of self-regulation which was so much favoured by the Robens Report, and adopted in the HSWA 1974. Three sets of complementary regulations were originally envisaged, dealing with the hazards of pressure, flammable materials and toxic materials. The last two of these proposed regulations now seem to have been abandoned.

These proposals, which are now almost a collector's item, were followed in 1984 by another consultative document[13] containing proposals for Regulations and two Approved Codes of Practice, one for Pressure Systems, the other for Transportable Gas Containers. These are still being considered by the different interests involved, 13 years after the publication of the first consultative document!

2.5.2 The Safety Representatives and Safety Committee Regulations 1977 (S.I. 1977, No. 500)

These regulations which were published with an Approved Code of Practice and HSE guidance note arise from section 2(4) of the HSWA 1974, under which recognised trade unions may appoint safety representatives from among the employees. The names of these nominees (who should have appropriate employment experience) and of the groups they represent must be must be notified to the employer. Their functions include:

● Representing employees in consultation with the employer and with inspectors on HS matters and receiving information from inspectors;

- Investigating work-related hazards and accidents;
- Investigating employees' complaints about HS conditions at work;
- Inspecting the scenes of notifiable accidents and relevant documents;
- Attending safety committee meetings.

Safety representatives may take time off with pay for training in and carrying out these functions, which, however, impose no duties on them. If no safety committee [19.1.9] exists, the employer shall establish one within three months if properly requested to do so by two or more safety representatives. The safety committee should promote cooperation between employers and employees on HS matters and provide a forum where employees can discuss them. Although a safety committee in no way relieves management of its safety responsibilities, it may usefully complement the work of professional safety personnel.

2.5.3 The Classification and Labelling of Explosives Regulations 1983 (S.I. 1983, No. 1140)

These regulations are seen by HSC[14] as a first step in updating the Explosives Act 1875. They should ensure that the hazards of particular explosives received by any user are clearly stated on the label.

2.5.4 The Classification, Packaging and Labelling of Dangerous Substances Regulations (CPLR) 1984 (S.I. 1984, No. 1244)

These regulations ensure that the hazards of most dangerous substances received in packages are marked on the label. The regulations require such substances to be classified and labelled in special but different ways depending on whether they are being 'supplied' or 'conveyed by road'. Since most packaged substances are supplied by road, both types of labels with their own marking systems have generally to be used. The supply provisions of CPLR stem from EC directives and the conveyance provisions from UN recommendations. Only the hazard warnings which apply to supply and which are generally marked on the labels are discussed here. 'Supply' has a wide meaning, including:

Transfer from a factory workhouse or other place of work and its curtilage to another place of work, whether or not in the same ownership.

Exemptions from CPLR are mostly substances whose packaging and labelling are covered by other regulations. These include those whose sole hazard is radioactivity and dangerous liquids or liquefied gases delivered in road or rail tankers or by pipeline, human and animal food, cosmetics, medicines, drugs and disease-producing micro-organisms. Substances similarly exempted from the supply provisions include explosives, goods in transit through the UK, compressed, liquefied or dissolved gases except in aerosol containers, and pesticides. The regulations have been amended twice[15].

The names of the 'dangerous substances' to which CPLR apply are given in an Approved List[16], some details of which are given in Appendix F.

Other regulations dealing with the transport, labelling and transport of dangerous substances include:

- The Petroleum Compressed Gases Order 1930;
- The Gas Cylinder (Conveyance) Regulations 1959 (S.I. 1959, No. 1919);
- The Dangerous Substances (Conveyance by Road and Tank Containers) Regulations 1981 (S.I. 1981, No. 1059);
- The Conveyance of Dangerous Substances in Packages Regulations 1986 (S.I. 1986, No. 1951).

2.5.5 The Reporting of Injuries, Diseases and Dangerous Occurrences Regulations (RIDDOR) 1985 (S.I. 1985, No. 2023)

These regulations came into force in 1986, replacing the Notification of Accidents and Dangerous Occurrences Regulations 1980 (NADOR). They require notification of all fatal accidents, major injuries, accidents causing more than three days off work, certain specified diseases and specified dangerous occurrences at work, as well as accidents involving flammable gases. Their obligations on managements are discussed in 19.2.4 and Appendix L. Two HSE advisory leaflets and a guide are available.

2.5.6 The Control of Substances Hazardous to Health (COSHH) Regulations 1988 (S.I. 1988, No. 1657)

These regulations which result from the EC Council Directive No. 80/1107/EEC came into force in October 1989 and are issued with two Approved Codes of Practice, 'Control of Substances Hazardous to Health' and 'Control of Carcinogenic Substances'. Equivalent legislation has been passed in the other EC countries.

The regulations completely abolish 31 orders and sets of regulations now in force, revoke or repeal 21 others, and repeal seven Sections of the Factories Act 1961. They should change our attitudes about personal exposure to potentially harmful substances and result in better monitoring and record keeping (particularly of the history of employees' exposure to hazardous substances). Some details of the regulations are given in Appendix G.

The following persons, trades and processes will be especially affected by these regulations:

- Users of substances hazardous to health in circumstances such as spraying which are most likely to involve high exposure levels;
- Users and handlers of highly toxic substances and carcinogens;
- 'Dusty trades' including ceramics, refractories, quarrying, foundries and metal manufacturing and finishing;
- Processes which generate appreciable quantities of substances hazardous to health, e.g. welding, cutting, grinding, milling or sieving.

The measures called for under these regulations are discussed further in 7.8.

2.5.7 The Electricity at Work Regulations 1989 (S.I. 1989, No. 635)

These regulations supersede The Electricity (Factories Act) Special Regulations 1908 and 1944 in controlling the use of electricity in factories. They include the requirement of permits-to-work on live high-voltage electrical circuits. The regulations extend to the use of electricity in flammable atmospheres [6.2], for which The British Approvals Service for Electrical Equipment in Flammable Atmospheres (BASEEFA) was set up in 1968. BASEEFA is controlled by the HSE and is responsible for certifying equipment for such duties to the appropriate British Standard.

2.5.8 The Noise at Work Regulations 1989 (S.I. 1989 No. 1780)

These regulations call for action by employers, employees and machine suppliers to protect people at work from noise-induced hearing damage. They require employers to assess noise levels in problem areas and reduce these as far as practicable, to inform those who may be exposed to noise levels above 85 dB(A) and provide ear protection for them and maintain machinery and protective equipment so that noise exposure remains at safe levels. Hearing protection is discussed further in 22.7.

2.6 The law and public inquiries into major accidents

We consider here the type of public inquiry which followed the Flixborough disaster [4], and will probably follow the next one in the UK. Like many others, although not always for the same reasons, the writer has serious misgivings about the effectiveness of such inquiries.

If an industrial disaster were to occur tomorrow, a public inquiry would probably be called for by the HSC under Section 14 of HSWA 1974 and would follow the Health and Safety Inquiries (Procedure) Regulations 1975[17]. The Commission would appoint a 'person' and assessors (who would constitute the court) to hold an inquiry, giving appropriate notice of its date, time and place to all persons entitled to appear.

The inquiry would have two main objectives: (a) to discover the causes of the disaster; and (b) to decide who was responsible for causing it.

2.6.1 The court of inquiry

The appointed person, who would be chairman of the court, would probably be a leading QC. The court's other members might be professors in the technical disciplines most involved, and a representative of the HSC. The court's main job would be to hear the evidence, reach its conclusions and write its report.

The appointed person would hold the inquiry in public except where:

- A Minister of the Crown rules that it would be against the interests of national security to allow the evidence to be given in public; or
- As a result of application made to him, he considers that the evidence to be given may disclose information relating to a trade secret.

2.6.2 'Interested parties', counsel and expert witnesses

A preliminary hearing would be held at which interested parties and their counsel (usually barristers), who would represent them in court, would be identified. The parties might be the operating company, its parent company, licensors, designers, contractors, trade unions, local authorities and Her Majesty's Factory Inspectorate (HMFI).

Both the court and the main parties would choose their own technical advisers (who might be rival firms of professional consultants). They, together with experts from HSE, would investigate the causes of the disaster before the court hearing started and later serve as expert witnesses in court.

2.6.3 Professor Ubbelohde's criticisms

Important criticisms of this type of inquiry were voiced by Professor Ubbelohde in 1974[18]. Some extracts follow:

> There are special features of every major accident . . . Promptitude in the scientific study of accidents is often hindered by established procedures of inquiry. One cause of delay is the age-old desire to find 'someone to blame' when things have gone wrong . . . Legal problems of allocating responsibility often overshadow scientific problems of how such accidents occur . . . Financial consequences of legal liability can become so oppressive that the whole tempo and management of accident inquiries (particularly while they are sub-judice) may acquire almost a nightmare quality. This concern with liability can distort and obscure basic scientific questions of how and why accidents occur. . . . Scientists would stress that their inquiries must be concluded without any emotional or other pressures on them . . . British practice seems no better than elsewhere. It can even be termed wasteful, antiquarian and lopsided, since full detailed scientific study often cannot take place until the scent is cold.
>
> Whatever form of organisation is chosen, to be effective, an officially established scientific inquiry should run side by side with any judicial inquiry, not after it. It should aim to have all the information collected and presented in an organised way by a small team of experienced scientific assessors. The professional independence and scientific integrity of these assessors must be properly protected. Their task should be to collect facts scientifically and their submissions would be examined and freely discussed by a body of scientists.

2.6.4 Further criticisms

Counsel are at least as much concerned to prevent the party they represent from being blamed as to uncover hazards which contributed to the disaster. They are so used to working in this way that they are unaware of their own blind spots and limitations. In technical matters they are largely dependent on the views of hand-picked experts whose 'Proof of Evidence' they have rehearsed and vetted. Who has ever heard of an expert witness giving evidence which might render his or her client responsible for the accident? The two objectives of the inquiry are thus in conflict.

Legal training and procedures (examination and cross-examination of witnesses) may have some value in establishing whether or not a witness is telling the truth. The procedures are, however, highly formal, and courts can only work on the evidence with which they are presented. They cannot crawl under the car and examine the under-chassis. They can only deal with what the mechanic or expert who has crawled under the car told them in evidence that he found.

Courts thus work in an artificial atmosphere of pre-selected facts and evidence over whose selection they may have rather limited control. They are seldom sufficiently familiar with the subject with which they are dealing to appreciate the relevance of every aspect.

Because of the heavy financial burden of daily legal representation at an inquiry, most parties want to see the whole thing over as soon as possible. In spite of this, some inquiries are very lengthy.

In the (usual) case of disagreement between experts, a court would probably prefer the views of its own advisers. This might be justified on the grounds that only its own advisers are genuinely impartial, but it is liable to lead to complaints from some parties that the court itself is biassed.

Even the court's technical advisors may find themselves in a delicate position. The court has only a transitory life and will be out of business as soon as the inquiry is over, whereas interested parties are usually powerful multinational companies with more staying power.

Another weakness of public inquiries is the unaccountability of courts themselves. Once the court has produced its report, it becomes 'defunct', though its members are alive and well. The court cannot be sued if its findings are in error. There is usually no provision for re-opening such a public inquiry, and even if there were, most parties would be against it because of the cost.

2.6.5 Hazards of unrecognised causes

Thus there are many reasons why the court of a public inquiry may fail to unearth the real causes of a disaster. This can lead to serious dangers, since the party who has wittingly or unwittingly covered up a hazard is likely to be the one most affected if he or she has other installations where it remains dormant, and could cause a further similar accident:

- If the hazard has been consciously covered up, the party has to adopt a double standard, denying publicly that such a hazard could exist, yet alerting those within his or her own organisation to be prepared for it. Thus one employee could be sacked for admitting the hazard's existence, and another for failing to take precautions against it!
- If it has not been recognised or if senior officials in the party's organisation have been persuaded by their own PR exercise that the hazard does not exist, it is likely that it will strike again.

2.6.6 What needs to be done?

An interesting paper by Dr Mecklenburgh[19] developed Professor Ubbelohde's thesis further. It suggested that the technical investigation of the causes of a disaster should be as complete as possible and reach definite conclusions before legal liability is dealt with. Thus there would be separate and consecutive inquiries to cover the technical and legal aspects, the first conducted by technical experts, the second by lawyers.

There are some like Professor Davidson[20] (a member of the Flixborough court) who consider that technical experts cannot be trusted to carry out a technical inquiry without the help of lawyers. Yet in the case of aircraft accidents, where the regulations covering investigations are more advanced[21], technical causes are investigated by technical experts unaided by lawyers. The views of Mr Wilkinson, former Chief Inspector of Accidents in the Department of Transport, are quoted below[22]:

> The UK approach to aircraft accident investigation has been tested and tried over many years and it works. This does not mean that it cannot be improved upon. The proof that the system has achieved general public acceptance is the fact that public inquiries and review boards are rare occurrences.
>
> I should now make it quite clear that, in my considered opinion and from bitter personal experience, there is no place whatever for lawyers in the investigation of aircraft accidents. Some superficially attractive arguments are regularly put forward, saying, in effect, that manufacturers, operators and airworthiness authorities, to name but a few, would, due to commercial pressures, not respond in a responsible manner to criticisms and recommendations without the immense pressures that are applied to them both collectively and individually after a major accident by lawyers acting for interested parties. This is nonsense, in fact, the reverse is much closer to reality. People who manufacture, operate and certify modern aircraft are generally responsible professionals who are concerned to give of their best.

There are, of course, important differences between disasters in air transport and those in the process industries. It is doubtful, for instance, whether it would be possible for a small country such as the UK to keep a permanent group of experts with sufficiently wide experience to be able to investigate the causes of all major accidents in all branches of the process industries. Nevertheless, HSE have gone a long way in developing such a capacity. They have investigated several serious accidents in the process industries (such as the explosion at King's Lynn in 1976 [5.2]) and published their findings[23], with which there has been little criticism. A wider pool of expertise for such investigations might be organised on an international or regional basis, e.g. within the EC.

Should the UK again experience a process disaster on the scale of Flixborough, it is hoped that the HSE, with the help where necessary of outside experts, will be allowed to complete the technical investigations and report them to the public before questions of legal liability are examined.

2.6.7 Could a Parliamentary Commission of Inquiry be used?

Those not happy with the idea of leaving the investigation of the causes of major industrial disasters entirely to technical people (even when they are employed by the HSC) might like to know the origins of our Public Inquiry system. It started after the Marconi scandal of Edwardian times. The Public Inquiry was invented as a device to replace Parliamentary Commissions of Inquiry for handling contentious political issues on which MPs were so sharply divided (on party political lines) that consensus was impossible[24].

This should not apply to the essentially technical issues of major accident causation. Members of Parliamentary Commissions, like lawyers, may not have the technical expertise to investigate the causes of process disasters, and they would thus be still largely dependent on outside help. Yet they seem to be more accountable to the public than the present courts of public inquiries. Parliamentary Commissions are certainly used in other countries to investigate the causes of industrial disasters. This was done in Italy after the Seveso disaster[25] [5.3], and no doubt could be done in the UK if this was agreed to be the best course.

2.7 The role of standards[26]

Modern industry operates with the aid of a complex network of standards (including codes of practice). Most of these have no legal status, although many form the bases of contracts, while non-compliance with important standards [2.9] would weaken the litigant's case in a lawsuit. The future safety of any plant depends on the choice and use of appropriate engineering standards which should be quoted in contracts between owners, designers and plant contractors. Those responsible for supervising design and construction on behalf of the owners must be aware of these requirements and be sufficiently familiar with the standards concerned to detect departures from them at an early stage. They must also have the necessary authority to be able to enforce compliance with the standards [23.5].

The aims of standardisation are defined as[27]:

- Overall economy in terms of human effort, materials, power, etc. in the production and exchange of goods;
- Protection of consumer interests through adequate and consistent quality of goods and services;
- Safety, health and protection of life;
- Provision of a means of expression and communication among all interested parties.

These aims are interdependent, but where health and safety are concerned, it is seldom possible to adopt the most economic solution (except in the broadest sense to society as a whole). Although some standards, such as BS 2092, 'Specification for industrial eye protection', are primarily concerned with safety and health, many others are crucial to safety.

Standards may be classified according to their subject, level and aspect, examples of which are given in Table 2.1.

Some of the earliest standards were units of measurement. Most countries now use 'SI' units [3.1.1]. This simplifies the harmonisation of

Table 2.1 Classification of standards

Subject	Level	Aspect
Engineering	International	Terminology
Transport	Regional	Specification
Building	National	Sampling
Food	Industrial	Inspection
Agriculture	Company	Tests and Analyses
Textiles		Limitation of variety
Chemicals		Grading
Information		Code of practice
Science		Packaging
Education		Transport
Health		Safety

national standards and the creation of international ones, but due to the reluctance of the USA to abandon inches, pounds and gallons, many international standards are drawn up in two versions, one using SI units, the other American ones.

One group of standards reduces manufacturing costs by standardising the dimensions of manufactured items such as nuts, bolts and pipes. Another group covers material specifications and gives chemical compositions and other properties of materials specified in design standards.

Design standards form another group of engineering standards. These give detailed design procedures, e.g. for pressure vessels, with recommendations on manufacture, inspection and testing.

Application standards form another group. They give the main characteristics of the equipment, e.g. a pump or motor, specify the ancillary equipment needed, and the operating conditions for which the equipment is designed.

Some national standards bodies produce codes of practice on a range of subjects such as tower cranes and machine guards. These give general guidance, design parameters and modes of operation, and refer to other standards for further details.

2.8 Levels of standards

Standards are next briefly discussed by level.

2.8.1 International standards

These, apart from those in the electrical field which started in 1908, are produced by the International Standards Organisation which was set up in Geneva in 1946. It attempts to harmonise national standards and to assist developing countries to establish their own. Despite the growing number of international standards, their coverage is thinner than the national standards of major industrial countries. There are also often long delays before international standards can be agreed. Since an international standard on any subject cannot be established without the agreement of the countries which have already produced their own standards on it, the standard usually has to wait until some countries give way and bring their national standards into line with others.

2.8.2 Regional standards

Regional standards committees have been established in various areas, often characterised by the use of a common language. They aim to encourage trade in the area by establishing common national standards within it.

2.8.3 National standards

These are produced by national standards bodies which have technical divisions responsible for different areas of technology. They work through specialist committees which draft new national standards and revise old ones.

Some national standards can restrict competition from countries with different ones. The UK, however, does not insist that all goods and equipment, whether made in the UK or imported, should comply with a British standard. Critical plant items such as pressure vessels which do not comply with the appropriate British standard may be imported so long as they conform to a reputable foreign standard carrying an equivalent assurance of safety and reliability.

2.8.4 Industrial standards

These cover particular industries and are produced mainly by professional bodies within those industries. They are specially important in the USA, many such standards being used internationally. The bodies producing these standards are agents of the American National Standards Institute, and include:

- ASTM   American Society for Testing and Materials;
- ASME   American Society of Mechanical Engineers;
- API   American Petroleum Institute;
- FPA   Fire Protection Association;
- ACGIH American Conference of Governmental Industrial Hygienists.

In the UK, there has been an unfortunate tendency for different organisations with a common interest in any subject each to produce and publish their own standard, all differing in matters of detail. As an example, codes covering the handling and use of liquid petroleum gases have been published by the HSE, the Home Office, the Institute of Petroleum, the Liquid Petroleum Gas Industry Technical Association, the Fire Protection Association, and by ICI in conjunction with RoSPA. Not all of these are periodically amended to keep them up to date. Such organisations should be encouraged to pool their efforts to produce a joint standard or code which is regularly updated.

When modern industrial technologies are transferred to other countries, particularly Third World ones which have little industry, appropriate standards are often lacking in the recipient country. The safety problems which result are discussed in Chapter 23.

2.8.5 Company Standards[28]

Company standards supplement national and industrial standards and incorporate the company's own experience. A company which pioneers a new technology is obliged to develop its own standards for it. These are often the forerunners of standards at industrial, national and international levels. Most companies have their own standards and an organisation for creating them and keeping them up to date. Most company standards are of three types:

- Buying standards, e.g. for raw materials, plant and equipment which it buys, and staff and labour whom it hires;
- Internal standards for design, operation, safety and general practice within the company;
- Standards and specifications for its own products.

Company standards are an important management tool, and play a key role in determining the company's competitiveness and safety record. If the standard contains valuable commercial information, it will be treated as confidential.

Company buying standards usually refer first to any accepted international, national or industrial standards on which they are based (e.g. ASME, BS, DIN or ISO), and then detail the deviations and additions to the standard which the company requires.

Some company standards are important for the safety of process plant. It is easier to develop effective company standards in large companies than in small ones. Some large companies have developed extensive codes of practice in safety matters for their own use.

2.9 Safety standards and codes of practice

While most standards have some significance for safety, some deal specifically with safety topics, e.g. protective headgear, scaffolding, insulation and colour coding of electrical wiring. Many of these are codes of practice. Several have been published by the ILO dealing with safe working practices in entire industrial fields.

Under HSWA 1974, standards issued by the British Standards Institution and professional and industrial bodies are treated as 'guidance literature'.

In 1982 HSC issued a guidance note on standards relating to safety[29]. This proposed that HSE should become more involved in BSI technical committees in cases where it expects to make use of the resulting standards. It also discussed legal backing for some British standards, which might take any of three forms (listed in order of diminishing status):

1. 'Application of a standard by regulation', i.e. the standard becomes mandatory in all circumstances;
2. 'Approval of a standard under Section 16 of HSWA';
3. 'Reference to standards in the course of guidance' (i.e. in HSE Guidance Notes).

Other British standards would carry less legal weight. A list of Codes of Practice approved under Section 14 of HSWA is given in Appendix H. Some of these have been prepared by the HSE in support of recent legislation. Others are British Standards, and some are from other sources.

HSE have issued a list[30] of some 625 British Standards which are significant to health and safety. A selection of those which apply to the process industries are given in Appendix H.

Despite HSE's list of 625 British Standards, there are still no adequate British standards for some important features of process plant, e.g. for the design of pressure relief systems for oil and petrochemical plant. For these we usually rely on appropriate foreign standards, particularly American ones.

It is impossible to list here all relevant British, American and European standards. Lees[3] gives extensive lists of standards in use in 1979, although some of these (such as those issued by ICI in conjunction with RoSPA) are no longer published for general use.

Despite the proliferation of standards, it is essential for professional people to be familiar with and up to date in the standards in their own fields. There is a need for special courses to update professionals in the standards they have to use.

References

1. Lasok, D. and Bridge, J. W., *Introduction to the Law and Institutions of the European Communities*, 3rd edn, Butterworths, London (1982)
2. Van Atta, F. A., 'Federal standards on occupational safety and health', in *Safety and Accident Prevention in Chemical Operations*, edited by Fawcett, H. H. and Wood, W. S., 2nd edn, Wiley-Interscience, New York (1982)
3. Lees, F. P., *Loss Prevention in the Process Industries*, Butterworths, London (1980)
4. Goodman, M. J. (ed.), *Encyclopedia of Health and Safety at Work. Law and Practice*, Sweet and Maxwell, London (loose leaf with frequent updates)
5. HSC Consultative Document, *The Control of Substances Hazardous to Health: Draft Regulations and Draft Approved Codes of Practice*, HMSO, London (1984)
6. Lord Robens's Committee, *Health and Safety at Work*, HMSO, London (1972)
7. HSC Advisory Committee on Major Hazards, *First, Second and Third Reports*, HMSO, London (1976, 1979 and 1984)
8. Morgan, P., *Historical Review of Major Hazard Legislation*, paper presented at Major Hazards Summer School 1986, International Business Communications Ltd, Byfleet (1986)
9. Lees, F. R. (ed.), *Safety Cases*, Butterworths, London (1989)
10. 'The Flixborough disaster, Report of the Court of Inquiry', HMSO, London (1975)
11. BS 5500:1985, *Unfired fusion welded pressure vessels*
12. HSC Consultative Document: *Proposals for New Legislation for Pressurised Systems*, HMSO, London, (1977)
13. HSC Consultative Document: *Proposed Pressure Systems and Transportable Gas Containers Regulations and Approved Codes of Practice*, HSE, Bootle (1984)

14. *Health and Safety Commission Report 1984–1985*, HMSO, London
15. HSC Consultative Paper, *Proposed Second Amendment of CPLR 1984*, second edition of Approved List and revision of ACOP on Classification and Labelling, HSE (1987)
16. HSC, *Information approved for the classification, packaging and labelling of dangerous substances*, HMSO, London (1984)
17. S.I. 1975, No. 335, as amended by SI 1976, No. 1246, *Health and Safety Inquiries (Procedure) Regulations 1975*, HMSO, London
18. Ubbelohde, A. R., Letter to *Financial Times*, 4 December (1974)
19. Mecklenburgh, J. C., 'The investigation of major process disasters', *The Chemical Engineer*, August (1977)
20. Davidson, J., 'On public inquiries', *The Chemical Engineer*, June (1984)
21. S.I. 1983, No. 551, *The Civil Aviation (Investigation of Accidents) Regulations 1983*, HMSO, London
22. Wilkinson, G. C., 'UK aircraft accident investigation procedures', *Air Law*, Kluwer, Deventer, The Netherlands, IX, No. 1, (1984)
23. HSE, *The Explosion at Dow Chemical Factory, King's Lynn, 27 June 1976*, HMSO, London (1977)
24. *Report of the Royal Commission on Tribunals of Inquiry.* (chairman, Lord Justice Salmon), HMSO, London (1966)
25. *The official report of the Italian parliamentary commission of inquiry into the Seveso disaster, translated by HSE,* HSE, Bootle (about 1984)
26. King, R. W., 'The role of standards in the safe transfer of technology to developing countries', paper given at ILO Symposium on *Safety, Health and Working Conditions in the Transfer of Technology to Developing Countries*, ILO, Geneva (1981)
27. ISO, *The Aims and Principles of Standardisation*, Geneva (1972)
28. British Standards Institution, *Guide to the Preparation of a Company Standards Manual*, London (1979)
29. HSC Consultative Document, *Reference to Standards in Safety at Work*, HMSO, London (1982)
30. HSE, *Standards Significant to Health and Safety at Work*, HSE (1985)

Note: A list of HSC Approved Codes of Practice (ACOPs) is given in Appendix G.

Chapter 3

Meanings and misconceptions

People's reactions to particular words and symbols differ widely, and are not always what their users intended. This problem, which can be serious enough when only English is used, becomes more acute when two or more languages are involved. Mistranslations may be amusing [23.4.1] but they lead to misunderstandings which can cause accidents. (That's why the Tower of Babel was never finished!)

Here units and general nomenclature are discussed first [3.1], followed by health and safety terms [3.2]. Section 3.3 discusses misconceptions, particularly those which sometimes permeate an entire organisation and lead to disasters.

3.1 Units and nomenclature

3.1.1 Units

Most of us use different systems of units at various times. This, with the conversions between them, can be a source of misunderstandings and possible accidents. Writers seldom define their units completely, and assume that their readers will automatically know which system they are using. Although this is true for most readers, a minority usually get it wrong. Examples are Imperial and American gallons (6 American gallons = 5 Imperial gallons), long tons (2240 pounds) and short tons (2000 pounds) and ounces of different kinds (avoirdupois, US fluid and troy). Furthermore, engineers and others who are alert to discrepancies when using familiar units easily lose this critical faculty when working in unfamiliar ones.

While the UK has been in the process of changing from Imperial units to metric ones for the past 30 years, this has coincided with the introduction (starting in 1960) of a special version of the metric system, *Le Système International d'Unités* (SI units), in which several hitherto widely used metric units have no place[1]. In it calories have given away to joules (J) and kg/cm^2 to pascals (Pa). The wholesale adoption of the SI system (as indeed that of metrification generally) within industry and commerce has still a long way to go. Thus British Gas still measures gas in cubic feet and sells it by the therm, while the latest (1988) edition of the *Handbook of Chemistry and Physics* continues to give most thermal data in calories.

An advantage claimed for the SI is that it has only one unit for each kind of quantity[2]. This, however, has proved so restrictive that several non-SI units (e.g. minutes, hours, days, degrees Celsius, litres, tonnes, bars) (but not calories!) are used with it to make the system workable[1] in various fields. The prefixes and their symbols (representing various powers of ten) which must be added as factors to base SI units to convert them to larger or smaller units can also be sources of error, as BS 5555[1] admits:

4.4 Errors in calculations can be avoided more easily if all quantities are expressed in SI units, prefixes being replaced by powers of 10.

In this book a minimum of formulae and mathematics are used. Most numerical quantities which are taken from other sources are quoted and where necessary calculated in the units quoted. With the occasional exception of joules (for which the writer still prefers calories and kilocalories (cals and kcals) for sensible and latent heats) SI units and others allowed in conjunction with them[1] are used in examples given here for the first time.

Two pitfalls in quoting pressures should be noted:

1. While many British and Americans use psi to mean gauge pressures, only writing psia for absolute pressures, many Europeans quote pressures in bars or pascals (1 bar = 10^5 Pa) to mean absolute pressures, in conformity with the historical origin of these units. Thus when we quote gauge pressures in bars, we are well advised to to write 'bar g.'). We should also check what others mean when they quote bars or pascals without qualifying them.

   It is curious that the Flixborough Report [4.1] quoted plant pressures before the disaster as kg/cm^2 only, when some company personnel understood this to mean gauge pressure while others understood it to mean absolute.

2. A second pitfall is that some units have approximately the same value, yet the differences are large enough to cause hazards if they are taken as identical in critical cases, e.g. the setting of pressure relief valves. Thus one standard atmosphere = 1.01325 bar = 1.0333 kg/cm^2.

3.1.2 Nomenclature

This section is mainly intended for non-UK readers. The abbreviation HS is frequently used for 'health and safety' in a general sense and persons responsible on a full-time basis for specified aspects of safety are referred to as safety professionals, despite wide differences in their training, backgrounds and responsibilities. Most large companies and 'works' engaged in the process industries employ several safety professionals, with a safety manager in charge. The overall responsibility for HS should lie with a named director of the company[3] [19.1].

Some technical terms used here where British and American usage sometimes differs are given in Table 3.1.

Table 3.1 Some technical terms

Term	Description
Mixer	Used for complete mixing equipment of various kinds
Stirrer	Only used for rotary shaft stirrers
Spectacle plate	An 8-shaped plate permanently installed between flanged joints to isolate equipment (see Figure J.1)
Line blind	Circular plate with stub temporarily inserted between flanged joints to isolate equipment (see Figure J.1)
Mild steel	Used for carbon steel except when speaking of high, low or medium carbon steel
'P&V' valve (pressure and vent)	Weight-loaded valve used to maintain a low positive pressure in tanks and equipment by controlling flow of inert gas entering and venting (referred to in the USA as 'conservation vent')
Interlock	A device which prevents a valve, switch, etc. from being actuated while another valve, switch, etc. is in a certain mode
Trip	An automatic device, such as a combination of a pressure-sensitive switch and a three-way solenoid in the air supply to a pneumatic valve controlling a process stream, which opens or closes the valve when the temperature, pressure, level or some other variable reaches a pre-set value

Other preferred terms are 'supervisor' rather than foreman or chargehand, 'non-return valve' rather than check valve, 'bund' rather than dike, 'flametrap' rather than flame arrester, 'generator' rather than dynamo, electrical 'earth' rather than ground, 'branch' (on a vessel) rather than nozzle, 'bursting disc' rather than rupture disc.

3.2 Meanings of health and safety terms used

The definitions given here are mainly derived from the writer's earlier book[4] and from a short IChemE guide[5].

3.2.1 Hazard

Hazard is a very important concept. Put simply by Heinrich[6]:

A hazard is a condition with the potential of causing injury or damage.

The pursuit of safety is largely a matter of identifying hazards, eliminating them where possible or otherwise protecting against their consequences. The IChemE guide[5] defines **chemical hazards, major hazards** and **hazardous substances**.

Often two hazards need to be present simultaneously to cause a major accident. The law has not always recognised this, and much time has been wasted arguing which of two hazards was the 'proximate cause' of an accident[7].

3.2.2 Accident

The many meanings of this common word are the subject of two reviews[8,9]. The writer prefers the definition given by Heinrich[8] with the word 'sudden' added. This definition includes 'near misses':

An accident is a [sudden] unplanned event which has a probability of causing personal injury or property damage.

The inclusion of the word 'sudden' differentiates accidents from slower forms of deterioration such as corrosion or exposure to low levels of airborne asbestos. But sudden total failure following a long period of deterioration might be classed as an accident.

The word is also commonly used to mean an accidental injury[10,11], when it is often qualified by adjectives such as 'fatal', 'notifiable', 'reportable'.

3.2.3 Injury

The word injury as used here, without qualification, refers only to physical injury to a person caused by an industrial accident, which by the above definition is sudden. Other types of injury are qualified.

Sudden injuries caused by accidents need to be distinguished from disabilities caused by industrial diseases over long periods.

3.2.4 Industrial disease

Following Hunter[11], the term 'industrial disease' or 'occupational disease' is used here for any type of ill-health arising from conditions at work.

The border-line between an accident such as 'gassing ' caused by brief exposure to a high concentration of a toxic gas and a disease caused by chronic exposure to low concentrations of the same gas is often difficult to draw.

A number of industrial diseases are notifiable in Britain, with an obligation to report all cases of them to the appropriate health authority[13]. Many are furthermore 'prescribed'. This means that those suffering from them who have worked in particular occupations for a minimum period are entitled to claim compensation from the state.

3.2.5 Damage

Accidental damage applies to things rather than people and is usually measured in money terms. Damage may be to plant and equipment (part of the fixed capital) and/or to materials in process (part of the working capital).

Damage control consists of recording all accidents, including near-misses, analysing their causes (hazards) and working out and implementing steps to remove or reduce them[14]. These activities help to reduce injuries, since many accidents which damage things also injure people.

3.2.6 Loss

This term is used in insurance and refers to the costs of injuries and damage.

Consequential loss is additional to direct financial loss and refers to that caused by the interruption of production during the period of repair following an accident.

3.2.7 Disaster

This word generally refers to a major accident or natural event and/or a serious epidemic of an occupational or natural disease which results in the death of a number of people. The word is a subjective one, and refers more to the degree of shock and suffering rather than the size of the event. Western has produced a classification of natural and man-made disasters[15]. Turner's study[16] of man-made disasters is discussed in 3.3.

3.2.8 Probability

Probability is a mathematical term having a value between 0 and 1, where 0 represents complete impossibility and 1 represents absolute certainty[17].

Probability may also be quoted as a probable frequency (the number of specified events occurring in unit time). It is vital to the study of **risk** and **reliability** [14].

3.2.9 Risk

This word is often used with different meanings, occasionally being used to mean hazard. I prefer the definition given by the Chartered Insurance Institute[7]:

Risk is the mathematical probability of a specified undesired event occurring, in specified circumstances or within a specified period.

Other meanings of risk are sometimes used in insurance, e.g. the subject insured (a dancer's legs) or the eventuality insured against (having twins).

Within our chosen meaning of risk as a probability, further contrasting types of risk have to be considered.

- **Pure risk** or **speculative risk**[7];
- **Individual risk** or **societal risk**[17]

A **pure risk** is one where the only possibility is loss or breaking even, whereas a **speculative risk** is one where there is a possibility of gain or loss[7]. Many risks found in industry which at first sight appear as pure risks, are found when we look closer to be speculative ones. The risk of head injuries to workers not wearing protective hats in an area where overhead construction work is going on might appear to be a pure risk, until we realise that by taking the risk someone has saved the cost of the hats.

Estimates of the value of human life are based on the consideration of speculative risks which people are prepared to take with their own lives and those of others[3].

Total elimination of risk is never possible. The questions of what degree of risk is acceptable, how much money should be spent in reducing them and how different risks should be balanced have led to two professional disciplines, **risk analysis** and **reliability engineering**[17] [14].

Societal risk, **individual risk** and **risk analysis** are discussed in Chapter 14.

3.2.10 Safe

The best definition[18] the writer has found is:

> A thing is provisionally categorised as safe if its risks are deemed known and in the light of that knowledge judged to be acceptable.

Safety case has a special meaning in connection with the CIMAH Regulations [2.5.2].

3.2.11 Other terms

Several other specialised terms relating to explosions, fires, toxicity, release and dispersion of hazardous substances, hazard studies, reliability and risk criteria are defined in the IChemE guide[5]. These and others are explained as needed in the text.

3.3 Misconceptions and disasters

This section develops some of Dr Barry Turner's ideas[16] and shows how they can be applied to safety in the process industries. Examples are given in Chapters 4 and 5.

3.3.1 Learning from past accidents

Do we learn as much as we should from past accidents? Is the necessary information available? Or are we swamped with so much information that we cannot find the right bits? Technology can help. The proverbial needle in a haystack is easier to find if we have a metal detector. Similarly, a computer can search information for us quickly when it is stored in a well-indexed data bank. Yet the 'information explosion' can overburden our minds and hinder learning. As one writer put it, 'The more an organism learns, the more it has to learn to keep itself going'[19].

Nevertheless we expect to be able to learn from disasters, for as Robert Stephenson said in 1856[20]:

> Nothing was so instructive to the younger Members of the Profession, as records of accidents in large works . . . A faithful account of those accidents, and of the means by which the consequences were met, was really more valuable than a description of the most successful works.

The older Engineers derived their most useful store of experience from the observation of those casualties which had occurred to their own and other works, and it was most important that they should be faithfully recorded in the archives of the Institution.

In spite of this, there are often obstacles which prevent us gaining more than a superficial knowledge of the causes of major accidents in the process industries. Major accidents tend to destroy or obscure much of the physical evidence. But perhaps more important is the fear in some directors' minds of publicising information which might reveal some negligence on the part of their companies. Commercial confidentiality may be invoked as a (legal) reason for concealing it[21]. Thus sometimes only the more obvious hazards of major accidents are revealed.

More fruitful lessons in hazard identification can often be learnt from the investigation of lesser accidents, and even 'near-misses', where there is less concern about being blamed and a greater readiness to examine and admit possible causes. This is a good argument for the system known as 'damage control'[14] [3.2.5].

3.3.2 The ingredients of disasters

Turner[16] concludes from a study of several man-made disasters that:

Large scale disasters need time, resources and organisation if they are to occur. . . . Since these conditions are most unlikely to be met solely as a result of a concaternation of random events, we can almost suggest that simple accidents can be readily arranged, but that disasters require much more organising ability!

Usually a combination of two or more different types of hazard is necessary. This was well expressed by Sir Geoffrey De Havilland and Mr P. B. Walker[22]:

There is a modern trend which is steadily changing the overall character of investigations, though without affecting the basic principle. Accidents are becoming less and less attributable to a single cause, more to a number of contributory factors. This is due to the skill of the designers in anticipating trouble, but it means that when trouble does occur, it is inevitably complicated. . . . This trend can make nonsense of public inquiries and of lawsuits where allocation of responsibility is attempted.

Turner proposed the following general formula or equation for man-made disasters:

'Disaster equals energy plus misinformation.'

Energy here may be the kinetic energy of a train, the potential energy of water stored in a dam or the chemical energy of combustible materials and explosives in a process plant or in storage.

Examination of several major process plant disasters [4, 5] shows clear evidence of multiple causation. Two types of hazard are usually involved:

1. The hazardous material or agent itself, its nature and its quantity, which determine the type and maximum extent of the damage it can do, if it is

unleashed. I call this the **inherent hazard**. It corresponds to Turner's 'energy', although we need to add 'toxic substances' as an alternative to energy in his equation.

2. The mechanism by which the inherent hazard becomes unleashed. This generally consists of one or more inadequacies in the means of control and containment. These I term **initiating hazards**. Since the means of control and containment should be known, the initiating hazards generally correspond to misinformation in Turner's equation.

3.3.3 Misinformation

Misinformation may have several causes:

- The information is completely unknown at the time (e.g. the carcinogenic properties of vinyl chloride in the 1950s).
- The information is known to some people, but not known to or properly appreciated by those in need of it at the time. The dissemination of this information may be inhibited if the senior management have misconceptions about it. These are then termed **organisational misconceptions**.
- The information needed is about fast-moving events as they unfold. Here misinformation may arise through failure of people to respond quickly enough, or through inadequacies in the means of perception or the channels of communication. An example is an air collision or near-miss, the avoidance of which depends on rapid communication between the pilots of both aircraft and a ground flight controller.
- Information which is misunderstood, e.g. due to language problems, or because it is too complicated, or because it contains apparent contradictions or illogiicalities.

Even, however, when all these types of misinformation are included, there still appears to be something missing from the right-hand side of Turner's equation. In its crudest form, this consists of deliberate actions in defiance of information known at the time. In many cases this defiance of information involves taking a 'speculative risk' [3.2.9]. A person may be praised or advanced if he 'gets away with it' and perish or be pilloried if he fails. If the pressures, economic, social or emotional, are strong enough he may feel that he has no option but to take the risk of ignoring information on which even his life may depend. In other cases it may involve a death wish on the part of the individual, or it may be an act of wilful sabotage. A person may also act in defiance of information which he knows to be right, because of some uncontrollable desire (e.g. for sex, nicotine, alcohol or drugs).

In yet other cases, although the information lies in the person's unconscious mind, he cannot recall it to his consciousness when it is needed, and behaves as though he were misinformed. This happens to all of us when our minds are overloaded, or because of some 'hang-up' or mental blockage (often caused by an inner conflict).

Defiance of information may thus be due to economic, sociological or psychological causes, and a number of disasters had such origins. Behind many of them there is a human drama. As the American humorist Artemas

Ward put it, 'It ain't so much the things we don't know that gets us into trouble. It's the things we know that ain't so'. I would therefore modify Turner's equation to read:

'Disaster equals energy and/or toxic substances plus misinformation or rejection of information.'

The **initiating hazard** is then often the result of misinformation, or the rejection of correct information.

3.3.4 The hazard of over-centralised management control

The need for standardised procedures tends to lead to rather centralised control, with hierarchical management structures. In the case of a transnational enterprise, with subsidiary companies operating similar plants in different countries, the procedures may be worked out by the parent company, which passes them on as established wisdom to its foreign subsidiaries. Personnel from the parent company who visit the subsidiary company to 'advise' on this or that are often imbued with status symbols reminiscent of religious orders:

In holy orders, social distinction and the symbols of such distinction become so interwoven that the vestments themselves of the priest inspire awe and wonder. Who knows whether the priest is a holy man? And who knows whether the majestic scholar, wending his way in mortarboard, hood and gown to the chapel on Convocation Day, is a wise man? What we do see is a drama of hierarchy wherein rank is infused with a principle of hierarchical order[23].

Thus while there is a real need for strictly standardised procedures, the type of organisation to which this gives rise may make it more difficult to correct organisational misconceptions within itself.

Four of the five major accidents discussed in Chapters 4 and 5 occurred within the foreign operating subsidiaries of transnational companies. Organisational misconceptions appear less likely to occur on the home pitch of a transnational company, because of the greater knowledge, experience and self-reliance of its technical staff compared with those of its subsidiary companies, who are less expected to think for themselves.

3.3.5 The incubation of disasters

Turner has shown that man-made disasters do not come 'out of the blue' in an organised situation, be it a railway system, a mine or a chemical plant, but usually after a lapse of time which may range from a few hours to several years after one or more departures from normality have occurred. He reached this conclusion after studying a number of official reports of disasters investigated by UK authorities. This time lapse represents the **incubation period**, during which action could have been taken to prevent the disaster, had the danger been recognised by the organisation concerned.

This may be a very sensitive matter and it often needs a character of independent mind within the organisation to perceive and warn of the hazard – in short a whistle-blower[24]. To respond properly to such warnings and deal fairly with the whistle-blower is a severe test for any management [19].

Turner's concept of the incubation period is given in the following passage:

> We began by imagining a situation where a group or community possessed of sufficiently accurate information about their surroundings to enable them to construct precautionary measures which successfully warded off known dangers, to provide us with a 'notationally normal' starting point for the development of disasters. From this starting point, for each disaster or large-scale accident which emerges, we have suggested there is an 'incubation period' before the disaster which begins when the first of the ambiguous or unnoticed events which will eventually accumulate to provoke the disaster occurs, moving the community away from the notationally normal starting point. Large-scale disasters rarely develop instantaneously, and the incubation period provides time for the resources of energy, materials and manpower which are to produce the disaster to be covertly and inadvertently assembled.

The incubation period for a disaster on a safely designed, built and managed process plant may begin when some misguided action is taken which goes against the original norms and reduces the integrity of the plant, although by itself it could hardly cause a disaster.

The mistake might then be recognised and corrected, and the integrity of the plant restored, so that resulting accidents are averted. If not, some minor accident may follow without the mistake being recognised. When the damage has been repaired and the plant is started again, its integrity remains impaired, since the hazard resulting from the mistake is still present.

Now a second mistake may be made, which is also not recognised. A further hazard is introduced which reduces the integrity of the plant still further. One or two further accidents due to one or both of these hazards may follow. After each the damage is repaired and the plant restarted, although the mistakes and the hazards which resulted from them remain unrecognised and uncorrected. Finally, when a certain combination of conditions are reached, both hazards are activated, and disaster occurs. The incubation period has lasted from the time of the first mistake through a succession of minor accidents until the disaster.

It is possible that the first mistake was made in the initial design of the plant. Here one could date the incubation period either from the time the mistake was made or from the time the plant was first started up.

Having discussed Turner's general theory of the origin of a disaster and of the opportunities available (during its incubation period) for preventing it, some examples are examined in the next two chapters. They discuss the technical causes of some past disasters in the process industries, the incubation periods which preceded the disasters, and the organisational misconceptions which led to them. They also extend the lessons drawn from the reports of official inquiries. The mistakes or departures from norms which occurred were, in hindsight, quite elementary ones and not caused by ignorance of abstruse information.

References

1. BS 5555: 1981/ISO 1000–1981, *Specification for SI units and recommendations for the use of their multiples and of certain other units*
2. Weast, R. C. (editor in chief), *Handbook of Chemistry and Physics*, 1st student edition, CRC Press, Boca Raton, Florida, (1988)
3. Lord Robens Committee, *Health and Safety at Work*, HMSO, London (1972) and 'The Health and Safety at Work etc. Act 1974'
4. King, R. W. and Majid, J., *Industrial Hazard and Safety Handbook*, Butterworths, London (1980)
5. The Institution of Chemical Engineers, *Nomenclature for Hazard and Risk Assessment in the Process Industries*, Rugby (1985)
6. Heinrich, W. R., *Industrial Accident Prevention*, 4th edn, McGraw-Hill, New York (1968)
7. CII Tuition Service, *Elements of Insurance*, Chartered Insurance Institute, London (1974)
8. Crane, N. C., 'Just what is an accident?', *Industrial Safety*, **23**, No. 3, 10 (1977)
9. The National Institute of Industrial Psychology, *A Review of the Industrial Accident Research Literature*, HMSO, London (1972)
10. The Factories Act 1961, Section 80, Part V, HMSO, London
11. The International Labour Office, *International Recommendations on Labour Statistics*, Geneva (1976)
12. Hunter, D., *The Diseases of Occupations*, English Universities Press, London (1957)
13. S.I. 1980, No. 377, *The Social Security (Industrial Injuries) (Prescribed Diseases) Regulations 1980*, HMSO, London
14. Bird, F. E. and Germain, G. L., *Damage Control*, American Management Association, New York (1966)
15. Western, K. A., *The Epidemiology of Natural and Man-made Disasters: the present state of the art*, Academic Diploma in Tropical Public Health, London University (1972)
16. Turner, B. A., *Man-made Disasters*, Taylor and Francis, London (1978)
17. McCormick, N. J., *Reliability and Risk Analysis*, Academic Press, New York (1981)
18. The Council for Science and Society, *The Acceptability of Risks*, Barry Rose (Publishers) Ltd, London (1977)
19. Dewey, J., *The Night Country*, Garstone Press, New York (1974)
20. Stephenson, R. quoted in *Engineering Progress through Trouble*, edited by Whyte, R. R., The Institution of Mechanical Engineers, London (1975)
21. HSC Discussion Document, *Access to Health and Safety Information by Members of the Public*, HMSO, London (1985)
22. De Havilland, G. and Walker, P. B, 'The Comet failure' in *Engineering Progress through Trouble*, edited by Whyte, R. R., The Institution of Mechanical Engineers, London (1975)
23. Duncan, H. D., *Communication and Social Order*, Oxford University Press, Oxford (1968)
24. The Council for Science and Society, (report of a working party), *Superstar Technologies*, Barry Rose (Publishers) Ltd, London (1976)

Chapter 4

Flixborough and its lessons

The largest peacetime explosion in Britain (equivalent to about 30 t of TNT) occurred at 16.53 hours on Saturday 1 June 1974 above the chemical works of Nypro (UK) Ltd, near Scunthorpe. It followed the rupture of a 20-inch pipe and the escape of about 80 t of hot liquid cyclohexane at $\simeq$ 155°C and 8 bar g. from the cyclohexane oxidation plant. It wrecked the works and main office block (fortunately unoccupied), and damaged property within a radius of 5 km. Twenty-eight men working on the site were killed and hundreds of people outside it were injured. Nypro was jointly owned, 55% by Dutch State Mines (DSM) who provided its know-how and top management, and 45% by the National Coal Board (NCB). Its main product was caprolactam, the raw material for Nylon 6.

A public inquiry into the cause of the disaster under the chairmanship of Roger Parker QC sat for 70 days and published its report[1] (subsequently referred to as 'the Report') in April 1975. There was widespread criticism of the Report in the technical press and even more[2] after my explanation[3] [4.5] was published. An article in *The Engineer*[2] which reviewed the inquiry concluded thus:

> Perhaps a good way of starting along the new path might be to declare the conclusion as reached in the Flixborough inquiry 'non-proven' and either leave it at that or re-open the inquiry. . . . The probable outcome of a re-opened inquiry is that it would come to an altogether different conclusion.

This chapter discusses the disaster and its causes. As well as the cause stated in the Report, two important contributory ones which emerged later are also given here. One[3,4] published in September 1975 is discussed only briefly [4.5]. The other and no less important cause [4.6] is not (at the time of writing) widely known. Lessons to be learnt both from the disaster and the subsequent public inquiry are discussed here [4.7].

4.1 Process description and normal start-up[1,2]

The process used in this plant had been developed by DSM and was widely used. Stamicarbon, DSM's engineering subsidiary, provided the process design. Sim-Chem Ltd was the main contractor. The plant started in 1973, producing a mixture of cyclohexanol and cyclohexanone by the liquid-phase air oxidation of cyclohexane under pressure at 155–160°C. Pressures in this chapter are quoted in kg/cm^2 as in the Report rather than in SI units used elsewhere in this book. All control, indicating and recording instruments referred to here were located in the plant control room which was destroyed.

Figure 4.1 shows the plant as it was designed and built, and Figure 4.2 as it was before the disaster. During operation 250–300 m^3 per hour of cold cyclohexane were pumped into column C2521 where it scrubbed and cooled gas leaving C2544. Liquid from C2521 was pumped through a water separator S2522 to the top of column C2544 (containing 40 m^3 of 2-inch ceramic rings), where it contacted the same gas leaving the reactors, cooling it and becoming heated to about 155°C. The operating temperature in C2544 was kept high enough to prevent water vapour in the gas from condensing in it.

Liquid leaving C2544 at 155–160°C and 8.5–9 kg/cm^2 flowed in cascade through (originally) six stirred reactors R1 to R6, each holding 33 m^3 of liquid which flowed between them through 28-inch flexible stainless steel bellows connected by flanged joints to branches on opposite sides of each reactor. These took up differential thermal expansion between the reactors. In some plants large 'U' bends are used rather than bellows.

Compressed air entered the base of each reactor and bubbled through the liquid, a small percentage of which was converted via organic peroxides to cyclohexanol and cyclohexanone with smaller amounts of organic acids. The gas leaving the reactors was mainly nitrogen with some unconverted oxygen, oxides of carbon, cyclohexane and water vapour.

Liquid leaving R6 passed first through the after-reactor R2529, which provided further time for peroxides to decompose. It was then washed first with aqueous caustic soda to remove the acids and finally with water. The washed cyclohexane layer containing the product cyclohexanol and cyclohexanone in solution was separated in a distillation unit from which unconverted cyclohexane was returned to C2521.

Gas leaving C2521 passed through a chilled scrubber C2522, at the outlet of which its pressure was controlled by the valve PV which was actuated by a control instrument. The gas then passed to the flare stack. The pressure was normally controlled at 9.5 kg/cm^2 abs. during both start-up and operation, although by at least one account[5] it had been reduced by Nypro to 8.5 kg/cm^2 abs. after the failure of R5 in March 1974. The pressure of the gas leaving the reactors was, however, bound to rise by at least 0.5 kg/cm^2 above the control setting when air was introduced because of the pressure drop during its passage through the three columns and pipework. Neither of these points were mentioned in the Report, which also failed to state whether the plant pressures which it quoted in kg/cm^2 were in gauge or absolute units [3.1.1]. These pressures are subsequently quoted here as kg/cm^2 (?).

During start-up the cyclohexane was heated by the reboiler E2521, the steam to which was controlled by an instrument which controlled the temperature of the cyclohexane leaving E2521 by actuating the air-operated valve TV.

To restart the plant after a short shut-down, when the reactors were full

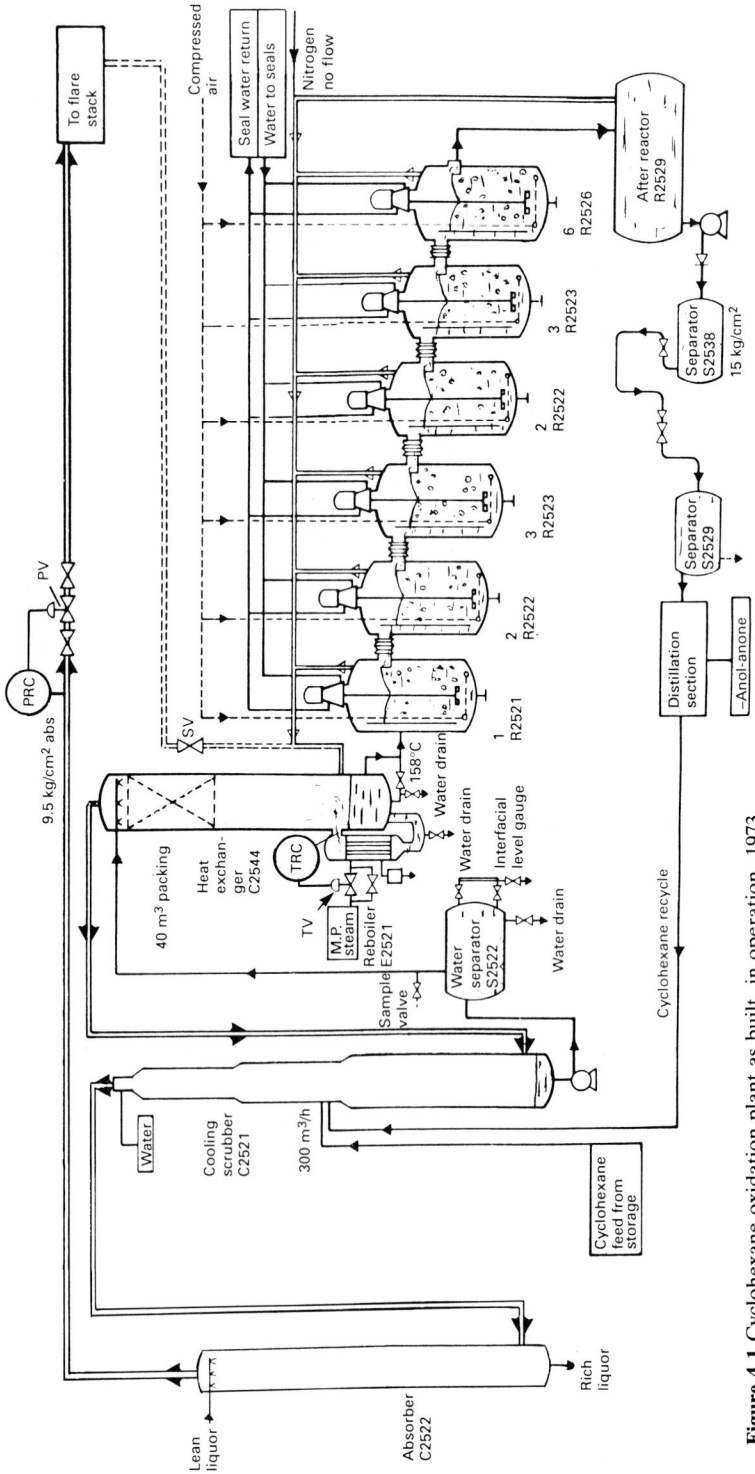

Figure 4.1 Cyclohexane oxidation plant as built, in operation, 1973

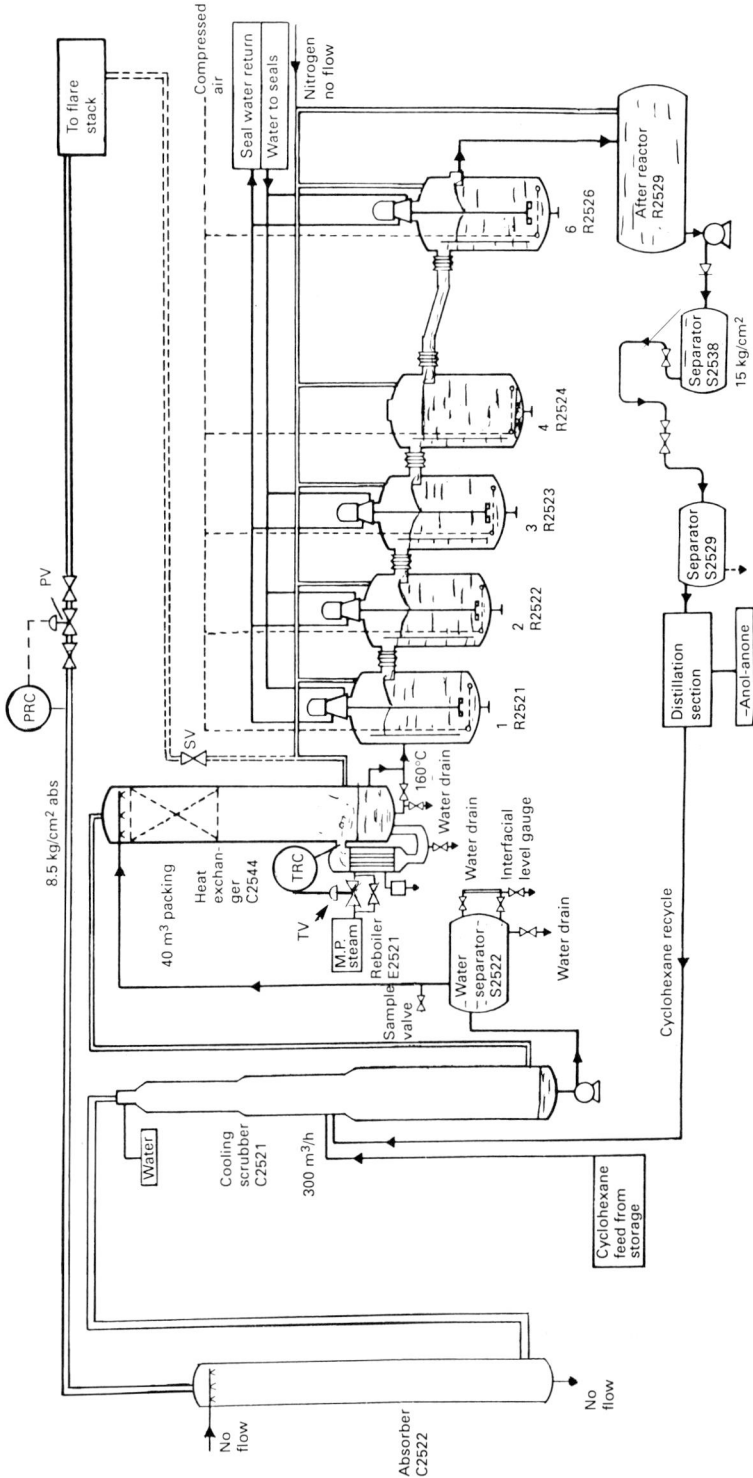

Figure 4.2 Cyclohexane oxidation plant as modified, in hot circulation, 1 June 1974

of cyclohexane, PV would be set (at $9.5\,kg/cm^2$ abs. before the failure of R5), and nitrogen would be admitted to the plant until a pressure of $4.5\,kg/cm^2$ abs. was reached. The cyclohexane feed pump and the reactor stirrers would then be started and the flow controller set to give about $125\,m^3$ of cyclohexane through the system. Steam would then be applied to E2521, and its temperature controller set at 160°C. Provided E2521 and C2544 (including its packing) were free of water at the start of circulation, the plant pressure and the temperature of R6 would gradually rise and then steady out over a period of four to five hours to about $9.25\,kg/cm^2$ abs. and 155°C. Air and catalyst were then normally introduced, although these could be delayed and hot-circulation continued if there were any hitches. The rest of the start-up procedure is not described since the explosion occurred during extended hot-circulation before air was introduced.

4.2 Conditions during start-up on 1 June 1974[1,2]

Two plant changes had taken place between the time of the initial start-up in 1973 and the disaster.

1. Towards the end of 1973 the drive of the stirrer in R4 failed and the stirrer was removed. Operation was continued with five stirred reactors and R4 unstirred.
2. On 27 March 1973, the plant was shut down hurriedly when a leak occurred in R5 and a large split was found in its side (Figure 4.3, based on Plate 7 of the Report). The reactor was emptied and removed and replaced by a hurriedly made 20-inch SS pipe, in the shape of a dog-leg, with two welded mitre joints in it. Its ends were flanged to join the flanged ends of the bellows left attached to R4 and R6 when R5 was removed.

This pipe, and the two bellows attached to it, are subsequently referred to as the 'by-pass'. This was inherently weak both because the compressive forces caused by the bellows produced a strong bending moment on both mitred joints and because each bellows was now only anchored at one end, thus allowing the whole assembly to squirm.

The plant was restarted at the beginning of April, and ran with the by-pass in position till the end of May. It was then shut down to repair a leak on the sight glass of S2522, from which the cyclohexane was first displaced by water into C2544 and the reactors. Most of the water left in C2544 was subsequently drained, but any water entering R1 could not be drained and would have stayed there until the plant was restarted. After repairing the leak the plant was pressurised to $4\,kg/cm^2$ (?) with nitrogen.

At 05.00 on Saturday 1 June, circulation of cyclohexane was restarted and steam was reapplied to E2521. The plant pressure rose abnormally rapidly. By 06.00 it had reached $8.5\,kg/cm^2$ (?) when the temperature in R1 had only reached 110°C and the other reactors were successively cooler. Another leak was then found. Circulation was stopped and the steam was shut off. The leak was cured after the shift change at 07.00. The pressure had then fallen to about $4.5\,kg/cm^2$ (?) without venting.

Figure 4.3 Damaged reactor R2525

Circulation was restarted at 09.00 and steam was reapplied to E2521, with the temperature controller set to maintain the outlet temperature of E2521 at 160°C.

By 11.30 to 12.00, the pressure had reached 9.1 to 9.2 kg/cm^2 (?) (0.3 to 0.4 above that required). The block valve on the off-gas line (which had been closed earlier to conserve nitrogen) was opened slightly until the pressure fell to $8.8\,kg/cm^2$ (?), and was then closed again. The plant warm-up continued without venting to the end of the shift at 15.00.

According to Hewitt[6], the supervisor of the morning shift, the pressure and temperature had steadied out by about 13.30 and the conditions by 15.00 at the shift change when he handed over to Simpson (who was killed on the following shift) were:

- Cyclohexane circulation rate: $125\,m^3/hour$;
- System pressure: $8\,kg/cm^2$ (?) with off-gas block valve shut;
- Reactor temperatures: about $155°C$ with steam by-pass valve shut.

At 16.52 the by-pass failed. Both bellows had ruptured, and the 20-inch pipe lay jack-knifed (at one of its mitred joints) on the ground below. Cyclohexane had escaped at high velocity through the two opposing 28 inch openings, mixing rapidly with the surrounding air to form a very large flammable cloud which ignited and exploded within 50 seconds of the failure.

4.3 Possible causes for the failure of the by-pass assembly

Early in the inquiry DSM's counsel Leggat produced a list of possible causes[7]:

1. Failure of the 20-inch pipe resulting from a hypothetical but relatively small pressure rise;
2. Failure of a different (8-inch) pipe (followed by an initial fire or explosion) before the failure of the 20-inch pipe;
3. Prior failure of some other part of the system;
4. Explosion in the air-line to the reactors.

Much of the inquiry was spent in examining and rejecting hypothesis (2), known as the 8-inch pipe theory. Causes (1), (3) and (4) received little attention in the court hearings although Leggat suggested several possible causes for a pressure rise (1):

(a) Entry of high-pressure nitrogen into the system due to some instrument malfunction;
(b) Entry of a slug of water into the system;
(c) Temperature rise in the system due to the use of excessive steam on the reboiler of C2544;
(d) A leaking tube in the reboiler of C2544 causing both actions (b) and (c);
(e) Explosion of peroxides formed in the process;
(f) Air in the reactor system which might cause a local explosion.

4.4 The court's views on the immediate cause of the disaster

The Report concluded that disaster was caused by the failure of the inadequate by-pass installed in March 1974. It also stated (without explanation):

§225 (vi) On 1 June the assembly was subjected to conditions of temperature and pressure more severe than any which had previously prevailed.
§88(a) The unusually fast rise in pressure during the early hours of 1 June remained unexplained.

and

§88(f) During the course of our investigations the possibility of a sudden rise in pressure during the final shift due to some internal incident was considered. We were able to exclude all of the possible internal incidents suggested.

Professor Newland had put forward a theory of 'dynamic squirm' to explain the failure of the by-pass[1]. Stated simply, the squirming of the bellows imparted a dynamic load that jack-knifed the mitred pipe-joint. This theory however still called for a small rise in pressure over the $8.8 \, kg/cm^2$ (?) that would have been present at 15.00 on 1 June even had this been gauge pressure. (A larger pressure rise would, of course, have been needed on his theory had this pressure been in absolute units.) Newland's theory, though accepted by the court and its technical advisers, Cremer and Warner, was not verified experimentally.

The Report admitted (§128) that Newland's theory was a rather unlikely cause which 'would readily be displaced if some greater probability to account for the rupture could be found'.

4.5 The pressure rises and their cause

My paper published in September 1975[3] showed that 100 litres or so of water (which from the recent history of the plant would have remained in the reactor system on 1 June) would have caused the fast rise in pressure at about 06.00 (which the Report noted but could not explain), and an even faster one at about the time of the disaster[3,4]. This stems from the fact (well known in physical chemistry) that a physical mixture of water and a hydrocarbon liquid such as cyclohexane or petrol boils at a considerably lower temperature than the boiling point of either liquid alone. Thus far from being unexpected, it would in the circumstances have required a minor miracle to have prevented these pressure rises from occurring.

Without repeating the calculations given in the paper, this showed that the water left in the packing of C2644 after the shut-down would have been vaporised with cyclohexane during hot-circulation, leading to the fast pressure rise observed at about 06.00. Later in the day when hot-circulation was continued this water, plus any present in the first three (stirred) reactors, would have been carried in suspension in cyclohexane into the fourth reactor from which the stirrer had been removed, and would have settled as a pool at the bottom. Continued circulation of hot cyclohexane at 155°C through the reactors would have heated the interface between the pool of water and the cyclohexane above it until it reached a temperature of about 145°C. At this point boiling would have started at the interface which was unstable above that temperature, followed by rapid mixing of water and cyclohexane, rapid boiling of the mixture and a fast, almost explosive rise in pressure of about $1 \, kg/cm^2$.

Four conditions (all present in the plant at the time) were necessary for this to happen:

- An unstirred reactor in which water can settle;
- A system pressure lower than the sums of the vapour pressures of water and cyclohexane at the mean temperature in the unstirred reactor;
- Enough water in the system to form a pool in the unstirred reactor;
- Continued circulation of hot cyclohexane prior to oxidation for long enough to allow the water–cyclohexane interface to reach the temperature at which it becomes unstable.

The hazard is discussed more generally in 6.4.6 under 'physical explosions', in a later paper[8] and in an oil company's safety booklet[9].

The main concern for anyone who may be faced with a similar hazard is to ensure that the first and second of these conditions cannot arise simultaneously. A quantitative analysis of the consequences when all four conditions are met requires knowledge of several factors which include:

- The minimum amount of water present;
- The latent and specific heats of water and cyclohexane (or other hydrocarbon);
- The mutual solubilities of water and the hydrocarbon present;
- The volumes of the system filled with gas and liquid.

The expected pressure rise, while sudden and violent, would probably not have ruptured a well-designed and constructed plant, but was clearly sufficient to destroy the faulty by-pass assembly. Had the reasons for these pressure rises been appreciated by the court at an early stage of the inquiry, this could have been concluded many weeks earlier.

4.5.1 Why the stirrer was removed from R4

The main purpose of the stirrers in R1 to R6 was to mix air intimately with the liquid, but they also prevented any water present in the reaction system at start-up from settling out during hot-circulation. The removal of the stirrer from R4 was only casually mentioned in evidence at the inquiry and the Report attached no significance to it. It had a curious history.

The state of emergency declared in Britain on 13 November 1973 following the miners' overtime ban included strict measures to save electricity[10]. It was then decided to try running the plant without using the stirrers, since the air was well dispersed as it entered the reactors. The plant ran satisfactorily in this fashion at a reduced throughput for several weeks. In January 1974, with the restoration of normal power supplies, the stirrers were restarted.

The drive of the stirrer of R4 was then found to be damaged. Having operated for several weeks with unstirred reactors, it was decided to remove the stirrer and continue operation without it. This appeared to work satisfactorily, and at the time of the disaster the removal of stirrers from all reactors was being considered. This would have been safe if the reactor pressures had been high enough to prevent vaporisation of the water/cyclohexane mixture during hot-circulation at start-up.

4.6 What caused the earlier failure of R5?

On the evening of 27 March 1974, while the plant was running, cyclohexane was found to be leaking from R4. The plant was shut down hurriedly, thus narrowly averting the bursting of the reactor. Had this happened, the Flixborough disaster might have occurred two months earlier. (The main task of any inquiry which followed it would then have been to examine why the reactor failed!)

R5 (like the other reactors) had a half-inch mild steel shell with a one eighth-inch stainless steel liner. It was externally lagged and clad with aluminium for weather protection. While the plant was being depressurised and shut down, water was sprayed over the leak to reduce fire risks, and the lagging was removed. A seven-foot long, roughly vertical crack was then found in the reactor shell (Figure 4.3, based on Plate 7 of the Report). It ran close to the annular reinforcing pad which was welded to the shell around one of the two 28-inch flanged openings. There was a shorter crack in the stainless steel lining.

The vessel was removed and the 20-inch dog-legged by-pass pipe was made and installed in its place. The reactors had been designed and tested individually (without the bellows) to a pressure of $16.5\,kg/cm^2\,g$. There was no apparent reason for one to fail at half this pressure.

Specimens of the shell through which the crack passed were cut out for metallurgical examination by DSM in Holland. These were cut from the ends of the crack which was still growing as the plant was shut down and not from near the middle, where it might have been expected to start. The metallurgical report, dated 3 May 1974, did not reach Nypro till several days after the disaster. It concluded that the crack was caused by (nitrate) intergranular stress-corrosion cracking of the carbon steel shell. It noted that the scale on the shell near the crack contained 0.75% of nitrate. This it attributed to cooling water which had been sprinkled over the reactor casing for some weeks in November 1973 for fire prevention when a valve above the reactor was leaking.

There was widespread scepticism about this report. Any nitrates found in the specimens were more likely to have come from water sprayed over the reactor while it was being depressurised than from water applied earlier before the crack started. Nitrate stress corrosion is only found in highly stressed metal near its yield point[11]. Although this applied to the ends of the spreading crack, there was no apparent reason for such high stresses where the crack started, nor was there any evidence of nitrate stress corrosion at this point. At the close of the inquiry, there was still no general agreement on the cause of the crack. Nevertheless, the court accepted DSM's report without seeking or hearing other evidence[1]:

§212 The cracked Reactor R2525 initiated the sequence of events which led to disaster. Examination of the crack by expert metallurgists showed that the crack had been caused by nitrate stress corrosion.

Important information about the design of the reactors had, however, been uncovered by HMFI's pressure vessel inspector, Glynne Evans, before the inquiry finished. He has informed me (by private communication):

That the vessel drawing specified a maximum thrust on the 28-inch nozzle stubs of 9 tonnes, whereas by calculation and experiment it was seen that there was a thrust of 38 tonnes at normal operating pressure. The crack ran longitudinally down the vessel, right at the toe of the weld at the doubling plate. No other so-called stress-corrosion cracks were found in this vessel that was sprayed by nitrate contaminated water, but cracks were discovered in identical places at other vessels that had never been sprayed with contaminated water.

The reactors were thus not designed for the thrust forces of the bellows, whose effect on a reactor was, as Evans put it, 'like driving a two-inch nail into an inflated truck tyre'.

4.7 Organisational misconceptions

Here, following Turner's ideas as developed in 3.3, I consider the organisational misconceptions (OMs) which made the disaster possible, and the incubation period leading up to it during which it might could have been averted.

The Report appreciated that the disaster did not come 'out of the blue' but considered it arose solely from a mistake made two months earlier.

§225(i) The scene was set for disaster at Flixborough when, at the end of March 1974, one of the reactors in the cyclohexane oxidation train on the plant was removed owing to the development of a leak, and the gap between the flanking reactors bridged by an inadequately supported by-pass assembly consisting of a 20-inch dog-leg pipe between two expansion bellows.

§225(iv) The integrity of a well designed and constructed plant was thereby destroyed and, although no-one was aware of it, disaster might have occurred at any time thereafter.

Both these statements are unfortunately belied by the information given in 4.6, which points to much earlier origins for the disaster. The first organisational misconception (OM1) in fact occurred during the design of the plant. Crudely stated it might read thus:

Unrestrained pressurised bellows pieces impose no significant thrust forces on the pipes or equipment to which they are connected.

The incubation period of the disaster dated from this point in the plant design.

A second organisational misconception (OM2) occurred in November 1973 when it was decided to operate the plant with the stirrers stopped. It might be stated simply thus:

There is no risk of a sudden eruption if you heat a hydrocarbon liquid and pass it through an unstirred vessel which contains some water.

(An exact statement of OM2 would be more complex, as the conditions for an 'eruption' [4.5] show.) The hazard arising from this misconception was 'frozen into the plant' in January 1974 when the stirrer of R4 was removed.

Organisational misconception (OM)

○ Incident or event

⊗ Incident linked to OM

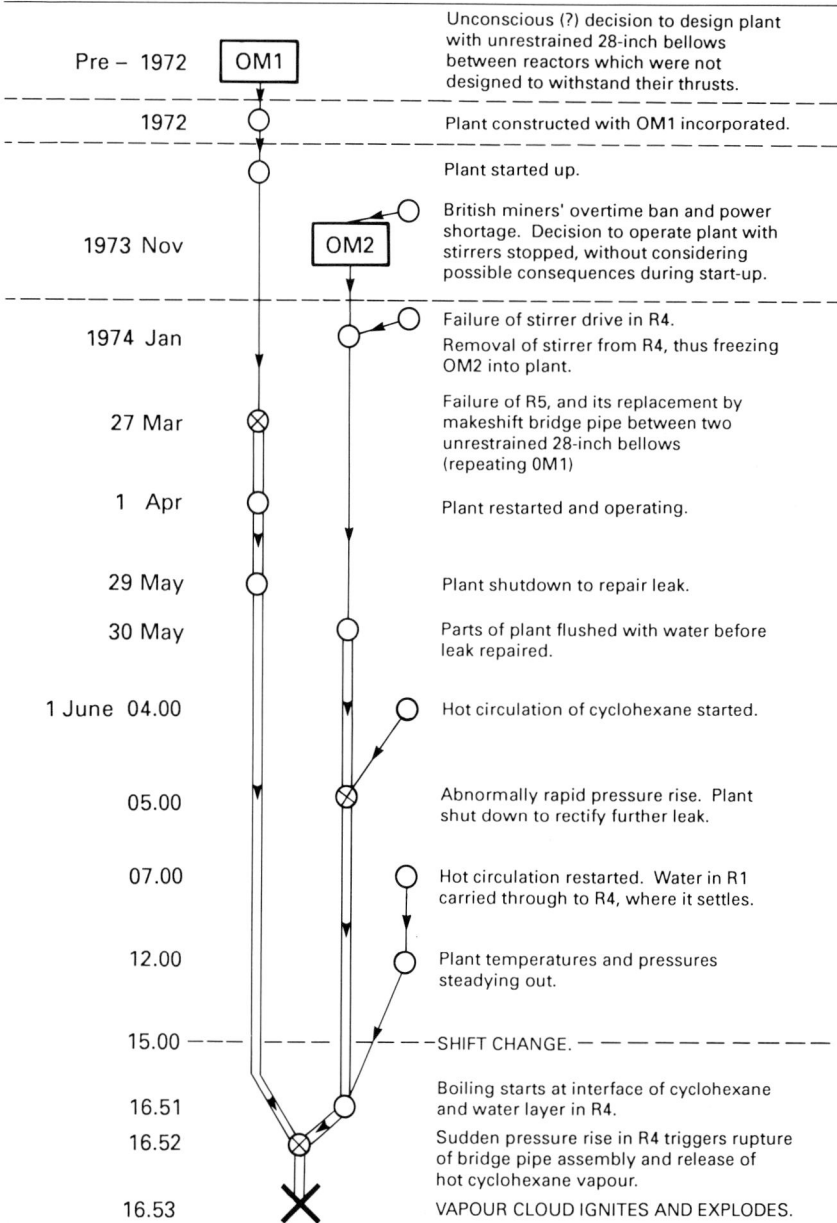

Pre – 1972	OM1	Unconscious (?) decision to design plant with unrestrained 28-inch bellows between reactors which were not designed to withstand their thrusts.
1972	○	Plant constructed with OM1 incorporated.
	○	Plant started up.
1973 Nov	OM2	British miners' overtime ban and power shortage. Decision to operate plant with stirrers stopped, without considering possible consequences during start-up.
1974 Jan		Failure of stirrer drive in R4. Removal of stirrer from R4, thus freezing OM2 into plant.
27 Mar	⊗	Failure of R5, and its replacement by makeshift bridge pipe between two unrestrained 28-inch bellows (repeating OM1)
1 Apr	○	Plant restarted and operating.
29 May	○	Plant shutdown to repair leak.
30 May	○	Parts of plant flushed with water before leak repaired.
1 June 04.00	○	Hot circulation of cyclohexane started.
05.00	⊗	Abnormally rapid pressure rise. Plant shut down to rectify further leak.
07.00	○	Hot circulation restarted. Water in R1 carried through to R4, where it settles.
12.00	○	Plant temperatures and pressures steadying out.
15.00		SHIFT CHANGE.
16.51	○	Boiling starts at interface of cyclohexane and water layer in R4.
16.52	⊗	Sudden pressure rise in R4 triggers rupture of bridge pipe assembly and release of hot cyclohexane vapour.
16.53	✕	VAPOUR CLOUD IGNITES AND EXPLODES.

Figure 4.4 Chronological chart of organisational misconceptions and incidents leading to disaster (time scale is non-uniform)

The failure of R5 in March 1974 followed from OM1. This was repeated in the 'design' (or lack of design) of the by-pass assembly. The disaster was caused by OM1, probably acting in combination with OM2.

A chronological chart of the incubation period showing the OMs and significant accidents within this period is shown in Figure 4.4. The disaster could have been averted at any time during this period had the OMs been appreciated and the appropriate (but expensive!) action taken to eliminate the faults caused by them from the plant.

OM1 seems to have originated in the design organisation. It may be that the reactor specification was made with the intention of connecting the reactors with large pipe U bends, before a decision was taken to use bellows instead in order to give a more compact layout. Had the cost of constructing reactors which would adequately withstand the thrust from the bellows been considered, these might never have been used. The error seems to have passed unnoticed and uncorrected by the engineers of several companies involved in the plant design and construction. Following this high-level mistake in the design organisation, R5 failed in service in April 1974. Another mistake – the inadequately supported by-pass, also resulting from OM1 – then occurred at a lower organisational level (i.e. in Nypro's maintenance department). This happened under stressful circumstances and led to the disaster. This illustrates the point made in 3.3.4 that once an OM exists at a high level in the organisation (e.g. in the parent company), it is prone to be perpetuated at a lower level (e.g. the subsidiary company), where it tends to be regarded uncritically as 'established wisdom' and is therefore more difficult to eradicate.

The question also needs to be asked why the court failed to recognise and point out this error, which makes nonsense of paragraphs 212 and 225(v) of its Report.

4.8 Lessons to be learnt

The chain of causation of the disaster just revealed suggests several general lessons for technical staff, company management as well as for tribunals of major accident inquiries. Some of the lessons are prompted by *Superstar Technologies*[12], which examined the problems of managing and monitoring industrial processes with high hazard potential.

The lessons which follow are put as questions which those addressed should ask themselves and answer honestly before evaluating their answers.

4.8.1 Questions for technical staff

- Has a mistake of which I am aware already been made which could prove to be the seed of a future accident or disaster?
- If so, how can I bring it to the attention of management so that it is properly investigated and corrected?
- Having brought the mistake to management's attention, how confident am I that it will be dealt with and rectified?

- Before doing something novel (such as dispensing with a stirrer) with a view to saving costs or improving the efficiency of a process, have I considered the safety consequences of this action, or could it cause an accident in a way which is not immediately obvious?

4.8.2 Questions for managements

- Are there adequately trained and experienced staff in all technical roles relevant to our company's activities? Are our competent technical staff overstretched because of rapid expansion or other reasons?
- Do we take properly into account the inherent hazardousness of our processes when deciding on the number and quality of technical staff needed to operate and monitor them?
- Quite honestly, do we tend to promote 'Yes-men' who can be relied on to agree with us at the expense of better-trained and more experienced operators who are sometimes more critical of our edicts and decisions?
- How do we treat our staff who question, particularly on safety grounds, the wisdom of decisions which we have already made or approved?
- Have we the courage to unscramble such decisions when it is subsequently found that they could lead to a major accident, particularly when the cost of unscrambling them is high?
- Do we have adequate staff and procedures for comprehensive monitoring of hidden hazards resulting from design, modifications and maintenance?
- Are our operating companies, particularly subsidiaries, starved of high-quality technical staff?
- Is there open exchange of technical information and consultation on safety aspects between our process licensors and designers, plant contractors and equipment manufacturers?
- Do we have updated and properly used plant manuals which include checklists of features (such as stirrers) which are essential for safety as well as functional reasons, so that if the latter are found to be dispensable, the safety reasons are not overlooked?

4.8.3 Questions for accident inquiry tribunals

Problems in the organisation of investigations of major accidents were discussed in 2.6. The following questions were framed with legally dominated inquiries of the Flixborough type primarily in mind:

- Is the sole purpose of our investigations to discover the causes of the accident, or are we also attempting to determine who was responsible for it?
- Are there 'interested parties' who feel they stand to lose if certain causes for the accident are established by our inquiries?
- What is the prime purpose of those representing 'interested parties' during our investigations, to protect them from being blamed or to uncover the causes of the accident? Are these purposes always compatible?
- Do we have adequate and appropriate expertise for independent investigation of all possible causes of the accident?

- Do we do give proper and balanced attention in our investigations to all these possible causes or do we accept statements from 'interested parties' that some causes can be ruled out without further investigation on our part?
- How do we handle improbable theories put forward by counsel and expert witnesses on behalf of 'interested parties' which would, if proven, 'let them off the hook'?
- Do we become so preoccupied with the pursuit and refutation of such theories that we fail to do justice to other and more probable ones?
- Are there consultants or other experts in the employ of 'interested parties' who have special knowledge and information which might help us if they were free to give it, but who are prevented from doing so either by their employers or their counsel or by the inquiry procedures followed?
- Are we subject to political or other pressures to avoid blaming certain interested parties for the accident?
- Are technical advisers employed by us for the purpose of the inquiry anxious not to offend certain interested parties by pursuing lines of investigation which could cause them trouble?
- Are we accountable for the conclusions we reach and publish, and if so, to whom?
- What mechanism, if any, is there for reopening the inquiry or revising our conclusions if it is subsequently found that they are incomplete or incorrect?

References

1. 'The Flixborough disaster, Report of the Court of Inquiry', HMSO, London (1975)
2. 'The Engineer Report – The lessons to be learned from the Flixborough enquiry', The Engineer, Morgan Grampian, London, 11 December (1975)
3. King, R. W., 'Flixborough – The role of water re-examined', Process Engineering, Morgan Grampian, London, pp. 69 to 73, September (1975)
4. King, R. W., 'A mechanism for a transient pressure rise', paper at Nottingham symposium, The Technical Lessons of Flixborough, The Institution of Chemical Engineers, Rugby (1975)
5. Cremer and Warner Report No. 2 to the Flixborough court of inquiry, 'General' (1974)
6. Hewitt, D. W., during examination by Jupp, K. G. (QC), Daily transcripts, Flixborough Court of Inquiry, day 17, p. 47, by courtesy of W. B. Gurney & Sons, Official Shorthand Writers (1974)
7. Leggat, A. P. (QC), Daily transcripts, Flixborough Court of Inquiry, day 8, pp. 74 and 75, ibid.
8. King, R. W., 'Latent superheat – a hazard of two-phase liquid systems', paper in I. Chem. E. Symposium Series No. 49, The Institution of Chemical Engineers, Rugby (1977)
9. AMOCO, Booklet No.1, Hazard of water, 5th edn, AMOCO, Chicago (1971)
10. Leading article, 'Electricity chiefs call for further power cuts', The Times, 7 December (1973)
11. HSE Technical Data Note 53/2, 'Nitrate stress corrosion of mild steel', HMSO, London (1976)
12. The Council for Science and Society, (report of a working party), Superstar Technologies, Barry Rose (Publishers) Ltd, London (1976)

Chapter 5

Four other major accidents

This chapter examines four other major accidents in recent years in the process industries:

1. The explosion in Shell's Pernis oil refinery in Holland on 20 January 1968[1];
2. The explosion at Dow Chemical Company's factory at King's Lynn, 27 June 1976[2];
3. The unintended formation and release of a few kg of the super-toxin 2,3,7,8-tetrachlorodibenzoparadioxin (TCDD) at Icmesa Chemical Company's works at Seveso on 10 July 1976, causing widespread illness and lasting environmental damage[3];
4. The release of some 30–40 t of toxic methyl isocyanate vapour at Union Carbide India Ltd's pesticide factory at Bhopal during the night of 2/3 December 1984, causing over 2000 deaths and affecting over 50 000 people[4].

A common theme runs through all four accidents. All resulted from a combination of technically simple hazards which had been created by conscious decisions of technically trained staff, usually those responsible for production operations. In few cases would they have lacked the knowledge to appreciate the consequences of their decisions had they thought them through. One must then conclude that generally they failed to project their decisions through to their likely consequences, and face up to them. Why was this so?

- Were the persons too busy and overstretched in their jobs?
- Did they lack the ability to think their decisions through?
- Were they inhibited by mental blockages caused by preoccupation with other matters, e.g. production targets and the odium of failing to meet them?

Such failures were collectively termed 'organisational misconceptions' in 3.3. Between the time of their occurrence and the disaster, several hours, days, weeks or months, referred to in 3.3.5 as the 'incubation period'[5], had elapsed. During this time the consequences of the decision might have been appreciated and averted had someone had the guts and authority to do so.

The general lesson is that every decision should be projected forward. Every decision maker should be conscious of the fact that the present is possibly some stage in the incubation period of a disaster, and that the fatal decision(s) might already have been taken. A special lesson for managements is that they must ensure that their production and engineering staff have the ability, knowledge, time and support to think their decisions through, and that they are not subject to (usually management created) pressures which inhibit them from doing so, or from acting on their conclusions. In plants where the potential for disasters is high, there is a need for competent staff with no production responsibilities to act as 'technical long-stops'. Their task should be to monitor the consequences of technical decisions taken by line managers and veto them where necessary, even when this leads to lost production.

5.1 The explosion at Shell's Pernis refinery in 1968[1]

A large aerial explosion occurred above Shell's Pernis oil refinery near Rotterdam at 04.23 on Saturday, 20 January 1968. Human fatalities and injuries were relatively light (two dead, nine hospital cases and 76 persons slightly injured), but the damage to the world's largest refinery was severe and estimated at US $28 million. Much neighbouring property including buildings in nearby Rotterdam was also damaged.

5.1.1 The inquiry and its findings

An inquiry was held by a group consisting of four senior police officers, seven officers of the factory inspectorate and the head of Shell's safety inspection division. Its sole task was to establish the technical cause of the explosion. Its report[1] was published within two months of the disaster. Its summary stated:

> The inquiry showed that the cause is to be found in a storage tank for oil slops which had steam heating near the bottom. Owing to the cold weather during the previous two weeks, the steam heating was in operation. As a result of difficulties in breaking down an emulsion during the preceding days in a desalting installation for crude oil there had been a large amount of waste oil, partly in the form of a water-in-oil emulsion with a high water content, which was among the liquids present in the tank concerned. There was a vigorous boiling effect in the tank, as a result of which a large quantity of hydrocarbons were expelled into the air within a short period of time. The presence of a very light, variable wind caused a large cloud to form, consisting of an explosive mixture of air and mist from these hydrocarbons. This explosive cloud was ignited by a source which cannot be established with certainty and exploded violently.

An Annex to the report concluded that 50 to 100 t of hydrocarbons must have been involved in the explosion.

The term 'slops' in an oil refinery refers to oil, often emulsified, which has been recovered, mainly in oil–water separators, from refinery effluents. The slops tank, no. 402, was 9 m high with a diameter of 15.2 m and contained four steam coils in its base with a heating surface area of 36 m². Steam had been connected to the coils on 9 January.

The slops at that time contained unusually large quantities of a crude oil–water emulsion from a crude-oil desalting unit. Most crude oils contain inorganic chlorides which have to be removed before distillation to prevent severe corrosion in the distillation units [11.7.4]. This was done in a continuous unit by mixing clean water into the crude-oil to dissolve the salts present, thus forming a water-in-oil emulsion. The emulsion was separated into oil and water layers by high voltage, the oil going to the crude distillation units and the water layer to the effluent oil separators.

Because of the stability of the emulsion formed with the Bachaquero crude oil then in use, the desalting unit could not separate it properly, and discharged a considerable amount of water-in-oil emulsion with a high water content to three settling tanks. As these in turn failed to break the emulsion, it passed into tank 402. Here the desperate remedy of heating the tank contents was being tried to an attempt to break the emulsion. The hazard of doing this has long been recognised by oil companies[7].

The only alternatives to this were to process untreated crude oil in the distillation units, with risk of severe corrosion and other hazards, or to shut down one or more crude distillation units and reduce refinery throughput.

Tank 402 had been full for two days and contained about 1500 m³ of slops. The steam to the coils was controlled by a valve which was found to be partly open after the explosion. The slops had a specific gravity of 0.82, and were estimated to have an initial boiling point of 60°C and to contain about 30% of light oil fractions.

The report concluded:

The explosion was in all probability the result of the ignition of a large cloud consisting of air and a large quantity of hydrocarbon mist. The formation of this cloud was due to the weather conditions, namely a very light variable wind, after there had been an eruption of hydrocarbons from tank 402 in a very short space of time.

1. The contents of tank 402 consisted mainly of a mixture of hydrocarbons with a relatively low initial boiling point. In addition there was a relatively large quantity of a stable water-in-oil emulsion in the tank.

2. As the tank was heated by means of a steam coil, the emulsion at the bottom of the tank was able to reach a relatively high temperature, while the oil layer above it remained considerably cooler.

3. At a temperature of approximately 100°C the emulsion layer split into a water layer and a relatively heavy oil which therefore also had a temperature of approximately 100°C.

4. The moment the temperature of the light oil layer reached the initial boiling point, vapour developed as a result of vapour formation at the unstable interface between the hot and the cooler oil layer, and this caused rapid mixing of the hot oil layer with the rest of the contents of the tank. The vapour thus formed led to the overflowing of the tank and the subsequent eruption of hydrocarbon mist.

Little can be said with certainty about the way the explosive cloud ignited. The front of the explosive mist-cloud had moved approximately 100 m from its source when the explosion occurred.

5.1.2 Comment

It is clear that a considerable temperature difference had developed between the lighter and more volatile oil in the top of the tank and the heavier water-containing layer in the bottom while the tank contents were being heated. The contents of the tank as a whole were then in a thermodynamically unstable condition. The sensible heat of the hot layer in the lower part of the tank was sufficient to vaporise a considerable proportion of lighter and cooler hydrocarbon above it when the contents mixed, causing a physical explosion [6.4.3] inside the tank. When its contents then overflowed, the hydrostatic pressure inside it would have been reduced, thereby accelerating the boiling process. The sequence of events is illustrated diagrammatically in Figures 5.1(a) to (d).

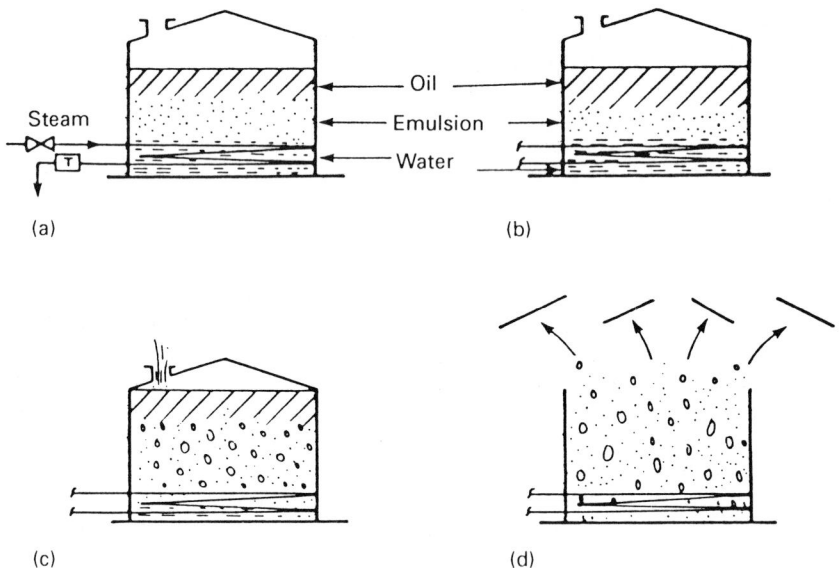

Figure 5.1 Sequence of events in Pernis disaster, 1968: (a) Tank filled with slops and steam turned on to heating coil; (b) emulsion partly separates as tank contents warm up; (c) water layer reaches its boiling point with cooker oil layer above it; (d) water–oil mixture boils explosively and erupts

The phenomenon, discussed elsewhere in general terms[8], was similar to the 'missing link' [4.5] in the explanation of the Flixborough disaster.

The accident followed from the decision to heat the lower part of an unstirred tank containing an oil with volatile components to a temperature above its initial boiling point. The hazard was accentuated by the presence of a considerable amount of partly emulsified water in the tank.

The die was set for disaster when the steam supply was connected to tank 402 on 9 January, thus giving an incubation period or lead time of 10 days. Had the mistake been recognised then and the steam supply to the tank disconnected, the disaster would have been averted. The resulting loss of refinery production would have cost Shell far less than the disaster.

5.2 The explosion at Dow Chemical Company's factory at King's Lynn, 27 June 1976

This explosion, in which one man was killed and considerable damage caused, was investigated by HMFI. Its report[2], on which this section is based, uses both metric and Imperial units [3.3.1] which are retained here.

5.2.1 General background

Dow Chemical Company, on whose works the explosion occurred, is a pioneer in health and safety practices in the chemical industry. Its well-known guide[9] on the identification and quantification of fire and explosion hazards is discussed in 12.1. Success in the design and operation of process plant has something in common with that of a ship or aeroplane. Safety factors and features must be adequate and balanced and based on a sound analysis of the hazards. The task is very demanding, and even the most skilful can make mistakes.

The material which exploded was a dinitrobenzene derivative, 3,5 dinitro-2-toluamide, $C_8H_7N_3O_5$, known in the trade as 'Zoalene' and used as a poultry food additive. It had been manufactured at King's Lynn in the 1960s, when a vacuum tray dryer was used to dry the product. Its production in the UK ceased in 1970, since when it was imported from France and Spain.

Early in 1975 the stocks at King's Lynn were found to be below the minimum specified purity of 98%. It was hence decided to re-dry about 85 t of the imported material at King's Lynn. Instead of the tray dryer used in the 1960s, an available double-coned glass-lined steel dryer was used. This had a jacket for steam or cooling water and was rotated on a hollow horizontal shaft through which steam and vacuum connections passed. It normally operated under a vacuum of about 27 inches of mercury (0.10 bar abs.), with steam at 45 psig (3.1 bar g.) in the jacket and rotated slowly for two hours or more until the material was dry. Steam would then be turned off, the vacuum released, and water passed through the jacket for about 40 minutes to cool the material before discharging it into drums.

5.2.2 Events leading up to the explosion

After drying several batches in June and July 1975, the plant was shut down. It was restarted in June 1976. A problem arose because of a cake-like residue left adhering to the inside of the dryer after discharging most of the dried material. After about every third batch this residue was removed by half-filling the dryer with water, and then rotating and heating it until the sides were clean. More Zoalene was then added to make up the total solid charge to between 1200 and 1400 kg. The dryer was closed and evacuated, rotation started, and steam turned onto the jacket.

This operation differed from that used for most batches in that far more water was present initially. This had to be evaporated before the solid could be dried. After drying nine batches normally, the dryer was charged with a slurry made from the residues from the previous batches. The day shift on Friday 25 June then charged 350 kg of material from kegs, making a total charge of about 1450 kg of solids.

Drying was started at 15.00 hours on the same day, and continued until 14.00 hours on Saturday 26 June. It was then shut down, the steam supply shut off and the vacuum broken with air. The lid of the charging opening was removed and the charge inspected. The lid was then replaced and clamped loosely to keep the charge dry. It was left lying in the hot stationary dryer until it could be emptied by the day shift on Monday morning. The dryer was not cooled by passing water through the jacket at the end of drying on Saturday as would have been done if the contents were to be discharged that day.

5.2.3 The explosion

Nothing abnormal was noticed until 17.07 on Sunday 27 June, when the shift foreman heard a hissing noise and saw white smoke emerging from the dryer room. He sounded the fire alarm before the explosion occurred at 17.10. A member of the works fire-fighting team who was on his way to the team's assembly point was killed by the blast.

The explosion was equivalent to that of 100–130 kg of TNT. It was later deduced that before the explosion, the Zoalene in the dryer had been decomposing at an accelerating rate accompanied by a fast temperature rise, which led to local temperatures in excess of 800°C. The final explosion was a detonation [9.1]. Calculations indicated that the failure of the dryer took place in two stages. First, the lid, which was held only by a hinge and a single clamp, would have failed at about 7 bar g. The dryer itself would have failed at about 50 bar g. (static loading) although the shock wave at detonation would have generated much higher pressures.

5.2.4 The cause of the explosion

The basic reason for the explosion lay in the chemical instability of the material, a substituted dinitrotoluene. Nearly all monocyclic dinitro aromatic compounds are liable to explode, with the release of considerable amounts of energy and gaseous combustion products, although they are less powerful explosives than TNT. One, dinitrophenol, as well as its salts, falls under the legal definition of an explosive [9.9.1][10].

Like most aromatic dinitro compounds, Zoalene would probably be difficult to detonate at room temperature, but would be virtually bound to explode if heated in a closed container to a certain threshold temperature. Zoalene had been subjected many times previously to temperatures up to 130°C during drying, without exploding.

Most energy-rich compounds decompose even at ambient temperature at a very slow rate with the evolution of heat. If the quantity is small and the initial temperature is low, this leads to a steady equilibrium state where the small amount of heat evolved is easily lost to the surroundings. But once a critical combination of quantity, initial temperature and surrounding temperature is exceeded, heat can no longer be lost as fast as it is generated. A runaway reaction with rising temperature then begins slowly and gradually accelerates, often leading to an explosion. This phenomenon which was analysed by Frank-Kamenetskii[11] is discussed in 8.7.

Tests, discussed in 9.6, have been devised and standardised for evaluating the stability of materials such as Zoalene. The methods in use in the UK at the time, while confirming that Zoalene is an unstable heat-sensitive material, would not have indicated that a mass of 1.4 t, when heated to about 130°C, was eventually liable to explode. Dow Chemical Company had recently developed a more sensitive technique known as accelerating rate calorimetry in the USA. Although this would have apparently pinpointed the hazards of drying Zoalene, this had not yet been done as the technique was new and a number of unstable compounds manufactured by Dow were waiting their turn to be tested.

Another important point is that the stability of potentially explosive compounds is critically sensitive to certain impurities, particularly acids. Many explosions occurred in the early manufacture of nitrocellulose which were eventually traced to the presence of acidic impurities. One would expect that Zoalene, an amide, would be partly hydrolysed when heated with water to the ammonium salt of 2,3-dinitro-2-toluic acid. This would lose ammonia under the influence of heat and vacuum, leaving free acid. The conditions during the drying of the batch which exploded thus seemed to be conducive to the formation of free acid.

It is hard to avoid the impression that a long accident-free record in drying Zoalene had dulled the appreciation of Dow's management to the fact that they were dealing with an unstable explosive material. Economic motives may have played a role. The profit from drying about £1 million worth of off-specification material which might otherwise have had to be scrapped was probably insufficient to justify the installation of safe drying equipment.

While more powerful and unstable explosives are handled and dried in the explosives industry, the dryers are designed and located on the principle of minimising injury and damage should one explode. A strict operating routine is also laid down from which deviations are not allowed.

The history of the drying operation up to the time of the accident reveals a number of changes which increased the risk of an explosion:

1. When production of Zoalene in the UK ceased but drying of imported material was started, Dow's management at King's Lynn no longer had control over its production or the quality of the product. The fact that imported material was below specification and required treatment to put it right might have been a warning of trouble ahead. Water may not have been the only impurity.
2. The original vacuum tray dryers, although more primitive, were safer than the double-cone dryer, particularly when this was stationary. Thus the material in the tray dryers would have been present in shallow layers, which would have reduced the tendency to form a critical mass, whereas when it was lying stationary in the cone dryer it was present in one single heap. Even if there had been a runaway reaction in the tray dryer, it had panels designed to blow out in the event of an explosion. These would have probably have released the contents of the dryer before any substantial risc in pressure could develop. At worst, there

might have been a deflagration. The double-cone dryer had more of the characteristics of a bomb.

3. Although batches were usually cooled in the cone dryer before being discharged, this appears to have been regarded as an operating convenience rather than a safety requirement. Thus the batch which exploded was simply left in the hot dryer on Saturday afternoon in the expectation that it would be cool enough to empty by Monday morning.

4. The recovery of residues stuck to the inside of the dryer by re-slurrying with water and evaporation of this water in the dryer were both new features which arose from the use of the double-cone dryer. When this had been used for drying Zoalene the previous year, the slurry had first been filtered, and only the solid cake was put in the dryer.

While it is hard to draw a line here between the risks which were acceptable and those which were not, had all the decisions taken been carefully considered at the time the probability of an explosion would have been apparent.

5.2.5 The conclusions of HSE's report

The report concludes that the following matters should have received closer attention:

1. The incorporation, in the case of certain batches, of re-washed material likely to contain a higher proportion of impurities;
2. The length of time during which batches were retained in the vessel after processing;
3. The procedure for cooling the material;
4. The instrumentation in respect of temperature and moisture content.

It recommended that:

Manufacturers and users of substances that are liable to exothermic reaction (and in particular nitro-aromatic compounds) in any condition of process or of use, should re-appraise the explosive potential of those substances in near adiabatic conditions. Attention is drawn to the value of ARC [Accelerating Rate Calorimetry] for the purpose of such investigations. It is also important that those who are concerned with the processing of such materials should ensure that the known and likely effects of their decomposition are taken into account at the design stage, and further that the implications of variations from normal practices are fully understood by all concerned before such variations are made.

5.2.6 Were there organisational misconceptions?

While a combination of factors caused the explosion, all might be attributed to a single OM stated thus:

The precautions and routines established for processing potentially explosive materials may be waived for temporary and short-term operations for which improvised equipment and procedures are permissible.

The incubation period should probably be dated from June 1975 (thus lasting a year) when it was decided to re-dry off-specification material at King's Lynn in a dryer which was unsuited to the risks involved.

5.3 The 'Dioxin' release at Seveso on 10 July 1976[3, 12]

This accident has influenced European thinking on the control of chemical hazards and led to the 'Seveso' Directive (82/501/EEC) which is implemented in the UK by the CIMAH Regulations [2.4.2]. This account is based on *The Superpoison*, by Margerison, Wallace and Hallenstein[3], on the report of the Italian Parliamentary Commission of Inquiry[12], and on a more recent paper by Cardillo and Girelli[13].

5.3.1 The accident

At 12.37 hours on Saturday 9 July 1976 a bursting disc fitted in the vent line of the reactor producing 2,3,5 trichlorophenol (TCP) at the chemical factory of Icmesa ruptured because of internal overpressure. Icmesa was an Italian company which, since 1969, was owned by Givaudan, which itself was part of the (Swiss) Hoffmann La Roche group. The factory was at Meda, on the outskirts of Seveso, a town of about 17 000 inhabitants about 20 km from Milan.

The bursting disc discharged a cloud of about two tonnes of hot chemicals directly into the open air, on a hot summer's day with a light northerly breeze. Heavy rain fell soon after the discharge, bringing the contents of the cloud down to earth over a largely urban area.

Unfortunately the cloud contained a small quantity, later estimated at 2 kg, of one of the most toxic compounds known, 2,3,7,8 tetrachlorodibenzoparadioxin, commonly known as TCDD or simply dioxin.

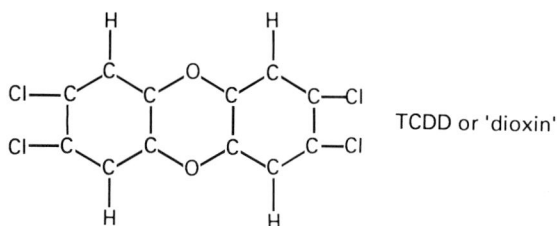

TCDD or 'dioxin'

An area of about ten square miles (which included some 40 factories) was contaminated, and by August at least 730 people had been evacuated.

The fatal dose of TCDD for the average man is less than 0.1 mg. It can enter the human body by ingestion, inhalation and through the skin, leading in mild cases to chloracne. Its symptoms include cysts and pustules, grey or brown staining of the skin, and purple urine. TCDD causes liver and kidney damage, and birth defects in children born to mothers exposed during pregnancy. It is a stable slow-acting poison, insoluble in water, and very difficult to destroy.

Over 700 local inhabitants were affected by the poison and many animals died. A considerable area of agricultural land was rendered unusable for many years.

5.3.2 TCP production and dioxin formation

TCP is a chemical intermediate in the manufacture of hexachlorophene, an anti-bactericidal agent made by Givaudan, and a hormone type weed-killer known as 2,4,5-T.

TCP is generally made by reacting tetrachlorobenzene (TCB) with an excess of caustic soda in the presence of a solvent, in the temperature range 160–200°C, preferably 170–180°C. This reaction, which is strongly exothermic, produces the sodium salt of TCP and sodium chloride and water as by-products.

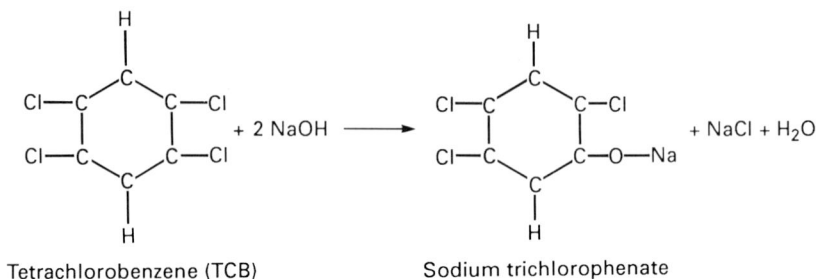

Tetrachlorobenzene (TCB)   Sodium trichlorophenate

The reaction is generally carried out in a stirred vessel, with a jacket or internal coil for heating or cooling and an overhead condenser for condensing solvent vapour and water formed (Figure 5.2).

Condensed solvent could either be returned to the reactor to cool its contents during the reaction or collected separately and recovered at the end of it. While the reaction needed cooling during its middle stages to remove the large amount of heat generated, the temperature was raised in the final stage to drive it as near to completion as possible.

Several variations on the process were employed by different companies, using different solvents and operating pressures which depended on the boiling point of the solvent. In a process developed in West Germany, methanol was used as solvent. While this was cheap, the process required an operating pressure of about 20 bar g. When a higher-boiling solvent such as ethylene glycol was used, the reaction could be carried out at atmospheric pressure, but vacuum was needed to recover the solvent by distillation after the reaction.

When a solvent in which water was insoluble was used it was easy to separate the condensed water and remove it during the reaction, so that only solvent was returned to the reaction mixture. This removal of water favoured the reaction and enabled it to proceed nearer to completion. But the solvent had to be one (such as an alcohol) in which sodium hydroxide is soluble. As a compromise, Givaudan used a mixture of two solvents, ethylene glycol (BP 197°C) in which sodium hydroxide is soluble, and orthodichlorobenzene (BP 181°C) or xylene (BP 142°C) in which water is

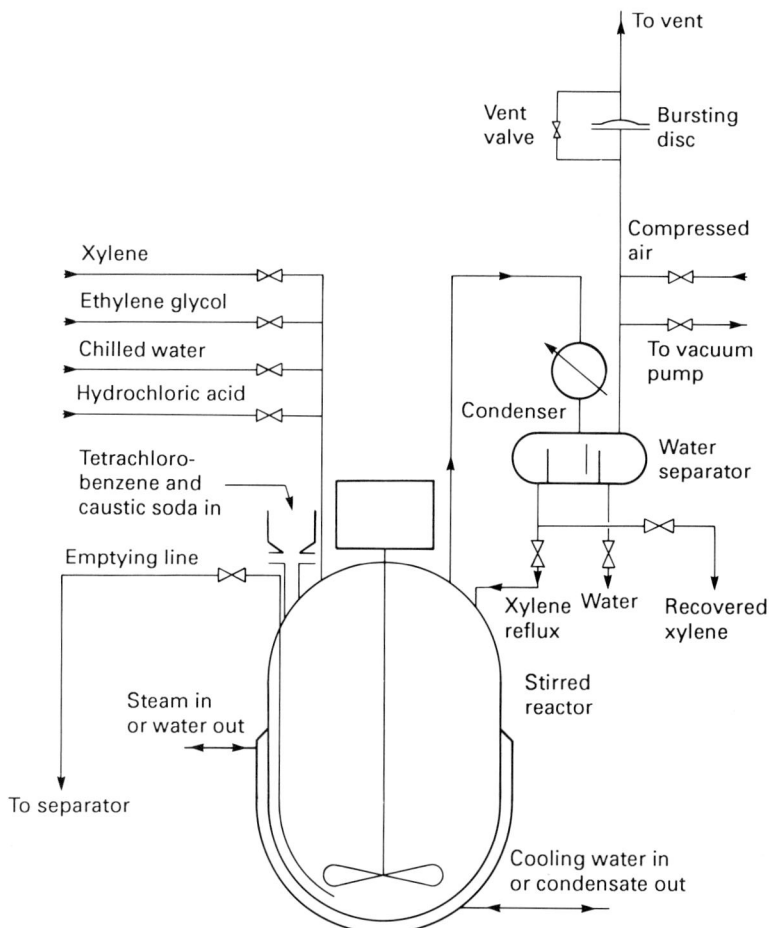

Figure 5.2 TCP production plant at Meda (diagrammatic)

insoluble. A mixture of ethylene glycol and xylene was in use at Meda at the time of the accident. Whichever solvent was used, most of the sodium chloride formed was thrown out of solution, causing operating difficulties by blocking lines and valves and interfering with reactor stirring.

At the end of the reaction, part of the solvent was first recovered from the reaction mixture by distillation. The reactor contents were then diluted with water and acidified to separate the molten TCP product (melting point 60–68°C) and dissolve the sodium chloride formed:

$$C_6H_2Cl_3ONa + HCl \rightarrow C_6H_2Cl_3OH + NaCl$$

In some processes this was done in a different vessel, in others in the same vessel used for the main reaction.

The history of TCP production was beset from the outset by cases of chloracne among the workers, for which dioxin was found to be responsible by Dr K. H. Schulz in 1957[14]. Traces of dioxin (about 25 ppm) in the TCP are formed as a by-product by the condensation of two molecules of sodium trichlorophenate and elimination of sodium chloride. This reaction, like that of TCP production itself, is exothermic and the quantity of dioxin formed increases at higher temperatures.

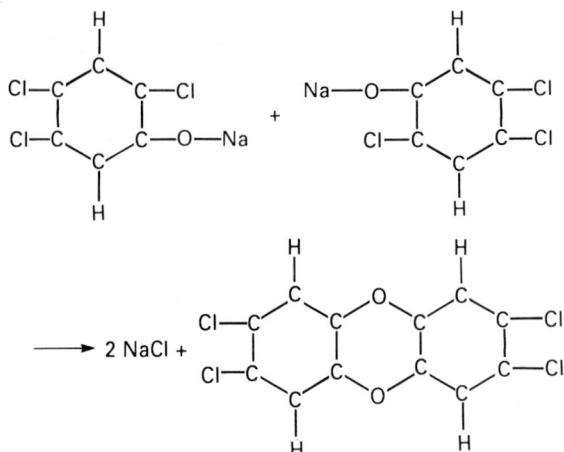

5.3.3 Previous incidents at TCP plants and remedial measures

A list of known incidents involving escapes of dioxin from TCP plants which resulted in illness among workers is given in Table 5.1. Several of these escapes resulted from runaway reactions which led to high temperatures and pressures, the opening of relief valves, and sometimes the failure of joint gaskets. After such escapes it proved extremely difficult to decontaminate the process buildings and workers' clothing.

Table 5.1 Incidents causing human exposure to dioxin before 1976[12]

Year	Location	Company	Number of cases
1949	Nitro, W. Virginia, USA	Monsanto	117
1952/1953	Hamburg, W. Germany	Boehringer	37
1953	Ludwigshagen, W. Germany	BASF	55
1953 to 1971	Grenoble, France	Rhone-Poulenc	97
1963	Amsterdam, Holland	Philips-Duphar	30
1964	Midland, Michigan, USA	Dow Chemical	60
1964	Neuratovice, Czechoslovakia	Spolana	72
1968	Bolsover, Derbyshire, UK	Coalite	79
1972	Linz, Austria	Chemie-Werk	50

Dioxin present in TCP also caused trouble in the final products (e.g. disabilities caused by exposure to dioxin in the 2,4,5-T used as a chemical defoliant during the Vietnam war). To overcome this problem Dow and some other chemical companies reduced the dioxin content of their TCP to less than 1 ppm by a special purification process, as well as adopting special hygiene measures in TCP production.

5.3.4 TCP production at Meda

The TCP plant at Meda used a mixture of ethylene glycol and xylene as solvent, and operated at atmospheric pressure except for distillation of glycol, which was done under vacuum. Trials in which several blockage problems were encountered were carried out in 1970/1971. There was little production for several years, but in 1974/1975 the plant was modified and production was expanded to meet Givaudan's needs for its US and Swiss factories. These needs had become urgent because many manufacturers had abandoned TCP production on account of its health hazards.

The reactor was heated by medium-pressure steam which had a maximum temperature of 190°C. After distilling off the xylene at the end of a batch, as much as possible of the glycol was distilled directly from the reactor under vacuum. The contents of the reactor were then quenched with cold water and acidified in the reactor itself. They were then transferred by compressed air into another vessel where they were separated into a water layer containing dissolved salt and glycol, and a molten TCP layer (MP *ca* 68°C). The unit was protected against overpressure during product transfer by a bursting disc fitted in the vent line from the condenser. This relieved directly to the atmosphere and was designed to rupture at 3.6 bar g.

5.3.5 Operation of the Givaudan plant in 1976

The TCP plant was operated for five days a week on three shifts, with two men per shift. In principle, a new batch of TCP was started every morning at 06.00 when the new shift arrived, and was completed before the night shift left at the same time the following morning. Because of minor operating problems and breakdowns, each batch tended to start progressively later during the week, until on Friday there was not enough time to complete the fifth batch of the week. It was then often decided to start a batch later on Friday and interrupt it part-way through, sometimes after distilling off the solvents, and before adding water and neutralising and washing the crude TCP product. The last shift then had instructions to shut off the steam to the reactor jacket and apply cooling water to it, and leave the water running until work restarted on Monday morning. This caused considerable delays on Monday morning as the contents of the reactor were now solid and had to be melted by applying steam to the jacket for several hours before it was safe to start the stirrer and continue work on the batch.

Several months before the accident, the Friday night shift workers were instructed that if a batch were unfinished at 06.00 on Saturday morning they were to shut off the steam heating but not to turn on the cooling water, so that the reactor cooled only slowly over the weekend. The contents were then sufficiently fluid on Monday morning for work to be started quickly, and time and money to be saved. Thirty-four batches had been interrupted over a weekend since production started in 1975, half of these after the decision was taken not to leave the cooling water running. On all but four of these occasions, the reaction had been completed and quenched with water before the night shift left.

5.3.6 The final batch

On Friday 9 July the fifth run of the week could not be started until late in the afternoon. By the time the night shift started, the reactor had been loaded with its charge, which consisted of:

Ethylene glycol	3235 kg
Xylene	609 kg
Tetrachlorobenzene	2000 kg
Flake caustic soda	1100 kg

The contents were heated to reaction temperature with steam at about 12 bar g. (190°C). The reaction was apparently completed during the night, and all of the xylene and about 500 kg of glycol were distilled off. At 04.45 the steam valve to the heating coil was closed and the stirring was continued for 15 minutes before the stirrer was stopped. The vacuum was turned off and the night shift then went home.

During Saturday morning, an exothermic runaway reaction developed unnoticed in the reactor, accompanied by considerable rises in temperature and pressure. This caused the bursting disc to rupture at 12.37 hours, with the discharge of a cloud of vapour and fine droplets high into the air. Some calculations indicated that a temperature of 400°C was needed in the reactor to produce a pressure high enough to rupture the bursting disc. A sample of material taken from the top of the reactor showed a dioxin content of 3500 ppm.

5.3.7 The cause of the runaway reaction

At the inquiry various theories were advanced to explain the runaway reaction. The most probable was that the reaction between tetrachlorobenzene and caustic soda had not finished when the reactor was shut down, and continued slowly in the unstirred and uncooled reactor, liberating heat with accelerating rises in temperature and pressure. This is consistent with an experimental study made some time after the disaster by Cardillo and Girelli in which they prepared a mixture of the same composition as that left in the reactor at 04.45 on 10 July, and subjected it to sensitive thermal analysis[13]. They found that 'a slow moderate exothermic process is noticeable at 180°C after about 4 hours'. This was much lower than the temperature of 230°C which was previously reported to be necessary for such a runaway reaction to start.

Other theories included:

- The exothermic condensation of glycol with elimination of water;
- That air entering the reactor reacted with material in it thus developing a hot spot which promoted other exothermic reactions;
- That hydrochloric acid was either accidentally or deliberately added to the reactor on Saturday morning.

5.3.8 Mistakes which led to the disaster

Two crucial mistakes whose possible consequences had not been appreciated (i.e. organisational misconceptions [3.3]) clearly played key roles in the disaster.

1. *The reliance on a pressure relief device (bursting disc) which discharged directly to the atmosphere to cope with any runaway reaction which might develop.* This caused both the pressure and temperature in the reactor to rise considerably when a runaway reaction occurred, leading to a very large increase in dioxin formation.

    It furthermore ensured that when the pressure built up and the bursting disc broke, much of the contents of the reactor were rapidly discharged high into the air and spread over a much larger area than would have happened if a vent valve on the reaction system had been open.

    This reliance was probably not a deliberate decision but rather one taken by default. The bursting disc was a safety device to protect the reaction system from overpressure from whatever cause and was probably installed to meet some code requirement. Had the consequences of a runaway reaction been considered properly, any gas/vapour passing the bursting disc (and a vent valve which by-passed it) should have been piped to a 'blow-down' vessel and condenser large enough to condense and contain all vapour released by the reaction.

    It appears that this mistake had persisted for nearly a year, from the time the TCP plant was modified in 1974/1975.
2. *The act of allowing a mixture of (hot) chemicals which can react further in a closed, unstirred and uncooled vessel, virtually unattended over a weekend.* This again may not have been a deliberate act, but it resulted from the dislocation of a batch-production programme which was geared to a particular shift system. The possible consequences of leaving an unfinished and still reactive batch unattended had not been foreseen.

    This mistake seemed to have persisted for several weeks, since four interrupted batches had already been left uncooled over a weekend before water and acid were added to quench the reaction and neutralise the caustic soda present.

    Had these mistakes been recognised and their consequences thought through in time, they could surely have been rectified. Thus the disaster was preventable.

5.4 The Bhopal disaster in December 1984[4]

This account of the world's worst industrial disaster is based on a detailed article in *India Today*[4] and on information researched by Granada Television for an award-winning documentary[15] which its director Laurie Flynn kindly made available to me.

Some 2500 Indians were killed and 100 000 injured by the release of 26 t of highly toxic methyl isocyanate (MIC) vapour into the midnight air of Bhopal, a city of more than 700 000 inhabitants which lies in the centre of India. The plant was located in the north-east of the city where it was surrounded by commercial premises and workers' quarters (Figure 5.3).

The following problems discussed elsewhere in this book featured in the disaster:

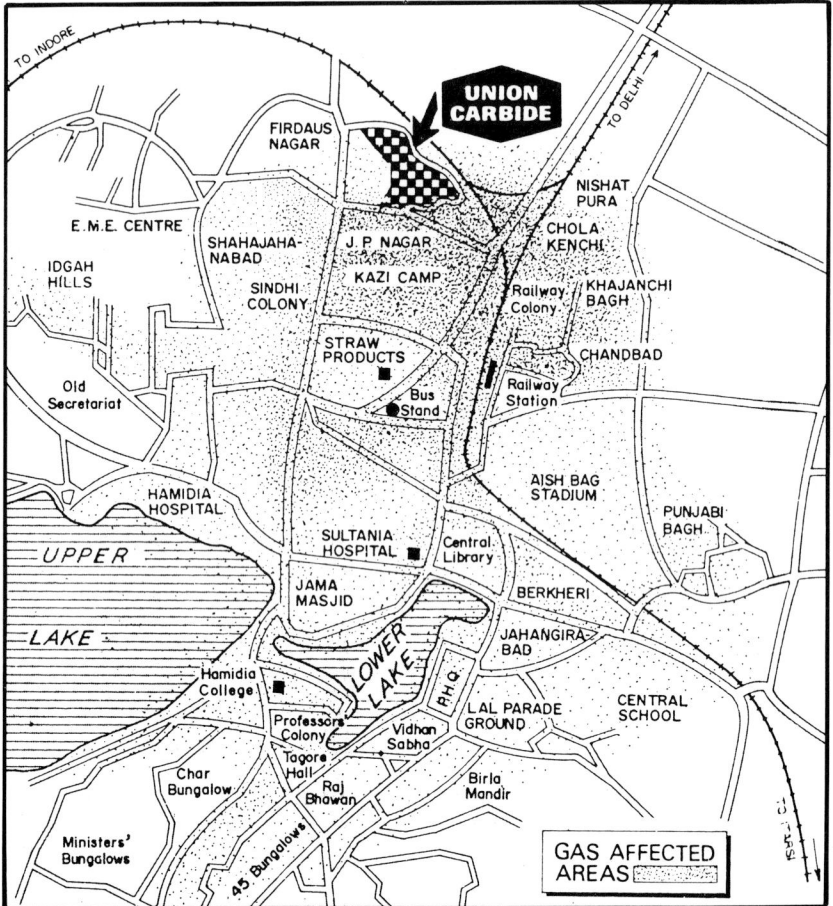

Figure 5.3 Map of Bhopal. (By B. K. Sharma, courtesy *India Today*)

- The siting of major hazard installations [16.6.1];
- Toxic vapours [7];
- Spontaneous exothermic polymerisation [8.6.4];
- Process protective systems [15];
- Staff training [21] and selection [19];
- Emergency procedures [20.3];
- Metal corrosion [11];
- Safety in technology transfer to Third World countries [23.5].

5.4.1 Background[4]

Most insecticides, while considered vital to food production, are toxic to humans to some degree. Their sales are often short-lived as more effective and safer ones are invented.

Union Carbide India Ltd (UCIL) started its Agricultural Products Division in 1966/1967 in Bombay. This moved in 1968 to Bhopal, where a formulation plant was set up for the insecticide carbaryl, under the trade name Sevin. At first the concentrate was imported from Union Carbide Corporation (UCC) in America, only grinding, blending and packing being done at Bhopal.

Sevin, Temik and MIC, their common intermediate, were made by UCC at a plant at Institute, West Virginia. MIC, itself highly toxic, was made by a route involving several other toxic gases and volatile liquids. Some properties of MIC are given in Table 5.2[16].

Table 5.2 Properties of methyl isocyanate

Formula	$CH_3-N=C=O$
Molecular weight	57.05
Boiling point at 760 mm Hg	40°C
Liquid density at 20°C	0.96
Vapour density (air = 1)	2.0
Heat of combustion, gross	19 770 J/g
Occupational Exposure Limit in USA (1980)[17]	0.02 ppm or
and in many other countries[15]	0.05 mg/m^3
Lower flammability limit in air, % by volume	5.3
Upper flammability limit in air, % by volume	26

Toxicity	The vapour is extremely toxic, attacking the skin and the mucous membranes of the eyes and respiratory system.
Reactivity	MIC is very reactive. It polymerises readily in the presence of various catalysts, including chlorides of iron, to form the cyclic trimer, 1,3,5-trimethyl isocyanurate. It reacts with water to form carbon dioxide, methylamine and other compounds. Both reactions are strongly exothermic, proceeding slowly below 20°C, but can become violent at higher temperatures and generate enough heat to vaporise most of the MIC

UCC made plans for the manufacture of 5000 t/yr of Sevin and Temik at Bhopal in two phases. In the first phase, which came into operation in 1977, the MIC was imported from the USA in stainless steel drums, and reacted with locally produced chemicals to make Sevin and Temik. In the second phase, which went into production in February 1980, MIC was also made at Bhopal.

The MIC plant, which started production in 1980, included the following stages (Figure 5.4):

- Production of phosgene from chlorine, and carbon monoxide produced on site:

  $CO + Cl_2 \rightarrow COCl_2$

- Production of methylcarbamoyl chloride from phosgene and monomethylamine, using chloroform as reaction solvent:

  $COCl_2 + CH_3NH_2 \rightarrow CH_3NHCOCl + HCl$

- Pyrolysis of methylcarbamoyl chloride:

  $CH_3NHCOCl + heat \rightarrow CH_3NCO + HCl$

- Fractional distillation of crude MIC in a 45-plate column at high reflux ratio, to remove chloroform and unconverted methylcarbamoyl chloride which was recycled to the pyrolysis stage.

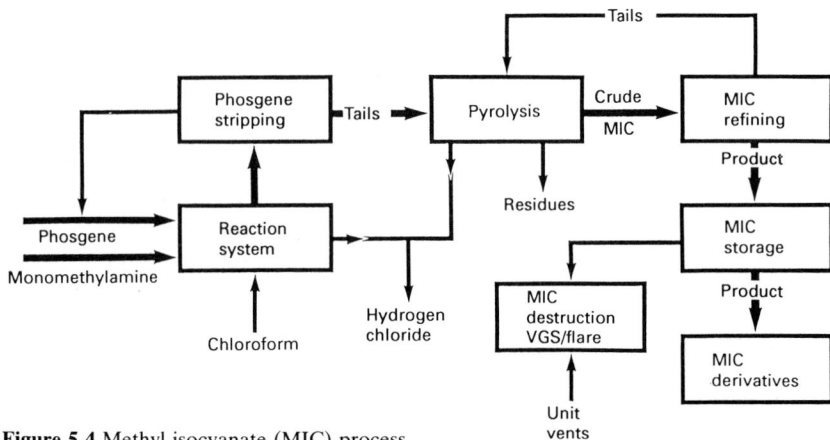

Figure 5.4 Methyl isocyanate (MIC) process

Chlorine, monomethylamine and chloroform were brought by road or rail from other parts of India and stored on site.

UCC's specification for MIC included a maximum chloroform content of 0.5%, but because of difficulties in operating the MIC column it was sometimes higher.

5.4.2 Storage of MIC and safety features

MIC was stored in two horizontal stainless steel pressure vessels referred to as 610, 611. These were each of $57 m^3$ nominal capacity, designed for full vacuum to 2.8 bar g. at 121°C (Figure 5.5). The grade 604 stainless steel employed, while resistant to rusting and oxidising acids, would be rapidly attacked by hydrochloric acid, to form chlorides of iron and other metals [11.6.1]. These vessels were covered with earth mounds with concrete decks to protect against accidental impact, external fire and for thermal insulation. This insulation, while useful in preventing heat from entering the vessels when the contents were refrigerated, also prevented escape of heat generated by polymerisation of MIC if the refrigeration system was out of action [8.2.1, 8.6.4].

A third similar pressure vessel, 619, was installed to receive off-specification material which could either be returned for reprocessing or destroyed in the vent gas scrubber (VGS).

A 30-ton refrigeration system with a heat exchanger through which MIC was circulated was provided to maintain its storage temperature at 0°C or lower, in order to retard, control and minimise polymerisation.

Instrumentation of each MIC storage vessel included:
- A temperature indicator and high-temperature alarm;
- A pressure indicator/controller to regulate the pressure by admitting nitrogen or venting vapour to the VGS or flare;
- A liquid level indicator/alarm for high and low level.

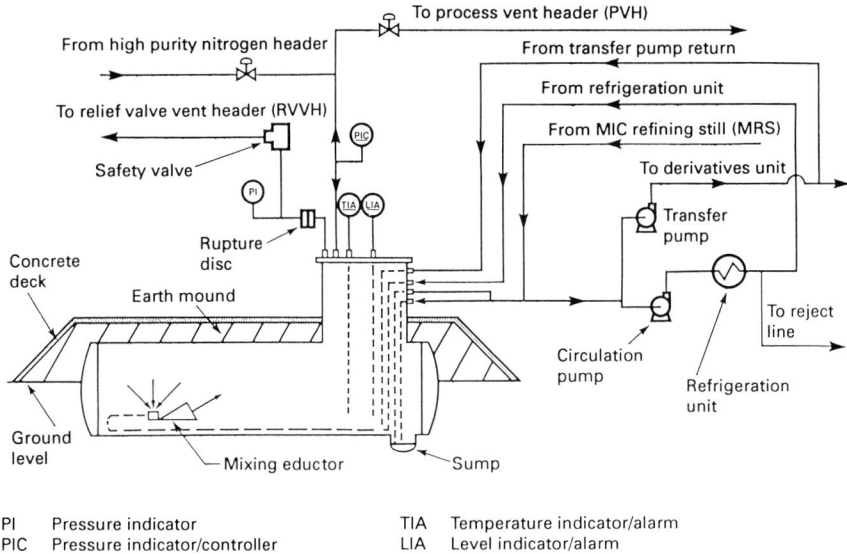

PI   Pressure indicator   TIA   Temperature indicator/alarm
PIC   Pressure indicator/controller   LIA   Level indicator/alarm

Figure 5.5 MIC storage vessel 610. (Courtesy Laurie Flynn and Granada Television)

The pressure in the vessels was controlled between the limits 0.14 and 1.7 bar g. This minimised nitrogen consumption and loss of MIC vapour. A pressure relief system fitted with a bursting disc and a relief valve in series was set to open if the pressure in a tank reached 2.8 bar g., discharging the vapour to the relief vent header.

The vent gas scrubber (VGS) was a 1.68 m diameter packed column in which gases entering were contacted with circulating caustic soda solution, the strength of which could be maintained by adding 50% caustic from storage tanks. Gas leaving the top of the VGS discharged through a stack to the air at a height of 33 m. Instrumentation included a caustic solution flow indicator/alarm which would automatically start the spare caustic circulation pump in case of low flow. Under emergency conditions, the VGS could neutralise 4 t of MIC in the first 30 minutes, provided the circulating caustic solution was fresh. After that its capacity would be reduced to below 2 t/h because there were no means of cooling the circulating soda.

A 30 m high flare tower equipped with a flame-front generator and wind shield was primarily intended to burn vent gases from the carbon monoxide unit and methylamine from a relief valve. It was situated some distance from the MIC storage tanks.

Both the process vent header and the relief vent header were connected to the VGS and the flare line and could be routed to either. Fixed fire-water monitors were provided which could be used to provide a water curtain round sensitive areas and knock down some toxic gases and vapours.

An artist's impression of these safety features with notes on their condition at the time of the disaster is given in Figure 5.6.

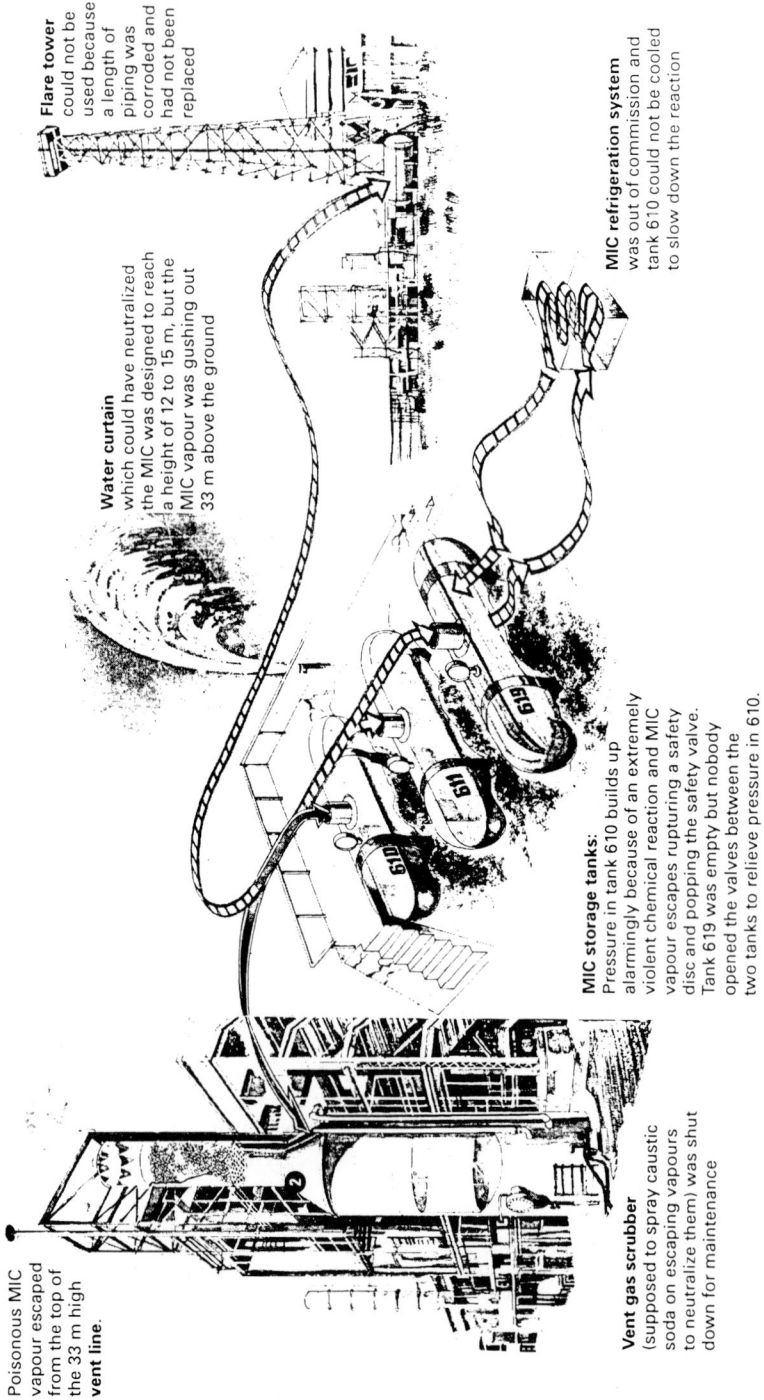

Flare tower could not be used because a length of piping was corroded and had not been replaced

MIC refrigeration system was out of commission and tank 610 could not be cooled to slow down the reaction

Water curtain which could have neutralized the MIC was designed to reach a height of 12 to 15 m, but the MIC vapour was gushing out 33 m above the ground

MIC storage tanks: Pressure in tank 610 builds up alarmingly because of an extremely violent chemical reaction and MIC vapour escapes rupturing a safety disc and popping the safety valve. Tank 619 was empty but nobody opened the valves between the two tanks to relieve pressure in 610.

Vent gas scrubber (supposed to spray caustic soda on escaping vapours to neutralize them) was shut down for maintenance

Poisonous MIC vapour escaped from the top of the 33 m high **vent line.**

Figure 5.6 State of safety features of MIC plant at Bhopal at time of disaster. (Drawing by Itu Chaudhuri, courtesy *India Today*)

5.4.3 The period before the disaster

Following the start-up of the MIC plant, there were several accidents and minor escapes of toxic vapours. A maintenance fitter was fatally injured at the end of 1981 through exposure to a leak of phosgene. Six weeks later 16 workers were injured by another leak of toxic gas. Further accidents occurred in October and December 1982 and in February 1983, and morale fell.

The plant was surveyed in May 1982 by a team of Union Carbide's American safety experts who were concerned by the extent of corrosion and warned of a possible escape of toxic gases.

Articles started to appear in the local press in September 1982 warning the people of Bhopal of the threat which the factory posed to their lives.

Besides these fears about plant safety, there were now worries about the health hazards of Sevin and Temik (which many farmers could not afford in any case). Sales and production dropped sharply. Expenditure was cut to the bone and many of the best staff left. UCC was reported to be considering dismantling the plant and selling it to Indonesia.

The refrigeration system which cooled the MIC in 610 and 611 was taken out of service in June 1984, and its refrigerant removed. It is not clear who made that decision or whether he was aware of its likely consequences.

A company safety survey by six health and safety experts on UCC's MIC II Unit at Institute was made during the week starting 9 July 1984. An extract of their report, a copy of which reached the management at Bhopal on 19 September, follows:

MAJOR CONCERNS
SM1   *MIC Storage Tank Runaway Reaction*
There is a concern that a runaway reaction could occur in one of the MIC Unit Storage tanks [vessels] and that response to such a situation would not be timely or effective enough to prevent catastrophic failure of the tank.

These (in my words) were the main reasons for their concern:

1. The vessels were being used for relatively long-term storage, and not being sampled regularly, so that there was more chance of contamination going undetected.
2. The refrigeration system (on the American plant) used brine rather than chloroform as coolant, so that if there was a leak, water could enter the MIC and react with it.
3. The vessels had been contaminated with water in the past as a result of leaking condensers and coolers in the MIC production unit.
4. Catalytic materials could conceivably enter the vessels from the flare system.

Their recommendations included one that the vessels be sampled at least daily, and that the seriousness of water contamination be emphasised to all operating personnel.

The team were also worried that flame failure alarms (thermocouples) on the flare tower were not working, with the danger of MIC vapour being released through the flare without being burnt.

Had the team then visited and inspected the Bhopal plant, history might have been different.

MIC production at Bhopal ceased on 22 October 1984, when the last MIC which entered 610 was impure and contained between 12% and 16% of chloroform. This contaminated MIC should have been routed to 619 and not 610. The contents of 610 were not mixed after 19 October. On 23 October the VGS was shut down and put on 'stand-by', apparently as an economy measure. At this time or soon after, the flare tower went out of service because of corrosion.

By Sunday 2 December, the day of the disaster, 610 contained 41 t of (impure) MIC. This had not been stirred or sampled for nearly six weeks, nor had its temperature been logged. The high-temperature alarm had been deactivated when the refrigeration system was shut down. The pressure on Saturday 1 December was apparently recorded as 0.14 bar g. (2 psig), and the same pressure was reported at 22.20 hours the next day.

5.4.4 The disaster

At 23.00 hours on Sunday 2 December 1985 a rather high pressure (0.7 bar g.) was noticed on the pressure indicator of 610 by the control room operator. One and a quarter hours later this had risen rapidly to over the top of the scale (3.8 bar g.). The vessel was hot, the concrete above it was cracking and MIC vapour was screeching through the relief valve. An attempt was made to start caustic circulation through the VGS. At 01.00 hours on Monday the toxic gas alarm was sounded by an operator in the derivatives area and firewater monitors were turned on and directed to the VGS stack to try to knock down escaping MIC vapour. At 02.00 hours the relief valve of 610 (set at 2.8 bar g.) reseated and MIC emission ceased. 610 and its associated pipework were later found to be intact and pressure tight. Four hours later the temperature in the caustic accumulator of the VGS was about 60°C, thus indicating that some circulation of caustic had occurred.

5.4.5 Quantity of MIC vapour released

Estimates made from mass and energy balances indicated that about 36 t of material had escaped from 610, of which about 25 t were MIC vapour, the rest being entrained liquids and solids formed from MIC. The average pressure in 610 during the two-hour period of the MIC escape was estimated to have been 12.2 bar g. As this was well above the hydrostatic test pressure of 4.1 bar g., it is fortunate that the vessel did not rupture. A peak temperature of 200°C in 610 appears to have been reached after the relief valve closed.

5.4.6 Initiation and nature of the reaction in 610

Two possible explanations for the initiation of the reaction soon emerged.

1. A gradually accelerating runaway polymerisation reaction started in 610 some time after the refrigeration system was decommissioned, and its contents had warmed up to the ambient temperature. Such runaway

reactions [8.6.4] were involved in the accidents at King's Lynn [5.2] and Seveso [5.3]. The critical temperature for any material above which one can start falls as the mass, thermal insulation and surrounding temperature increase. In this case the runaway reaction might well have started at 15–20°C.

The interval of five months between the shut-down of the refrigeration system and the disaster is not, however, easy to explain. Unfortunately we have no record of the temperature of 610 since June. At 0°C and below, both the polymerisation and hydrolysis rates of MIC are probably very low, even with metal chlorides present. Little reaction would have occurred until sufficient heat had entered the vessel from the ground to raise its temperature to about 15°C. Once the temperature in 610 reached that of the ground (15–20°C), the reaction, though still very slow, probably generated heat faster than it could be lost, and accelerated as the temperature rose. The vessel also contained metal chlorides (resulting from HCl corrosion of the stainless steel equipment), which catalyse the polymerisation. These could have lain dormant for a considerable period in the base of the vessel before becoming activated.

2. The polymerisation was initiated by the entry of water into 610, either through a leaking valve or as a result of sabotage, shortly before the disaster.

This explanation has received wide publicity.

Chemical analysis of the solids left in 610 might be expected to throw light on the question. A summary of the analysis by UCC of such a sample is given in Table 5.3.

Table 5.3 Analysis of core sample of solids in 610

Constituent	% weight
1,3,5-trimethyl isocyanurate (MIC trimer)	40–55
1,3-dimethyl isocyanurate	13–20
Methylamine hydrochlorides	7–10
Methylbiurets	7–14
Dihydro-1,3,5-trimethyl-1,3,5-triazine-2,4-dione	5–7
Methylureas	3–6
Iron, chromium and nickel salts	0.18–0.26
Chloroform	0.4–1.5
Water[a] up to	2

[a] The water may have entered the sample after the disaster

The major constituent, MIC trimer, would be expected from the metal chloride catalysed polymerisation of MIC. The small amount of chloroform resulted from chloroform impurity in the MIC.

The formation of methylamine hydrochlorides required hydrogen chloride. The UCC investigators thought the hydrogen chloride was formed by hydrolysis of chloroform in the MIC, although it could have been formed from methylcarbamoyl chloride or even phosgene had they been present in the MIC in 610. (Sir Frederick Warner [4] and Dr Luxon, health and safety director of the Royal Institute of Chemistry, were reported as saying that 'the victims suffered from phosgene poisoning'[18].)

The total amount of iron, chromium and nickel in the salts found amounted to approximately 9 kg. Their ratio was approximately the same as in 304 stainless steel, indicating chloride corrosion of 610 (and/or other equipment) as their source. Some of the metal salts were probably present before December, and some formed during the runaway reaction in 610.

The analysis shows a number of organic compounds which might have been formed directly from MIC, or indirectly by 'cooking' of its trimer after the relief valve had closed. The UCC team concluded from their analyses of the residues that:

> 1000 to 2000 lb of water and 1500 to 3000 lb of chloroform were required.

Far less water would, however, have been needed if the hydrochlorides had been formed from methylcarbamoyl chloride rather than chloroform. Dr S. Varadarajan, a senior Indian government scientist, stated[18] that:

> Just half a kilogram of water entered the underground methyl isocyanate tank, triggering a runaway reaction.

A pipe-washing operation was suggested as the source of water. UCC team's comment on this reads:

> Water could have been introduced inadvertently or deliberately directly into the tank through the process vent line, nitrogen line or other piping. Records indicate that the safety valve discharge piping to the relief valve vent header from four MIC process filters was being washed shortly before the incident. Oral discussions indicate that a slip blind was not used to isolate the piping being washed. However, entry of water into Tank 610 from this washing in the MIC unit would have required simultaneous leaks through several reportedly closed valves, which is highly improbable.

Whether or not water entered 610 after the MIC plant was shut down, the shutting down of the refrigeration unit in June was clearly a major factor in enabling the runaway reaction to occur.

5.4.7 Emergency measures

According to Wally Schaffer, chairman of the UCC workers' safety committee at Institute, there was no emergency or evacuation plan for people outside the plant at Bhopal, as there was at Institute. The toxic gas alarm at Bhopal was not activated until MIC had been escaping for at least an hour, and half of it had escaped. Even then it was turned down after a few minutes, when it could only be heard by UCIL workers. Criticisms have been raised[4] that more might have been done to introduce fresh caustic into the VGS so as to have made it more effective. There were, however, few personnel present at the time of the disaster as most plants were shut down. They were working in high concentrations of MIC vapour with apparently only 'half-hour masks' for protection[19].

One desperate measure might have averted the worst effects of the disaster once the MIC emission had started. This would have been to set fire to the vapour escaping from the vent stack. Rockets or a machine gun firing incendiary bullets might have done this! Such ignition might, however, have caused a vapour cloud explosion [10.5.4] leading to an escape of stored chlorine or phosgene.

5.4.8 Mistakes or organisational misconceptions

Two major mistakes, expressed here as 'organisational misconceptions', lay at the root of the disaster:

1. That it was safe to build a plant storing large quantities of very toxic gases and volatile liquids close to a densely populated area. This seems to have arisen about ten years before the disaster.
2. That it was safe to shut down the refrigeration system designed to allow MIC to be stored safety. This happened 5 to 6 months before the disaster.

Union Carbide have been criticised on several other points which, though serious in themselves, appear less important than these two. They include:

• A safer route for the manufacture of MIC or of the insecticides themselves could have been chosen.
• The MIC storage capacity should have been much smaller.
• The storage vessels should have been made strong enough to contain a runaway reaction if it occurred.
• The VGS should have been large enough to cope with the maximum rate at which MIC vapour could have been released in a runaway reaction.
• The pressure relief system should have been designed for two-phase flow of vapour and liquid and not for vapour only.

Most of these criticisms lie in the grey area between what is 'practicable' and what is 'reasonably practicable' [2]. Compliance with them would have added significantly to the project cost, and might well have rendered it a non-starter in the first place. Such questions are now studied under 'risk-benefit analysis' [14.7]. Many other Third World chemical companies would have followed the same practices as Union Carbide on these points, and would probably have got away with them.

Had someone deliberately tried to organise the disaster, he could hardly have done better! The words of Turner[5] quoted in 3.3.2 are particularly apt.

The first mistake probably had a profound effect on the morale, quality and competence of the technical management at UCIL's Bhopal works. 'Good company men' would obviously be expected to subscribe to the misconception. The effect on their morale when they could no longer accept it would have been shattering.

This was well expressed in the TV programme by engineer Pareek, who was responsible for safety for one and a half years at Bhopal:

Interviewer: 'Were inexperienced people being brought in?'
Pareek: 'Yes, er, you know in the industry the news gets around that a

plant is not healthy, so nobody who was worth anything would like to come and join that plant'.

The lead time or 'incubation period' [3.3.5] during which the first mistake persisted thus lasted about ten years from 1974, when the siting of the MIC plant was decided. From then on, although many came to recognise this as a mistake, it became increasingly difficult and expensive to rectify. One might as well have tried to move the city as the plant. One hopes a better site would have been found for the plant in Indonesia!

References

1. Ministry of Social Affairs and Public Health, *Report concerning an inquiry into the causes of the explosion on 20th January, 1968 at the premises of Shell Nederland Raffinaderij in Pernis*. State Publishing House, The Hague, Holland (1968)
2. HSE, *The Explosion at Dow Chemical Factory, King's Lynn, 27 June 1976*, HMSO, London (1977)
3. Margerison, T., Wallace, M. and Hallenstein, D., *The Superpoison*, Macmillan, London (1981)
4. 'Bhopal – city of death', *India Today*, pp. 6 to 25, 31 December (1984)
5. Turner, B. A., *Man-made Disasters*, Taylor and Francis, London (1978)
6. Davenport, J. A., 'A study of vapour cloud incidents', Paper 24, *Eleventh Loss Prevention Symposium*, 83rd National Meeting, Am. Inst. Chem. Engrs, New York (March 1977)
7. AMOCO, Booklet No.1, *Hazard of Water*, 5th edn, AMOCO, Chicago (1964)
8. King, R. W., 'Latent superheat – A hazard of two phase liquid systems', I.Chem. E. Symposium Series No. 49, *Chemical Process Hazards with Special Reference to Plant Design VI*, Rugby (1977)
9. Dow Chemical Company, *Fire and Explosion Index – Hazard Classification Guide*, 5th edn, Midland, Mich. (1980)
10. Watts, H. E., *The Law Relating to Explosives*, Charles Griffin, London (1954)
11. Frank-Kamenetskii, D. A., *Diffusion and Heat Exchange in Chemical Kinetics*, Translated by J. P. Appleton, Plenum Press, New York (1969)
12. *The official report of the Italian parliamentary commission of inquiry into the Seveso disaster, translated by HSE*, HSE, Bootle (about 1984)
13. Cardillo, P., and Girelli, A., 'The Seveso runaway reaction: A thermoanalytical study', paper in I. Chem. E. Symposium Series No. 68, *Runaway Reactions, Unstable Products and Combustible Powders*, Rugby (1981)
14. Schulz, K. H., *Arch. Klin. Exp. Derm.*, **206**, 540 (1957)
15. Flynn, L., and Smithson, J., *The Betrayal of Bhopal*, 'World in Action' TV documentary, available as video cassette from Granada Television, London
16. Morel, C., Gendre, M., Limasset, J. C. and Cavigneaux, A., *Isocyanate de méthyle – Fiche toxicologique no. 162*, INRS, Paris, (1982)
17. The International Labour Office, *Occupational Exposure Limits for airborne toxic substances – a tabular compilation of values from selected countries*, 2nd edn, Geneva (1980)
18. 'Water leak caused fatal chain reaction says Bhopal expert', *The Guardian*, Saturday 5 January (1985)
19. Bhatia, S. and McKie, R., 'Time bomb at Bhopal', *The Observer*, 9 December (1984)

Chapter 6

Electrical and other physical hazards

This chapter discusses electrical and other physical hazards of a non-mechanical nature which can to a large extent be controlled during design. General electrical hazards are first discussed in 6.1 followed by electrical ignition hazards in 6.2. Static electricity and lightning are discussed in 6.3. Section 6.4 deals with several physical hazards involving liquids.

6.1 General electrical hazards

Marshall[1] makes the following general points and gives several case histories of electrical accidents in the chemical industry.

Design

- All wiring in process plant should be piped.
- For indoor duties where condensing steam or corrosive vapours may be present, galvanised rather than black enamel conduit is recommended.
- Totally enclosed motors, though larger and more expensive than screen-protected and drip-proof types, are much safer in areas where combustible or corrosive materials are being handled.

Installation and maintenance

- The retaining bolts or set screws of all angle bends and inspection plates must be in place and screwed up tightly.
- Conduits must be protected from damage, e.g. from glancing blows by fork-lift trucks.
- All spare conduit holes in distribution boards and fuse boxes must be sealed.
- Temporary connections made in an emergency to maintain production should be replaced by permanent ones, on the same day if possible, and should never be allowed to become semi-permanent.
- Before replacing a blown fuse the electrician should investigate and rectify the cause of its failure.
- The insulation of all electrical equipment should be periodically tested and the results recorded.

- For belt-driven equipment where the position of the motor requires occasional adjustment, the connection to the terminal box should be through flexible metal conduit. This requires very regular inspection to ensure that neither it nor the cables are damaged.

Operation

- Motors including those of ventilation fans should not be left running when plant is shut down and unattended.

6.1.1 Shock and flash burns[2]

To produce electric shock a current has to pass through part of the human body. The degree of shock depends on:

- The current flowing;
- The time for which it flows through the body;
- The type of current (d.c. or a.c.) and its frequency;
- The parts of the body in the path of the current. The heart is the most vulnerable part of the body and a current flowing from the left arm to a leg is more likely to be fatal than one flowing from one leg to the other.

Unfortunately, a.c. electricity of mains frequency (50 or 60 Hz) is more dangerous than either d.c. or high-frequency a.c. electricity of the same voltage. Mild electric shock is felt as stabbing pains in the parts of the body in contact with the conductor. Severe shock causes muscles to contract and involuntary movements, including the gripping of a conductor more tightly, or jumping and falling. High voltage 50–60 Hz a.c., however, causes violent muscular contractions which may throw the victim clear of the circuit.

In the worst cases electric shock affects the heart, preventing it from pumping blood, thus depriving the brain of oxygen and leading to death unless help is given swiftly. As the electrical resistance of the dry skin is much higher than the internal resistance of the body, the degree of shock much depends on whether the parts of the body in contact with the conductor are wet or dry. For mains frequency a.c. electricity, the minimum voltages needed to produce various effects on the human body with a wet skin are roughly as follows:

Feeling	10 volts
Severe pain	20
'Hold-on'	25
Death	45

Normally the shock from contact with 220 volt a.c. supply is only fatal if the skin is wet. The use of insulating gloves and footwear is important where there is any risk of contact with live electrical circuits.

Welders are at risk from open-circuit voltages on arc welding sets while changing electrodes or setting up work. Such shocks, though usually mild, can be severe if the welder is sweating profusely. Welders must be trained to keep their bodies insulated both from the work and from the metal electrode and holder.

Rescue and treatment of electric shock victims must be speedy. If the victim is still in contact with a live conductor, the electric supply should be switched off if at all possible before approaching him. If this cannot be done at once, and the voltage is below 500, the rescuer may drag the victim away from the conductor by his clothes providing they are dry, or push him away by a piece of dry wood or plastic.

Successful treatment of a shock victim whose heart has stopped beating requires prompt artificial respiration and heart massage. As his heart is liable to cease beating again for some time after revival, he needs to be accompanied and watched by a trained first-aider while being taken to hospital.

Electric burns which are deep and slow to heal can be caused by opening switches, removing fuses from energised circuits, or shorting cables. It is important that only qualified electricians should install and maintain electrical equipment, that other persons be excluded from areas where there is a risk of shock or burns, and that electrical faults be promptly reported and rectified. Temporary electrical wiring and cables should only be allowed under a written permit system for a strictly limited period.

6.1.2 Unintended opening and closing of electrical circuits

Many serious accidents have come about through motors (e.g. of pumps or stirrers) being started on plant which was under maintenance. The use of a permit system [18] with motor switches locked and/or fuses removed is necessary. The unintended stopping and tripping-out of running motors through overload devices, as well as from total power failure, can also cause accidents as well as spoiled production. The effects of all such possible stoppages need to be examined by HAZOP studies [16.5.5] when the plant is designed. Protection is achieved by trips and interlocks [15.7] and by operating procedures [20.2] worked out to deal with each case.

6.2 Electrical ignition hazards

Standard industrial electric installations are not designed to prevent sparks and hot spots which might ignite a flammable gas or vapour. Special equipment and precautions are needed when electrical equipment is used in areas where such materials may be present. Decisions on these depend on a complex network of standards (discussed by Lees[3]) and have to be taken when the plant is designed. They are usually the responsibility of electrical specialists, with guidance from process engineers and others on the nature and extent of the fire and explosion risks in different areas. Thereafter, once the plant is operating, care is needed to ensure that extensions and replacements of electrical items comply with the established system, that it is not violated by process modifications, and, above all, that the electrical installation is regularly inspected and properly maintained.

Only broad principles are discussed here. The following aspects have to be considered:

- The gases and vapours liable to be present and their ease of ignition by: (a) electrical sparks, (b) hot surfaces. (Both affect the design and type of electrical equipment chosen.)
- The probability of a flammable concentration of gas or vapour being present wherever electrical equipment is used;
- Methods of protection and limits of their applicability;
- Specifications for 'flameproof', 'intrinsically safe' and other types of electrical installations for use in different categories of hazardous areas.

6.2.1 Grouping of flammable gases and vapours for ease of ignition

The flammability of gases and vapours is discussed in 10.2. Following BS 5345[4], flammable gases and vapours are divided into four groups (called apparatus sub-groups) – I, IIA, IIB and IIC – according to their ease of ignition by electrical sparks[5] [10.2.4]. Methane, the only member of group I, has a minimum ignition energy of 0.29 mJ. Most flammable gases and vapours belong to group IIA, but some such as diethyl ether, ethylene and ethylene oxide belong to group IIB. Hydrogen, the only member of group IIC, has a minimum ignition energy of 0.019 mJ. Acetylene and carbon disulphide, which has a minimum ignition energy of 0.01–0.02 mJ, are excluded from all groups.

BS 4683[6] assigns flammable gases and vapours to six temperature (T) classes according to their ignition temperatures. This determines the maximum surface temperature for electrical equipment (e.g. the casing of a motor) which may be allowed in a particular atmosphere.

Class T1, which allows a maximum surface temperature of 450°C, includes hydrogen, methane, propane, benzene, methanol and other compounds with high ignition temperatures. At the other extreme, carbon disulphide is restricted to class T5, which allows a maximum surface temperature of only 85°C.

Somewhat different groupings of flammable gases and vapours apply when other forms of protection such as intrinsically safe circuits are used.

6.2.2 Hazardous zones or areas

BS 5345[4] divides a site where flammable gases and vapours are stored or processed into three zones and one area:

- Zone 0 is one in which an explosive gas–air mixture is present continuously or for long periods (e.g. the gas space in a fixed roof tank containing petrol);
- Zone 1 is one in which an explosive gas–air mixture is likely to occur in normal operation;
- Zone 2 is one in which an explosive gas–air mixture is not likely to occur in normal operation, and if it does, will only exist for a short time;
- A non-hazardous area is one in which an explosive gas–air mixture is not expected to be present in quantities such as require special precautions to be taken for the construction and use of electrical equipment.

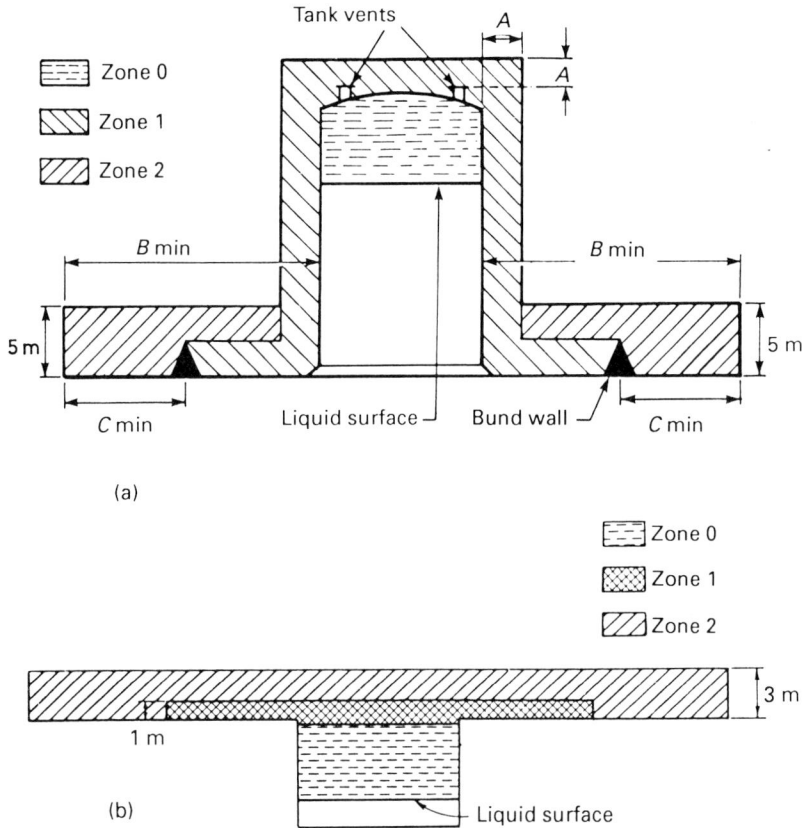

Figure 6.1 Hazardous area classification (a) for fixed roof tank and (b) for open-topped oil/water separator. (Courtesy ICI/RoSPA)

While this zonal classification is common to international, national and company standards, its interpretation largely depends on the code of practice followed[7]. Examples from the ICI/RoSPA code are shown in Figure 6.1. The economic savings in being able to use standard (non-flameproof) electrical equipment are such that various means are used to establish non-hazardous 'islands' (e.g. control rooms), within areas classed as zone 2 which require flameproof or otherwise protected equipment. The most commonly adopted method is to fit air-locks at the entrance/exit to the control room, etc., and maintain a slight positive air pressure inside it by supplying it with air from a non-hazardous area by a fan and ducting. This may be needed both to reduce explosion risks and to prevent the entry of toxic vapours.

The system of zoning has prevented the outbreak of many fires resulting from small leaks of flammable fluids. It is, however, doubtful if it has prevented the ignition of many massive escapes of flammable vapours (such as those listed in Appendices D and E). These usually spread into non-hazardous areas where normal industrial electrical equipment is used.

6.2.3 Methods of protection

The principal methods of protection used are:

- *Segregation*, i.e. keeping all electrical installations well away from flammable gases and vapours. This may require considerable ingenuity, but is necessary with substances such as acetylene and carbon disulphide.
- *Flameproof enclosures* are the most widely practised method. The switch, motor or other item which is liable to spark is contained in a strong enclosure which air can enter or leave only via a very narrow gap (e.g. between two machined surfaces). This prevents any flame or explosion inside the enclosure from breaking it or propagating into the surrounding atmosphere. The engineering specifications for flameproof enclosures become more stringent on passing from zone 2 to zone 0, and from apparatus class I to apparatus class IIC.
- *Intrinsically safe systems*[8] apply only to very low power applications such as communications and some instrument circuits. Their principle is that the energy of any spark produced should be less than the minimum ignition energy of any flammable gas or vapour present. As with flameproof enclosures, a system which is accepted for one gas or vapour may not be accepted for another.
- *Pressurising and purging* was explained in 6.2.2.

Besides these four methods, several others[3,9] which are mainly of historical interest are still in use on older plants.

6.2.4 Specifying electrical equipment for hazardous areas

Besides the normal electrical specifications for the equipment, the 'zone' where it is to be used, the apparatus class for flameproof equipment and the temperature class of the gases and vapours liable to be present must be supplied to the manufacturer. Similar information must be given for intrinsically safe systems. The ranges of flameproof and intrinsically safe equipment are much smaller than for normal industrial equipment. They are mainly used in the oil, gas and chemical industries.

6.2.5 Inspection and maintenance of protected equipment

The importance of maintaining the integrity of flameproof equipment must be appreciated, especially by maintenance personnel. When the cover plate or a cable connection of a flameproof enclosure is removed it must be correctly replaced. The whole principle of flameproofing depends on the maintenance of extremely small gaps between the enclosing box and covers, cable conduit, etc. In the same way, the integrity of a pressurised control room is lost if the pressurising fan is switched off or the ducting bringing air to the fan from a non-hazardous area is removed or leaking. If the ignition protection is to be more than a cosmetic exercise, it needs regular inspection by a competent person who understands the principles on which it is based, has the equipment to check critical gaps, knows what faults to look for and has the authority to ensure that his or her advice is followed promptly.

6.3 Static electricity (including lightning)

There used to be a joke in fire protection circles that whenever an investigator could not discover what started a fire, he put it down to static electricity. (In the same vein, metallurgists were accused of attributing every metal failure they could not explain to stress corrosion.) Today the assessment of static risks is made quantitatively by calculation and measurement where needed, and experts prescribe appropriate precautions[10]. Much of this should be done when the installation is designed. The discussion here is mainly qualitative and intended to help the non-expert.

Static electricity is generated whenever two different materials in contact are separated or rubbed together. Eventually an electric field powerful enough to break down the resistance of the air-gap separating the objects, different potentials may be built up. This breakdown strength varies with the gap and for flat surfaces is about 30 kV per centimetre. Although the main hazard of static electricity is ignition, it can produce small shocks in humans and false readings of sensitive instruments and cause dusts to deposit in unwanted places and powders to agglomerate, thus preventing their free flow in hoppers, etc. Some electronic systems are very sensitive to static electricity and special precautions are needed to prevent the accumulation of static charges in computer rooms.

Static electricity is used in some processes – for removing dusts from gases, for paint spraying and in modern printing.

The current flowing in systems where static electricity is continuously being formed by the movement or flow of material is low, between 1 and 1000 nA for liquids flowing in pipelines and between 10 and 100 000 nA for powders leaving a grinder. Potentials of 10 to 40 kV can readily build up unless the charge is conducted away from the material as fast as it is formed.

The maximum energy of a spark from an object carrying a charge to an earthed conductor is given by:

$$E = \frac{CV^2}{2}$$

where E = energy, J (joules); C = capacity of object, F (farads); V = potential, V (volts).

If the charged object is a conductor, nearly all of this energy is released in the spark, which is heard as a 'crack', but if the object has a low conductivity much of the energy is spent in overcoming its resistance.

The capacitance of a man is about 250 pF, that of a metal drum about 150 pF, and of a tin lid about 10 pF. Thus the maximum energy of a spark from an unearthed metal drum with a potential of 60 kV into which a powder is flowing, to a nearby earthed conductor, is about 0.2 mJ. This compares with the following spark energies required to ignite various materials:

Flammable dust clouds	5 mJ upwards
Mixtures of flammable vapours in air	0.01 to 1 mJ
Sensitive explosives	0.001 mJ upwards

The energy needed to ignite various flammable material/air mixtures is lowest when they contain about enough air for complete combustion. The ignition energy for a dust cloud increases with the particle size of the dust, while that for gases and vapours follows their grouping in BS 5345 [10.2.4].

A dust cloud generally requires a more powerful spark to ignite it than a vapour/air mixture because of the extra heat required to raise the solid particles to their ignition temperature. Besides spark discharges, corona discharges (of low energy) which can only ignite the most sensitive materials can occur from points on charged objects. These are accompanied by a faint glow and sometimes a hissing sound. If a stream of a non-conducting powder flows into the earthed metal container, the electrical energy may be released as a harmless corona discharge as it enters the container. The same can apply when a stream of non-conducting liquid enters a tank from the top. But if the powder or liquid strikes a metal object separated from the bin or tank by some non-conducting material, that object may build up sufficient electrical energy to give an intermittent spark discharge with sufficient energy to ignite any flammable vapour present, or even the powder itself.

A brush discharge is a third type of static discharge which occurs between discrete areas on the surface of a non-conductor and a conductor. This has little energy except when it involves a sheet of a high-resistivity non-conductor with large opposing charges on its two sides. It is then known as a propagating brush discharge and can be dangerous.

Three conditions are needed for static electricity to produce a fire or explosion:

1. A flammable vapour/air or powder/air mixture must be present;
2. An electric charge must have built up, generally on a conducting object insulated from surrounding ones, with sufficient potential to discharge as a spark to a nearby object (usually an earthed one);
3. The spark must have sufficient energy to ignite the surrounding flammable mixture.

To prevent ignition by static means eliminating one or more of these conditions. The principal methods of doing this are by:

1. Earthing (or bonding) of stationary conductive equipment;
2. Increasing the conductance of insulating floors, footwear, wheels and tyres for personnel and trucks, etc.;
3. Increasing the conductivity of non-conductors by incorporating conducting additives, surface layers and films and by humidification of the atmosphere;
4. Increasing the conductivity of the atmosphere by ionisation.

6.3.1 Earthing and bonding

'Bonding' is used here in the electrical sense of connecting the objects to be bonded by an electrical conductor, usually metallic. It is done to eliminate potential differences between objects, whereas earthing eliminates a potential difference between an object and the ground. Neither are effective when the objects themselves are non-conductors. Earthing, which is generally the preferred choice, may also serve as protection against lightning and for electrical circuits, but it must be effective. The earth connections used for electrical circuit protection are usually adequate for static electricity provided they are not disconnected when the circuit is moved or changed.

In areas where flammable materials may be present, all conducting equipment which is separated by insulating materials with a resistance to earth of more than one megohm needs to be identified and earthed. (A lower resistance is needed where sensitive explosives and unstable chemicals are handled.) Normal plant construction generally provides adequate earthing, but for equipment mounted on rubber anti-vibration mountings, special earthing wires may be needed. These need to be checked after the equipment is maintained, moved or painted. Similarly, when adhesive or other insulating joints are made in machines and structures, special means such as conducting pads or wires must be used to ensure that they are electrically bonded and/or earthed. For transfer of flammable liquids, antistatic hose which has a metal fibre running through it is available. Dust-collecting bags with similar protection can also be obtained.

Expert advice on earthing is required for equipment which is insulated from the earth as a result of cathodic protection.

6.3.2 Protecting personnel and equipment on wheels

The resistance of the human body is low enough for it to readily conduct static charges. These it can accumulate in various ways if it is insulated from the earth, e.g. by walking across a pile carpet. Such charges on a person can damage sensitive electronic components, cause personal shocks with reflex actions which lead to accidents, and even discharge as sparks capable of igniting flammable mixtures. The resistance of most footwear worn in contact with clean floors of wood, cork or concrete is normally low enough to prevent accumulation of dangerous static charges by the body, but carpets, some synthetic floor coverings and some rubber and plastic footwear have high resistance and promote body accumulation of static charges. Personnel resistance monitors[10] can be used to check the resistance of footwear in use. Where necessary, specially warranted antistatic footwear whose moderate electrical resistance is still high enough to protect the user from shock caused by contact with mains electricity is available[11]. This protects against static ignition of most gases and vapours, but not against that of explosives. Where these are handled, conductive flooring and footwear of much lower resistance are needed[11, 12]. Trolleys, etc. are protected by tyres of rubber which contains a conducting additive[13].

6.3.3 Increasing the conductivity of high-resistance materials

These materials fall broadly into two classes, organic and inorganic. Of the former, most natural products are sufficiently conducting to prevent the build-up of static charges providing the atmosphere is not very dry. Some synthetic plastics, rubbers and fibres have very high resistances. These are available with antistatic additives incorporated into them to increase their conductivity[13]. Of the inorganic class, ceramics and glass normally have very high resistance. This can be reduced by applying surface-conducting films, which is usually done during manufacture.

In a humid atmosphere with a relative humidity of 65% or more, a thin conductive moisture film is formed on the surface of most materials. While this can be achieved (at a cost) by air conditioning, such moisture contents are too high for optimum comfort.

6.3.4 Increasing the conductivity of the atmosphere

The conductivity of the atmosphere can be increased locally by special devices which produce ions in it. Some of these are electrically operated, while others depend on the use of a radioactive source. Both methods bring potential hazards.

6.3.5 Hazards and precautions with gas discharges

When clean flammable gases (free of dusts and liquid droplets) discharge into the air (e.g. through vents and leaks) without forming solid particles or liquid droplets as they expand, they seldom acquire sufficient charge to ignite. Escapes of hydrogen from high-pressure plants (such as ammonia) are an exception and sometimes ignite, burning with an almost invisible flame. But when droplets or particles are contained in or formed from the gas which is discharging, these can acquire sufficient charge to ignite flammable vapours which are present. The release of carbon dioxide from liquid-filled cylinders usually produces particles of solid carbon dioxide and ice. This caused at least two disastrous explosions, one in West Germany in 1953 when an underground petrol store was being officially opened and a carbon dioxide extinguishing system was being demonstrated. Most of the assembled VIPs were killed! Carbon dioxide should not be used for this purpose unless special precautions have been taken to prevent the formation of solid particles in the discharge. The discharge of most liquefied gases is also accompanied by charged liquid droplets, while steam discharges such as 'steam curtains' used for the emergency dispersion of flammable vapour escapes in oil refineries and petrochemical plants can produce charged droplets which might cause ignition. One safeguard lies in ensuring that all conductors near the point of gas or steam discharge are earthed. Techniques intended for the inerting of flammable atmospheres should be checked by an expert on static hazards before they are adopted.

6.3.6 Hazards and precautions with flammable liquids

The resistance of many flammable liquids including ethanol, acetone and crude oils is sufficiently low to prevent risk of their developing dangerous static charges when they are stored in conducting and earthed containers. But flammable distilled hydrocarbons, including gasolines, naphthas and toluene, have high resistance which enables dangerous static charges to develop during storage even when their containers are conducting and earthed (unless they contain an antistatic additive).

Even low-resistance liquids can accumulate dangerous charges when they are in contact with conducting and earthed pipes, tanks and equipment. This applies to liquid droplets produced when a jet of falling liquid breaks up in air.

Proper containers which are earthed or stand on an earthed metal surface should always be used for the small-scale handling of flammable liquids. When pouring any flammable liquid, both containers as well as any hoses and nozzles used should be bonded and earthed (Figure 6.2). Such liquid containers should be of metal, except for sizes smaller than 2 litres, where plastic ones may be used[3].

Nozzle in contact
with container

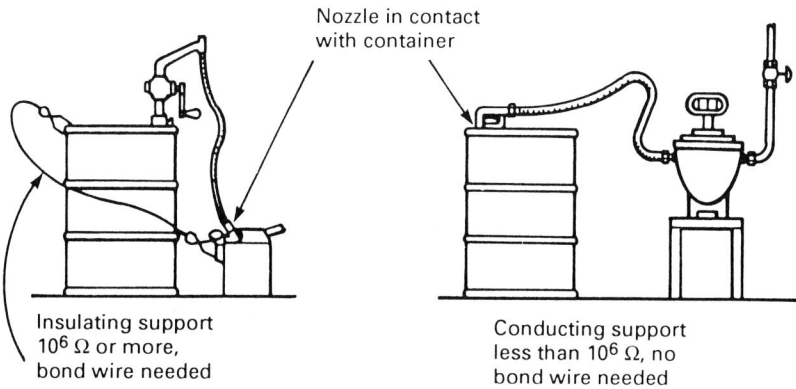

Insulating support
$10^6 \, \Omega$ or more,
bond wire needed

Conducting support
less than $10^6 \, \Omega$, no
bond wire needed

Figure 6.2 Examples of bonding and earthing

Liquids flowing in pipelines produce electrical charges at rates which increase rapidly with velocity. The presence of a second liquid phase (such as water) and constrictions such as valves and filters increase the rate of charge generation. Liquids falling freely into a tank also acquire considerable charges. The presence of a flammable atmosphere in the vessel or tank should be prevented if at all possible, e.g. by the use of a floating roof tank or by inert gas blanketing [10.2.2]. Other important precautions are:

1. *Earthing and bonding*. Storage tanks above ground for flammable liquids should be earthed with an uninsulated wire which is easily inspected for damage. Transfer systems, road and rail tankers and any drums or cans into which the liquid is to be transferred should be earthed. Strong spring-loaded earthing clips which make metal contact through any surface layer of rust or paint should be used. They should be fixed before the hose is connected and not removed until after it has been disconnected.

2. *Position of liquid inlets*. Flammable liquids should enter a tank or vessel through a bottom inlet or dipleg to avoid free fall.

3. *Flow velocity*. For high-resistance liquids such as gasoline, kerosene, toluene and other white oils and light hydrocarbons, the velocity in the pipe should be restricted to 7 m/s if no second phase is present and to about 1 m/s if a second phase is present. A company guide[14] gives detailed advice.

4. *Restrictions*. Valves and filters, etc. which cause restrictions should be located as far as possible from the inlet to the tank or vessel receiving the liquid.
5. *Antistatic additives*. Special additives which are effective at concentrations of 2 to 10 ppm are available for reducing the resistivity of liquids These are normal ingredients of aviation kerosenes and some flammable hydrocarbon liquids.
6. *Mixing and stirring*. Mixing of liquids in vessels containing a flammable atmosphere should be done with low-speed stirrers which are fully submersed in the lower part of the vessel.
7. *Gauging and sampling*. All metallic parts of gauging and sampling equipment used with flammable liquids should be connected to the tank or earthed. Plastic and other cords and dipsticks of low conductivity should not be used. When a low-conductivity flammable liquid has been pumped into or mixed in a tank it should not be gauged or sampled for ten minutes after pumping or mixing even when it is free of water. If it is liable to contain suspended water, it should be allowed to stand for 30 minutes before gauging or sampling.
8. *Tank cleaning with high-pressure jets or steam*. All metal parts of cleaning equipment used for tanks which may contain a flammable atmosphere should be in good electrical contact and earthed. Steam should not be introduced into such tanks until their atmosphere has been rendered non-explosive, e.g. by purging with nitrogen. Gauging and sampling should be avoided during cleaning and until all mist has settled. Monitoring for static charge generation during cleaning is advisable.
9. *Addition of powders to flammable liquids*. The addition of powders into flammable liquids in tanks or vessels can cause static charges capable of igniting the vapour/air mixture, particularly if the conductivity of the powder and/or liquid is low. Several fires are thought to have started in this way. The powder should preferably be admitted by a short earthed conducting chute or pipe extending to the liquid surface, but expert advice on a safe and practical method for particular situations may be needed.

6.3.7 Combustible powder handling

Although dust clouds require more powerful sparks than vapour/air mixtures to ignite them, most operations in which powders are produced or handled are powerful generators of static charges. Containers into which combustible powders flow should be made of conducting materials and earthed. Most industrial systems handling combustible powders in air have explosion vents of large cross section. They also often have an explosion-suppression system which releases an inerting gas into the system when the start of an explosive pressure rise is detected.

6.3.8 Removable non-metallic liners in containers

When removable non-metallic liners are used in containers used for materials wet with flammable solvents or for dry combustible powders, there is a danger of a propagating dust discharge which could ignite the vapour or powder when the liner is removed. The container should be conductive and liners with a surface resistivity of less than 10^{11} ohms should be used.

6.3.9 Machine earthing and other precautions

Care is needed to ensure that machinery is earthed and has sufficient conductance to prevent build-up of static electricity. Shafts and belting may need special brush or point collectors, and belts may be given a dressing which makes them conducting. The main point of danger is where the belt leaves the pulley wheel. The same problem arises where rolls of paper and other materials are unwound, and collecting points may sometimes be introduced close to the gap. Care is needed to prevent the collecting points themselves causing sparks which jump the gap.

6.3.10 Handling explosive and sensitive flammable materials and special situations

Specialist advice should be taken before starting operations involving materials with ignition energies below 0.2 mJ such as carbon disulphide and some explosive and unstable compounds. Such advice is also needed for electrostatic dust removal, electrostatic printing and electrostatic paint spraying.

The garments worn by personnel handling sensitive materials should have surface resistivities below 5×10^{10} ohms. Natural fibres such as cotton, flax or cotton are satisfactory at relative humidities of 65% or more and temperatures of about 20°C, but many synthetic fibres with higher surface resistivity require a special antistatic finish which usually needs to be reapplied every time the garment is washed.

6.3.11 Detection of static charges

While sophisticated methods such as electrostatic voltmeters[10] are available, a simple neon tube like that used for testing car sparking plugs provides a good indication, although care must be taken that its glow is not caused by stray currents induced by electrical circuits. Even simpler is a small bundle of dry cotton or silk fibres which are repelled from one another when placed near a static charge[15].

6.3.12 Lightning[15]

Lightning results from the large-scale discharge of atmospheric static electricity with very large voltage differences and currents of 1000 amps or more. It has caused several major fires of oil storage tanks and struck many tall buildings, including the Empire State Building, which is reported to have been struck over one thousand times. The principal items requiring protection in the UK (where thunderstorms are fairly infrequent) are storage tanks and tall buildings and structures. Protection for tank farms aims to provide sufficient points with conducting paths to earth to disperse any charge harmlessly. In providing lightning protection for storage tanks the following points are important.

1. The tanks and any other large metal objects must be well earthed.
2. Two or three earths are required, one or more of which should be a metal bar or tube sunk 2 m in most soils and at least twice this depth in sand or gravel.
3. The points of conductors must be high enough to be well clear of flammable vapours, to prevent a fire occurring from an arc at the point of contact.
4. Conductors must provide a continuous path to earth without sharp angles or bends.
5. Conductors should be well separated from metal which may provide an alternate path to earth.
6. Conductors must not pass close to flammable materials nor to places where people may be present.
7. Conductors should be of heavy copper, and be regularly inspected for corrosion, secure fixing and incidental damage.
8. Conductors should not be removed or interfered with except by a person authorised by the competent authority (i.e. the chief electrical engineer).

6.4 Physical hazards involving liquids

The following hazards are discussed in this section:

1. The thermal expansion of liquids in closed systems;
2. The freezing of liquids in pipes, valves, etc.;
3. Physical explosions which involve two phases, one or both of which are liquids and which are initially at different temperatures;
4. 'Turnover' of liquids at their boiling point in tanks and vessels (a serious hazard in the cryogenic storage of liquefied gases);
5. 'Boilover' which sometimes occurs during oil tank fires;
6. Irregular boiling of superheated liquids characterised as 'bumping'.

Other physical hazards, water-hammer, steam-hammer and cavitation are discussed in 15.3.2. While the hazards described here are deceptively simple, their elimination can be difficult and some 'solutions' cause other hazards as bad or worse than the one eliminated.

6.4.1 Thermal expansion of liquids

The problem arises when the temperature of a closed liquid-filled system increases. It is found:

1. In hot water systems;
2. In pipes and pipelines containing toxic and flammable liquids and liquefied gases;
3. In storage and transport containers filled with similar liquids and liquefied gases.

The coefficient of thermal expansion by volume $\times 10^{-3}$ of common pipe materials at 20°C ranges from about 0.03 for steels and cast iron to 0.07 for aluminium, and from 0.2 to 0.4 for polyethylenes. Figures for liquids on the same basis range from 0.21 for water to 1.12 for methanol, 1.6 for pentane and 2 or more for butane and propane. Since liquids have a very limited compressibility, very high pressures can be reached in closed liquid-filled systems.

1. *Hot water systems*
When water or any liquid is heated in a closed system which is not connected to an elevated vented expansion tank provision must be made for its thermal expansion. This is commonly done by connecting an expansion chamber containing an impervious rubber bag filled with air or nitrogen to the pipework. Some means are then needed of checking that the bag contains sufficient gas and is not punctured.

2. *Pipes and pipework containing hazardous liquids and liquefied gases*
Many pipes and pipelines carrying liquids and liquefied gases have to be provided with valves at both ends. If these are both closed when the pipe is full of liquid, and its temperature later rises, the pressure generated may distort or rupture the pipe, or blow a gasket in a flanged joint or the packing of a valve. This can occur if the pipe is exposed to hot sunlight, or if the temperature of the liquid in the pipe when the valves are closed is below ambient.

The usual solution[16] is to fit small pressure relief valves to pipe sections which are liable to be 'boxed in'. Provided the valve closes once the overpressure is relieved, the quantity of liquid released is small. But if the pipe contains scale which lodges in the valve and prevents it from reseating, its entire contents may be released.

In the case of LPGs, for relatively short pipes within the battery limits of a plant, the relief valves are often arranged to discharge onto the ground where the LPG soon evaporates. If the relief valve fails to reseat, this is revealed by ice on the downstream side of the valve. The pipe then has to be drained of liquid and the relief valve cleaned or replaced. But if the pipe is a large one, say 8 inches or more in diameter and 100 or more metres long, several tonnes of LPG may escape if the relief valve fails to reseat. In such cases the relief valve should discharge to a properly designed relief system, with an adequately sized liquid catch pot and a tall vent or flare stack [15.5.1].

Another solution is to route the discharge from the relief valve to part of the system at a lower pressure, e.g. a pressure storage vessel. In this case it is difficult to detect whether the relief valve has reseated or not, while an alternate means of disposal becomes necessary if the receiving vessel has to be emptied and isolated while the rest of the system is still in use.

Another solution is to establish a strict operating procedure whereby the line is partially drained of liquid before the second valve is closed. This needs close liaison between those responsible at each end of the line.

3. *Storage and transport containers for liquids and liquefied gases*
It is vital to leave a certain percentage of the volume of such containers free of liquid. This applies to drums and bottles containing liquids and especially to cylinders and road and rail tankers filled with liquefied gases.

Filling ratios for liquefied gas containers of different sizes, for different liquefied gases, and for use under different climatic conditions are given in BS 5355[17]. Cylinder filling plants and filling stations for road and rail tankers must be provided with positive means for ensuring that the specified filling ratio is not exceeded. Clear operating procedures must be established and those responsible for filling must be carefully trained and supervised. The Spanish camping site disaster in July 1978 where there was heavy loss of life was attributed to overfilling of a road tanker with liquid propylene[3].

6.4.2 Pipes and equipment containing liquids liable to solidify

The unintended formation of solids in pipes and equipment containing fluids is always troublesome and can be very hazardous, particularly when the fluid is water which expands when it freezes. Perhaps the worst case was the Fézin disaster of 1966[3]. This resulted from the freezing of water in a drain valve under an LPG storage sphere when both LPG and water were passing through the valve, thus preventing it from being closed. Since then, the design and operating procedures for the pressure storage of liquefied flammable gases for bulk LPG storage installations has received much attention[18]. Although different authorities have different views[19] on the safest way to drain settled water, all agree that it is essential to have two valves in series. The first valve nearest the storage vessel should be the larger of the two and of a type which can be quickly closed if the downstream valve becomes frozen in the open position. When draining water the first valve is opened wide and the flow is controlled by the second valve. Some authorities now insist on having a separate drain pot below the sphere, with a valve between it and the sphere which has to be closed whenever water is drained from the pot. In a cold climate there can then be a hazard of water freezing in the pot and bursting it.

Many hydrocarbon gases form hydrates which solidify at temperatures above 0°C. This tendency increases with the pressure, and can cause blockages in gas pipelines and valves[9]. Such pipelines must be thoroughly dried before hydrocarbon gas is introduced under pressure, and the gas must be dried to a low dew-point. Some hydrocarbon liquids solidify at temperatures above 0°C (benzene at 5.5°C, cyclohexane at 6.5°C). Many substances which are solid at the ambient temperature are moved by pipe in their molten state. For pipes carrying water and other non-flammable liquids, a common solution to the freezing problem is to 'trace' exposed pipes, valves and fittings with electrical heating tape and lag them. For some fuel oil pipes, steam tracing is used. The use of hot tracing, however, accentuates the problem of thermal expansion in 'boxed-in' pipes [6.4.1], while only flameproof electrical equipment may be used in zones where flammable gases and vapours may be present. Care must also be taken to determine the correct quantity and resistance of the tape to be used, and to safeguard against short circuits and shocks. For water pipes in which water is usually flowing and where only short lengths are exposed, it is sufficient in many climates to ensure that water is either kept continuously flowing in cold weather or that the pipe is completely drained when frost is forecast.

A serious hazard arises when part of a fire water system is frozen. I saw a large warehouse opposite a hotel where I was staying in Amsterdam burn down at breakfast-time before the fire brigade could pump water from a frozen canal.

Many accidents have been caused by trying to clear blocked pipes by partly dismantling them and/or rodding them through while the contents were still under pressure. Specific instruction and warnings are needed to suit different situations. Special techniques are sometimes needed.

6.4.3 Physical explosions or eruptions

I use the term 'physical explosion' for the sudden vaporisation of a liquid in a closed or partly closed system. The term 'eruption' is sometimes more appropriate when a blast wave is not involved. The phenomenon occurs when a liquid (usually water) suddenly comes into contact with a hot solid or another liquid, or when the energy contained in a superheated liquid system is suddenly released.

Marshall[20] uses the term physical explosion for what I call a 'mechanical explosion', namely the blast wave which follows the sudden rupture of a boiler or pressure vessel. Others might include electrical and even nuclear explosions as physical explosions. An important point to be noted is that the pressure rise resulting from a physical explosion is usually too fast for any standard pressure relief device to handle [15.3.2].

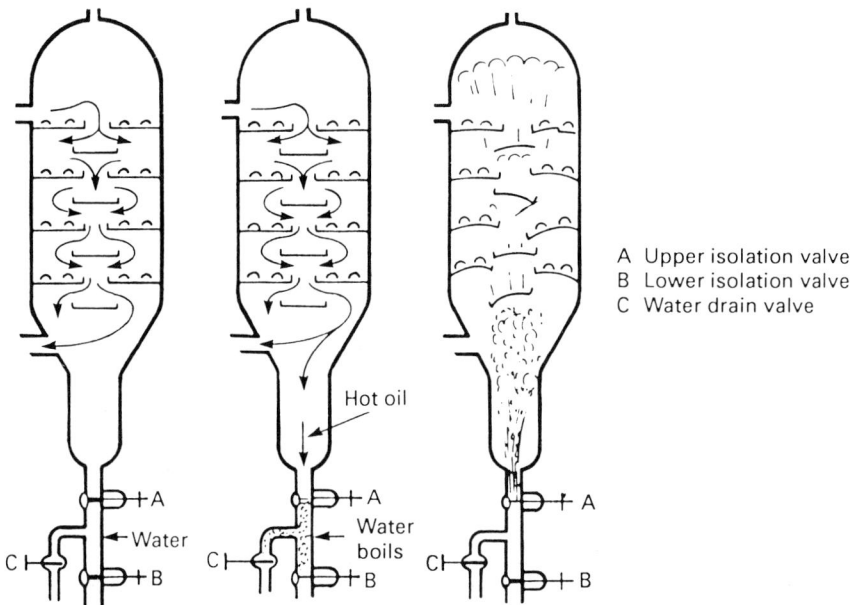

A Upper isolation valve
B Lower isolation valve
C Water drain valve

Figure 6.3 Physical explosion caused by mixing water and hot oil inside oil distillation column. The connection from valve C to the pipe joining A and B is too far above B, so that too much water is left undrained

A serious physical explosion in which seven men were killed and several injured occurred on 4 November 1975 at a steel plant at Scunthorpe[21]. This happened when molten steel was being poured into a special transport container into which water had entered through a leak in the cooling system. Molten steel was scattered over a wide area. The largest recorded physical explosion occurred during the Krakatoa eruption in 1883, when the sea poured into a submarine crater containing millions of tons of molten lava.

Physical explosions have frequently occurred during the start-up of distillation columns (especially vacuum columns) of oil refineries, when hot oil has come into contact with a pocket of water lying in a low point of the column or its associated pipework[22]. It is important to eliminate such low points as far as possible during design, and to ensure that the remaining ones are provided with drain valves at the lowest point of the pocket to remove water during start-up. An example of this problem is shown in Figure 6.3.

It is also possible for a physical explosion to occur in an unstirred tank or vessel containing water and a hydrocarbon layer above it, when one or other of these layers is heated. The phenomenon starts with boiling at the interface between the two liquid layers. This occurs at a temperature lower than the boiling point of the lower boiling liquid. The initial boiling rapidly mixes the two liquids which boil together at this lower temperature, thereby releasing a good deal of stored energy. I have shown elsewhere[23] [4.5] how this phenomenon probably triggered a pressure rise causing the rupture (a mechanical explosion) of the by-pass assembly at Flixborough, which in turn was followed by a large vapour cloud explosion.. The Pernis disaster discussed in 5.1 was another example of a physical explosion (or eruption) which led to a vapour cloud explosion. Another fire and explosion which apparently resulted from an internal physical explosion occurred on a plant in New Zealand in September 1974[24].

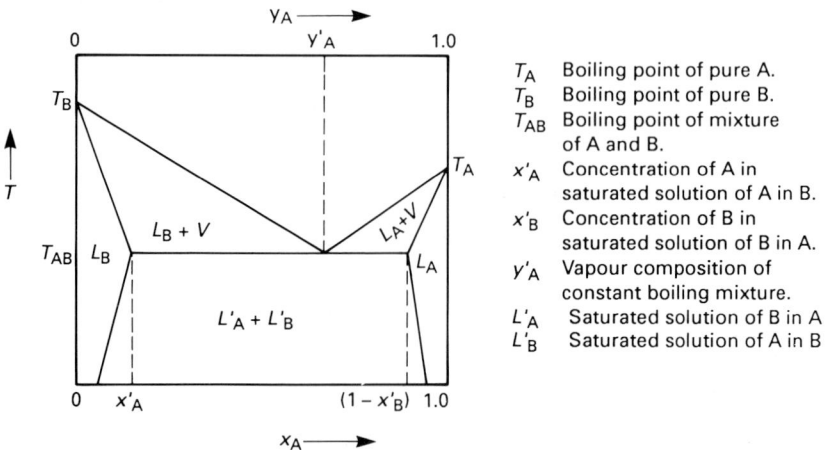

Figure 6.4 Liquid–vapour phase diagram at constant pressure for two liquids A and B of limited mutual solubility

The number of accidents of this type suggests that the hazard is still not widely recognised. The most important lesson is to avoid heating a tank or vessel containing two immiscible liquids unless one is sure that they are well stirred.

Figure 6.4 is a generalised temperature-composition or phase diagram for two nearly immiscible liquids at constant pressure. When in contact with each other these boil at a lower temperature than that of either liquid alone.

6.4.4 Rollover of LNG in cryogenic storage tanks

Rollover of liquids which are on the point of boiling is accompanied by a sudden increase in pressure and evolution of vapour and can occur in fields as different as jam making and the storage of liquefied natural gas. Its hazards are only discussed here in the last context[25], although the same principles apply generally. At the start of a rollover the entire contents of a tank or vessel may be at their boiling point, which is higher at the bottom of the tank because of the head of liquid above it. This situation is only possible when the composition of the tank contents is non-uniform and in such a way that the density of the liquid at the bottom would be higher than that at the top when both were measured at the same temperature. Due to its higher temperature, the density of the liquid at the bottom may then become less than that of the liquid at the top, causing the entire contents to roll over. As the hotter liquid rises, its hydrostatic pressure and boiling point are reduced so that it boils violently as sensible heat is converted to latent heat.

Gases are frequently stored as refrigerated liquids in insulated storage tanks or vessels at pressures only slightly above atmospheric. The word 'cryogenic' is used when the storage temperature is below −110°C. The surface of the liquid is in equilibrium with its own vapour. Refrigeration is often achieved by compressing and condensing the gas and returning it as liquid to the tank. If liquid of slightly higher density enters the tank and forms a layer in the bottom, its temperature can rise substantially before its density falls sufficiently for it to rise through the colder liquid above it. This is a rollover. It is accompanied by rapid turnover of the liquid contents, evolution of vapour and a sudden rise in pressure, the extent of which depends partly on the depth of the liquid. Refrigerated liquid storage installations must therefore be designed to withstand the maximum pressure rise which could result from a rollover. Other features which should be incorporated to reduce this hazard include[25]:

- *Top (splash) filling*. This relieves superheat in the incoming liquid. If the incoming liquid were superheated and of higher density than the bulk liquid, it would retain its superheat if it entered the tank at the bottom, where the hydrostatic pressure would prevent it from boiling;
- *Provision for circulating liquid from the bottom of the tank to the top*;
- *Provision of means for detecting a superheated state in the tank and predicting the incidence of a rollover*. This requires temperature monitoring at several levels in the tank and may also need the liquid density and/or composition to be monitored at several levels.

In the case of LNG which is predominantly methane the constituents which raise its density are ethane, propane and nitrogen. While the first two raise its boiling point, the last lowers it.

6.4.5 Boilover or slopover in oil tank fires[26]

In many oil tank fires a sudden eruption occurs when the fire has burnt for some time and consumed a considerable proportion of the oil. These cause large flames and the scattering of burning oil over a wide area, with the spread of fire to other tanks, injury to firemen, and the loss of firefighting equipment. Boilover is explained thus in the NFPA Manual[27]:

> Boilover occurs when the residues from the surface burning become denser than the unburnt oil and sink below the surface to form a hot layer which progresses downward much faster than the regression of the liquid surface. When this hot layer, called a 'heat wave', reaches water or a water-in-oil emulsion in the bottom of the tank, the water is superheated and subsequently boils almost explosively, overflowing the tank. Oils subject to boilover must have components having a wide range of boiling points, including both light ends and viscous residues. These characteristics are present in most crude oils and can be produced in synthetic mixtures.

According to this explanation, a boilover is the result of a physical explosion [6.4.3]. But since a 'heat wave' depends on the density of the hot surface layer exceeding that of the cooler oil below it, boilovers have features in common with rollovers [6.4.4].

This prompts the question 'Is water in an oil tank essential to a boilover?' At the start of an oil tank fire the surface layer loses light ends while its density falls initially because of the increase in its temperature. As the middle oil fractions evaporate its density rises again. The temperature of the lower layer meanwhile rises slowly by conduction while its density falls. By the time the densities of the two layers are again equal, the hot upper layer consists mainly of heavy ends, while the lower layer still retains most of its light ends. Mixing of the two layers could then lead to a boilover caused by the sudden vaporisation of some of these light ends.

Because of the risk of water causing a boilover, special care is needed when using water jets on oil tank fires. Jets should be used to cool the walls of tanks and other objects exposed to flames and radiation, but foam and water fog should be used for fighting the fire. According to Lees[3], the use of vertical strips of a temperature-sensitive paint on the wall of the tank can warn a trained fireman of the imminence of a boilover.

Where possible, the oil (together with any water) in the burning tank should be pumped out into an empty tank. It is sometimes preferable to let a tank fire burn itself out rather than than risk a boilover by applying water to it. Boilovers featured in some of the large losses summarised in Appendix E.

6.4.6 Irregular boiling of superheated liquids

'Bumping' and eruptions similar to physical explosions and rollovers sometimes occur when organic liquids or alkaline solutions are distilled or evaporated in clean industrial equipment and laboratory glassware. This is pronounced under vacuum. In the absence of bubbles, liquids are capable of sustaining several degrees of superheat before boiling commences, which then occurs with considerable violence, sometimes ejecting liquid from the container, sometimes rupturing it. The problem is cured by providing a source of bubbles in the liquid, e.g. pieces of a porous solid, or by arranging a small flow of inert gas through fine holes at the bottom of the boiling liquid. Laboratory workers should never point the mouth of a test tube which is being heated at anyone. The presence of circular stains on the ceilings of many a chemistry laboratory provides mute evidence of this hazard!

References

1. Marshall, G., 'Electrical fire hazards in the chemical industry', *Loss Prevention Bulletin 039*, The Institution of Chemical Engineers, Rugby (1981)
2. Dalziel, C. F., 'Effects of electric current on man', *Electrical Engineering* (February 1941)
3. Lees, F. P., *Loss Prevention in the Process Industries*, Butterworths, London (1980)
4. BS 5345 (12 parts), *Code of Practice for the selection, installation and maintenance of electrical apparatus for use in potentially explosive atmospheres*
5. Factory Mutual Corporation, *Handbook of Industrial Loss Prevention*, 2nd edn, McGraw-Hill, New York (1967)
6. BS 4683 (4 parts), *Specification for electrical apparatus for explosive atmospheres*
7. ICI Ltd-RoSPA, *IS/91 Electrical installations in flammable atmospheres*, RoSPA, Birmingham (1972)
8. Redding, R. J., *Intrinsic Safety*, McGraw-Hill, Maidenhead (1971)
9. King, R. W. and Majid, J., *Industrial Hazard and Safety Handbook*, Butterworths, London (1980)
10. BS 5958, *Code of practice for control of undesirable static electricity*
    Part 1: 1980, 'General considerations',
    Part 2: 1983, 'Recommendations for particular industrial installations'
11. BS 5451: 1977, *Specification for electrically conducting and antistatic rubber footwear*
12. BS 3187: 1978, *Specification for electrically conducting rubber flooring*
13. BS 2050: 1978, *Specification for electrical resistance of conducting and antistatic products made from flexible polymeric material*
14. Shell Chemical Co. Ltd, *Safety in Fuel Handling*, Shell, London (1963)
15. Underdown, G. W., *Practical Fire Precautions*, 2nd edn, Gower Press, Aldershot (1979)
16. API RP:520 *Recommended practice for the design and installation of pressure relieving devices in refineries*, American Petroleum Institute, New York (Part 1, 3rd edn, 1967, Part 2, 2nd edn,1963)
17. BS 5355: 1976, *Specification for filling ratios and developed pressures for liquefiable and permanent gases*
18. HSE, *The Storage of Liquefied Petroleum Gas at Factories*, HMSO, London (1973)
19. Anon., 'Safety in design of plants handling liquefied light hydrocarbons', *Loss Prevention Bulletin 042*,The Institution of Chemical Engineers, Rugby, (1981)
20. Marshall, V. C., 'Dust explosions and fire balls', paper given at Major Hazards Summer School, Cambridge, organised by IBC Technical Services Ltd, London (1986)
21. HSE, *The Explosion at Appleby-Frodingham Steel Works, Scunthorpe on 4th November 1975*, HMSO, London
22. The American Oil Company, *Hazard of Water*, 5th edn, AMOCO, Chicago (1964)
23. King, R. W., 'Flixborough – The role of water re-examined', *Process Engineering*, Morgan Grampian, London, pp. 69 to 73, September (1975)

24. *Report of the commission of inquiry into the explosion and fire which occurred at the factory of Chemical Manufacturing Company Ltd on 26th September 1974*, Government Printer, Wellington, New Zealand (1975)
25. Drake, E. M., 'LNG rollover – update', *Hydrocarbon Processing*, January 1976, pp. 119-122
26. Vervalin, C. H., *Fire Protection Manual for Hydrocarbon Processing Plants*, Gulf Publishing, Houston, Texas (Volume 1, 1984, Volume 2, 1982)
27. NFPA 30, *The Flammable and Combustible Liquids Code*, National Fire Protection Association, Boston, Mass. (1984)

Chapter 7

Health hazards of industrial substances

Substances are considered toxic if they have some adverse effect on the human body, although as Paracelsus (1493–1541) said, this all depends on the size of the dose – anything is a poison if you take enough of it. Thus while very small quantities of elements such as manganese, fluorine and iodine are essential to health, they are toxic in larger doses. Calciferol [50-14-6] or Vitamin D_2, which is essential to development in all mammals, is lethal in higher doses and is used in formulation with warfarin as rat poison[1].

Bitter experience has shown that persistent exposure of workers to relatively low levels of many industrial substances, particularly chemicals, can produce chronic disease, leading to serious disability or premature death. Many acute health hazards of the last century, e.g. silica, yellow phosphorus and lead used in the manufacture of cutlery, matches and paint, respectively, have been overcome by the use of safer materials. In recent decades their place has been taken by other hazards, some, new ones such as vinyl chloride, others, old ones in new guises, such as asbestos and chromates. It seems inevitable that new problems will continue to appear as old ones are solved.

Most readers, particularly those who have worked in production or research in the chemical and allied industries, know some of these problems only too well. I knew at least four former colleagues who worked with known carcinogens and died in youth or middle age from cancers – a professor of organic chemistry specialising in higher aromatics, a research chemist in a dyestuffs company, another working for a company making asbestos brake linings, and a physicist working on nuclear-powered submarines. I was also in Abadan in the 1940s when 40 refinery workers died from TEL poisoning [1.2.1]. Those like me who are in their seventies and enjoy good health are the lucky ones.

Thanks to the many man-made chemicals in use today, toxicology has become an advanced scientific discipline. In spite of continuous research for several decades, our present knowledge on the extent of ill-health caused by exposure to harmful chemicals and their vapours is still very limited [1.7]. Most of us rarely see more than the tip of the iceberg. Disease usually follows low-level occupational exposure to a variety of harmful substances over many years. Their effects are chronic rather than acute, and may be confused with those arising from other causes inherent in the worker's general health or life-style.

Several occupations in specific industries where workers have been exposed to particular hazardous process materials (mainly carcinogens) show appallingly high fatal incident frequency rates. A selection of the worst ones reported by Kletz[2] is given in Table 7.1. The work of Hamilton and Hardy[3] in the USA and Hunter[4] in Britain gives a clear and balanced picture.

Table 7.1 Fatal incident frequency rates in particular occupations (FIFR is defined as fatalities per 10^8 working hours, which approximately equals the number of deaths in a group of 1000 employees during their working lives) (courtesy Kletz[2])

Occupation	Cause of fatality	FIFR
Coal carbonising	Bronchitis and cancer of the bronchus	140
Viscose spinners (aged 45–64)	Coronary heart disease (excess)	150
Asbestos workers		
Males (smokers)	Cancer of the lung	115
Females (smokers)	Cancer of the lung	205
Rubber mill workers	Cancer of the bladder	325
Mustard gas manufacture (Japan 1929–1945)	Cancer of the bronchus	520
Cadmium workers	Cancer of the prostate	700
Amosite asbestos factory	Asbestosis	205
Nickel workers (employed before 1925)	Cancer of the nasal sinus	330
	Cancer of the lung	775
β-Naphthylamine manufacture	Cancer of the bladder	1200

While toxicology aims to study the effects of substances on the human body, living animals, which are less articulate than humans and enjoy fewer legal rights, are mainly used for experiment. Records of previous chemical exposure of workers also provide clues.

A very large amount of biological testing and health screening is now necessary in many countries before a new chemical is manufactured. In the USA these requirements are quite specific under the Toxic Substances Control Act[5]. In spite of this, a study made in 1984 showed that there was no toxicity data for nearly 80% of the chemicals used in commercial products and processes in the USA[6]. In the UK the obligation of employers to examine and protect workers against the health hazards of substances used in industry is implied under 'General duties' of Sections 2, 5, and 6 of HSWA 1974.

Current thinking in the EC is reflected by Council Directive No. 80/1107/EEC. This has led in the UK to the Control of Substances Hazardous to Health Regulations 1988[7] (COSHH) [2.5.6]. Other EC countries now have similar legislation. The measures called for under COSHH are discussed in 7.8.

While COSHH revokes most of the previous piecemeal legislation for the protection of workers' health in particular industries, it follows, extends and complements other recent regulations such as the Classification, Packaging and Labelling of Dangerous Substances Regulations[8]

(CPLR). Thus to the questions, 'How do we know if a substance is harmful to health, and in what way can it be harmful?' the first answer is: 'Read the label!' (Whether this always gives adequate warning, e.g. as in the case of tobacco, is debatable.) Besides protecting the health of workers, COSHH aims to extend our knowledge of how it is affected by the substances to which they are exposed at work.

Added to the fivefold classification of substances hazardous to health given in the supply provisions of CPLR (Table F.1), three overlapping classes of very dangerous substances listed in Part 1 of Schedule 1 of the first draft of the COSHH regulations[7a] are given in Table 7.2. Special precautions are needed in handling substances possessing these properties[9].

Table 7.2 Three classes and characteristic properties of substances with extreme health hazards (see Table F.1 for classification of harmful substances)

Class	Characteristic properties
Carcinogenic	A substance which if it is inhaled or ingested or if it penetrates the skin may induce cancer in man or increase its incidence
Teratogenic	A substance which if it is inhaled or ingested or if it penetrates the skin may involve a risk of subsequent non-hereditable birth defects in offspring
Mutagenic	A substance which if it is inhaled or ingested or if it penetrates the skin may involve a risk of hereditable genetic defects

Carcinogenicity concerns the ability of the material to produce cancer in human beings. Studies have suggested that well over 50% of human cancer has its origin in occupational and other man-made environments. According to data available in 1980, about 2000 chemicals were then suspected carcinogens[9]. Among the methods of testing used, the use of bacteria has proved to correlate well with the results of human exposure, and to be speedy and economic. Up-to-date lists of chemicals for which there is substantial evidence of carcinogenicity, and specific information about them, are prepared by the Carcinogen Assessment Group of the (US) Environmental Protection Agency.

Teratology is concerned with the birth of abnormal offspring when the material is administered to pregnant females. Teratogenic effects are produced by a variety of mechanisms. Their study is based mainly on administering the test material to the pregnant animal during a critical period of pregnancy.

Mutagenicity is concerned with the ability of the material to produce genetic changes. It is closely related to carcinogenicity and teratology. While difficult to investigate, it is of immense importance to the whole human race, since genetic changes once made cannot be undone, and are passed on from one generation to the next. Since our genetic make-up is the finely balanced result of long aeons of natural selection, any accidental genetic change resulting from chemical exposure is likely to be for the worse.

Before discussing the health hazards of various substances further, we first consider the roles of the professionals who deal with them.

7.1 Occupational health professionals

Professionals trained in several disciplines are employed to deal with health problems arising from exposure to harmful substances at work. Some are in government service, some are employed by industry, some work in universities and research institutions and some as consultants. While the professionals listed below have their own special fields of responsibility, because of their limited numbers there is a good deal of overlap in their work. Four types of professional are most involved with day-to-day problems in industry:

- The occupational physician,
- The occupational hygienist,
- The occupational nurse
- The dermatologist.

Professionals involved in more basic research include:

- The toxicologist,
- The epidemiologist.

Ergonomists and psychologists are also sometimes involved.

7.1.1 The occupational physician[10]

Prior to 1973, there were some 1300 appointed factory doctors in the UK. Most were general practitioners who carried out various statutory duties such as:
- Examination of young people first entering employment;
- Regular examination and in some cases certification and medical supervision of workers in processes covered by regulations;
- Investigation of gassing incidents and industrial diseases.

This system changed in 1973 with the formation of the Employment Medical Advisory Service (EMAS), under the Department of Employment. This is staffed by doctors with appropriate qualifications. Their services are provided free except for statutory medical examinations. They have powers to investigate and advise on occupational health problems on their own initiative and at the request of employers, employees, trade unions, factory inspectors and others.

Doctors employed within industry are known as occupational physicians. Their recommended duties include[10]:

1. Responsibility for the health of the whole enterprise;
2. Concern with the working environment and its health hazards;
3. Accessibility to workers for individual consultations and investigation of work-related hazards arising from them;
4. Education and advice for management and employees on work-related health problems, especially those arising from new processes, materials and equipment;
5. In a comprehensive health team, the occupational physician should be the leader and coordinator, while recognising the mastery of other specialists such as occupational hygienists in their own fields.

The occupational physician is often employed on a part-time basis, making regular visits to the establishment, which often has a well-equipped medical centre with a full-time trained nurse in charge of it. He is involved to a major extent with medical examination of workers and to a lesser extent with treatment. While he should not duplicate or take over the role of the general practitioner, skin diseases, other ailments and minor injuries arising from the working environment can often be best treated in the works medical centre. Under the COSHH regulations he will normally be the 'appointed doctor'. He will be expected to have overall control of occupational health records and health surveillance procedures called for under the regulations. He will also be responsible for medical surveillance of employees liable to be exposed to substances listed in Appendix 1 of the general ACOP published with the Regulations (Table G.2). The duties involved in medical surveillance are discussed in 7.8.1.

Medical examinations should be based on practical needs as well as statutory requirements. They fall into two groups, pre-employment examination and routine examination. Medical examinations are most needed for those whose work carries a specific health hazard.

7.1.2 The occupational hygienist

The British Occupational Hygiene Society, which is open to anyone with an interest in the subject, has about 1500 members, most of whom are professionally involved in it. The Institute of Occupational Hygienists, which has an entrance examination, has a somewhat smaller membership. Several universities and technical colleges have courses in occupational hygiene. About 1000 British occupational hygienists are understood to be employed in industry. Some large companies have well-established occupational hygiene departments whose duties include the monitoring of hazardous substances in the working atmosphere [7.7]. In medium-size and smaller companies with no occupational hygienist, this work may be done by a staff chemist or safety professional, or by one of several consulting firms which offer industrial hygiene services on a fee-paying basis.

In the USA the profession received considerable stimulus with the passing of the Occupational Safety and Health Act in 1970. The COSHH Regulations should provide a similar stimulus in the UK. The American Industrial Hygiene Association has given clear definitions of the profession and its practitioners, including their training[11] which should equip them:

1. To recognise the environmental factors and stresses associated with work and work operations and to understand their effect on Man and his well-being;
2. To evaluate, on the basis of experience and with the aid of quantitative measurement techniques, the magnitude of these stresses in terms of ability to impair Man's health and well-being;
3. To prescribe methods to eliminate, control or reduce such stresses where necessary to alleviate their effects.

The stresses which concern the industrial hygienist are chemical, physical, biological and ergonomic, and thus basically include all the health hazards considered in this book.

To evaluate health hazards in the working environment first requires detailed standards on tolerable levels of such hazards, and second, means of measuring the levels at which they are present. The standards with which we are mainly concerned in this chapter are the Occupational Exposure Limits for airborne substances given in HSE's Guidance Note 40[12] and discussed further in 7.5. Strategies for such monitoring are suggested in HSE's Guidance Note 42[13] and discussed further in 7.7. This work of the occupational hygienist should complement that of the occupational physician in monitoring the health of the workers.

As well as this monitoring task, the occupational hygienist is trained to recommend remedies to problems which are revealed in the course of it. Such remedies are basically of three types:

- Finding a less hazardous substitute for the harmful substance which causes the problem;
- Employing suitable means for reducing the intensity of the hazard in the working area. Typical examples are the use of local exhaust ventilation, and the substitution of a canned or diaphragm type pump for one with a packed gland when handling highly toxic materials;
- Providing personal protective devices such as dust masks and respirators and ensuring that these are worn. This is a remedy of last resort.

7.1.3 The occupational nurse[14]

Trained nurses are employed in many manufacturing establishments. They take on many of the duties of the occupational physician, both in medical examinations (e.g. of sight and hearing) and in the treatment of some injuries and diseases, particularly those of the skin. Most manufacturing enterprises have a first-aid centre or medical department with a full-time occupational nurse in charge during day working hours. Emergency first-aid treatment needed during shift work when the nurse is not present is given by trained voluntary first-aid workers, as required by the Health and Safety (First-Aid) Regulations[15]. When nurses are available to treat persons who are injured or suffering from exposure to harmful substances, they are usually in the best position to handle reports required under RIDDOR[16] [2.5.5] and other regulations, and to ensure that all relevant details provided by the injured or affected person, especially in regard to causation, are correct. In doing this they often become involved in occupational hygiene work. Many of the responsibilities of medical record keeping, which are more demanding under COSHH, fall on the shoulders of the occupational nurse.

7.1.4 The dermatologist[17]

The dermatologist is primarily concerned with diagnosing, treating and preventing skin diseases, many of which are caused by allergic reactions to particular substances. Many skin disorders have psychological causes and a good dermatologist has to be a bit of a psychologist as well. In diagnosing allergens that cause skin complaints in particular workers, the dermatologist can be assisted by observations made by other workers, supervisors, nurses, physicians and safety professionals.

7.1.5 The toxicologist[18]

The toxicologist is generally a research scientist specialising in the poisonous effects of substances on the human body. HSC's Advisory Committee On Toxic Substances (ACTS), which represents the views of British toxicologists, advises HSE on the subject. Methods used by toxicologists for assessing the toxicity of substances used or produced in the process industries include:

- Epidemiological studies [7.1.6],
- Experiments on animals,
- Experiments with micro-organisms.

Rats are mainly used for testing the toxicity of chemicals, despite protests from animal welfare organisations. Strict standardisation of several factors in such tests has been established by the EC to give statistically valid results when a group of animals is exposed to a chemical. These factors include route of admission (oral, dermal, inhalation), strain of rat, sex, age, temperature, diet, housing, season and time of observation. The toxicity of different substances and their effects on the human body are discussed in 7.3.

7.1.6 The epidemiologist[10]

The epidemiologist tries to diagnose the causes of outbreaks of diseases from the activities and environments of the population exposed to them. He is well versed in the use of statistics and computers. He is involved in the preparing questionnaires for surveys involving large numbers of people, and in the analysis of the data obtained from them. While every occupational physician and most occupational nurses like to don an epidemiologist's hat from time to time, full-time epidemiologists are more concerned with the frequency and distribution of diseases in defined populations than with day-to-day medical practice.

Epidemiologists have played an important role in bringing to light the health hazards of several materials which were previously thought to present little danger.

7.2 How harmful substances attack us

To produce a harmful effect a substance has either to attack an external surface of the body and/or to enter it in some way before attacking an internal organ. Three modes of entry need to be considered in the industrial context: by swallowing, through the skin, and by breathing. (Other modes used medically are by injection into a vein, muscle, or under the skin.) Once a toxic substance has entered the blood-stream it is transported to most internal organs, one or more of which may be attacked.

7.2.1 External attack

Some corrosive liquids and solids, particularly acids and alkalis and strong oxidising agents, attack the skin directly, causing inflammation and sometimes destroying living cells. Many organic liquids, especially solvents, dissolve and remove natural oils and fats and so render the skin more vulnerable to injury. Inflammation is a biological response to tissue injury in which the flow of blood increases and particular blood cells migrate to the affected area. It is one of our immune mechanisms for the repair of tissue damage[19]. This response can become hypersensitive after repeated exposure to small traces of many substances, e.g. ethylene diamine. The result is allergic dermatitis[17]. Sometimes the action of strong light on the skin exposed to the chemical is needed to complete the process, known then as photo-sensitisation. Creosote is a common example of this.

The skin is a sensitive indicator not only of symptoms of external attack but also of emotional reactions, which may result in pallor or blushing. Some allergies respond to psychiatric treatment. The development of allergic symptoms in one or two workers out of a number sharing the same environment can create special problems. The first step is to identify the material responsible (the allergen). This may involve special 'patch' tests under the supervision of a dermatologist[17]. If total protection against it cannot be guaranteed, it is advisable to transfer the worker to another job where he or she will no longer be exposed to the allergen.

The eyes are even more sensitive to harmful substances than the skin, and are much affected by sensory irritants in the atmosphere – fine dusts, and gases such as ammonia and sulphur dioxide. Because the eyes are so vital, there are many situations where eye protection [22.6] is essential. Gloves [22.3] and special clothes [22.2] to protect the skin are also needed in many of these situations.

A large number of substances can enter the body and attack it internally without affecting the eyes and skin.

7.2.2 Ingestion (swallowing)

Of the three methods of entering the body, swallowing is the most easily controlled, since eating and drinking are deliberate acts which should not be necessary at the workplace. Entry of many toxic materials into the body by ingestion is also less dangerous than by other means, since oral toxicity is in many cases lower than toxicity from inhalation or skin penetration[6]. The provision of adequate lockers, changing rooms, washrooms, canteens and restrooms and the enforcement of strict standards of hygiene and housekeeping are, however, essential in places where toxic materials are handled or processed. A total ban on eating, chewing, sucking, drinking and smoking in the workplace may be necessary when toxic or harmful materials are present. The dangers of putting fingers, pencils, etc. on which toxic dust has settled into the mouth should be pointed out, as such actions are often done unwittingly. The sucking by mouth on the ends of flexible tubes to start syphoning liquids from carboys and other containers should be banned, and proper equipment for the safe and easy transfer of liquids should be provided. Labels, small brushes and sewing thread should not have to be licked, especially when toxic substances are around. In laboratories, the use of mouth pipettes for measuring and dispensing solutions and of laboratory beakers for tea and other drinks should be prohibited and proper alternatives provided.

7.2.3 The dermal route (entry through the skin)

Several toxic liquids and their vapours pass readily through the skin into the bloodstream. These include tetraethyl lead, carbon disulphide, hydrogen cyanide, aromatic hydrocarbons such as benzene and toluene, nitro-compounds such as nitrobenzene, trinitrotoluene, and nitropropane, aromatic amines such as aniline, polychlorinated hydrocarbons such as trichloroethylene, organic phosphates, nicotine, and several insecticides and herbicides.

Outer clothing of PVC-impregnated fabrics with a nylon or terylene base offers good protection against most accidental chemical splashes, with the exception of some organic solvents and hydrofluoric acid [22.2.5]. Supposedly impervious outer clothing only aggravates the problem, however, if the liquid penetrates it and is absorbed on underclothing. Where this is suspected, particularly in a worker who has collapsed, one of the first actions should be to remove his clothing, wash the affected area of his body with soap and warm water, and dress or wrap him in clean underclothing or pyjamas. If he is taken to hospital by ambulance while still wearing contaminated clothing, he may be dead on arrival. It is important to inform suppliers of protective clothing of the chemicals and harmful substances against which protection is needed, and secure their help and advice.

Health and safety precautions should include listing all liquids used which are absorbed through the skin, warning workers of the dangers of their contact with skin and clothing, and providing showers or suitable washing facilities and clean clothing for anyone affected.

7.2.4 Inhalation (entry by breathing)

The main route taken by toxic substances into the human body is by inhaling. The human respiratory system (Figure 7.1) consists of an upper part, the nose, mouth and larynx, and a lower part, the trachea, bronchus, bronchioles and alveoli. The last two are the gas-exchange components of the lungs, and have a light, porous and spongy structure. The alveoli, of which there are a very large number, are blind cup-like pouches with thin elastic walls through which gases and vapours can pass readily into and out of the bloodstream. The system, which has a capacity of 5–50 l/min of air, is an easy route for contaminants to enter the body. Most gases and vapours reach the alveoli, but ones like ammonia and hydrogen chloride which are very soluble in water are absorbed before they get so far.

Whether particles will enter the respiratory system, how far they are likely to penetrate, and whether they will be trapped depends largely on their size. Particles whose mean diameters are greater than 50 μm do not usually enter it. Those with diameters greater than 10 μm are deposited in

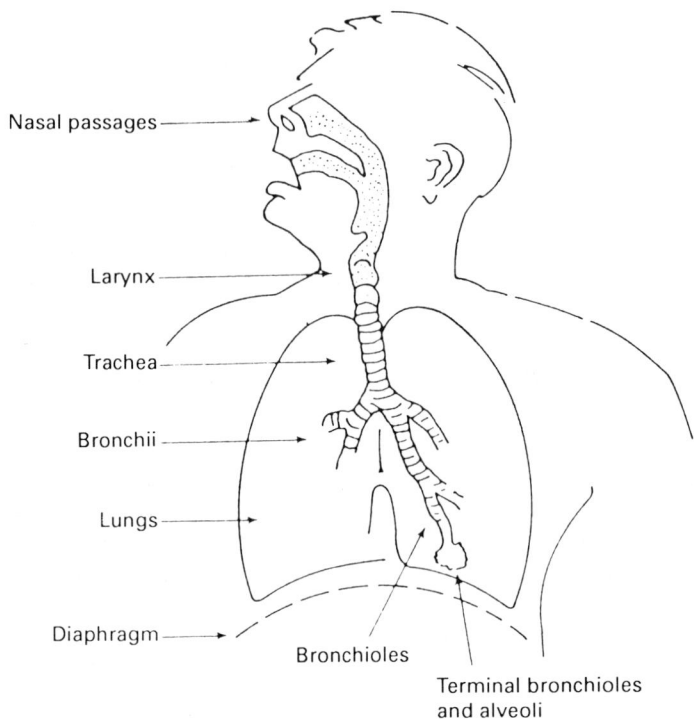

Figure 7.1 The human respiratory system

the upper respiratory tract. Those in the range 2–10 μm are deposited progressively in the trachea, bronchi and bronchioles. Only those smaller than 1–2 μm reach the alveoli. These are generally invisible to the naked eye, and are the most dangerous. Fibres with diameters of 3 μm and less and lengths of up to 50 μm can enter the lung; those longer than 10 μm tend to be trapped there. Fibres with diameters less than 1.5 μm and lengths up to 8 μm have the greatest biological activity. Asbestos was once the worst offender. Now that glass fibres with the same physical characteristics are becoming common, they may present an equal hazard.

7.3 Effects on body organs

Organs frequently affected are the respiratory system itself, the blood and bone marrow, the liver, the kidneys, and the brain and central nervous system. Some chemicals have specific effects on organs such as the bladder, the pleural cavity, the gastrointestinal tract, the prostrate, the nasal cavities and larynx.

7.3.1 The respiratory system

The natural protection of the respiratory system includes:

- A wet filter formed by the fine hairs (cilia) of the nose and trachea. Particles trapped there are washed out by mucus which can be eliminated by sneezing, blowing the nose, spitting or swallowing.
- Involuntary nervous contraction of the bronchioles which is triggered by various air contaminants and restricts breathing. This results in a struggle for breath and escape into cleaner air. Some subjects become hypersensitive to pollens and chemicals such as diisocyanates and fine spray containing platinum salts, which produce symptoms of asthma and hay fever.

Contaminants which primarily affect the respiratory system fall into five groups:

- Reactive gases such as ammonia, chlorine and sulphur dioxide; these irritate the upper respiratory system, swell the walls of the airways and cause involuntary flight reactions.
- Substances such as isocyanates, chromates and some wood dusts which cause bronchial restriction, and to which sensitive persons may become allergic.
- Most mineral dusts; some such as iron oxide appear merely to block the nasal passages and reduce breathing capacity; others such as silica and asbestos cause harmful and irreversible changes in the lung structure.
- Some contaminants which include blue asbestos, chromium and arsenic compounds and some complex hydrocarbons found in tars lead in time to the growth of malignant tumours in the respiratory system.
- Asphyxiant gases such as carbon dioxide and nitrogen which are all but inert; when present in high concentrations these reduce the oxygen concentration of the atmosphere to levels below those needed for normal work (about 18%) or life itself (about 10%).

7.3.2 The blood and bone marrow

The blood which is the internal transport medium for the whole body and the bone marrow which produces the red and white blood cells as well as platelets which initiate blood clotting following a wound are attacked by carbon monoxide, lead compounds, benzene and TNT, among other compounds.

7.3.3 The liver

The liver is the main detoxifying organ in the body. Most chemicals entering the body reach it and are converted there to other compounds which may be more or less toxic than the original ones. Most are more soluble in water than the original compounds, and end up in bile which is removed in the urine or faeces. Although the liver is fairly resistant to toxins, massive doses will inflame it, causing hepatitis and jaundice. Chlorinated hydrocarbons including insecticides such as DDT damage the liver, and vinyl chloride monomer causes a malignant liver tumour known as angiosarcoma.

7.3.4 The kidneys

The kidneys maintain the required levels of water, salt and pH in the body. They are damaged by compounds of mercury, cadmium and lead and by some organic compounds which include carbon tetrachloride.

7.3.5 The brain and central nervous system

The brain and central nervous system are affected by a number of organic compounds which include ethanol, nicotine and other alkaloids, diethyl ether and other anaesthetics, and a range of volatile solvents which are mild narcotics. The vapours of mercury and many volatile organo-metallic compounds also affect the central nervous system. The effects of many compounds on the central nervous system are fairly specific, producing hallucinations, exuberance, depression, tension, sedation, etc. Most of these symptoms can also have psychological causes.

7.4 Units and classes of toxicity

Table F.1 gives the five toxicity classes of substances harmful to health which are listed in the supply provisions of the CPL regulations. The units of toxicity used are discussed next.

7.4.1 Units of toxicity

The main unit is the lethal dose for rats which is determined under standard laboratory conditions [7.1.5]. It is usually expressed as LD_{50}, defined as that dose administered orally or by skin absorption which will cause the death of 50% of the test group within a 14-day observation period[20]. It is generally expressed as mg or μg of substance per kg of body weight of the animal (mg or μg/kg). All test conditions must be standardised, with proper controls. A number of experimental animals of as uniform a population as possible must be sacrificed in the test, since individual responses vary considerably. The results must be subjected to rigorous statistical analysis. The LD_{50} unit was developed in 1927 for the standardisation of drugs. It is now used for virtually all chemicals and has legislative backing.

Since LD_{50} is based on oral or dermal entry into the body, another unit, the lethal concentration LC_{50}, is used for airborne materials which are inhaled. This is defined as the concentration of airborne material, the four-hour inhalation of which results in the death of 50% of the test group within a 14-day observation period[20].

7.4.2 Classes of toxicity

Criteria for the classifications 'very toxic', 'toxic' and 'harmful' used in CPLR (Table F.1) are given in Table F.2. More extensive toxicity ratings of substances were given by Hodge and Sterner[21], who placed them in six classes, according to their probable lethal human dose taken orally (Table 7.3).

Table 7.3 Toxicity classes (Hodge and Sterner[21])

Class	Toxicity		Probable oral lethal dose (human) for 70 kg person
6	Super-toxic	<5 mg/kg	A taste (less than 7 drops)
5	Extremely toxic	5–50 mg/kg	Between 7 drops and 1 tsp
4	Very toxic	50–500 mg/kg	Between 1 tsp and 1 oz
3	Moderately toxic	500–5000 mg/kg	Between 1 oz and 1 pint
2	Slightly toxic	5–15 g/kg	Between one pint and one quart
1	Practically non-toxic	15 g/kg	More than one quart

Chemicals in the super-toxic class are included in the 'very toxic' class shown in Table F.2. The CIMAH Regulations[22] [2.4.2] also recognise such 'super-toxins', and list a number of them.

7.4.3 The NFPA health hazard rating[23]

The criteria for NFPA's health hazard ratings of chemicals are given in Appendix B along with those for reactivity and flammability. The main purpose of these ratings is as a guide for fire-fighting, where exposures are short (from a few seconds up to an hour). The resulting health hazards arise both from the properties of the chemicals themselves and from those of their combustion products. The hazard rating for any chemical indicates the nature and degree of protection required by those exposed to its spillages and fires.

7.5 Occupational Exposure Limits (OELs)

While LD_{50} and LC_{50} values are derived from experiment, OELs (a more arbitrary set of values) have been established and published in various countries for the control of several hundred airborne substances in the working atmosphere. They are the maximum concentrations in air of the substances which should not be exceeded in the breathing zone of workers. OELs are given for gases, vapours, liquid droplets and solid particles, but not for substances which are hazardous solely through their radioactive or pathogenic properties.

Published OELs apply to single substances only. In the absence of other data the effect of two or more toxic gases/vapours is estimated by adding the fraction:

$$\frac{\text{(concentration ppm)}}{\text{OEL (ppm)}}$$

for each of them. If the sum of these fractions exceeds unity, the OEL for the mixture is assumed to be exceeded. But since some toxic/irritant substances tend to neutralise each other while others interact to produce more serious results than the sum of their individual effects (synergism), expert advice is often needed when dealing with two or more toxic/irritant airborne substances present at the same time.

As more has been learnt about the toxicities of some substances, their OELs have had to be reduced. The OEL for vinyl chloride which was set at 500 ppm in 1962 has subsequently been reduced to 3 ppm on an annual basis, since it was discovered in 1974 to be the cause of a rare liver cancer. Other sharp reductions have been found necessary for benzene, acrylonitrile, carbon tetrachloride and many other compounds, and more are expected to follow as knowledge increases. Compliance with a published OEL is thus no guarantee that the atmosphere is harmless, and the concentrations of all unnatural airborne contaminants should be reduced as far as is reasonably practicable.

Limits for gases and vapours are specified both as ppm by volume and as mg per m^3 in air, but limits for solids and liquid droplets can only be given as mg per m^3.

7.5.1 OELs for the UK

OELs for the UK are set, published and revised annually by the HSE[12]. Until 1984 they were based on those recommended and published in the USA by the American Conference of Governmental Industrial Hygienists and known as Threshold Limit Values (TLVs). OELs in the UK fall into two categories:

1. *Control Limits.* These are given in regulations, ACOPs, EC directives or have been agreed by HSC. They are judged to be 'reasonably practicable' for all work activities in the UK. Failure to comply with Control Limits may result in enforcement action.
2. *Recommended Limits.* These are limits recommended by HSE, usually on advice from HSC's Advisory Committee on Toxic Substances. They provide realistic criteria for plant design and engineering and control of exposure, and assist in the selection and use of personal protective equipment.

'Maximum exposure limit' (MEL) and 'occupational exposure standard' (OES), both used in the COSHH regulations, are new terms for 'Control Limit' and 'Recommended Limit' respectively. The latest ideas on toxic substances which may be reflected in future Guidance Notes EH 40 are given in the *Toxic Substances Bulletin*, published by HSE.

The number of substances and groups of similar compounds for which both categories of limits are given has risen steadily and Guidance Note EH 40/89 contains over 700 entries. Table 7.4 lists the substances assigned maximum exposure limits in Schedule 1 of COSHH. This corresponds to the Control Limits listed in EH 40/89. Asbestos, coal dust and lead which are subject to other regulations are not included in Table 7.4.

OELs for toxic substances which feature in Part 2 of Schedule 3 of the CIMAH regulations are given in Table 7.8.

HSE Guidance Note EH 40/89 gives two sets of values for many substances – for long-term (usually 8-hour) and short-term (normally 10-minute) exposures. Specific short-term exposure limits are given for substances for which brief exposure may cause acute effects. Both are expressed as time-weighted averaged concentrations (TWAs) over the period specified. For substances for which no short-term limit is listed,

Table 7.4 List of substances assigned maximum exposure limits

Substance	Long-term max. exposure limit (8-hour TWA value)		Short-term max. exposure limit (10-minute TWA value)		Notes
	ppm	mg/m^3	ppm	mg/m^3	
Acrylonitrile	2	4	–	–	Sk
Arsenic and compounds, except arsine and lead arsenate (as As)	–	0.2	–	–	
1,3 Butadiene	10	–	–	–	
Cadmium and cadmium compounds, except cadmium oxide fume and cadmium sulphide pigments (as Cd)	–	0.05	–	–	
Cadmium oxide fume (as Cd)		0.05	–	0.05	
Cadmium sulphide pigments respirable dusts (as Cd)		0.04	–	–	
Carbon disulphide	10	30	–	–	Sk
Dichloromethane	100	350	–	–	
2,2'-Dichloro-4,4' methylene dianiline (MbOCA)	–	0.005	–	–	Sk
2-Ethoxy ethanol	10	37	–	–	Sk
2-Ethoxy ethyl acetate	10	54	–	–	Sk
Ethylene dibromide	1	8	–	–	
Ethylene oxide	5	10	–	–	
Formaldehyde	2	2.5	2	2.5	
Hydrogen cyanide	–	–	10	10	Sk
Isocyanates, all (as –NCO)	–	0.02	–	0.07	
Man-made mineral fibre	–	5	–	–	
1-Methoxypropan-2-ol	100	360	–	–	
2-Methoxyethanol	5	16	–	–	Sk
2-Methoxyethyl acetate	5	24	–	–	Sk
Rubber fume[a]	–	0.75	–	–	
Rubber process dust	–	8	–	–	
Styrene	100	420	250	1050	
1,1,1-Trichloroethane	350	1900	450	2450	
Trichloroethylene	100	535	150	802	Sk
Vinyl chloride[b]	7	–	–	–	
Vinylidene chloride	10	40	–	–	
Wood dust (hard wood)	–	5	–	–	

[a] Limit relates to cyclohexane-soluble material.
[b] Also subject to an overriding annual maximum exposure limit of 3 ppm.

HSE recommends that a figure of three times the long-term limit, averaged over 10 minutes, be used as a guideline for short-term excursions.

With the exception of the annual Control Limit for vinyl chloride monomer, and the Recommended Limit for cotton dust, all the limits relate to personal exposure measurements [7.7.7] rather than background levels.

The airborne substances listed enter the body mainly through normal respiration, but some (marked 'Sk' in the Table) can also enter through the skin. There are no sharp dividing lines between 'safe' and 'dangerous' concentrations, and for some substances there are no apparent thresholds below which adverse effects do not occur.

The substances listed in Table 7.4 show long-term limits ranging from 0.005 to 1900 mg/m^3. The stringency of the precautions needed in handling these different substances thus varies considerably. While the limits for one substance may be met without difficulty by general dilution ventilation, even the best local exhaust ventilation will not meet the more stringent limits of some others, which may require handling by remote control in totally enclosed systems (e.g. 'power fluidics' [15.7]). While some airborne vapours may be safely discharged from building and ventilation vents to the outside atmosphere, others must be removed by adsorption, neutralisation or incineration to prevent unacceptable environmental pollution.

Dusts require special consideration[24, 25] since only the smaller particles are liable to be inhaled. As an approximation it is often assumed that half of an airborne dust is respirable. Some substances present in minute concentrations which are well below the published OELs can cause allergic reactions in persons who have become sensitised to them[10]. The absence of a substance from the list does not imply that it is harmless, since the toxicities of a high proportion of the substances found in the process industries have not yet been fully assessed. All substances should therefore be handled with care.

7.5.2 OELs in other countries

In 1980 the ILO published the second edition of its report on the OELs for airborne toxic substances adopted in 18 countries[26]. This contains a review of the methodologies adopted by these countries, a list of 1116 substances with the OELs adopted for them, and sections on particulate matter and carcinogens showing how they are dealt with by the 18 countries. The report provides data for the following countries: Australia, Belgium, Bulgaria, Czechoslovakia, Finland, East Germany, West Germany, Hungary, Italy, Japan, Netherlands, Poland, Romania, Sweden, Switzerland, the USSR, the USA and Yugoslavia as well as the Council of Europe. UK figures were not shown separately as the US figures then applied in the UK. While there are considerable differences between the OELs adopted by various countries, the limits for any substance at any one time in most industrialised non-communist countries usually lie within a fairly narrow band. The figures for Sweden are generally lower, i.e. a quarter to a half of these figures, while those for the USSR are even lower, particularly for organic compounds. Some of these differences have been discussed by Fawcett[9].

7.6 Sources of exposure to airborne substances hazardous to health

Sources of exposure are classed by Carson and Mumford[6] as:

- *Periodic emissions* which arise from the need to open or enter the 'system' occasionally, for example, during sampling, cleaning, batch additions, bulk tank-car loading, line breaking, etc. Periodic emissions tend to be large and include both anticipated events and unplanned releases, in which human error may be a factor.

- *Fugitive emissions* are small but continuous escapes from normally closed sources; 15–20% of total volatile organic chemical emissions are fugitive. They occur from dynamic seals such as valve stems and pump or agitator shafts and from static seals such as flange gaskets.

The consequences of these two types of emission may be different. A periodic emission which results in a high local airborne concentration of a toxic substance may cause acute effects on an exposed person, which qualify as an accidental injury [3.2.3]. Fugitive emissions of toxic substances may produce airborne concentrations high enough to eventually cause occupational diseases [3.2.4] in those chronically exposed to them, but are unlikely to produce the acute symptoms associated with an accidental injury.

7.6.1 Periodic emissions

Several causes of emissions, mostly of this type listed by Carson and Mumford, are given below under the headings of spillage, leakage, unintended venting, failure of item at normal working pressure and failure of item due to excessive pressure. Some present fire as well as toxic hazards and several are discussed elsewhere in this book. The list should be of value in keeping accident records and identifying and controlling the hazards responsible for them. Causes which could appear under more than one heading are given only under the first one. The list includes contributory causes of all the major accidents discussed in Chapters 4 and 5.

Causes of spillages

Overflow, backing-up, blowback, air-lock, vapour-lock;
Excess pressure, wrong routing, loss of vacuum;
Vessel damaged, tilted, collapsed, vibrated, overstirred;
Overloading of open channel/conveyor;
Poor isolation, drains or doors open, flanges uncovered;
Failure of control or major service;
Surging, priming, foaming, puking, spitting;
Condensed products in vapour, change in normal discharge;
Malicious intent, vandalism.

Causes of leakage

Broken, damaged or badly fitted pipe, vessel, instrument, glass, gasket, gland, seal, flange, joint or seam-weld;
Internal leaks, overpressure of pipe or vessel;
Deterioration of bursting disc (pinholing).

Causes of unintended venting

Evaporation through open line, drain, cover;
Relief valves leaking, bursting discs blown, lutes blown;

Valve stuck, scrubber overloaded, ejector failure;
Dust formation, escape and accumulation;
Equipment failed/out of service (e.g. scrubbers, flares).

Causes of item failures at normal working pressure

Inadequate design, materials, construction, support, operation, inspection
   or maintenance;
Deterioration due to corrosion, erosion or fatigue
Mechanical impact.

Causes of failure due to excessive pressure

Overfilling, overpressurising or drawing a vacuum;
Overheating or undercooling;
Internal release of chemical energy;
Exposure to fire or other source of external heating (e.g. radiation).

Most periodic emissions can be avoided by careful design and operation. This means providing drains and vents which lead to safe disposal systems from equipment which may have to be opened, together with inert gas or steam purges and sometimes solvents to remove residual material in the equipment before it is opened. This also applies to hose connections to road and rail tankers, etc. Reliable and quick-closing valves can be provided as close as possible to the flanges or connections that have to be broken.

Batch additions of liquids should be made from calibrated vessels through pipes and valves which drain completely into the receiving vessels. Batch additions of solids should also be made as far as possible from enclosed hoppers through enclosed chutes and suitable valves, although this often presents problems. Some low-melting, soft and putty-like materials (such as sodium metal) can be added to stirred batch reactors by extrusion through a die, which is simpler than melting them in a separate melter and then adding them as liquids. The operating procedures must ensure that the design features provided are properly used, and in cases where they present problems, that they are modified until the problem is overcome. Operators, especially those new to the work, must also be fully informed by management of the toxic and other hazards of the materials with which they are dealing [21.1.3]. Awareness is easy in the case of a pyrophoric liquid which bursts into flames as soon as it enters the air. Appreciation is far more difficult for toxic, odourless and relatively inert substances which produce no immediate symptoms in persons exposed to them.

In spite of all these precautions, occasions sometimes arise, usually during maintenance or cleaning operations, when there are risks of personal exposure from periodic emissions of hazardous substances. For such operations, it is essential that those liable to be exposed are properly protected by appropriate and well-fitting breathing apparatus and clothing, and that persons not so protected are excluded from the danger area.

7.6.2 Fugitive emissions

Carson and Mumford[6] also provide indicative data on the rates of fugitive emissions from different sources. A selection of these emission rates for liquids, vapours and gases is given in Table 7.5.

Table 7.5 Emission rates for fluids from various equipment

Equipment	Emission rate (mg/s)
Pump shaft seals	
Regular packing without external lube sealant	140
Regular packing with lantern ring oil-injection	14
Grafoil packing	14
Single mechanical seals	1.7
Double mechanical seals	0.006
Bellows seal with auxiliary packing, diaphragm pump (double), canned pump	Nil
Valve stems (excluding pressure relief valves)	
Rising stem valves with regular packing	
Rating ⩽ 300 lb	1.7
Rating > 300 lb	0.03
Non-rising stem valves	0.005
Pressure relief valves, average release for valve vented to closed system without upstream protection by bursting disc	2.8
Compressors	
Reciprocating	
Single rod packing	45
Double rod packing	3.6
Rotary	
Labyrinth seal	45
Mechanical seal	As for pumps
Liquid film seal	0.006
Stirrer seals ≃ Pump seal data × shaft speed (m/s) ÷ 1.91	
Piping and flange connections	
Open ended pipe	0.63
Flange and asbestos gasket	
Rating 150–300 lb	0.056
Rating > 300 lb	0.056
Threaded connection	0.056
Welded connection	Nil

It is clear from this table that by far the largest fugitive emissions come from the glands of conventional seals on pump and compressor shafts and piston rods, followed by the seals of rising stem valves. Actual emission rates from these sources depend on further factors such as shaft speed, diameter, straightness, roundness, and surface condition, bearing alignment, internal pressure, fluid viscosity, out of balance and radial loads, vibration and, above all, on the state of maintenance. On plants handling hazardous fluids shaft seals should be of high quality or eliminated entirely.

Table 7.6 Dust emission rates from various equipment

Type of equipment/handling	Emission rate (mg/s)
Vibratory screens	
Open top	5.5 times top surface area in m^2
Closed top with open access port	
6-inch dia. port	0.11
8-inch dia. port	0.21
12-inch dia. port	0.44
Closed cover – no ports	Nil
Bag dumping	
Manual slitting and dumping	3
Semi-automatic (enclosed dumping but manual bag entry/removal)	
Fully automatic and negative pressure	Nil
Bagging machines (filling)	
No ventilation	1.5
Local ventilation	0.01
Total enclosure and negative pressure	Nil

The same authors also give typical dust emission rates from various equipment. These are reproduced in Table 7.6 and refer to total dust emissions. Values should be halved for an estimate of the respirable dust release. While again the actual figures depend on many additional factors, those given provide a useful starting point for estimating airborne concentrations of hazardous materials for initial design and for surveys when plant is running.

Under COSHH, occupational hygienists are more involved in the design of installations (especially ventilation) where substances hazardous to health are handled. Maximum allowable emission rates from plant and equipment should be included in their design specifications.

7.7 Monitoring the working environment for toxic substances

Advice on this difficult subject which will assist UK employers in meeting their duties under COSHH[7] is given in HSE's Guidance Note EH 42[13]. Similar advice is available in other countries from their occupational health authorities, e.g. OSHA (Occupational Health and Safety Administration) and NIOSH (National Institute for Occupational Safety and Health) in the USA.

Such monitoring involves an initial assessment followed by carefully planned programmes of air-sampling and analysis. The purposes of the sampling programmes discussed in EH 42 are:

1. To estimate personal exposure to hazardous substances, e.g. to investigate compliance with OELs;
2. To investigate the effectiveness of engineering and process control measures.

Sampling is also undertaken for epidemiological investigations and for environmental monitoring outside the working environment.

The measurement of airborne concentrations of any contaminant is complicated by the many variables present. These include the type, number and position of emission sources and their rates, the dispersion of contaminant in the air, and ambient conditions including air velocity, which is specially important for outdoor operations. Variations in personal exposure may occur within shifts, between shifts, between processes and between individuals, and between people doing the same job (such as filling bags) at the same time and place, particularly when the way they work affects the emission. A structured approach to the problem is thus needed, which incorporates the following stages in sequence:

1. An initial assessment,
2. A preliminary survey,
3. A detailed survey,
4. Routine monitoring.

If the initial assessment shows that the health hazard is negligible, it is not usually necessary to go further. Preliminary and detailed surveys are exploratory, self-contained and concluded once the prevailing conditions have been established. They may be needed on start-up of a new process, on setting of a new (lower) exposure limit, when the process of control measures have been modified or when unusual operations are to be carried out.

The sampling equipment chosen should match the analytical techniques used and the relevant OELs. For personal exposure monitoring the sampling equipment should, where possible, be worn on the person and the sample taken at a consistent position within his or her breathing zone.

7.7.1 Initial assessment

This can only be made by a trained and experienced person, who needs the following information:

- The substances present at the workplace, raw materials, intermediates, products, contaminants, auxiliary chemicals;
- The airborne form of the substance (dust, fume, aerosol or vapour);
- The health hazards of the substances, and whether combinations of non-hazardous ones could be dangerous;
- Whether people could become exposed to these substances through inhalation, ingestion or skin absorption;
- Where and when exposure is likely to occur;
- Which groups or individuals are most likely to be exposed;
- The likely pattern and duration of exposure;
- What information is available on similar exposures elsewhere.

Simple qualitative tests involving the use of dust lamps and smoke pellets are useful in the initial assessment. In some cases it will be necessary to sample the air and analyse it for possible contaminants.

7.7.2 Preliminary survey

Once people likely to be at risk have been identified, their exposures should be measured by sampling, analytical and control techniques as discussed in 7.7.7. Personal air-sampling techniques should be used initially, complemented where required by static sampling, e.g. to provide further information on the efficiency of engineering controls. The preliminary survey should provide basic quantitative information on the efficiency of process and engineering controls and on the likely extent of workers' exposure to harmful substances. It may reveal shortcomings in preventative measures, work procedures and operator training.

7.7.3 Detailed survey

This is needed where the preliminary survey does not give an adequate assessment of the extent and pattern of exposure, e.g. when:

- The results of the preliminary survey are very variable;
- Many people are at risk of excessive exposure;
- Personal sampling results are near the appropriate OELs and the cost of additional control measures cannot be justified without further evidence on the extent of the risk. EH 42[13] states that 'detailed surveys may typically take 3 to 10 days to complete in a workplace with between 25 and 200 employees'.

7.7.4 Routine monitoring

After preliminary/detailed surveys and any indicated remedial action has been carried out, further routine monitoring may be needed to check that control measures remain effective, and to reveal trends in exposure patterns. The frequency of monitoring may vary from monthly, where highly toxic or carcinogenic substances are involved, to less than yearly when exposures appear well controlled. The sampling methodology should be well planned to enable results to be analysed statistically.

7.7.5 Sampling strategies

Because of the wide variations possible in exposure patterns, a number of samples are usually needed to reduce errors, and a carefully planned programme is required. EH 42 discusses this in detail and suggests a three-level strategy. The first level employs relatively unsophisticated sampling equipment and experimental techniques and is often suitable for the preliminary survey. The second level should aim to provide accurate measurements of TWA exposures which can be related to the appropriate 10-minute and 8-hour TWA limits, and is more suitable for detailed surveys and routine monitoring. A third-level strategy is occasionally needed if, despite taking all reasonably practicable control measures, personal exposures remain close to the relevant OEL. This usually requires a highly sophisticated sampling programme and rigorous statistical analysis of the results[27].

7.7.6 Interpretation of results

This again is discussed in detail in EH 42. A distinction must be made between substances for which a Control Limit has been adopted by HSC, those for which a Recommended Limit is listed and those for which no exposure limits are listed in EH 40. In the last case it is suggested that employers should make their own assessment of the health risk and set tentative limits for their own use, after consulting their employees. EH 42 gives practical guidelines on the actions to be taken when the results of sampling programmes bear particular relations to the relevant OELs. These are summarised in Table 7.7.

Table 7.7 Guidelines on recommended actions following a personal exposure sampling programme

Results	Action
A. Preliminary survey; first-/second-level strategy	
< 0.1 limit	Normally none, if exposure is as low as reasonably practicable.
0.1 to 1.5 limit	Investigate process/control measures; make detailed survey.
Some results > 1.5 limit	Investigate; improve control measures, provide respiratory protection to workers with high exposure until improvements made and confirmed by survey.
B. Detailed survey; second-/third-level strategy	
< 0.25 limit	Normally none, if exposure is as low as reasonably practicable.
< 1.25 limit	Investigate and improve control measures; repeat survey using more refined techniques to improve accuracy.
Mean < 0.5 limit, all results < limit	Consider routine monitoring and suitable frequency.
Mean > 0.5 limit some results > limit	Investigate and improve control measures and repeat survey; consider routine monitoring.
Routine monitoring; second-/third-level survey	
All results	Check individual values, mean, 'range', etc. for reliability of compliance with OEL; consider need for corrective action.
Results significantly different from previous survey	Consider need for corrective action, detailed survey or revised basis for survey.

7.7.7 Air sampling and analysis[28]

Being specialised subjects, only broad principles are discussed here. Most precise methods depend on drawing a known volume of contaminated air through an apparatus which may contain a liquid or solid reagent, an adsorbent (such as charcoal or silica gel), or a filter (used mainly for solids). The equipment consists of a sampling head, a pump and a means of measuring the volume of air drawn through the apparatus. For several contaminants the total quantity of one or more present is determined by the extent of a colour change in a specific reagent. With apparatus using filters and adsorbents, the quantity and composition of the trapped contaminant is determined by removing the filter or adsorbent tube and analysing its contents.

Sampling of air for gas and vapour contaminants is fairly straightforward, since all gases and vapours, once mixed with air, are uniformly distributed, and their concentrations are not altered when air is drawn through the sampling tube. Sampling of air for finely divided solids and liquids is more difficult as these tend to be deposited in the sampling tube. Filters for solid contaminants are therefore incorporated into the sampling head itself (Figure 7.2).

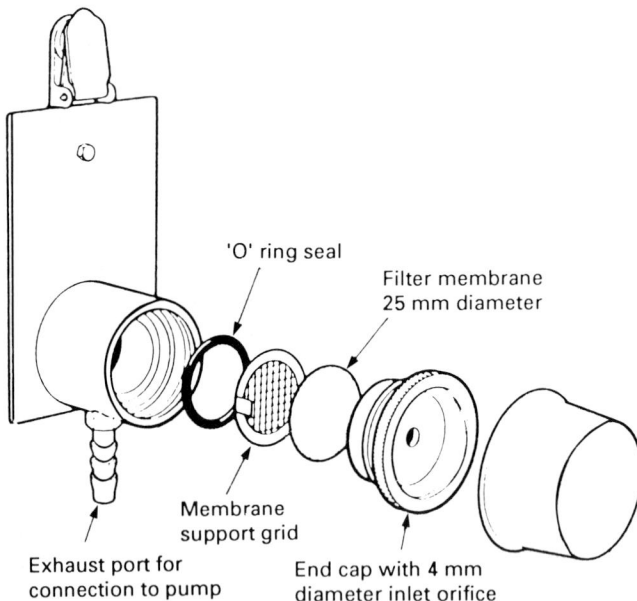

Figure 7.2 Orifice-type sampling head used for airborne lead. (From HSE Guidance Note EH28)

OELs for many airborne solids are quoted on two bases, 'total dust' and 'respirable dust'. The sampling apparatus for respirable dust contains a special elutriator in the sampling head before the filter. This separates the larger particles which would generally be stopped in the nose and upper respiratory tract and allows only the smaller particles (which normally reach the lungs) to pass through.

Some devices designed for use as personal dosemeters depend on the principle of diffusion and adsorption, and require no air pump. An adsorbent, normally charcoal, is contained in the middle of a short glass tube which has plugs of porous material sealed in one or both ends. Vapours present in the air diffuse through the plug at a rate proportional to their concentration in air and are trapped by the adsorbent. The tubes are sealed by caps which are removed immediately before use and replaced immediately afterwards. The adsorbent is later analysed for specific compounds. The main use of these devices, which are not considered as accurate as those using a measured volume of air, is in preliminary surveys of personal exposure [7.7.2]. The tube is fitted to a small holder which is clipped to the lapel or breast pocket of the wearer.

Other monitoring instruments based on physical principles require no measurement of air volume. These have detecting heads which are exposed to the air. Examples are:

- Electrochemical measurement of oxygen, carbon monoxide and hydrogen concentrations in air;
- Flammable gas monitors which use an electrically heated combustion catalyst (pellistors);
- Instruments which depend on absorption of light of a particular wavelength.

Fixed and portable instruments of all types are used. Portable instruments operating on physical principles are employed for checking the atmosphere in confined spaces before anyone is allowed to enter [18.8.1]. The measuring device with batteries is contained in a small case to which a sample probe and a rubber squeeze bulb are connected by flexible tubes. Readings are taken from a dial on the outside of the case. Although easy to operate, these instruments can go wrong, and need to be calibrated and adjusted by trained persons. Typical faults are caused by:

- Air leak between instrument and sample probe;
- Poisoning of catalyst;
- Flat battery.

For accurate measurements such as those required in detailed surveys and routine monitoring, compact lightweight personal sampling equipment which incorporates a low-flow, battery-operated air pump is preferred. Such pumps, which can be carried in the worker's pocket, give reasonably constant flow rates of 20 to 200 ml/min over an 8-hour period. The pump is connected by flexible tube to a sample head attached to the worker's overalls, etc., in a consistent position close to his or her breathing zone (Figure 7.3). A range of standard tubes through which the air is drawn and which contain solid reagents are available, both as 'short-term' and 'long-term' tubes. Tubes containing reagents which react with one of 30 or more specific contaminants to produce colour changes which are later measured are available. These are generally preferred to liquid reagents in bubblers for personal sampling. For gases and vapours for which no suitable reagent tubes are available, standard adsorbent tubes containing charcoal, porous polymers or silica gel are used. Details of methods recommended by HSE are published in their MHDS series[29], an up-to-date list of which is given in EH 40[12]. The following UK firms who

Figure 7.3 Personal air sampling equipment. (Courtesy Draeger)

are members of the Industrial Safety (Protective Equipment) Manufacturers' Association are listed in their 1988 reference book[30] as suppliers of
gas-detection equipment:

- Draeger Ltd, The Willows, Mark Road, Hemel Hempstead, Herts HP2
  7BW. Tel: 0442 3542
- MSA (Britain) Ltd, East Shawhead, Coatbridge, Scotland ML5 4TD.
  Tel: 0236 24966
- Sabre Safety Ltd, 225 Ash Road, Aldershot, Hants GU12 4DD. Tel:
  0252 31661
- 3M United Kingdom plc, PO Box No. 1, Bracknell, Berks RG12 1JU.
  Tel: 0344 426727

7.7.8 Detection of toxic/irritant gases and vapours by odour

Odour can often give a useful warning of an emission. Hydrogen sulphide
can be detected by smell by most freshly exposed persons at concentrations
well below its OEL, but exposure for only a few minutes paralyses the
olfactory nerves so that the sense of smell is lost. Common irritant gases
such as chlorine, sulphur dioxide and ammonia can be detected by about
50% of people exposed to them below their OELs. Some toxic and
asphyxiant gases such as carbon monoxide and nitrogen are odourless.

Individuals vary greatly in their sensitivity to smell, from four times below the mean to four times above it. Lynsky summarises a published study which compared the odour threshold with OELs (TLVs) in the 1982 US list for 214 substances[31]. This placed them in five classes ranging from 'A', for which 90% of distracted persons perceived a warning at the 8-hour OEL concentration, to 'E' for which less than 10% of attentive persons can detect the substance at the 8-hour OEL concentration. It also gives 'odour safety factors', i.e. the ratio of the 8-hour OEL to the odour thresholds for the 214 substances. Substances in category A include acrylates, amyl acetate, hydrogen sulphide, mercaptans and trimethylamine. Those in category E include carbon tetrachloride, chloroform, ethylene dichloride, ethylene oxide, methyl isocyanate, nickel carbonyl, nitropropane, toluene diisocyanate and vinyl chloride. The odour thresholds and safety factors for a number of common toxic organic chemicals listed in Part 2 of Schedule 3 of the CIMAH regulations are compared with their 8-hour OELs in Table 7.8.

Because of the wide variations in odour, safety factors and people's sense of smell, and the frequent dulling of this sense on long exposure to particular vapours, the apparent absence of odour should never be taken as

Table 7.8 Toxic substances of Group 2, Schedule 3 of CIMAH regulations

Substance	Quantity (tonnes)	OEL TWA (ppm) 8 hour	OEL TWA (ppm) 10 min	Odour threshold (ppm)	Odour safety factor
Acetone cyanohydrin	200	–	–	–	–
Acrolein	200	0.1	0.3	0.61	0.16
Acrylonitrile	200	2	–	0.12	17
Allyl alcohol	200	2	4	1.8	1.1
Allylamine	200	–	–	–	–
Ammonia	500	25	35	4.8	5.2
Bromine	500	0.1	0.3	2.0	0.05
Carbon disulphide	200	10	–	92	0.11
Chlorine	50	1	3	3.2	0.31
Ethylene dibromide	50	1	–	–	–
Ethyleneimine	50	0.5	–	1.5	0.32
Formaldehyde (concentration $\geq 90\%$)	50	2	2	0.83	2.4
Hydrogen chloride (liquefied gas)	250	5	5	0.77	6.5
Hydrogen cyanide	20	10	10	0.58	17
Hydrogen fluoride	50	3	6	0.042	71
Hydrogen sulphide	50	10	15	0.0081	1200
Methyl bromide	200	15	–	–	–
Nitrogen oxides	50	3	5	0.39	7.8
Phosgene	20	0.1	–	0.90	0.11
Propyleneimine	50	–	–	–	–
Sulphur dioxide	1 000	2	5	1.1	1.7
Tetraethyl lead	50	0.10 mg/m^3		–	–
Tetramethyl lead	50	0.15 mg/m^3		–	–

evidence that an atmosphere is safe to breathe and enter. Smell is no substitute for scientific monitoring. While smell may suggest the need to monitor the air for one chemical, there may be a far greater danger from another present which has no odour.

7.8 Substances hazardous to health, and the law

Numerous regulations about the use and handling of toxic substances in particular industries have been wholly or partly revoked and superseded by the COSHH regulations [2.5.6]. Only some aspects of these and the CIMAH regulations [2.4.2] are considered here.

7.8.1 Scope and practical effects of the COSHH Regulations[7] (see Appendix G)

The scope of COSHH includes prohibitions, assessment, control of exposure, monitoring exposure at the workplace, health surveillance, as well as information, instruction and training. The substances whose use or import are prohibited under COSHH are those formerly prohibited under other regulations which COSHH will replace.

An assessment by a competent person [7.7.1] must be made before work which may involve exposure of any employee to a substance hazardous to health is started. This should enable the precautions needed to comply with the regulations to be identified. The assessment required is related to the risk.

Exposure should preferably be prevented by elimination or substitution by a less hazardous substance. Where this is not reasonably practicable, adequate engineering controls over materials, plant and processes should be exercised. Personal protective equipment should only be used when adequate engineering controls are not reasonably practicable. Substances to which MELs have been assigned require the strictest safeguards. Those to which OESs are assigned have next priority. While exposure should not exceed OES concentrations, in cases where it does, control may still be considered adequate if the employer has identified the cause of non-compliance, and is taking steps to reduce exposure to below the OES as soon as reasonably practicable.

Employers are required to set their own working control standards for harmful substances to which neither MELs nor OESs have been assigned, on the basis of all relevant information.

Situations where the use of personal protective equipment may be necessary include emergencies caused by plant failure, and during routine maintenance.

Engineering control measures include the use of totally enclosed processes and handling systems, those which limit the generation of harmful airborne substances, local exhaust ventilation with or without partial enclosure, adequate general ventilation. and the provision of safe storage and disposal facilities. A thorough discussion of such measures, particularly ventilation as applied to the control of carcinogens in the American workplace, is given by Feiner[32].

Personnel control measures include reduction of numbers exposed, exclusion of non-essential access, reduction in period of exposure, cleaning walls and surfaces, provision of suitable personal protective equipment, prohibition of eating, drinking and smoking, and the provision of adequate facilities for washing, changing and storing clothing and laundering contaminated clothing.

Procedures for emergencies resulting from loss of containment or control should be established which give maximum employee protection against exposure to substances with health hazards.

Respiratory protective equipment (RPE) must be properly selected for its purpose, of an HSE-approved type or to a HSE-approved standard, and matched to the job and the wearer. Eye protection should comply with the Protection of Eyes Regulations 1974 and BS 2092.

Special protection is particularly needed by maintenance workers when isolating equipment for maintenance, as discussed in 18.6.1. Special respiratory apparatus for escape [22.8.5] is needed for those who might become trapped in a cloud of toxic gas in works with significant inventories of toxic gases and volatile liquids. The types of protection available for the eyes, respiratory system and skin and criteria for their selection are discussed in Chapter 22.

Both employers and employees should ensure that the control measures provided are properly used or applied, and procedures should be established for regular inspection and maintenance or other remedial action. Records of examinations, tests and repairs should be kept for at least five years.

Monitoring exposure at the workplace, like assessment, should be related to the risk. The methods and strategy recommended have been discussed in 7.7.

Health surveillance of workers, where appropriate, should be given to provide early detection of ill effects resulting from exposure to substances hazardous to health. In such cases health records should be kept for each worker involved. Like other measures, the level of health surveillance will depend on the nature and degree of the risk. Substances and processes considered 'appropriate' are listed in Schedule 5 [Appendix G.2]. They are those for which health surveillance was called for under former regulations now revoked by COSHH. Health surveillance will generally be supervised by an employment medical adviser (EMA) or an appointed doctor (AD), and may lead to the suspension of employees from work in which their health is adversely affected.

Health surveillance has four objectives: to protect the health of employees, to assist in evaluating control measures, to collect and use data for detecting and evaluating health hazards and to assess the immunological status of employees in work activities involving micro-organisms.

Health surveillance procedures include biological monitoring, biological effect monitoring, medical surveillance including clinical examinations, enquiries about symptoms, inspection by a responsible person, review of records and occupational history.

Employees are entitled to access to their health records.

Information, instruction and training of persons who may be exposed to substances hazardous to health is another important feature of the COSHH regulations, although most of this is straightforward and self-evident.

7.8.2 Toxic substances and the CIMAH Regulations 1984[22]

Schedule 1 provides 'Indicative Criteria' for toxic and very toxic substances which applies to Regulation 4. This elaborates on the criteria given in Table 7.3.

Two groups of toxic substances to which Regulations 7 to 12 (briefly summarised in 2.4.2) apply are listed in Schedule 3.

Group 1 is a list of the most toxic substances. It contains 32 substances and groups of compounds for which a minimum quantity of 1 kg applies, eight substances with a minimum quantity of 10 kg, 54 substances with a minimum quantity of 100 kg, one substance with a minimum quantity of 500 kg and two substances with a minimum quantity of 1 t.

The 31 substances to which 1 kg applies are all organic and include benzidene [23.3.3], fluoroacetic acid (used as a rodenticide by professional operators) and related compounds, 2-naphthylamine (now banned) and 2,3,7,8-tetrachlorodibenzo-para-dioxin, the active agent of the Seveso disaster [5.3].

The eight entries to which 10 kg applies include five inorganic ones: arsine, beryllium (powders and compounds), hydrogen selenide, nickel carbonyl, oxygen difluoride and selenium hexafluoride. Nickel carbonyl is an intermediate in the purification of nickel. Beryllium is used in atomic reactors and spacecraft, and in some copper and aluminium alloys.

The 54 entries to which 100 kg applies include five inorganic ones, arsenic trioxide and related compounds, cobalt and nickel powders and compounds, phosphine and tellurium hexafluoride. Many of the organic entries are herbicides or pesticides[1]. They include Aldicarb [5.4], Parathion, a phosphorus-based insecticide used for fruit trees, and Crimidine (LD$_{50}$ 1.25 mg/kg) and Warfarin, both rat poisons, the latter an anticoagulant.

One of the two entries to which 1 t applies is methyl isocyanate [5.4].

Several very toxic pesticides do not appear in Schedule 3 at all, e.g. strychnine sulphate [60-41-3], (LD$_{50}$ 5 mg/kg), used for mole control, and brodifacoum [56073-10-0], (LD$_{50}$ 0.4 mg/kg), another anticoagulant and very potent rat poison. The list of pesticides in group 1 of Schedule 3 will need frequent updating as new ones are developed.

Group 2 is a list of 23 common toxic industrial chemicals to which a minimum quantity of 1 t or more applies. The chemicals and their minimum quantities are listed in Table 7.8. This also gives the short- and long-term TWA OELs for the UK in ppm in cases for which they are available. (The fact that no OELs are published in the UK for four of these 23 common toxic chemicals reinforces the need to minimise exposure to all potentially harmful substances, whether their OELs are published or not.) The lowest actual quantity which applies to any of those listed is 20 t. Some of these are gases under ambient conditions, while the rest are volatile liquids.

7.9 Treatment of affected persons

This is mainly a medical question which is outside the scope of this book. The relationship between safety management and occupational health professionals, particularly as applying in the USA, is discussed by Kilian[33] and Murphy[14] (an occupational nursing consultant) in Fawcett and Wood's book.

Most works or factories in the process industries, being 'establishments presenting special or unusual hazards', are required to have their own 'first-aid room', with an adequate number of trained 'occupational first-aiders' even though there may be fewer than 400 employees working there[34].

The first-aid room is often part of a works medical centre, which is equipped to enable medical examinations to be carried out, or in larger organisations, a department of occupational medicine. This should have a positive role in promoting health and fitness among workers as well as treating injuries and sickness. It should have close working relations through EMAS[35] with the local health service and hospitals, and an up-to-date knowledge of specialised facilities, such as a burn centre, and skills, such as a retinal surgery service, to which patients can be referred without delay in emergencies. It should also keep up-to-date records both of workers' health and of all materials used on the premises to which they may be exposed.

Small factories which do not present 'special or unusual hazards' are almost all required to have at least one first-aid box or cupboard kept under the charge of a person trained in first-aid. Minimum standards for first-aid boxes and for first-aid training are given in the regulations and clarified by the Approved Code of Practice[36] and Guidance Notes[37] prepared for them.

First-aid requirements are specially critical in factories where toxic or corrosive substances are present. A recently published manual provides details of first-aid treatment for exposure to nearly 500 chemicals[38]. Oxygen resuscitation equipment, a stomach pump, emergency showers, special washing facilities, and antidotes to the toxic materials encountered are often required, as well as more common items such as stretchers, warm emergency clothing and blankets.

7.10 How does one decide if a disease is occupational?

While this important question is mainly one for health specialists and lies outside the scope of this book, a short but comprehensive guide (which reflects official American thinking) is published by NIOSH[39]. It includes:

- Data for a number of specific disease-producing agents, exposure standards, and a list of occupations with possible exposure to them;
- Considerations of medical, personal, family and occupational history;
- Clinical evaluation, signs and symptoms of occupational disease and laboratory tests;
- Epidemiological data, industrial hygiene sampling and data evaluation;
- Aggravation of pre-existing conditions and its legal ramifications;
- Sample questionnaires for evaluating respiratory symptoms:
- Qualifications of medical and industrial hygiene personnel.

References

1. 'Poisons, economic' in volume 18, *Encyclopaedia of Chemical Technology*, edited by Kirk, R. E. and Othmer, D. F., 3rd edn, Wiley, New York (*ca* 1983)
2. Kletz, T. A., 'The application of hazard analysis to risks to the public at large', *World Cong. Chem. Eng. Amsterdam*, Elsevier, Amsterdam (1976)
3. Hamilton, A., and Hardy, H. L., *Industrial Toxicology*, 3rd edn, Publishing Sciences Group, Acton, Mass. (1974)
4. Hunter, D., *The Diseases of Occupations*, 6th edn, Hodder and Stoughton, London (1978)
5. P.L.94-469, *The Toxic Substances Control Act*, Environmental Protection Agency, Washington DC (1976)
6. Carson, P. A. and Mumford, C. J., 'Industrial health hazards, Part 1: Sources of exposure to substances hazardous to health', *Loss Prevention Bulletin 067*, The Institution of Chemical Engineers, Rugby (1985)
7. S.I. 1988, No. 1657, *The Control of Substances Hazardous to Health Regulations 1988* and Approved Codes of Practice, *Control of Substances Hazardous to Health* and *Control of Carcinogenic Substances*, HMSO, London
7a. HSC Consultative Document, *The Control of Substances Hazardous to Health: Draft Regulations and Draft Approved Codes of Practice*, 1st edn, HMSO, London (1984)
8. S.I. 1984, No. 1244, *The Classification, Packaging and Labelling of Dangerous Substances Regulations 1984*, HMSO, London
9. Fawcett, H. H., 'Toxicity versus hazard' in *Safety and Accident Prevention in Chemical Operations*, edited by Fawcett, H. H. and Wood, W. S., 2nd edn, Wiley-Interscience, New York (1982)
10. Tyrer, F. H. and Lee, K., *A Synopsis of Occupational Medicine*, 2nd edn, John Wright, Bristol (1985)
11. *American Industrial Hygiene Association Journal*, **20**, 428–430 (October 1959)
12. HSE, Guidance Note EH 40/89, *Occupational Exposure Limits, 1987*, HMSO, London (1989)
13. HSE, Guidance Note EH 42, *Monitoring Strategies for Toxic Substances*, HMSO, London (1984)
14. Murphy, A. J., 'The role of an occupational health nurse in a chemical surveillance program' in *Safety and Accident Prevention in Chemical Operations*, edited by Fawcett, H. H. and Wood, W. S., 2nd edn, Wiley-Interscience, New York (1982)
15. S.I. 1981, No. 917, *The Health and Safety (First-Aid) Regulations 1981*, HMSO, London
16. S.I. 1985, No. 2023, *The Reporting of Injuries, Diseases and Dangerous Occurrences Regulations (RIDDOR)*, HMSO, London
17. Seville, R., *Dermatological Nursing and Therapy*, Blackwell Scientific, London (1981)
18. Williams, P. L. and Burson, J. L., *Industrial Toxicology*, VanNostrand, New York (1985)
19. Atherley, G. R. C., *Occupational Health and Safety Concepts*, Applied Science Publishers, London (1978)
20. The Institution of Chemical Engineers, *Nomenclature for Hazard and Risk Assessment in the Process Industries*, Rugby (1985)
21. Hodge, H. C. and Sterner, J. H., 'Tabulation of toxicity classes', *Am. Ind. Hyg. Assoc. Quart.*, **10**(4), 93 (December 1949)
22. S.I. 1984, No. 1902, *The Control of Industrial Major Accident Hazards Regulations 1984 (CIMAH)*, HMSO, London
23. National Fire Protection Association, *Standard 704 M, Identification systems for fire hazards of materials*, NFPA, Boston, Mass. (1975)
24. HSE, Guidance Note EH 44, *Dust in the workplace: general principles of protection*, HMSO, London (1984)
25. MDHS 14, *General method for the gravimetric determination of respirable and total dust*, HMSO, London

26. The International Labour office, *Occupational Exposure Limits for Airborne Toxic Substances – A Tabular Compilation of Values from Selected Countries*, 2nd edn, Geneva (1984)

27. Davies, O. L. (ed.), *The Design and Analysis of Industrial Experiments*, Oliver and Boyd, London (1954)

28. Lee, G. L., 'Sampling: principles, methods, apparatus, surveys', in *Occupational Hygiene*, edited by Waldon, H. A. and Harrington, J. M., Blackwell Scientific, London (1980)

29. *Methods for the determination of hazardous substances* (MDH Sseries, about 40 titles), HMSO, London

30. *Reference Book of Protective Equipment*, Industrial Safety (Protective Equipment) Manufacturers Association, 7th edn ISPEMA, London (1987)

31. Lynskey, P. J., 'Odour as an aid to chemical safety: odour thresholds compared with threshold limit values', *Loss Prevention Bulletin 060*, The Institution of Chemical Engineers, Rugby (1984)

32. Feiner, B., 'Control of workplace carcinogens', in Sax, I., *Cancer Causing Chemicals*, Van Nostrand Reinhold, New York (1981)

33. Kilian, D. J., 'The relationship between safety management and occupational health programs', in *Safety and Accident Prevention in Chemical Operations*, edited by Fawcett, H. H. and Wood, W. S., 2nd edn, Wiley-Interscience, New York (1982)

34. S.I. 1981, No. 917, *The Health and Safety (First-Aid) Regulations 1981*, HMSO, London

35. *Employment Medical Service Act 1972 – Guide to the Service*. HMSO, London

36. HSC, *Approved Code of Practice for the Health and Safety (First-Aid) Regulations 1981*, HMSO, London

37. HSE, *Guidance Notes for Health and Safety (First-Aid) Regulations, 1981*, HMSO, London (1981)

38. Lefevre, M. J., *First Aid Manual for Chemical Accidents*, Englishedition translated by Becker, E. I., Van Nostrand Reinhold, New York (1984)

39. Anon. *Occupation and Disease – A Guide for Decision-making*, NIOSH Center for Disease Control, Rockville, Maryland (1976)

Chapter 8

Chemical reaction hazards

The reactivity of chemical elements, compounds and free radicals is the very essence of chemistry, which most readers will have studied, and whose memories this chapter may help to jog. Explosivity, flammability and corrosivity are only touched on in this chapter, as the subjects are dealt with in Chapters 9–11.

8.1 Reactivities of the elements and structural groupings

The reactivity of an element is closely related to its position in the Periodic Table (Table 8.1), which is determined by its atomic number[1].

Table 8.1 The Periodic Table (up to Uranium: non-metals shown in italics)

Group	1A	2A	3B	4B	5B	6B	7B	←	8B	→	1B	2B	3A	4A	5A	6A	7A	8A
Period	1																	2
I	H																	*He*
	Light metals														*Non-metals*			
	3	4											5	6	7	8	9	10
II	Li	Be											*B*	*C*	*N*	*O*	*F*	*Ne*
	11	12											13	14	15	16	17	18
III	Na	Mg											Al	*Si*	*P*	*S*	*Cl*	*Ar*
							Transition metals											
	19	20	21	22	23	24	25	26	27	28	29	30	31	32	33	34	35	36
IV	K	Ca	Sc	Ti	V	Cr	Mn	Fe	Co	Ni	Cu	Zn	Ga	Ge	*As*	Se	*Br*	*Kr*
	37	38	39	40	41	42	43	44	45	46	47	48	49	50	51	52	53	54
V	Rb	Sr	Y	Zr	Nb	Mo	Tc	Ru	Rh	Pd	Ag	Cd	In	Sn	Sb	*Te*	*I*	*Xe*
	55	56	57	72	73	74	75	76	77	78	79	80	81	82	83	84	85	86
VI	Cs	Ba	a	Hf	Ta	W	Re	Os	Ir	Pt	Au	Hg	Tl	Pb	Bi	Po	*At*	*Rn*
	87	88	89	90	91	92												
VII	Fr	Ra	Ac	Th	Pa	U												

a The 15 rare earth elements fill this gap

The Periodic Table consists of seven periods (each representing a principal quantum number). No. I contains two elements, Nos II and III each contain eight, IV and V each contain 18, VI contains 32, and VII is incomplete and contains only a few radioactive elements which lie outside the scope of this book. The elements are also treated as belonging to eight groups which correspond to the eight elements each in periods II and III. Periods IV, V and VI each contain two groups of eight elements ('A' series) and seven elements ('B' series) and some additional elements. The properties of the first two and the last six elements in periods IV, V and VI have some resemblances to those in similar positions in periods II and III and are treated as belonging to the 'A' series of that group. The others are treated as the 'B' series of groups 1 to 7, apart from three elements in the middle of each period, which are treated as part of group 8B, and the 15 rare earths of period VI which are assigned to group 3B.

The position of an element in the Periodic Table, in particular its group, governs the number of electrons in the outer shell of the atom. An atom with a complete outer shell of electrons is extremely stable and will be one of the inert gases – helium, neon, argon, krypton or xenon.

8.1.1 Reactivity of metals

The elements of group 1 (1A) are the alkali metals – lithium, sodium, potassium, rubidium and caesium. Their atoms have a single electron in their outer shells, which they part with very readily to form positive ions. Because of this they react violently with water and most non-metals. The affinities of the alkali metals for most non-metals generally increase from lithium to caesium, although the heats evolved per unit weight of these elements in such reactions are generally in the reverse order and greatest for the lightest element. The same applies to the metals of the next two groups.

The atoms of group 2 (2A) elements have two electrons in their outer shells, which they also lose readily. They are the alkali earth metals: beryllium, magnesium, calcium, strontium and barium, which also react vigorously but less violently with water. The atoms of group 3 elements have three electrons in their outer shells and are also active, but besides shedding electrons to form positive ions, they may also achieve stability by sharing their electrons with other atoms. The first member, boron, is a reactive non-metal, followed by aluminium, a metal, and the less common metals scandium, yttrium and the rare earths.

Most of the elements in periods IV, V and VI are metals, and apart from the first three of each period and the rare earths, they are more or less stable in air and are used in engineering. They include the older metals – zinc, copper tin, iron, lead, bismuth, mercury, gold, silver and platinum – and the newer ones – titanium, vanadium, chromium, manganese, cobalt, nickel, molybdenum, palladium, cadmium, tantalum, tungsten, osmium and irridium. They show a limited range of reactivity among themselves, from titanium, the most active, through tantalum and tungsten, to the noble metals gold and platinum. Some indication of the activities of metals and non-metals is given by the their 'standard electrode potentials'[2]. When

Table 8.2 Electromotive series of elements

A. *Metals and metalloids*
Li Rb K   Sr Ba Ca Na Mg Al Be U   Mn Zn Cr Ga Fe Cd In Co Ni Sn Pb
H_2 Sb Bi As Cu Po Te Ag Hg Pb Pd Pt   Au

B. *Non-metals*
F_2 Cl_2 Br_2 I_2 O_2 S   Te

placed in order of decreasing potential they form an 'Electromotive Series', shown in Table 8.2[3].

The orders are somewhat ambiguous, since some elements have more than one position in the series, depending on the valency of the ion formed. Hydrogen is treated here as a metal, as are arsenic and tellurium. Metals higher in the series than hydrogen react with dilute hydrochloric and sulphuric acids, displacing hydrogen. A metal will displace any other metal lower than it in the series from a solution of its salts. When two metals in contact are immersed in an aqueous medium, the one higher in the series is attacked by galvanic corrosion, and the one lower in the series is protected. This is the principle of cathodic protection[4] [11.2].

The series also determines the ease of reduction of metallic oxides, the ease of corrosion in air, the action of water on metals, the solubility and stability of hydroxides and carbonates, the behaviour of nitrates on heating, and the occurrence of the metals in their free state in nature.

As regards corrosion, the alkali and alkaline earth metals rust rapidly in air, sometimes burning. Metals down to copper rust comparatively easily in air, although several metals such as aluminium, beryllium and chromium form rather resistant oxide layers on the surface, which protect the underlying metal from further attack. The alkali and alkaline earth metals displace hydrogen from water, even in the cold, with the evolution of much heat. Magnesium and succeeding metals, except those at the bottom of the list, will displace hydrogen from steam. The properties of individual metals are, however, greatly modified by alloying with others and by heat treatment; thus alloys are available with specific corrosion resistance against various media and conditions.

Many of the metals and their compounds in the middle of periods IV, V and VI are active catalysts for particular reactions. Thus although they do not react themselves, they have the effect of increasing the activities of other elements and compounds.

8.1.2 Reactivity of non-metals

The elements of group 7 (7B) are the halogens, fluorine, chlorine, bromine and iodine, whose atoms have outer shells with one electron short of a full set. They react strongly with atoms with surplus electrons, to form negative ions. This activity decreases from fluorine to iodine. Atoms which need two electrons to complete their outer shells (oxygen, sulphur, selenium, tellurium and the radio-active metal polonium) are rather less active. Besides the inert gases, there are only fifteen non-metals, of which some, arsenic and tellurium, are border-line cases and sometimes behave as metals.

Carbon which forms strong covalent bonds with other carbon atoms, as well as with hydrogen, oxygen, sulphur, nitrogen, phosphorus and the halogens, is the cornerstone of organic chemistry. Single carbon–carbon bonds are less reactive than double bonds which, in turn, are less reactive than triple bonds, although six-membered aromatic rings with alternate single and double bonds have comparable stabilities to chains with single bonds.

While some elements are much more reactive than others, their reactivity usually only manifests itself when one element is brought into contact with other elements and compounds. The greatest activity occurs when elements with opposite and complementary properties are brought together, e.g. caesium and fluorine. In the same way, compounds with opposite and complementary properties show the greatest activity when they mix, e.g. acids with bases and oxidising agents with reducing agents.

8.1.3 Reactivity of some structural groupings

The activity of structural groupings is too large a subject to cover in this book, but three broad and overlapping types of structural groups with high hazard potential are discussed:

- Those which confer instability to the compounds in which they are present, often rendering them explosive. A number of these are listed in Table 9.2.
- Those which render the compound liable to polymerise, with the evolution of heat. These are unsaturated linkages between carbon and

Table 8.3 Structural groupings liable to cause exothermic polymerisations

Type	Formula	Examples
Vinyl	$-CH=CH_2$	Styrene, vinyl chloride, ethyl acrylate
Conjugated double bonds with carbon, nitrogen and oxygen atoms	$-CH=CH-CH=CH-$ $-CH=CH-CH=O$ $-CH=CH-C=N$	Butadiene, isoprene, chloroprene, cyclopentadiene Acrolein, crotonaldehyde Acrylonitrile
Adjacent double bonds	$-CH=C=O$ $-N=C=O$	Ketene Methyl isocyanate, toluene di-isocyanate
Three-membered rings	$\overset{\displaystyle O}{\overset{\displaystyle \diagup\diagdown}{-CH\!-\!CH_2}}$ $\overset{\displaystyle NH}{\overset{\displaystyle \diagup\diagdown}{-CH\!-\!CH_2}}$	Ethylene and propylene oxide, epichlorhydrin Ethylene imine
Aldehydes	$-CH=O$	Acetaldehyde, butyraldehyde

carbon, carbon and oxygen, and carbon and nitrogen atoms, as well as three-membered rings containing these atoms. Structural groupings of some of the more common industrial chemicals which exhibit this tendency are given with examples in Table 8.3. Hazardous polymerisations involving some of them are discussed in 8.6.4.

- Those which render the compound liable to attack by atmospheric oxygen with the formation of peroxides, hydroperoxides and other compounds containing the –O–O– grouping. A list of these are given in Table 8.4 and their hazards are discussed in 8.5.1.

8.2 Reaction rate

The rates of chemical reactions often bear little relation to the heat or energy which they generate. Some very fast ones (e.g. ionic reactions in aqueous solution) produce little or no energy or heat, while others which develop a great deal of heat proceed very slowly in the absence of a catalyst. Reaction rates are studied under chemical kinetics and reaction power under thermodynamics, both branches of physical chemistry.

The simplest chemical reactions for study are uncatalysed, homogeneous ones, which take place entirely in a a single gas or liquid phase.

While many ionic reactions in aqueous solutions fulfil these conditions, they are in most cases almost instantaneous, and considerably faster than the speed with which their insoluble products come out of solution. These reactions seldom present serious hazards in themselves. Some which are important in the context of corrosion are discussed in 11.3.

Non-ionic reactions of this simple type usually involve one, two or three molecules, which may be identical or different. The rates at which such reactions proceed depend on the concentrations of the reactants, the temperature and reaction medium, and, in the case of gas reactions, on the pressure. This dependence on reactant concentrations is not a straightforward one. Many reactions proceed in stages, some faster than others. The overall reaction rate is then determined by the rate of the slowest stage, whose equation may be quite different from that of the overall reaction. Many reactions, including those of combustion, are *free radical chain reactions*.

An example of these is the gas phase dealkylation of toluene by hydrogen. The overall forward reaction is:

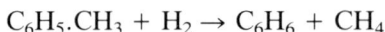

$$C_6H_5.CH_3 + H_2 \rightarrow C_6H_6 + CH_4$$

If this were a straightforward bimolecular reaction, the rate would be proportional to the product of the concentrations or partial pressures of toluene and hydrogen (strictly speaking their 'activities'). The rate, however, is actually proportional to the product of the toluene concentration and the square root of the hydrogen concentration. Without going into detail, this rate agrees with the following mechanism:

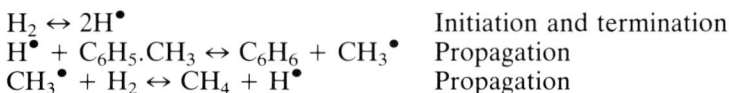

$H_2 \leftrightarrow 2H^\bullet$	Initiation and termination
$H^\bullet + C_6H_5.CH_3 \leftrightarrow C_6H_6 + CH_3{}^\bullet$	Propagation
$CH_3{}^\bullet + H_2 \leftrightarrow CH_4 + H^\bullet$	Propagation

Note: $^\bullet$ signifies a free radical.

The rate r_f for this forward reaction is given by the equation:

$$r_f = k_2[C_6H_5.CH_3][H_2]^{\frac{1}{2}}$$

This shows that the rate of disappearance of toluene equals a constant k_2 *times* the concentration of toluene *times* the square root of the concentration of hydrogen. Since the reaction is reversible, there will be a similar equation for the reverse reaction.

Reactions can have various 'orders', from zero to 3, the order being the sum of the power factors in the rate equation. Thus the simple bimolecular reaction would be a second-order reaction, whereas the actual chain reaction shown above has an order of 1.5. While chemical kinetics is a complicated subject, the designer of a reaction system should know the order of the reaction and the rate constants for the forward (and reverse) reactions over the range of temperatures which might be employed. He or she should also have similar information for any side reactions.

Complications arise when more than one phase is present, and one or more molecular species has to move from one phase to another before or after reacting. The degree of mixing, the interfacial area available and the diffusion rates of the molecules all then have to be taken into account. Where solids are involved, their surface areas are also important.

Catalysis is another very important but complicating factor, since most industrial reactors operate with catalysts which may increase reaction rates one hundredfold or more, and thus promote one desired reaction over other competing ones. Some reactions are autocatalytic, being catalysed by their own products, so that after a sluggish start the reaction rate gradually gathers speed, even though the temperature is maintained constant.

In spite of all these complications, there is a characteristic rate constant for every reaction under every condition.

8.2.1 Reaction rate and temperature

The rates of most chemical reactions, particularly homogeneous ones, i.e. those taking place in a single liquid or gas phase, increase exponentially with temperature, doubling or trebling with every 10°C temperature increase (Figure 8.1(a)). There are exceptions to this, such as the third-order gas-phase oxidation of nitric oxide to nitrogen peroxide:

$$2NO + O_2 \rightarrow 2NO_2$$

In this the reaction rate decreases slowly with temperature (Figure 8.1(b)).

Some catalytic reactions are controlled by the rate of adsorption of one of the reactants on the catalyst surface, which may decrease at elevated temperatures. Their rate may first increase with temperature almost exponentially, then rise more slowly to a peak, giving an S-shaped curve (Figure 8.1(c)).

The rates of biochemical reactions which involve living organisms (fungi, yeasts, bacteria) also show a maximum at the optimum temperature of the organism, but fall off to zero at slightly higher temperatures, which kill the organism (Figure 8.1(d)). Moulds and fungi generally die at temperatures of 50°C and higher, but bacteria are somewhat more resistant and survive up to a temperature of 70°C.

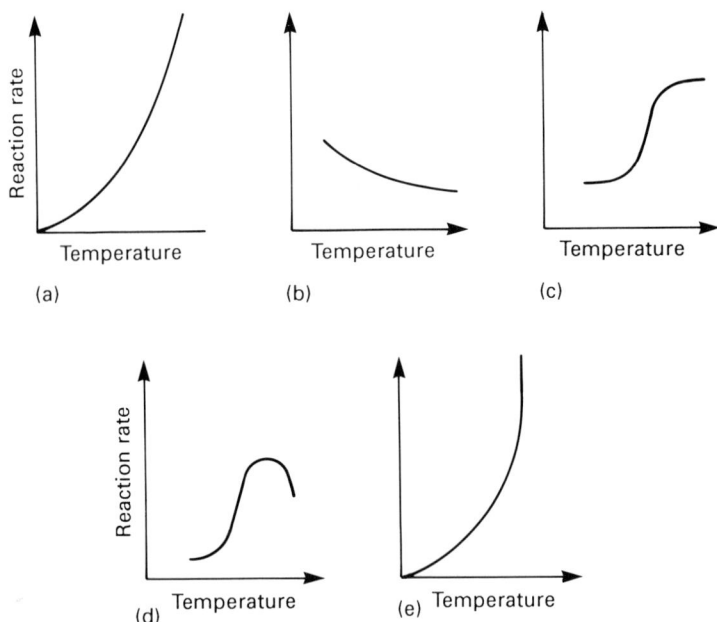

Figure 8.1 Dependence of reaction rate on temperature: (a) most homogeneous reactions in single fluid phase; (b) third-order gas-phase oxidation of nitric oxide; (c) catalytic reaction controlled by adsorption at solid surface; (d) biochemical reactions; (e) decomposition culminating in explosion

The rates of some decomposition reactions initially increase exponentially with temperature until their character changes to an explosion (Figure 8.1(e)).

8.3 The power of reactions

The main clue to the possible violence of any reaction lies in the heat liberated, the temperature that may be reached and the volume and nature of any gases and vapours formed. Thermodynamics can tell us:

1. How much heat is given out or absorbed in the reaction, and
2. Whether it is equilibrium-limited or can proceed virtually to completion.

The first question requires a heat balance to be drawn up between the starting materials and end products. This requires information on heats of reaction, specific heats, latent heats if there is a change of state, as well as work done on or by the reaction system, e.g. through expansion, compression or electrolysis. The heat of reaction is calculated from the heats of formation of the reactants and products. The second question is answered by the equilibrium constant for the reaction at the temperature in question. This is governed by the 'free energy change' of the reaction which can be calculated from the free energies of formation of the reactants and products. The thermodynamic data required can generally be found in the chemical literature[6-8], or else estimated by special methods[5, 9]. Broadly speaking, any reaction (of which there are many) that can lead to a rise in temperature of 300°C and/or the production of a significant amount of gas or vapour may pose a significant hazard[10]. This is discussed further in 8.6.6 and the use of computer programs for screening compounds for chemical stability is discussed in 9.5.1.

8.4 Inorganic reactions

Most of the more powerful inorganic reactions come under the headings of:

- Oxidation-reduction reactions,
- Acid-base reactions,
- Hydrations and hydrolyses.

8.4.1 Oxidation-reduction reactions

Reactions between the many oxidising and reducing agents are usually powerful and sometimes explosive. Oxidising agents include oxygen, chlorine, nitric, chloric and hypochloric acids and their salts and hydrogen peroxide. Reducing agents include hydrogen, carbon, most metals and several non-metals, sulphur dioxide, sulphites, nitrites, ammonia and hydrazine.

Examples of oxidation-reduction reactions carried out on an industrial scale include:

- The smelting of many oxide ores of metals such as haematite with coke,
  $Fe_2O_3 + 2C \rightarrow 2Fe + CO + CO_2$
- The production of sulphur dioxide by burning sulphur in air,
  $S + O_2 \rightarrow SO_2$
- The production of phosgene by reacting carbon monoxide with chlorine,
  $CO + Cl_2 \rightarrow COCl_2$
- The 'Thermit' welding process in which a mixture of powdered aluminium and ferric oxide is heated locally to initiate a strong reaction to give aluminium oxide and molten iron or steel. This reaction, which can be started by a powerful blow, can be a hazard when a rusty steel roof (e.g. of an oil storage tank) has been painted with aluminium paint.

8.4.2 Acid-base reactions

These result whenever a 'base' such as a metal oxide or hydroxide or ammonia reacts with an acid to form a salt. Reactions between 'strong' acids such as sulphuric and hydrochloric and 'strong' bases such as sodium hydroxide are very violent when the reactants are concentrated. Industrial examples are the production of ammonium nitrate and sulphate for fertilisers and the pickling of steel wire and sheet to remove oxide layers before applying protective coatings.

8.4.3 Hydrations and hydrolyses

Several acids and bases are formed industrially by the combination of oxides with water, e.g. the production of sulphuric acid,

$$SO_3 + H_2O \rightarrow H_2SO_4$$

and the slaking of lime,

$$CaO + H_2O \rightarrow Ca(OH)_2$$

Many anhydrous salts add water to form crystalline hydrates while titanium chloride is hydrolysed by water to give titanium dioxide pigment,

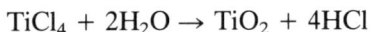

$$TiCl_4 + 2H_2O \rightarrow TiO_2 + 4HCl$$

8.5 Some hazardous organic reactions and processes

As with inorganic reactions, there are powerful oxidation-reduction reactions (which are mostly thought of as oxidations, reductions and hydrogenations), acid-base reactions involving both organic and inorganic acids, bases and anhydrides, and hydration reactions involving organic acid anhydrides and acid chlorides. There are also important reactions involving the formation and use of organo-metallic compounds, and a great variety of organic syntheses which have little parallel in inorganic chemistry. Two principal difference between organic and inorganic reactions are:

- There is virtually no upper limit to the temperature at which inorganic reactions are carried out, and many ore-reduction processes take place at temperatures above 1000°C. Most organic reactions, however, have to be carried out at temperatures ranging from ambient to 250°C, due to the thermal instability of the materials. A few carefully controlled reactions such as ethylene production, with very short contact times (one second or less), are, however, carried out at temperatures of 800°C or more.
- There is nearly always a fire hazard with organic reactions, which are mostly undertaken within totally enclosed plant and pipework, often under pressure. Most inorganic reactions are free from serious fire hazards, and many of them are carried out in plant which is open to the atmosphere.

Certain structural groupings liable to cause instability or polymerisation have been touched on. The polymerisation of some aldehydes (e.g. acetaldehyde, butyraldehyde), when catalysed by small amounts of alkalis, is followed by the elimination of water and further evolution of heat if the reaction mixture is not well cooled. The overall process is one of condensation.

Many compounds containing two or more different reactive atoms or groups such as (active) hydrogen, chlorine, hydroxy-, amino-, nitro-, etc. undergo intermolecular or intramolecular (i.e. between two or more molecules) condensation reactions with the elimination of water, hydrogen chloride, ammonium chloride or nitrogen. Sometimes these condensations are accompanied by the release of considerable amounts of heat.

Among the most hazardous organic reactions[10, 12, 13] both on a laboratory and an industrial scale are oxidations, halogenations and nitrations (all of which are very exothermic). Polymerisation and copolymerisation reactions and condensations of the Friedel–Crafts, Claisen and Cannizaro types can be hazardous if temperature control is lost (e.g. through the breaking of a thermocouple). Other potentially very hazardous reactions are those involving acetylene (Reppe chemistry) and organo-metallic compounds of lithium, zinc and aluminium, and other metals high in the electromotive series. These are too specialised to be discussed here. Reactions which generally impose fewer safety problems include esterifications (other than nitration), reduction, amination, hydrolysis, hydrogenation (apart from the hazards of hydrogen under high pressure) and alkylation.

8.5.1 Peroxide formation

Many organic compounds which contain an 'active' hydrogen atom directly attached to carbon can react with oxygen when in contact with air at ambient temperature, to form peroxides and other 'peroxy-' compounds which contain the –C–O–O– grouping. These reactions involve the formation of free radicals which react with oxygen molecules through a chain reaction [8.2.1]. Most peroxy- compounds are unstable and many are explosive [9.3]. Table 8.4 gives a list of organic structural groupings with active hydrogen atoms which form peroxides with oxygen[14, 15].

Peroxy compounds are very 'labile'[11] (liable to spontaneous change). Thus the first one formed (usually a hydroperoxide) may react further to form a peroxide which may then form a dimer or trimer or change in some other way. Since most peroxides are less volatile than the original compound, they are left behind when it evaporates. There have been many small laboratory explosions after (diethyl) ether which readily forms a peroxide with air has been allowed to evaporate in a dish. Sodium, potassium and other alkali metals readily form peroxides in air as do their alkoxides, amides and organo-metallic compounds. Bretherick[14] quotes three lists of compounds which readily form peroxides, according to their hazards:

List A, giving examples of compounds which form explosive peroxides while in storage, include diisopropyl ether, divinylacetylene, vinylidene chloride, potassium and sodium amide. Review of stocks and testing for peroxide content by given tested procedures at 3-monthly intervals is recommended, together with safe disposal of any peroxidic samples.

List B, giving examples of liquids where a degree of concentration is necessary before hazardous levels of peroxides will develop, includes several common solvents containingone ether function (diethyl ether, ethyl vinyl ether, tetrahydrofuran), or two either functions (p-dioxane, 1,1-diethoxyethane, the dimethyl ethers of ethylene glycol or 'diethylene glycol'), the secondary alcohols 2-propanol and 3-butanol, as well as the susceptible hydrocarbons propyne, butadiyne, dicyclopentadiene, cyclo-hexene and tetra- and deca-hydronaphthalenes. Checking stocks at 12-monthly intervals, with peroxidic samples being discarded or repurified, is recommended here.

Table 8.4 Structural groupings containing a peroxidisable hydrogen atom[14, 15]

Grouping	Examples
$>C-O-$ $\quad\|$ $\quad H$	Acetals, ethers, oxygen heterocycles
$-CH_2$ $\qquad >C-$ $-CH_2 \quad\|$ $\qquad H$	Isopropyl compounds, decahydronaphthalenes
$>C=C-C-$ $\qquad\|$ $\qquad H$	Allyl compounds
$>C=C-X$ $\quad\|$ $\quad H$	Haloalkenes
$>C=C-$ $\quad H$	Other vinyl compounds (monomeric esters, ethers, etc.)
$>C=C-C=C<$ $\quad\| \qquad \|$ $\quad H \qquad H$	Dienes
$>C=C-C\equiv C-$ $\quad\|$ $\quad H$	Vinyl acetylenes
$-C-C-Ar$ $\quad\|$ $\quad H$	Cumenes, tetrahydronaphthalenes, styrenes
$-C=O$ $\quad\|$ $\quad H$	Aldehydes
$-C-N-C<$ $\quad\| \| \quad\|$ $\quad O \quad H$	N-alkyl-amides or -ureas, lactams

Note: The products of the last two types readily degrade and do not usually accumulate to a dangerous level

List C contains peroxidisable monomers, where the presence of peroxide may initiate exothermic polymerisation of the bulk of material. Precautions and procedures for storage and use of monomers with or without the presence of inhibitors are discussed in detail. Examples cited are acrylic acid, acrylonitrile, butadiene, 2-chlorobutadiene, chlorotrifluoroethylene, methyl methacrylate, styrene, tetrafluoroethylene, vinyl acetate, vinylacetylene, vinyl chloride, vinylidene chloride and vinylpyridine.

Where a peroxide-containing material has to be distilled, sufficient non-volatile mineral oil should be mixed with it before distillation to keep the peroxide concentration at a safe low level at which there is no danger of an explosion or violent reaction.

Serious explosions due to peroxides have been reported in vessels in which diethyl ether, butadiene and dihydrofurane had evaporated.

Small amounts of antioxidants such as phenols and amines often have to be incorporated into organic liquids and solids to inhibit peroxide formation. The free radicals involved in peroxide formation are trapped by these antioxidants, thereby terminating the chain reaction at an early stage. The antioxidant is eventually consumed. Thus it is necessary to check the antioxidant concentration regularly when materials liable to form peroxides are stored, and to 'top-up' with further antioxidant where necessary. In many cases (e.g. with monomers for synthetic rubber and plastics) the antioxidant has to be removed before using the material in the process. The amount of antioxidant-free material should then be kept to a minimum, and any remaining when the plant is shut down should be blended back with inhibited material in the main storage.

Peroxides are involved in many self-heating phenomena of organic solids, such as soya beans, rape seed, oiled fibres and spent grains from brewing, during drying, storage and transport. The problems and prevention of fire in these situations are discussed in 8.7.

8.5.2 Oxidation processes

A variety of oxidizing agents are used including oxygen, air, hot dilute nitric acid, hydrogen peroxide, hypochlorous acid, permanganates and chromates. Several valuable chemical intermediates are extensively made by the selective oxidation of hydrocarbons with air or oxygen. Three types of hazard are encountered with these processes:

- General hazards of handling volatile flammable liquids;
- Hazards resulting from the deliberate mixing of hydrocarbons and oxygen inside the plant;
- Several of the materials handled (e.g. ethylene oxide, cumene hydroperoxide, acrylonitrile, acrolein) have special hazards.

All these processes are continuous and the main reactions which are highly exothermic are accompanied by even more exothermic side reactions leading to normal combustion products, water and carbon dioxide. Large amounts of heat must be removed from the reactors.

All plants have distillation columns and other equipment for separating the products from the crude oxidation mixture and usually for recycling unconverted hydrocarbons to the reactors. Special instrumentation is used to maintain the gas compositions in the reactors outside the explosive range and to detect fires and hot spots in them.

These potentially very hazardous processes fall into two types, liquid phase and vapour phase, some of the more important being shown in Table 8.5.

Liquid phase processes
In these processes air or oxygen is bubbled into the liquid hydrocarbon (or a solvent) in which the reaction takes place, usually via a peroxide intermediate which is in some cases isolated before reacting further. Temperatures range from 100 to 225°C and pressures of 5–10 bar g. are usually necessary to maintain the liquid state. Cooling is generally achieved by evaporation of part of the hydrocarbon which is condensed and

Table 8.5 Some major hydrocarbon oxidation processes

Feed	Product(s)	Main final products
Liquid phase processes		
Cyclohexane	Cyclohexanol/cyclohexanone	Caprolactam (nylon) solvents
Cumene	Phenol/acetone	Resins, acrylics
Paraxylene	Terephthalic acid	Polyester fibres
Light naphtha	Acetic acid	Vinyl acetate
Toluene	Benzoic acid	Terephthalic acid
Ethylene	Acetaldehyde	Ethyl hexanol
Ethyl benzene	Ethyl benzene hydroperoxide	Styrene (rubbers and plastics)
Vapour phase processes		
Ethylene	Ethylene oxide	Antifreeze
Butenes or benzene	Maleic anhydride	Unsaturated polyester resins
Butenes	Butadiene	Rubbers
o-Xylene or naphthalene	Phthalic anhydride	Plasticisers for PVC
Propylene	Acrolein	Glycerol, acrylics
Propylene + ammonia	Acrylonitrile	ABS resins
Ethylene + HCl	Ethylene dichloride	PVC

scrubbed from the 'off-gases' and returned to the reactors, which generally contain large inventories of liquid hydrocarbons at temperatures above their atmospheric boiling points. Thus in addition to very high fire loads there is usually the potential for the massive escape of a flash-vaporising flammable liquid followed by a vapour cloud explosion (VCE) [10.5.4]. Control of this hazard depends on preventing a major emission. Steps which may be taken include:

- Careful plant siting and layout in relation to prevailing wind, ignition sources, concentrations of people, traffic and valuable property (e.g. other plants);
- Reducing the inventories of superheated flammable liquids in the reactors as far as possible;
- Employing very high standards of design, construction, inspection, maintenance and operation, including hazard and operability studies [16.5.5];
- Reducing the number of flanged joints in the train of reactors, e.g. by replacing several reactors in series with a single vessel;
- Separating vessels in a reaction train where feasible with automatic quick-closing valves to isolate the contents of each vessel in an emergency;
- Special precautions to prevent the unintentional accumulation and concentration of unstable peroxides in any part of the plant.

Not surprisingly, several major fires and explosions have occurred on these plants. Probably the worst was the Flixborough explosion of 1974[16] [4]. This, however, happened while the plant was being started up, and before air was introduced into the reactors and neither peroxides nor oxygen were involved.

There is little danger of combustion occurring in the main body of cyclohexane oxidation reactors during normal operation when air is dispersed as small bubbles in the liquid. Fires have, however, occurred at the point where compressed air is introduced into such reactors[17], during start-up following a short shut-down. They were accompanied by darkening of the crude liquid product and fortunately did not lead to the escape of flammable materials. This hazard was eliminated by reducing the compressed air temperature to 100°C before it entered the reactors, and by ensuring that the compressed air line sloped continuously downwards from its control valve until the compressed air entered the reaction liquid.

Vapour phase processes
In these a mixture of the hydrocarbon with air or oxygen is reacted on the surface of a solid catalyst. The processes shown in Table 8.5 include two in which other reactants are introduced as well, ammonia in the *ammoxidation* process for acrylonitrile, and hydrogen chloride in the *hydrochlorination* process for ethylene dichloride. The vapour phase processes employ temperatures from 250 to 600°C. The major hazard is explosion of the mixed feed if its composition reaches an unsafe level. All processes rely greatly on instrumentation and automatic devices to prevent this happening. The reactors of most processes are essentially tubular heat exchangers with the tubes packed with catalyst. Special measures are needed to prevent preferential flow in some tubes at the expense of others. Fluidised-bed reactors which avoid this problem and reduce the explosion hazard have been developed, but with limited success, for some of these processes. The heat is removed by special heat transfer liquids or molten salts outside the tubes and used in waste-heat boilers to raise steam. Some plants use a molten mixture of sodium/potassium nitrate/nitrite as a heat transfer medium. This is a highly oxidising material which generally causes a fire or explosion if it contacts hot organic material. There have been several reactor explosions following leakage between the hydrocarbon/air stream inside the tubes and molten salt outside. Their effects can be minimised by good reactor design, and tube corrosion should be monitored by regular analysis of the salt for metals.

Special material hazards
While several intermediates, products or by-products of oxidation processes constitute major hazards, only two with explosion risks, cumene hydroperoxide (CHP) and ethylene oxide (EO), are discussed here. Of the others, acrolein and acrylonitrile pose serious health risks and all create fire, and to some extent, explosion risks.

CHP, which is produced in solution in cumene, is concentrated by evaporation before conversion to phenol and acetone. It is a very unstable material and can in certain conditions decompose at an accelerating rate and then explode. This can happen through contamination with mineral acids, through overheating, or simply through allowing CHP solutions to stand at normal process temperatures without adequate mixing or heat removal. Such an explosion in a process vessel could release its contents and lead to a VCE.

EO is very volatile, has wide flammability limits, polymerises readily and is both highly reactive and toxic. The concentrated vapour can also explode [9.5.3]. There have been a number of severe accidents on EO plants, some due to explosions with air or oxygen in the oxidation reactors and some to explosions of EO alone in the purification column.

Conclusions

Hydrocarbon oxidation processes which are increasingly used by major transnational companies to make many of the chemical intermediates of synthetic fibres, rubbers, plastics and paints have high potential for major explosions. Their employment makes special demands on managements and on sound engineering, particularly instrumentation.

8.5.3 Halogenation processes

There are two main types of halogenation processes, addition to double bonds and substitution of hydrogen with elimination of hydrogen halide. Reactions with fluorine are the most violent, more so generally than with oxygen, and extreme measures such as dilution with nitrogen are needed to control them.

Chlorinations are also very vigorous, and besides the direct use of chlorine, a variety of chlorinating agents are used including hydrogen chloride (for addition to a double bond), sodium hypochlorite, phosgene, thionyl chloride, sulphuryl chloride and chlorides of phosphorus. Oxy-chlorination processes which use a mixture of hydrogen chloride with air or oxygen with a catalyst in place of chlorine are now also common, particularly for the manufacture of ethylene dichloride from ethylene as a stage in the manufacture of vinyl chloride. Most hydrocarbon gases burn in chlorine, in which they exhibit upper and lower flammability limits. While measures similar to those used for oxidation reactions are needed in chlorination reactions to control reaction temperature and avoid explosive mixtures, the high corrosivity of chlorine and hydrogen chloride in the presence of water and the toxicity of chlorine and many chlorine compounds add to the overall hazard. First-class engineering, instrumentation, materials of construction, operator training and protective and emergency measures are needed for these plants.

When introducing a chlorine-using process on any site where chlorine is not already produced it is worth considering the installation of a special chlorine production plant. This may be less hazardous than importing chlorine in road or rail tank cars and storing it on site.

Bromination reactions pose similar though less acute hazards. They are less common and are usually carried out on a smaller scale than chlorinations. Iodinations are again less vigorous and less common.

Some halogenated organic compounds such as vinyl chloride and methyl iodide are carcinogens, and require special measures for air monitoring and personal protection.

8.5.4 Nitration processes

In nitration processes two basic types of product are produced. Nitrate esters which contain the nitrate group $-O-NO_2$ are formed from compounds containing hydroxyl groups, such as cellulose, glycerine, ethylene glycol and pentaerythritol. Nitro compounds containing the nitro group $-NO_2$ linked directly to a carbon atom are formed from paraffin hydrocarbons such as methane and propane. and from aromatic compounds such as benzene, toluene and phenol. More highly nitrated compounds such as TNT, nitroglycerine and guncotton are manufactured as explosives, whereas less nitrated compounds with lower nitrogen contents have industrial uses – nitrocellulose lacquers, dyestuff and other chemical intermediates, and nitroparaffin solvents.

In the production of nitrate esters and nitro-aromatic compounds a mixture of concentrated nitric and sulphuric acids is used at the lowest possible temperature, to make all except the least nitrated products. The sulphuric acid serves to 'mop up' the water formed as a by-product and thus drive the reaction nearer to completion. These reactions present at least two major hazards:

- The instability of the products, which decompose violently above a critical temperature. Efficient cooling and temperature control of the reactions are hence vital, and emergency arrangements for dumping the contents of a reactor (into a large volume of ice and water) may be needed.
- Exothermic oxidative side-reactions in which the nitric acid is reduced to the very toxic gas, nitric oxide. These are promoted by the presence of water and other impurities in the starting material, e.g. metal fragments and jute in cotton waste. When such side reactions 'take over' in the batch nitration of cellulose the reactor is said to be 'on fire' because of the copious production of brown fumes with which the exhaust ventilation system is often unable to cope. There is then a serious health as well as explosion hazard, as the writer recalls from wartime work in an explosives factory, when it was sometimes impossible to see from one side to the other of a nitration room because of the fumes. In addition to the damage to the respiratory system, the teeth of workers in such nitration rooms suffer permanent damage.

Added to these hazards are those of human exposure to acid burns, and to the vapours of some nitration products, e.g. nitroglycerine, which causes severe headache for those freshly exposed and a fall in blood pressure. There is also a danger to workers of explosive compounds lodging in their clothing or hair or under their fingernails [9.8.6].

The danger of destructive explosions in batch nitrators for glycerine, in which a tonne or more of nitroglycerine is produced per batch, has been very much reduced by the development of simple short contact-time continuous nitrators[18].

The nitro-paraffin solvents are mostly made in a continuous vapour phase process[19], in some cases using nitrogen peroxide and oxygen as the nitrating medium, at temperatures of 150–300°C and pressures of about 10 bar g. These processes have resemblances to vapour phase oxidation processes, and have similar hazards.

8.6 Reactivity as a process hazard

As a potential process hazard, the chemical reactivity of any substance should be considered in the following contexts:

- Its reactivity with elements and compounds with which it is required to react in the process;
- Its reactivity with atmospheric oxygen;
- Its reactivity with water;
- Its reactivity with itself, i.e. its propensity to polymerise, condense, decompose or explode;
- Its reactivity with other materials with which it may come in contact unintentionally in process, storage or transport;
- Its reactivity with materials of construction, i.e. its corrosivity [11].

The hazards of exothermic reactions occur in several of these contexts, particularly in storage of compounds which tend to polymerise or

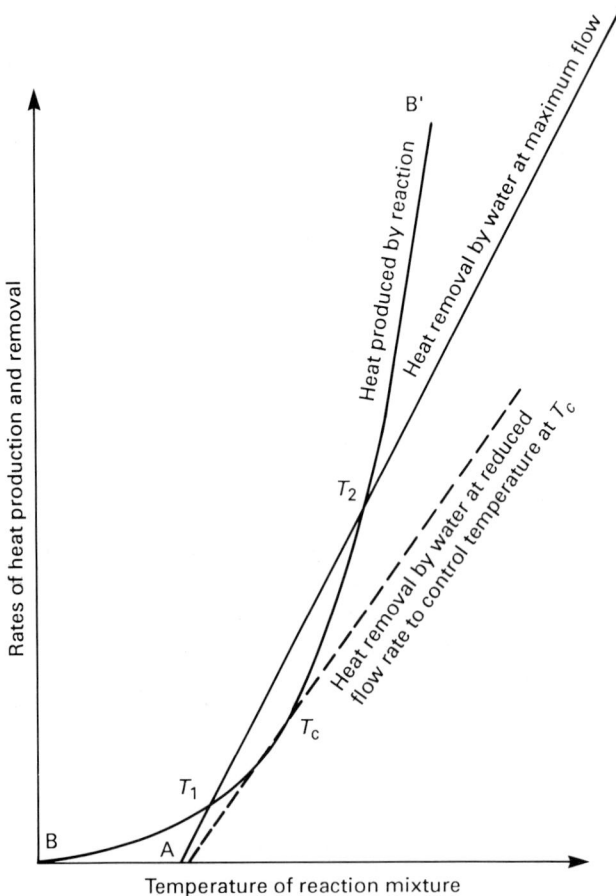

Figure 8.2 Rates of heat generation and removed from reactor with homogeneous phase exothermic reaction

decompose, and in process reactors themselves[12]. The rates of most reactions increase rapidly with temperature [8.2.1], leading to the danger of their getting out of control, with large rises in temperature and pressure and 'loss of containment' of the process materials.

The hazard is shown in its simplest form in Figure 8.2, where both the rates of heat production and removal from a water-cooled and stirred reactor are plotted against the temperature of the reaction mixture.

If the flow of cooling water is constant (and at its maximum value) the rate of heat removal increases roughly in proportion to the temperature difference between the incoming water and the reaction mixture, giving the line A–A'. The rate of heat generated by the reaction, on the other hand, increases exponentially with the temperature, and is shown by the curve B–B'. This crosses the line A–A' at a lower temperature T_1 and a higher temperature T_2. At temperatures below T_2, the (maximum) rate of heat removal is greater than the rate of heat evolution, and the temperature will drop to T_1. The temperature can be controlled at some level T_c below T_2 by adjusting the flow of cooling water, but at the temperature T_1 the maximum water flow is required. Operation at this temperature is very stable, since if the temperature is displaced above or below it, it will return to this value without any adjustment in the cooling water rate. But once the temperature rises above T_2, the rate of heat evolution exceeds the (maximum) rate of heat removal. The temperature will then continue to rise, and the reaction goes out of control. The design should therefore provide a substantial difference between T_2 and T_c, to allow an adequate margin of safety.

The curves are not fixed for all time when the plant is designed. The line A–A' will be lowered if the cooling surfaces become fouled and heat transfer coefficients fall, or if the water pressure falls. The curve B–B' may be raised by the presence of catalytic impurities or by an increase in the quantity of reacting material in the reactor.

In a very similar way, we arrive at the concept of a 'critical mass' of any material liable to 'self-heating' in storage, above which there will be an uncontrolled runaway reaction[20]. This is based simply on the fact that the rate of heat evolution for a material undergoing a (slow) exothermic reaction will be proportional to the mass or volume of the material, which increases with the cube of a linear dimension, while the rate of heat removal is proportion to its surface area, and thus increases with the square of a linear dimension. In dealing with large piles of solids of low thermal conductivity there is also a greater tendency for hot spots to arise in the interior, where the reaction rate, although still low, may get out of control first. Once this has happened the runaway reaction will spread throughout most of the pile. This critical mass concept, elaborated by Frank-Kamenetskii[20], has been at the root of some major accidents in the process industries [5.2, 5.4]. It is discussed further in 8.7 in connection with the self-heating of materials.

8.6.1 Reactivity between reactants in processes

This must be carefully studied when the reaction system is designed, both from the thermodynamic and the kinetic aspects[21]. The information is vital to the design of the process in sizing heat exchangers and determining heating and cooling requirements, not to mention safety.

From the safety viewpoint, the main thing to know is whether the reaction is strongly exothermic, moderately exothermic, mildly exothermic, thermally neutral or endothermic. These expressions are quantified in Table 8.6 in terms of the heat given out or absorbed by the reaction, as J/g of total material fed to the reactor, including reactants, solvents and diluents.

Table 8.6 Degrees of reaction heat

Category	Heat (joules) given out per gram of total reactants, solvents and diluents
Extremely exothermic	$\geqslant 3000$
Strongly exothermic	$\geqslant 1200$ and < 3000
Moderately exothermic	$\geqslant 600$ and < 1200
Mildly exothermic	$\geqslant 200$ and < 600
Thermally neutral	$\geqslant -200$ and < 200
Endothermic	< -200

An experienced chemist or chemical engineer can usually make a fair assessment of which of these categories applies by considering the overall chemical equation of the reaction. Reactions involving direct oxidation of hydrocarbons by air or oxygen, chlorination reactions, and ethylene polymerisation without diluents are extremely exothermic. Nitration reactions, and polymerisation of propylene, styrene and butadiene are strongly exothermic. Most condensation and polymerisation reactions of compounds with molecular weights from 60 to 200 are moderately or mildly exothermic. Reactions between aqueous solutions of inorganic salts to form precipitates and esterification reactions between organic acids and alcohols (in the absence of strong dehydrating agents) are usually thermally neutral. The cracking and dehydrogenation of hydrocarbons, and the reduction of most metallic oxides to metals, are very endothermic reactions which require the application of large amounts of heat or energy.

The application of such thermochemical information to evaluating the inherent hazardousness of process units is discussed later in Chapter 12 in connection with the Dow[22] and Mond[23] 'hazard indices'.

Exothermic reactions are usually least difficult to control in continuous processes involving only gases and liquids. In such cases the inventory of at least one of the reactants in the reaction system is usually fairly low, and its supply can be quickly interrupted in case of trouble. Heat exchange surfaces in the reaction system are also usually clean, and it is not difficult to design satisfactory emergency systems for relief of excess pressure or for dumping liquids whose reactions have got out of hand. Even here, however, there is a danger, particularly when starting up, that the reaction fails to start when expected, but suddenly 'takes off' when the reactor is nearly full of unreacted material. This applies particularly to autocatalytic reactions, many of which depend on the formation of free radicals.

Exothermic reactions are most difficult to control in batch processes where the entire charge of reactants is added at the start of the batch, and where both liquids and solids are present. Here the inventory of reacting material is high at the start of a batch, and the presence of solids may foul heat exchange surfaces and block valves and pipes including pressure relief and blowdown systems. The problem can often be alleviated by carrying the reaction out in a refluxing solvent or inert liquid which boils (usually under pressure) at the desired reaction temperature. Cases have, however, been known when the pipe returning the condensed solvent to the reactor has become blocked with solid, the solvent has all boiled off, and a runaway reaction has developed with disastrous consequences [5.3].

Equilibrium limitations on the reaction are usually first calculated from published free energies of formation, and checked experimentally where necessary. They have an important bearing on the pressure and temperature chosen for the reaction, on the relative amounts of reactants employed, and on measures taken to drive an equilibrium-limited reaction nearer to completion, e.g. by removing one of the products of the reaction such as water as it is formed.

Kinetic aspects which must be studied experimentally affect the reaction temperature and time (hence the size of the reactor), the pressure in the case of gas reactions, the degree of agitation in the case of reactions involving more than one phase, and the catalyst used (if any). They also often affect the relative yields of wanted product and by-products.

8.6.2 Reactivity with atmospheric oxygen

Most hazards caused by reactivity with atmospheric oxygen are dealt with elsewhere in this book. Problems arising from oxidative self-heating are discussed in 8.7. The hazards of flammable process materials are covered in Chapter 10, and problems of rusting and corrosion are described in Chapter 11. There are also many cases where atmospheric oxygen has to be excluded from process plant and storage.

In most continuous organic chemical reactors which operate under pressure, air is automatically excluded, except where it is deliberately introduced as oxidant for a reaction. In some cases more stringent measures are taken, not merely to prevent air entering plant while it is running but also to remove it from the plant before starting up, and to remove oxygen from materials entering the process. This includes the use of oxygen scavengers such as sodium nitrite, sodium sulphite, sulphur dioxide, hydrazine and tertiary butyl catechol. These are generally used in aqueous solution. Such cases include:

- Removal of dissolved oxygen from boiler feed water to reduce corrosion of the boiler tubes and hence the risk of boiler explosions;
- Prevention of peroxide formation [8.5.1] with many organic compounds;
- Protection of oxygen-sensitive materials such as photographic developers, organo-metallic compounds, alkali and alkaline earth metals and their hydrides, yellow phosphorus, titanium trichloride, powdered

metals such as zirconium, titanium, uranium and plutonium, and even iron and lead, and special polymerisation catalysts;
• Prevention of aerobic fermentations in alcoholic beverages.

The adventitious formation of pyrophoric materials, such as iron sulphide on the inside of crude oil carriers and storage tanks, presents another hazard of reactivity with oxygen. This needs to be anticipated and prevented as far as possible, e.g. by applying an oil- and hydrogen sulphide-resistant coating to the inside of tanks liable to contain sour crudes. Otherwise it may be necessary to remove all flammable gases and vapours before air is admitted to the tank. Pyrophoric iron sulphide usually reacts harmlessly with atmospheric oxygen to form ferrous sulphate provided it is kept wet. Where problems of this sort arise, the help of an experienced chemist is needed.

8.6.3 Reactivity with water

Many chemicals react violently with water, which is seldom far away, and is widely used in process plants for cooling and cleaning. A short list of these, with the gases formed, is given in Table 8.7[24].

Table 8.7 Materials which react strongly with water

Material and state (s, l or g)	Action	Gas, etc. liberated
Calcium (s)	Moderate	Hydrogen
Lithium (s)	ditto	ditto
Sodium (s)	Vigorous	ditto
Potassium (s)	Explodes	ditto
Calcium hydride (s)	Vigorous	ditto
Lithium hydride (s)	ditto	ditto
Aluminium alkyls (l)	ditto	Alkanes
Aluminium alkyl halides (l)	ditto	ditto
Zinc alkyls (l)	ditto	ditto
Calcium carbide (s)	Moderate	Acetylene
Calcium phosphide (s)	ditto	Phosphine
Aluminium phosphide (s)	ditto	ditto
Fluorine (g)	Vigorous	Oxygen and ozone
Sodium peroxide (s)	Moderate	Oxygen
Aluminium chloride (s)	Vigorous	Steam and acid fumes
Phosphorus pentoxide (s)	ditto	ditto
Sulphur trioxide (s)	ditto	ditto
Acetyl chloride (l)	ditto	ditto
Phosphorus trichloride (l)	ditto	ditto
Silicon tetrachloride (l)	ditto	ditto
Sulphuric acid (l)	ditto	ditto
Thionyl chloride (l)	ditto	ditto
Titanium tetrachloride (l)	ditto	ditto
Calcium oxide (s)	ditto	Steam
Sodium hydroxide (s)	Moderate	ditto
Potassium hydroxide (s)	ditto	ditto
Activated alumina (s)	ditto	ditto
Activated silica (s)	ditto	ditto
Activated molecular sieves	ditto	ditto

These materials should only be handled by trained operators, wearing appropriate protection. They must always be stored in sealed watertight containers and kept in a dry place which is not subject to flooding. Plant in which they are used must be carefully designed and operated to prevent any possibility of water entering the process accidentally, e.g. through leaking heat exchangers.

8.6.4 Self-reacting compounds (mainly monomers)

As explosive materials are discussed in Chapter 9, the main compounds considered here are organic monomers, which are liable to polymerise spontaneously with evolution of heat, especially in storage. The same principles may be applied to compounds which are liable to condense, isomerise, transform or decompose without explosion in some other way, such as the α- and β-forms of sulphur trioxide, selenium, and antimony (of which there is an explosive form!). Monomers include styrene and substituted styrenes, vinyl chloride, acrylonitrile, butadiene, isoprene and cyclopentadiene, as well as methyl isocyanate whose runaway reaction caused the Bhopal disaster of 1984 [5.4]. Most of these polymerise spontaneously at low rates at ambient temperature even in the absence of a catalyst. All, however, polymerise much faster in the presence of a catalyst, such as a peroxide formed by the action of atmospheric oxygen on the material itself.

Liquid cyclopentadiene, chloroprene (the monomer of Neoprene rubber), butadiene and isoprene all form dimers at room temperatures, the reaction rates increasing rapidly with temperature. The uncatalysed dimerisation rates for cyclopentadiene, butadiene and isoprene at various temperatures are shown in Figure 8.3[25]. Although the rates for butadiene

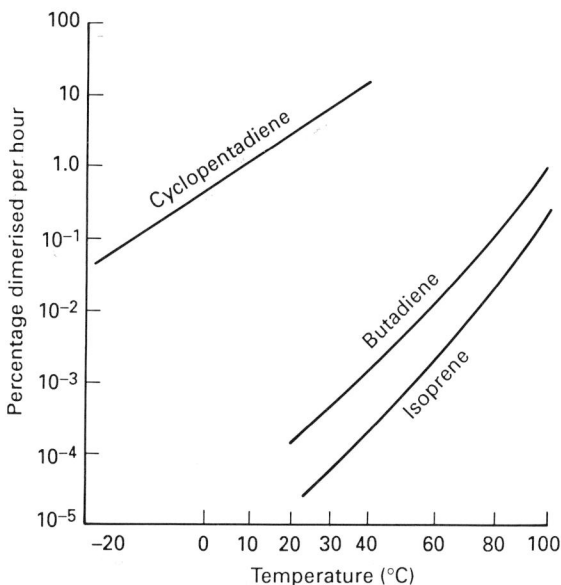

Figure 8.3 Rates of uncatalysed dimerisation of three-diene hydrocarbons versus temperature

and isoprene are low enough for them to be stored for limited periods at ambient temperature, the rate for pure cyclopentadiene is so high that it is normally not stored as such, but produced as it is needed by thermal depolymerisation of its dimer.

These reactions evolve considerable amounts of heat and their rates increase exponentially as the temperature rises. As explained earlier, there is usually a critical temperature and/or a critical mass for such materials in storage above which disaster may lie imminent. In the case of monomers stored as liquids, the temperature of the unpolymerised material will rise until it reaches its boiling point, and then boil vigorously as the rest of it polymerises. With large storage tanks and pressurised spheres, the flow of escaping vapour may be so large that it is impossible to contain or dispose of it safely. Some monomers such as cyclopentadiene polymerise so rapidly at ambient temperature, even in the total absence of a catalyst, that their storage at ambient temperature is impossible.

The writer recalls a 10 m by 2.5 m (dia.) horizontal tank containing C_5 hydrocarbons rich in cyclopentadiene which had been produced 35 years ago as a large-scale experiment. This polymerised exothermally and continued to boil vigorously for three days, causing a serious hazard in the works, despite all efforts to cool it by spraying water over the tank. A similar incident which occurred in road transport is mentioned in 10.5.5.

Most monomers must be stored out of contact with air, with an inhibitor added which will prevent polymerisation and/or destroy any peroxide catalysts that may have formed [8.5.1]. Inhibitors commonly used are hydroquinone and its monomethyl ether, tertiary butyl catechol and phenothiazine.

Inhibition of many polymerisations (e.g. of acrylonitrile and vinyl acetate) is complicated by the fact that the inhibitor only works effectively in the presence of some oxygen[26]. Some monomers with inhibitors added are stored more safely in a tank which air can enter than under a nitrogen blanket. Oxygen is, however, rigorously excluded from butadiene and chloroprene during the last stages of their manufacture by using an oxygen-scavenging compound such as sodium nitrite. Tertiary butyl catechol is added in storage of butadiene to protect against leakage of oxygen into it with the formation of butadiene peroxide. A dangerous situation arises if inhibitor is lost from a monomer during storage. This can happen through separation of the inhibitor from the product or by reduction in its efficiency.

Since inhibitors are far less volatile than the monomers they protect, the vapour in a tank of stored monomer is free of inhibitor, and liquid condensing from it may form solid polymer. This has caused blockages in vents and pressure relief devices on some monomer storage tanks and vessels.

Loss of inhibitor efficiency can occur either through lack of oxygen as described above or through contact with so much oxygen that all the inhibitor has reacted. Stored monomers thus need to be periodically sampled and analysed to check the effectiveness of inhibition, which must be adjusted when necessary.

Bond[26] reports seven accidents resulting from spontaneous polymerisation of monomers in storage in which the following were involved: styrene,

Table 8.8 Adiabatic bulk storage of monomers: calculation of percentage polymerised when liquid boils or bursting disc blows

Monomer	Acrylo-nitrile	Styrene	Methyl styrene	Vinyl pyridine	Butadiene
Molecular wt	53	104	118	105	54
Boiling point (°C)	77	145	165	158	(81)
Heat of polymerisation (J/g)	1410	680	293	718	1352
Specific heat (av.) (J/g °C)	2.27	1.81	1.89	1.81	2.48
Polymerisation at BP (%)	8.5	30	86	35	(10.1)
Final state:					
Vapour (%)	70	48	7	46	
Polymer (%)	30	52	93	54	

methyl methacrylate, acrylonitrile, acrolein, vinyl chloride and acrylic acid.

Table 8.8 shows the results of calculations made to show the amount of polymer which would have been formed from five monomers stored in bulk at an initial temperature of 25°C, by the time they reached their boiling points. Adiabatic storage with no heat loss to the surroundings was assumed. The Table also shows what percentages of the initial material would escape as vapour and be left as polymer, when all liquid had either polymerised or vaporised. Four of the monomers are assumed to be stored at atmospheric pressure in vented tanks. The fifth, butadiene, is assumed to be stored in a pressure sphere with a bursting disc set at 10 bar g. This calculation shows the percentage of polymer formed by the time the bursting disc blows.

8.6.5 Reactivity with adventitious materials

The reactions with which one is mainly concerned here are those which might follow from some incident involving loss of containment, which would increase its consequences. Some types of chemicals which react strongly with each other have been considered already, and care must be taken when planning their storage to eliminate the risk of their coming into contact with each other as the result of an accident. Thus oxygen cylinders should always be stored well away from acetylene and LPG cylinders and other fuels. Drums of sodium chlorate, sodium nitrate and other oxidising materials should also be stored at some distance from organic materials and powdered metals. Chlorine and acids should be kept far apart from alkalis, alkaline solutions, light metals, carbides, hydrides and other materials with which they might react with dangerous consequences, e.g. fire, explosion or the formation of more toxic materials. This applies to storage in drums and portable containers as well as in tanks, tank farms and silos.

One of the most critical operations where the hazards of incompatible chemicals are at their greatest is in bulk loading of road and rail tank-cars.

A useful tabular guide to the compatibility of 24 types of chemicals was published by McConnaughey[27] and is reproduced as Table 23.4 of Lees's book[13]. This shows unsafe combinations of chemicals for adjacent loading at a glance, and should be considered when planning bulk loading facilities.

Drainage systems need careful study to ensure that spillages of incompatible materials do not mix and react in them to produce toxic or flammable vapours, perhaps some distance from the original spillages. With certain materials one has to consider whether they should ever be allowed to enter a drain at all, particularly a public one. If the answer is 'No', proper means of containing them and disposing of them safely must be provided[28, 29]. One must also consider the accidental formation and mixing of substances which have adverse catalytic effects on materials in process or storage. Peroxides have already been mentioned as polymerisation catalysts. The corrosion products of several metals, e.g. iron and manganese, catalyse the decomposition of unstable compounds such as hydrogen peroxide and hypochlorites.

Materials, particularly liquids which react violently with each other, are sometimes used as auxiliary materials in the same process, although they are never intended to come into contact with each other. As an example, a solution of caustic soda is used to remove acid impurities from the products of hydrocarbon alkylation and acid-treating processes. Both plants and storage units must be carefully designed to ensure that mutually antagonistic materials can never mix inadvertently. This requires critical examination by an experienced chemist. The biggest problems often lie in process plant and storage depots which are not recognised as 'chemical', but which contain mutually antagonistic materials. Here there may be no qualified chemists on the staff to appreciate the hazards.

8.6.6 Practical reactivity screening in chemical manufacture

A system of reactivity evaluation set up for a UK company making more than 30 bulk organic chemicals by batch processes in which over 150 stages are involved is described by Coates and Riddell of Sterling Organics[30]. This followed a plant incident involving a reaction (decomposition of a diazonium salt) in which a large volume of gas was evolved. After proceeding smoothly for several years, a batch 'turned nasty', with rapid pressure build-up in a vessel. Investigations showed the reaction only 'misbehaved' when the main reactant was impure.

The system involved the setting up of a Hazard Evaluation Laboratory under a Hazard Evaluation Manager, whose task included the screening of both materials and processes for uncontrollable exothermic behaviour.

Of the many materials and processes used in the works the following received priority for study:

1. All plant processes where any doubt existed regarding the reaction controllability, or which had in the past shown evidence of variable exothermic behaviour;
2. Residues from distillations or products/residues which were subjected to temperatures in excess of 250°C;

3. Raw materials, intermediates and products which contain functional groupings known from experience to have potential for instability;
4. All plant processes involving nitrations, or strongly oxidising conditions, and processes with high gas evolution.

Material evaluation

Most materials tested are typical plant samples. If it is suspected that the material can detonate or deflagrate violently, it is first tested by HSE. Their typical tests include:

- Home Office mortar test, to indicate whether the material exhibits moderate detonation properties;
- Detonation/propagation test with thermal ignition in a sealed tube;
- Home Office pressure–time test;
- Ballistic-torpedo friction test.

After HSE have cleared that the material is safe for hazard evaluation, the company's testing programme is started. The tests and testing equipment are as simple as possible, and designed for routine use by technicians. The following tests are carried out in sequence, results being assessed after each before deciding whether to proceed to the next:

- *Deflagration test*
- *Simple exotherm test.* This measures the lowest temperature T_e at which the material exhibits exothermic behaviour. After carrying out the test on fresh material, it is repeated on a fresh sample which has been subjected for 24 hours to a temperature 20°C above its normal operating temperature. Compounds which show no exothermic behaviour are not tested further, but where the maximum temperature to which the material is intended to be subjected is within 100°C of T_e, the material is considered to be potentially hazardous, and subjected to the next test.
- *Adiabatic exotherm test.* This is a refinement on the last test and may give values of T_e some 20-30°C lower. Plant material must not be subjected to temperatures closer than 50°C to the value of T_e given by this test unless it satisfies the next test.
- *Long-term hot storage test.* This gives the lowest temperature at which the material will undergo self-heating under isothermal conditions. The storage period is determined by the maximum holding time in the plant. The test is first carried out at a temperature 50°C higher than the maximum operating temperature proposed, and no further tests are done if no self-heating occurs. But if self-heating is found, the test is repeated on fresh material at progressively lower temperatures until self-heating is absent. The plant operating conditions are then adjusted to give a safe margin between the maximum operating temperature and the minimum self-heating temperature.
- *Drying tests.* These are carried out on materials which are dried in contact with air to determine whether the value of T_e determined by the adiabatic exotherm test is affected by exposure of the material to air under drying conditions. The details of the test are adapted to the type of dryer used or proposed. The method and conditions of drying may have to be altered to ensure that the material will not decompose during drying.

Reaction evaluation

Most of the reactions are normally carried out batch-wise on a plant scale, one or more materials being added progressively as the reaction proceeds. With exothermic reactions there is often a danger that the reaction fails to start or proceeds slowly on the first addition when expected, leading to an accumulation of reactants in the mixture. This may then suddenly react too rapidly to be controlled.

The full exothermic potential of a reaction is first assessed by carrying it out in a stirred Dewar vessel without cooling. This allows it to be classed as one of three types:

1. Non-exothermic
2. Exothermic but not runaway
3. Runaway.

Further tests are required when the reaction is of type (3). The possibility of reactions normally of types (1) and (2) degenerating into type (3) as a result of process malfunction or maloperation also have to be considered and sometimes checked experimentally.

The thermal stability not only of the materials used or produced in the reaction but also of the reaction mixture at various stages may require to be tested by the simple exotherm test. If the samples contain materials which are liable to evaporate during the test, they should be contained in sealed tubes fitted with bursting discs.

The heat of reaction and the rate of heat evolution of exothermic reactions are determined by carrying out the reaction in a laboratory calorimeter equipped with an automatic temperature recorder. Expert interpretation of the temperature record shows the point in the reaction where an irreversibly runaway reaction starts, and allows temperatures to be set for the operation of alarms, quenching devices, etc. on a plant scale.

The possible effects of unplanned reactions or conditions are determined by carrying out a series of experiments in Dewar flasks.

Where runaway reactions might occur, routines are established for the gradual or stepwise addition of reactants to the reaction mixture, and for checking that the reaction is proceeding as planned before more reactant is added. In some cases this may necessitate special investment, e.g. in microprocessors to control the reaction on the plant scale and in sampling and testing facilities to check the composition of the reaction mixture during the reaction before more material is added.

The information gathered from these laboratory hazard evaluations, valuable though it is in setting safe parameters for plant design and operation, can seldom enable all possible incidents on a full-scale plant to be foreseen. Factors which appear insignificant in a laboratory reaction may become of critical importance on a plant, where conditions of stirring, surface/volume ratios and materials of construction may be totally different. Thus the conclusions from laboratory hazard evaluations often require to be confirmed and extended by further work on a pilot plant, specially built or adapted for the reaction in question. This tends to set the design and operating conditions for the full-scale plant. Once these have been established for a potentially hazardous process they should not be altered without submitting the proposed change to the type of procedure discussed under 'Hazard and Operability Studies' in 16.5.5, where it is examined by a team of staff with experience of all aspects involved, led by a specialist in such studies. Where doubts still exist, the Hazard Evaluation Manager would be consulted. He may then decide on the need for further experiments to study the effects of the proposed change.

8.7 Self-heating hazards of solids

Several organic materials, mainly of natural origin, which are processed in bulk to make foods, textiles, paper, etc. are liable to oxidative self-heating which may lead to combustion when they are dried, stored and transported. Examples are given later in 8.7.3. Similar problems are encountered in the sometimes violent decomposition of unstable materials such as ammonium nitrate, 'high-strength' calcium hypochlorite, organic peroxides and various nitro-compounds, which is usually preceded by a period of self-heating without atmospheric oxygen.

To put the first problem in perspective, about 8×10 t per year of atmospheric carbon dioxide are continuously being converted into organic matter by natural photosynthesis. Some of this gets burnt, some eaten, some used temporarily by industry, and some buried until it gets dug up again. But probably the bulk of it simply 'decays' at temperatures only a little above ambient, ultimately reverting to carbon dioxide and water. This process 'has to happen', for without decay the cycle of life would long since have ground to a halt. The temperature at which decay takes place is bound to be above ambient, since although decay is much slower than combustion, it uses the same amount of oxygen and produces the same amounts of carbon dioxide, water and heat.

There is fortunately a gap of a few hundred degrees Celsius between the temperature ranges in which decay and combustion take place, so that although decay raises the temperature of the matter involved, it seldom starts to burn. But when this gap is bridged, either by the decay process taking place at abnormally high temperatures or by combustion starting at unusually low ones, decay leads directly to combustion.

We are concerned to prevent our artefacts won from nature from going up in smoke and also usually try to arrest their decay for as long as possible. In spite of this, they do deteriorate in time. In so doing they are sometimes liable to 'self-heat' to the point where they burst into flames. A book by Bowes[31] based mainly on work at the Fire Research Station provides much practical information and underlying theory about this problem.

8.7.1 Mechanisms for self-heating of organic materials

Self-heating of organic materials can take place through various exothermic reactions, most of which depend on atmospheric oxidation. There are also exothermic degradation processes which proceed in the absence of free oxygen. Sufficient has been learnt about most self-heating processes for measures to be taken to prevent their leading to ignition, even when they are not fully understood, as in the case of the farmer's haystack.

It is useful first to gain a general appreciation of the amounts of heat liberated during both aerobic and anaerobic decay processes. As a simple illustration, the theoretical dehydration of softwood (treated as cellulose) to charcoal (carbon) is considered, and compared with the combustion of the wood itself and of the charcoal produced.

Starting with one mol of dry cellulose (taken as 162 g for $C_6H_{10}O_5$), the processes of combustion and of dehydration with their (net) thermal effects are roughly as follows:

- Combustion:
  $$C_6H_{10}O_5 + 3O_2 \rightarrow 6CO_2 + 5H_2O \qquad + 3185 \, kJ$$
  $$6C + 3O_2 \rightarrow 6CO_2 \qquad + 2410 \, kJ$$
- Hence carbonisation (dehydration):
  $$C_6H_{10}O_5 \rightarrow 6C + 3H_2O \qquad + 780 \, kJ$$

This shows that the simple dehydration of cellulose, if it could be achieved without side reactions, would be a highly exothermic process, yielding 24% of its total net heat of combustion. Another way of looking at this would be to say that even very wet cellulose containing 60% water could in theory be converted isothermally to solid carbon and water vapour without the need for any external heat.

In fact, heat must be supplied during the first stages of charcoal production, when methanol, acetic acid and other combustible liquids are formed by destructive distillation, as well as solids with lower hydrogen and oxygen contents than the original cellulose. The final stages of charcoal production from partly carbonised solids are, however, exothermic and thermally self-sustaining, and raise the charcoal to temperatures at which it readily ignites if air is admitted. Clean timber does not present a self-heating hazard, although at least one fire has been reported where a steam pipe passed snugly through a hole in a thick beam.

There are two types of mechanism for the oxidative self-heating of organic matter:

- Microbiological processes, which occur at usual ambient temperatures in the presence of moisture and nutrients. These can only raise the temperature of the material to the maximum which the organism can withstand, which is still well below usual ignition temperatures.
- Purely chemical processes, of which those involving the initial formation of a peroxide or hydroperoxide are the norm.

Microbiological processes

Two different types of microbiological organism which are present everywhere, aerobic moulds and bacteria, soon get to work. Moulds operate most effectively at a temperature of about 40°C, but are mostly killed at 50°C. In the case of hay they require a moisture content of at least 25% to operate effectively. Bacteria require a somewhat higher moisture content of 40%, and work most effectively at a temperature of about 60°C but die at 75°C.

Parallel with these aerobic processes, anaerobic ones play a part in oxygen-depleted zones of the material. In the case of cellulose, these involve hydrolysis to sugars and fermentation to carbon dioxide, methane, ethanol, glycerol or lactic acid, all exothermic processes.

The generation of impure methane (marsh gas) from decaying vegetation proceeds fast in shallow lakes in tropical countries. On reaching the surface it may undergo partial combustion as faint flashes of a 'cool flame' known as 'Will-o-the-Wisp' [10.2.3]. Marsh gas is not always so harmless. It is recorded that a paddle steamer on an inland lake in Uganda was destroyed by fire resulting from the release of trapped methane by the ship's paddle wheels!

Chemical oxidation

The first step in peroxidation processes is the addition of two oxygen atoms to a C–H bond or an R–C radical. These are 'free-radical chain reactions' which proceed autocatalytically with the accumulation of peroxide bodies. Slow oxidation at moderate temperatures is accompanied by evolution of 5–15% of the final oxidation products, CO, CO_2 and H_2O. Almost all the oxygen consumed at a given temperature can, however, be recovered as these oxidation products by heating to a higher temperature. This suggests that once sufficient oxygen has been absorbed, further self-heating may take place in its absence. The rate constants for peroxidation processes increase rapidly with temperature, but the actual rates in piles of organic matter are restricted by the diffusion of oxygen through the outer layer of the pile and by the availability of reactive sites which oxygen molecules can reach. Nevertheless, these peroxidation processes, and the thermal decomposition of peroxides which succeed them, are sometimes enough to raise the temperature at one or more places in the pile to ignition point.

Besides peroxidation, there are other mechanisms which account for the rapid ignition of fresh surfaces of cool charred hay or activated charcoal with atmospheric oxygen. These may involve the exothermic adsorption of oxygen or water or the presence of chemically active centres on the fresh surfaces. Such self-heating is often prevented by atmospheric weathering of the cooled material spread in thin layers for a few days before bulk storage or shipment.

The effect of moisture on self-heating

Materials which self-heat to the point of ignition are often wet. This is surprising, considering the additional heat which self-heating has to supply to dry the material before it can ignite. Thus wet haystacks are more liable to fire than dry ones, while wood charcoal produced for fuel in tropical countries[32] has to be stored under cover, due to frequent fires which break out in material which has been exposed to rain. It is, however, clear from the thermal data on cellulose quoted earlier that there is no shortage of chemical energy to evaporate a fair quantity of water, provided this energy can be released.

In other cases the risk of self-heating occurring and leading to fire is greatest if the material has been 'over-dried'. This hazard can be controlled by the design of the dryer, and by the use of instruments which monitor the moisture content of the product. While the influence of moisture on self-heating is rather ambiguous there seems to be an optimum range of moisture contents corresponding to a relative atmospheric humidity of 25–50%, for minimum fire risk through self-heating.

8.7.2 Effects of pile size, initial temperature and time

The critical importance of the mass or size of the pile of material for self-heating has been mentioned earlier [8.6]. For the simplest case of a material undergoing self-heating, the following heat balance equation can be written:

$$dq/dt = kQV\phi T - k'A(T - T_0)$$
$$= 0 \text{ for a steady state condition}$$

where dq/dt is the rate of heat accumulation of the reacting mixture,
Q is the heat of reaction per unit mass of reactant consumed,
V is the volume of material reacting,
A is the surface area of the volume reacting,
T is the internal temperature,
T_0 is the external temperature,
$\phi T = r$ = reaction rate = mass of reactant consumed per unit volume, which is taken as a function of the internal temperature T,
k and k' are constants.

The first term on the right-hand side is the rate of heat production and the second term the rate of heat loss to the surroundings. In most cases the reaction rate r increases with T much faster than $(T - T_0)$.

At sufficiently low values of T_0 or V/A the equation will have a real root in T and a steady state will be set up, with a corresponding internal temperature. But if T_0 or V/A are raised, eventually a condition will be reached where the equation has no real roots, and a steady state condition is no longer possible. The temperature will then rise until the pile ignites.

Although material in storage in real piles, silos, bins, bales and other containers (all subsequently referred to as piles) which is subject to an oxidative self-heating process presents a more complicated picture, the concepts of a mutually dependent critical temperature and mass or size still apply. The dependence of critical size and temperature is discussed in a paper by Boddington, Gray and Walker[33]. As important, however, as the temperature T_0 of the surroundings is the initial temperature at which the pile is made or the bin etc. is filled. This may be quite high if the material comes direct from a dryer. It is then often necessary to cool it to a lower temperature before it is safe to store.

Three factors which arise with self-heating materials complicate the above picture:

- Most self-heating materials have a cellular, granular or fibrous structure with a very low thermal conductivity. Thus when self-heating occurs, the temperature in the interior of the pile is much higher than the surface temperature, which may be closer to the outside temperature than the average temperature of the pile.
- The oxygen concentration in oxidative self-heating processes which depends on the diffusion of atmospheric oxygen into the pile will be

highest at the surface and decrease inside with distance from the surface. This factor acts in opposition to the first one.

- The time taken for a self-heating pile to reach a point of rapid temperature rise and ignition may be considerable (i.e. several days). The slow rate of oxygen transfer into the pile, the initial build-up of peroxidic bodies and the need to dry out the pile make the overall process a lengthy one.

Two illustrations of the concepts of critical mass and critical temperature are taken from Bowes's book[31]. Table 8.9 shows the results of ignition tests on mixed hardwood sawdust, samples of which were placed to various depths on a thermostatted hot-plate. Several tests were done for each depth of sawdust at different surface temperatures, and the minimum temperature at which ignition occurred was found for each one. The time to reach ignition was also recorded.

Table 8.9 Ignition tests on mixed hardwood sawdust on hot surface

Depth of layer (mm)	Ignition temperature (°C)	Time to ignition (min)
5	355	9
10	320	24
20	290	50
25	280	105

Table 8.10 Critical ambient temperatures for self-ignition of piles of activated carbon

Critical temperature (°C)		40	60	80	100
Length of shortest side (m)	Rectangle	1.3	0.55	0.21	0.09
	Cube	1.8	0.73	0.28	0.14

Table 8.10 shows the critical ambient temperatures for rectangular and cubical packages of a typical batch of activated charcoal.

Bowes provides mathematical equations with constants derived from laboratory experiments on particular materials to predict their behaviour in real situations. These are too specialised to be given here. He also discusses computer modelling of self-heating processes. The underlying theory is largely based on the work of Frank-Kamenetskii[20].

8.7.3 Auto-ignition hazards

Fires sometimes occur in industrial driers, even when the highest temperatures appear to be well below the ignition temperatures of the material being dried. Self-heating may occur in driers in the following situations:

- In deposits of dust on ledges and heated surfaces;
- In material left in the dryer when it is shut down;
- In bodies of immobile material which 'get stuck' in the dryer;
- In bins or hoppers into which the dried material is discharged without cooling.

Serious fires have also happened in spray dryers, particularly in the production of milk powders[34-36].

'Lagging fires' occur in the lagging of hot pipes which have become impregnated with oils and fats, particularly drying oils, even when the material of the lagging is non-combustible, such as rock-wool and fibre glass.

Fresh charcoal is rather subject to self-heating and auto-ignition, as are fresh charcoal surfaces exposed by crushing, grinding, etc. The risk of auto-ignition is greatly reduced by weathering the cooled charcoal for eight days in shallow layers. Chemically active charcoals sometimes require to be packed with a polythene liner or cover which is relatively impervious to oxygen.

Coals vary considerably in their proneness to self-heating, but the size of many piles increases the risk of auto-ignition. The presence of a high proportion of fines in piles of broken coal increases the risk. The self-heating tendency increases with the softness of the coal, on passing from anthracite through bituminous coals to lignites. Pyrites increases the tendency. As with charcoal, fresh coal surfaces are the most active. Serious problems are encountered when pulverised coal has to be stored, particularly when it has been dried in an oxygen-free atmosphere.

Timber presents little hazard of self-heating. Sawdust, wood shavings and chips sometimes present a hazard, particularly if they contain a high proportion of resin, or if contaminated by oil, especially drying oils. The same applies to fibre board. Clean paper presents little hazard, but fires have occurred in transport containers tightly packed with rolls of hot, freshly-dried board made from recycled newsprint and corrugated boxes.

Cotton linters present a fairly low hazard, although fires sometimes occur in steam-heated dryers of cotton waste. Piles of oily rags present an auto-ignition hazard which, like sawdust and wood shavings, are easily controlled by good house-keeping.

Raw and solvent-extracted natural fibres are seldom a problem. Oiled fibres present some hazard when highly unsaturated oils such as linseed and tung are used, but there is far less danger with non-drying oils and solid vegetable fats.

Oil seeds such as rape which contain easily oxidisable oils have occasionally ignited as a result of auto-oxidation, but the presence of natural anti-oxidants in the seeds generally protects them. Oil seed cake and meal and soya meal which have been subjected to strong mechanical action present some hazard, especially if they are filled into silos or bins at high initial temperatures. Fish meal is also rather liable to self-heating and ignition in storage.

Cereal grains if stored wet or infested with insects are liable to self-heating, but there has been little evidence of ignition, except possibly for wet oats in silos. Dried spent grains from breweries and distilleries are liable to self-heating and ignition, especially if dried to less than 5% moisture. As a precaution, these should be cooled below 38°C before bulk storage.

In most cases careful attention to the five following points will reduce the risks of auto-ignition in storage and transport to an acceptable level:

- Control of the initial moisture content within limits, which need to be determined for each material;
- Protection of the material from rain and other sources of moisture during storage and transport;
- Limitation of the temperature of the material entering the storage or transport container;
- Limitation of the size of the piles or containers to one known to be safe for the material and the conditions of storage, and ensuring free air circulation around the piles or containers during storage or transport;
- Temperature monitoring inside large piles or containers of combustible materials.

Buried hot or warm objects such as inspection lamps within a pile or container have been implicated as another cause of self-heating and fire. It only requires one lamp inside a large silo to initiate the chain of destruction!

The same precautions apply to unstable materials liable to self-heating in the absence of atmospheric oxygen.

With materials specially prone to oxidative self-heating for which these precautions are not sufficient the access of oxygen to the material may have to be restricted or prevented, e.g. by enclosing the materials in plastic film.

8.8 Reactive substances and CIMAH Regulations 7 to 12[37]

Group 3 of Schedule 3 of these regulations [2.4.2], given in Table 8.11, is a list of highly reactive substances with the minimum quantities of each to whcih Regulations 7 to 12 apply.

Most of these substances are peroxy-compounds [8.5.1]. Their main hazards are explosive decomposition and fire, and of catalysing exothermic runaway polymerisations of monomers. They are also liable to react violently with many other chemicals.

3,3,6,6,9,9-Hexamethyl-1,2,4,5-tetroxacyclononane is one of the least known of several cyclic trimeric peroxides, with powerful and sensitive explosive properties. Of the other substances, acetylene, ammonium nitrate, ethylene oxide, hydrogen and sodium chlorate present hazards of fire and explosion as well as of reactivity, and are discussed in Chapters 9 and 10 as well as in this chapter. The footnote to Table 8.11 suggests that ammonium nitrate and sodium chlorate have safe states which are exempt from the Regulations.

Propylene oxide is less sensitive and more easily transported than ethylene oxide, but it still needs to be handled with some care[38]. The main hazard of ethyl nitrate is explosion, but its use appears to be limited. The lower nitroparaffins which have similar hazards and are more commonly used are not included in the list.

Table 8.11 Schedule 3, Group 4 of CIMAH Regulations – Explosive substances (for the application of Regulations 7 to 12)

Substance	Quantity (t)	Number	
		CAS	EEC
Acetylene	50	74-86-2	601-015-00-0
Ammonium nitrate[a]	5000	6484-52-2	
2,2-Bis (*tert*-butyl-peroxy) butane (concentration ≥ 70%)	50	2167-23-0	
1,1-Bis (*tert*-butyl-peroxy) cyclohexane (concentration ≥ 80%)	50	3006-86-8	
tert-Butyl peroxyacetate (concentration ≥ 70%)	50	107-71-1	
tert-Butyl peroxyisobutyrate (concentration ≥ 80%)	50	109-13-7	
tert-Butyl peroxy isopropyl carbonate (concentration ≥ 80%)	50	2372-21-6	
tert-Butyl peroxymaleate (concentration ≥ 80%)	50	1931-63-0	
tert-Butyl-peroxypivalate (concentration ≥ 77%)	50	927-07-1	
Dibenzyl peroxydicarbonate (concentration ≥ 90%)	50	2144-45-8	
Di-*sec*-butyl peroxydicarbonate (concentration ≥ 80%)	50	19910-65-7	
Diethyl peroxydicarbonate (concentration ≥ 30%)	50	14666-78-5	
2,2-Dihydroperoxypropane (concentration ≥ 30%)	50	2614-76-8	
Di-isobutyryl peroxide (concentration ≥ 50%)	50	3437-84-1	
Di-*n*-propyl peroxydicarbonate (concentration ≥ 80%)	50	16066-38-9	
Ethylene oxide	50	75-21-0	603-023-00-X
Ethyl nitrate	50	625-58-1	007-007-00-9
3,3,6,6,9,9-Hexamethyl-1,2,4,5-tetroxacyclononane (concentration ≥ 75%)	50	22397-33-7	
Hydrogen	50	1333-74-0	001-001-00-9
Methyl ethyl ketone peroxide (concentration ≥ 60%)	50	1338-23-4	
Methyl isobutyl ketone peroxide (concentration ≥ 60%)	50	37206-20-5	
Peracetic acid (concentration ≥ 60%)	50	79-21-0	607-094-00-8
Propylene oxide	50	75-56-9	603-055-00-8
Sodium chlorate[a]	50	7775-09-9	017-005-00-9

[a] Where this substance is in a state which gives it properties capable of creating a major accident hazard

The list also does not include many common reactive chemicals such as sodium, fluorine, hydrogen peroxide, sodium hypochlorite, hydrazine, sulphur trioxide, phosphorus trichloride and the lower aluminium alkyls. It is, however, expected to be extended when the Regulations are amended, when it will probably include sulphur trioxide. The fact that a chemical site may have none of the substances on this list is no proof that it does not contain dangerous quantities of highly reactive chemicals.

References

1. Huheey, J. E., *Inorganic Chemistry*, 3rd edn, Harper & Row, London (1983)
2. Lewis, G. M. and Randall, M., *Thermodynamics*, McGraw-Hill, Maidenhead (1923)
3. Hodgman, C. D. (ed.) *Handbook of Chemistry and Physics*, 32nd edn, Chemical Rubber Publishing Co., Cleveland, Ohio (1950)
4. Uhlig, H. H., *Corrosion and Corrosion Control*, 3rd edn, Wiley-Interscience, New York (1985)
5. Coffee, R. D., 'Chemical stability' in *Safety and Accident Prevention in Chemical Operations*, 2nd edn, edited by Fawcett, H. H. and Wood, W. S., Wiley-Interscience, New York (1982)
6. National Bureau of Standards, *Selected Values of Properties of Hydrocarbons, Circular 461*, Washington, DC: US Govt Printing Office (1947)
7. National Bureau of Standards, *Selected Values of Chemical Thermodynamic Properties, Circular 500* (1952) and *Circular 270-3* (1969) Washington, DC: US Govt Printing Office
8. Rossini, F. D. (ed.), *Selected Values of Physical and Thermodynamic Properties of Hydrocarbons and Related Compounds. API Research Project 44*, Pittsburgh, Pa (1952) and Williams C. C. (ed.) *ibid* Texas (1975)
9. Wenner, R. R., *Thermochemical Calculations*, McGraw-Hill, New York (1941)
10. NFPA *Manual of Hazardous Chemical Reactions, 491M-1975*NFPA, Quincy, Mass. (1975)
11. Swern, D. (ed.), *Organic Peroxides*, Wiley-Interscience, New York, vol. 1 (1970), vol. 2 (1971), vol. 3 (1972)
12. Austin, G. T., 'Hazards of commercial chemical operations' and 'Hazards of commercial chemical reactors', in *Safety and Accident Prevention in Chemical Operations*, 2nd edn, edited by Fawcett, H. H. and Wood, W. S., Wiley-Interscience, New York (1982)
13. Lees, F. P., *Loss Prevention in the Process Industries*, Butterworths, London (1980)
14. Bretherick, L., *Bretherick's Handbook of Reactive Chemical Hazards*, 4th edn, Butterworths, London (1990)
15. *Recognition and Handling of Peroxidisable Compounds*, Data Sheet 655, National Safety Council, Chicago (1976)
16. *The Flixborough disaster, Report of the Court of Inquiry*, HMSO, London (1975)
17. Alexander, J. M., 'The hazard of gas phase oxidation in liquid phase air oxidation processes', paper in I. Chem. E. Symposium series 39a, *Chemical Process Hazards with special reference to plant design – V*, Rugby (1974)
18. Kletz, T. A., *Cheaper and Safer Plants*, The Institution of Chemical Engineers, Rugby (1984)
19. Waddams, A. L., *Chemicals from Petroleum*, 4th edn, John Murray, London (1978)
20. Frank-Kamenetskii, D. A., *Diffusion and Heat Exchange in Chemical Kinetics*, Translated by J. P. Appleton, Plenum Press, New York (1969)
21. Stull, D. R., 'Identifying chemical reaction hazards' in *Loss Prevention* **4**, 16 (1970) and, 'Linking thermodynamics and kinetics to predict real chemical hazards', in *Loss Prevention* **7**, 67 (1973)
22. Dow Chemical Company, *Fire and Explosion Index – Hazard Classification Guide*, 5th edn, Midland, Mich. (1980)
23. ICI Mond Division, *The Mond Index*, 2nd edn ICI PLC, Northwich, Cheshire (1985)
24. King, R. W. and Magid, J., *Industrial Hazard and Safety Handbook*, Butterworths, London (1980)
25. Kirk, R. E. and Othmer, D. F., *Encyclopaedia of Chemical Technology*, 3rd edn Wiley-Interscience, New York (*ca* 1979)
26. Bond, J., 'Violent polymerisations', *Loss Prevention Bulletin 065*, The Institution of Chemical Engineers, Rugby (1985)
27. McConnaughey, W. E. *et al*, 'Hazardous materials transportation', *Chemical Engineering Progress*, **66**(2), 57 (1970)

28. Fawcett, H. H., 'Chemical wastes' in *Safety and Accident Prevention in Chemical Operations*, 2nd edn, edited by Fawcett, H. H. and Wood, W. S., Wiley-Interscience, New York (1982)

29. Ross, R. D., 'Disposal of hazardous materials' *ibid.*

30. Coates, C. F. and Riddell, W., 'A system of hazard evaluation for a medium size manufacture of bulk organic chemicals' Paper in I. Chem. E. Symposium Series 68, *Runaway Reactions, Unstable Products and Combustible Powders*, Rugby (1981)

31. Bowes, P. C., *Self-heating: evaluating and controlling the hazards*, HMSO, London (1984)

32. Uhart, E., *Potential Charcoal Development in Uganda*, United Nations Development Programme, Vienna (1975)

33. Boddington, T., Gray, P., and Walker, I. K., 'Runaway reactions and thermal explosion theory', paper in I. Chem. E. Symposium series 68, *Runaway Reactions, Unstable Products and Combustible Powders*, Rugby (1981)

34. Duane, T. C. and Synnot, E. C., 'Effect of some physical properties of milk powders on minimum ignition temperatures', *ibid.*

35. Gibson, N. and Schofield, F., 'Fire and explosion hazards in spray dryers', in *Chemical Process Hazards*, vol. 6, The Institution of Chemical Engineers, Rugby (1977)

36. The Institution of Chemical Engineers, *User Guide to Fire and Explosion Hazards in the Drying of Particulate Materials*, Rugby (1977)

37. S.I. 1984, No. 1902, *The Control of Industrial Major Accident Hazards Regulations 1984 (CIMAH)*, HMSO, London

38. Gait, A. J., 'Propylene oxide', in *Propylene and its Industrial Derivatives*, edited by E. G. Hancock, Ernest Benn, London (1973)

Chapter 9

Explosion hazards of process materials

'An explosion,' as one writer put it, 'is, like an elephant, difficult to define, but easily recognised when you are confronted with one'. Here we are concerned with the potential of substances (which may be solids, liquids or gases) to decompose explosively in the absence of air. The explosivity of mixtures of flammable gases, vapours, solid particles and liquid droplets with air is discussed in Chapter 10. Physical explosions are covered in 6.4.3.

All chemical explosions have three things in common:

- They are very fast;
- They give out heat;
- They make a loud noise.

Most chemical explosions produce large volumes of hot gases, but there are exceptions to this, such as silver acetylide, which produces only solid silver, carbon, heat and a sharp 'crack'[1]:

$$Ag_2C_2 \rightarrow 2Ag + 2C$$

The explosion of some compounds such as nitrogen iodide is initiated by a minor impulse or shock, while that of others such as ammonium nitrate requires much energy to initiate.

9.1 Explosive deflagrations and detonations

Chemical explosions fall into two classes, *explosive deflagrations* and *detonations*.

- A deflagration is caused by a chemical reaction which passes through the deflagrating material at well below sonic velocity. It normally only develops an appreciable pressure if it is confined. Most deflagrations are thus not explosions, but very fast fires. The burnt products of a deflagration move in the direction opposite to the combustion wave, which is considered to propagate by heat transfer through the deflagrating material. An *explosive deflagration* produces an appreciable *blast wave*, with the potential to do damage. Explosives which normally deflagrate are termed propellants or low explosives. Cordite, a blend of nitrocellulose and nitroglycerine, is an example. It develops its energy gradually as the shell or bullet is accelerated through the gun-barrel.

- A *detonation* is caused by a very rapid chemical reaction which passes through the exploding material at speeds of 1 to 10 km/s – well in excess of sonic velocity. High pressures are developed, and the products of combustion move in the same direction as the wave. A detonation is a *shock wave* accompanied by a chemical reaction which sustains it. Explosives which normally detonate are termed high explosives, and have high shattering power even when unconfined. Trinitrotoluene (TNT) which is used for filling shells is an example.

The terms in italic type are defined in an Institution of Chemical Engineers publication[2].

A detonation produces much higher pressures and greater devastation at close range than a deflagration in which the same amount of energy is released. At a greater distance, however, where the shock wave of a detonation has slowed to sonic velocity and degenerated into a blast wave, the destructive effects of the two types of explosion, which are still considerable, become comparable[3]. Whether a material will detonate or deflagrate on ignition depends on the material itself, on its quantity, whether and how it is confined, and how it is ignited. Many substances which deflagrate when weakly ignited will detonate under sufficiently strong ignition. A deflagration wave can in some circumstances accelerate spontaneously to a detonation.

Many explosives will burn harmlessly if spread in thin layers and not confined. It is thus most important when designing plant for potentially explosive materials, to minimise inventories at every stage, and avoid confinement as far as possible. The alternative of making the plant strong enough to withstand detonations is seldom practical.

9.2 Industrial chemicals with explosive potential

While explosives are manufactured on a large scale for military purposes, blasting, fireworks and rockets, etc., many compounds which might and sometimes do explode are made in the chemical industry for entirely different purposes. They may possess other desirable properties in themselves, or they may be intermediates in the manufacture of other products.

Some examples are shown in Table 9.1. These include a few compounds which readily release oxygen and form explosive mixtures with organic matter and other fuels, such as the nitrates and chlorates of sodium, potassium and ammonium, hydrogen peroxide and nitric acid. The fuels include porous or finely divided solids such as charcoal, coal dust, sugar, sawdust, sulphur and aluminium powder, and combustible liquids such as fuel oil, ethanol and hydrazine.

The explosion hazards of materials which fall within the scope of the Explosives Act of 1875 are very strictly controlled[1] during manufacture, transport and storage. This has not always been so with compounds made

Table 9.1 Some industrial chemicals with explosive potential

Chemical	Actual use	Intermediate for
Acetylene	Gas welding and cutting	Vinyl chloride, acrylics, perchloroethylene
Ammonium nitrate	Fertilisers, explosive ingredient	Nitrous oxide
tert-Butyl hydroperoxide	Polymerisation initiator	
Ethylene oxide	Fumigant, rocket fuel	Detergents, paints, antigreeze, etc.
1,3-Dinitrobenzene	Explosive ingredient	Dyestuffs
Nitromethane	Solvent, underwater explosive, corrosion inhibitor	Chloropicrin
Picrin acid	Yellow dye	
Trinitrobutyl-toluene	Perfumery ingredient	
Vinyl acetylene		Neoprene rubber
Oxidants		
Hydrogen peroxide	Bleaching agent, antiseptic	Organic peroxides
Manganese dioxide	Batteries	Manganese compounds, ferromanganese
Nitric acid	Descaling concrete	Nitrates, nitro compounds
Nitrogen peroxide	Rocket propellant	Nitromethane
Potassium chlorate	Weedkiller, explosive ingredient	Potassium perchlorate
Potassium nitrate	Explosive ingredient, fertiliser	

for entirely different purposes. Several serious explosions have occurred in the manufacture, transport and use of ammonium nitrate and other industrial compounds and their mixtures (Figure 9.1). Hence there is always a need to screen industrial chemicals and their mixtures for potential explosivity, and to examine those with this tendency for destructive power and sensitivity to heat, friction, impact and other forms of initiation. Only then can safe conditions for their manufacture, storage, transport and use be established.

9.3 Structural groups which confer instability

Table 9.2, due to Coffee[4], shows 17 structural groupings, nine of which contain nitrogen, which confer instability to the compounds in which they are present, often rendering them explosive. These groupings are known as 'plosophors'. The list is far from exhaustive, and a more extensive one containing 42 structural groupings which confer instability is given by Bretherick[8]. Some of these are rarely seen outside research laboratories.

Figure 9.1 Crater and destruction caused by ammonium nitrate explosion at Oppau, Germany, in 1921 (Badische Anilin und Soda Fabrik)

Table 9.2 Structural groupings indicative of potential instability[4]

Acetylide	$-C\equiv C-METAL$
Amine oxide	$\equiv N^+-O^-$
Azide	$-N=N^+=N^-$
Chlorate	$-ClO_3$
Diazo	$-N=N-$
Diazonium	$(-N\equiv N)^+X^-$
Fulminate	$-(C\equiv N\rightarrow O)$
N-Haloamine	$-N\overset{\displaystyle Cl}{\underset{\displaystyle X}{\big\langle}}$
Hydroperoxide	$-O-O-H$
Hypohalite	$-O-X$
Nitrate	$-O-NO_2$
Nitrite	$-O-NO$
Nitro	$-NO_2$
Ozonide	$-O\underset{O}{\overset{\displaystyle O}{\diagup\!\!\diagdown}}O-$
Peracid	$-C-O-O-H$ (with $\parallel$ O below C)
Perchlorate	$-ClO_4$
Peroxide	$-O-O-$

Some more complex groupings which create explosive tendencies in organic compounds (including salts of organic bases) are[5]:

Primary nitramine	$-NH-NO_2$
Secondary nitramine	$>N-NO_2$
Nitroso	$-N=O$
Diazosulphide	$-N=N-S-N=N-$
Picrates	$[C_6H_2(NO_2)_3.O]'$
Iodates	$[IO_3]'$

Acetylene and acetylenic compounds are very sensitive to heat, shock and abrasion, and many are explosive.

Molecular nitrogen, ammonia and amino groups are in themselves very stable, but the compounds formed when ammonia or an amino group reacts with an oxidising agent are among the most unstable; they include lead azide and mercury fulminate which are used in detonators[1]:

Detonators

Mercury fulminate $C=N-O-Hg-O-N=C$

Lead azide
$$\begin{matrix} N \diagdown \\ \| \quad N-Pb-N \\ N \diagup \end{matrix} \begin{matrix} \diagup N \\ \diagdown \| \\ \diagdown N \end{matrix}$$

Mercury fulminate is also a health hazard to those working with it[6]. Most azides are explosive and very sensitive.

Hydrazine vapour, $H_2N=NH_2$, is shock sensitive and explodes violently, and most substituted hydrazines and hydrazones can form explosive compounds.

Nitrate and nitro groups introduce oxygen at a high energy level into organic compounds and provide the main basis for the explosives industry. The explosivity of such compounds depends largely on their oxygen balance and the $C:N$ ratio in the compound. Mono-nuclear aromatic compounds with three nitro groups are common explosives. Most with two nitro groups will explode at mildly elevated temperatures. Some have been used as explosive ingredients, and some have other industrial uses (Table 9.1). All dinitro-aromatic compounds should be treated as potentially explosive and thermally sensitive.

Most aromatic compounds with only one nitro group are fairly stable at ambient temperatures but liable to decompose explosively at higher temperatures. Thus explosions of (mono) nitrotoluenes have occurred during their distillation when heated to high temperatures, or when held at moderate temperatures for long periods. Harris and others of ICI Dyestuffs division have shown that such decomposition on a large scale is preceded by an induction period which increases as the temperature is lowered[9]. At 190°C decomposition occurred after 80 days.

Of the nitro-paraffins, nitro-methane, which is used both as an explosive and a solvent, has been involved in serious tank car explosions in the USA. Nitro-ethane is also explosive, though less sensitive than nitro-methane, but becomes more so with the addition of small amounts of amines.

Peroxides and peroxy-compounds, also discussed in 8.5.1, form a very large subject, due to their great number and variety. Bretherick classes peroxides in three main groups, inorganic, organic, and organo-mineral[8].

Most peroxides are hazardous and some are explosive. Rather more than 65 organic peroxides are made and used commercially, the total consumption in 1978 in the USA being about 13 000 t. This does not include production of peroxides as intermediates (sometimes only in dilute solution) in the manufacture of bulk chemicals such as phenol and cyclohexanol by direct oxidation processes. Their main use is as free radical initiators for industrial polymerisations. They are also used as bleaching agents for edible oils and flour, disinfectants, fungicides, cross-linking agents for thermoplastics, curing agents for resin and rubbers, shrink-proofing agents for wool, in the manufacture of epoxy resins, and as the active ingredient in some pharmaceuticals. Only peroxides which can be produced, shipped and utilised with a reasonable degree of safety are made and sold commercially. While the hazards of peroxides should be well known to those making and using them, the risks posed by those made inadvertently by reaction of various substances with atmospheric oxygen [8.5.1 and 8.7] are more insidious.

Although several hazard-classification systems have been developed for organic peroxides, none has been accepted commercially. The Factory Mutual Engineering Corporation classifies organic peroxides according to their fire and explosion hazard into five classes, and gives safeguards for their storage and use[10]. Those at the extremes of the range are:

Class I. These present a high explosion hazard through easily initiated, rapid explosive decomposition. This group may include peroxides that are relatively safe under highly controlled temperatures or in a liquid solution, where loss of temperature control or crystallisation from solution can result in severe explosive decomposition.

Class V. These present a low or negligible fire hazard, and with them combustible packing materials may present a greater hazard than the peroxide itself.

Most of the 'Highly reactive substances' listed in Group 3 of Schedule 3 of the CIMAH Regulations [8.8] are organic peroxy-compounds, and come under the scope of Regulations 7 to 10 for quantities of 50 t or more[11].

Of the inorganic peroxides, hydrogen peroxide which is made in grades ranging from 3% to 98% concentration (in water), decomposes in all concentrations with the evolution of heat and oxygen in the presence of small amount of catalytic impurities such as platinum, manganese dioxide and alkalis. In concentrations of 86% and higher, the liquid will detonate with a high-energy ignition source, but many mixtures of lower-grade hydrogen peroxide with organic compounds, which occur when hydrogen peroxide is used as an oxidising agent in a chemical reaction, are explosive[12].

Another danger of hydrogen peroxide lies in the use of acetone or other ketones as a solvent for reactions in which it is used, since they themselves form peroxides which readily transform into crystalline dimers and trimers, which are explosive and very sensitive[7,8].

Other inorganic peroxides are made and used on a large scale in household washing powders, but they seldom appear to be involved in explosions. Some little-used organo-mineral peroxides are very dangerous.

The importance of a thorough literature search can hardly be over-stressed before using, making or storing unfamiliar chemicals. Thus two serious explosions of propargyl bromide, an acetylenic compound, occurred in the 1960s in the USA. Reports in the Russian literature that it was thermally sensitive to detonation had apparently been missed[4].

9.4 Preliminary screening of materials for explosivity

Table 9.2 provides a useful first checklist for materials of known composition. On finding such a grouping in the material being screened, the literature should be consulted, starting perhaps with Bretherick[8]. If any compound contains two such groupings, the probability of it proving explosive is strong.

Two simple but effective preliminary tests are first quoted. These can be applied to materials whose composition is not known, provided effective precautions are taken[5].

The first, which should be done before all others, is to take about one-tenth of a gram of the material and drop it onto a hot plate at above 300°C. If it goes *pop, bang, snap* or *crackle*, one is warned of possible trouble. If the sample decomposes or chars the test is continued until no further decomposition is apparent.

A test for instability in liquids is to place about 50 ml in a small beaker located behind suitable barricades. A small 50 watt coil of heating wire, attached by flexible cable to a plug and socket switch in a safe location, is then placed in the liquid. The observer having retired to a safe observation point, the current is switched on and left on until the sample has completely evaporated, decomposed or burnt. The beaker should not be approached again until it is cool and the electricity has been disconnected. This simple test, while not infallible, has detonated compounds whose explosivity more sophisticated tests had failed to reveal[5].

9.5 Thermochemical screening

When any substance explodes, a considerable amount of heat is usually released, partly in the form of pressure waves and partly as a rise in temperature of the explosion products. The power of the explosion of the substance is related to this quantity, known as its heat of decomposition (ΔH_d). Values of ΔH_d and of ΔH_c, the heat of combustion, are quoted here in MJ/kg 'net' (as recommended by Lees[3], i.e. assuming the water formed to be in the gaseous state) at 25°C and atmospheric pressure. The reader is warned that values of ΔH_d used by some writers appear to be 'gross' figures (i.e. assuming the water formed to be in the liquid state). Thermodynamicists (unfortunately for me) treat the heat of decomposition as negative when heat is given out and positive when heat has to be supplied when a substance decomposes. Here, to avoid confusion when comparing different heats of decomposition, I reverse this convention and treat ΔH_d when heat is given out during decomposition as positive (except in Figure 9.2 which is based on Stull's correlation).

Knowledge of ΔH_d should provide some guide to the explosive potential of any substance. Thus we might (tentatively) expect any substance capable of exploding in the absence of air to have at least a certain minimum value of ΔH_d.

Closely linked with the heat of decomposition is the oxygen deficiency in the molecule compared with that needed to convert all carbon and hydrogen in it to carbon dioxide and water (and to convert any other elements to their combustion products – oxides, carbonates, etc.). For compounds and mixtures containing only carbon, hydrogen, nitrogen and sufficient oxygen to convert all carbon and hydrogen to carbon dioxide and water, the heats of decomposition and combustion should be identical, apart from the effects of small amounts of carbon monoxide, ammonia, oxides of nitrogen and hydrocarbons in the decomposition products. But for compounds and mixtures with a deficiency of oxygen, the heat of decomposition is always less than the heat of combustion. By mixing a material such as nitrocellulose which has an oxygen deficit with nitroglycerine which has an oxygen excess, both the heat of decomposition and the explosive power are increased. The same applies when TNT is mixed with ammonium nitrate.

A theoretical check should also be made on the volume of gases which would be released by the decomposition of unit weight of the compound. This together with the theoretical heat release provides a rough appreciation of its destructive potential. Thus whilst copper and silver acetylides both decompose explosively with a sharp crack, their decomposition products, carbon and a metal, are both solid, and their main hazard is probably as a source of ignition or initiation for some other flammable or explosive material. Likewise the detonators mercuric fulminate and lead azide produce only small volumes of gases per unit weight, compared with explosives such as TNT and cordite.

9.5.1 The ASTM 'CHETAH' computer program

The 'CHETAH' (CHEmical Thermodynamic And energy Hazard evaluation) program, developed by the American Society for Testing and Materials, and first published in 1974[13], is claimed to enable one to predict the maximum value of ΔH_d for any compound or mixture of compounds whose chemical formula and structure are known. The program also estimates three other quantities[3]:

- (the heat of combustion ΔH_c) – (the heat of decomposition ΔH_d) (both taken here as positive when energy is released);
- the oxygen balance (which defines whether the substance contains more or less oxygen than that required for complete combustion);
- the 'y' criterion (proportional to the molecular weight of the substance and the square of its heat of decomposition).

The probability of the substance exploding under the impact of shock was found to depend on all these four quantities, the most important being the heat of decomposition. As a very rough guide, only those substances whose heats of decomposition ΔH_d exceed 0.7 kcal/g (2.9 MJ/kg) are likely to explode when subjected to mild heat or shock. This figure is about 10%

more than the amount of heat required to bring cold water to the boil and convert it into steam.

CHETAH can be applied to a mixture of chemicals in a process such as a chemical reactor to characterise the hazard potential of the system as high, medium or low by calculating the heat of reaction ΔH_r or decomposition ΔH_d:

High $\Delta H_d > 2.9\,\mathrm{MJ/kg}$
Medium $2.9 > \Delta H_d > 1.25\,\mathrm{MJ/kg}$
Low $\Delta H_d < 1.25\,\mathrm{MJ/kg}$

Materials with medium or high heats of decomposition should always be treated with great caution.

Stull[14] suggested that a better correlation of the probability of any substance exploding is given by plotting ΔH_d vs the difference $(\Delta H_c - \Delta H_d)$ for a number of potentially explosive substances. This is discussed in 9.5.3.

9.5.2 On estimating heats of decomposition

The heat given out when a substance explodes is difficult to measure and depends on the composition of the explosion products. If one knows this composition and the heat of formation of the substance, its heat of (explosive) decomposition can be estimated.

The heats of decomposition and compositions of the explosion products of commercial explosives have been well studied and reported. But when a substance which is normally thought of as safe explodes during storage or shipment, there may be considerable uncertainty about its heat of (explosive) decomposition, since nobody has had the opportunity to collect and analyse the explosion products. One can in such cases only estimate a range of values for ΔH_d, the highest value being sometimes many times higher than the lowest.

Most explosive substances contain only the elements carbon, hydrogen, nitrogen and oxygen, while their main decomposition products are solid carbon and the gases carbon monoxide, carbon dioxide, hydrogen, nitrogen, oxygen and water vapour. The heats of formation of the elements carbon, hydrogen, oxygen and nitrogen under standard conditions (usually at 25°C) are by definition zero. The heats of formation at 25°C in megajoules per kg-atom of contained oxygen of the main combustion products are -110.58 for CO, -196.94 for $\frac{1}{2}(CO_2)$ and -242.01 for H_2O as vapour. Some other gases such as ammonia, oxides of nitrogen and methane are also usually formed. Of these, methane and ammonia have negative heats of formation while both nitrous and nitric oxides have positive heats of formation. Thus the presence of methane or ammonia in decomposition products rather than elemental carbon, hydrogen and nitrogen raises the heat of decomposition. Methane and ammonia both decompose into their elements at high temperatures. They may be expected to be formed during the initial stages of many explosions and then decompose if a high enough temperature is reached during the explosion. Significant concentrations of methane and ammonia in explosion products thus tend to be characteristic of weak explosions.

Free oxygen is only likely to be present in the decomposition products when the amount of oxygen contained in the substance exceeds that needed to convert all carbon and hydrogen present to carbon dioxide and water vapour. In the usual case, where less oxygen is present, one should consider the proportions in which it is converted to carbon dioxide, carbon monoxide and water vapour. These proportions can have a significant effect on the calculated heats of decomposition. A substance giving a decomposition product containing a given amount of carbon, hydrogen and oxygen in the form of carbon and water vapour will have a considerably higher heat of decomposition than if it is in the form of carbon monoxide and hydrogen.

Thus although the heat released when a chemical compound explodes depends (like its heat of combustion in air) on its chemical formula and heat of formation, it also depends on the composition of the combustion products. For compounds which rarely explode this composition is difficult to predict. Many compounds normally thought of as safe and whose heats of decomposition are relatively low are known to explode when sensitised by particular impurities (such as metal oxides) which can affect the mechanism of the explosion and the composition of the explosion products, thus giving a heat of decomposition near to the upper end of its possible range.

Here in estimating the heats of decomposition of several potentially explosive compounds I have used two methods, one which gives low (and in some cases normal) values, the other tending to give high values which may only apply in special circumstances.

Assumptions for estimating lower heats of decomposition
1. Combined oxygen in the substance first forms carbon monoxide until all carbon present is thus accounted for;
2. Any remaining oxygen first converts the carbon monoxide to carbon dioxide;
3. When all carbon present has been thus accounted for, the hydrogen present forms water with the remaining oxygen.

Assumptions for estimating upper heats of decomposition
4. Combined oxygen in the substance first forms water vapour until all hydrogen present is thus accounted for;
5. Any remaining oxygen first converts carbon present to carbon monoxide;
6. Any oxygen still remaining converts carbon monoxide to carbon dioxide.

While assumption (4) may appear rather extreme, it provides some allowance for increases in the heat of decomposition caused by the presence of methane and ammonia in the combustion products.

9.5.3 Heats of decomposition for particular compounds

Table 9.3 shows ten common industrial compounds with explosive potential including three well-known explosives, nitroglycerine, cellulose

Table 9.3 Oxygen deficiency and net heats of decomposition and combustion for known and possible explosives

Compound	State	Empirical formula C H N O				Atoms in CP*	Oxygen deficit atoms	%	Heats (MJ/kg) ΔH_c		ΔH_d	$[\Delta H_c - \Delta H_d]$
Glycerol trinitrate	l	3	5	3	9	8.5	−0.5	−5.9	6.31		6.28	0.03
Cellulose nitrate 13.4% N	s	6	7	3	11	15.5	4.5	29.0	9.13		4.06	5.07
Trinitrotoluene	s	7	5	3	6	16.5	9.5	63.6	14.65		4.69	9.96
2,4-Dinitrotoluene	s	7	6	2	4	17.0	13	76.5	18.89	upper	4.35	14.54
										lower	2.18	16.71
4-Nitrotoluene	s	7	7	1	2	17.5	15.5	88.6	26.03	upper	3.26	22.77
										lower	1.34	24.69
Nitromethane	l	1	3	1	2	3.5	1.5	42.8	10.89	upper	5.07	5.82
										lower	4.47	6.42
2-Nitropropane	l	3	7	1	2	9.5	7.5	78.9	20.75	upper	3.40	17.35
										lower	0.44	20.31
Ammonium nitrate	s	0	4	1	3	2	−1	−50	0.96		0.96	0
Acetylene	g	2	2	0	0	5	5	100	48.33		8.74	39.59
Ethylene oxide	g	2	4	0	1	6	5	83.3	26.75	upper	3.04	23.71
										lower	0.06	26.69

* CP = combustion products

trinitrate, and trinitrotoluene (TNT). Several appeared in Table 9.1 and were discussed in 9.2 and 9.3. The table shows the formulae of the compounds, the number of atoms of oxygen in the combustion products, and the theoretical deficiency (or excess) of oxygen in the molecule for complete combustion. It also shows the net heats of combustion and decomposition and the differences between these two quantities. For compounds whose heats of decomposition were uncertain, the figures given here were estimated using the assumptions given in 9.5.2. Upper and lower values are given for five compounds but in the cases of acetylene and ammonium nitrate these values coincide and only a single figure is given. Figure 9.2 shows the values of ΔH_d plotted against $[\Delta H_c - \Delta H_d]$ for all ten compounds, including ranges for the five. An arrow points from high to low probability. By sketching contours of equal probability at right angles to this arrow, acetylene (ΔH_d 8.74 MJ/kg), dinitrotoluene (ΔH_d 2.18–4.47 MJ/kg) and ammonium nitrate (ΔH_d 0.96 MJ/kg) would all have about the same probability of exploding. This chart is a fair guide to the explosive probability of the compounds shown and is a far better criterion of this than ΔH_d alone.

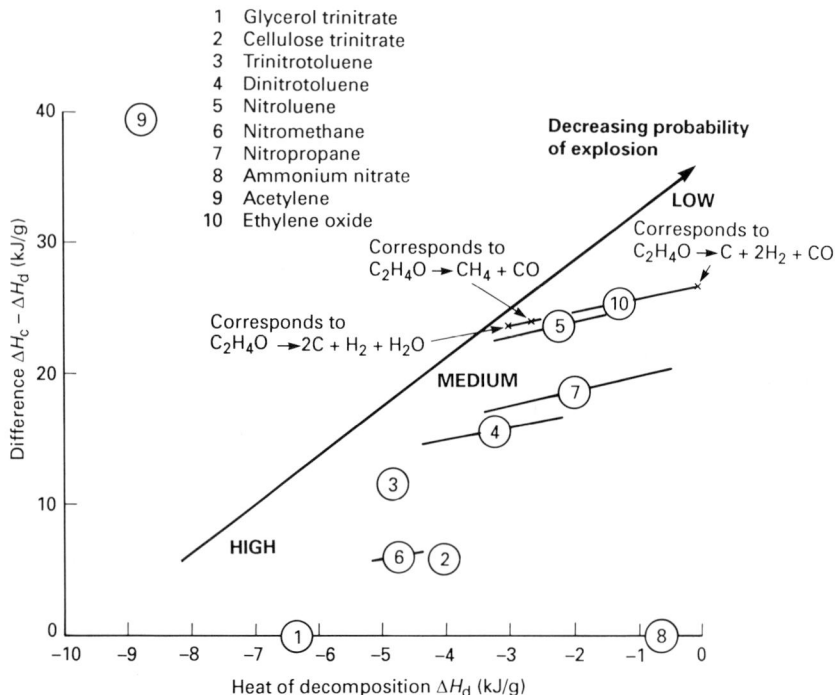

Figure 9.2 Probability correlation for CHETAH programme showing values for substances listed in Table 9.3

2,4-Dinitrotoluene (DNT) with an estimated range for ΔH_d of 2.18–4.35 MJ/kg decomposes at 250°C, the decomposition becoming self-sustaining at 280°C. There is a published report of an explosion of DNT which had mistakenly been held at 210°C in a pipeline. A maximum handling temperature of 150°C has been recommended for it[8].

4-Nitrotoluene (ΔH_d 1.34–3.26 MJ/kg) is made and used on a large scale commercially. There have been a few reports of explosions in vacuum distillation stills used to separate and purify mononitrotoluenes although most of these were attributed to higher nitrotoluenes or other unstable compounds (aci-nitro salts) derived from nitrotoluenes[8].

Nitromethane (ΔH_d 4.47–5.07 MJ/kg), which is used both as an explosive and a commercial solvent, is dangerous to handle because it can readily be detonated by shock, high temperatures, the sudden application of gas pressure and forced high-velocity flow through restrictions. Nitropropane (ΔH_d 0.44–3.4 MJ/kg) appears to be generally safe to handle although it is susceptible to thermal decomposition, particularly in the presence of metal oxides.

Ammonium nitrate when heated carefully in a test tube to 170°C decomposes quietly into nitrous oxide and water vapour, giving out some heat at the same time. But if the temperature rises above 250°C, the decomposition may become explosive, and oxygen and nitrogen are formed[8]. The two equations with the heats of decomposition liberated in each case are:

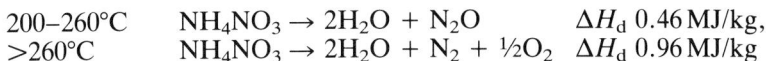

200–260°C $NH_4NO_3 \rightarrow 2H_2O + N_2O$ ΔH_d 0.46 MJ/kg,
>260°C $NH_4NO_3 \rightarrow 2H_2O + N_2 + \frac{1}{2}O_2$ ΔH_d 0.96 MJ/kg

There have been a number of ammonium nitrate fires which have burnt freely, liberating oxygen. Although it is practically impossible to detonate ammonium nitrate on a small scale, some of the worst accidental chemical explosions have occurred with ammonium nitrate, with heavy loss of life. Substantial confinement (as at the bottom of a large pile), which allows high temperatures to develop over a period of time as a result of slow decomposition starting at ambient temperature, has been an important factor in some of these explosions. As with other materials which decompose very slowly even at room temperature, the critical mass concept applies to ammonium nitrate[8] [8.6.4 and 8.7].

Another factor contributing to some ammonium nitrate explosions has been the presence of a small amount of hydrocarbon applied to the surface of the 'prills', to reduce their hygroscopicity and caking tendency, and retard their solution when applied to the soil as a fertiliser[3].

The addition of 7.5% w. of carbon to ammonium nitrate increases the heat of decomposition to 3.64 MJ/kg:

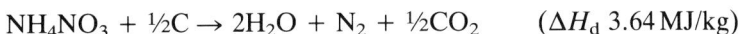

$NH_4NO_3 + \frac{1}{2}C \rightarrow 2H_2O + N_2 + \frac{1}{2}CO_2$ (ΔH_d 3.64 MJ/kg)

The explosive decomposition of acetylene may be represented by:

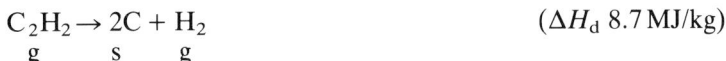

$\underset{g}{C_2H_2} \rightarrow \underset{s}{2C} + \underset{g}{H_2}$ (ΔH_d 8.7 MJ/kg)

(It is a surprise to find that acetylene, which contains no oxygen, has a higher heat of decomposition than TNT.)

The explosive decomposition of gaseous acetylene in the absence of air can be initiated by heat or shock at pressures of 2 bar g. and higher, and can escalate into a destructive detonation in pipelines. In the UK acetylene comes under the control of the Explosives Inspectorate when handled at pressures above 24 psig[1] (1.655 bar g.).

The explosive decomposition of ethylene oxide under clean conditions can be represented by:

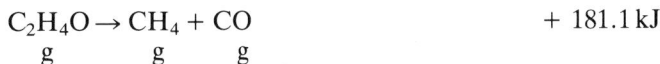

$\underset{g}{C_2H_4O} \rightarrow \underset{g}{CH_4} + \underset{g}{CO}$ + 181.1 kJ

rather than by:

$\underset{g}{C_2H_4O} \rightarrow \underset{s}{C} + \underset{g}{2H_2} + \underset{g}{CO}$ +   2.5 kJ
                              +   g

From its molecular weight of 44, its heat of decomposition in the vapour state is 2.68 MJ/kg, compared to the upper estimate in Table 9.3 of 3.04 MJ/kg. Had liquid ethylene oxide been able to explode, the heat of decomposition (with methane again formed) would have been 2.01 MJ/kg. The case of ethylene oxide demonstrates the strong effect of methane in the decomposition products in raising ΔH_d to the point where explosion is possible.

While ethylene oxide vapour will explode if ignited, it is stable in the liquid state provided no polymerisation catalysts such as iron oxide are present. But if the liquid polymerises rapidly, the heat evolved will vaporise unpolymerised material, causing an explosion hazard.

Sodium chlorate is another common chemical with a small explosion hazard in itself which is enormously increased by the addition of small amounts of organic matter.

The values of ΔH_d given here are mainly estimated from heats of combustion given in the *CRC Handbook of Chemistry and Physics*[15]. The steps involve:

- Calculation of heats of formation from published heats of combustion;
- Calculation of net heats of combustion (with water formed in the vapour state);
- Estimation of lower and upper values of heats of decomposition using the assumptions stated.

For compounds whose heats of formation or combustion are unknown, it is often possible to estimate them from bond energies between the various atoms[16, 17].

9.6 Stability and sensitivity tests

Tests for thermal and shock stability of potentially explosive substances are needed to establish safe processing conditions. Thermal stability is important for operations such as drying, reaction, evaporation, distillation and other processes involving elevated temperatures. Shock stability is important for size reduction, pumping, blending and transport.

Unfortunately, the stability of many of these materials depends on their concentration, purity and the nature of the impurities present, as well as on their temperature and physical environment. These tests are therefore not infallible. Sometimes an impurity may be present in an industrial compound which considerably reduces its stability. In the manufacture of guncotton, for example, the crude nitration product, after separation and washing to remove spent acids, still contains unstable esters. These are removed by repeated boiling, first with water and then with very dilute alkali. If this is not done thoroughly, or if traces of unstabilised material by-pass this stage, the guncotton is liable to explode during subsequent drying. Similarly, the shock sensitivity of many explosives (e.g. nitroglycerine) depends on whether they are in the liquid or solid state. The solid is less stable than the liquid. Thus stability tests need to be carried out not only on pure materials and commercial products, but also on samples representative of their condition in the processes concerned. This applies to all potentially explosive substances irrespective of their purpose.

Likewise, when considering the process conditions which the tests are intended to validate, a generous allowance should be made for 'upset conditions', with temperatures, etc. outside the normal operating range.

Needless to say, processes for hazardous materials should have special safety instrumentation and emergency devices (such as 'dump tanks') to counter hazardous upset conditions.

9.6.1 Thermal stability

Apart from the crude screening tests already mentioned [9.4], a number of special tests are used. These include the use of modified melting point apparatus[5] and thermal analysers which can be employed for differential thermal analysis (DTA) using standard procedures[18, 19]. Since these work at atmospheric pressure and may fail to detect exotherms on volatile materials, confinement tests which use thermal stability bombs have also been developed[4, 20]. These tests also measure the pressure rise in a confined space.

Other more refined methods include differential scanning calorimetry (DSC) and accelerating rate calorimetry (ARC). These are carried out in the UK by the Royal Armament Research and Development Establishment, EM2 Branch, on behalf of HM Inspector of Explosives[3, 21]. During the investigation of the explosion discussed in 5.2 the material showed exotherms by DSC in the temperature range 248–274°C which had previously been missed by DTA [5.2.5]. The theory and methods developed for thermal analysis, and their limitations, are discussed by Daniels[22].

9.6.2 Shock sensitivity

Thermally sensitive materials which decompose exothermally are also sensitive to shock or friction, although the stimulus needed may be severe. Tests range from a simple hammer test, through drop weight impact tests[23] (Figure 9.3), to more severe ones which use explosives for initiation, such as card gap tests[24, 25].

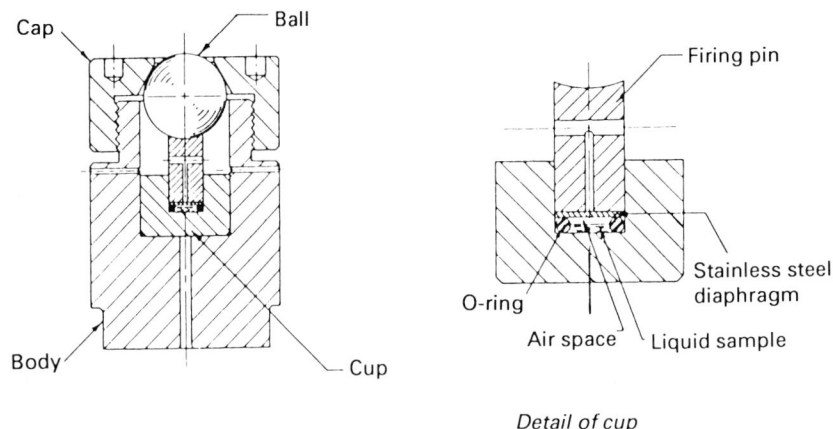

Detail of cup

Figure 9.3 Sample holder of Olin drop-weight impact tester. (From Coffee[4], courtesy Wiley-Interscience)

A number of other tests, including the Cartridge Case test for discriminating between detonating and deflagrating explosives and a Bonfire test to investigate the behaviour of the bulk material in a fire, are discussed by Lees[3] and Connor[20]. Lees stresses that the object of carrying out reactivity or explosivity tests should be to obtain a positive result, using extreme conditions where necessary. Less extreme conditions under which it might be handled in practice are then tested.

9.7 Classification of materials with explosive potential

Coffee has proposed classifying such materials by their behaviour in three basic but different stability tests[4]:

- Theoretical computation by CHETAH;
- Thermal analysis by DTA;
- Impact sensitivity by drop weight.

Taking results as either positive or negative, eight combinations are possible. Further examination reduces these to three groups representing low, medium and high potential hazards. These are shown in Table 9.4.

Table 9.4 Classification of explosive hazard of materials

Combination number	Heat of reaction	Thermal stability	Impact sensitivity	Overall rating rating
1	−	−	−	Stable and incapable of
2	−	−	+	high energy release
3	−	+	−	
4	+	−	−	Capable of high energy
5	+	+	−	release but hard to initiate
6	+	−	+	Capable of high energy
7	−	+	+	release and sensitive
8	+	+	+	thermally and mechanically

Code

Heat of reaction, $-\Delta H_d$ (CHETAH)	−	<0.7 kcal/g
	+	>0.7 kcal/g
Thermal stability (DTA)	−	no exotherm
	+	exotherm
Impact sensitivity	−	insensitive at 550 in-lb (solids) or 100 kg-cm (liquids)
	+	sensitive at 550 in-lb or 100 kg-cm

9.8 Explosions of industrial chemicals outside the explosives industry

A few examples of major explosions during the manufacture, storage and transport of chemicals for uses other than as explosives follow under the headings of the chemicals involved. One good example, the explosion of a dinitro aromatic compound in a dryer, was discussed in 5.2. A large number of minor explosions have occurred in the chemical industry, generally during the first years of manufacture of some unstable compound the hazards of which were not fully appreciated. Thus the celebrated Dr Schwartz wrote at the turn of the century about 'Chemico-Technical Factories and Colour Works'[26]:

> The progress of chemical reactions on a working scale must not be excluded from inspection, proceed they never so smoothly *in the laboratory*. When working *on the large scale*, quite different and dangerous reactions may arise. Hence the risk of insuring such establishments should not be accepted until a summer and winter working season has passed without an accident from fire and explosion.

Elsewhere the same author in discussing acetylene which was then coming into wide use for illumination states 'between 1897 and 1900, the number of explosions amounted to 32, 17.5, 5.4 and 2.2 per thousand users of the gas'.

9.8.1 Ammonium nitrate

Although Schwartz had warned that 'ammonium nitrate explodes by percussion or at 70°C'[26], several large explosions of ammonium nitrate made primarily for use as a fertiliser occurred between 1918 and 1947. The two most destructive incidents (both described in more detail by Lees[3]) were at Oppau, Germany, in 1921 when 430 people were killed and at Texas City in 1947 when 552 people were killed and over 3000 injured.

At Oppau two explosions in close sequence occurred at the BASF works where some 4500 t of a mixture of ammonium nitrate and sulphate were stored, when blasting powder was being used to break up piles of the material which had become caked. The explosion created a crater 75 m in diameter and 15 m deep, destroyed the works and 1000 nearby houses (Figure 9.1), while the air pressure wave caused considerable damage in Frankfurt more than 80 km away.

The Texas City explosion involved two ships in the harbour with cargoes of bagged ammonium nitrate. The first explosion occurred about one hour after a fire was reported (and could not be extinguished) on a ship carrying 2300 t of ammonium nitrate. The explosion set fire to a cargo of sulphur which the second ship moored 200 m away was carrying as well as ammonium nitrate, which exploded 16 hours after the first explosion. The explosions caused fires in many tanks in nearby oil refineries, some of which burned for nearly a week, and caused extensive damage to warehouses, business premises and residential property.

9.8.2 Vinyl acetylene

Vinyl acetylene (C_4H_4) is formed by dimerisation (the combination of two molecules) of acetylene and also in high-temperature cracking and dehydrogenation processes as a by-product of butadiene. Like acetylene, it is a high-energy compound which can decompose explosively into its elements without the involvement of oxygen. It also polymerises readily with the evolution of heat, which can initiate an explosion of the unpolymerised material. For many years it was produced from acetylene as the first step in the manufacture of chloroprene (2-chlorobutadiene), from which the oil- and chemical-resistant rubber Neoprene is made. Being an easily liquefiable gas like butane and butadiene, it is generally handled as a liquid, either refrigerated or under pressure. Two accidents are discussed here, one involving vinyl acetylene produced from acetylene, the other the separation of butadiene from vinyl acetylene and other C_4 hydrocarbons.

On 25 August 1965 several explosions occurred at Dupont's Louisville works (known as 'Rubber Town') on the banks of the Ohio river, where about 200 t of vinyl acetylene produced there from acetylene was stored as a liquid. Twelve people were killed (most by the first explosion) and much of the plant was wrecked. A company investigation[27] showed that the first explosion was caused by the mechanical failure and overheating of a compressor which circulated vinyl acetylene gas. The subsequent explosions which occurred over a period of eight hours were initiated by flying metal fragments and fires started by the first explosion, and by transmission through pipelines.

On the evening of 23 October 1969 a butadiene purification column on one of Union Carbide's butadiene plants at Texas City exploded[28]. This column, which operated at pressure of 2–3 bar g., separated butadiene (as overhead product) from 2-butenes and vinyl acetylene left in the butadiene-rich stream from which most other impurities had been removed by extractive distillation with a selective solvent. The boiling points at atmospheric pressure of these hydrocarbons (°C) are:

butadiene	−4.6
trans butene	0.8
cis-butene-2	3.7
vinyl acetylene	5.0

Although vinyl acetylene has the highest boiling point, so that its highest concentration might be expected to be at the bottom of the column, it actually accumulates several trays higher because of the 'non-ideality' of the mixture.

At 09.00 a compressor was running hot and had to be shut down to change a valve. As this was not expected to take long, instead of shutting the whole plant down the feed was stopped but not the steam to the reboiler, and the column was put on 'total reflux' (to save time when the plant was restarted). A laboratory analysis on a sample from the base of the column at 11.00 showed 36.9% vinyl acetylene, which probably remained unchanged until 19.23, when the column exploded, the upper part rising several feet before falling on its side (Figure 9.4). A cloud of vapour was released and ignited. Fortunately no one was seriously injured. Examination of the column showed the explosion to have centred on the fourteenth tray (from the bottom).

Computer studies showed that a maximum vinyl acetylene concentration of 57–60% would have been expected on the tenth tray. Investigations showed that vinyl acetylene in most concentrations in butadiene would exhibit an exotherm starting at 135–140°C and culminate in an explosion.

Figure 9.4 Base of butadiene distillation column after internal explosion of vinyl acetylene. (Courtesy Gulf Publishing Company)

Such temperatures were just possible in the reboiler tubes. The effects of several possible initiators for the explosion were investigated. Of these, only sodium nitrate lowered the threshold temperature for the exotherm by 20°C. Solids containing 13% of sodium nitrate were found in the reboiler of the column. It had been formed by the oxidation of sodium nitrite a solution of which (following established wisdom) was continuously circulated through butadiene-distillation columns to scavenge traces of oxygen and prevent the formation of butadiene peroxide.

It was concluded that the explosion of vinyl acetylene had probably been triggered by a thermal polymerisation assisted by sodium nitrate.

The following preventative actions were taken for future operation:

- The steam temperature in the reboilers of similar columns was limited to 105°C;
- A high-temperature alarm set at 55°C was installed in the base of each similar column;
- Continuous analysers of vapour from the base and tray 10 of each column were installed and equipped with an automatic alarm and trip to shut the steam valve to the reboiler if the vinyl acetylene concentration exceeded 30%;
- Operation of these columns at total reflux was no longer permitted;

● Sodium nitrite was eliminated as an oxygen scavenger and replaced by an organic inhibitor.

The following practice which the writer initiated on another butadiene plant is worth mentioning. A vinyl acetylene-rich sidestream is removed continuously from a tray in the column where its concentration peaks. This passes to a selective hydrogenation unit where it is converted to butadiene and butenes and the treated stream is returned to the butadiene plant.

9.8.3 Cumene hydroperoxide[29]

One of three phenol plants at Philadelphia, USA, was destroyed on 9 February 1982 by an explosion when 25 000 gallons of cumene hydroperoxide in a tank were being heated by steam. Following this, safe temperature limits were set for heating this unstable peroxide, and arrangements were made for the rapid release of vapour from similar tanks [8.5.1].

9.8.4 Ethylene oxide[30]

On an ethanolamine plant in Kentucky in April 1962 an explosion occurred when ammonia feed to the plant inadvertently entered the ethylene oxide feed vessel and reacted violently. One man was killed and 12 were injured.

9.8.5 Hydrazine derivative[30]

Process equipment in which a hydrazine derivative was being made exploded in February 1961 on Olin Matheson's plant at Lake Charles, Louisiana. Hydrazine, like acetylene, is an unstable compound which can decompose explosively on its own when heated. One man died from injuries.

9.8.6 Two mini-explosions

Schwartz[26] reports the case of a nitroglycerine plant worker who had returned home and was lighting a cigar when an explosion occurred which stripped the flesh from his thumb and forefinger. It was caused by a few milligrams of nitroglycerine trapped under his thumbnail.

Walking past a gas-separation plant which was shut down for maintenance in the late 1950s, the writer was surprised to hear a series of sharp 'pops'. These were explosions of small deposits of copper acetylide which had formed on the outside of copper instrument lines inside the lagging. They resulted from small leaks over a long period of gases containing traces of acetylene. The lines were very cold during plant operation, but the explosions occurred during maintenance when the lagging was removed, the lines warmed up to room temperature, and fitters were working on the plant. They, not surprisingly, downed tools until all the explosive deposits were destroyed. After that the copper lines were replaced by stainless steel.

9.9 Features of the explosives industry and the Explosives Acts of 1875 and 1923[1]

The manufacture of substances intended for use as or in explosives has special features, particularly where safety is concerned. From its earliest days it has required a licence from the government and been subject to strict control and inspection.

The Explosives Acts of 1875 and 1923, and subsequent supporting legislation, apply principally to the operation of private and public companies engaged in the manufacture of explosives and pyrotechnics intended for non-military purposes (e.g. blasting, demolition, sporting ammunition, fireworks, flares and distress signals). The manufacture of explosives for military purposes has hitherto been carried out mainly in government factories. These are exempt from the provisions of the Explosives Acts, although subject to similar controls.

Explosives are defined in Section 3 of of the Explosives Act 1975 as:

(1) Gunpowder, nitroglycerine, dynamite, guncotton, blasting powders, fulminate of mercury or the fulminates of other metals, coloured flares, and every other substance whether similar to those above-mentioned or not, used or manufactured with a view to produce a practical effect by explosion or a pyrotechnic effect; and

(2) Fog signals, fireworks, fuses, rockets, percussion caps, detonators, cartridges, ammunition of all descriptions, and every adaptation or preparation of an explosive as above defined.

While the manufacture of the substances defined under (1) is a branch of the process industries (as defined in this book's introduction), the manufacture of the products defined under (2), like that of rubber tyres and plastic buckets, lies outside their scope.

The Explosives Acts lay down strict requirements for all safety aspects of this specialised industry. Critical processes and operating conditions are closely defined. Process and storage units are required to be in separate buildings or underground, and minimum distances between them are stated. Buildings with the greatest risk of explosion are designated as 'Danger Buildings' and detailed requirements are given for their construction and bunding. Strict limits are placed on the inventories of explosive materials and the number of persons present in any 'Danger Building'. Materials produced at each stage are subjected to analytical and stability tests which they must satisfy before they may be used in the next stage. All possible sources of ignition and explosion initiation such as electrical installations, static electricity and smoking are closely controlled.

The Explosives Act of 1875 also led to the establishment of the Explosives Inspectorate, now part of HSE, which exercises close control of safety aspects under the Act. A unique band of inspectors quaintly known as 'Danger Building Visitors' used to police the most hazardous operations. Other inspectors sample and test the raw materials, intermediate and final products, and thus provide an independent check on their quality and stability.

Safety *and security* are both top priorities in everything connected with explosives. Factories and depots are obvious targets for criminals, sabotage and enemy action. While the industry enjoys an enviable safety record, its products, manufacturing methods and hazards do not receive the same degree of publicity and open discussion as other process industries. Thus the Major Hazards Committee which was set up to assist HSC in the wake of the Flixborough disaster did not discuss installations subject to the control of the Explosives Acts in its reports. Similarly, the NIHHS Regulations 1982 [2.4.1] excludes substances to which the Explosives Act 1875 applies from its schedules of hazardous substances which require notification.

Explosive materials intended for use as such lie outside the scope of Bretherick's handbook[8]. Modern examples are cyclotrimethylamine trinitramine (RDX), ethylenediamine dinitramine (EDNA), pentaerythritol tetranitrate (PETN), ethylene glycol dinitrate, tetranitroaniline (tetryl) and trinitroresorcinol (styphnic acid). These and other explosives are, however, included in Group 4 of Schedule 3 of the CIMAH regulations, given later in Table 9.5.

9.9.1 Legal classification of explosive substances

The Act specifies seven classes of explosives. The first five which are discussed here are explosive substances. The last two are explosive products.

Class 1 – Gunpowder class
This consists exclusively of gunpowder, a mechanical mixture of saltpetre (65–75%), sulphur (10–20%) and charcoal (10–15%). Saltpetre is usually potassium nitrate but sometimes sodium nitrate. Once widely used for many purposes, the main uses of gunpowder today are for fireworks, fuses and igniters, and to a limited extent for blasting.

Class 2 – Nitrate-mixture class
This means any preparation other than gunpowder formed by mixing a nitrate with any form of carbon or non-explosive carbonaceous material, whether sulphur is added or not. Ammonium nitrate is the usual nitrate in this class of explosive, and fuel oil the usual carbonaceous material. Explosives containing perchlorates are included in this class if they do not fit other classes. These explosives are mainly used for blasting.

Class 3 – Nitro-compounds class
This is a very wide class and includes all compounds which are themselves explosive or which can form explosive compounds with metals, and are produced by the action of nitric acid (with or without sulphuric acid) on a carbonaceous substance. (The term includes nitrate esters and nitramines made by the action of nitric acid on organic compounds, as well nitro compounds which contain the $-NO_2$ group directly linked to a carbon atom.)

Class 3 includes mixed explosives containing ammonium nitrate, sodium nitrate, sodium chloride and other substances in addition to 'nitro-compounds' made as described above. This class has two divisions:

Division 1 comprises compounds or mixtures consisting wholly or partly of nitro-glycerine or some other liquid 'nitro-compound'.

Nitroglycerine itself as made industrially is a pale yellow oil which solidifies below 12°C, the solid being particularly sensitive. Figure 9.5 shows a flow sheet of a continuous glycerine nitrator and nitroglycerine separator, the latter being located in a special room surrounded by a blast-resistant bund.

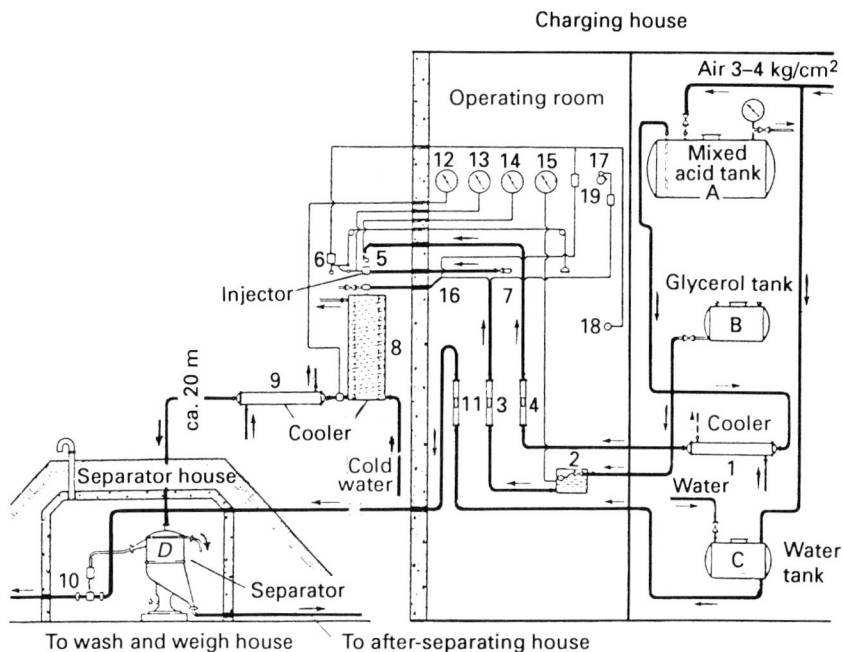

Figure 9.5 Flow sheet of nitro glycerine plant. (From *Encyclopaedia of Explosives,* edited by S. M. Kaye, vol. 8, fig. 53, published by US Army Armament and Research Division, Large Caliber Weapons Systems Laboratory, Dover, New Jersey, 1978)

Nitroglycerine vaporises at ambient temperatures, is absorbed through the skin, produces severe headache to those not regularly exposed to its vapour, and has been used to reduce blood pressure. Nitroglycerine is not easily ignited, but on sudden heating it explodes at about 215°C. It is about as sensitive to shock as mercury fulminate and there have been several explosions during its manufacture, storage and transport. Because of this it is rarely stored or used alone. Its stability is however greatly increased when it is incorporated with nitrocellulose (in blasting gelatine and in the propellant cordite) and with diatomaceous earth (in dynamite).

Division 2 comprises all 'nitro-compounds' as defined above which are not included in the first division. They clearly include cellulose nitrate of high nitrogen content which is used alone for military demolition (as 'guncotton slabs and primers'), TNT, picric acid and most trinitro-aromatic compounds which are used as explosives.

An explosion which occurred in a nitrocellulose dryer in a Dutch factory and its subsequent investigation are described by Hartgerink[31].

Nitrocellulose of explosive grade when made from cotton and known as guncotton has a nitrogen content of about 13%, but when made from paper it has a nitrogen content of about 12.3%. Nitrocellulose of 12.3% nitrogen content made from cotton is known as collodion. This is not regarded as explosive when mixed uniformly with at least one third of its weight of water, and it is used in ether solution for the coating of photographic plates and in surgery. Nitrocellulose of lower nitrogen content (9 to 11%) is used in the manufacture of varnishes, paints and celluloid.

Trinitrotoluene (TNT) is a yellow solid which melts at about 78°C. It is insoluble in water but dissolves in acetone, toluene, etc. TNT reacts with alkalis to form explosive salts and is very stable and not easily affected by shock. It begins to decompose at 150°C and decomposes rapidly at about 310°C. TNT is used for filling shells and bombs, both alone and mixed with ammonium nitrate.

Class 4 – Chlorate mixture class
This includes any explosive containing a chlorate.

Class 5 – Fulminate class
The term fulminate is not confined to its strict chemical meaning, but means any compound or mixture, whether included in the foregoing classes or not, which through its susceptibility to detonation is suitable for employment in percussion caps or other appliances for developing detonation, or which from its extreme sensibility to explosion and great instability is especially dangerous. (The term detonation as used in the Act appears to include explosive deflagration as discussed in 9.1.)

9.9.2 Prohibited explosives

There are only two prohibited explosives. These consist of mixtures of sulphur with any chlorate and mixtures of phosphorus with any chlorate, **for use in fireworks**. These mixtures are very sensitive to friction or percussion and explode with great violence, although they are allowed in detonators.

9.9.3 Exemptions of certain substances from the definition of an explosive

Exemptions apply both to explosive substances and products. Only the former are considered here:

Collodion cotton is not subject to the provisions of the Explosives Act when in solution in a solvent or in various specified mixtures;
Picric acid, picrates, dinitro-phenol and dinitro-phenolates under certain conditions.

These compounds which have other uses besides explosives are exempted when mixed with at least half their weight of water or when other specified conditions are met.

Acetylene, whose uses are mainly industrial, is exempted under the following conditions (which are quoted in full):

(a) When acetylene is subjected to a pressure not exceeding 9 lb per square inch above that of the atmosphere.
(b) (i) When acetylene is being used in the production of organic compounds, and the pressure does not exceed three hundred pounds per square inch above that of the atmosphere, and subject to the conditions approved by the Secretary of State, and
    (ii) When acetylene is subjected to pressure not exceeding twenty-two pounds per square inch above that of the atmosphere, and is being used in admixture with air or oxygen, and subject to the conditions approved by the Secretary of State.
(c) When acetylene is contained in a homogeneous porous substance with or without acetone subject to the conditions laid down in Order of Secretary of State No. 9, dated the 23rd June 1919[1].
(d) Acetylene in admixture with oil-gas, in accordance with the Orders of Secretary of State Nos. 5 and 5(a), dated 28th March 1898 and 29th September 1905 respectively[1].

The Acetylene (Exemption) Order 1977, No. 1798 exempts acetylene in liquid or compressed form or when mixed with air or oxygen from being deemed an explosive within the meaning of the Explosives Act 1875 in the case of acetylene contained in steel cylinders which have been made, filled and tested in accordance with the provisions of Order in Council (No. 30) dated 2 February 1937.

Ammonium nitrate impregnated with oil is excluded from certain provisions of the Explosives Act, but is subject to a licence from the Secretary of State by virtue of Statutory Instrument 1958, No. 416, The Ammonium Nitrate in Oil Exemption Order, 1958.

Despite the vintage of the Explosive Acts, the definitions of explosive substances contained in them are sufficiently wide to allow many which were unknown when the Acts were drafted to be brought within their scope.

9.10 Explosives and CIMAH Regulations 7–12[10]

Group 4 of Schedule 3 of the CIMAH Regulations [2.4.2] is a list of 'Explosive substances' to which Regulations 7 to 12 apply, and is given in Table 9.5. Schedule 1 which applies to Regulation 4 provides the 'Indicative Criterion' that:

Explosive substances are those which may explode under the effect of a flame or which are more sensitive to shocks or friction than dinitrobenzene.

The exemption under Regulation 3 of any 'factory, magazine or store licensed under the Explosives Act 1875' from the provisions of the CIMAH Regulations clearly excludes all substances 'manufactured with a view to produce a practical effect by explosion or a pyrotechnic effect'. From this it is appears that the only explosive substances to which the CIMAH

Regulations apply in the UK are those which are made entirely for other purposes. Of these, acetylene is already covered by extensions made many years ago to the Explosives Acts.

Table 9.5 comprises thirty chemical compounds which include many whose main use is to 'produce a practical effect by explosion or a pyrotechnic effect'. Of dinitro-aromatic compounds, several of which are made industrially for non-explosive purposes, it only includes the salts of dinitro phenol, which are already covered by the Explosives Acts. Several others, including Dinex [131-89-5], 2,6-dinitro-2-methylphenol which is used as a molluscicide to control water snails, and Zoalene a chick feed additive [5.2], are not included, although they have undoubted explosion hazards, e.g. during drying. Other notable omissions are the lower nitro-paraffins, nitromethane and nitroethane and acetylene derivatives

Table 9.5 Schedule 3, Group 4 of CIMAH Regulations – Explosive substances (for the application of Regulations 7 to 12)

Substance	Quantity (t)	CAS number	EEC number
Barium azide	50	18810-58-7	
Bis (2,4,6-trinitrophenyl)-amine	50	131-73-7	612-018-00-1
Chlorotrinitrobenzene	50	28260-61-9	610-004-00-X
Cellulose nitrate (containing >12.6% nitrogen	100	9004-70-0	603-037-00-6
Cyclotetramethylene-tetramine	50	2691-41-0	
Cyclotetramethylene-trinitramine	50	121-82-4	
Diazodinitrophenol	10	7008-81-3	
Diethylene glycol dinitrate	10	693-21-0	603-033-00-4
Dinitrophenol, salts	50		609-017-00-3
Ethylene glycol dinitrate	10	628-96-6	603-032-00-9
1-Guanyl-4-nitrosamino-guanyl-1-tetrazine	10	109-27-3	
2,2',4,4',6,6'-Hexanitro-stilbene	50	20062-22-0	
Hydrazine nitrate	50	13464-97-6	
Lead azide	50	13424-46-9	082-003-00-7
Lead styphnate	50	15245-44-0	609-019-00-4
Mercury fulminate	10	20820-45-5 and 628-86-4	080-005-00-2
N-Methyl-N,2,4,6-tetranitro aniline	50	479-45-8	612-017-00-6
Nitroglycerine	10	55-63-0	603-034-00-X
Pentaerythritol tetranitrate	50	78-11-5	603-035-00-5
Picric acid	50	88-89-1	609-009-00-9
Sodium picrate	50	831-52-7	
Styphnic acid	50	82-71-3	609-018-00-9
1,3,5,-Triamino-2,4,6-trinitrobenzene	50	3058-38-6	
Trinitroaniline	50	26952-42-1	
2,4,6-Trinitroanisole	50	606-35-9	609-011-00-0
Trinitrobenzene	50	25377-32-6	609-005-00-8
Trinitrobenzoic acid	50	35860-50-5 and 129-66-8	
Trinitrocresol	50	28905-71-7	609-012-00-6
2,4,6-Trinitrophenetole	50	4732-14-3	
2,4,6-Trinitrotoluene	50	118-96-7	609-008-00-4

such as vinyl acetylene[28, 32] [9.8.2] and propargyl bromide, all of which are produced on an industrial scale for non-explosive purposes, but which have been the cause of several serious explosion incidents.

Thus while the list is interesting as a reference list of very explosive materials, the number of installations in the UK which may be affected by it is probably very small.

References

1. Watts, H. E., *The Law Relating to Explosives*, Charles Griffin London (1954)
2. The Institution of Chemical Engineers, *Nomenclature for Hazard and Risk Assessment in the Process Industries*, Rugby (1985)
3. Lees, F. P., *Loss Prevention in the Process Industries*, Butterworths, London (1980)
4. Coffee, R. D., 'Chemical stability' in *Safety and Accident Prevention in Chemical Operations*, edited by Fawcett, H. H. and Wood, W. S., 2nd edn, Wiley-Interscience, New York (1982)
5. Snyder, J. S., 'Testing reactions and materials for safety', *ibid.*
6. Hunter, D., *The Diseases of Occupations*, 6th edn, Hodder and Stoughton, London (1978)
7. Swern, D. (ed.), *Organic Peroxides*, Wiley-Interscience, New York, vol. 1 (1970), vol. 2 (1971), vol. 3 (1972)
8. Bretherick, L., *Bretherick's Handbook of Reactive Chemical Hazards*, 4th edn, Butterworths, London (1990)
9. Harris, G. F. P., Harrison, N. and MacDermott, P. E., 'Hazards of the distillation of mononitrotoluenes', paper in I.Chem. E. Symposium series 68, *Runaway Reactions, Unstable Products and Combustible Powders*, Rugby (1981)
10. *Loss prevention data sheet 7-80, Organic Peroxides*, Factory Mutual Engineering Corporation, 1151 Boston-Providence Turnpike, Norwood, Mass. 06062 (1972)
11. S.I. 1984 No.1902, *The Control of Industrial Major Accident Hazards Regulations 1984 (CIMAH)*, HMSO, London
12. Kirchner, J. R., 'Hydrogen peroxide', in vol. 13, *Encyclopaedia of Chemical Technology*, edited by Kirk, R. E. and Othmer, D. F., 3rd edn, Wiley-Interscience, New York (ca 1979)
13. *CHETAH – The ASTM Chemical Thermodynamic and Energy Release Evaluation Program*, DS51, American Society for Testing and Materials, Philadelphia, Pa (1974)
14. Stull, D. R., 'Identification of reaction hazards', *Loss Prevention*, **4**, 16 (1970)
15. Weast, R. C. (editor) *CRC Handbook of Chemistry and Physics*, 1st student edn, CRC Press Inc., Boca Raton, Florida (1988)
16. Wenner, R. R., *Thermochemical calculations*, McGraw-Hill, London (1941)
17. Benson, S. W., *Thermochemical Kinetics*, Wiley, New York (1968)
18. ASTM E537-76, *Assessing the Thermal Stability of Chemicals by Methods of Differential Thermal Analysis*. American Society for Testing and Materials, Philadelphia, Pa (1974)
19. ASTM E487-76, *Test for Constant-Temperature Stability of Chemical Materials*. American Society for Testing and Materials, Philadelphia, Pa (1974)
20. ASTM 476-73, *Test for Thermal Instability of Confined Condensed Phase Systems*. American Society for Testing and Materials, Philadelphia, Pa (1974)
21. Connor, J., 'Explosion risks of unstable substances – Test methods employed by EM2 (Home Office) Branch, Royal Armament Research and Development Establishment', In *Loss Prevention and Safety Promotion in the Process Industries*, edited by Buschmann, C. H., Elsevier, Amsterdam (1974)
22. Daniels, T., *Thermal Analysis*, Kogan Page, London (1973)
23. ASTM D2540-70, *Test for Drop Weight Sensitivity of Liquid Monopropellants*, American Society for Testing and Materials, Philadelphia, Pa (1974)

24. ASTM D2539, *Test for Shock Sensitivity of Liquid Monopropellants by the Card Gap Test*. American Society for Testing and Materials, Philadelphia, Pa (1974)
25. ASTM D2541, *Test for Critical Diameter and Detonation Velocity of Liquid Monopropellants*. American Society for Testing and Materials, Philadelphia, Pa (1974)
26. Von Schwartz, *Fire and Explosion Risks* (first English edition) Charles Griffin, London (1918)
27. Armistead, J. G., 'Polymerisation explosion at Dupont' in Vervalin, C. H., *Fire Protection Manual for Hydrocarbon Processing Plants*, 3rd edn, Gulf Publishing, Houston (1985)
28. Griffith, S. and Keish, R. G., 'Butadiene plant explodes', *ibid*.
29. Garrison, W. G., *100 Large Losses – A thirty-year review of property damage losses in the hydrocarbon-chemical industries*, 11th edn, Marsh and McLennan Protection Consultants, 222 South Riverside Plaza, Chicago, Illinois 60606 (1988)
30. Vervalin, C. H., *Fire Protection Manual for Hydrocarbon Processing Plants*, 3rd edn, Gulf Publishing, Houston (1985)
31. Hartgerink, J. W., 'Investigation into the cause of an explosion during the drying of porous nitrocellulose powder', paper in Symposium on Chemical Process Hazards with special reference to Plant Design, I. Chem. E. Symposium series No 39a, Rugby (1974)
32. Carver, F. W. S., Hards, D. J. and Hunt, K., 'The explosibility of some unsaturated C_4 hydrocarbon fractions', *ibid*.

Chapter 10

Flammability, fires and explosions involving air

The word flammable is used here rather than 'inflammable' (which could be misunderstood for 'non-flammable'). Flammability simply means the ease with which a material burns in air (or occasionally in some other stated gas such as oxygen). It applies to gases, liquids and solids.

In the process industries we are most concerned with the flammability of gases, liquids and their vapours, and of dispersed dusts and liquid droplets, all of which we refer to as fuels. The result of ignition can be a fire or explosion or sometimes both. Accidental fires and explosions of flammable air mixtures often follow the escape of combustible materials, which may result from an uncontrolled reaction or an explosion within the process or storage plant [8, 9]. Despite the differences in their destructive effects, fires and explosions of fuel/air mixtures are both treated in this chapter.

The flammability of plastics, textiles and building materials is a major concern in homes, offices, stores, schools, hotels, etc. Most of these materials are products of the process industries, but their flammability in service, which lies outside the scope of this book, is treated in a recent handbook[1]. The constructional materials used in the process industries are mostly of low flammability, although the plant and structures used often require fire protection [15.1].

We assume that readers are:

- Familiar with combustion and the concept of the fire triangle with its three essential components, a fuel, oxygen, and heat or an ignition source (Figure 10.1);
- Aware that flammable liquids must generally vaporise before they can ignite;
- Aware that mixtures of flammable gases and air only ignite readily when their composition lies within a certain range – the flammability limits.

Flammable gases and vapours of flammable liquids are in many ways more dangerous than the liquids themselves. This is because they are invisible, cannot easily be contained, and may travel considerable distances before they reach an ignition source and ignite. The flame then usually follows the gas or vapour back to its source – either a leak or the exposed surface of a flammable liquid. Flammable liquids should therefore not be kept or handled in open containers.

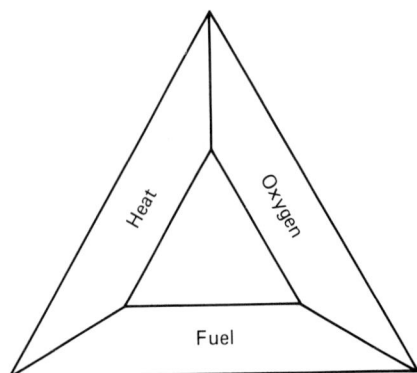

Figure 10.1 The fire triangle

Although the word combustible applies to all materials which will burn (including flammable ones), the National Fire Protection Association (NFPA) (of the USA) classifies liquids as either flammable *or* combustible, depending on their flash-point[2].

10.1 NFPA flammability classification of materials

The NFPA hazard rating system[3], the main features of which are given in Appendix B, recognises four degrees of flammability.

NFPA Flammability degree 4
Materials which will rapidly or completely vaporise at atmospheric pressure and normal ambient temperature or which are readily dispersed in air, and which will burn readily. This degree includes:

- Gaseous materials;
- Cryogenic materials;
- Class IA flammable liquids (see Table 10.1);
- Materials which on account of their physical form or environmental conditions can form explosive mixtures with air and which are readily

Table 10.1 Classification of combustible liquids[2]

Classification	Flash-point (°C)	Boiling point (°C)
Flammable liquids:		
Class IA[a]	<22.8	<37.8
Class IB[a]	<22.8	≥37.8
Class IC	≥22.8 and <37.8	
Combustible Class II	≥37.8 and <60	
Combustible Class IIIA	≥60 and <93.4	
Combustible Class IIIB	≥93.4	

[a] Class IA and IB liquids rate as 'Highly Flammable' under UK regulations[14]

dispersed in air, such as dusts of combustible solids and mists of flammable or combustible liquid droplets.

NFPA Flammability degree 3
Liquids and solids that can be ignited under almost ambient temperature conditions. Materials in this degree produce hazardous atmospheres in air under almost all ambient temperatures or, though unaffected by ambient temperatures, are readily ignited under almost all conditions. This degree includes:

- Class IB and IC flammable liquids (see Table 10.1);
- Solid materials in the form of coarse dusts which may burn rapidly but which do not generally form explosive atmospheres with air;
- Solid materials in a fibrous or shredded form which may burn rapidly and create flash fire hazards such as cotton, sisal and hemp;
- Solids which burn with extreme rapidity, usually by reason of self-contained oxygen (e.g. dry nitrocellulose);

NFPA Flammability degree 2
Materials that must be moderately heated or exposed to relatively high ambient temperatures before ignition can occur. Materials in this degree would not under normal conditions form hazardous atmospheres with air, but under high ambient temperatures or under moderate heating may release vapour in sufficient quantities to produce hazardous atmospheres with air. This degree includes:

- Liquids having a flash-point above 37.8°C, but not exceeding 93.4°C (Class II and IIIA combustible liquids, see Table 10.1);
- Solids and semi-solids which readily give off flammable vapours.

NFPA Flammability degree 1
Materials that must be preheated before ignition can occur. Materials in this degree require considerable preheating under all ambient conditions before ignition and combustion can occur. This degree includes:

- Materials which will burn in air when exposed to a temperature of 815°C for a period of 5 minutes or less;
- Liquids, solids and semi-solids having a flash-point above 93.4°C (Class IIIB flammable liquids);

This degree includes most ordinary combustible materials.

NFPA Flammability degree 0
Materials that will not burn. This degree should include any material that will not burn in air when exposed to a temperature of 815°C for a period of 5 minutes.

The highest degree of hazard (no. 4) is reserved for those materials whose mixtures with air at normal ambient temperatures could explode for reasons either of flash-point or particle size.

The next degree (no. 3) includes two classes of materials which are usually thought of as very hazardous, those which ignite spontaneously in air (which makes determination of their flash-points impossible) and solids such as dry nitrocellulose which (because of oxygen contained in them) burn extremely rapidly in air. Materials of both of these classes, however, have higher scores on the NFPA *reactivity* rating than most materials with a flammability rating of 4.

Although combustible solids of low volatility such as coal have low flammability ratings when free of dust, their flammability ratings increase to degree 3 or even 4 when they are pulverised. This reflects the fact that combustible solids are much easier to ignite when finely divided. They may then even ignite spontaneously when deposited in thick layers [8.7]. Very fine dusts of coal and most other combustible materials can explode when dispersed in air as thick clouds if a strong ignition source is present[5]. The main criterion affecting the explosiveness of a dust cloud is its particle size. Particles larger than 200 μm present little explosion risk, but those in the range 10–50 μm can cause powerful explosions [10.4].

10.2 Parameters of flammability (mainly for gases and vapours)

There is no single parameter by which the flammability of different materials can be compared. Several need to be considered. The flash-point is specially important for liquids. The following list includes properties (nos 8 to 11) which, while affecting flammability to some degree, are mainly important for containing and fighting fires:

1. Flammability limits (for gases and vapours);
2. Flash-point (for liquids and low-melting solids);
3. Autoignition temperature (mainly for gases and vapours);
4. Ignition energy (for gases, vapours, dust clouds and mists);
5. Burning velocity (mainly for gases and vapours);
6. Heat of combustion (for all materials);
7. Oxygen requirements for complete combustion;
8. Specific gravity relative to:
   - Air, for gases and vapours,
   - Water, for liquids;
9. Solubility in water (mainly for liquids);
10. Melting point and softening point for solids;
11. Viscosity (mainly for liquids and low-melting solids);
12. Carbon/hydrogen ratio (mainly for hydrocarbons).

10.2.1 Flammability limits in air[6, 7]

For every flammable gas or vapour there is a range of concentrations (% by volume) in air within which ignition can occur. The stoichiometric mixture which contains just enough oxygen for complete combustion usually lies near the geometric mean of this flammable range. The flammability limits which define this range should be determined using standardised apparatus and conditions to eliminate variations due to temperature, dimensions of the apparatus and spark strength[8]. The flammability range widens, as would be expected with increasing temperature, pressure and oxygen concentration. The ratio of the upper to the lower flammability limit of most stable hydrocarbons and organic compounds is between 3:1 to 5:1. Carbon disulphide, hydrogen, acetylene and ethylene oxide, however, have exceptionally wide flammability ranges. The upper flammability limits of acetylene[9] and ethylene oxide[10] which can decompose explosively under the action of a spark in the complete absence of air are 100%. For hydrogen the limits are 4% and 75%.

All mixtures of gases and vapour with air which are within the flammable range are liable to explode on ignition, particularly if the mixture is partly or wholly confined. Such explosions are generally deflagrations [9.1] which can lead to pressure rises up to eight times the original pressure. Accidental explosions of this type are all too common. Gas-fired boilers and furnaces are specially prone to this hazard, e.g. when gas is admitted to a cold furnace and ignition is delayed.

Some flammable gases and vapours when mixed with air within a still narrower concentration range can detonate when strongly ignited. Such detonations [9.1] can cause peak pressures up to twenty times the original pressure and create shock waves which are far more damaging than the blast waves caused by explosive deflagrations[6]. Fortunately they seem to be fairly rare, except when the mixture is closely confined as in a pipeline, or when a large quantity of flammable mixture is involved, and the character of the combustion process changes progressively:

Deflagration → Explosive deflagration → Detonation

Gases and vapours whose mixtures with air can detonate include hydrogen, acetylene, and diethyl ether. Their flammability and detonability limits in air are given in Table 10.2. The detonability limits are less well

Table 10.2 Flammability and detonability limits for some gases and vapours

Gas or vapour	% volume in air			
	Flammability		Detonability	
	Lower	Upper	Lower	Upper
Acetylene	2.5	100	4.2	50
Hydrogen	4.0	75	18.3	59
Diethyl ether	1.85	36.5	2.8	4.5

established than the flammability limits. It has even been suggested that if a sufficiently powerful initiator were employed, the limits would be practically identical[6]. Mixtures of a number of unlikely compounds with air have been claimed to detonate, even carbon tetrachloride, a non-flammable liquid, formerly used in 'vaporising-liquid' fire extinguishers![6]

It is often essential to know the concentration of flammable gas or vapour in a given atmosphere (tank, vessel, open air, etc.), and to have some warning if this exceeds a certain figure. Fixed and portable instruments which give these figures as a percentage of the lower flammable limit are discussed in 18.7.1.

10.2.2 Flash-point

The flash-point of a liquid is the lowest temperature at which air saturated with its vapour will ignite. This depends both on its lower flammability limit and its vapour pressure. For a series of similar compounds, it can be related to the boiling point. The term 'flash-point' only applies to air-stable liquids and solids which melt before they can ignite. It has no relevance to gases, nor does it apply to spontaneously inflammable compounds such as zinc and aluminium alkyls.

Several methods of test are used by different authorities. All involve heating the liquid slowly in contact with air, applying an ignition source at intervals above its surface, and noting the lowest temperature at which the vapour ignites (flash-point) or the liquid catches fire (fire-point). The methods are of two types, the 'closed cup' for liquids of low flash-point and the 'open cup' for liquids of high flash-point. Closed cup methods tend to give somewhat lower results than open cup methods, but differences between methods of the same type appear to be small.

Three methods of test widely used in the petroleum industry are the 'Tag' Closed Tester[11] for low flash-point liquids (<200°F) the Pensky–Martin Closed Tester[12] for high flash-point liquids (≥200°F) and the Cleveland Open Cup test[13] which is also used for high flash-point liquids. BS 3900: Part 9[14] describes an apparatus for determining the flash-points of paint, petroleum and related products in the range 5–110°C. This has several alternative closed cups which enable the flash-point as specified in various national standards to be determined.

Liquids which can burn are grouped in Table 10.1 on the basis of their flash-points.

The relation between temperature, partial pressure, flammable limits and flash-point is shown in Figure 10.2[6]. This shows that there is an upper as well as a lower equilibrium temperature at which air saturated with the vapour of the liquid will ignite. The lower temperature which is the flash-point corresponds to the lower flammability limit, while the upper temperature corresponds to the upper flammable limit. This is important for the safety of fixed roof tanks vented to the atmosphere in which flammable liquids are stored. An explosion is possible in the free space above the liquid if the temperature there lies between the lower and upper equilibrium temperatures. To avoid this risk as well as save liquid through 'breathing losses', floating roofs for such liquids should be used wherever possible.

Another possibility is to use a nitrogen or other inert gas 'blanket' over the liquid in fixed roof tanks, with automatic pressure-operated gas admittance and relief valves. A selection of liquids whose upper and lower equilibrium temperatures straddles the typical ambient temperature of 15°C is given in Table 10.3, together with their flammability limits in air and equilibrium temperatures.

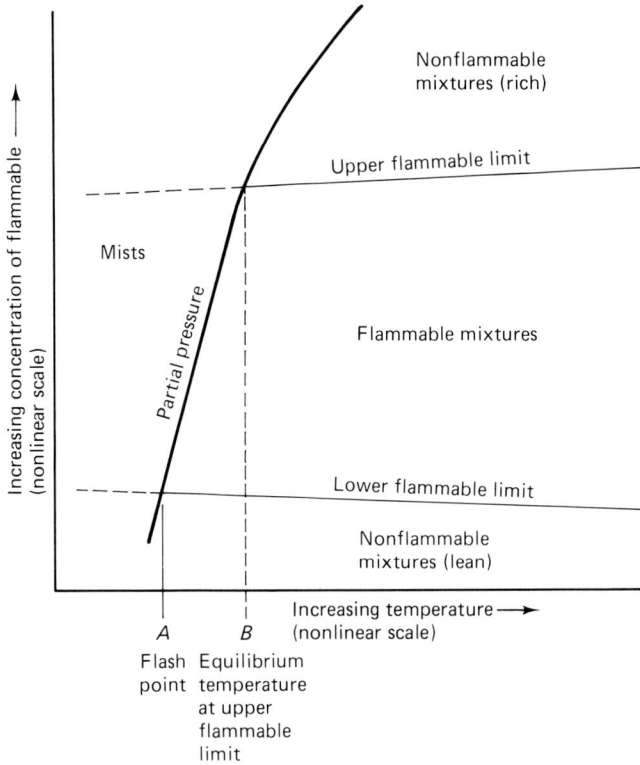

Figure 10.2 Relationship between temperature, partial pressure, flammability limits and flash-point

Table 10.3 Liquids posing vapour space explosion hazard in tank storage

Liquid	Flammability limits (% vol.)		Equilibrium temperatures (°C)	
	Lower	Upper	Lower (flash-point)	Upper
Carbon disulphide	1.3	50	−30	26
Cyclohexane	1.3	7.8	−20	15
Benzene	1.4	8.0	−11	16
Methyl ethyl ketone	1.9	10	−10	21
n-Heptane	1.1	6.7	−7	27
Toluene	1.2	7.1	4	40
Methanol	6.7	40	11	40
Ethanol	3.3	19	13	40
Isopropanol	2.2	12	13	38

10.2.3 Autoignition temperature

This is the lowest temperature at which a flammable gas or vapour–air mixture will ignite from its own heat or from contact with a heated surface without the need for a spark or flame. It is sensitive to the purity of the material and the nature and area of the hot surface. The autoignition temperature may be reduced by 100°C or more for surfaces which are lagged or contaminated by dust[5]. Reproducible results are only obtained when a standard method of test is used[15]. Autoignition temperatures vary widely, from 100°C for carbon disulphide, and 170°C for di-ethyl ether to 538°C for methane, 560°C for hydrogen and 630°C for ammonia[16]. Ignition temperatures of dust clouds have also been measured. These vary from 190°C for sulphur to over 700°C for coal-tar pitch[4].

A special feature of the autoignition of many organic vapours, including hydrocarbons, alcohols, ethers, aldehydes and acids, is their tendency to form 'cool flames' at temperatures several hundred degrees below their normal autoignition temperatures[17]. These emit a pale blue light which is visible only in the dark. This is a form of incomplete combustion which usually involves the formation of an unstable peroxide and its decomposition to an aldehyde. A list of structural groupings which render compounds specially prone to peroxide formation and cool flames was given in Table 8.5. The main hazard of cool flames lies in their potential for transition into true combustion. Further hazards are that the products of cool flames are usually less stable, more reactive and more toxic than the original compounds.

Special care is needed with processes where materials of low ignition temperature are present to eliminate hot surfaces, and to thoroughly insulate and effectively seal off any which cannot be eliminated. Electrical motors and other equipment for use in atmospheres where flammable gases and vapours and fine dusts of combustible materials can be present must be carefully chosen, and of a temperature class[18] [6.2.1] which is well below the lowest ignition temperature of any materials present. Such surfaces must also be kept clean and free from dust deposits which hinder their cooling and allow higher surface temperatures to be reached.

Other common ignition sources for flammable vapours and dust clouds are welding torches, matches, the hot flues and external walls of furnaces, electrical heaters and broken light bulbs, and bearings and seals which run hot through poor alignment or lack of lubrication.

10.2.4 Ignition energy

Ignition of any flammable gas–air, vapour–air or dust–air mixture requires a minimum amount of energy. This is of prime importance for ignition by electric sparks, whether from electrical circuits or static electricity. The ignition energy of any flammable material in air depends on the material itself and on the proportion in which it is present, and is at a minimum near the stoichiometric mixture. The minimum ignition energies of most flammable gases and vapours lie between 0.1 and 0.3 millijoules (mJ), but carbon disulphide, hydrogen, acetylene and ethylene oxide have abnormally low ignition energies – between 0.01 and 0.02 mJ. These substances pose special problems not only for the design and protection of electrical circuits in potentially flammable atmospheres[19-21] [6.2] but also in the prevention of sparks from static electricity[22] [6.3]. Leaks of hydrogen under pressure are specially prone to ignite spontaneously through static electricity. The fact that these flames are practically invisible adds to the danger.

10.2.5 Burning velocity

This is the velocity at which a gaseous fuel–air mixture issuing from a burner 'burns back' on itself. Like the previous properties described, it varies with the gas concentration in air, the dimensions of the apparatus, and the conditions of the test, all of which must be standardised. The figure usually quoted for different fuel gases is the 'maximum fundamental burning velocity in air at atmospheric temperature and pressure'. Actual flame speeds from industrial burners and in unintentional fires in which gas turbulence plays a role may be much higher. Nevertheless, the burning velocity of a gas or vapour is the main factor which determines the stability of its flame.

When a jet of gas escapes into the atmosphere through an opening and ignites without impinging on any solid object the flame is most readily blown off the opening (and hence extinguished) if the burning velocity is low, whereas a high burning velocity ensures that the flame 'sits' on the burner. The likelihood that any gas–air or vapour–air mixture will explode increases with its burning velocity.

The burning velocities of the lighter paraffin and aromatic hydrocarbons lie in the range 35–50 cm/s, but those of olefins are higher (ethylene 69 cm/s), while those of acetylene (173 cm/s) and hydrogen (320 cm/s) are exceptionally high.

10.2.6 Heat of combustion

The heat of combustion is related to the total heat liberated and to the maximum flame temperature of a material. Heats of combustion are generally quoted on a gross basis, to include the heat given out in condensing any water formed, although for our purpose a net basis is better. For liquids and solids they are usually quoted on a weight basis, and for gases on a volume basis. Most hydrocarbons have heats of combustion ranging from 42 000 to 55 000 kJ/kg. Hydrogen has the highest value of 143 000 kJ/kg (due to its low atomic weight).

The presence of oxygen, nitrogen and sulphur atoms in an organic compound lowers its heat of combustion. Methanol, for example, has only half the heat of combustion of most hydrocarbons, while that of carbon disulphide is about one third. While carbon disulphide is very easily ignited, the flame has a low temperature and gives out little heat. The possibility of forming an explosive vapour cloud by the flashing of hot process liquid which is escaping from a pressurised source is less with non-hydrocarbons than with hydrocarbons [12.4.1]. This is due to the higher heats of combustion and lower latent heats of vaporisation of hydrocarbons.

10.2.7 Oxygen requirements for complete combustion

The speed with which a flame can spread through a material depends to a large extent on the speed with which the oxygen required for complete combustion can reach the flame. Hence materials with lower oxygen requirements may be expected to burn faster, other things being equal. Materials with the lowest oxygen requirements are generally those which already contain a fair proportion of the oxygen needed, e.g. nitro-compounds and peroxides, and they do in fact burn faster than most others. This is a criterion to be borne in mind when comparing the flammability of different materials.

10.2.8 Specific gravity (SG)

The SG of gases and vapours is expressed relative to air and that for liquids and solids relative to water. The former is important in determining whether and to what extent a gas or vapour which has escaped into relatively open surroundings will rise and disperse rapidly, or descend to and remain for some time at ground level. But if the escape is into a closed building, a gas of low SG may present as great an explosion hazard as one of high SG.

When a liquefied gas of low SG (i.e <1.0) such as ammonia or methane escapes its evaporation may cool the air with which it mixes so much that the resulting gas–air mixture is denser than the surrounding air, and thus tends to remain near ground level. Escapes of such liquefied gases can thus form dangerous gas clouds which persist for a considerable time at ground level. Even the cooling which takes place when a compressed gas escapes can have the same effect. Thus the fact that a gas which escapes is significantly lighter than air is no guarantee that it will always disperse rapidly upwards.

The SG of a flammable liquid which is insoluble in water is the main factor in determining whether water can be used to fight fires of spilled material, or whether it is more likely to spread the fire, as well as transferring flammable materials into drains and effluents. Water jets should not be applied to fires of hydrocarbon liquids of low flash-point (SG <1.0), although water mist supplied by atomising nozzles is used on such fires by trained fire-fighters. In the case of compounds such as nitrobenzene and chlorobenzene, which are denser than water and insoluble in it, a pool fire is easily extinguished by covering it carefully with water.

10.2.9 Solubility in water

Fires and spillages of materials which are soluble in water can generally be dealt with effectively by water unless their presence at low concentrations in drains and effluents is unacceptable on toxicological or other grounds. This depends on local circumstances as well as on the toxicity of the materials themselves. Thus the discharge of even small quantities of cyanides into a first-class salmon river would be unacceptable. The recent poisoning of the Rhine, which after treatment provides most of Holland's drinking water, has proved a costly mistake for the firm responsible.

10.2.10 Melting point or softening point (for solids)

The nature of a solid fire is completely altered if the material melts before burning. The act of melting generally spreads and accelerates the fire. While crystalline materials usually have well-defined melting points, pitches and thermoplastics first soften and then form viscous liquids whose viscosity remains high over a considerable temperature range. Such burning, hot, sticky semi-solids present a serious hazard to fire-fighters.

10.2.11 Viscosity (for liquids and low-melting solids)

The viscosity of a liquid affects its behaviour in a fire in much the same way as the melting point affects the behaviour of a solid. Thus a viscous liquid emerging from a ruptured tank will move more slowly than one of low viscosity. This may at first be thought to make the low-viscosity liquid more dangerous in the event of a fire. Nevertheless, the high-viscosity liquid has special dangers to fire-fighters and others who may come into contact with it. This is clear from considering the effects of 'Napalm', a war weapon consisting basically of petrol whose viscosity has been increased a hundredfold or more by incorporating soap into it.

10.2.12 Hydrogen/carbon ratio (for hydrocarbons)

The ratio of hydrogen to carbon atoms in hydrocarbons range from a little less than 1:1 for polynuclear aromatics to 4:1 for methane. Although this may have little direct effect on whether the material will ignite, it has some influence on the spread of fire, caused by differences in the heat radiation from the flame. This depends on the colour, luminosity and smokiness of the flame. The lower the H:C ratio, the smokier and more luminous will the flame become.

The luminosity and the smokiness of the flame tend to work in opposite directions. While more of the heat in a luminous flame is radiated sideways and less is carried upwards in the combustion products compared with a non-luminous one, much of the radiated heat from a smoky flame is absorbed by the carbon particles in the smoke, which screen other objects from the radiation. For any fire situation, there is thus an intermediate C:H ratio which gives the maximum rate of heat absorption, and hence of fire-spread, for objects on the same level or lower than the flame.

From the point of view of fire-fighting, the smoky flame is the more hazardous, both in restricting vision and increasing respiratory hazards.

10.3 Flammability and CIMAH Regulations 7–12[23]

The CIMAH Regulations [2.4.2], in common with those of other EC countries, give limits to the inventories of various flammable materials in manufacturing installations above which Regulations 7 to 12 apply. The limits are given in Schedule 3 under *Group 5 – Flammable substances*. The definitions of the Flammable substances are given in Schedule 1. The limits and definitions follow:

(i)   *Flammable gases*: limit 200 t
      Substances which in the gaseous state at normal pressure and mixed
      with air become flammable and the boiling point of which at normal
      pressure is 20°C or below;
(ii) *Highly flammable liquids*: limit 50 000 t
      Substances which have a flash-point lower than 21°C and the boiling
      point of which at normal pressures is above 20°C;
(iii) *Flammable liquids*: limit 200 t
      Substances which have a flash-point lower than 55°C and which
      remain liquid under pressure, where particular processing condi-
      tions, such as high pressure and high temperature, may create
      major accident hazards.

The reason for the low limit in (iii) lies in the possible processing
conditions involved – where an escape could lead to the sudden formation
of a large flammable or explosive vapour cloud. The definition of
flammable liquids in (iii) includes highly flammable liquids as defined in
(ii).

10.4 Flammable dusts and explosive dust clouds[5, 6]

The ease of ignition and flammability of combustible materials increases
greatly when they are finely divided. When mixed with and suspended in
air as dust clouds, most finely divided combustible solids are capable of
powerful explosions, which have caused many serious accidents. Some of
the worst, on a par with the Flixborough disaster, have been in starch
plants and grain silos in the USA. Many serious mine explosions in the UK
which were first attributed to methane were later realised to have been
caused mainly by coal dust[5]. The terms dust and powder are sometimes
used interchangeably, although BS 2955: 1958 defines a powder as a
material with particles less than 1000 μm in diameter, and dusts as those
with particle diameters less than 76 μm[24]. The word dust usually implies
valueless airborne material which has found a resting place, whereas
powders may be valuable. Here the word dust is used to include powders.
   Several authorities have published lists of materials which are liable to
form explosive dust clouds[5, 25]. Table 10.4 gives some of the more common
ones.
   As in the case of flammable gases and vapours, there are lower and
upper dust concentrations in air within which an explosion is possible and
outside which none is likely to occur, although the upper limit is rather
ill-defined. The lower limit ranges from about 20 mg/l for plastics such as
polyethylene, polystyrene and urea formaldehyde resins to 100 mg/l for
activated carbon, with aluminium, magnesium, coal, sulphur, sugar and
flour having intermediate values. In appearance an explosive dust cloud
resembles a very thick fog. Dust explosions can raise the pressure inside
confined spaces by as much as eight times with cellulose acetate,
magnesium and sugar, but only five to six times with sulphur, TNT and
phthalic anhydride[5]. The size of the particles is, however, more important
than their composition in determining the force of a dust explosion and
whether one will occur. The water and incombustible content of the dust

Table 10.4 Common powders liable to form explosive dust clouds

Type	Example
Carbon	Coal, peat, charcoal, coke, lampblack
Fertilisers	Bone meal, fish meal, dried blood
Food products and by-products	Starches, sugars, flour, cocoa, powdered milk, grain dust
Metal powders	Aluminium, magnesium, zinc, iron
Natural resins, waxes	Shellac, rosin, gum sodium resinate, soap powder, waxes
Plastics, thermoplastic	Nylon, PE, PP, PS, PVA, PVC etc., cellulose esters and ethers
Plastics, thermosetting	Melamine-, phenol-, urea-formaldehyde, epoxy and polyester resins, etc.
Spices, beverages and natural insecticides	Cinnamon, pepper, gentian, tea dust, coffee and cocoa, pyrethrum
Wood, paper and tanning materials	Wood flour, wood dust, cellulose, cork, bark dust, wood extract
Miscellaneous	Hard rubber, phthalic anhydride, sulphur, TNT, tobacco

are also important. The most severe explosions occur with particle diameters between 10 and 50 µm. Coarser particle with diameters greater than 200 µm present far less explosion risk. Dust explosions normally behave as deflagrations and not as detonations, and are treated as deflagrations in planning protective measures. The flame speeds in dust explosions are high, comparable to those in gas deflagrations. The NFPA rate the explosivity of combustible dusts according to their maximum rate of pressure rise[26] (Table 10.5).

This rating is important in relating the vent area of vessels subject to a dust explosion risk to the vessel volume and relief pressure.

The ignition sources for dust explosions are similar to those of fires, and include flames, radiant heat and hot surfaces, mechanical sparks, self-heating, static electricity and electrical circuits. The minimum ignition temperatures for dust clouds often differ considerably from those for the same material in layers, being in some cases higher and in others lower[5]. They range from 190°C for sulphur to over 700°C for coal tar pitch. The differences are connected with the self-heating tendencies of the dusts [8.7]. If the plant in which a dust cloud is present has a hot surface such as a steam pipe on which dust can settle, slow combustion can begin in the

Table 10.5 NFPA explosion rating of combustible dusts

	Maximum rate of pressure rise	
	lbf/in^2 s	kN/m^2 s
Class St.-1	<7 300	<50 000
Class St.-2	7 300–22 000	50 000–150 000
Class St.-3	>22 000	>150 000

settled layer at a much lower temperature than the ignition temperature of the dust cloud. This local fire can then initiate a dust cloud explosion.

The minimum ignition energies of dust clouds are about one hundred times higher than those of most flammable vapour–air mixtures, ranging from 15 mJ for sulphur to 60 mJ for bituminous coal.

Dust explosions fall broadly into two categories, those which occur in buildings and other places where the dust is present adventitiously, often having settled on floors, girders and on top of equipment, shelves and tables, etc., and those which occur inside process equipment and storage bins and silos where the 'dust' is a powder, sometimes deliberately suspended in air for pneumatic conveying or drying.

10.4.1 Explosions due to adventitious dust in buildings, etc.

The first type can be prevented by careful design of the plant to prevent the escape of dust into the general atmosphere, and by careful housekeeping and regular removal of dust by proper cleaning methods. For these there is a legal requirement in Section 31 of the Factories Act 1961, which includes the following:

(1) Where, in connection with any grinding, sieving, or other process giving rise to dust, there may escape dust of such a character and to such an extent as to be liable to explode on ignition, all practical steps shall be taken to prevent such an explosion, by enclosure of the plant used in the process, and by removal and prevention of accumulation of any dust that may escape in spite of the enclosure, and by exclusion or containment of possible sources of ignition.

Extreme measures are sometimes needed to prevent the accumulation of combustible dusts. It must be realised that many combustible dusts are liable to spontaneous self-heating and auto-combustion, which can lead to dispersion and explosion. The chances of this occurring are increased by the size of the pile and the longer it remains [8.7.2]. Dusty operations should where possible be segregated, totally enclosed, and operated under a slight negative pressure with air extraction and dust collection equipment to prevent leakage of dust into the general work area. Floor openings should be kept to a minimum and openings for pipes and ducts should be sealed. Frequent cleaning and inspection are essential, using cleaning methods which remove dust effectively and do not merely transfer the hazard to somewhere else. The use of compressed air jets for cleaning should be banned. Special vigilance is needed to prevent dust accumulation on horizontal surfaces such as girders and shelves which are above eye level, and small portable mirrors with telescopic handles may be needed for inspection. Buildings themselves with high dust explosion risks such as flour mills and plastics plants should be constructed with hinged windows or blow-out wall panels which will fall in a predetermined area, to prevent structural collapse.

Portable fire extinguishers and hose reels with atomising nozzles should be available, rather than types which give solid jets which stir up dust. All potential ignition causes should where possible be eliminated and otherwise carefully controlled. These sources include open flames, welding, sparks from friction, hand tools, grinding, and static electricity, overheated bearings and smoking. Non-ferrous tools and truck wheels are recommended to reduce sparking risks, although it is doubtful whether the sparks arising from them could ignite a dust cloud. Only approved dust-tight electrical motors, wiring and equipment should be used. Control measures may require the use of increased humidity (although this sometimes increases the risk of self-heating) and inert gas.

A sequence of events in a serious industrial dust explosion has been described by Lees as follows[6]:

> A primary explosion occurs in an item of the plant. The explosion protection is not adequate to prevent the flame issuing from the plant, due either to rupture of the plant or to poor explosion venting. The air disturbance disperses the dust in the work room and causes a secondary explosion. The quantity of dust in the secondary explosion often exceeds that in the primary one. Moreover the building in which the secondary explosion occurs may be weaker than the plant itself. The secondary explosion is thus often more destructive than the first one. In some cases the primary explosion also occurs in the open and disturbs dust deposits and this causes a secondary explosion. In other cases the primary explosion occurs in one unit of the plant and the explosion propagates within the plant to other units.

10.4.2 Dust explosions inside plant and equipment

Section 31 of the Factories Act 1961 requires plant to be designed to prevent such explosions:

> (2) Where there is present in any such plant used in any such process as aforesaid dust of such a character and to such an extent as to be liable to explode on ignition, then, unless the plant is so constructed as to withstand the pressure likely to be produced by any such explosion, all practicable steps shall be taken to restrict the spread and effects of such an explosion by the provision, in connection with the plant, of chokes, baffles and vents, or other equally effective appliances.

The types of plant or equipment liable to dust cloud explosions include:

- Milling, grinding, pulverising, disintegrating and stamping machines;
- Kilns, pneumatic driers, rotary drum dryers, spray dryers and fluidised-bed dryers;
- Screens, classifiers, bag filters and dust collectors;
- Conveyors and elevators of various types;
- Cyclones and settling chambers;
- Storage bins and silos.

The prevention and control of dust explosions in such equipment should be fully considered at the design stage. Sometimes it is economically feasible to prevent the possibility of a dust explosion completely, e.g.

- By avoidance of dust;
- By the use of an inert gas in place of air;
- By the use of wet instead of dry methods of processing.

Where such steps are impossible, suitable protective measures should be taken, such as:

- Making the plant strong enough to withstand a dust explosion;
- Placing the plant out of doors where an explosion will cause no injury or further damage;
- The use of explosion-relief panels;
- The use of explosion-suppression systems.

10.4.3 Prevention of dust-explosions inside plant

Wherever practicable, plant which does not produce dust clouds or which minimises their size should be used. For conveying dusts and fine powders, drag-link type conveyors which convey the material in a solid mass are preferred to bucket elevators, screw-, pneumatic- and vibro-conveyors. The cross section of the return legs of drag-link conveyors can be much smaller than the forward legs if the plates or links fold up. Wet processes for grinding, pulverising and disintegrating should where feasible be employed.

Wet dust washers (e.g. Venturi type), which are in most cases just as effective and economic as dry ones, can often be used when the dust or powder is not required in a dry state. Wet type collectors are obligatory for grinding magnesium and its alloys[27].

An inert gas system using nitrogen, carbon dioxide or the gaseous combustion products of a fuel, although generally more expensive than air, is sometimes used in its place. When used in dryers, it is generally necessary to condense the water vapour from the exhaust gas which is recycled through the dryer. Sometimes the use of inert gas in drying improves the quality of the product enough to offset its cost. When an inert gas system is used, it is not necessary to exclude air entirely in order to prevent dust explosions. With many organic materials, reduction of the oxygen content to 10% is sufficient, although some metals require the oxygen content to be reduced to 2%.

Magnetic or pneumatic separators should be used to remove spark producing foreign objects from the material being processed[26].

10.4.4 Protection against dust explosions inside plant[5, 6, 26, 28]

The design of protective features against dust explosions in cases where their prevention cannot be guaranteed is a specialised field for which experts may need to be consulted. Only rather general points are given here.

Air-classified rod mills and other mechanical equipment where air–dust mixtures are usually present are normally built sufficiently strongly to withstand dust explosions, although the inlet and outlet ductings should be specially checked. Large equipment such as cyclones and bag filters which cannot be built to withstand the pressure of a dust explosion should be located separately from the rest of the plant in an area where access is restricted, e.g. on a roof.

The spread of an explosion in a dust-handling system can generally be restricted by the use of chokes, e.g. a rotary star valve between a hopper and a bin below, or a horizontal screw conveyor with a flight removed and a stationary segment baffle in its place which seals the air space while allowing the powder to pass under it[28]. Plants should in any case be designed to minimise both the volume in which a dust explosion can occur and the quantity of combustible dust present.

The ease of ignition, speed and force of a dust explosion are for most materials critically dependent on the moisture content of the dust, the control of which can do much to reduce the hazard. This is usually required (as in the case of dryers) for economic or other reasons. For many materials which absorb water, such as flour, it is possible to install a relatively cheap control instrument which operates on the wet bulb principle, to regulate the steam or fuel supply to the dryer[29]. This will not only control the moisture content at the desired level but usually pays for itself in a few weeks in fuel savings, as well as reducing the risk of ignition from static electricity.

Spaces inside equipment which may contain dust suspensions are best illuminated through armoured glass panels by external lamps.

Protection in the form of bursting panels, explosion doors, and explosion-suppression systems is often required.

As a rough guide, a relief area of $0.2\,m^2$ per cubic metre of plant volume for organic dusts, and twice this for metal powders, is needed for bursting panels and explosion doors[28]. They should vent to the open air in an area to which personnel access is restricted[6, 26, 28].

Explosion-suppression systems are based on the principle that a dust explosion is not instantaneous, but is preceded by a slower initial rise in pressure lasting for perhaps 10 or 15 ms. A fast-acting pressure-sensitive element forming part of the explosion-suppression system detects this initial pressure rise and transmits an electrical signal to one or more suppressors. Each contains a cartridge filled with compressed inert gas or vaporising liquid or dry powder, and a detonator. The principle is illustrated in Figure 10.3. Suppressors should release sufficient inert gas to rapidly dilutte the oxygen in the dust cloud to below the minimum explosive concentration.

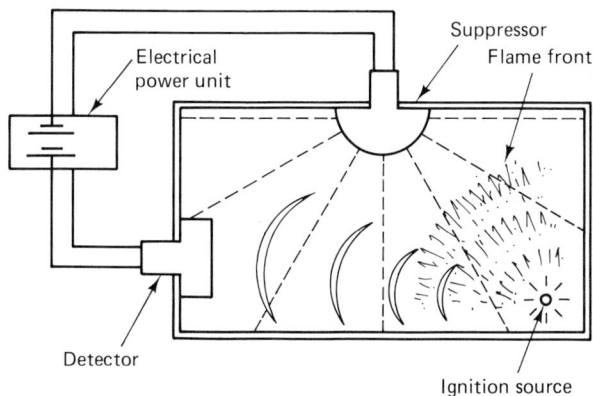

Figure 10.3 Basic explosion-suppression system

The release of the suppressant even in the absence of an explosion will, in many cases, over-pressurise the equipment, so that an explosion relief panel, door or disc is still required. This may be opened or broken by a detonator operated by the same detector that initiates the suppressor.

All joints, inspection doors, slide valves, etc. on plant handling flammable dusts should be dust-tight, with suitable flanged joints and packing.

10.4.5 Dust fires

Lees classifies dust fires into two types, flaming and smouldering fires[6]. The first type can only be sustained if sufficient volatile material is evolved from the dust, or when carbon monoxide formed by incomplete smouldering combustion in the dust burns on its surface. Smouldering combustion requires a minimum depth of dust, which can in some cases be as little as 2 mm. It is sometimes the culmination of a self-heating process [8.7.1], and can proceed very slowly for a long time with no obvious effects such as smoke and smell. The principal hazard of a smouldering dust fire is that of disturbing it and causing a dust explosion. There are also hazards of igniting other combustible materials and of the evolution of toxic combustion products (such as carbon monoxide). The detection and extinction of smouldering dust fires are often difficult, especially when they occur in confined spaces such as a ship's hold. In such cases the atmosphere should be monitored and the necessary breathing apparatus [22.8.6] worn by persons who have to enter the space to deal with the fire. A particular hazard here is subsidence due to hollows resulting from combustion. In storage hoppers, ships' holds and other transport containers regularly used for dusty materials where there is a significant risk of smouldering combustion it is recommended to monitor the temperature at several points inside by the use of thermocouples and temperature indicator/ alarms. Smouldering fires which have been apparently extinguished, e.g. by excluding atmospheric oxygen, must be thoroughly cooled by water or other means to prevent re-ignition when the means of extinction has been removed. Sometimes the charred material left after a smouldering dust fire has been extinguished and cooled is pyrophoric and ignites again on contact with air.

10.5 Liquid and vapour fires and aerial explosions

The types of incident discussed here include pool fires, flash fires, fireballs, vapour cloud explosions and BLEVEs. They are the outcome of primary hazards discussed elsewhere in this book and usually require the services of the fire brigade and the undertaker rather than the safety specialist. Their avoidance, however, calls for hazard control, influencing its forms, pattern and quantitative aspects. These include protective features, the possible damage caused to neighbouring installations, and possible 'domino' effects.

Quantitative treatment and extensive references up to 1979 on all types of incident are given by Lees[6]. Practical aspects are discussed by Vervalin[30] and others. More recent references are given in the following subsections.

The definitions which follow are from an I. Chem. E. booklet[31]:

Pool fire: the combustion of material evaporating from a layer of liquid at the base of the fire;

Flash fire: the combustion of a flammable vapour and air mixture in which flame passes through that mixture at less than sonic velocity such that negligible damaging overpressure is developed;

Fireball: a fire, burning sufficiently rapidly for the burning mass to rise into the air as a cloud or ball;

Vapour cloud explosion (VCE): the preferred term for an explosion in the open air made up of a mixture of a flammable vapour or gas with air;

BLEVE (boiling liquid expanding vapour explosion): used to describe the sudden rupture of a vessel system containing liquefied flammable gas under pressure due to fire impingement; the pressure burst and the flashing of the liquid to vapour creates a blast wave and potential missile damage, and immediate ignition of the expanding fuel–air mixture leads to intense combustion creating a fireball.

This definition of a BLEVE, which restricts the fuel to a liquefied flammable gas, the cause of rupture to an existing fire, and the consequence of ignition of the vapour to a fireball (which presumably excludes a VCE) seems rather narrow. The writer prefers Roberts's definition of a BLEVE as 'the sudden release from containment of any liquid at a temperature where its vapour pressure exceeds atmospheric pressure'[32]. The term BLEVE originated in the USA following several explosions of LPG road and rail tank cars which were involved in fires.

Many serious fires in the process industries have followed massive escapes of liquefied flammable gases, some of which 'flashes' leaving the rest as liquid boiling on the ground, etc. If ignition occurs, any of the following are possible: a low-level flash fire, a large pool fire of the remaining liquid, a fireball, or a vapour cloud explosion. If sufficient fuel is available, all can occur in the same incident.

Whether the vapour burns predominantly as a flash fire, a fireball or explodes depends on the wind, the geometry of the vapour cloud, the degree to which it has mixed with air at the time of ignition, and the degree to which it is confined by neighbouring plant and buildings. Most flammable vapours are heavier than air. Unless released at high velocity, they tend to slump to the ground and spread like a pancake, burning as a flash fire if then ignited. But if vapour is released more rapidly, e.g. from a ruptured LPG tank car, and ignited before it has had time to spread, the fire is likely to be more local and concentrated, rising as a fireball. If vaporising liquid under pressure is released through a nozzle, mixing so rapidly with air as to produce a large volume of flammable mixture within the flammable limits, the conditions are more favourable for an explosion. These considerations are discussed further under BLEVEs [10.5.5].

10.5.1 Pool fires

Common examples are tank fires and fires of spillages of flammable and combustible liquids and low-melting solids on the ground and on water. The burning rate of a pool fire is the linear rate of evaporation of liquid under the fire, typically in the range 0.7–1.0 cm/min. Pool fires may be contained or uncontained. They may involve liquids of high or low flash-point, which may be soluble or insoluble in water. The first priority with an uncontained pool fire is to cut off the flow of fuel to it. For this, remotely operated or automatic isolation valves on bottom pipe connections of tanks and large vessels containing flammable liquids are needed, and the valves themselves need fire protection. A pool of burning liquid with a high flash-point can generally be extinguished with water, but a pool fire of a liquid of low flash-point needs to be blanketed by foam or dry chemical.

Considerations of pool fires affect the layout and design of liquid storage installations and their bunds. They also determine the preparation, slope and drainage of ground surfaces below and adjacent to process plant and storage vessels, fixed fire protection including fire-proofing of structures and vessels, and the firefighting facilities needed.

A serious hazard of fighting hydrocarbon (particularly crude oil) tank fires with water is 'boil-over' or 'slop-over' of burning liquid [6.4.5].

The size of many modern oil storage tanks (over $100 000 \, \text{m}^3$) and supertankers has increased the scale of pool fires enormously. A common cause of tank fires is overfilling, one of several dangers of storage tanks stressed by Kletz[33]. Another is the ignition by lightning [6.3.12] of vapour that has accumulated above the deck of a floating roof tank. Some essential points for the storage of flammable liquids which are stressed in an American loss-information bulletin include[34]:

- Generous spacing between individual tanks and between storage and process areas is imperative.
- Separate bunds for large tanks are essential.
- Piping, valves and flanges within bunds should be kept to a minimum and buried as far as possible, and no fittings liable to fail quickly when exposed to fire should be allowed within bunds.
- No pumps should be installed inside bunds.
- Great care is needed in transferring hot liquid from a burning tank to another tank to avoid vaporising material in the second tank.
- Regular gas checks should be made in the vapour space above floating roof tanks.
- Plastic materials should be avoided for the construction and insulation of large tanks.
- High-level alarms should be installed on tanks filled from pipelines and tied-in to the pump station/office controlling the filling. High-level trips which shut down the filling pump are desirable. Filling operations should always be closely watched.
- A permit system for any maintenance, especially hot work near tanks, must be enforced. Good housekeeping on tank roofs and around tanks must be enforced.
- Adequate fire-water supply and storage capacity, with spare pumps with diesel engine drives and ring mains, should be provided. Adequate road access for fire appliances to all hydrants must be ensured.
- Foam dams with approved delivery equipment should be provided on floating roof tanks. Sub-surface foam injection is recommended for large fixed roof tanks.

- Sufficient foam for use in a major emergency should be kept on the premises.
- Tanks for flammable liquids should be sited on ground lower than that of adjacent occupancy. Roadways lower than bunds, pipe racks, and drainage trenches should be designed to protect those using them from burning or flammable materials.

10.5.2 Flash fires

A flash fire often follows the escape of a heavier-than-air flammable gas or vapour at low velocity and low level into still air. By the time this reaches a source of ignition, the flammable gas or vapour will have spread like a pancake and may have covered a considerable area. The fire flashes through the flammable mixture lying between the more concentrated fuel gas/vapour close to the ground and the air above it, mixing unburnt gas and air as it passes between them. Burning is usually complete in a few seconds.

Flash fires are usually thought of as large-scale phenomena, but the ignition of the vapour of a low-boiling solvent from an open container in a workroom by an electric fire on the floor several metres away is also a flash fire.

A flash fire occurred at about 09.00 hours at a petrochemical works where the writer was working in the late 1950s. The material was cold liquid ethylene which had escaped from a line normally carrying excess ethylene gas at ambient temperature and low pressure from a gas-separation plant to a gasholder. The failure of a control instrument had caused liquid ethylene to enter the line, which contracted and fractured spilling vaporising ethylene onto the ground. This spread over an area of a few thousand square metres, being visible as a low-lying mist. Some day workers who were arriving, mostly on foot and one on a bicycle, had entered the affected area, seeing only a normal morning mist, before some of them realised the danger. One, E. (later President of the Institution of Chemical Engineers), was in the process of placing a warning barrier across the road when the gas reached the works boiler house and ignited. Those on foot who managed to stay on their feet, including E., during the few seconds of the fire were badly burnt on the lower parts of their bodies. The man on a bicycle fell to the ground and was killed by the flames which enveloped him. A small explosion which did little damage occurred near the point of ethylene escape where the gas–air mixture was confined by plant and buildings.

A more serious flash fire occurred in Abadan refinery in about 1941. It resulted from the overloading of a knockout vessel, direct contact condenser and vent stack which formed part of the emergency blowdown system for four thermal reforming units. A cloud of condensing hydrocarbon vapour emerged from the stack, sank to ground level and moved slowly through the adjacent refinery area on a very light breeze. Regular refinery workers who saw it coming moved quickly cross-wind out of its path. A large gang of contract labourers with donkeys and panniers laden with bricks were less fortunate. They fled downwind of the cloud, which overtook and enveloped them before igniting and killing most of them.

The cause of the overloading of the blowdown system was quite unique. This system was provided to dispose of the hot contents of the tubes of any one of the reforming furnaces if a tube failed. Their internal pressure and temperature were about 100 bar g. and 550°C. The operating instructions in the event of a tube failure (which was immediately evident from smoke and flames from the furnace) were to shut the main fuel valve to the furnace, stop the feed pump, open the valve from the tubes to the blowdown system, then wait until the pressure in the furnace tubes fell below the pressure in the steam main (about 15 bar g.) before opening a valve to admit steam to the furnace tubes. After a tube failure in one of the furnaces (a not-uncommon event in wartime, when spare tubes were scarce), the European shift operating supervisor, after opening the blowdown valve, opened the steam valve too soon, so that hot hydrocarbon vapour entered the steam main. All furnaces had oil-fired burners, with atomising steam supplied by the same main that now contained hydrocarbon vapour. Smoke and flames next appeared from the furnace adjacent to the one being blown down. The supervisor, thinking that a tube in this had also split, repeated the same incorrect blowdown procedure. This was followed by flames from the other two furnaces which he then also blew down. This completely overloaded the knockout vessel and condenser, which were only designed to deal at most with two simultaneous emergency furnace blowdowns.

The disaster was clearly caused by misconceptions [3.3.3] on the part of the operating supervisor in a situation which required cool nerves and thinking.

Once a substantial escape of flammable vapour has occurred, one's personal safety demands speedy evacuation, with an eye on the wind for its probable path. If there are adjacent sources of ignition which can be extinguished, or prevented from entering the area covered by vapour, this should be done, but it is usually too late to shut down fired furnaces. If actually caught in a flash fire at ground-level, it seems important to keep on one's feet, or even climb out of the fire if this is possible during the few seconds that it lasts.

Fireballs and VCEs which are discussed next are often preceded by flash fires.

10.5.3 Fireballs

Fireballs usually quickly follow BLEVES involving LPG road and rail tankers. Most fireballs have started with the sudden release and ignition of a flammable vapour and liquid droplets at a low level. Flame around the core of relatively concentrated vapour heats it until it has sufficient buoyancy to rise. The initial fire at low level may cause even more casualties and damage than radiation from the fireball after it has formed and risen.

Some general characteristics of fireballs are:

- Viewed from a distance, they appear as roughly circular balls of fire, rising steadily, with a clear lower boundary. If at the same time there is a pool fire beneath it, this may appear to join the fireball by a 'stalk' of fire and smoke. They can be of considerable size, those quoted by Marshall[35] having diameters ranging from 35 to 300 m.

- They burn very rapidly, combustion being usually complete within 5–20 seconds, depending on the quantity of fuel present. Combustion rates of several tons of fuel per second are usual.
- Heat radiation is very intense, and capable of causing casualties and igniting combustible materials at considerable distances.

A review of fireballs by Marshall[35] includes a number of references and a list of 11 major fireballs between 1970 and 1979, all but one in the USA. One involved butane, five LPG (presumably mixed butane/propane), three propane and two vinyl chloride monomer (VCM). The quantities of these spilled materials ranged from 2 to 435 t.

Besides these large fireballs, small ones have been produced experimentally by bursting portable LPG containers in the open air[36]. Whether a fireball is formed depends, among other things, on the speed with which the flammable vapour is released and ignited. Fireballs often occur when the roof of a large building whose contents are burning collapses, releasing a quantity of flammable vapour formed from combustible materials in the building by the heat[37] (see Figure 10.4).

Figure 10.4 Fireball following collapse of roof of burning building (Warrington Guardian)

One of the worst fireball incidents was the liquefied propylene tank car disaster at the crowded Spanish holiday camp of San Carlos de la Rapita on 11 July 1978[6, 35]. This followed a BLEVE and resulted in the deaths of 210 lightly clad holiday-makers. Many of the casualties resulted from a shower of propylene droplets accompanied by flash and pool fires. The effects of radiation from fireballs has varied widely, some producing surprisingly few radiation casualties. Marshall cites firemen in uniform only 30 m from bursting LPG rail tankers who survived, while a fireball containing about 10 t of propane killed one man at 90 m and severely injured three others at 140 m.

While the conditions for a fireball are similar to those for a VCE [10.5.4], the most appropriate personal protection is different. With a fireball which has no blast effects, people are better protected from its radiation if they are in a building or under cover. The duration of a fireball, while short, is much longer than that of an explosion, giving those exposed to one a few seconds to take cover. With a VCE, on the other hand, where roofs and entire buildings are liable to collapse (unless they are of blast-proof construction), people are probably better off in the open (although they may be struck by missiles), since the human body is better able to survive nearby explosions than most buildings. Fortunately neither fireballs nor VCEs usually occur without a degree of warning in which to escape or take cover.

10.5.4 Vapour cloud explosions (VCEs)

Two or more disastrous VCEs, each causing property damage of over US$10 million, now usually occur every year world-wide[31, 38-41]. Appendix D gives the numbers of VCEs of different severities with the types of installation where they occurred up to 1983, with comments. Appendix E gives a list of property damage losses in excess of US$10 million (1988 values) in the oil/chemical industries for the 30-year period 1958–1987, with the plants where they occurred and the materials involved. At least half of these were caused by vapour cloud explosions and many of the others by BLEVEs with fireballs. Two VCEs, the Flixborough disaster of 1974 [4] and the Pernis disaster of 1968 [5.1] were discussed earlier.

Until the advent of large-scale oil and petrochemical plant in the 1940s it was thought that a flammable air-vapour/gas mixture could only explode if it was confined (e.g. in a tank, building, container or the cylinder of a car engine). This belief appeared to be shattered by several large explosions outside oil and petrochemical plants, for which the term 'unconfined vapour cloud explosion' was coined. Following more recent studies, many experts now think that the presence of adjacent plant structures, columns and buildings which provide obstacles to air movement are needed for flame velocities of 100 m/s which are typical of an explosion. At least one VCE, that at Port Hudson, Missouri, on 9 December 1970, when a pipeline fracture created a cloud of propane in an open area with a low building density, casts doubt on this view[6, 32]. The adjective 'unconfined' has, however, now been generally dropped from the term.

We are more concerned here with the prevention of VCEs than with theories about their propagation. In a plant where a VCE is possible, the damage it could cause has to be taken into account in its layout, design and insurance [12.4], in the protection provided, and in emergency planning [20.3].

The following features characterise most recorded VCEs:

1. They were explosive deflagrations and not detonations [9.1], and developed from fires.
2. They were usually preceded by the release of a flammable liquid or liquefied gas which had been contained under pressure at a temperature considerably higher than its atmospheric boiling point. Massive releases of natural gas at pressures of 60 bar g. from fractured 24-inch pipes and larger have also caused them. A few VCEs have followed smaller releases of ethylene from high-pressure polymerisation plants.
3. Most VCEs have involved the presence of at least 5 t and usually 10 t of flammable vapour in the cloud.
4. The materials most commonly involved have been light hydrocarbons. Other materials occasionally involved have been vinyl chloride, ethylene oxide, hydrogen and isopropanol.
5. At a distance the damage follows a pattern which can be modelled on the explosion of an equivalent amount of TNT at the centre of the VCE, although there may be greater directional effects caused by the asymmetry of the cloud[41]. Nearer the centre of the explosion, the damage, while far less than expected from the TNT model, is nevertheless usually sufficient to cause 100% property damage.

Regarding (2), the release of a vaporising liquid at high velocity into the open air produces a far more extensive and persistent explosive cloud than the release of gas, vapour or liquid only. A flammable gas or vapour issuing at high velocity from a jet usually entrains enough air to reduce its concentration to below the lower flammability limit [10.2.1] before it has travelled far, while a jet of flammable liquid whose temperature is below its boiling point seldom produces enough vapour for a VCE[42]. Of course, if a very large and rapid release of flammable gas takes place and the wind velocity is low, there just may not be enough air round the point of release to dilute it to a safe concentration. Also if the liquid released is much colder than the ground (e.g. refrigerated LNG or LPG), it may evaporate fast enough to form a large explosive cloud with air, although a fire is more probable.

Apart from the large scale and appropriate conditions of many hydrocarbon processes, there are two other reasons for the preponderance of light hydrocarbons in VCEs [12.2.1]. One is their higher heat of combustion compared with compounds containing chlorine, oxygen, nitrogen and other elements. The other is their low heat of vaporisation compared to their specific heat. Thus when a liquid hydrocarbon contained under pressure at a temperature, say, 50°C above its atmospheric boiling point is released a higher proportion of it will vaporise than would in the case of a flammable compound containing oxygen, such as ethanol.

Summary of VCE sources
The following types of installation, etc. should be considered as possible sources of VCEs.

- Storage and transport of 10 t or more of liquefied flammable gases, both under pressure and refrigerated;
- Large-scale pipeline transport of flammable gases in liquid form;
- Large-scale pipeline transport of flammable gases such as natural gas and ethylene at pressures above 35 bar g.;
- Tanks containing crude oils and other oils having flash-points below 60°C if fitted with steam coils or other means of heating;
- Plants containing substantial quantities of a flammable gas at pressures above 35 bar g.;
- Plants containing substantial quantities of liquefied flammable gases under any conditions;
- Plants containing substantial quantities of liquid hydrocarbons with five to nine carbon atoms per molecule under pressure at temperatures above their atmospheric boiling points;
- Plants containing substantial quantities of flammable non-hydrocarbon liquids with flash-points below 40°C under pressure, at temperatures above their atmospheric boiling points.

10.5.5 Boiling liquid expanding vapour explosions (BLEVEs)

The abbreviation BLEVE rhymes with 'heavy' and is used both as a noun and a verb ('the rail tanker BLEVEd'). The cause of most BLEVEs was the failure of a vessel containing a flammable liquid under pressure at a temperature above its atmospheric boiling point. Sometimes the failure is confined to a small split, but once one develops, the stresses on adjacent parts of the shell increase greatly, and the vessel frequently bursts. Since the circumferential stresses in the shell of a cylindrical vessel are usually twice the longitudinal ones, such vessels tend to fail by splitting on or adjacent to a circumferential seam. This is accompanied by a loud bang (the BLEVE). Most of the liquid contents then escape, partly flash-vaporising and forming a dense cloud of vapour and liquid droplets.

If the vessel contains LPG and failure is the result of fire, the cloud will ignite immediately, rising and burning rapidly as a fireball. If the failure is due to causes other than fire, and ignition is delayed until the edge of the cloud reaches a furnace or engine, etc., the cloud may in the meantime have mixed with sufficient air to cause an explosion (VCE), or to have slumped to the ground and spread to produce a flash fire. There are no hard and fast dividing lines between these three possibilities, which describe rather the dominant nature of the fire/explosion.

When the vessel splits, the remaining liquid in the ends of the two halves vaporises rapidly until its temperature has dropped to its atmospheric boiling point. Unless the two halves of the vessel are very securely fixed, one or both of them will be propelled violently by jet action. The initial thrust is readily calculated as the (gauge) pressure in the vessel at rupture multiplied by its cross section at the split. One or both parts of a horizontally mounted cylindrical pressure vessel (for storage or transport) will be propelled horizontally, while the top half of a sphere which has split on a horizontal seam, or the upper part of a vertically mounted cylindrical vessel, will shoot upwards.

Large parts of vessels may be thrown considerable distances when a

BLEVE occurs. Part of a 2 m diameter and 13 m long 'bullet' tank was reported to have been thrown 1200 m by the LPG incident at Mexico City in 1984 [C] before it destroyed two houses[31]. Most of a 20-ton vertical absorber column, 16 m high and 2.6 m dia., was propelled over 1000 m by propane at 14 bar g. when it struck and toppled a high-voltage transmission cable at Romeoville, Illinois, in July 1984[43].

The explosion energy of a BLEVE has two sources[32, 41]:

- The pressure of the vapour in the container, which is immediately available when the container ruptures, producing a blast wave [10.5];
- The vapour created by flashing of the superheated liquid left in the ends of the container.

The commonest cause of BLEVEs is fire, usually resulting from smaller leaks from the vessel which later BLEVEd. Most pressure vessels are fitted with pressure relief valves which open well before the stresses in its walls reach a dangerous level [15.2] at its operating temperature. Flame impingement on the upper part of the vessel shell which is above the level of liquid inside it may, however, raise its temperature sufficiently to reduce its strength to the point where failure occurs. This has happened in most major BLEVEs, including a large LPG storage sphere at Fézin in France on 4 January 1966 (Figure 10.5), which caused 18 deaths, 81 injuries and extensive damage[5, 43].

Five VCEs are known to have occurred up to 1977 during pressure storage of C_3 and C_4 hydrocarbons in spheres and bullets, and seven in

Figure 10.5 The Feyzin explosion (United Press International)

transport of these materials in road and rail tankers[40] [D]. Most of these were in the USA. Four of these transport incidents occurred as a result of rail tank car collisions, and all but one of these happened in shunting or 'humping' operations in rail yards. Some commentators have blamed the frequency of these incidents on the poor state of maintenance of US railways and rolling stock. While some of these incidents were better described as fireballs than VCEs, most and probably all of them included a BLEVE. Several BLEVEs occurred in the major loss incidents from 1978 to 1987 listed in Appendix E, two of which are referred to in this section.

To avoid a BLEVE, it is vital to cool the outside of pressure vessels containing flammable liquids and liquefied flammable gases if they are exposed to fire. Permanent water sprays for storage vessels of this kind are now generally mandatory. This is not, of course, possible with road and rail tankers, and whether these can be water cooled when involved in a fire usually depends on how quickly the local fire service can respond.

Accidental fires are not the only cause of BLEVEs. Reports indicate that the LPG road tanker which failed at the Spanish holiday camp may have been overfilled with liquid propylene, which expanded from the heat of the sun[32]. A BLEVE listed in Appendix D occurred during the transport of acrolein, a highly toxic as well as flammable liquid with a boiling point of 52°C. This resulted from its spontaneous exothermic polymerisation [8.6.4]. A similar BLEVE was narrowly avoided in the UK, when a low-boiling monomer being carried by road tanker started to polymerise exothermically while *en route* via the Dartford tunnel to Middlesbrough[37]. Emergency services were alerted and it was fortunately possible to cool the contents sufficiently by spraying the tanker with water to prevent a BLEVE.

BLEVEs are not always confined to flammable materials, and when they are, they are not always followed by fires. Other serious consequences may follow, including:

- Inhalation of toxic gases such as chlorine, acrolein;
- Frostbite from being showered by droplets of propane or flakes of solid carbon dioxide;
- Asphyxiation by the last two and other heavier than air gases which may displace atmospheric oxygen from working areas;
- Scalding by hot steam from boiler explosions which by Roberts's definition are also BLEVEs.

References

1. Schultz, N., *Fire and Flammability Handbook*, Van Nostrand Reinhold, New York (1986)
2. NFPA Standard No. 30, *Flammable and Combustible Liquids Code*, National Fire Protection Association, Quincy, Mass. 02269 (1977)
3. NFPA Standard No. 704M, *Identification of the Fire Hazards of Materials*, National Fire Protection Association, Quincy, Mass. 02269 (1977)
4. S.I.1972, No.917 as modified by S.I.1978, No.209, *The Highly Flammable Liquids and Liquefied Petroleum Gases Regulations 1972*, HMSO, London
5. Palmer, K. N., *Dust Explosions and Fires*, Chapman and Hall, London (1973)

6. Lees, F. P., *Loss Prevention in the Process Industries*, Butterworths, London (1980)
7. Wood, W. S., 'Safe handling of flammable and combustible materials', in *Safety and Accident Prevention in Chemical Operations*, edited by Fawcett, H. H. and Wood, W. S., 2nd edn, Wiley-Interscience, New York (1982)
8. Zabetakis, M. G., *Flammability characteristics of gases and vapours*, Bulletin 627, Bureau of Mines, Washington, DC (1965)
9. Miller, S. A., *Acetylene – Its Properties, Manufacture and Use*, Benn, London (1964)
10. Chemical Industry Safety and Health Council, *Code of Practice for Chemicals with Major Hazards – Ethylene Oxide*, London (1975)
11. ASTM D 56, *Standard Method of Test for Flash Point by Tag Closed Cup Tester*, American Society for Testing and Materials, Philadelphia, Pa (1979)
12. ASTM D 93, *Standard Method of Test for Flash Point by Pensky-Martens Closed Tester*, American Society for Testing and Materials, Philadelphia, Pa (1979)
13. ASTM D 92, *Standard Method of Test for Flash Points by Cleveland Open Cup*, American Society for Testing and Materials, Philadelphia, Pa (1979)
14. BS 3900: Part 9: 1986, *Determination of flashpoint (closed cup equilibrium method)*, British Standards Institution, Milton Keynes
15. BS 4056: 1966, *Method of test for ignition temperature of gases and vapours*, British Standards Institution, Milton Keynes
16. NFPA Standard No. 325M, *Fire Hazard Properties of Flammable Liquids, Gases and Volatile Solids*, National Fire Protection Association, Quincy, Mass. (1969)
17. Coffee, R. D., 'Cool flames' in *Safety and Accident Prevention in Chemical Operations*, edited by Fawcett, H. H. and Wood, W. S., 2nd edn, Wiley-Interscience, New York (1982)
18. BS 5501, *Electrical apparatus for potentially explosive atmospheres*, British Standards Institution, Milton Keynes
19. Redding, R. J., *Intrinsic Safety – The safe use of electronics in hazardous locations*, McGraw-Hill, Maidenhead (1971)
20. Fordham Cooper, W., *Electrical Safety Engineering*, 2nd edn, Butterworths, London (1986)
21. RoSPA-ICI Ltd, *Electrical installations in flammable atmospheres, Engineering Codes and Regulations*, RoSPA, Birmingham (1972)
22. Haase, H., *Electrostatic Hazards, their Evaluation and Control*, Verlag Chemie, New York (1977)
23. S.I. 1984, No. 1902, *The Control of Industrial Major Accident Hazards Regulations 1984 (CIMAH)*, HMSO, London
24. BS 2955: 1958, *Glossary of terms related to powders*
25. National Safety Council, *Accident Prevention Manual for Industrial Operations*, 7th edn, NSC, Chicago, 1131 (1974)
26. NFPA Standard No. 68, *Explosion Venting Guide*, National Fire Protection Association, Quincy, Mass. (1974)
27. S.R. & O. 1946, No. 2107, *The Magnesium (Grinding of Castings and other Articles) Special Regulations 1946*
28. HSE Booklet 22, *Dust Explosions in Factories*, HMSO, London (1970)
29. King, R. W. and Magid, J., *Industrial Hazard and Safety Handbook*, Butterworths, London (1980)
30. Vervalin, C. H., *Fire Protection Manual for Hydrocarbon Processing Plants*, 3rd edn, Gulf Publishing, Houston, Texas (Volume 1, 1985, Volume 2, 1982)
31. *Nomenclature for Hazard and Risk Assessment in the Process Industries*, The Institution of Chemical Engineers, Rugby (1985)
32. Roberts, A. F., 'Vapour Cloud Explosions and BLEVES', paper given at Major Hazards Summer School, Cambridge, organised by IBC Technical Services Ltd, London (1986)
33. Kletz, T. A., *What Went Wrong?*, Gulf Publishing, Houston, Texas (1985)
34. Oil Insurance Association, *Loss Information Bulletin No 400-1, Tank Fires*, Industrial Risk Insurers, Hartford, Connecticut (1974)

35. Marshall, V. C., 'Dust explosions and fire balls', paper given at Major Hazards Summer School, Cambridge, organised by IBC Technical Services Ltd, London (1986)
36. Roberts, A. F., 'The effect of conditions prior to loss of containment on fireball behaviour', *I. Chem. E. Symposium Series No. 71*, Rugby (1982)
37. *Report by HM Chief Inspector of Factories 1985*, HMSO, London
38. Gugan, K., *Unconfined Vapour Cloud Explosions*, I. Chem. E./George Godwin, London (1979)
39. Strethlow, R. A., 'Unconfined vapour cloud explosions – an overview', paper in *Fourteenth Symposium on Combustion*, Combustion Institute, Pittsburgh, Pa (1973)
40. Davenport, J. A., 'A study of vapour cloud incidents', paper in *AIChE Loss Prevention Symposium*, Houston, Texas (March 1977)
41. Lewis, D., 'The blast, radiation and interactions with people, control rooms and surroundings', paper given at Major Hazards Summer School, Cambridge, organised by IBC Technical Services Ltd, London (1986)
42. Kletz, T. A., 'Some of the wider questions raised by Flixborough', paper given at Nottingham symposium, *The Technical Lessons of Flixborough*, The Institution of Chemical Engineers, Rugby (1975)
43. Garrison, W. G., *100 Large Losses – A thirty-year review of property damage losses in the hydrocarbon-chemical industries*, 11th edn, Marsh and McLennan Protection Consultants, 222 South Riverside Plaza, Chicago, Illinois 60606 (1988)

Chapter 11

Corrosion hazards and control

The corrosion of metals, particularly steel and ferrous alloys, is of great economic importance. Its cost in the USA and Europe is estimated at 4% of gross national product[1]. Corrosion proceeds slowly, and is usually more the concern of engineers, chemists and metallurgists than of safety professionals, yet it has caused catastrophic failures with heavy loss of life.

This chapter deals mainly with metal corrosion, followed by a few notes on the corrosion and deterioration of non-metallic constructional materials. The action of corrosive chemicals on the human body [7.2.1] is not considered here.

For many corrosive environments, several alternate metals might be considered, ranging from very expensive ones such as tantalum to the cheapest, unprotected mild steel. Notes on the costs of corrosion resistant metals are given in 11.6.4. Occasionally one encounters a process for which no satisfactory materials of construction have been found, or whose operation is restricted to conditions imposed by limitations of the materials available.

Corrosion tends to be concentrated in certain zones of the equipment where conditions of temperature and concentration favour it.

The harmful results of corrosion include:

1. Mechanical explosions of boilers which were once common. Now, with better understanding of corrosion causes, improved methods of control, and statutory inspection and maintenance, they are infrequent[2].
2. Sudden failures of pipes, pressure vessels and other equipment containing toxic or flammable materials. An example was the rupture of an old pipeline carrying liquid propane near Port Hudson, Missouri, in 1970 which caused a massive explosion[3].
3. Small leaks of dangerous fluids past closed valves or through shaft seals, joints, pinholes in pipes, etc. The Bhopal disaster of 1985 may have been started by a small leak of water past a closed valve[4] [5.4].
4. Failure of instrument components, e.g. thermocouples, pressure and level gauges and transmitters, upon which the safe operation of a plant depends. The corrosion failure of a pressure switch on the suction of a vinyl chloride compressor resulted in a calamitous explosion[5].
5. Production of solid corrosion products which restrict flow, prevent valves from closing or reduce heat transfer. A fire which destroyed

three large LPG storage tanks at Ras Tanura in 1962 was attributed to corrosion products which prevented a valve from closing[5].

6. Production of corrosion products which contaminate process materials or have undesirable catalytic effects on them. Lead [1.2.1] is now prohibited in food and beverage plant[6], while copper deteriorates soaps, margarine and synthetic rubbers.

7. Production of explosive, pyrophoric and oxygen-consuming corrosion products. Copper acetylide, a sensitive explosive, is formed on copper tubes and brass fittings in contact with moist gases containing acetylene. Pyrophoric iron sulphide, which also deoxygenates air without combustion, is formed inside steel tanks used for sour crude oils.

8. Production of flammable and toxic gases (such as hydrogen, hydrogen sulphide, phosphine and arsine) inside vessels and closed spaces as a result of the acid corrosion of impure metals.

9. Failure of foundations and underground structures caused by spillages of acids and corrosive chemicals.

10. Failure of rubber and plastic hoses, gaskets and gland seals through chemical or solvent action.

Fatigue, erosion, cavitation, and high stresses are also common causes of metal failure. The combined effect of chemical and physical influences acting together is often far greater than the sum of their individual effects. Several such combinations are recognised hazards, e.g. stress corrosion, fatigue corrosion, erosion–corrosion and electro-chemical corrosion.

A survey of 685 service failures in process equipment over the period 1968–1971 showed that 55% were due to corrosion (Table 11.1) and 45% to mechanical causes[7].

Table 11.1 Causes of corrosion failure of process equipment[7]

Cause	%
Cavitation★	0.5
Cold wall	0.7
Cracking, corrosion fatigue★	2.7
Cracking, stress corrosion★	23.8
Crevice	1.6
Demetallification	1.1
End grain	0.7
Erosion–corrosion★	6.9
Fretting★	0.5
Galvanic	0.7
General	27.6
Graphitisation	0.2
High temperature	2.4
Hot wall	0.2
Hydrogen blistering	0.2
Hydrogen embrittlement	0.7
Hydrogen grooving	0.5
Intergranular	10.1
Pitting	14.4
Weld corrosion	4.5

Headings marked ★ in Table 11.1 are combinations of chemical and mechanical effects. Galvanic corrosion lies at the root of many of the causes listed.

Corrosion reduces the reliability and integrity of plant, which is rarely fully restored to its original condition after maintenance.

While there is often a complete solution to a corrosion problem, this may not be acceptable economically. The capital investment in a process plant is often depreciated over 10 years. Any benefit gained by operating it for longer is an unplanned bonus, which helps to pay for those closed down prematurely. Hazard control in process plant can, without undue cynicism, be compared to a sailing race, the prize going to the skipper who sails just close enough to the wind. This analogy is particularly apt where corrosion is concerned.

The monitoring of plant for corrosion, both on a continuous basis using automatic recording instruments and by qualified inspectors during plant shutdowns, are dealt with under inspection and maintenance [17]. These are essential to prevent the plant integrity being gradually 'eaten away' by corrosion.

11.1 Acceptable corrosion rates

This is an important question for the designer. Corrosion of metals is reported in the case of uniform attack as mm penetration per year (mm/yr), mils per year (1 mil = 0.001 inch), and grams per square metre per day (g/m² day); g/m² day is multiplied by (0.365/density) to give mm/yr. Corrosion is unfortunately often non-uniform, as in pitting corrosion, crevice corrosion and inter-granular corrosion. As well as the overall rate, the designer needs to know what type of attack to expect, and to make additional allowance where this is non-uniform. Acceptable rates depend on the functions of the items concerned. For mirrors, standard weights and electrical resistances, virtually no corrosion can be allowed. For a pressure vessel requiring a wall thickness of 25 mm in the absence of corrosion, a corrosion allowance of 3 mm would provide a working life of 10 years with a uniform rate of corrosion of 0.3 mm/yr. If such pressure vessels were connected by stainless steel bellows with a wall thickness of 1 mm, of which only 0.1 mm could be allowed for wastage, their allowable corrosion rate would be only 0.01 mm/yr.

Factors other than the integrity of the equipment may limit the allowable corrosion rate. Where the colour of the product is important (e.g. thermoplastics) or metal contamination may be harmful (e.g. foodstuffs), it may be necessary to choose a stainless steel with a very low corrosion rate even though that expected from mild steel would otherwise be acceptable.

Uhlig gives the following classification of allowable (uniform) corrosion rates for various duties[1]:

1. <0.15 mm/yr – for critical duties, e.g. valve seats, pump shafts, impellers and springs;
2. 0.15 to 1.5 mm/yr – for most tanks, piping, valve bodies and bolt heads.;
3. >1.5 mm/yr – seldom satisfactory.

Corrosion rates feature in the Dow and Mond Hazard Indices [12]. Rates of less than 0.1 mm/yr make no contribution to the indices, but rates above 1 mm/yr make a serious contribution, especially if they are combined with erosion effects and/or local pitting. Rates of 0.1 mm/yr are not, however, always acceptable, e.g. for thin linings, thin-walled equipment items and heat transfer surfaces where small leaks could have serious consequences.

Pitting can be allowed for if the degree (shallow or deep) and the pitting factor are known. This is the ratio of the deepest metal penetration to the average penetration as determined by weight loss of the specimen. Other forms of non-uniform corrosion are difficult to allow for, and must as far as possible be prevented by the choice of metal, the use of suitable coatings, modifying the corrosive environment or the design of the item, or a combination of these means. In cases where non-uniform corrosion cannot be eliminated, methods of monitoring must be employed which enable it to be detected at an early stage.

11.2 Galvanic corrosion

Most forms of metal corrosion in aqueous media can be explained by the theory of galvanic corrosion (named after Luigi Galvani), which also suggests ways of controlling it. When two dissimilar metals which are in electrical contact are immersed in an aqueous solution, atoms of the more active or negative one (the anode, see Table 8.2) tend to lose electrons and dissolve, forming positive ions. The electrons flow as an electrical current to the less active or more positive metal (the cathode) which attracts dissolved ions of any more positive metals, as well as hydrogen ions. These ions are discharged on reaching the cathode if their metal (hydrogen included) is more electropositive than the anode metal. Metal ions on being discharged plate out as metal on the surface of the cathode, whereas hydrogen ions may form bubbles of hydrogen gas after reaching it. In the presence of dissolved oxygen, hydrogen ions combine with it to form water as they are discharged.

If the cathode and anode are separated and connected through an external circuit, positive current (a fictional concept) is said to flow from the cathode through the circuit to the anode (whereas in fact electrons flow in the reverse direction).

The two metals, the electrical connection between them, and the solution form a galvanic cell. A cell with a zinc anode and iron cathode at which hydrogen ions are being converted to water is shown in Figure 11.1.

The rate at which the anode dissolves is proportional to the current flowing. The voltage of the cell is at a maximum when no current flows, and decreases as the current increases. The (maximum) 'open circuit' voltage of a cell is greater, the wider the separation between the two metals in the electromotive series. The behaviour of galvanic cells also depends on many other factors, including the acidity or alkalinity of the solution (its pH), the concentrations of various ions in the solution, the concentrations of dissolved oxygen and other gases in the solution, the formation of insoluble corrosion products, and the presence of 'complexing agents' which combine with various metal ions.

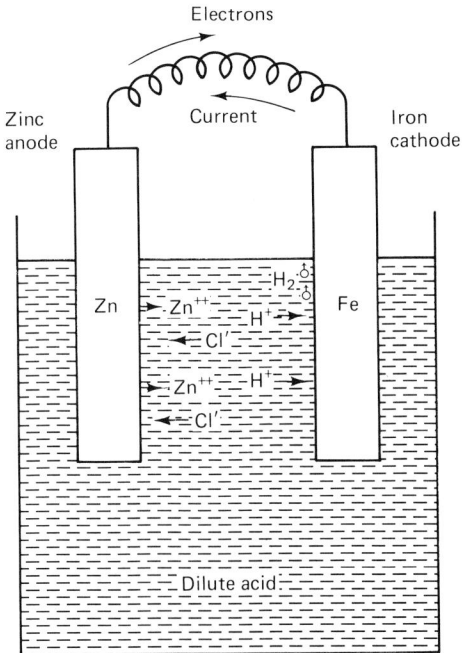

Figure 11.1 A simple galvanic cell

While corrosion of the above type which is dependent on the presence of hydrogen ions proceeds fastest in acid solutions, i.e. where their concentration is high, some metals, such as aluminium, zinc and chromium, whose oxides can behave as acids as well as bases, react with hydroxyl ions and dissolve to form negative ions (anions) in alkaline solutions.

Examples of cells with two dissimilar metals, where one dissolves at the anode to form positive ions, include copper and iron pipes connected together in domestic water systems, and bronze tubes in a steel shell in a heat exchanger. In such cases it is always the more negative metal which corrodes. This may be reduced by inserting some electrically insulating material between the two metals at the point of contact.

With galvanised steel, the zinc corrodes preferentially. Similarly, 'sacrificial' anodes of magnesium or zinc may be attached to steel items exposed to an aqueous environment to protect them. Taking this a stage further, a separate electrode of high corrosion resistance (e.g. a 'non-consumable' platinised-titanium anode) may be placed in the same aqueous medium as the metal (usually steel) item which it is intended to protect, and an external voltage applied between the steel and the anode to oppose the natural voltage of the cell, thus preventing a current from flowing. This is the principle of 'cathodic protection' which is widely applied, e.g. to ships' hulls, tanks and pipelines.

Dissimilar metal cells are also found on a micro-scale as the result of electrically conducting impurities on a metal surface. Galvanic cells are also formed between cold-worked metal in contact with annealed areas of the same metal. Another class of galvanic cell in which the anode and cathode are of identical materials is the concentration cell, of which there are two types. The first is the salt concentration cell, which is found where there is a difference in the concentration of dissolved salt containing ions of the metal exposed, at different parts of the exposed surface. Here the metal in contact with the lower concentration of ions is the anode and dissolves preferentially.

The second and more important type of concentration cell is the differential aeration cell. An example is two iron electrodes in dilute sodium chloride solution, the electrolyte round one being well aerated, that round the other being unaerated. The difference in oxygen concentration causes a current to flow from the cathode, which is well aerated, to the less aerated anode where corrosion occurs. The discharge of hydrogen ions at the cathode leaves the solution there short of them and increases the concentration of hydroxyl ions, making the solution slightly alkaline. The hydroxyl ions on meeting ferrous ions formed at the anode undergo a secondary reaction, precipitating ferrous hydroxide whose saturated solution has a pH of 9.5. This settles over the anodic area and reacts with any free oxygen present forming rust (hydrated ferric oxide) and creating a stagnant oxygen-depleted zone below it. Iron thus continues to dissolve in the oxygen-depleted zone below the deposit, while hydrogen ions continue to be discharged and form water in the more aerated areas.

Differential aeration cells cause pitting corrosion of steel, stainless steels, aluminium, nickel and other so-called passive metals, especially when they are exposed to salt water (Figure 11.2). Other typical forms of

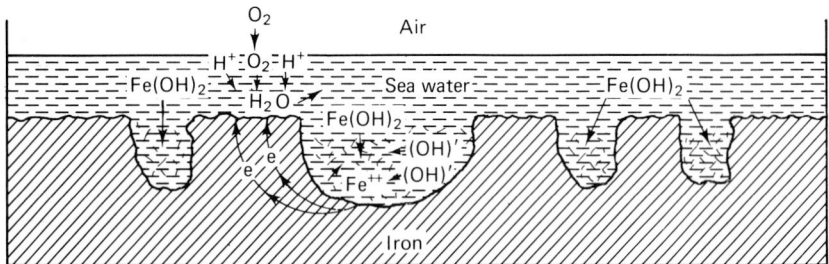

Figure 11.2 Pitting corrosion caused by differential aeration

corrosion caused by differential aeration are water-line corrosion which occurs on partly immersed metal surfaces, just below the water-line (Figure 11.3), and crevice corrosion which causes damage in threaded connections and other crevices where the oxygen concentration is lower than elsewhere.

Temperature differences between different parts of the same aqueous medium which is contained within a tube or shell of the same metal can cause e.m.f. differences between hot and cold parts, thus setting up a thermal type of galvanic corrosion cell (found in the corrosion of heat exchangers and boilers).

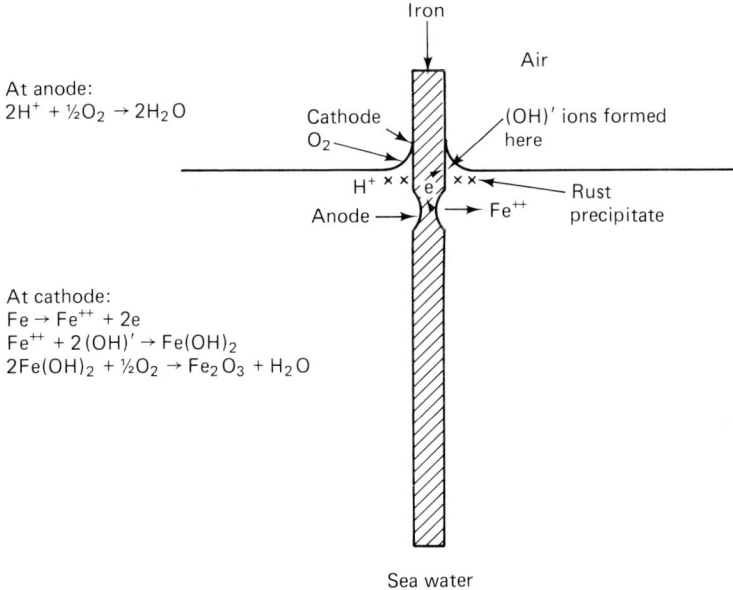

At anode:
$2H^+ + \frac{1}{2}O_2 \rightarrow 2H_2O$

Cathode
O_2

$(OH)'$ ions formed
here

Iron

Air

H^+ × × e × ×

Anode →

Fe^{++}

Rust
precipitate

At cathode:
$Fe \rightarrow Fe^{++} + 2e$
$Fe^{++} + 2(OH)' \rightarrow Fe(OH)_2$
$2Fe(OH)_2 + \frac{1}{2}O_2 \rightarrow Fe_2O_3 + H_2O$

Sea water

Figure 11.3 Water-line corrosion

11.3 Corrosion of iron and steel in aqueous media

The corrosion of iron and steel in aqueous media takes place as a network of short-circuited galvanic cells on the metal surface. The solution of iron to form ferrous ions and electrons at the anode is rapid in most media provided the electrons can escape. In deaerated solutions the cathodic reaction is:

$$e + H^+ \rightarrow \frac{1}{2}H_2$$

This is fast in acids, but very slow in neutral or alkaline solutions. In the presence of air dissolved oxygen reacts with hydrogen atoms adsorbed on the iron surface:

$$2H + \frac{1}{2}O_2 \rightarrow H_2O$$

This proceeds as fast as oxygen reaches the metal surface (often hindered by a barrier of ferrous hydroxide).

Two kinds of variables affect the corrosion of iron and steel in aqueous media. First, there are the environmental variables such as pH, oxygen partial pressure, temperature, liquid velocity relative to the metal, and the concentrations of dissolved substances in the water. Second, there are factors inherent in the object exposed to the medium, its composition and structure, its previous history and heat treatment, surface condition, any bonding to other metals, and mechanical stresses. The whole subject is very complex. Several variables – oxygen partial pressure, temperature and salt concentration – first increase corrosion rates to a peak when they are raised from a low to an intermediate level. A further increase in the level of the variable then causes a marked decline in corrosion rates. The reasons for such behaviour will not be discussed here.

Despite this complexity, there is a wide range of conditions, both of the aqueous medium and of the exposed metal, over which the steady corrosion rates of unprotected iron and steel settle after a few days' exposure within the fairly narrow range of 0.07 to 0.11 mm/yr. This applies to relatively still water which is non-scaling with regard to calcium and magnesium carbonates, with a pH range of 4 to 10 and a temperature range of 10–25°C, in contact with air. It applies to most types of iron and low-alloy steels. Superimposed on this background are the trends caused by the common variables which are discussed next. The effects of bacteria, hydrogen and other variables are discussed in 11.4.

11.3.1 Air and oxygen

In completely deaerated water the corrosion rate is negligible (less than 0.005 mm/yr). This increases as the partial pressure of dissolved oxygen in the water increases up to about half an atmosphere. It then decreases sharply as the oxygen partial pressure rises further. This does not occur in salt water where corrosion rate increases steadily with oxygen partial pressure.

11.3.2 Salt concentration (NaCl)

The corrosion rate roughly doubles on passing from fresh water to sea-water (*ca.* 3%NaCl), but falls back to its original value as the salt content is increased to 10%. Dissolved salt increases the pitting tendency.

11.3.3 Temperature

In a system open to the air, the corrosion rate roughly doubles as the temperature increases from 15°C to 80°C. It then falls to well below its original value as the water reaches its boiling point, and dissolved oxygen is expelled.

11.3.4 Water velocity

On passing from still water to about 0.3 m/s in an open system, the corrosion rate may increase by a factor of two or three, but a further increase in velocity in the absence of salt reduces the corrosion rate to below its original value, until quite high velocities are reached. Erosion and cavitation then begin to play a part, and penetration increases, sometimes spectacularly.

11.3.5 Calcium and magnesium carbonates

Many natural waters contain calcium and magnesium carbonates kept in solution by dissolved carbon dioxide. When this is lost, e.g. by exposure to air, heat, or the addition of alkalis, a protective scale layer forms on exposed metal surfaces which greatly reduces the corrosion rate. Waters are often deliberately treated so that they have a slight scaling tendency, but unless this is carefully controlled, there is a risk that a thick layer will form leading to restrictions in flow, etc.

11.3.6 pH and acids

In the absence of weak acids (e.g. acetic, carbonic), the corrosion rate is practically unaffected by pH over the range 4–10, but rises rapidly at lower pH values, when surface deposits of corrosion products dissolve giving greater oxygen access to the surface. Hydrogen is also formed at low pH values. Corrosion rates fall at pH values above 10 but may rise again at pH values above 13. Weak acids in which solid corrosion products are soluble increase the lower end of the pH range from 4 to 5 or 6. The preferred pH for corrosion control is usually about 8.

11.3.7 Metal coupling

Coupling of iron or steel to a more noble metal increases its rate of corrosion within a critical distance from the metal junction. This distance may be as little as 5 mm in soft drinking water and as much as 500 mm in sea-water. Coupling to a less noble metal similarly reduces corrosion.

11.3.8 Varieties of iron and steel

As the corrosion rate of iron or steel in natural waters, including sea-water over the pH range 4–10, is controlled by the rate of oxygen diffusion to the surface it is practically unaffected by the type of iron or steel. This should therefore be chosen on other grounds, e.g. price and mechanical properties. In the acid (pH <4) and extreme alkaline (pH >13.5) ranges, however, where corrosion is faster and proceeds with evolution of hydrogen, a relatively pure iron corrodes at a much lower rate than iron or steel with higher contents of carbon, nitrogen, sulphur, phosphorus and other impurities.

Cast iron water pipe may last much longer than steel for two reasons:

- It is made with a greater wall thickness;
- In the case of grey or 'ductile' cast iron containing spheroidal graphite, the corrosion products tend to cement the residual graphite flakes together, thus allowing the completely corroded pipe to continue functioning for many years so long as it is not disturbed.

11.3.9 Metal composition

Although this has little effect on the corrosion rates of iron and steel in natural waters and soils, it becomes important in other conditions. The addition of small amounts (0.1–1.0%) of chromium, copper or nickel markedly reduces the rate of atmospheric corrosion. An increase in the carbon content of a steel may cause a slight increase in its corrosion rate in sea-water, although not in fresh water.

Stainless steels (containing >12% Cr), high-silicon irons and high-nickel alloys give a very marked reduction in corrosion rates over low-alloy iron and steel for most ordinary applications, but are often only economical in special applications (such as chemical duties).

Sulphur and phosphorus increase corrosion rates in acid solutions, but manganese counteracts the effect of sulphur.

Heat treatment has little effect on corrosion in the usual environments in which oxygen diffusion is controlling, but in handling acid oil-well brines, marked local corrosion often occurs near welds.

11.4 Other types of metal corrosion

The types of corrosion just discussed were dominated by galvanic action. Other types, some of which are shown schematically in Figure 11.4, are now considered.

11.4.1 Corrosion–erosion

This combination of corrosion and erosion results from a liquid or gas flowing at high velocity, with or without suspended solids or liquid droplets, and can lead to the rapid failure of valves and other equipment. A valve stem subject to erosion by steam or gas often appears to have been 'wire-drawn'. The use of harder materials, corrosion inhibitors, and cyclones and settlers to exclude hard gritty particles from the flowing liquid all reduce erosion. Water droplets can be excluded from steam turbines by raising the inlet temperature. Sometimes the system has to be redesigned to reduce the energy loss through the affected item. The pressure drop is spread by placing other resistances to flow (e.g. a length of smaller diameter pipe) in the fluid path. Liquid erosion is often caused by cavitation.

11.4.2 Cavitation–erosion

Cavitation results from the collapse of vapour bubbles formed within a liquid behind a fast-moving metal surface, or in liquid chokes and valves where sonic velocities are reached, often accompanied by a noise like grinding shingle. An example was the failure of an automatic by-pass valve on a minimum flow return line from the discharge to the suction of a firewater pump serving a large works [16.6.9]. It is sometimes found with pump rotors and impellers as a series of small but deep pits (Figure 11.5), which can lead to the break-up of fast-moving parts. Here it can be avoided by limiting the speed of the impeller and/or increasing the liquid head at the pump suction. Materials most resistant to cavitation are hard, have good fatigue and corrosion resistance, small grain size, and are able to work harden under repeated stressing[8].

11.4.3 Fretting–corrosion

This occurs between two surfaces in contact, one or both of which are metal, and subject to slight slip. It is often caused by vibration and leads to

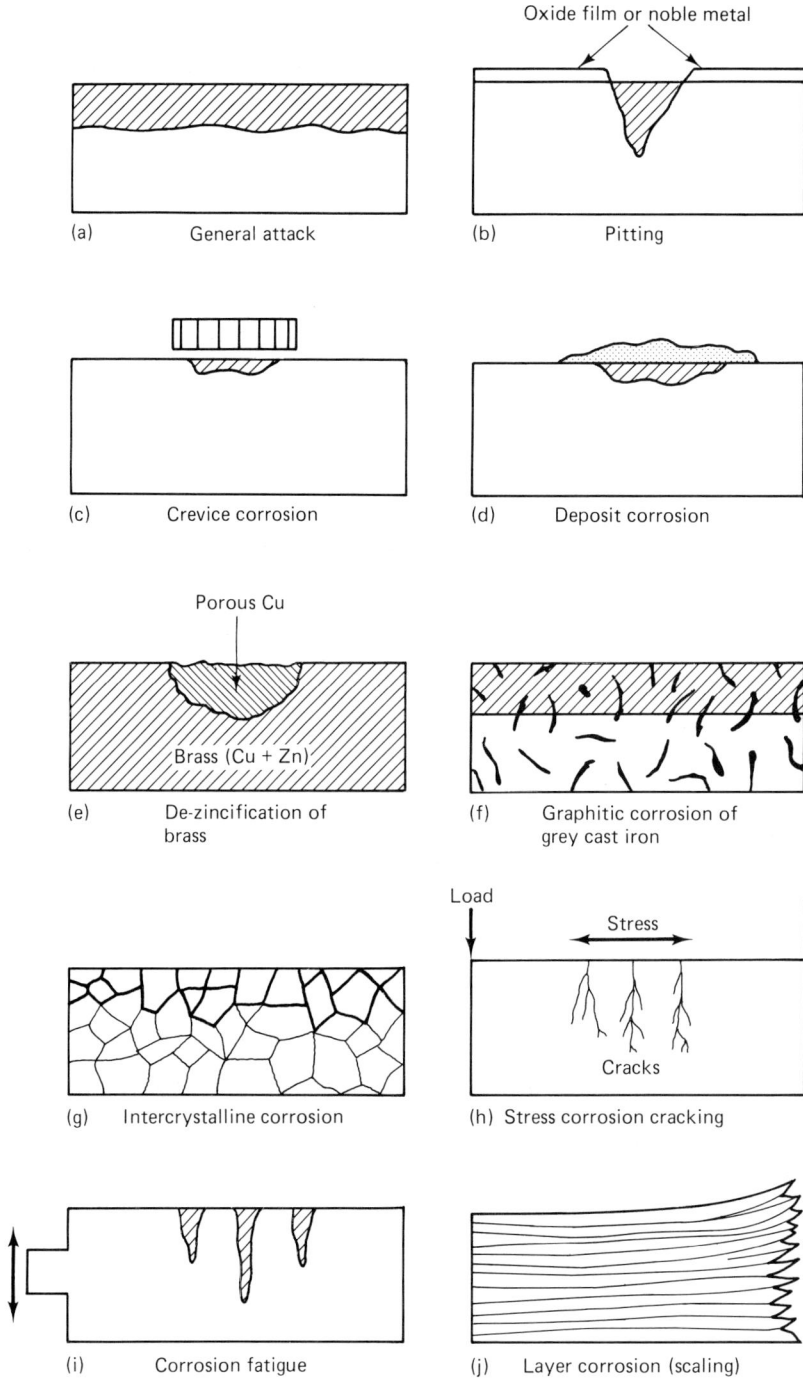

Figure 11.4 Schematic illustration of different types of corrosion (courtesy Gösta Wranglin and Institut för Metallsbrydd, Stockholm)

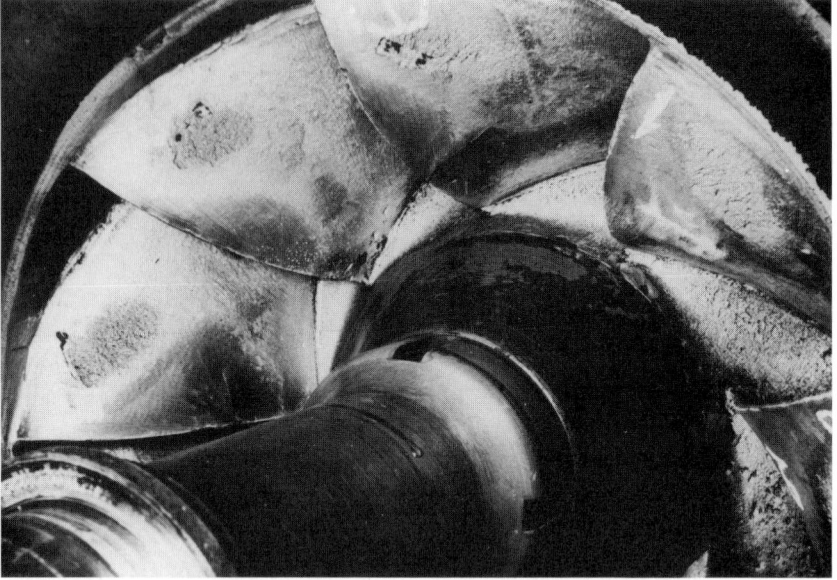

Figure 11.5 Cavitation damage to pump rotor (courtesy the Director, National Engineering Laboratory)

pitting which is evident when the corrosion debris has been removed. It has caused failures of suspension springs, bolt and rivet heads, and parts of vibrating machinery. Stationary ball bearings under static load have been destroyed by fretting caused by vibration [13.3.2]. Fretting–corrosion of steel requires oxygen and is practically eliminated in a nitrogen atmosphere.

11.4.4 Dezincification of zinc-containing alloys

Dezincification is a disguised hazard which can occur when zinc-containing alloys are exposed to aqueous media at high temperatures under stagnant conditions, especially if the medium is acid and the surface covered with a layer of porous scale. Apart from tarnishing, the item may appear undamaged. The area affected is porous, and has lost its ductility and much of its strength. Alloys containing less than 15% of zinc, or with small additions of tin, arsenic, antimony or phosphorus, are less subject to dezincification and corrosion inhibitors in the water also help[1]. Copper alloys containing aluminium are subject to a similar form of corrosion, the aluminium dissolving preferentially.

11.4.5 Intergranular corrosion

This occurs at the grain boundaries of a metal, with rapid loss of strength and ductility. The grain boundaries are anodic and the grains cathodic. It occurs with 18-8 stainless steels which have been heated to temperatures of

500–800°C, and then exposed to a corrosive environment. Annealing such items at 1100°C for one hour for every inch of metal thickness, followed by rapid water quenching, usually prevents it. For items which have to be welded and cannot be properly heated afterwards, alloys stabilised with titanium or columbium and of low carbon content should be used. Copper aluminium alloys which have not been properly heat treated are also subject to this form of attack.

Non-electrochemical grain boundary attack can occur when nickel is heated in an atmosphere containing sulphur.

11.4.6 Stress-corrosion cracking

This occurs when a metal surface is simultaneously exposed to a high static stress and a specific corrosive environment. The stress may be residual (arising from cold working or heat treatment), or externally applied. Stresses developed during fabrication can largely be relieved by heat treatment.

Most structural metals, particularly heavily loaded high-strength steels, are subject to stress-corrosion cracking in some environment such as solutions of inorganic chlorides, alkalis and nitrates. The observed cracks may be intergranular or transgranular, depending on the metal and the environment and they concentrate the stresses, thus starting other forms of failure, (fatigue [13.1.1] and brittle failure [13.1.3]). Stress-corrosion cracking was common in riveted steam boilers where it was known as 'caustic embrittlement'.

11.4.7 Corrosion fatigue

This is analogous to stress corrosion and occurs when a dynamically loaded metal component is subjected to repeated tensile stresses in a corrosive environment. These may be well below the critical stress needed to cause failure. A number of branching cracks appear over the affected surface. Complete separation eventually occurs either by brittle fracture [13.1.3] at ambient or low temperatures or by yielding or shear [13.1.5] at higher temperatures.

11.4.8 Hydrogen attack

This takes various forms, *embrittlement*, *blistering* and *cracking*, the last resembling stress-corrosion cracking. It results from the diffusion of atomic hydrogen through steel and other metals, forming molecular hydrogen on reaching voids or imperfections in the metal. If this happens at a large number of small centres, it causes embrittlement, but if it occurs at a small number of voids, it may generate sufficient pressure to raise blisters or cause cracks.

Atomic hydrogen is formed at the cathodic areas of galvanic corrosion cells before it combines to form molecules. It can also be formed when cathodic protection is applied to metals, and at high temperatures by the dissociation of hydrogen molecules. Sulphur and arsenic compounds promote and accentuate hydrogen attack, which is described as sulphide cracking when hydrogen sulphide is responsible.

Hydrogen damage can occur at high temperatures and high partial pressures, e.g. in catalytic reformers, and in plant and equipment handling natural gas and oil-brines containing hydrogen sulphide. Chromium and molybdenum increase the resistance of steels to such attack.

11.4.9 Scaling and high-temperature oxidation

These become pronounced with steel at temperatures above 500°C, but its resistance at higher temperatures can be improved by the addition of chromium as shown in Table 11.2[1].

Table 11.2 Chromium contents of steels for various maximum temperatures (in air)

Cr in Cr–Fe alloy (%)	Max. temperature (°C) in air
Nil	500
4–6	650
9	750
13	750–800
17	850–900
27	1050–1100

Silicon, nickel, aluminium and yttrium also improve the high-temperature resistance of steel alloys and extend their use up to 1350°C. Nickel alloys containing 20% chromium have good oxidation resistance and mechanical properties up to 1150°C and are used for supporting furnace tubes and for gas turbine blades.

11.4.10 Carburisation of ethylene-cracking tubes

The austenitic 18/8 and 25/20 tubes of ethylene-cracking furnaces which operate at temperatures of 700–900°C suffer rapid internal attack (carburisation) if the hydrocarbon feedstock is sulphur-free. This is prevented by the addition of small amounts of sulphur compounds to the feedstock.

11.4.11 Sulphate-reducing bacteria

These bacteria which are present in many soils and air-free natural waters cause rapid corrosion of steel, particularly buried pipelines, oil-well casings and pipe from deep water wells, at pHs of 5.5 to 8.5. They reduce inorganic sulphates to sulphides in the presence of hydrogen or organic matter, and are assisted by an iron surface, cathodic areas of which supply atomic hydrogen. The bacteria use this to reduce sulphate ions, producing ferrous hydroxide and sulphide. Chlorination, aeration and specific bactericides are used for control.

These bacteria must be eliminated from water injected into oil-wells to assist oil recovery, to prevent the production of hydrogen sulphide which converts 'sweet' reservoirs into sour ones.

11.4.12 Atmospheric corrosion or rusting

The rate of rusting depends on the air humidity and temperature and its content of sulphur dioxide, sea-salt and other local contaminants. In an industrial area the rate may be more than a hundred times the rate in a desert or arctic region. Iron and steel generally need protection by surface coatings, and aluminium alloys by anodising. Wrought iron (which contains a small quantity of silicate slag) has better resistance to atmospheric corrosion than steel. An inscribed iron pillar erected by a Hindu king near Delhi has remained 'rust-free' for 1500 years (Figure 11.6). Its surface is covered with a thin film of magnetite (Fe_3O_4). Factors which have contributed to its preservation are the dry and until recently unpolluted climate, its low sulphur and high phosphorus content, and its large mass and heat capacity which reduce diurnal temperature fluctuations.

Figure 11.6 A 1500 year old iron pillar at Delhi

Handrails which are exposed to rain sometimes fail from corrosion from the unprotected inner surface which become anodic as a result of oxygen depletion [11.3.1]. Insulated pipes for steam and process fluids suffer corrosion where they are exposed to leaks and dripping water, particularly if the insulation contains soluble salts which can lead to stress corrosion.

11.4.13 Zinc embrittlement of stainless steels[9]

Highly stressed stainless steel at temperatures of 750°C and above are liable to fail rapidly in contact with small amounts of molten zinc. This could happen in a plant fire if molten zinc from hot galvanised sheet dropped onto a red-hot stainless steel pipe.

11.5 Passivation

Some metals high in the electromotive series, e.g. chromium, nickel, molybdenum, titanium and zirconium, normally corrode at very low rates and are said to be passive. Although not fully understood, passivity is generally consistent with the formation of a very thin, tough, adherent and almost invisible surface oxide film. Passivity can break down in certain chemical environments, e.g. chloride solutions and organic acids.

According to a wider definition, lead in sulphuric acid, magnesium in water and iron in inhibited hydrochloric acid are also passive. Here the passivity is due to a surface layer of an insoluble salt.

11.5.1 Passivation of iron and steel

Iron and steel are rendered passive by exposure to oxygen above a certain partial pressure and to aqueous solutions of some oxidising compounds. Of these, sodium chromate is added in small amounts to circulating cooling water for process plants and power stations, and nitrites are used to protect petroleum product pipelines and petrol storage tanks in which small amounts of water are present. Other passivators which are only effective in the presence of atmospheric oxygen include alkalis and alkaline salts such as caustic soda, sodium phosphate, sodium silicate and sodium borate and neutral and slightly acid solutions of sodium benzoate, sodium cinnamate and sodium polyphosphate.

Passivation may also be effected by applying a small voltage between the metal and the liquid in contact with it and making the metal the anode. This is anodic protection, the reverse of cathodic protection [11.2], and is sometimes used to protect steel tanks containing sulphuric and nitric acids

11.6 Corrosion-resistant metals and alloys

The passive properties of chromium, nickel, etc. are also conferred on their alloys, giving a wide range of stainless steels, high nickel and other alloys which have excellent corrosion resistance in specific environments. Their cost [11.6.4] increases with their corrosion resistance. This encourages the use of non-metals such as chemical glassware, graphite, plastics, and steel with special linings such as glass, rubber and PTFE. Corrosion data for metals and non-metals in various environments are given in 'Perry'[10].

11.6.1 Stainless steels

Stainless steels are iron-based, with 10–30% of chromium, 0–22% of nickel, and minor amounts of carbon, columbium, copper, molybdenum, selenium, tantalum and titanium. Several are produced in wrought and cast form and widely used in the process industries, although they generally lack resistance to hydrochloric acid and reducing acids. There are three groups of stainless steels:

- *Martensitic alloys* contain 12–20% of chromium with other additives and are used in mildly corrosive environments. Welding is by manual arc, with preheating to 200°C, and post-weld treatment at 650–750°C to prevent cold cracking.
- *Ferritic alloys* contains 15–30% Cr with low carbon content. Their corrosion resistance is good except against reducing acids such as HCl, although they are sensitive to intercrystalline corrosion. Welding requirements are similar to martensitic alloys although they tend to enbrittlement on welding.
- *Austenitic stainless steels* which contain 16–22% Cr, 6–22% Ni, with low carbon are the most corrosion resistant, although their resistance to chloride ions and reducing acids is generally poor. Some have been developed for high-temperature service (e.g. tubes of ethylene-cracking furnaces), and others containing 2.5–3.5% Mo for very corrosive duties at lower temperatures. They are readily welded by manual arc, tungsten inert gas (TIG) and metal inert gas (MIG). Ti, Nb or Ta are added to prevent intergranular corrosion of field-welded joints.

11.6.2 Other ferrous alloys

There are two groups of 'ferrous' alloys (in which iron is not a major constituent), with higher corrosion resistance than stainless steels: *medium alloys* and *high alloys*.

- *Medium alloys* contain 30–44% Ni+Co, 13–30% Cr and other elements. They include Carpenter 20, Incoloys which resist sulphuric acid over a wide range of concentrations and temperatures, and Hastelloy G which is used in wet-process phosphoric acid where fluorides are present. They have limited resistance against hydrochloric acid.
- *High alloys* contain 55–80% Ni and various amounts of Cr, Mo, Fe and other elements. They include Hastelloy B which resists hydrochloric acid over a range of concentrations and temperatures but lacks resistance against nitric acid and oxidising salts. High-nickel alloys are sensitive to contamination during welding, especially from S, Pb and Zn. They should be stress-free before welding, which is similar to that of stainless steels, preferably with argon shielding.

11.6.3 Non-ferrous metals and alloys

Nickel is used to handle and concentrate alkaline solutions. Monel 400 (67% Ni, 30% Cu) resists alkalis and hydrofluoric acid, and moderately oxidising and reducing environments.

Aluminium resists atmospheric conditions, industrial fumes, fresh and brackish waters, concentrated nitric and sulphuric acids, but corrodes rapidly in caustic solutions.

Copper resists sea-water, alkalis and solvents but is rapidly attacked by oxidising acids. Brasses, though used for domestic water fittings, are little used for chemical duties. Bronzes, especially aluminium and silicon bronzes, have better resistance. Cupronickels (10–30 Ni) are the most resistant of copper alloys and used in heat exchangers on sea-water duties. Copper, however, is a deleterious contaminant in many processes. This limits its use and that of its alloys.

Lead, often with the addition of antimony or tellurium, was once widely used in corrosive environments containing sulphate, carbonate and phosphate ions, which form thin protective coatings on its surface. Its toxicity, low mechanical strength and poor temperature resistance now greatly limit its use in the process industries.

Titanium is strong and resists nitric acid better than stainless steels, and it also resists sea-water, wet (but not dry) chlorine, oxidising solutions hot or cold, and hypochlorites. It does not resist aqueous hydrogen fluoride, fluorine, moderately concentrated sulphuric and hydrochloric acids with no oxidising agent present, oxalic, formic and anhydrous acetic acids, boiling calcium chloride >55%, hot concentrated alkalis and dilute alkalis containing hydrogen peroxide. Small additions (0.1%) of palladium and platinum greatly improve its resistance against hydrochloric and sulphuric acids.

Titanium reacts with both atmospheric oxygen and nitrogen above 600°C. TIG welding with careful argon protection is needed and fabrication can be difficult.

Zirconium has similar although generally superior corrosion resistance to titanium. It resists:

- Alkalis of all concentrations up to their boiling point, and fused alkalis;
- Hydrochloric acid of all concentrations up to the boiling point;
- Nitric acid of all concentrations up to the boiling point;
- Sulphuric acid <70%, boiling;
- Phosphoric acid <55%, boiling;
- Boiling formic, acetic, lactic, and citric acids.

Zirconium does not resist oxidising metal chlorides (e.g. $FeCl_3$, $CuCl_2$), hydrofluoric and fluosilicic acids, wet chlorine, oxygen, nitrogen and hydrogen at elevated temperatures, aqua regia and boiling trichloracetic and oxalic acids. TIG welding is used with argon shielding of all zirconium surfaces above 400°C.

Tantalum has even better resistance than zirconium to oxidising and reducing acids and it also resists wet or dry chlorine, aqua regia, oxidising metal chlorides hot or cold, and lactic, oxalic and acetic acid. It does not resist alkalis, hydrofluoric acid and fluorides, fuming sulphuric acid, oxygen, nitrogen and hydrogen at elevated temperatures, and methanol solutions of hydrogen chloride. Welding of tantalum is similar to that of zirconium although otherwise it is easily fabricated.

11.6.4 Note on metal costs

Equipment and piping account for between 25% and 50% of the erected cost of most process plant, while the use of special corrosion-resistant metals is generally limited to a minor proportion of heat-exchange surfaces, vessel linings, pumps and pipes. Thus the high cost of some metals is seldom prohibitive when there is a real need. In comparing metal costs their densities and strengths must be considered. Metal prices, while sometimes steady for several months, are subject to sudden and violent fluctuations, and at the time of writing the price of nickel is rocketing. Some metal densities and bulk prices typical of the period 1985–1988 are given in Table 11.3. Prices of some metals and alloys in semi-finished form per cubic metre are given in Table 11.4. While SS plate is shown as five to six times as expensive as MS plate, the cost of a SS vessel or heat exchanger will be only two and a half to three times that of an equivalent MS one.

Table 11.3 Some metal densities and typical costs, 1985–1988

Metal	Density t/m^3	Cost in bulk (£)	
		per t	per m^3
Steel	7.8	160	1 250
Aluminium	2.7	1 200	3 250
Lead	11.3	350	3 960
Zinc	7.1	650	4 620
Copper	8.8	1 400	12 300
Titanium	4.5	5 800	25 500
Chromium	7.2	4 500	32 500
Nickel	8.8	6 000	53 000
Zirconium	6.4	Not published	
Tantalum	16.6	Not published	

Table 11.4 Some typical costs of semi-finished metal products

Metal or alloy	Cost (£)	
	per t	per m^3
Mild steel plate	350	2 750
18%Cr 9%Ni SS plate	2 000	16 000
Lead sheet	700	8 000
Copper tubes	3 000	26 000
Titanium sheet	15 000	68 000
Nickel sheet	11 000	97 000

Prices of zirconium and tantalum are seldom published but are usually well above those of titanium and nickel.

For some very corrosive duties which stainless steels cannot satisfy, the choice often lies between a high-nickel alloy and titanium or one of its alloys, which have become increasingly competitive over the past 25 years.

11.7 Examples of industrial corrosion problems

Corrosion is a complex subject and, without thorough knowledge of the resistance of metals and alloys under operating conditions, oil and chemical companies quickly run into trouble. The following examples are based on first-hand experience.

11.7.1 A mild steel sulphuric acid tank

Mild steel is corroded quickly by sulphuric acid in the concentration range 5–60% but becomes passive at higher concentrations, and it is used to store 90–98% sulphuric acid.

A water hose was used to clean down the conical roof of a tank containing 95% acid, without it being realised that the roof had corroded internally and had at least one hole. Next morning the tank was in two pieces with a clean cut running round it, just above the acid level.

Water had trickled down inside the wall and formed a layer above the denser acid. Diffusion then produced a narrow zone of hot, dilute and highly corrosive acid. The incident drew attention to the corrosion of the roof of this and similar tanks by dilute acid formed by condensation of acid fumes underneath the roof and absorption of atmospheric water vapour by the condensed acid.

11.7.2 An overseas herbicide plant

A herbicide plant built in the 1960s corroded so fast during commissioning that it could not be handed over to the client in running order. This was due to the use of stainless steel for handling dilute hydrochloric acid.

The more expensive alloys Hastelloy B or titanium with 0.1% Pd could have been used for heat exchange surfaces and difficult parts, while less expensive non-metals, chemical glassware, polypropylene or ABS (acrylonitrile-butadiene-styrene polymer), or steel lined with rubber, glass or PTFE could have been used for other items.

11.7.3 A new petrochemical process

A plant was built by Shell in the late 1960s in southern France, using a new and revolutionary process based on iodine for the dehydrogenation of butane to butadiene. Many millions of pounds had been spent on development. In spite of this the plant could not be commissioned, apparently because of intractable corrosion problems. The project was abandoned and the large investment written off.

11.7.4 Overhead condensers of crude oil distillation columns

Rapid corrosion was occurring in the overhead condensers and associated pumps, pipework and vessels of the primary flash and atmospheric columns of the main crude oil distillation units at Abadan refinery in the 1940s. One serious fire followed a pipe leak caused by this corrosion. The writer had the job of investigating the problem.

The primary flash columns operated under moderate pressure and removed LPG gases and low-boiling liquid hydrocarbons. They had tubular water-cooled condensers with mild steel shells and 'Admiralty' tubes.

The atmospheric columns operated at atmospheric pressure with light naphtha as overhead product, and various sidestreams. They had direct-contact condensers made of mild steel in which the vapour was condensed by contact with cooling water in short columns packed with ceramic rings.

A simplified flow-scheme of the distillation units and their crude oil supply is shown in Figure 11.7.

The corrosion of the condensers of the primary columns was caused mainly by dilute hydrochloric acid, which was sometimes found in them. This was traced to hydrolysis in the columns of inorganic chlorides present in the crude oil. These were only found in the oil from two of ten wells in one of the three oilfields which supplied the refinery by pipeline. They only occurred when the production rates from these wells exceeded a critical figure. The problem was solved by:

1. Setting maximum production rates for each well, checked by regular analyses of the crude oil for inorganic chlorides;
2. Injecting a small controlled amount of ammonia into each primary column to maintain the pH of water separated from the condensed liquid at about 8. Excess of ammonia which would have attacked the copper alloy condenser tubes had to be avoided.

The corrosion of the overhead systems of the atmospheric columns was found to have different causes and resulted from the combined action of:

1. Dissolved oxygen in the cooling water;
2. Small amounts of hydrogen sulphide formed by 'cracking' organic sulphur compounds present in the crude oil in the fired heaters of the columns.

As a result, a layer of elemental sulphur was found (to my amazement!) inside the domed tops of the direct-contact condensers.

The problem was solved by lining the shells of the overhead condensers and water separators with a corrosion-resistant cement, and replacing mild steel parts which could not be lined with stainless steel.

11.8 Notes on 'corrosion' of non-metals

As much care is needed in choosing and maintaining non-metallic constructional materials as with metals. The deterioration of plastics and other non-metallic materials which are susceptible to swelling, crazing, cracking, softening, etc. is essentially physiochemical rather than electrochemical. Thus it is seldom possible to evaluate the chemical resistance of non-metals by weight loss or surface penetration alone.

A range of coatings and linings have been developed to protect steels from corrosion by media ranging from rainwater to concentrated hydrochloric acid. These are often used in conjunction with cathodic

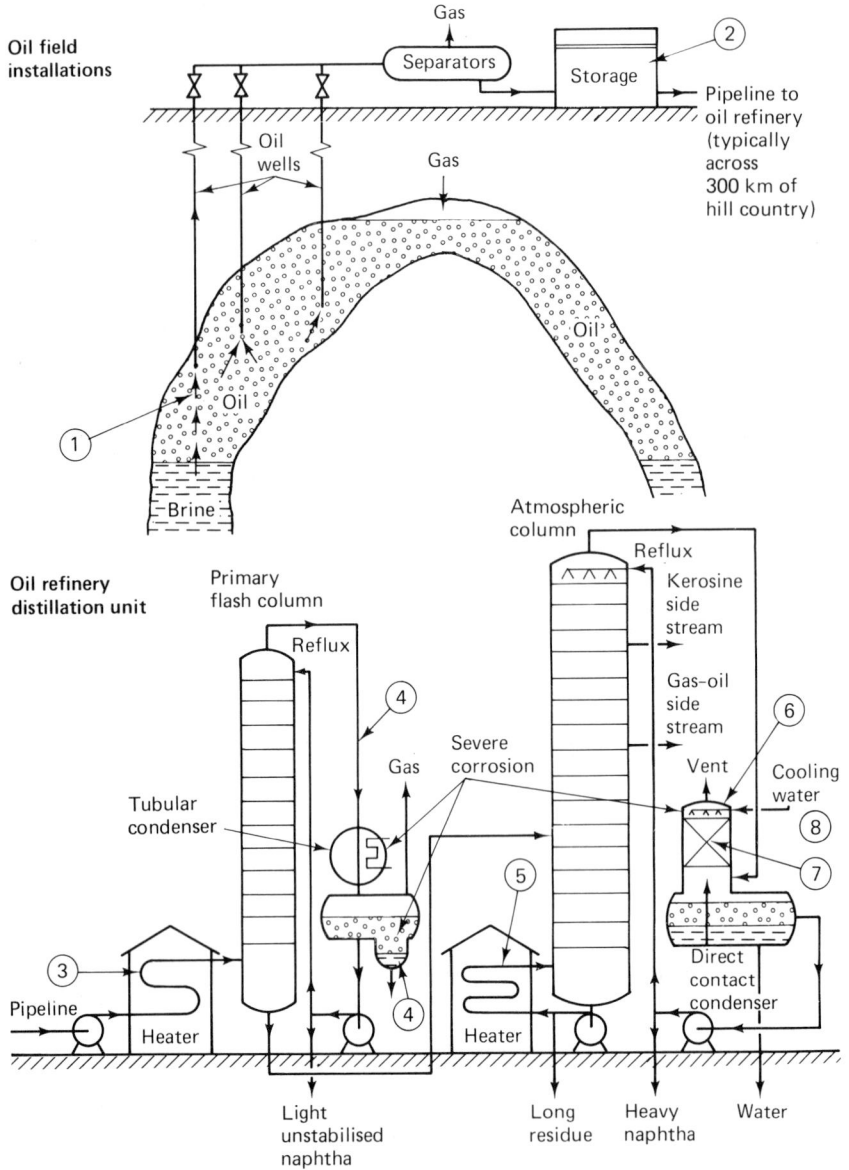

Causes of condenser corrosion
1. Brine drawn up into oil well at high oil production rate.
2. Oil always contains sulphur compounds, sometimes
   inorganic chlorides and water.
3. $MgCl_2$ hydrolysed here.
4. HCl found here.
5. H_2S formed here (from sulphur compounds in oil).
6. Sulphur found here.
7. $H_2S + O_2$.
8. Dissolved oxygen in cooling water.

Figure 11.7 Simplified flow scheme of crude oil production and distillation

protection. Electrical methods of detecting pinholes and other flaws in coatings and linings are widely used. The failure of a lining is then apparent well before the lined shell or pipe fails.

Some problems of non-metallic plant constructional materials follow.

11.8.1 Concrete foundations

These need to be protected both against salts and bacteria in the soil and the effects of spilt process fluids, particularly acids. In the mid-1950s the structure of a synthetic alcohol plant in which sulphuric acid was used nearly collapsed because the foundations had disintegrated through the action of acid which had spilt over a period of years from leaking pump glands. This could have been avoided had acid-resistant flooring and drains been provided in the area at risk.

11.8.2 Flange gaskets and jointing materials

A surprising number of leaks occur through the gaskets of flanged pipe joints of operating plant. Because of this, specialist contractors provide a service of sealing leaks from the outside, without shutting down the plant [17.1.4]. Yet a wide range of jointing materials are available from which ones with good resistance to almost any process fluid can be chosen. The use of spirally wound gaskets reduces the danger of a sector of gasket material between two bolts being blown out.

The main causes of such leaks are probably mechanical, e.g. stresses in the pipework and joints (particularly due to differential thermal expansion), badly aligned flanges, improperly tensioned bolts, and the use of unsuitable flanges.

11.8.3 Rubber and plastic hoses

Serious accidents occur through the failure of flexible hoses of various kinds used to transfer fluids between storage tanks and vessels and transport vehicles, and for various unplanned and emergency operations. Many of these accidents result from mechanical causes – improperly secured hose connections and damaged hoses. Some are due to chemical deterioration of the hose, and the use of rubbers and plastics for fluids to which they are not resistant. It is therefore important:

- To provide secure storage for hoses which are not in immediate use;
- To check all hoses regularly;
- To have a system for identifying hoses of different materials;
- To have clear instructions on which types of hoses may and may not be used for the various fluids present.

There is often a tendency in an emergency operation to grab the nearest hose to empty a process vessel, etc. without considering whether it is resistant to the fluid. This may work at the time, but may cause the hose to deteriorate seriously before it is required for its normal duty.

11.8.4 Chemical glassware

Chemical glassware is widely used for small industrial-scale operations and for pilot plant and bench-scale work. Its good corrosion resistance and transparency give it a special appeal. It rarely fails through corrosion since its limitations against hydrogen fluoride and strong alkalis are well known, and early signs of deterioration are usually visible. Unfortunately it is all too easy to overlook its fragility and low tensile strength. In the only serious chemical accident for which the writer felt some personal responsibility, a friend and junior colleague (who since died), was badly burnt in a fire following the escape of a few litres of benzene from a pilot plant distillation unit constructed of chemical glassware. Hazards arising from its fragility always need to be carefully considered before it is chosen and used.

References

1. Uhlig, H. H., and Revie, W. R., *Corrosion and Corrosion Control*, 3rd edn, Wiley, New York (1985)
2. Frey, D., 'Case histories of corrosion in industrial boilers', *Materials Protection and Performance*, **20** (2), 49 (1981)
3. Davenport, J. A., 'A study of vapour cloud incidents', *Eleventh Loss Prevention Symposium*, New York, American Institution of Chemical Engineers (1977)
4. 'City of death' in *India Today*, 31 December (1984)
5. Lees, F. P., *Loss Prevention in the Process Industries*, Butterworths, London (1980)
6. Hunter, D., *The Diseases of Occupations*, 6th edn, Hodder and Stoughton, London (1978)
7. Collins, J. A. and Monack, M. L., 'Stress-corrosion cracking in the chemical process industry', *Materials Protection and Performance*, **12**, 11 (1973)
8. King, R. W. and Magid, J., *Industrial Hazard and Safety Handbook*, Butterworths, London (1980)
9. HSE, Guidance Note PM 13, *Zinc embrittlement of austenitic stainless steel*, HMSO, London (1976)
10. Perry, R. H. and Chilton, C. H., *Chemical Engineers' Handbook*, 5th edn, McGraw-Hill, New York (1973)

Chapter 12

Fire and explosion hazard rating of process plant

The hazards of substances (which range from those of salt to nitroglycerine) and the plants which process them differ widely in degree and type. Plants with high hazard potential need greater spacing, greater care in their design and operation, and more safety and protective features than less hazardous ones. Several methods are used to rate the fire and explosion hazards of process plants. The results have important financial implications. Faced with choosing one of several processes with different capital and operating costs, and different degrees of hazard, managements need to know how these will affect the project costs.

Dow Chemical Company's 'Fire and Explosion Index' (F&EI) Hazard Classification Guide', now in its sixth edition, is probably the best-known method. The degree of hazard of each unit of a process plant is rated as a single number. The third edition[1], published as a manual by the American Institute of Chemical Engineers[2], forms the basis for the more complex later ones[3,4] and for the Mond index[5-7] which was developed by ICI plc. Both the third edition of Dow's guide (chosen for its comparative simplicity) and the second edition of the Mond Index[6] are described and discussed here. Later editions of the Dow guide use different numerical bases for the index (Table 12.1) and include credit factors for various loss-control measures, estimates of maximum probable property damage

Table 12.1 Dow Fire and Explosion Index Range and Degree of Hazard

Degree of hazard	Index range			
	3rd edn (1973)	4th edn (1976)	5th edn (1980)	6th edn (1987)
(Mild)		0–20		Not
Light	20–40	1–50	1–60	applicable
(Moderately heavy)	60–75			
Moderate	40–60	51–81	61–96	
Intermediate		82–107	97–126	
Heavy	75–90	108–133	128–156	
(Extreme)	90 and up			
Severe		134–up	159–up	

and other damage parameters. Those using the Dow guide occasionally should keep to the same edition to prevent confusion. The fourth edition is described by Lees[4]. The methods can be used as an aid to feasibility studies, process design, layout and mechanical design and can be applied at any subsequent stage in the plant's life.

To use the Dow guide one needs a process flow diagram of the plant, a plot plan, a copy of the guide and a good understanding of the process. To apply the Mond index, one also needs detailed cost data for the installed equipment, pipework, buildings and structures, a drawing compass and a calculator (or computer and program).

This chapter concludes with a section [12.4] on estimating the potential loss from vapour cloud explosions from projected and existing plants.

12.1 The Dow Fire and Explosion Hazard Index, 3rd edn[1, 2]

The description given here is necessarily abbreviated and several finer points have had to be omitted. It nevertheless gives all the essential steps with a worked example, including the safety features called for in the design.

The first step is to divide the plant into units. Here a unit is defined as a part of a plant which can be readily and logically characterised as a separate entity. One should start with the units as defined by the process design and, if possible, split them into smaller ones which do not overlap significantly. Figure 12.1 is a very simplified flow diagram of a hypothetical plant making alcohols from propane oxidation which is divided into six units (Table 12.6). Recycled propane vapour plus make-up propane from pressure storage is mixed with a small proportion of compressed air to give a mixture well above the upper explosive limit before entering a vapour phase reactor at 315°C and about 22 bar g. The oxygen and part of the propane react exothermally to give a mixture of alcohols, aldehydes and ketones, with nitrogen and unreacted propane. The hot gases leave the reactor at 538°C and, after cooling, pass through an absorber where water is used to recover the alcohols, aldehydes and ketones. These are separated from the water in a stripping column and passed to three columns in series where acetaldehyde, acetone and methanol are separated as overhead products. The gas leaving the absorber and consisting principally of nitrogen and propane passes to a second absorber where most of the propane is removed by circulating absorption oil. Propane is recovered from the absorption oil in a pressurised stripping column and recycled as vapour to the preheater. Vapour phase hydrocarbon oxidation processes were discussed in 8.5.2.

The F&EI for each unit is evaluated from the following factors:

1. The material factor for the unit, MF,
2. Special material hazards factor,
3. General process hazards factor,
4. Special process hazards factor.

(In later editions of the guide (1) and (2) are combined and the MF is a function of three NFPA hazard ratings[8].)

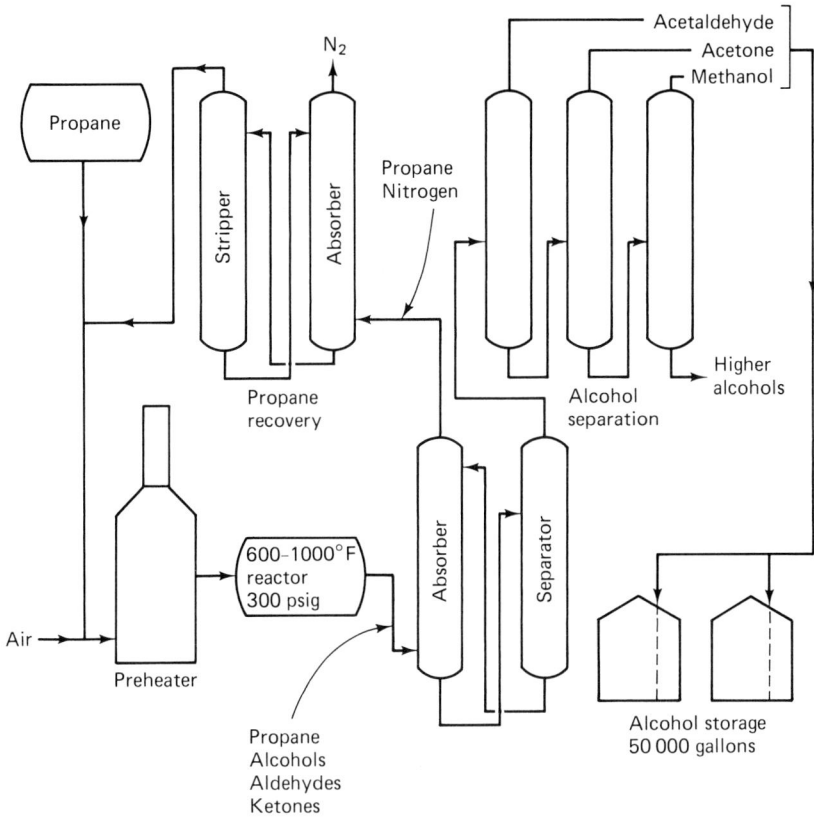

Figure 12.1 Flow diagram of hypothetical alcohols from propane plant

12.1.1 The Material Factor (MF)

This is a number (generally from 1 to about 60) which denotes the intensity of energy release from the most hazardous material or mixture of materials present in significant quantity in the unit. MF is calculated as:

$$\text{MF} = \frac{\Delta H_c}{2326}$$

where ΔH_c = net heat of combustion (kJ/kg).

For combinations of reactive materials such as oxidising and reducing ones, the heat of reaction is used instead of the heat of combustion. MFs for several materials are given in Table 12.2.

12.1.2 Special material hazards

An additional percentage as given in Table 12.3 must be added to the MF of materials which have certain hazardous properties, provided these have not already been taken into account in determining the MF. Where ranges are given, judgement is needed to decide the percentage to be added.

Table 12.2 Material Factors for common chemicals

Compound	MF	Compound	MF
Acetone	12.3	Isopropyl alcohol	13.1
Acetylene	20.7	Methyl chloride	5.5
1,3-Butadiene	19.2	2-Nitrotoluene	11.2
Carbon disulphide	6.1	Potassium perchlorate	0.0
Chlorine dioxide	0.7	Stearic acid	15.7
Diethyl ether	14.7	Styrene	17.4
n-Dinitrobenzene	7.2	Sulphur	4.0
Dowtherm A	14.0	Triethyl aluminium	18.9
Ethyl acetate	10.1	Urea	3.9
Ethylene oxide	11.7	Vinyl chloride	8.0
Hydrogen	51.6	Xylene	17.6

Table 12.3 Percentages to be added to MF for special material hazards

Special property of material	Percentage to be added to MF
A. Oxidising	0 to 20
B. Reactive with water to produce combustible gas	0 to 30
C. Subject to spontaneous heating	30
D. Subject to spontaneous polymerisation	50 to 75
E. Subject to explosive decomposition	125
F. Subject to detonation	150
G. Other unusual hazardous properties	0 to 150

Table 12.4 Percentages to be added to MF for general process hazards

Process characteristic	Percentage to be added to MF
A. Handling and physical change only	0 to 50
B. Continuous reactions	0 to 30
C. Batch reactions	25 to 50
D. Multiplicity of reactions in same equipment	0 to 50

12.1.3 General process hazards

A further percentage must be added to the MF for each of the applicable process characteristics given in Table 12.4.

12.1.4 Special process hazards

A further percentage must be added to the MF for each of the applicable special process characteristics given in Table 12.5.

Table 12.5 Percentages to be added to MF for special process hazards

Special process characteristic	Percentage to be added to MF
A. Low pressure	0 to 100
B. Operation in or near explosive range	0 to 150
C. Low temperature	15 to 25
D. High temperature. Use one only	
1. Above the flash-point	20
2. Above the boiling point	25
3. Above the auto-ignition temperature	35
E. High pressure	
1. 17 to 200 bar g.	30
2. Above 200 bar g.	60
F. Processes or reactions difficult to control	50 to 100
G. Dust or mist explosion hazard	30 to 60
H. Greater than average explosion hazard	60 to 100
I. Large quantities of combustible or flammable liquids in unit (use one only)	
1. 7.5 to 22.5 m^3	40 to 55
2. 22.5 to 75 m^3	55 to 100
3. 75 to 190 m^3	75 to 100
4. More than 190 m^3	100 or more
J. Other unusual process hazards	0 to 20

Table 12.6 Units of alcohols from propane plant with estimated F&EIs

	Unit	F&EI	Degree of hazard
1.	Propane storage	56.3	Moderate
2.	Reaction area	76.5	Heavy
3.	Alcohol absorber	28.0	Light
4.	Alcohol separation	15.5	Mild
5.	Propane recovery	39.0	Light
6.	Alcohol storage	27.8	Light

12.1.5 Evaluation of the F&E indices

The indices for the six units of the hypothetical alcohols plant are evaluated on standard calculation sheets which are combined in Figure 12.2. The results are summarised in Table 12.6.

The reaction unit has the highest F&EI followed by the propane storage unit which is assumed to consist of pressure spheres. The only units whose F&EIs are affected by the plant scale are propane storage, propane recovery and alcohol storage. The F&EIs of all other units would be the same for a pilot plant as for a large commercial one (a possible weakness in the method).

12.1.6 Selection of preventative and protective features

Protective systems may be considered as passive, i.e. systems without moving parts such as blast-walls and knockout vessels, or active, i.e. systems with moving parts such as valves and switches. Systems without

FIRE AND EXPLOSION INDEX	NAME			
LOCATION				JOB NUM
PLANT Alcohol				CHARGE
UNIT		Propane storage		Reaction area
MATERIALS		Propane		Propane
CATALYSTS		None		None
REACTIONS		None		$C_3H_8 + O_2 \rightarrow CH_3OH$ $CH_3COCH_3 +$
SOLVENTS		None		None
1. MATERIAL FACTOR FOR:		Propane	19.9	Propane
2. SPECIAL MATERIAL HAZARDS	% factor suggested	% factor used	19.9	% factor used
A. Oxidizing materials	0-20			
B. Reacts with water to produce combustible gas	0-30			
C. Subject to spontaneous heating	30			
D. Subject to rapid spontaneous polymerization	50-75			
E. Subject to explosive decomposition	125			
F. Subject to detonation	150			
G. Other				
Add percentages A–G for special material hazard (SMH) total		0		0
((100 + SMH total)/100) × (material factor) = subtotal no. 2 ⟶			20	0.2 ⟶
3. GENERAL PROCESS HAZARDS				
A. Handling and physical changes only	0-50	25		
B. Continuous reactions	25-50			50
C. Batch reactions	25-60·			
D. Multiplicity of reactions in same equipment	0-50			
Add percentages A–D for general process (GP) total		25		50
((100 + GP total)/100 × (subtotal no. 2) = subtotal no. 3 ⟶			25	⟶
4. SPECIAL PROCESS HAZARDS				
A. Low pressure (below 15 psia)	0-100			
B. Operation in or near explosion range	0-150			100
C. Low temperature: 1. Carbon steel 50° to −20°F	15			
2. Below −20°F	25			
D. High temperature (use one only)				
1. Above flash point	10-20			
2. Above boiling point	25	25		25
3. Above autoignition point	35			
E. High pressure: 1. 250-3000 psig	30			30
2. Above 3000 psig	60			
F. Processes or reactions difficult to control	50-100			
G. Dust or mist hazard	30-60			
H. Greater than average explosion hazard	60-100			
I. Large quantities of combustible liquids (use one only)				
1. 2000-6000 gallons	40-55			
2. 6000-20,000 gallons	55-75			
3. 20,000-50,000 gallons	75-100			
4. Above 50,000 gallons	100+	100		
J. Other				
Add percentages A–J for special process (SP) total		125		155
((100 + SP total)/100) × (subtotal no. 3) = F & EI ⟶			56.3	⟶

Figure 12.2 Combined Dow F&EI calculation sheets for units of alcohols plant (courtesy Dow Chemical Co.)

	DATE		
Alcohol absorber	Alcohol separation	Alcohol storage	Propane recovery

MATERIALS AND PROCESS

Alcohol absorber		Alcohol separation		Alcohol storage		Propane recovery	
Reactor products Propane		Alcohols, ketones, aldehydes		Alcohol		Propane	
None		None		None		None	
None		None		None		None	
Water		None		None		Absorption oil	
Propane	19.9	Methanol	8.6	Methanol	8.6	Propane	19.9
% factor used	19.9	% factor used		% factor used		% factor used	
0		0		0		0	
0.2 ———→	20	0.2 ———→	8.6	0.2 ———→	8.6	0.2 ———→	20
0		0		25		0	
0		0		25		0	
———→	20	———→	8.6	———→	10.7	———→	20
				50			
10				10		10	
		25					
30						30	
		55				55	
				100			
40		80		160		95	
———→	28	———→	15.5	———→	27.8	———→	39

moving parts give fewest problems, but all need regular maintenance and testing, which must be budgeted for. The simplest ones are usually the best. When many rarely used protective devices are installed, some readily fall into disrepair and are not available when a real need arises. The checklists which follow are intended as guides for selection and not imperatives to be followed blindly.

Dow's preventative and protective features are grouped in three categories:

1. Basic features needed in any plant with an F&E hazard;
2. Features whose application depends on the F&EI of the unit (Table 12.7);
3. Special preventative features (Table 12.8) which relate to specific hazards listed in Tables 12.3 to 12.5.

Many of the same features appear in both Tables 12.7 and 12.8. The method used here in selecting them is explained in 12.7.

Table 12.7 Preventative and protective features related to F&E hazard

Legend	*Priority*
Feature optional	1
Feature suggested	2
Feature recommended	3
Feature required	4

Feature		*Priority for calculated F&EI*					
		0 to 20	20 to 40	40 to 60	60 to 75	75 to 90	over 90
A. Fireproofing	[16.6.5]	1	2	2	3	4	4
B. Water spray	[16.6.9]						
a. Directional		1	2	3	3	4	4
b. Area		1	2	3	3	4	4
c. Curtain		1	1	2	2	2	4
C. Special instrumentation[a]	[15.7]						
a. Temperature		1	2	3	3	4	4
b. Pressure		1	2	3	3	3	4
c. Flow control		1	2	3	4	4	4
D. Dump/blowdown/spill control	[15.5]	1	1	2	3	3	4
E. Internal explosion relief and/or suppression	[10.4.4]	1	2	3	3	4	4
F. Combustible gas monitors	[15.7]						
a. Signal alarm		1	1	2	3	3	4
b. Actuate equipment		1	1	2	2	3	4
G. Remote operation	[15.8]	1	1	2	3	3	4
H. Building ventilation	[16.6.5]	see Dow code					
I. Building explosion relief	[16.6.5]	see Dow code					
J. Diking	[16.6.5]	1	4	4	4	4	4
K. Dust explosion control	[10.4]	see Dow Guide					
L. Blast and barrier walls/separation	[16.6.5]	1	1	2	3	4	4

Table 12.8 Safety features for particular hazards (based on Table VII of Dow guide[1,2])

Hazard type	Protective features
2. Special material hazards	
A. Oxidising materials [8.5.2]	Keep separate from combustible material and store in fireproof area
B. Reacts with water to produce combustible gas [8.6.3]	Protect from water and ignition sources, ventilate gas formed and comply with appropriate electrical code
C. Subject to spontaneous heating [8.7]	Provide cooling
D. Subject to spontaneous polymerisation [8.6.4]	Provide polymerisation inhibitor system, cooling and over-pressure relief
E. Subject to explosive	Design equipment to contain or safely relieve explosion; provide temperature and/or pressure control if effective and consult recognised authority
3. General process hazards	
A. Handling and physical change only	For loading and unloading, provide excess flow and remotely operated valves, alarms for inadequate electrical grounding, purge procedures for vessels and lines, special fire-extinguishing systems and ventilation
B. Continuous reactions	Prevent reactant unbalance by appropriate instruments or other method; provide safe over-pressure relief, over/under-temperature alarm or automatic shutdown; consider measures to keep out hazardous impurities
C. Batch reactions	Same as (3B) with special procedures/instruments to avoid hazards during turnaround
D. Several reactions in same equipment	Same as (3C) with positive separation of reactants when not required
4. Special process hazards	
A. Low pressure	Provide instrument interlocks to avoid hazardous pressure range and alarms to indicate approach to hazardous condition, e.g. pressure, oxygen concentration
B. Operation in or near explosive range	Design equipment to contain or safely relieve explosion; consider explosion suppression, dilution or inerting to avoid explosive range and back-up instrumentation for process control
C. Low temperature	Provide special vent or dump systems
D. High temperature	Arrange to minimise flow of flammables; consider flammable gas monitors for alarm, shutdown or deluge actuation; provide special vent or dump system
E. High pressure (> 17 bar g)	Provide quick-acting and safe vent/dump systems and remote or instrument operated valves to minimise consequences of line or equipment failure; consider flammable gas monitors for alarm, shutdown or deluge actuation
F. Processes or reactions difficult to control	Design equipment to contain or safely vent worst situation; consider dump, vent and quench systems
G. Dust hazard	Same as 4B

1. **Basic preventative and protective features**
A. Adequate water supply for fire protection [16.6.9];
B. Structural design of vessels, piping, structural steel [16.6.5];
C. Overpressure relief devices [15.2–15.4];
D. Corrosion resistance and/or allowances [11.1];
E. Segregation of reactive materials in process lines and equipment [8.6.1];
F. Electrical equipment grounding [6.3.1];
G. Safe location of auxiliary electrical gear [6.2];
H. Normal protection against utility loss (alternate electrical feeder, spare instrument air compressor, etc.) [16.6.6];
I. Compliance with various applicable codes [2.9];
J. Fail-safe instrumentation [15.7.2];
K. Access to area for emergency vehicles and exits for personnel evacuation [16.6.2];
L. Drainage to handle probable spills plus fire-fighting water [16.6.7];
M. Insulation of hot surfaces that reach 80% of autoignition temperature of any flammable in the area [10.2 3];
N. Adherence to the appropriate electrical codes [6.1];
O. Limitation of glass devices in flammable or hazardous service [15.8.4];
P. Building and equipment layout appropriate to hazard [16.6.2];
Q. Protection of pipe racks, instrument cable trays and supports from fire exposure [16.6.5];
R. Provision of accessible block valves at battery limits [16.6.3];
S. Protection of cooling tower(s) [16.6.6];
T. Protection of fired equipment against accidental explosion and fire [10.5];
U. Special precautions for critical rotating equipment [13.3];
V. Appropriate construction of control rooms and their separation/ isolation from control laboratories, electrical switchgear/transformers and potential sources of hydrocarbon release [16.6.5].

2. **Features dependent on F&EI of unit**
Table 12.7 gives minimum preventative and protective features which must be considered, but whose application depends on the probability and expected intensity of a fire or explosion as indicated by the F&EI calculated for the process unit. The features should be interpreted rather broadly, as explained in the guide and summarised as follows:

A. *Fire protection of structural supports*
This applies only when flammable liquid might be retained in the area, the fireproof rating depending on the expected depth of liquid retained. Drainage is, however, generally preferred to fireproofing of structures as a means of protection against pool fires.
B. *Water spray protection of equipment and area*
Water spray is mainly needed for protection against liquid and solid fires. Directional spray may be used to protect pumps, vessels and cable trays. Area spray is used in addition for F&EIs of 60 and over.
   Water curtains serve to reduce the movement of vapour clouds and may be installed between furnaces and potential points of vapour

release. Sprinkler piping, supports and nozzles are vulnerable to explosions and should be robustly constructed where this danger exists.

C. *Special instrumentation*

This covers instruments installed for safety rather than process control, e.g.

- Interlocks [15.7.1] between flow controllers which cause one to 'fail safe' if another flow fails,
- Analysers which initiate alarms or shut down equipment when a hazardous condition is approached,
- Remotely operated valves to stop the flow of flammable fluids in case of fire.

The need for such special instrumentation starts with F&EIs of 40, when it may be added to normal control loops [15.7]. At F&EIs of 60 or more, additional safety instrumentation which is quite separate from control instrumentation is needed, and at F&EIs of 75 or more, the installation of redundant safety instrumentation should be considered. Safety instrumentation needs careful selection to have most effect.

D. *Dump, blowdown or spill control*

This covers systems designed to remove hazardous materials quickly from danger spots in an emergency. It includes fixed deluge equipment to wash away spilled materials, pump-out equipment, sloped drainage, headers to vent and flare stacks and the injection of inert and 'short stop' materials to control 'runaway' reactions.

E. *Internal explosion*

This category covers techniques which prevent explosive mixtures from forming inside process equipment or – where this is impossible – those which relieve, contain or suppress the explosion and/or eliminate sources of ignition. Many of these techniques depend on the use of inert gases.

F. *Combustible gas monitors*

These may sound an alarm, actuate protective equipment such as sprinklers or ventilation fans or close valves on process lines. They should be located between points of possible release of flammable fluids and sources of ignition (e.g. furnaces) and in areas where natural ventilation is poor.

G. *Remote operation*

This is specially needed for operations which are too hazardous for personnel to be allowed in the same area. All process plants with control rooms depend to a large extent on remote operation.

H. *Building ventilation*

Three types are covered in the Dow guide, which refers to the appropriate NFPA standards:

- Removal of smoke and hot gases during a fire,
- Air change in areas where flammable liquids are handled,
- Removal of airborne dust and harmful vapours in areas where their formation cannot be entirely prevented.

I. *Building explosion relief*

Operations where there is an explosion risk should preferably be located in separate detached buildings. The area of explosion vents needed depends in the expected explosion intensity.

J. *Diking*
   This includes the sloping of ground at an angle of at least 2° away from tanks and other possible areas of spillage towards impounding basins capable of holding the largest spill which could occur. Where dike (bund) walls are provided round tanks their average height should not exceed 1.9 m. Bunded areas should be provided with means of draining water in such a way that flammable spills do not escape.

K. *Dust explosion control*
   This generally means providing explosion relief panels, the ratio of whose area to the volume relieved depends on the maximum expected rate of pressure rise. This in turn is largely dependent on the particle size of the dust [10.4].

L. *Blast and barrier walls and separation*
   Blast walls are used to confine damage from mishaps with operations involving high pressures, potentially explosive materials and mixtures, and the use or processing of explosives. They may also be used to protect sprinkler valves. Where space is limited barrier walls may be used to separate flammable storage areas from process units.

3. **Preventative features for specific hazards**
   These are given in Table 12.8.

12.1.7 Safety features for the example given

Table 12.9 shows the special features stipulated in Table 12.8 for each unit together with the corresponding feature and degree of priority indicated in Table 12.7. The features appropriate to the specific hazards of each unit were first selected from Table 12.8. Their priorities were then assessed

Table 12.9 Safety features for each unit with priorities

Unit	F&EI	Feature and priority	
		From Table 12.8 with 12.7	*From Table 12.7 only, units (1) and (2)*
1.	56.3	3A not applicable 4D C(3), F(2), J(4) 4I same + G(2)	A(2), B(3), D(2), E(3) L(2)
2.	76.5	3B C(4), D(3) 4B C(4), E(4) 4D C(4), F(3), J(4) 4E C(4), D(3), F(3), G(3)	possible A(4), B(4) and L(4)
3.	28.0	4D C(2), F(1), J(4) 4E C(2), D(1), F(1), G(1)	
4.	15.5	4D C(1), F(1), J(1) 4I same + G(1)	
5.	39.0	4D C(2), F(1), J(4) 4E C(2), D(1), F(1), G(1)	
6.	27.8	4I C(2), D(1), F(1), G(1)	

from Table 12.7 and the F&EI of the unit. For units with F&EIs of 40 or over, any other features in Table 12.7 which could apply are also listed with their priorities.

12.1.8 Some comments on the Dow guide, 3rd edn

The different protective and preventative measures suggested by the guide seem reasonable although their priorities are more questionable and their details still have to be decided. As examples, the fireproofing of supports for propane pressure-storage vessels and the use of water spray to prevent their tops from overheating in a fire would now be generally regarded as essential rather than merely 'suggested' (A2) or 'recommended' (B3). This is hardly surprising, since the guide's recommendations are based on combinations of rather general hazards and not on specific cases. Final decisions and details of the safety features needed require far more information than that given in the very sketchy flow diagram shown in Figure 12.1. Safety features must, moreover, be considered realistically. Most can themselves fail or go wrong and compete with other items for management attention.

The Dow guide is essentially a condensation of much experience which would fill several volumes. One risk in using it against which it warns lies in taking an oversimplified view and applying it without considering the details of the case. Its main virtue is that it enables a reasonable profile both of the hazards and of the safety features needed for a plant to be quickly drawn from preliminary process information. This is particularly important during preliminary studies when alternative investment opportunities are being considered. Here the method enables process hazards and the cost of safety features needed to combat them to be taken into account at the earliest stage. The guide is also a useful adjunct to other hazard studies, particularly in the checklists which it provides, as a project takes shape and its design proceeds.

12.2 The Mond Index[5, 6]

The 'Mond Fire, Explosion and Toxicity Index' was first presented by D. J. Lewis in 1979[5]. The second edition of the Mond Index which is discussed here was published as an 80-page booklet by the Explosion Hazards Section of the Mond Division of ICI[6]. It is developed from the third edition of Dow's F&EI guide and deals primarily with F&E hazards. Toxicity is considered only as a complicating factor. Like the Dow Index, it is a rapid hazard-assessment method for use on existing chemical plant, during process and plant development, and in plant design and layout[7, 9]. When used during development design and layout, it highlights features requiring further study, thus enabling problems to be recognised and often eliminated, with financial saving. Its use generally should lead to a rational expenditure on safety items.

The Mond Index extends the calculation procedures of the Dow Index to highlight particular hazards. Thus it provides separate indices for fire, internal explosion and aerial explosion potential, as well as an overall hazard rating. Only its general features are given here. Readers intending to use the Mond Index are advised to attend one of the short courses which are offered on it. A computerised version is also available for use on an IBM PC.

The first step, as with the Dow Index, is to divide the plant into units, and it is better to start with too many than too few. The next step is to determine the material factor B, which provides a numerical base for the indices. The base is then modified by many other considerations contained in the following sections:

1. Special material hazards, M;
2. General process hazards, P;
3. Special process hazards, S;
4. Quantity hazards, Q;
5. Layout hazards, L;
6. Acute health hazards, H.

Standard calculation sheets (Figure 12.3) are used. These list the conditions with hazard potential with suggested penalty ranges for each of the six hazard factors.

Most penalties are positive, but 'negative penalties' can also be applied, e.g. to gases which rise rapidly and to marginally flammable materials such as trichloroethylene. From the seven hazard factors for each unit, the overall or equivalent Dow index and the three special hazard ratings are calculated.

The equivalent Dow index, D, is given by the formula:

$$D = B(1 + M/100)(1 + P/100)(1 + [S + Q + L + T]/100)$$

The three special indices for fire, F, internal explosion, E, and aerial explosion, A, are first calculated for basic standard protection only. Offsetting factors for various forms of special protection are then applied, and final adjusted values of these three indices are then calculated. Finally an overall risk rating is calculated, before and after offsetting this for the special protective measures applied or considered.

12.2.1 Dominant material and material factor B

The dominant (key) material on which the material factor is based is next selected. It is defined as:

That compound or mixture in the unit which, due to its inherent properties and the quantity present, provides the greatest potential for an energy release by combustion, explosion or exothermic reaction.

The material factor, B, is in most cases the net heat of combustion of the material in air, expressed as thousands of Btu per pound (2326 kJ/kg). For reactive combinations of materials, the heat of reaction is used if it exceeds the heat of combustion. This material factor is practically the same as that given in the third edition of Dow's guide.

12.2.2 The six hazard factors

The first three, M, P and S, of these factors are elaborations of those given in the third edition of Dow's guide.

The **Special Material Hazards Factor (M)** is applied to take into account any special properties of the key material which may affect either the nature of the incident or the likelihood of its occurrence. Ten properties are listed, with appropriate penalties. They include any tendencies of the key material to act as an oxidant, to polymerise spontaneously, to decompose violently, to detonate, etc. One property designated (m) represents the mixing and dispersion characteristics of the material and also features in the aerial explosion index. The highest penalties recommended are for unstable materials which can deflagrate or detonate.

The **General Process Hazards Factor (P)** relates to the basic type of process or other operation being carried out in the unit. Six main types are listed, including material transfer, physical change only and various types of reaction with different characteristics.

The **Special Process Hazards Factor (S)** reflects 14 listed features of the process operation which increase the overall hazard beyond the basic levels already considered. These take account of operating temperature and pressure, corrosion, erosion, vibration, control problems, electrostatic hazards, etc.

S is evaluated on the assumption that the plant has an adequate control system for normal operation. Credits for more sophisticated safety features such as explosion suppression and combustible gas monitors are applied later.

The **Quantity Hazards Factor (Q)** represents the quantity of combustible, flammable, explosive or decomposable material in the unit which is treated as a separate factor in the Mond Index. It is related to the total quantity K in tonnes of such material in the unit. K also features in the fire index.

The **Layout Hazards Factor (L)** is another separate factor in the Mond Index. The normal working area N of the unit in square metres also features in the fire index, and is defined 'as the plan area of the structure associated with the unit, enlarged where necessary to include any pumps and associated equipment not within the plan area of the structure'.

The height H in metres above ground at which flammable materials are present in the unit also features in the aerial explosion index. L also includes factors for the relation of ventilation rates to flammable vapours which could escape, and 'domino effects' involving the spread of incidents from one unit to another.

The **Acute Health Hazards Factor (T)** is not intended to reflect health hazards as such, but rather the delay caused by the toxicity of escaping materials when tackling a developing or potential fire or explosion. The factor is the sum of penalties for skin effects and inhalation.

12.2.3 Calculation of indices

The equivalent Dow Index (third edition) whose formula was given earlier is not used for interpretive purposes but features in later calculations.

MOND INDEX 1985

SPECIAL PROCESS HAZARDS (Section 8)

1.LOW PRESSURE(BELOW 15 PSIA)	50 TO 150
2.HIGH PRESSURE	0 TO 160
3.LOW TEMP.:1.CARBON STEEL +10C TO -25C	0 TO 30
2.CARBON STEEL BELOW -25C	30 TO 100
3.OTHER MATERIALS	0 TO 100
4.HIGH TEMP.1.FLAMMABLE MATERIALS	0 TO 35
2.MATERIAL STRENGTH	0 TO 25
5.CORROSION & EROSION	0 TO 400
6.JOINT & PACKING LEAKAGES	0 TO 60
7.VIBRATION,LOAD CYCLING,ETC.	0 TO 100
8.PROCESSES/REACTIONS DIFFICULT TO CONTROL	20 TO 300
9.OPERATION IN OR NEAR FLAMMABLE RANGE	25 TO 450
10.GREATER THAN AVERAGE EXPLOSION HAZARD	40 TO 100
11.DUST OR MIST EXPLOSION HAZARD	30 TO 70
12.HIGH STRENGTH OXIDANTS	0 TO 400
13.PROCESS IGNITION SENSITIVITY	0 TO 100
14.ELECTROSTATIC HAZARDS	10 TO 200

SPECIAL PROCESS HAZARDS TOTAL

QUANTITY HAZARDS (Section 9)

MATERIAL TOTAL TONNES
QUANTITY FACTOR

LAYOUT HAZARDS (Section 10)

HEIGHT IN METRES
WORKING AREA IN SQUARE METRES

1.STRUCTURE DESIGN	0 TO 200
2.DOMINO EFFECT	0 TO 250
3.BELOW GROUND	50 TO 150
4.SURFACE DRAINAGE	0 TO 100
5.OTHER	50 TO 250

LAYOUT HAZARDS TOTAL

ACUTE HEALTH HAZARDS (Section 11)

1.SKIN EFFECTS	0 TO 50
2.INHALATION EFFECTS	0 TO 50

ACUTE HEALTH HAZARDS TOTAL

MOND INDEX 1985

LOCATION

PLANT

UNIT

MATERIALS

ADDITIONAL INFORMATION

COMMENT NUMBER

PRESSURE = psig   TEMPERATURE t= DEG.C
MATERIAL FACTOR (Section 5)

KEY MATERIAL OR MIXTURE :
FACTOR DETERMINED BY : B= RANGE
MATERIAL FACTOR

INITIAL   FACTOR   REVIEW

SPECIAL MATERIAL HAZARDS (Section 6)

1.OXIDISING MATERIALS	0 TO 20
2.GIVES COMBUSTIBLE GAS WITH WATER	0 TO 30
3.MIXING & DISPERSION CHARACTERISTICS	-60 TO 100
4.SUBJECT TO SPONTANEOUS HEATING	30 TO 250
5.MAY RAPIDLY SPONTANEOUSLY POLYMERISE	25 TO 75
6.IGNITION SENSITIVITY	-75 TO 150
7.SUBJECT TO EXPLOSIVE DECOMPOSITION	75 TO 125
8.SUBJECT TO GASEOUS DETONATION	0 TO 150
9.CONDENSED PHASE PROPERTIES	200 TO 1500
10.OTHER	0 TO 150

SPECIAL MATERIAL HAZARDS TOTAL

GENERAL PROCESS HAZARDS (Section 7)

1.HANDLING & PHYSICAL CHANGES ONLY	10 TO 50
2.REACTION CHARACTERISTICS	25 TO 50
3.BATCH REACTIONS	10 TO 60
4.MULTIPLICITY OF REACTIONS	25 TO 75
5.MATERIAL TRANSFER	0 TO 150
6.TRANSPORTABLE CONTAINERS	10 TO 100

GENERAL PROCESS HAZARDS TOTAL

MOND INDEX 1985 PAGE NO. 3
FILE NO.

OFFSETTING INDEX VALUES FOR SAFETY & PREVENTATIVE MEASURES

A. CONTAINMENT HAZARDS (Section 16.1)
1-PRESSURE VESSELS
2-NON-PRESSURE VERTICAL STORAGE TANKS
3-TRANSFER PIPELINES A)DESIGN STRESSES
B)JOINTS & PACKINGS
4-ADDITIONAL CONTAINMENT & BUNDS
5-LEAKAGE DETECTION & RESPONSE
6-EMERGENCY VENTING OR DUMPING

PRODUCT TOTAL OF CONTAINMENT FACTORS K1=

B. PROCESS CONTROL (Section 16.2)
1-ALARM SYSTEMS
2-EMERGENCY POWER SUPPLIES
3-PROCESS COOLING SYSTEMS
4-INERT GAS SYSTEMS
5-HAZARD STUDIES ACTIVITIES
6-SAFETY SHUTDOWN SYSTEMS
7-COMPUTER CONTROL
8-EXPLOSION/INCORRECT REACTOR PROTECTION
9-OPERATING INSTRUCTIONS
10-PLANT SUPERVISION

PRODUCT TOTAL OF PROCESS CONTROL FACTORS K2=

C. SAFETY ATTITUDE (Section 16.3)
1-MANAGMENT INVOLVEMENT
2-SAFETY TRAINING
3-MAINTENANCE & SAFETY PROCEDURES

PRODUCT TOTAL OF SAFETY ATTITUDE FACTORS K3=

D. FIRE PROTECTION (Section 17.1)
1-STRUCTURAL FIRE PROTECTION
2-FIRE WALLS,BARRIERS
3-EQUIPMENT FIRE PROTECTION

PRODUCT TOTAL OF FIRE PROTECTION FACTORS K4=

E. MATERIAL ISOLATION (Section 17.2)
1-VALVE SYSTEMS
2-VENTILATION

PRODUCT TOTAL OF MATERIAL ISOLATION FACTORS K5=

F. FIRE FIGHTING (Section 17.3)
1-FIRE ALARMS
2-HAND FIRE EXTINGUISHERS
3-WATER SUPPLY
4-WATER SPRAY OR MONITOR SYSTEMS
5-FOAM & INERTING INSTALLATIONS
6-FIRE BRIGADE ATTENDANCE
7-SITE CO-OPERATION IN FIRE FIGHTING
8-SMOKE VENTILATORS

PRODUCT TOTAL OF FIRE FIGHTING FACTORS K6=

MOND INDEX 1985 PAGE NO. 4
FILE NO.

EQUATIONS

EQUIVALENT DOW INDEX (for initial assessment and review)

D = B(1+M/100)(1+P/100)(1+(S+Q+L+T)/100))

FIRE INDEX

INITIAL ASSESSMENT AND REVIEW F= BK/N

OFFSET F*K1*K3*K5*K6

INTERNAL EXPLOSION INDEX

INITIAL ASSESSMENT AND REVIEW E=1+(M+P+S)/100

OFFSET E*K2*K3

AERIAL EXPLOSION INDEX

INITIAL ASSESSMENT AND REVIEW A=B(1+m/100)(1+p)(QHE/1000)(t+273)/300

OFFSET A*K1*K2*K3*K5

OVERALL RISK RATING

INITIAL ASSESSMENT AND REVIEW R=D(1+(.2E*SQUARE ROOT(AF)))

OFFSET R*K1*K2*K3*K4*K5*K6

INDICES COMPUTATION

INDEX	INITIAL VALUE	INITIAL CATEGORY	REVIEW VALUE	REVIEW CATEGORY	OFFSET VALUE	OFFSET CATEGORY
D						
F						
E						
A						
R						

Figure 12.3 Calculation sheets for Mond Index (courtesy Explosion Hazards Section, ICI Mond Division, Northwich)

The **Fire Index (F)** relates to the amount of flammable material in the unit, its energy release potential and the area of the unit and is given by:

$$F = B \times K/N$$

F values in the range 0–2 class as 'Light', 5–10 as 'Moderate' and 100–250 as 'Extreme'. The Fire Index is related to the Fire Load which would equal $2442 \times F$ in kJ/m^2 ($215 \times F$ in BTU/ft^2) if all the available combustible material were consumed. In practice often only 5–10% is consumed before the incident is controlled.

Typical fire durations are related to the fire load for four different materials, solid combustibles, heavy crude oil, flammable light liquids, and LPG/LNG, and decrease in that order. For a fire load of 100 000 Btu/ft^2 the fire duration ranges from 10 minutes to 2½ hours, and for a fire load of 1 million Btu/ft^2 from 40 minutes to one day.

The **Internal Explosion Index (E)** is a measure of the potential for explosion within the unit and is given by:

$$E = 1 + (M + P + S)/100$$

An internal explosion index of 0–1.5 is categorised as light, one of 2.5–4 as moderate and one above 6 as very high.

The **Aerial Explosion Index (A)** relates both to the risk and magnitude of a vapour cloud explosion [10.4.5] originating from a release of flammable material, usually present within the unit as a liquid at a temperature above its atmospheric boiling point. This index includes quantitative and quantitative factors, and is given by:

$$A = B(1 + m/100)(QHE/1000)(t + 273/300)(1 + p)$$

The **Overall Hazard Rating (R)** is used to compare units with different types of hazards, and is given by:

$$R = D (1 + [0.2 \times E \times (AF)^{\frac{1}{2}}])$$

An Overall Hazard Rating of 0–20 is categorised as light, 100–500 as moderate, 1100–2500 as high and over 12 500 as extreme.

12.2.4 Criteria and review

Ranges of the four indices for different degrees of hazard are given in Table 12.10.

The most important index is the overall hazard rating R. Experience from applying the full method to operating plants has shown that it is uncommon for a unit, after a complete assessment, to have an overall hazard (R) with a category rating greater than 'high'. It is therefore reasonable to assume that a unit assessed at this level can be operated in a satisfactory manner given full regard to the hazards indicated by the assessment. Offsetting usually reduces the overall hazard category by one or two levels and gives a clearer picture of the relative importance of the different protective measures which could be taken.

When the initial assessment is unfavourable, the estimates should be refined by the use of better data. The effects of possible changes in materials of construction, sizes and types of equipment and process

Table 12.10 Mond index ranges for various degrees of hazard

Potential hazard category	Fire F	Internal Explosion E	Aerial Explosion A	Overall Hazard R
Mild or light	0–2	0–1.5	0–10	0–20
Low	2–5	1.5–2.5	10–30	20–100
Moderate	5–10	2.5–4	30–100	100–500
High	10–30	4–6	100–400	500–2500
Very high	20–50	>6	400–1700	2500–12 500
Extreme	100–250		>1700	12 500–65 000
Very extreme	>250			>65 000

conditions, and reduction in inventory should also be considered. When changes thus indicated have been made and all factors have been reviewed, their new values are entered in the 'Reduced Value' column of the form with a note on the reason for the change. The final stage of the index calculations in which the hazards are reduced by applying special safety features and protective measures is done on the basis of these reduced values.

The scope for reductions in the indices by design changes is greatest before the design is finalised. On existing plants most improvements result from the incorporation of the safety features and preventative measures contained in the offsetting section. However, reducing inventories has a significant effect on fire potential and can usually be achieved on new and existing plants.

12.2.5 Index reduction by offsetting measures

Safety features and preventative measures may reduce the probability or magnitude of an incident (sometimes both). The booklet classifies them in these two groups and devotes 20 pages to discussing them. Factors are suggested by which the values of the appropriate index should be multiplied when a safety feature or preventative measure (which is additional to the basic standard) is introduced. Before such measures can be evaluated the basic standards which would apply to the design, construction, operation and personnel training have to be defined. As examples, the basic standard for pressure vessel design is taken as Pressure Vessel Construction Category 3 of BS 5500, and the basic standard for process control instrumentation is the minimum compatible with operation under normal design conditions (i.e. without alarms or trip systems).

Three broad categories of safety features and preventative measures which reduce the *probability* of an incident, and the symbols used for the product totals of their sub-factors, are:

A. Features which improve containment of process materials K_1
B. Features which improve the safety of process control K_2
C. Features which improve safety awareness of personnel K_3

There are several possibilities in each category. The factor for each category is the product of the suggested values for the features and measures which apply.

Three more broad categories of safety features and preventative measures are considered to reduce the *magnitude* of any incident. These are:

D. Fire protection K_4
E. Isolation of process materials K_5
F. Fire fighting K_6

The factor for each category is obtained in the same way as for the first group. Brief descriptions of the features and measures considered in each category are given in the calculation sheet. Where only the basic standards apply, a factor of 1 is used.

The factors K_1 to K_6 are calculated for the actual or proposed protective features. The offset indices are then obtained by multiplying the original (reduced) indices by the appropriate offsetting factors.

Offset Fire Index $= F \times K_1 \times K_3 \times K_5 \times K_6$
Offset Internal Explosion Index $= E \times K_2 \times K_3$
Offset Aerial Explosion Index $= A \times K_1 \times K_2 \times K_3 \times K_5$
Offset Risk Rating $= R \times K_1 \times K_2 \times K_3 \times K_4 \times K_5 \times K_6$

The benefits given by the protective features are assessed by comparing the degrees of hazard for the original and the offset indices. These benefits apply only when the protective hardware is maintained and is in proper working order and when the management procedures on which the benefits depend are followed. Neglect of either will cause the indices to revert to their original values. (The importance of maintaining special protective features was clearly demonstrated by the Bhopal disaster [5.4].)

12.3 Plant layout and unit hazard rating[7]

The layout of process installations is a complex task, even where there are no serious hazards. The site and layout chosen must satisfy commercial criteria as well as being socially and legally acceptable (in the eyes of the local planning authority assisted in the UK by advice from HSE and its Major Hazard Assessment Unit). While industrial planners and plant contractors deal with layout problems frequently, most personnel in the industry, including safety specialists, usually have to live with existing layouts. Risk analysis [14.3] is often required. The final layout is usually a compromise. Mecklenburgh in the I. Chem. E. guide *Plant Layout*[9] put it thus:

In general it is true to say that the most economical plant layout is that in which the spacing of the main equipment items is such that it minimises interconnecting pipework and structural steelwork. . . . As a general rule, as compact a layout as possible with all equipment at ground level is the first objective consistent with access and safety requirements.

'Consistency with safety requirements' is open to a wide range of interpretations. Absolute safety being impossible, one has to decide what degree of risk is acceptable[10]. If this were judged on purely economic grounds and insurance was 100% efficient, we would have reached a state where the layouts adopted minimised the overall costs, including capital, operation, maintenance and insurance. In practice only a few large enterprises (which often carry much of their own insurance) have the knowledge needed to achieve such optimum layouts.

Up to quite recently, safety distances were decided largely on the basis of personal judgement and company policy. As Yelland put it in 1966, 'The existing literature in this field is wide, diffused and often contradictory[11].' Minimum safety distances have, however, been specified for the storage of highly flammable liquids[12], liquid petroleum gases[13] and explosives[14] [9]. The Oil Insurance Association of Chicago has for many years published general guidelines and recommended spacings for plants and plant units in the oil, gas and petrochemical industries[15].

Before the Flixborough disaster (1974), however, the main safety criterion used in the layout of oil and chemical installations was the Institute of Petroleum's electrical safety code[16] [6.2]. This code has an important bearing on the distances of control rooms, offices and laboratories (usually located in 'Safe Areas') from pipes and equipment containing flammable materials. It does not, however, specify safe distances between one hazardous unit and another, nor does it relate the extent of hazardous areas to the amount of flammable material contained in them.

Since Flixborough and other more recent disasters in the process industries, far more attention has been given to plant layout and siting. Lees[4] deals with the subject in depth and gives a comprehensive bibliography on layout. Liston's chapter in Fawcett and Wood's book[17] is useful reading for those faced with layout problems. Figure 12.4(a) from this reference shows the cross-section of a recommended 'in-line' arrangement of process equipment, with the relative positions of roads, pipebridges, pumps and plant structures. This has roadways flanking the equipment on either side, giving excellent access for firefighting, etc. Next to the road on the left is a gantry-way with a crane on rails to handle heavy equipment for maintenance, with buried cooling water headers running below it. This provides useful space for tube bundles and other equipment to be stored or worked on during maintenance. Even where a gantry cannot be justified, an open strip between the road and the plant is recommended. Next to the right lies an area depressed by about 200 mm, where fractionating towers, heat-exchangers, vessels and other stationary process equipment are located. This area is divided by low walls (about 200 mm) running across it to isolate spills of process liquids which drain into special sewers. Process pipes run on raised racks to the right of the depressed area, thus allowing unhindered passage to persons under them. Air-cooled heat-exchangers are located above the pipe racks. Pumps and other rotating equipment are located in a line to the right of the pipe racks and adjacent to the road on the right flank of the in-line arrangement. Other components of the unit such as control rooms, furnaces and reactors may be located at either end of the in-line arrangement. Figure 12.4(b)

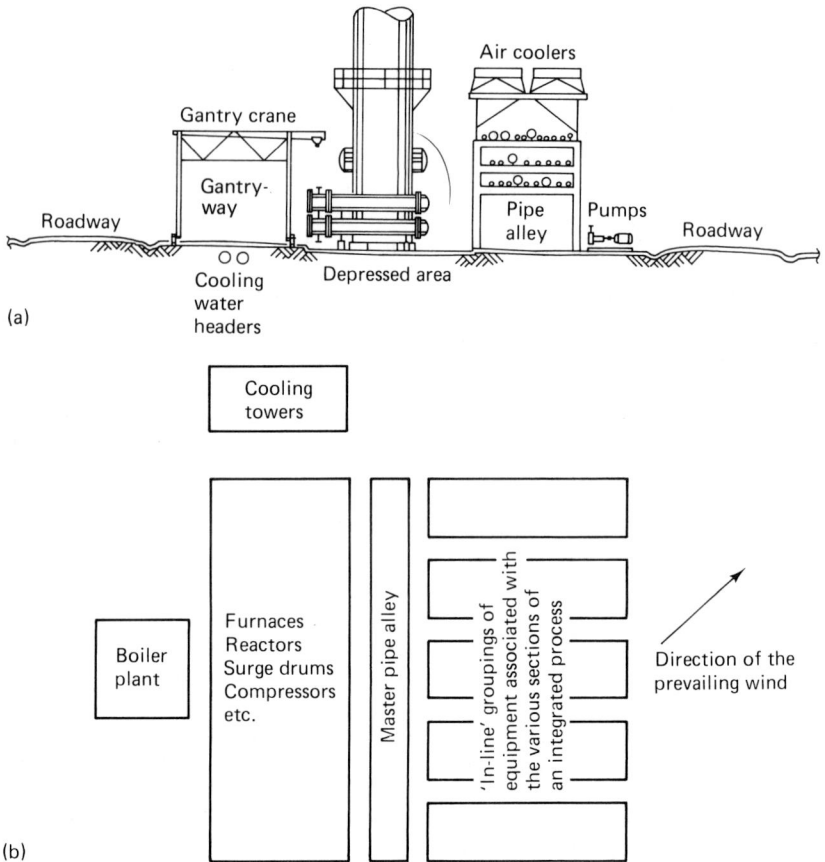

Figure 12.4 (a) Cross-section of in-line arrangement of process equipment. (b) Example of applying the in-line arrangement to integrated processing units (from Liston[17], courtesy Wiley-Interscience)

from the same reference shows an integrated plan of several 'in-line' arrangements of process units, with a main pipe-rack at one end separating them from furnaces, reactors, compressors and boiler plant, etc.

Few of these references give recommendations on the separation of plant and storage units of various degrees of hazard from one another and from boundaries, office and other buildings, etc. This can only be done if one has some numerical scale such as the Dow or Mond Index for the degree of hazard of units themselves. Dow's guides do not give explicit spacing recommendations, although some latter ones[3] allow the maximum probable property damage for any proposed layout to be estimated, and the layout to be revised to reduce excessive damage estimates.

12.3.1 Application of the Mond Index to plant layout[7]

After publication of the Mond Index its author, D. J. Lewis, gave a further paper on its application to plant layout and spacing problems[7]. The spacings proposed in this method are intended to ensure that incidents apart from disasters will produce only moderate damage to adjacent plant sites, with minimal effects outside the works boundary. The spacings needed to prevent knock-on effects arising from disasters such as major explosions, major tank froth-overs and catastrophic vessel failures are not considered to be reasonably practicable.

Lewis refers to the Overall Risk Rating before allowance for offsetting factors as R_1, and after allowance for offsetting as R_2. He first recommends the following basic principles for an optimum layout:

1. Roads should enter the plant from at least two points, preferably on opposite sides of the plant perimeter.
2. Emergency vehicles should have access to units with moderate and high fire risks from at least two directions.
3. Control rooms, amenity buildings, workshops, laboratories and offices should be close to the site perimeter, and next to units of mild or low risk to shield them from higher-risk units.
4. Pipebridges should not allow incidents to be transferred easily from one unit to another. Key pipebridges should be assessed for hazard potential.
5. High-risk units should be separated from each other by units of lower risk.
6. Ignition sources such as furnaces, electrical switchgear and flare stacks should be adequately separated from units.
7. Medium-risk units should not be placed next to populated buildings.
8. Units with high risk of toxic release should be well separated from populous buildings and areas both inside and outside the works perimeter.
9. Units with high values of the Aerial Explosion Index A_2 should be separated from plants or works boundaries by areas of low-risk activities with low population densities.
10. Pipebridges with medium to high overall risk rating should be placed where they are at least risk from transport accidents and incidents on tall process units.
11. Units previously treated separately for hazard ratings may be combined into larger ones, providing (a) the risks are similar, (b) potential losses are not excessive and (c) the reassessed rating R_2 is not excessive.
12. The process flow should be as logical as possible to minimise pipework.
13. Pipebridge routes should be chosen to minimise the spread of incidents and should allow as much control of flows as possible if there is an incident.
14. Storage units should be well separated from operational areas and as far as possible from roads and railways within the site.

These principles allow an initial layout to be prepared with a nominal inter-unit spacing of 10 m. This layout is related to the flow of materials and control and maintenance requirements. The basic pipebridge and access routes are marked on it and the nominal inter-unit distance is then replaced by distances chosen logically. After explaining how spacing distances are to be measured (e.g. whether from the bund wall of a storage tank or from the tank itself), Lewis gives five tables of spacing distances involving process and storage units.

In cases where process and other considerations do not allow the recommended spacing to be used, special means such as fixed water sprays should be considered to reduce the effects of heat radiation or the spread of fire or to control the escape of toxic or flammable materials.

After considering the effects of serious releases of toxic and flammable materials (over 5 t for the latter), and units of high Aerial Explosion Index A_2, Lewis recommends that these be located near the centre of a site.

A problem which can arise with Lewis's method is that the process units selected for the Mond Index tend to be small. Two or more may occupy the same structure and interpenetrate, with no clear boundary between them. Lewis therefore suggests that small units may be amalgamated into larger ones. His proposals on distances are consistent although his figures are open to debate. His paper contains no actual examples of the application of his method to plant layout.

12.4 Maximum probable property damage from vapour cloud explosions (VCEs)

At present about two or more VCEs occur world-wide every year. Because of their enormous destructive potential, insurers are concerned both to identify installations where one might occur and to estimate the maximum resulting loss. International Oil Insurers (IOI), London, have published EML (estimated maximum loss) 'Rules'[18], recently updated as 'guidelines'[19] for use by insurance surveyors. Industrial Risk Insurers (IRI), Chicago, have also published their own guide[20] which differs in several respects from the IOI rules, for estimating the 'maximum probable property damage' (MPPD). (The terms EML and MPPD appear to have identical meanings. Both refer to property losses including plant inventories, but exclude personal injuries and business interruption losses.)

CWA Information and Research Ltd of London offer computer programs for EMLs resulting from VCEs based on the IOI rules, and also for 'instantaneous fractional annual losses'[21] caused by fires and explosions on plants handling flammable materials.

The method outlined here is a fairly simple one (which the writer developed in the 1970s for an international insurance broker). It aims to combine the best features of the IOI and IRI methods and serves as an introduction to more elaborate procedures.

12.4.1 Flammable materials with VCE potential

Broadly speaking, all flammable gases and flammable liquids (flash point <40°C) are assumed capable of supporting a VCE, when present in appropriate quantity and condition. C_1 to C_4 hydrocarbons are classed as gases and most C_5 to C_9 hydrocarbons as flammable liquids.

Gases

Gases vary considerably in their fire and explosion hazards [10.2]. Some like ammonia, methyl chloride and carbon monoxide which have low heats of combustion are unlikely to support VCEs on their own, but may contribute when mixed with other gases. Some vapours such as acetaldehyde are very readily ignited, e.g. by a hot steam pipe. Butadiene, vinyl acetylene, formaldehyde, acetaldehyde, acrolein and ethylene oxide are capable of spontaneous exothermic polymerisation under appropriate conditions [8.6.4], while acetylene, vinyl acetylene and ethylene oxide can explode in the absence of air [9.2].

To qualify as a VCE hazard, a gas needs to be flammable and present in one of the following forms:

1. As gas under a pressure of 35 bar g. or more. This form is generally restricted to gases with critical temperatures below 20°C. Examples are hydrogen, carbon monoxide, methane and ethylene.
2. As refrigerated liquid. This form can cover the entire range of flammable gases from hydrogen to butane and ethylene oxide.
3. As liquid under pressure. This form is generally restricted to gases with critical temperatures above and boiling points below 20°C. Examples are C_3 and C_4 hydrocarbons, ammonia, ethylene oxide, ethyl chloride and vinyl chloride.

The gases involved most frequently in vapour cloud explosions are C_3 and C_4 hydrocarbons, present initially in form (3).

Liquids

To qualify as a VCE hazard, a flammable liquid must generate a large amount of vapour very rapidly when it is released. This usually means that it is initially under pressure and at a temperature at least 10°C higher than its atmospheric boiling point. Most hydrocarbon liquids at their boiling points under a pressure of 20 bar g. vaporise completely when the pressure is suddenly reduced to atmospheric. Their heats of combustion (gross) lie between 42 000 and 50 000 kJ/kg. For non-hydrocarbon organic liquids, the percentage vaporised under similar conditions lies between 25 and 85%, with heats of combustion between 10 000 and 40 000 kJ/kg. These differences partly explain why hydrocarbon liquids have been far more frequently involved in VCEs than non-hydrocarbons.

Liquid hydrocarbons which have been involved in VCEs include gasolines, intermediate light fractions and naphthas in the gasoline boiling range, the light ends of crude oils, aromatics and cyclohexane.

Organic liquids which could in theory support a VCE include the lower alcohols, aldehydes, ketones, esters, ethers, oxides, acid anhydrides, as well as some compounds containing chlorine, sulphur and nitrogen. Only a few of these have actually caused a VCE.

12.4.2 Sources of VCEs

While the usual point of escape has been a pipe or pipe fitting, the sources of the material escaping have generally been vessels, due to their much larger inventories compared to pipes, (except for pipelines). Sources are therefore grouped as:

1. Storage vessels and tanks;
2. Process vessels (including columns and reactors);
3. Pipelines (generally large or long).

Storage vessels and tanks

Pressurised storage vessels containing more than 10–20 t of liquefied flammable gases are nearly all potential sources of VCEs. In theory the entire inventory of a full vessel could escape and vaporise in the event of a rupture. Most liquid storage tanks under ambient conditions do not qualify except for:

- Storage tanks containing flammable liquids immiscible in water which have been fitted with steam coils or other means of heating [10.5.4];
- Storage tanks containing a material capable of spontaneous exothermic polymerisation causing part of the contents to boil [8.6.4];
- Cryogenic storage tanks containing liquefied flammable gases including LNG and LPGs. As with pressurised storage, the entire contents could escape and vaporise in the event of a rupture.

Provided the storage vessels and tanks are well isolated by automatic or remotely operated valves, the maximum amount of an escape is usually limited to the contents of the largest single tank or vessel whose contents qualify for VCE potential.

Process vessels and equipment

Two types must be considered, high-pressure gas-filled equipment and moderate to low-pressure equipment in which most of the flammable material is in the liquid state.

High-pressure gas-filled equipment is typical of hydrocrackers and HP hydrogenation processes, OXO plant reactors, and HP polyethylene, ammonia, synthetic methanol and ethanol plants. The inventory of flammable gas in a HP train is calculated on the assumption that the entire contents of the high-pressure system downstream of the compressor will escape in the event of rupture, unless proven means of automatically isolating parts of the system rapidly in case of rupture have been installed.

Process vessels containing liquid include feed and buffer vessels, reactors, settlers and separators, liquid–liquid extraction columns, gas-absorption columns, adsorbers and driers, distillation columns with associated reboilers, condensers, reflux drums and heat exchangers. In estimating the maximum amount of flashing liquid which could escape from any vessel in the event of rupture, it is assumed to be filled with liquid to its maximum normal operating level at the time of the escape. The corresponding volume of liquid, V_{max} is calculated from this level and the vessel dimensions, after deducting the volumes of vessel internals, packing, catalyst, etc.

The maximum weight of liquid is then given by:

$$W_L = d_L V_{max}$$

where W_L = maximum liquid inventory, t

$\quad\quad d_L$ = flammable liquid density under operating conditions

To judge the extent to which the contents of connected vessels contribute to the escape, the main process lines containing flammable liquids, both upstream and downstream of the hypothetical point of escape, need to be examined. Here it is assumed that (properly maintained) non-return valves downstream of the point of escape close when an escape occurs, as do automatic and remotely operated isolation valves designed to close in the event of an escape.

The position of branches on vessels connected to the pipe or vessel from which the escape occurs is important. Only the liquid contents of these vessels which lie above these branches need normally be considered, except where there are internal baffles or there is reason to expect considerable flashing and entrainment in the vessel. This is allowed for in 12.4.3.

The probability of escape from other parts of the process unit cannot be ignored. Pumps, stirrer glands, heat exchanger tubes and pipework are more likely to fail than the shells of pressure vessels.

12.4.3 Maximum plausible size of explosion

The maximum plausible size of explosion is described as tonnes of TNT equivalent, which is related to the quantity of flammable gas or vapour escaping by the equation:

$$W_e = W_V Hf/4652$$

where W_e = equivalent weight of TNT, t
$\quad\quad W_V$ = weight of flammable vapour released, t
$\quad\quad H$ = heat of combustion (gross) of flammable vapour, kJ/kg
$\quad\quad f$ = explosive yield factor
$\quad\quad 4652$ = heat of combustion of TNT, kJ/kg

The main problem lies in assessing a realistic value for f, which allows for the facts that part of the vapour will escape without ignition, and that part will burn before the rest explodes. For estimating MPPD, a value for f of 0.05 is assumed. This is based on historical studies of many VCEs. For escapes of flammable liquids at temperatures above their atmospheric boiling points, W_V is calculated as:

$$W_V = \frac{2W_L C_p(T_1 - T_2)}{H_V}$$

with a maximum value of $W_V = W_L$
where W_L = weight of liquid escaping, t
$\quad\quad C_p$ = specific heat at $(T_1 + T_2)/2$ of liquid escaping, kJ/kg
$\quad\quad T_1$ = temperature of liquid in plant
$\quad\quad T_2$ = atmospheric boiling point of liquid
$\quad\quad H_V$ = heat of vaporisation of liquid at T_2, kJ/kg

Here the weight of vapour flashing at the point of release is arbitrarily doubled to allow for the further evaporation of liquid spray in air.

For liquefied flammable gases with boiling points below 20°C:

$$W_V = W_L$$

Where it is necessary to take account of vapour released from a series of connected vessels by flashing of liquid inside them, the following expression is used:

$$W'_V = \frac{W'_L.C_p(T_1-T_2)}{H_V}$$

with a maximum value of $W'_V = W'_L$

W'_V and W'_L apply to the flammable liquid contents of other vessels connected to the vessel considered as the source of the vapour cloud, and which lie below the connecting nozzles of the vessels to which W'_L refers. The total vapour released is then given by:

$$W_{VT} = W_V + W'_V$$

12.4.4 Maximum probable property damage (MPPD)

Assessment of maximum probable property damage is based on the assumption that the vapour cloud explodes with its centre either at the point of release or at some other centre within the works boundary [12.4.5] where it might cause even greater damage. Here one needs an accurate plot plan of the works, detailed cost data for the installed equipment, pipework, buildings and structures, and for the inventories of process materials in process and storage, as well as a calculator and drawing compass.

Either the IOI[18, 19] or the IRI[20] method may be used for estimating property damage. The latter is more detailed although the IOI method appears adequate for most purposes considering the approximate nature of these estimates. It assumes an average property damage of 80% within a circle of radius R centred on the probable centre of the explosion, and an average property damage of 40% due to blast alone within a concentric annular area between radii R and $2R$. To the latter figure a further 40% is added for refinery and petrochemical plant and storage areas which contain significant inventories of flammable materials, to allow for damage caused by secondary fires. The radius R is related to the power of the explosion. Table 12.11, which has been adapted from the IOI method to match the present treatment, gives values of R and $2R$ for various TNT equivalence.

Having selected one or more potential sources of VCE emissions and estimated their TNT equivalence, circles with the radii indicated in Table 12.11 are drawn on the works plan with the sources as centres. The damage occurring within the inner circle and in the annulus for each source is then estimated and the two are added to give the total MPPD for each source. The MPPD is usually calculated on the basis that it includes the inventory of process materials. Whether it does so or not should be clearly stated. Where there are several potential sources within the same works, one is mainly interested in the source giving the highest MPPD.

The question of cloud drift before ignition is mainly of interest when there are areas with higher densities of property value within or adjacent to

Table 12.11 Assumed VCE damage in target areas

Explosion equivalence (t TNT)	Circle radii (m)	
	R, inner circle 80% damage	2R, outer circle 80% blast damage in annulus
5	80	160
10	100	200
20	125	250
30	145	290
40	160	320
50	170	340

the works perimeter than those near the source of escape. This should be apparent from the works plan with the property values marked on it.

12.4.5 Ignition sources, drift and other factors

While most large escapes of flammable vapour with VCE potential have reached an ignition source and burnt and/or exploded, several such escapes have occurred without igniting, and at least one cloud drifted 0.5 km before exploding. The further a cloud drifts, the more of its flammable vapour will have been diluted by air to below the lower flammable limit, thus reducing the explosion potential. (On the other hand, some limited drift in which more concentrated flammable vapour is diluted to below the upper explosive limit may increase the force of an explosion.) It is important to appreciate that there are some very large potential VCE sources each containing over 1000 t of liquefied flammable gases under pressure (such as C_3 and C_4 storage spheres). The contents of one of these if released suddenly could travel several kilometres in a light breeze (e.g. over an estuary) before exploding.

There are thus several reasons for considering possible points of ignition for any potential VCE source which is being studied. The ignition point is unlikely to become the centre of the explosion because only the edge of the flammable vapour cloud, which is usually very large and may cover an area the size of a football pitch, has to reach it for ignition to occur. The generally accepted ignition source of the Flixborough explosion was thought to have been the furnace of another plant about 100 m from the point of the vapour escape. Yet the probable centre of the explosion was much nearer the point of escape than to the furnace. If there are clearly identifiable permanent ignition sources within 150 m of a large potential VCE source, the cloud is unlikely to travel far before it ignites.

In making the study the name of the material being considered as a VCE source is marked on the plan, together with its ignition temperature in air [10.2.3]. Potential points of ignition for the source material are then considered and marked on the plan. Methane, ethane and propane with ignition temperatures over 450°C require a strong ignition source such as a furnace, an open flame (flare, burning pit, welding torch), a road vehicle or a high-voltage cable. C_5 to C_9 hydrocarbons except aromatics have ignition temperatures in air between 230 and 300°C, and could be ignited by high-pressure steam pipes, hot oil pipes and other hot objects.

MPPD studies should be made for all 'major hazard' installations with VCE potential. It is important that these be done at the design stage and that their implications are fully appreciated then.

References

1. Dow Chemical Company, *Fire and Explosion Safety and Loss Prevention Guide – Hazard Classification and Protection*, 3rd edn, Dow Chemical Company, Midland, Mich. (1972)
2. *Dow's Process Safety Guide*, American Institute of Chemical Engineers, New York (1973)
3. Dow Chemical Company, *Fire and Explosion Index – Hazard Classification Guide*, 4th edn (1976), 5th edn (1980), 6th edn (1987), Dow Chemical Company, Midland, Mich.
4. Lees, F. P., *Loss Prevention in the Process Industries*, Butterworths, London (1980)
5. Lewis, D. J., 'The Mond Fire, Explosion & Toxicity Index – A development of the Dow Index', paper presented at the AIChE Loss Prevention Symposium, Houston, 1–5 April 1979
6. ICI Mond Division, *The Mond Index*, 2nd edn, Imperial Chemical Industries plc, Explosion Hazards Section, Technical Department, Winnington, Northwich, Cheshire CW8 4DJ (1985)
7. Lewis, D. J., 'Application of the Mond Fire, Explosion & Toxicity Index to Plant Layout and Spacing Distances', *Loss Prevention*, **13**, 20 (1980)
8. NFPA Standard No. 325M, *Fire Hazard Properties of Flammable Liquids, Gases and Volatile Solids*, National Fire Protection Association, Quincy, Mass. (1969)
9. Mecklenburgh, J. C., *Plant Layout*, Leonard Hill Books, Aylesbury (1973)
10. Dransfield, P. B., Lowe, D. R. T. and Tyler, B. J., 'The problems involved in designing reliability into a process in its early research or development stages', *I.Chem.E.Symp. Ser. No. 66, 67, 1981*, Rugby (1981)
11. Yelland, A. E. J., 'Design Considerations when Assessing Safety in Chemical Plant', *The Chemical Engineer*, September (1966)
12. HSE, *Guidance Note CS/2, The storage of highly flammable liquids*, HMSO, London (1977)
13. HSE, *Guidance notes for the storage of liquefied petroleum gas at fixed installations*, HMSO, London
14. S.I. 1951, No. 1163, *Stores for Explosives Order*, HMSO, London
15. Oil Insurance Association, *Publication 631, General Recommendations for Spacing in Refineries, Petrochemical Plants, Gasoline Plants, Terminals, Oil Pump Stations and Offshore Properties*, Chicago, Illinois (1972)
16. The Institute of Petroleum, *Model Code of Safe Practice, Part 1. Electrical*, I.P., London (1965)
17. Liston, D. M., 'Safety aspects of site selection, plant layout and unit plot planning' in *Safety and Accident Prevention in Chemical Operations*, edited by Fawcett, H. H. and Wood, W. S., 2nd edn, Wiley-Interscience, New York (1982)
18. International Oil Insurers, *EML Rules*, London (1976)
19. International Oil Insurers, *The evaluation of estimated maximum loss from fire or explosion in oil, gas and petrochemical industries with reference to percussive unconfined vapour cloud explosion*, London (1985)
20. International Risk Insurers, *Vapor cloud loss potential estimation guide*, Chicago (1975)
21. Whitehouse, H. B., 'IFAL – a new risk analysis tool', *I. Chem. E. Symposium No 93*, Rugby (1985)

Chapter 13

Hardware hazards

This chapter deals first with the mechanical causes of metal failure, then with general machine hazards, and finally with the hazards of centrifuges, mixers, pumps and compressors.

13.1 Mechanical causes of metal failure

Metal failure is a complex subject. A survey of the causes of 685 metal failures in process equipment over the period 1968-1971 showed that 55% of the failures were caused by corrosion [11] and 45% by mechanical failure[1], of which the commonest cause was fatigue (Table 13.1).

The tensile or (more recently) the yield strength of a metal is its main mechanical property considered in design. This is used in conjunction with

Table 13.1 Causes of mechanical failure of process equipment

Cause	Percentage of failures
Fatigue	33.2
Overload	12.1
Abrasion, erosion or wear	12.1
Poor welds	9.8
Cracking, plating	6.9
Creep or stress-rupture	4.2
Overheating	4.2
Cracking, heat treatment	4.2
Defective material	3.6
Embrittlement, strain, age	2.9
Brittle fracture	2.7
Cracking, thermal	1.3
Cracking, weld	1.3
Leaking through defects	0.9
Warpage	0.9
Embrittlement, sigma	0.7
Blisters, plating	0.2
Brinelling	0.2
Cracking, liquid metal penetration	0.2
Galling	0.2
Impact	0.2

a largely empirical factor of safety. However, only a small proportion of failures are caused by inadequate tensile strength. This is mainly because most metals stressed beyond their yield point will undergo a large amount of plastic deformation before rupture occurs. In a well-designed structure this deformation will redistribute the stresses over other members, thereby making the structure less prone to tensile failure. Wise lists the principle mechanical modes of metal failure as[2]:

1. Fatigue,
2. Brittle fracture,
3. Creep,
4. Collapse due to buckling or general yielding,
5. Tearing or shear failure.

These are discussed next followed by an introduction to fracture mechanics.

13.1.1 Fatigue

Fatigue damage occurs locally on the surface of a metal which is subject to repeated stresses. The fatigue strength depends on both the geometry of the object, which may concentrate stresses, and its surface condition (roughness, corrosion, etc.). Fatigue may be initiated by corrosion pits or cracks on the metal surface. The fatigue strength of an object drops gradually with increasing number of stress applications to a limit which is usually reached after 10^6–10^7 cycles.

Although the causes of fatigue in rotating, reciprocating or vibrating machinery are easily recognised, they are less obvious in statically loaded vessels and pipes, where the following causes may be present[3]:

• Periodic temperature transients;
• Restrictions of expansion or contraction during normal temperature variations;
• Applications or fluctuations of pressure;
• Forced vibrations;
• Variations in external loads.

The fatigue strength of a component may be surprisingly low and bear little relationship to its tensile strength. Fatigue failure spreads gradually over the affected section, usually until another mode of failure, often a fast-spreading brittle fracture, supervenes, when the failure can be catastrophic (Figure 13.1). Fatigue is not always recognised as the initial or preceding stage of such failures which would not however have occurred in its absence. Some writers have estimated that over 80% of all metal failures are initially due to fatigue[2].

13.1.2 Creep

This is the gradual yielding of a continuously loaded object which occurs above a temperature typical of the metal which has limited use above that

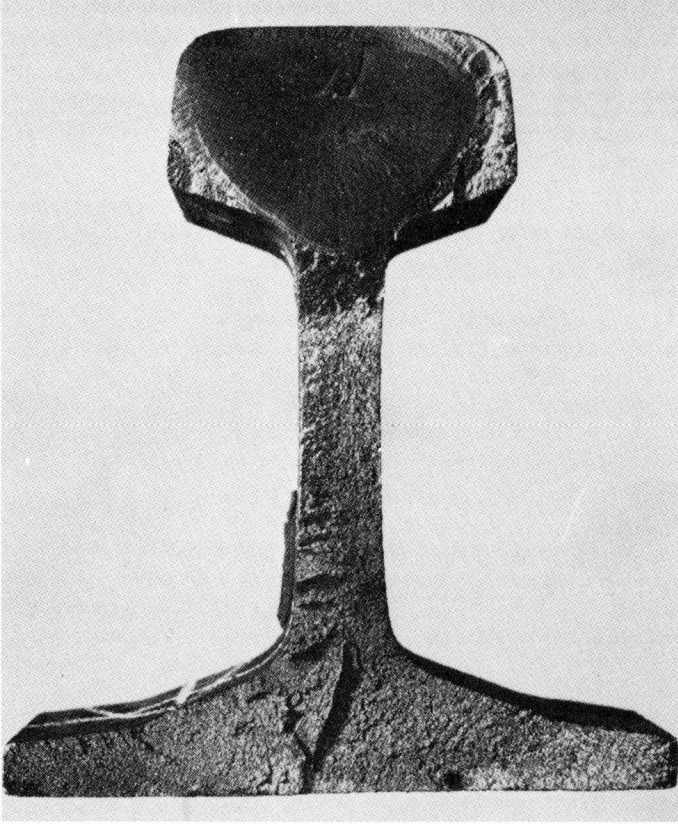

Figure 13.1 Fatigue flaw in head of rail leading to brittle fracture (from S. Wise, 'Why metals fail', in *Engineering Progress through Trouble*, by permission of the Council of the Institution of Mechanical Engineers)

temperature. Approximate creep threshold temperatures are:

Mild steel	400°C
Low-allow steels	500°C
Austenitic stainless steels	600°C

13.1.3 Brittle fracture

Brittle fracture applies to iron and steel and arises from their crystal structure (a body-centered cube). This allows two modes of fracture. The first is by extended slip along a plane at 45° to the principal tensile stress, while the second is by brittle cleavage on a different set of crystal planes at right angles to this stress. The first tends to be slow with much plastic flow, the second rapid and without yielding. A 'notch', meaning any change of section which affects the local stress distribution, is essential for brittle fracture to occur in steel. 'Notches' include keyways, circumferential grooves, holes, contour changes, scratches, threads and fatigue cracks, as well as the standard notches cut in metal specimens for impact testing.

Specimens of the same steel when subjected to a standard notched impact test at different temperatures behave differently at high and low temperatures. Above the 'transition temperature' the tests give high results and failure is by ductile fracture. Below this temperature failure is by brittle fracture and the energy required to break the specimen is reduced to one quarter or less (Figure 13.2).

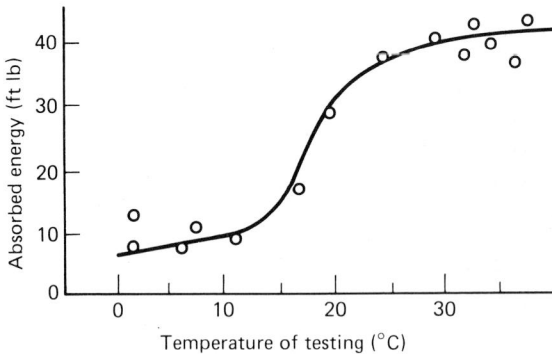

Figure 13.2 Effect of temperature on notched impact strength of steel (from S. Wise, 'Why metals fail', in *Engineering Progress through Trouble*, by permission of the Council of the Institution of Mechanical Engineers)

For plain carbon steels, the transition temperature goes up with the carbon content, from about 20°C for 0.25% carbon to 100°C for 0.5% carbon. BS 4360[4] makes it possible to specify structural steels with a guaranteed maximum transition temperature of −15°C or less. For lower temperatures alloy steels, particularly austenitic stainless steels, are used. Resistance to brittle fracture also depends on other factors besides the metal specification. It decreases as the thickness increases and also depends on the notch toughness, the extent of cracks and the degree of local embrittlement at tips of pre-existent defects. The UK pressure vessel design standard BS 5500 gives methods of selecting and testing alloy steels for minimum design temperatures down to −110°C[5].

Fatigue cracks which concentrate stresses at their tips commonly initiate brittle fracture. Other initiating sources are hydrogen cracks [11.4.8], cracks under welds in high-tensile steels, and shrinkage tears in weld metal and steel castings.

Brittle fracture was held responsible for several bridge and ship disasters, including the SS *Schenectady* at moorings in 1942 and several other Liberty Ships built in World War II. It was probably responsible for the failure of a steel LNG storage tank (at −155°C) at Cleveland, Ohio, in 1944, which resulted in a fire that killed 128 people and injured hundreds.

The pipework of gas-separation plants is subject to the combined hazard of low-temperature embrittlement and high stresses caused by thermal contraction. This combination can come about if a liquified gas such as ethylene at a low temperature passes into pipework designed to operate at a higher temperature, and has happened as a result of the failure of a liquid level controller. An escape of ethylene caused in this way took place on a UK petrochemical plant where the writer was working in the late 1950s [10.5.2]. It was followed by a flash fire which resulted in several deaths and injuries. The same dual hazard was probably responsible for the explosion at DSM's ethylene plant at Beek, Holland, in 1975, in which 14 people were killed[3].

13.1.4 Buckling

Long or slender columns under compression loads tend to buckle due to tensile yielding on one side of the column. This occurs when the length/radius ratio of the column exceeds a critical slenderness ratio. This ratio depends on the mechanical properties of the material, and generally lies between 120 and 150. Columns for which this ratio is exceeded become unstable when a critical load is reached. They then continue to yield until they fail without any further increase in the load. Parts of machines and structures which are normally under tension sometimes fail by buckling when they are exposed to compressive forces under conditions which were not anticipated in their design. Another cause of buckling is the weakening of load-bearing members caused by drilling holes to attach brackets, etc. or by removing or loosening essential parts of a structure. There have been several disastrous failures through buckling of 'falsework', hastily erected without proper design or engineering supervision, during the construction of large concrete structures[6]. To design structures which never buckle requires well-trained and experienced structural engineers who can anticipate unusual loadings (including earthquakes). To prevent weakening of structures caused by unauthorised modifications requires good overall supervision by a competent engineer.

13.1.5 Shear

Although the shear strengths of steels and other metals are well known to the designer, shear failures commonly occur through failure to appreciate the shear stresses to which bolts, rivets, pins, shafts and shaft couplings may be exposed. The problem is most acute when rotating machinery with considerable momentum is suddenly accelerated or decelerated, and on machinery where an obstruction can be met, e.g. mixers. Here shear-pins which fail deliberately when certain torques are exceeded are sometimes needed to prevent worse damage. Torque-limiting devices are frequently necessary, and fitters require training to ensure that bolts and nuts are not overtightened during construction and maintenance.

13.1.6 Fracture mechanics[2]

This relatively new engineering science started with the discovery that the brittle fracture of materials such as glass is due to the presence of many randomly distributed internal flaws. The strength of the material depends on the stresses at the ends of the flaws which in turn depend on the overall stressing and the size of the flaws. On applying the theory to metal fractures, it was found that the flaws were not distributed at random, but resulted from fatigue, hydrogen cracking or defects in manufacture.

The theory has led to two new concepts, 'stress intensity' at a notch or flaw and 'fracture toughness'. 'Stress intensity' K describes the intensification of stress in front of a crack in terms of its length:

$$K = \sigma(\pi a)^{\frac{1}{2}}$$

where σ is the nominal stress and a the crack half-length.

The 'fracture toughness' K_C is the stress intensity at fracture measured under carefully controlled conditions.

From this theory the critical size of cracks which can be accepted for a component operating under given conditions can be calculated. By extending the theory, the rate at which a (fatigue) crack will propagate can also be estimated. Since cracks can be measured by non-destructive testing methods [17.4], fracture mechanics provides a valuable aid to the inspection of metal components. It helps to establish acceptance limits, to decide on the maximum interval between inspections, and also to decide whether a component in which a flaw of a certain size has been detected may remain in service (e.g. while a replacement is obtained) and if so, for how long. The theory also helps in the design of metal components and in the development of improved engineering materials.

13.2 General hazards of moving machinery

Few process operators are exposed to the machine tool risks faced by metal and wood workers. Yet they often have to fill and empty mixers, centrifuges and the hoppers of extrusion and tableting machines by hand, and clear blockages of process materials close to moving parts. Although the complete automation of such machines reduces the exposure of process operatives, it can increase the exposure of maintenance workers and cleaners to machine hazards.

The following precautions apply to most moving machinery:

- One or more emergency stop buttons for the motor driving the machine should be installed within easy reach of the machine operators;
- Greasing points should be brought to safe locations by extension tubes;
- Where it is necessary to inspect moving parts of a machine, viewing windows should be incorporated into the guards;
- On machinery subject to automatic 'on–off' control (e.g. by level switches), a notice should be prominently displayed, e.g. **This machine is liable to start without warning**;
- A carefully considered decision must be taken about all electrically operated equipment (especially motors) as to whether it should restart automatically or only by human intervention when electricity is restored after an electricity failure.

Machine hazards are characterised generally by the type of motion:

- Rotary,
- Reciprocating and/or sliding,
- Oscillating,
- Complex.

(a)

(b)

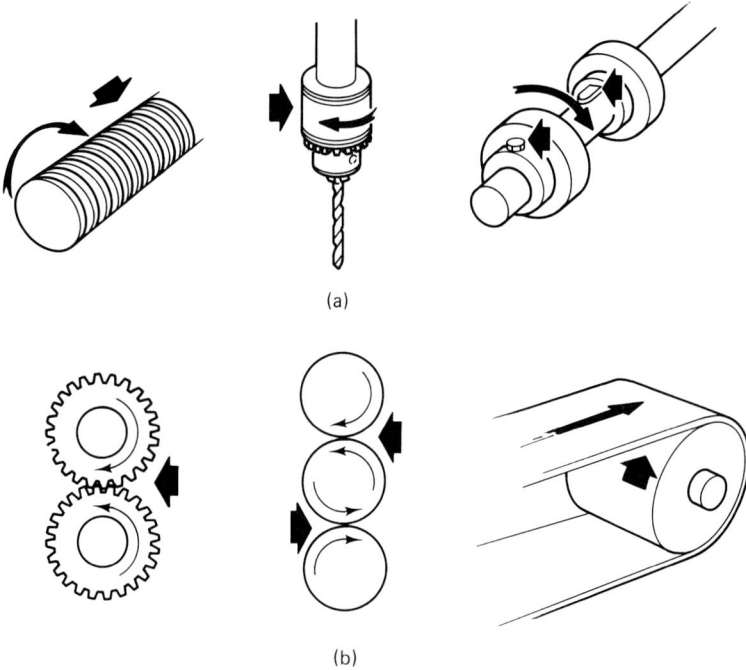

Figure 13.3 Hazards of rotary motion: (a) rotating parts operating alone, (b) in-running nips

Several types of hazard are caused by rotary motion (Figure 13.3), e.g. shafts and other rotating parts, in-running nips, and screw and worm mechanisms.

Noise and vibration are produced by most machines and special care is needed in their selection, mounting, maintenance and in measures to mitigate these nuisances. Excessive noise [22.7] is not only a health hazard, causing permanent hearing impairment, but is also a safety hazard in preventing people from properly hearing speech and audio signals of all kinds. Excessive vibration is also both a health and safety hazard, affecting the bodies of workers, moving objects resting on high ledges till they fall off, and causing mechanical damage and metal fatigue. Both subjects have received a great deal of specialised study in recent years and expertise is readily available to be drawn on. It is important that these problems receive adequate attention when plant is designed and equipment is ordered. The first cost of machines which satisfy acceptable criteria for noise and vibration is nearly always higher than that of machines which do not, but the direct and indirect costs of the problems to which these machines lead generally far outweigh the initial 'saving'.

13.2.1 Shafts and rotating parts

Many revolving shafts which appear deceptively smooth and harmless have projections such as key-heads, set-screws or cotter-pins. They can catch loose ties, scarves, tapes, torn overalls, hair, shoe-laces or string projecting from a pocket. They cause air currents which helps to trap the loose material, which, once lapped round, is quickly wound up, dragging the wearer into the machine, often with fatal results. Even more dangerous are rotating parts with arms and discontinuities such as open-arm pulleys, fan blades, spoked gear- and fly-wheels, centrifuges and mixers. Other dangerous rotating machinery includes rotating drum dryers, 'beaters' used in paper making, carding machines and cotton openers. All rotating shafts should be guarded and totally enclosed. If access to one is absolutely vital, e.g. for lubrication or monitoring vibration, a special permit [18.3] signed by the plant manager or other appropriate person should be issued to the named and trained person who is authorised to do the work. He or she should be responsible for removing and replacing the guard immediately before and after the work, and signing an appropriate record with the time and date when this is done.

13.2.2 In-running nips

These exist wherever two parallel shafts, rolls or wheels in close proximity rotate in opposite directions. Examples are found in steel rolling mills, calenders in the rubber and paper industries, and rotary printing machines. Another type of running nip is that between a belt, chain or moving fabric and a pulley-wheel, sprocket or roll. Nip points are also found between moving and stationary parts of machines. Workers must be protected from the dangers of trapped fingers and hands in nip points by proper guards. Gloves worn near to nip points increase the danger. Special care in the design of guards and in devising safe procedures is needed when the machine has to be cleaned while it is running.

13.2.3 Screw and worm mechanisms

The shearing action between the moving screw and the fixed parts of the machines poses a hazard. Examples are screw conveyors, mincing machines and helical blade mixers. These must be adequately guarded to protect operators, and shear-pins or other torque-limiting devices should be used to protect the machines from serious damage by hard objects.

13.2.4 Reciprocating and sliding motion

Examples include reciprocating pumps, compressors and hydraulic machinery. One danger lies when the moving part of a machine crosses a fixed part. Another occurs when a moving part extends beyond the baseline of the machine near to a wall or fixed object where a person standing there could be crushed. Such danger areas should be enclosed, e.g. by handrails, to prevent access.

13.2.5 Complex motions

Examples are cam-operated mechanisms which combine a sliding and rotary motion, and oscillating parts such as pendulums and crankshafts. These motions can be more dangerous than simple ones since they are less predictable.

13.2.6 Anti-vibration mountings and pipe connections

Rotary and other moving machinery, especially reciprocating compressors and batch centrifuges which may vibrate considerably, must be very securely mounted. Anti-vibration mountings are often required. Flexible pipe connections onto the vibrating item are equally necessary and must be adequate for the test pressure of the system. Pipework must be securely anchored in ways which allow for differential expansion.

13.2.7 Fitting, assembly and clearance between moving and stationary parts

Care must be taken to ensure that machines are correctly assembled during manufacture or after maintenance, that spare parts meet dimensional tolerances, that clearances between moving and stationary parts are correct and that everything is properly and securely fitted. Some machines need trial runs in their installed positions before starting or resuming their normal duties.

13.2.8 Guards, covers and doors

A wide range of guarding devices and principles exist. These include fixed guards, interlocking guards, trip devices, mechanical restraint devices, and feeding and take-off devices. Suitable guards should be purchased with the machine and not bought and attached later. Power-operated guards must 'fail safe' in case of power or circuit failure. Simplicity, reliability and ease of maintenance are prime considerations when choosing a guard system. BS 5304 provides a guide to their selection, installation and use[7]. A successful machine guard is one which allows operators to increase production by removing the fear of injury from their minds. The problem of preventing guards being removed from operating machinery mainly arises when the guard retards production or is constantly having to be removed to clean the machine, clear blockages or remove faulty work.

Enclosed process equipment containing moving parts often has doors or covers for charging, discharging or inspection. Strict precautions are then needed to ensure that the covers are fixed in position and the doors closed when the machine is operating. For this, interlocks are generally required.

13.2.9 Training and work permits for guarded machinery

Operators of guarded machinery must be taught the hazards of the machine before being allowed to operate it. Procedures are needed to ensure that the following points are observed:

- No guard may be adjusted or removed without the written authority of the responsible supervisor, and then only by a person responsible for maintaining the machine.
- No machine may be started unless the guard is in position and functioning properly.
- A written permit system as explained in Chapter 18 should be used before any maintenance is done on the machine to ensure that it is isolated from electrical power, other services and process materials.

- Employees working on or near process machines must be appropriately dressed, with no loose hair, pendants or other loose articles which could get caught by a moving part.
- Plant inspection and maintenance, discussed in Chapter 17, must include the guards themselves.

13.2.10 Ignition hazards of machinery

Machinery can give rise to sparks, static electricity and high surface temperatures which can ignite flammable gases, vapours and dusts present, both inside and outside the machine. If flammable materials are known or liable to be present, ignition hazards and precautions needed against them must be evaluated in the light of their properties [10.2]. Electrical ignition hazards are discussed in 6.2 and 6.3. Where fire and explosion risks exist inside machinery such as mixers and centrifuges, a continuous purge of inert gas through the equipment is generally needed to reduce the oxygen content to below 2–5%, depending on the hazard[8] [13.4.4].

A survey should be made of possible high surface temperatures outside machinery such as mechanical brakes, bearings, seals, couplings and gears. Heat-sensitive paints and crayons which undergo an irreversible colour change when a certain temperature is reached are useful for thermal monitoring[9].

13.3 Common hazards of rotary machines

The process industry employs many high-speed single-shaft rotary machines. Motor armatures, centrifuge bowls, stirrer, turbine and fan blades, and compressor and pump impellers are assembled on the shaft to form rotors. These rotate inside stationary casings, often with very little clearance. Most machines have rotary shaft seals and outboard bearings on at least one end of the shaft, which is driven via a coupling. Some machines have electro-magnetic or mechanical brakes.

High centrifugal forces are generated in the rotors which are subject to imbalance and resulting vibration. This vibration increases at and near critical speeds which coincide with a natural frequency of the rotor. Both the balance and the natural frequencies of a rotor are modified by process materials in contact with it. The operating speed of a rotor is often higher than its lowest critical speed which is passed through whenever it is started or stopped.

A serious hazard is that of the rotor breaking at speed, perhaps after hitting the casing because of severe imbalance, or of distortion from uneven expansion under heat, e.g. on the start-up of a stand-by unit, or as a result of sudden deceleration caused by a seized bearing. Unless the casing is very strongly constructed, fragments of the broken rotor may break it and fly into the operating area, causing further damage and possible injury, as well as releasing process materials.

Although this is rare in a well-designed and maintained plant, a number of points need expert attention.

13.3.1 Rotor balance and vibration

Unloaded rotors which run at moderate and high speeds should be statically and dynamically balanced. Machines such as centrifuges and mixers, etc. whose rotors are unbalanced by their loads must be designed to tolerate a certain degree of vibration and imbalance. Limits, albeit crude ones which depend on subjective judgement, should be set on vibration levels for the various rotary machines present. Brittle lacquers can be useful here [17.5.1]. If the limit is exceeded, the cause should be investigated at once. If it cannot be found and cured quickly while the machine is running, the machine should be shut down for investigation, despite the loss of production.

Operating and critical speeds
The maximum recommended safe operating speed should never be exceeded. A reliable speed-limiting device is often needed with variable-speed drives.

The critical speeds of the rotor, both loaded and unloaded, should be known. The operating speed should differ sufficiently from any critical speed to prevent excessive vibration (Figure 13.4). This is seldom difficult to achieve with fixed-speed drives, but presents problems when turbines and other variable speed drives are used.

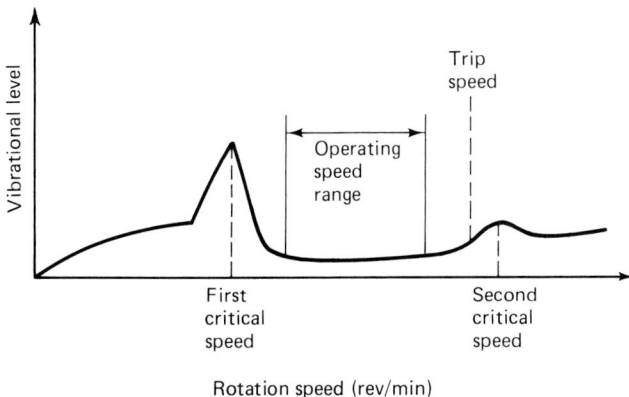

Figure 13.4 Vibrational response plot of rotor of centrifugal compressor

When the rotor has to pass through one or more critical speeds to reach the operating speed, the drive should have sufficient torque at the critical speeds for this transition to take place rapidly. The rotor should be as lightly loaded as possible when passing through a critical speed. There should be sufficient braking (in some cases provided by the process fluids) for critical speeds to be passed quickly when the machine is stopped.

A special vibrational problem can arise when two similar and adjacently mounted machines, e.g. rotary blowers, are operating in parallel on the same duty, but running at different speeds. This can happen when one or both machines are driven through variable-speed drives. If the difference between the speeds of the two machines corresponds to a natural frequency of the pipework and/or spring-loaded relief valves, these may be seriously damaged.

13.3.2 Bearings

Proper alignment and lubrication are essential, and care is needed to prevent grease or lubricating oil being contaminated or dissolved by process materials, or blown out of bearings by gas pressure. When ordering any rotary machine it is essential to ensure that its seals and bearings can cope with the most severe process conditions possible. As an example, a centrifuge which may have to operate in an inert gas atmosphere will need adequate shaft seals to conserve the gas and protect the bearing(s).

Suitable forms of lubricant monitoring, e.g. for temperature, pressure, flow or contaminant [17.5.4], should be used with alarms and/or trips for critical bearings. Bearing, seal and drive alignments should be checked under operating conditions, when thermal differential expansion may upset what seemed a perfect alignment when the equipment was cold. Bearings and their housings must be designed for all possible loads, including axial ones, to which they may be subjected.

Bearings may be damaged by vibration even at rest (see 'Fretting corrosion' [11.4.3]). It is good practice to provide independent anti-vibration mountings for stand-by units so that their bearings are protected from vibration from adjacent running units.

13.3.3 Shaft seals[10]

Rotary shaft seals present a major leak hazard and are responsible for a high percentage of fires as well as maintenance. Only the simplest treatment can be given here. Considerations of critical speed often require seals to be very compact. When the fluid to be sealed is a liquid, either packed glands or mechanical seals are employed. Where absolute freedom from leaks is vital, a 'canned pump' is used. Its motor and bearings are totally enclosed and exposed to the process liquid which acts as lubricant and coolant. When the fluid to be sealed is a gas, 'labyrinth seals' are generally used.

Packed glands need a small continuous leak to prevent overheating. Shafts must be perfectly straight and circular and central in the stuffing box. Their finish where they pass through the gland should be smooth, generally polished. Packing materials are available as coils, spirals and rings of cotton, asbestos, braided lead and copper, etc., impregnated with suitable lubricants, to fit various sizes of stuffing boxes.

Many glands have lantern rings to which a second fluid is supplied, part of which leaks into the machine (Figure 13.5). This serves to:

- Prevent solids in the process fluid from entering the gland;
- Prevent the escape of toxic, flammable or volatile liquids;
- Provide lubrication.

The choice, supply and control of the second liquid, which is bound to contaminate the process material, are critical.

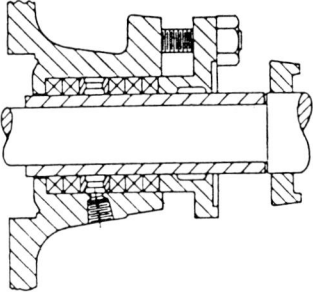

Figure 13.5 Packed stuffing box with lantern ring

Stuffing boxes are designed to allow adjustment and sometimes renewal of packing material while the machine is running in order to counteract wear and compaction of the packing. This is a potentially hazardous operation which may involve removing guards and the use of personal protective equipment. Safe procedures which take the particular hazards into account must be established for this work; only trained and authorised personnel should be allowed to do it.

A mechanical seal is a packaged assembly which forms a running seal between flat, precision-finished surfaces. It contains four basic elements, a rotating seal ring, a stationary seal ring, a spring-loaded section for maintaining seal-face contact, and static seals (Figure 13.6). Mechanical seals normally pass only very minute amounts of fluid, generally as vapour. They are generally dearer and more effective than packed glands. Their main disadvantage is that when they leak, the machine has to be shut down to adjust or replace the seal. There are several types which include balanced seals, and double seals where a second liquid at a higher pressure is introduced into the space between the seals, to prevent escape of process liquid.

Face-type seals are seldom used for gases because of heat-removal problems. Labyrinth seals, which consist of a number of sharp-edged discs making light contact between the rotating and stationary parts, are commonly used when slight leakage can be accepted. They provide

Static seals   Rotating seal ring   Spring

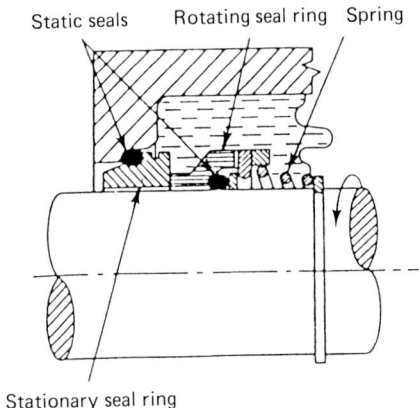

Stationary seal ring

Figure 13.6 Mechanical seal components

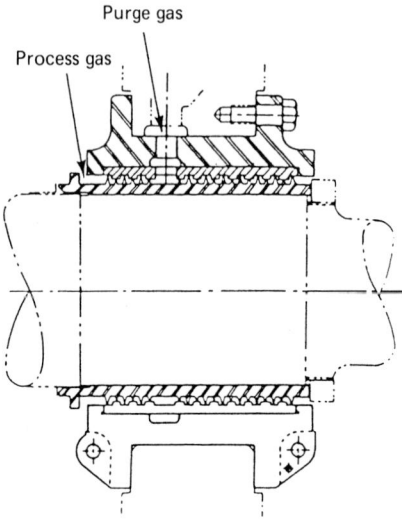

Figure 13.7 Purged labyrinth seal for centrifugal compressor

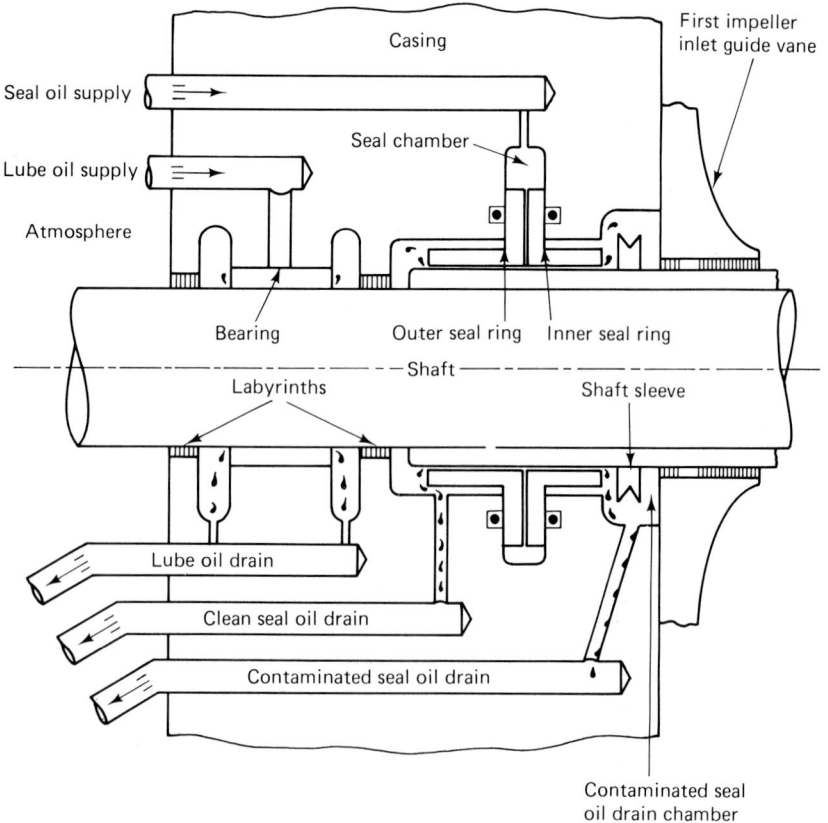

Figure 13.8 Oil-film shaft seal for centrifugal compresor

stepwise expansion of the gas to minimise overall leakage. In cases where no leakage of the process gas is permissible, a purge of some inert gas is introduced at an intermediate point in the seal (Figure 13.7), or oil film seals (Figure 13.8) are used.

All types of rotary seal need frequent inspection for leakage and heating. In critical cases suitably placed continuous gas detectors and temperature monitors may be required. Misalignment, corrosion and the intrusion of process solids into rotary shaft seals are common problems. Since perfect alignment between the drive and driven shafts is often difficult to achieve, flexible couplings are often used. Good alignment is nevertheless important. This should be re-checked after a machine has been installed, even if it has been checked before leaving the supplier.

Besides outward leakage of process fluid, inward leakage of air through the seal has often to be taken into account. This risk is obvious when the equipment is under vacuum but it can also occur with pumps and compressors where, despite slight positive inlet pressure, the pressure inside the shaft seal is negative. The seal may then overheat due to lack of liquid causing the leak to become worse. This impairs the performance of the pump and may stop it pumping altogether. It may also affect oxygen-sensitive materials, e.g. by forming dangerous peroxides. Where possible, the inlet pressure should be raised to prevent this problem. Otherwise a double mechanical seal or a gland with a lantern ring should be used, with a suitable purge fluid.

More specific hazards of centrifuges, mixers, pumps and compressors are considered next.

13.4 Centrifuges[8, 10]

The centrifugal force on any object constrained to move in a circle is proportional to the square of the rotational speed multiplied by the radius of rotation. Since the centrifugal force is ultimately limited only by the strength of the rotating bowl or whatever link keeps the object moving in a circle, it may reach many thousands of times that of gravity. Such forces are common in centrifuges which are widely used to separate mixtures of liquids and solids as well as immiscible liquids. A centrifuge generally consists of:

1. A rotating bowl to whose contents centrifugal force is applied;
2. A drive shaft;
3. Bearings for the drive shaft;
4. A casing or covers to segregate the separated products, and contain gases and vapours;
5. Seals (when it is important to prevent air from entering the casing, or vapour in the casing from escaping, or where the separation has to be carried out under pressure);
6. A shaft drive mechanism (usually an electric motor or turbine);
7. A braking mechanism, (either electromagnetic via the drive-motor or mechanical);
8. A frame to support and align these items.

Centrifuges may be classed as batch or continuous, or by the principle of separation, i.e. sedimentation or filtration (Figure 13.9). Sedimentation centrifuges have solid bowls while filtering centrifuges have perforated ones which are sometimes called baskets. These are usually lined with a wire screen and sometimes a cloth as well.

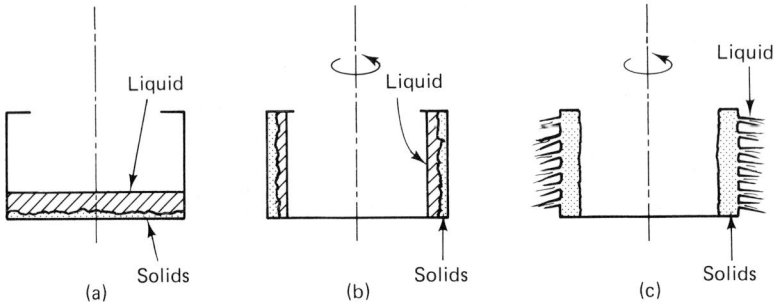

Figure 13.9 Principles of sedimentation and filtering centrifuges. (a) Bowl stationary, (b) sedimentation in rotating solid bowl, (c) filtering in rotating perforated bowl

Most centrifuges run above the first critical speed. Since their loads are liable to be unbalanced, very rugged bearings are needed, with some degree of damping in their mounting to accommodate the vibration as the critical speed is passed.

13.4.1 Batch and continuous centrifuges

Batch centrifuges are mainly of the filtering type, and are used to separate solids from liquids. They have to be stopped and the cover, usually hinged, opened to remove the separated solid, generally manually. There are also batch centrifuges of the sedimentation type, which are used to remove traces of solids and water from lubricating oils and other liquids. These only need to be stopped occasionally for emptying and cleaning.

Continuous centrifuges are provided with a mechanism for removing the separated solid without stopping the machine (although in some types the centrifuge is slowed down).

13.4.2 Continuous sedimentation centrifuges

These have solid shallow-angle conical bowls, with an inner lip or weir at the broader end, over which separated liquid flows. They are fed with slurry through a hollow shaft to near the middle of the bowl. A helical scroll which rotates at a speed slightly different to that of the bowl rakes the solid along the inside of the bowl, where liquid is drained, to the narrower end where it is discharged (Figure 13.10).

13.4.3 Continuous filtering centrifuges

Of these there are several types whose names characterise the mechanisms used to discharge the separated solid. They include ploughing, peeler and pusher centrifuges.

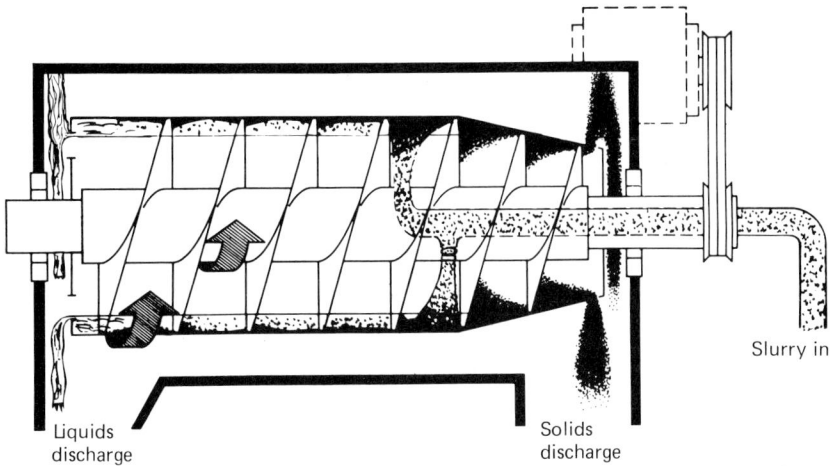

Figure 13.10 Sedimentation centrifuge with helical conveyor (courtesy Pennwalt Ltd)

Ploughing and peeler centrifuges

These are virtually automatic mechanical batch centrifuges, which do not stop to discharge. After building up sufficient solids in the bowl, the slurry feed is stopped and spinning is continued while further liquid is extracted, and the cake may be washed in the bowl with liquid introduced inside it. A mechanical knife is advanced into the inside of the cake, cutting it and discharging it via a chute through the casing, usually to a hopper or conveyer below it. Machines which slow down for removal of cake are known as ploughing centrifuges and those which do not are termed peeler centrifuges.

Pusher centrifuges

These have a mechanism which intermittently and at frequent intervals pushes the drained solid through the bowl to one end, where it is discharged through the casing (Figure 13.11).

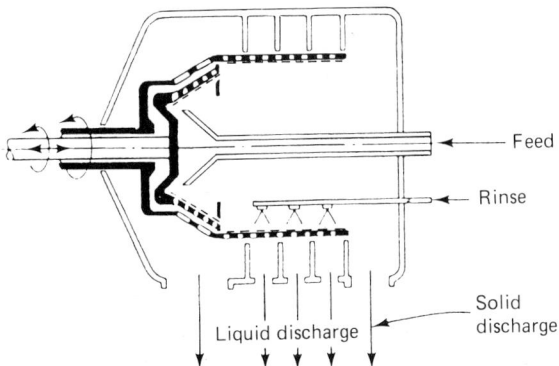

Figure 13.11 Single-stage pusher centrifuge

Other types
These employ vibration, oscillation and other means to move the solid along the bowl.

13.4.4 Hazards of centrifuges

Centrifuges have most of the hazards common to rotary machines [13.3] as well as other more specific ones such as:

- Fires and explosions from flammable solvents,.which may be ignited by mechanical friction or static electricity. The latter still persists after the centrifuge has stopped. Thus an operator who is insulated from the floor may carry a static charge. Special precautions are needed when flammable solvents are processed.
- Hazards to personnel through contact with or inhalation of toxic materials at low concentrations over long periods.
- Projectiles caused by mechanical failure resulting from basket imbalance, bearing failure, faulty assembly or corrosion.

Batch centrifuges which are constantly being stopped and started and opened for removal of solids present more serious explosion, toxic and mechanical hazards than continuous ones. Ploughing and peeling centrifuges tend to be more hazardous than pusher or other types, because of mechanical stresses, friction and imbalance set up when the knife cuts into the rapidly moving cake.

The 'windage' within the casing, caused by the rotating bowl, which acts as a centrifugal fan, causes problems in containing vapours, dust and inert gases. Thus a centrifuge without adequate shaft seals sucks in air where the shaft passes through the casing, and blows it out (laden with solvent vapour) with the solids discharged, making it difficult to prevent vapour and dust from escaping.

Other problems include:

- The loosening and opening of joints between the casing and ducts, etc. connected to it as a result of vibration;
- Bridging and blocking of the discharge ducting by semi-dry solids. This may lead to improvised means of clearing them, which result in gaps through which gas, vapour and dust pass readily.

Such problems should be carefully considered in relation to the fire, explosion and toxicity risks when the installation is designed and the possible use of an inert gas within the casing is being decided.

Fire and explosion risks
These are categorised as follows in the I. Chem. E. guide[8]:

- Lowest level – Non-ploughing centrifuges processing high boiling point solvents at temperatures well below their flash-points. No inert gas blanketing is required.
- Moderate level – Single ploughing centrifuges processing solvents at temperatures well below their flash-points, and single-non ploughing centrifuges handling solvents near or above their flash-points. Inert gas blanketing is necessary.

- High level – Ploughing centrifuges processing solvents at temperatures near or above their flash-points, multiple installations of moderate level type, or special risk situations. Inert gas blanketing with oxygen meters to confirm the presence of an inert gas atmosphere is necessary.

Risks from toxic and aggressive materials are similarly categorised in the guide. The main precautions recommended, in addition to regular inspection and maintenance, relate to ventilation and protective equipment for operators and general housekeeping. Monitoring of the working atmosphere is needed for many toxic materials, while special attention must be paid to design details which affect the escape of toxic materials, particularly as vapour or dust.

The same guide contains detailed mechanical specifications, which include materials of construction, installation, lubrication and special requirements for specific types of centrifuge. It also gives system specifications for flammable materials (with several illustrated examples of inert gas blanketing), a system specification for toxic and aggressive materials and much other information.

13.5 Mixers[11]

Mixers include all types of liquid stirrers, paste kneaders and solid blenders, as well as those used to promote the solution and reaction of gases in liquids. A wide range of mixers are used for various functions and materials, both for batch and continuous operation (Figure 13.12). Some are used solely for physical mixing or solution without chemical reaction, some in conjunction with heat transfer and drying operations, and some as essential features of chemical reaction systems. While mixing is a purely mechanical function, and subject to all the mechanical hazards associated with it, its purpose is usually to effect some change in the materials treated. It is thus also associated with process and material hazards.

Another I. Chem. E. guide[11] provides a detailed classification of mixers, and discusses their hazards generally and in relation to their types. Most of the hazards discussed in this book and all the mechanical ones discussed in this chapter are found in different mixing operations. Mixing (or the lack of it) has been important factors in many industrial accidents, including four out of the five major accidents discussed in Chapters 4 and 5.

13.5.1 Major accidents involving mixing or lack of it

The removal of a stirrer from one of the reactors was probably a factor in the causation of the Flixborough disaster [4.5].

The Pernis oil refinery explosion of 1968 [5.1] could hardly have occurred if the 'slops' tank had been provided with a stirrer and this had been operating while its contents were being heated. Both the explosion at Dow's factory at King's Lynn in 1976 [5.2] and the 'run-away' reaction which led to the Seveso disaster of 1976 [5.3.7] took place in equipment which had a mixing function which had been stopped for the weekend. This stoppage was crucial to both accidents.

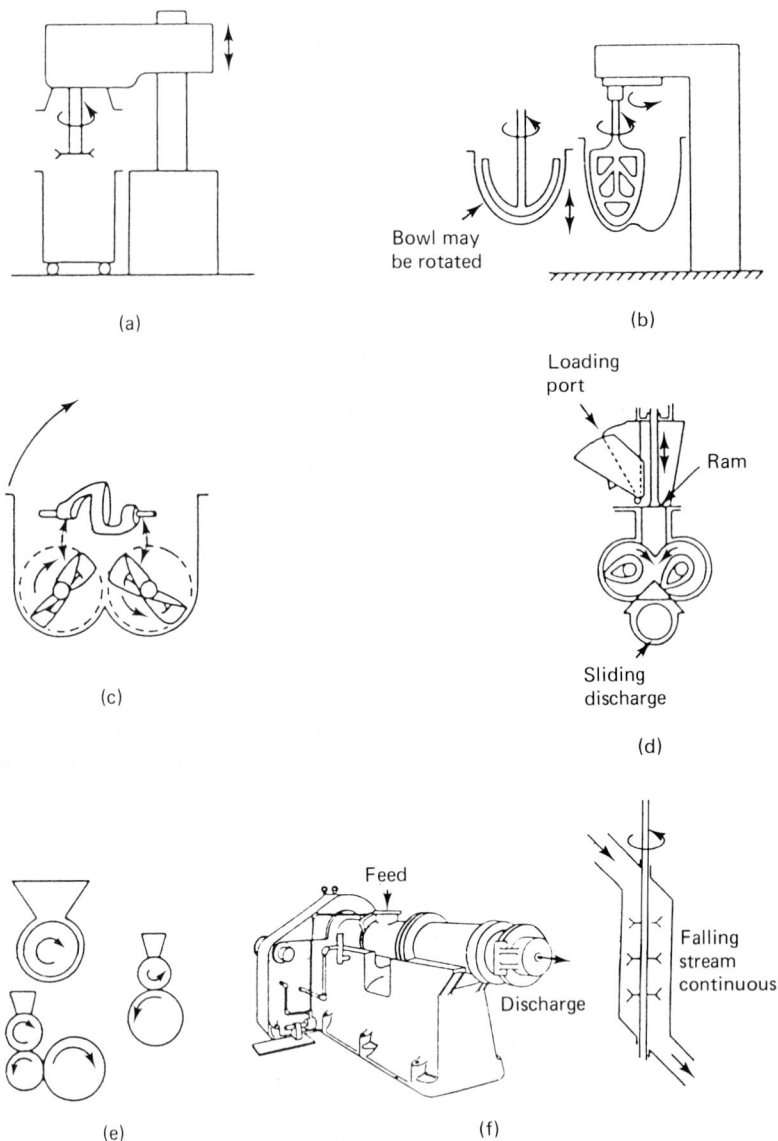

Figure 13.12 Types of mixer: (a) change can with high-speed impeller; (b) planetary or small clearance (also change can); (c) heavy paste tilting bowl (Z-blade); (d) internal (Banbury) mixer; (e) roll mills; (f) continuous mixer/extruder. (From *Guide to Safety in Mixing*[11], courtesy Institution of Chemical Engineers)

Several serious accidents have occurred through attempts to assist mechanical mixers by hand or by hand-held paddles, poles, etc. There is often a strong temptation to assist a mixer in this way when unmixed material accumulates, often as a tacky cake, in corners and stagnant zones behind baffles in the mixing vessel. As a general rule, the cover of a mixing vessel should never be opened while the mixing mechanism is operating and until its contents have come to rest. Many industrial mixers are provided with electrical interlocks to prevent the cover being opened while the mixer is working, and vice versa.

Fire and explosion hazards in mixers are similar to those in centrifuges, and the ignition risks and remedies are similar.

13.5.2 Example of mixing hazard

The following example of an unexpected 'process-mechanical' hazard involving mixing is one which the writer encountered on a pilot plant in the late 1950s.

This consisted of a stirred and jacketed 200-litre stainless steel batch reactor provided with an overhead condenser and product receiver, arrangements for introducing liquids and solids, and emptying, and for operating under vacuum or an inert gas blanket. The product, diethyl aluminium monochloride, is a liquid boiling at about 195°C, which bursts into flame immediately on contact with air and explodes on contact with water. It was then made in two reaction stages, from aluminium granules, ethyl chloride, and a catalyst in the first stage, with the addition of sodium in the second stage. At the end of the second stage the reactor contained roughly equal quantities of two liquid phases at a temperature of about 100°C. The lighter one was the impure product and the denser one consisted of a molten solid, sodium aluminium monoethyl trichloride (melting at about 60°C), with a good deal of suspended aluminium powder and sodium chloride.

The procedure for the final separation of pure product from by-products was first to distil off all the diethyl aluminium monochloride under vacuum with the stirrer operating, using jacket heat, and then drain the molten sodium aluminium monoethyl trichloride into an open-top drum. Unless the stirrer was running, the distillation was extremely slow and incomplete because the heat from the jacket had first to pass through the dense, molten-solid layer.

On one occasion the batch procedure had to be interrupted for the weekend before the product could be distilled off, and the plant was shut down and left under an inert gas blanket. On resuming on Monday, the reactor was cold and the sodium aluminium monoethyl trichloride had first to be melted by applying heat through the jacket. When this appeared from the temperature to be complete, the stirrer motor was started. It shuddered and revolved the stirrer about a quarter of a turn, then halted with sounds of distress from inside the reactor.

It was soon clear that the blades of the stirrer were embedded in a large block of unmelted solid which also contained the lower parts of a long thermopocket and an inlet tube for introducing ethyl chloride into the base of the reactor. Both of these were now wrapped around the stirrer, which could not be be used. We were now faced with the awkward problem of removing nearly 100 litres of the dangerous liquid from the reactor as well as the low-melting solid. Fortunately we managed this safely. The main lesson which the writer learnt from it was:

• When there is any possibility of a stirrer or mixer being hindered by solid process materials, make arrangements for it to be turned safely through one or two revolutions by hand before the motor is started. This facility should be built into the design. It is tempting Providence if one has to remove the guard round a coupling on the drive shaft in order to do it. Also ensure that a shear-pin or other torque-limiting device is incorporated into the drive.

13.6 Pumps

Most pumps used in the process industries are either of the rotary or reciprocating type. Of the former, centrifugal pumps predominate, but 'impulse' and positive displacement types are also used. All reciprocating pumps are of the positive displacement type.

Pumps seldom interact with processes in the same way as mixers. Their normal purpose is to move liquids by imparting mechanical energy to them. Even where a pump is used to mix the contents of a vessel, it is usually external to the vessel so that the functions of liquid circulation and mixing are separate. Unlike mixers and centrifuges, the working volumes of operating pumps are full of liquid, and there is little risk of internal explosion of mixtures of air and flammable vapours.

Yet despite the simple nature of their hazards, pumps are a common source of fires which start at leaking seals. While glandless canned pumps avoid this problem, they are expensive and unsuitable if the liquid contains solids.

Before starting a pump, all valves on the suction and discharge lines should be in the correct mode, and lines, joints and supports must be in good condition. These should be watched when the pump is started since this imposes a shock on the delivery line which has split gaskets and dislodged hoses. Hard, solid objects such as grit and the ends of welding rods cause serious damage to pumps and their valves. These are generally excluded during the start-up of a plant by temporary in-line strainers. Where this danger is persistent (e.g. when the pump is supplied by a long pipeline), permanent suction strainers or baskets are needed.

Each pump usually requires an individual pressure gauge on both the discharge and the inlet, with isolation valves or cocks to allow the gauge to be replaced without draining the system. The discharge gauge should be sufficiently damped to even out pulsations, and should be protected where necessary against overpressure. Remotely operated emergency isolation valves or other means of stopping the flow of liquid are needed when the consequences of a pump fire are serious. Pumps handling flammable liquids at temperatures approaching or above their flash-points should be located in well-ventilated areas, usually in the open air, and not under pipe racks, electrical and instrument cables, or other objects which would add to the fire risk.

Besides handling single, clean liquids, pumps often deal with mixtures of immiscible liquids, and suspensions of solids in liquids, generally termed slurries.

13.6.1 Slurry pumps

The design and operation of slurry pumps are more critical than those handling clean liquids. The slurry must not contain solid particles above a size which the pump can handle, and care must be taken to avoid conditions which cause agglomerations of particles into lumps, or the shearing of polymeric particles to produce a mass of fibres.

Suspended solids must not be allowed to enter shaft seals. Internal passages and valves must be free of restrictions and sharp changes in direction, and throttling-type valves can seldom be used. Special positive displacement pumps thus tend to be used for slurries, with variable-speed drives to regulate the flow. Interruptions to slurry flow often result in settling of solids in suction and discharge lines and in the pump itself. It is thus usually necessary to have an installed liquid flushing system, both for the shaft seals and to displace solids from the pump and lines and leave them full of liquid only when the pump is shut down.

13.6.2 Centrifugal and impulse (turbine type) pumps

Common features of these pumps are:

- Compactness,
- Non-pulsating flow,
- Discharge pressure dependent on fluid density without rising to dangerous levels against a closed discharge.

They are generally driven at constant speed by an electric motor when the flow is controlled by an automatically operated valve on the discharge line. When run against a closed discharge, they quickly develop heat which may vaporise the liquid, vapour-lock the pump and damage the seal and other parts. Many pump circuits require 'minimum flow return lines' which by-pass liquid from delivery to suction, but these in turn require cooling to prevent overheating.

Great care is needed in selecting and installing these pumps to ensure that they have sufficient suction head to prevent cavitation–corrosion of the impeller [11.4.2]. Means of venting gases from the pump casing and of ensuring that it is full of liquid when the pump is started must be provided. Pump motors must have adequate torque, not only at the operating speed but also for starting. Electric motors do not usually have sufficient torque to start a centrifugal or turbine pump on load. When the flow is controlled by an automatic flow-control valve on the pump delivery line this will generally be wide open when the pump is started. If the pump is then started with its discharge valve open, the motor will probably 'trip out'. Such pumps are usually started with the discharge valve closed. This is gradually opened when the pump reaches its operating speed, and the proper discharge pressure is reached. The discharge valve is only opened wide when the automatic control valve has taken over. It is also sometimes necessary to start a centrifugal pump with a closed discharge valve in order to achieve correct hydraulic balance.

In some cases, however, it is essential to choose a pump/motor/starter combination which will start on full load (with the discharge valve open). These cases include:

- Standby pumps which are required to start and come on-stream automatically;

- Pumps which are required to restart automatically on restoration of power after an electricity failure;
- Pumps subject to on–off control, e.g. through the operation of level switches.

In complex processes with many pumps running simultaneously, smooth pump operation is very dependent on proper choice of instrumentation (e.g. one-, two- and three-term controllers), and on their proper setting.

13.6.3 Positive displacement pumps

These include both rotary and reciprocating types, the latter including piston, diaphragm and plunger pumps, the last type being frequently used for metering liquid flows. Flow is controlled by varying the speed or the stroke, either manually or by plant instrumentation. Diaphragm pumps which are free of sealing problems are used for toxic or corrosive liquids which are difficult to seal.

All these pumps, if started against a closed discharge, develop very high pressures, stop the drive mechanism and are usually damaged. To prevent this, most are fitted with internal pressure relief valves which by-pass liquid from discharge to suction. Unlike centrifugal and impulse pumps, they are started with their discharge valves open. As their flows pulsate to a greater or lesser extent, compressed air- or gas-cushion chambers are often required on the pump discharge, particularly with reciprocating pumps, to even out the flow and reduce vibrations in the pipework. These chambers need to be adequately sized, with the connection between the pump and the chamber as short and large as practicable. They usually need to be provided with a diaphragm or bladder to hold the charge of air or gas and prevent it being lost by solution in the liquid being pumped. A set procedure must be laid down for checking the charge of gas or air. These chambers are specially prone to fatigue and proper allowance must be made for this in their design. (The writer had a lucky escape when an improvised cushion chamber of a small reciprocating pump burst in front of him.)

13.7 Compressors[10, 12]

Most compressors are of aerodynamic (axial flow or centrifugal), or positive displacement (reciprocating or rotary) types. Axial flow compressors which are only suitable for very large flows (inlet $200–100\,000\,m^3/min$) and discharge pressures up to $10\,bar\,g$. have limited application in the process industries.

Centrifugal compressors handle inlet flows from 20 to $200\,m^3/min$, and discharge pressures up to $400\,bar\,g$. and are used for process and refrigerant gases for specific applications where the gas composition, pressure, flow rate and temperature can be expected to remain in a fairly narrow range. Their main use is in the oil, gas and petrochemical industries.

Rotary positive displacement compressors handle inlet flows from 3 to 200 m³/min and discharge pressures up to 10 bar g. The oil-lubricated screw type are used for medium-size process and refrigeration duties, although they tend to be noisy and require special hoods or housing to protect the hearing of operators.

Reciprocating compressors are available for a wide range of flows (inlet 0 to 2000 m³/min) and discharge pressures up to 3500 bar. They are are used exclusively for very high pressures, and for duties at lower pressures where the flows are too low for other types of compressor. Although they are usually the most efficient type, they tend to be much larger than other types for the same duty, and require special provisions to protect pipework and supports from pulsations and vibration.

Compressors are large consumers of power, often several thousand kilowatts per machine. Drives include electric motors, gas and steam turbines for centrifugal machines and gas and steam engines for reciprocating machines. Electric motors are geared up to drive the faster revving centrifugal compressors, and geared down to drive reciprocating compressors.

Most electric motors used for driving compressors are of constant speed, induction type, with rather low torques while starting and coming up to speed. Due to the high cost of variable-speed drives, electrically driven compressors have generally to be run at constant speed, and must be provided with valves and other facilities to allow them to be brought up to working speed before any appreciable external load is applied. Large electric motors in hazardous areas have to be pressurised and cooled with uncontaminated air brought in from outside the plant area. The hot air has often to be discharged outside this area.

Gas turbines are well suited to driving centrifugal compressors. Their speeds match those required by compressors and can be varied to suit the loads. Both single-shaft and two-shaft machines are used. The latter type is the most flexible arrangement and the drive turbine usually has sufficient starting torque to start the compressor on load. In hazardous areas, however, the supply of clean air, the disposal of exhaust gases and the enclosure of turbines and combustion chambers to prevent them from becoming ignition sources sources pose serious problems.

Gas engines built integrally with large reciprocating compressors and which share the same crank-shaft and crank-case with them have long been used in the oil and gas industries. Smaller compressors are usually driven by separate engines. Engines and their ignition systems in hazardous areas must meet the codified flameproofing requirements.

Mechanical considerations limit the maximum compression ratio per stage to 5–10 for reciprocating, 4 for rotary positive, 3–4.5 for centrifugals and 1.5 for axial flow compressors. Multi-stage compression is thus common.

The high compressibility of gases and the temperature rise that accompanies compression necessitates cooling, during compression and between stages. Gases show wide variations in their temperature rise during compression. For adiabatic compression from 15°C, the discharge temperatures for a doubling of pressure amount to 108°C for argon, 80°C for nitrogen, 66°C for carbon dioxide and 36°C for butane.

Considerations of power loss, corrosion, polymerisation, decomposition of gas and lubricating oil, and differential expansion and loss of strength of parts of the compressor, limit the discharge temperatures of compressor stages to about 150°C (but lower for air compressors [17.1.5]).

The presence of liquid in the inlet gas is a hazard in most types of compressor, and can cause serious damage, particularly to reciprocating compressors, where the action of a piston backed by the energy of a large fly-wheel on an incompressible liquid in a limited volume can be disastrous. Any gas liable to contain slugs of liquid (e.g. from condensation) needs to pass first through liquid separators of adequate capacity, fitted with high-level trips and alarms, before entering any stage of a compressor.

Because of the high cost of process compressors and their drives, spare ones are seldom installed, and staff must be thoroughly trained in their operation, trouble spotting and maintenance before a plant is commissioned. They are not operated in isolation but in conjunction with a train of other equipment. Details of the many operating steps in which they may be involved must be worked out between senior operating and design staff while the plant is being designed, and written up as a draft operational manual or operability study, which may be invaluable for spotting hazards [16.5.5]. Most faults in compressor performance can be diagnosed from the normal instrument readings coupled with unusual sounds and appearances, but vibration and contaminant analysis and other condition-monitoring techniques may also be needed on complex high-cost compressors and drives [17.5].

13.7.1 Centrifugal compressors[13, 14]

Most centrifugal compressors are multistage machines with a rotor consisting of up to nine impellers or 'wheels', a balancing drum and a thrust disc mounted on a shaft (Figure 13.13). This rotates between seals and bearings within a stationary casing which contains diaphragms, diffusers and seals. While a range of standard impellers of different size and materials are available, their arrangement in any compressor is for a specific application for which the compressor is virtually 'tailor made'.

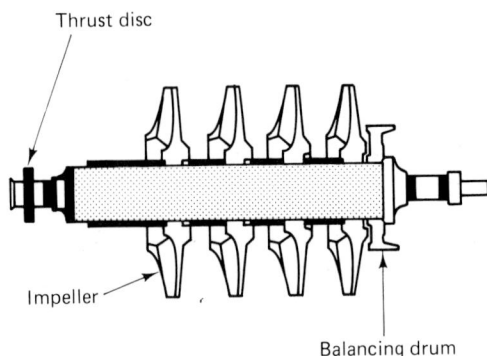

Figure 13.13 Rotor of centrifugal compressor

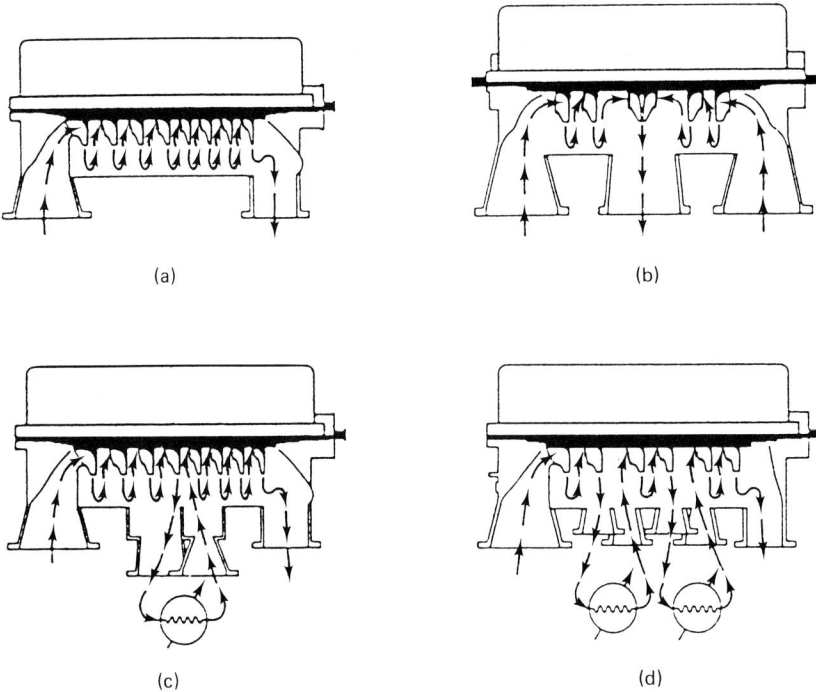

Figure 13.14 Flow arrangements for centrifugal compressors: (a) straight-through, (b) double-flow, (c) single intercooled, (d) double intercooled

Possible arrangements include – straight through with or without intercoolers, double-flow and side-load machines. Some typical arrangements are shown in Figure 13.14.

Over a limited speed range, the head developed by an impeller varies with the square of its rotational speed, and the inlet flow varies in proportion to it. Since the volume of a gas decreases as it passes through a compressor, the size of the impellers, particularly their blade width, becomes progressively smaller on passing through a machine. Three factors limit the speed of impellers:

- Stresses in the impeller;
- Vibration caused by approach to a critical speed;
- Aerodynamic limit caused by approach of the tip speed to the speed of sound in the gas.

Where corrosive gases (such as hydrogen sulphide) are present, special alloys of relatively low yield strength have to be employed, which limit the speed.

Vibration problems of centrifugal compressors are much accentuated when a critical rotor speed corresponding to a resonant frequency of the rotor system is reached. This has various critical speeds which increase by multiples of two. These can be calculated during design but should be checked by testing when the compressor is built. The operating speed range of a centrifugal compressor usually lies between the first and second critical speeds, and must be restricted to ensure that it does not approach either of them (Figure 13.4). The critical speeds of the driver must also be analysed to check that they are compatible with those of the compressor and that the combination is suitable for the specified operating range. Precautions must be taken when starting up and shutting down to ensure that the first critical speed is passed fairly quickly and without excessive vibration.

The flow stability is significantly reduced as the tip speed approaches the speed of sound, which depends on the molecular weight and other properties of the gas. These therefore have to be considered in determining the maximum shaft speed.

The characteristic curve of a single-stage centrifugal compressor is shown in Figure 13.15. When running at constant speed with inlet and outlet open to the atmosphere, the inlet gas flow Q and head H are represented by (1). On throttling the discharge flow slightly, the head jumps rapidly to (2), the stonewall point, with only a small reduction in gas flow. Between (1) and (2) the flow is somewhat unstable. On throttling the discharge further the head increases and the flow decreases, passing first through (3), the normal operating point, finally reaching (4), when the compressor starts to surge. The flow is then highly unstable, with gas passing backwards and forwards through the compressor, and the discharge pressure fluctuating wildly. This can severely damage the compressor, to prevent which an automatic surge-control device is needed. This opens a by-pass valve to recycle gas from discharge to suction before the surge point is reached. The compressor can operate only in a flow range lying between (2) and (4).

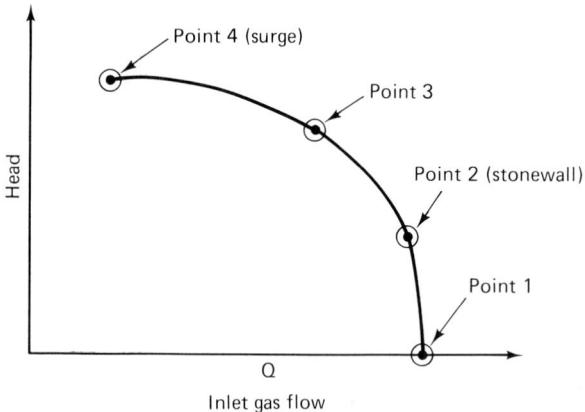

Figure 13.15 Characteristic curve for centrifugal compressor

In a multistage machine each impeller is designed for the inlet pressure, temperature and inlet flow reached at that stage with the gas in question. Suppose that the machine is driven at constant speed and the density of the gas increases while other conditions remain the same. The first compression stage is least affected, but its discharge pressure and density will be higher, which reduces the (volumetric) flow of gas to the second stage. The same effect is found to a more pronounced degree in the third and subsequent stages. The surge point may then be reached before the last stage. While this situation may be somewhat alleviated by by-passing a quantity of gas round the compressor from discharge to suction, this has the effect of moving the operating point of the first stage further to the right in Figure 13.15. In difficult cases it is impossible to prevent surge conditions in the final stage without reaching the stonewall point in the first stage, and the compressor becomes quite inoperable at that speed. Similar effects can occur if the pressure, temperature, flow rate, etc. were to alter. Flow stability could, however, be restored if the compressor speed could be reduced. But although the operational flexibility of a multistage machine is increased if its speed can be adjusted to compensate for changes in gas properties or operating conditions, arbitrary speed changes may lead to unstable flow conditions.

The capacity of constant-speed units may be controlled by a valve in the inlet piping or adjustable inlet vanes in the compressor. For variable-speed units the capacity is controlled over the operational speed range by the governor setting of the drive unit. The minimum process instrumentation needed is shown in Table 13.2.

Table 13.2 Minimum process instrumentation for centrifugal compressors (other than standard protection against failure of cooling water or bearings)

	Indicator	Alarm	Shutdown
Inlet pressure (each section	•		
Inlet temperature (each section)	•		
Outlet pressure (each section)	•		
Outlet temperature (each section)	•	•	
Reference gas pressure		•	•
Balancing-drum differential pressure	•	•	•
Buffer gas differential pressure	•	•	•
Total flow through compressor	•		
Recycle flow through compressor	•		
Gas molecular weight or S.G.	•		
Speed	•	•	

While labyrinth seals are used for the internal seals of centrifugal compressors, and for external seals on air, nitrogen and oxygen compressors where small leaks are permissible, it is generally necessary to use some positive oil-film shaft seal for other gases. This oil system which becomes contaminated by the gas must be kept separate from clean oil systems used for bearings, gears, etc.

13.7.2 Reciprocating compressors[15]

The essential features of most reciprocating compressors are still those of a bicycle pump:

- The cylinder or barrel of uniform internal cross section;
- The close-fitting piston that moves inside it;
- Inlet and outlet valves;
- The piston and drive mechanism which operates the piston in both directions.

Only the diaphragm compressor, which replaces the piston and part of the barrel by a flexible diaphragm, departs from these features. Reciprocating compressors may be single or double acting, compressing gas on one or both sides of the piston. As with centrifugal compressors, the effects of the gas temperature rise during compression set a limit (usually between 3 and 7) to the compression ratio which may be employed in a single stage. For higher compression ratios, multistage compression with intercooling between stages is used.

For constant-speed units, capacity control is achieved by suction valve unloading, clearance pockets, a combination of both or a by-pass. For variable-speed units, capacity control is achieved by speed control, supplemented as necessary by the previous methods. Where the compressor must be unloaded during start-up, proper provision must be made for this.

For high-pressure compressors forced lubrication is generally supplied both to the middle of the cylinder and to the packing case, which contains proprietary packing rings based on PTFE, lead, graphite or other materials. The piston rod passes through a compartment with its own stuffing box to prevent contaminated oil and gases from coming into contact with the crosshead, connecting rod and crankshaft. This compartment may be purged with inert gas, vented and drained to a spent oil receiver, and fitted with a removable cover to service the packing case.

The problems of lubricating air compressors are discussed in 17.1.5. Where compressed air for breathing apparatus is required, compressors with carbon piston rings or diaphragm-type compressors are recommended.

In addition to temperature and pressure indicators for gas, oil and water entering and leaving each stage or intercooler, pressure-lubricating systems should be provided with a pressure switch or other device to stop the compressor in the event of low oil pressure. Alarm or shutdown switches are also generally needed for:

- High gas discharge temperatures from each cylinder;
- High cooling water temperatures from each cylinder;
- High engine jacket temperatures for engine-driven compressors;
- High packing rod temperatures;
- Excessive vibration;
- Low oil level or drive failure on forced-feed lubricators;
- Leakage of starting air check valves in power cylinder heads;
- Low pressure or flow in frame-lubrication system;
- High liquid level in separators.

Instruments should be located for ease of testing, inspection and maintenance and freedom from vibration.

References

1. Collins, J. A., and Monack, M. L., 'Stress-corrosion cracking in the chemical process industry', *Materials Protection and Performance*, **12**, 11 (1973)
2. Wise, S., 'Why metals break' in Whyte, R. R. (ed.), *Engineering Progress through Trouble*, The Institution of Mechanical Engineers, London (1975)
3. Lees, F. P., *Loss Prevention in the Process Industries*, Butterworths, London (1980)
4. BS 4360: 1986, *Specification for weldable structural steels*
5. BS 5500: 1985, *Specification for unfired fusion welded pressure vessels*
6. HSE, *Final Report of the Advisory Committee on Falsework (The Bragg Report)*, HMSO, London (1976)
7. BS 5304: 1975, *Code of practice for safeguarding of machinery*
8. The Institution of Chemical Engineers, *User Guide for the Safe Operation of Centrifuges*, Rugby (1976)
9. Collacott, R. A., *Mechanical Fault Diagnosis and Condition Monitoring*, Chapman and Hall, London (1977)
10. Perry, R. H. and Chilton, C. H., *Chemical Engineers' Handbook*, 5th edn, McGraw-Hill, New York (1973)
11. *Guide to Safety in Mixing*, The Institution of Chemical Engineers, Rugby (1982)
12. Ludwig, E. E., *Applied Process Design for Chemical and Petrochemical Plants, Volume 3*, Gulf Publishing Company, Houston, Texas (1965)
13. API Standard 617, *Centrifugal compressors for general refinery service*, 3rd edn, American Petroleum Institute, New York (1973)
14. Lapina, R. P., *Estimating Centrifugal Compressor Performance*, Gulf Publishing Company, Houston, Texas (1965)
15. API Standard 618, *Reciprocating compressors for general refinery service*, 2nd edn, American Petroleum Institute, New York (1974)

Chapter 14

Reliability and risk analysis

Reliability and risk analysis are interrelated subjects which are important for safety. Reliability is defined here as[1]:

> The characteristic of an item expressed by the probability (R) that it will perform a required function under stated conditions for a stated operating period (t).

Many readers will be familiar with theories which attempt to relate the reliability of a system to that of its components. During recent decades some older ideas have been found inadequate, particularly in the mechanical field, and newer concepts of 'inherent reliability' and 'loading roughness' have been introduced. Modern ideas on reliability are outlined with the minimum of mathematics in 14.1 where they are discussed in the context of design, equipment selection and failure, maintenance, deliberate redundancy, safety and safety equipment. Some reliability data on instruments and equipment is given in 14.2.

Risk analysis is introduced in 14.3 and developed in the following sections. Risks to life are outlined in 14.7. Other related topics discussed elsewhere in this book include hazard indices [12] and hazard and operability studies [16.5.5].

14.1 Introduction to reliability

While a brief introduction to reliability theory is given here to assist the general reader, critical reliability calculations are best left to specialists. Reliability theory, particularly in its mechanical application, is discussed critically by Carter, with many practical examples and exercises[2]. Several mathematical models which apply to complex process systems are given by Lees[3].

14.1.1 Ideas and terminology

Reliability (R) is the probability, quoted as a number between 0 and 1, that an item will perform as intended for a given length of time, expressed as (t), under stated conditions. The time t should be comparable to the intended working life of the item, expressed as T, under the same conditions. R is calculated from the mean of a number of observations (e.g. of service times to failure) which represent a sample of a much larger 'total population' of possible observations of the same thing. Because of the frequent lack of data on which to calculate reliability, it is often described in subjective terms, without putting a number to it.

The confidence which one can place on a numerical value of reliability depends on the number of observations on which it is based. This is expressed as 'confidence limits' (i.e. upper and lower limits to the mean) at various levels, say, 60%, 90% or 99%. The true mean of all possible observations then stands a 60%, 90% or 99% chance of lying within these limits. The larger the number of observations, the narrower is the range between the lower and upper limits for any given confidence level. Probability tables and charts are available to relate the ratios of the upper and lower limits to the mean for any number of observations at various levels of confidence. The use of these tables or charts does, however, depend on the observations following a particular distribution or pattern. For operational reliability, 60% or 90% confidence limits may be acceptable, but higher confidence limits are usually called for when people's lives are at stake[4].

The 'spread' in the numerical values of the observations from which the mean is calculated is expressed as the 'standard deviation', which is discussed in 14.1.4. It is related to the confidence limits. To establish the numerical reliability of most items of process equipment with acceptable confidence limits requires a large number of tests under controlled conditions which are representative of the process. The amount of high-quality reliability data available which applies to any process equipment under particular conditions is often very limited. If we had to depend on it, we might have to conclude that there is no equipment available on which we can rely!

An 'item' may be any component, sub-system, system or equipment which can be considered and tested on its own.

A 'failure' occurs when an item becomes unable to fulfil its specified function. There are many types and shades of failure [14.1.8].

'Conditions' include the loading applied to the item (e.g. mechanical or electrical), the environment in which its function is performed (e.g. tropical, arctic, vibratory, or in contact with a process fluid of a particular composition, temperature and pressure). 'Conditions' also include the human beings who install, operate and maintain the item. Loading is seldom constant, but varies from some low value typical of shutdown or idling conditions, through a mean for normal operation to a maximum for upset or emergency conditions. Higher loads than this are considered here as 'misuse', even though the equipment may withstand them.

Two common criteria of how often an item breaks down are:

1. *Mean time between failures – MTBF*. This applies to repairable items. Thus if an item fails, say, four times during a period of use totalling 1000 hours and is repaired instantly each time, the mean time between failures would be 1000 ÷ 4 = 250 hours. In this case we are generally referring to otherwise reliable equipment which contains one or more components with much shorter lives (such as filters or mechanical seals) which require periodic replacement.

2. *Mean time to failure – MTTF*. This applies to non-repairable items and is the average time an item may be expected to function before failure, under stated conditions of environment and load.

The failure rate for a particular type of item, designated by λ, represents the fraction of such items operating in a particular environment which fail in unit time, which may be one hour, one day or one year, etc.

14.1.2 Change of reliability with time

Even when conditions remains roughly constant, the reliability of any item changes during its working life. From early reliability studies, mainly on electronic components which then depended on thermionic valves with short working lives, it was concluded that the working life of most items has three phases:

1. The early failure period, beginning at some stated time and during which the failure rate decreases rapidly. This is a sort of 'running-in' period when manufacturing flaws (e.g. in materials or assembly) and installation faults cause a small proportion of the items tested to fail. In process plant many early failures occur during the commissioning period, when the failed items are repaired or replaced.
2. The 'constant failure rate period' during which some of the surviving items fail apparently at random at a low but approximately uniform rate. This defines the working life of the item under the conditions specified. Despite the random nature of these failures, they all have definite causes which can usually be diagnosed. For non-repairable items in process plant, the constant failure rate period should exceed the anticipated lifetime of the plant.
3. The 'wear-out' period during which the surviving items (usually the great majority) fail at an increasing rate due to deterioration (e.g. as a result of wear, corrosion or fatigue).

Since the early days of computers which depended on thermionic valves the reliability of electronic equipment has increased spectacularly, so that modern integrated circuits on microchips might outlast the Pyramids!

A plot of failure rate against time (of failure-prone items) gives the classic 'bath-tub curve' of Figure 14.1(a). Many components of process equipment show this failure pattern.

An easily visualised example of an apparent 'constant failure rate' is provided by an underground lighting system consisting of a large number of identical bulbs wired in parallel, each of which is replaced as soon as it fails. When the system was new, most bulbs would have lasted for several months and then failed one by one within a shorter period. As time went on and all bulbs had been replaced several times, bulbs would fail more at less at random, giving a 'pseudo-constant failure rate'. The appearance of a constant failure rate is in fact due to a random and roughly constant replacement rate. If records of individual bulbs had been kept, it might have been found that no bulbs failed for the first three months of service although most failed in the fourth. Their true failure rate would then not

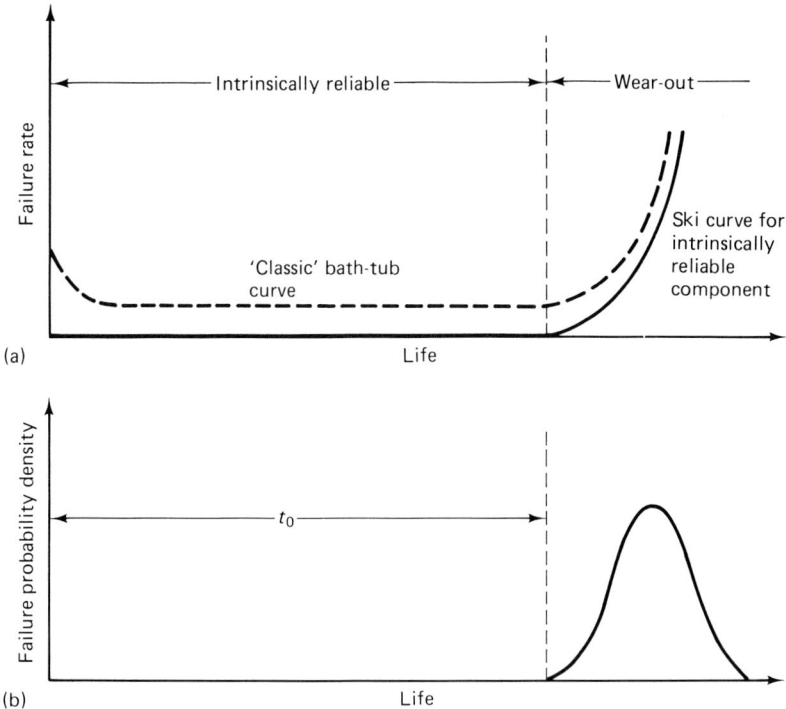

Figure 14.1 Failure-rate versus age curves: (a) the classic 'bath-tub' curve; (b) 'ski' curve for intrinsically reliable component

have shown the first two phases of the bath-tub curve at all, but would have been better represented by a 'ski curve' as shown in Figure 14.1(b).

The same appearance of a constant failure rate is commonly found with the instruments of a process plant which has been in service for several years, where a policy of repair maintenance [17.1] was adopted (for the instruments).

Carter and other reliability engineers in the mechanical field have long questioned the universal validity of the bath-tub curve, particularly its implication that they are dependent on components which are liable to fail at random at a low but steady rate for most of their working lives[2]. This fatalistic view would make air-travel rather unsafe. It would also make long production runs on complex single-stream continuous plant almost impossible, and there would be little point in scheduled preventative maintenance [17.1.1].

With improvements in engineering, manufacturing techniques and quality control it is now generally possible to produce components in which the first two phases of the bath-tub curve are replaced by a single one which is essentially free from failures, provided the loading and other conditions to which the component is subjected remain within specified limits. This is achieved for mass-produced components without incurring economic penalties. Indeed in most cases it is more profitable to produce equipment (like a quartz crystal watch) which can be guaranteed to remain failure-free for, say, a year's use, than lower-quality items which fail at random. The bath-tub curve is now becoming outmoded. Although the ski curve cannot always be achieved, it is the target for which most engineers strive. Three things should be noted about it:

1. The early failure period of the bath-tub curve can be virtually eliminated by improved quality control, proof testing and/or a period of test running before the item leaves the works or factory.
2. By providing a margin of safety in design which is adequate for all anticipated variations in loading and other conditions, and in the strength of the items themselves, the constant failure rate period can be replaced by a 'zero failure rate period' during which the item is said to be 'intrinsically reliable'. This is not done by applying large and arbitrarily selected margins of safety, but it requires detailed study of possible load and strength variations, wear and corrosion rates and mechanisms, and careful attention to all possible modes of failure.
3. Wear, corrosion and other forms of deterioration still impose a limit on the period of use during which components remain intrinsically reliable. This (like the learners' skis of the *ski évolutif* system) can be long, medium or short. At the end of this period the items are worn but by no means all worn out. Although now past their guarantee periods, and subject to occasional failures, they may continue to give useful service for much longer. The shape of this part of the curve can vary as shown in Figure 14.2, and is often typical of the mode of failure, e.g. stress-rupture, fatigue or corrosion.

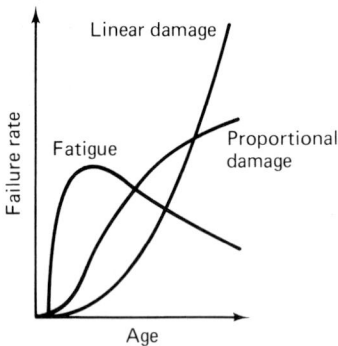

Figure 14.2 Failure-rate versus age curves for different failure modes during wear-out period

The policy to be adopted at the end of the intrinsically reliable period depends on considerations both of safety and of economics. It is sometimes recommended, as in the case of factory lighting [17.1.7], to scrap and replace all bulbs or other short-life items at the end of this period. While there may be an occasional failure within this period, it is usually difficult to attach any meaningful statistic to it.

The obvious advantages to the user of intrinsically reliable items have resulted in their becoming the norm for most machine components and many domestic articles. Such items are generally sold under guarantee for a certain period. Because of this, premature failures are quickly reported to the manufacturer, who takes the necessary action to prevent their recurrence.

14.1.3 Some simple theory

Reliability theory is usually presented on the basis of constant failure rate, λ. Reliability, R (a number between 0 and 1), is related to λ by the equation:

$$R = e^{-\lambda t} = e^{-t/T}$$

where e is the base of natural logarithms (2.718) and T is the MTBF or MTTF (whichever applies). λ is thus the reciprocal of T. This is the 'exponential law of reliability'.

The ratio t/MTBF or t/MTTF is important. When this equals one, whether the time is one minute or ten years, the reliability equals $1/e$, i.e. 0.368.

If the probability of malfunction or failure is F, and both F and R are expressed as decimals, then:

$$R + F = 1 \quad \text{and} \quad R = 1 - F$$

The reliability of any system depends on the individual reliabilities of its components. According to classical reliability theory, if n components operate in series so that the failure of any one would cause the system to fail, the system reliability equals the product of the reliabilities of each component:

System reliability $R_s = R_1 \times R_2 \times \ldots \times R_n$

This is called the Product Rule of Reliability. Although it is quite easy to visualise and demonstrate, it is now recognised that it only applies in certain fields (such as electronics) and other cases where the loading roughness [14.1.5] is low[2]. As an example of the rule, let us return to the underground lighting system, and suppose that this time six identical bulbs with a reliability of 0.9 over a period of 1000 hours, measured under the same conditions as those which will now apply to their use, are wired in series. If the bath-tub curve is applied to them and the bulbs were all in their 'constant failure rate' period, the reliability of the set of bulbs over the period of 1000 hours should equal $(0.9)^6 = 0.53$.

Now let's consider what happens if the voltage is subject to unpredictable surges which are large enough to cause some bulbs to fail. While the measured reliability of the bulbs will be different for this new condition, the reliability of the system of bulbs in series is now partly dependent on the frequency of voltage surges, and the product rule will no longer apply.

Such reservations on the product rule of reliability have been expressed thus by Carter[2]:

[According to the product rule] the components of any series system containing a large number of parts must have a very high degree of reliability. To achieve a reliability of 0.99 in a series system of 400 components requires a mean component reliability of 0.999975. [To measure a reliability of this magnitude with any degree of confidence would require millions of item test hours to be carried out.] Everyday experience suggests that this approach is over-simplified. For example 20 or more vehicles each containing 100 or more components may be held up at a traffic light, giving 200 000 or more components in series. Yet only rarely does the traffic fail to move when the lights change from red to green, and studies made on the supposed correlation between mechanical reliability and the number of components in series have failed to reveal any such correlation.

The same considerations apply to a complex continuous process plant containing several units in series with little or no intermediate storage. Success in fact depends on the weakest individual component of the system being capable of coping with the most severe loading or environment to be encountered. This is the old adage 'the strength of a chain is that of is weakest link'. Reliability theory thus has to 'recognise the fact of variability or scatter, both in the capacity or strength of the product and in the duty it will have to face'[2]. This is discussed further in 14.1.5.

Some good books[5] also tell us there is also a Product Rule of Unreliability which applies to systems consisting of similarly independent components in parallel. If the system unreliability F_s is defined as the probability that all components fail, it should equal the product of the probabilities of failure of each of these components, i.e.

$$F_s = F_1 \times F_2 \times \ldots \times F_n$$

or

$$(1 - R_s) = (1 - R_1)(1 - R_2) \times \ldots \times (1 - R_n)$$

From this it appears that very high reliabilities can be achieved from parallel systems employing redundant elements. In practice a number of qualifications also apply to this concept, and one needs to delve more deeply both into probability theory and into the physical realities of the system in question. Pope's precept is very apt here:

A little learning is a dangerous thing;
Drink deep, or taste not the Pierian spring:

Suppose now we have a system consisting of two electric light bulbs wired in parallel and connected in the first place to a constant voltage supply, each with a reliability of 0.9 (unreliability 0.1) for a period of 1000 hours. According to the Product Law, the probability of both failing within this time should be $(0.1^2) = 0.01$, and the reliability of the system should be 0.99. Once one fails, however, the system has changed to a single bulb with a reliability of 0.9, until the failed bulb is replaced by one with the same reliability. But if a bulb is instantly changed at the moment of failure, the chance of both failing simultaneously within the time t becomes vanishingly small – far less than 0.01. In either case the behaviour of an apparently simple system is far more complex than suggested by the product rule of unreliability.

It is also clear that if, as in the case of the six bulbs in series, there were voltage surges, both bulbs would be liable to fail together whenever a surge occurred, and the reliability of the two bulbs in parallel would then be little higher than that of a single bulb.

The reliability of even simple parallel systems is more complicated than that of series systems, and factors such as detectability of failure, repairability (or replaceability), repair time and load variations have to be taken into account.

For process plant, the need for spare or redundant equipment is more often governed by production than by safety requirements. In both cases, availability, a combined function of reliability and repairability, is at least as important as reliability itself. In spite of these complications, reliability theory has been developed and applied to the design of the most complex systems, and without it, Man could probably never have landed on the moon.

14.1.4 Confidence levels and distribution of test results

For many purposes where failure rates or other raw data are available it is unnecessary to calculate numerical reliabilities as such. But whether reliabilities are calculated or not, the confidence level (CL) of the data is important. The higher the confidence level required, the more items need to be tested. If 100 batches each with 100 identical items were tested for the same period of service and in the same conditions, the average number of failures per batch might average 10, with failures per batch ranging from four to 19, most batches having between eight and 13 failures. The reliability would then be 0.9. A plot of the number of failures per batch against the number of batches with that number of failures might give a frequency-distribution curve such as shown in Figure 14.3.

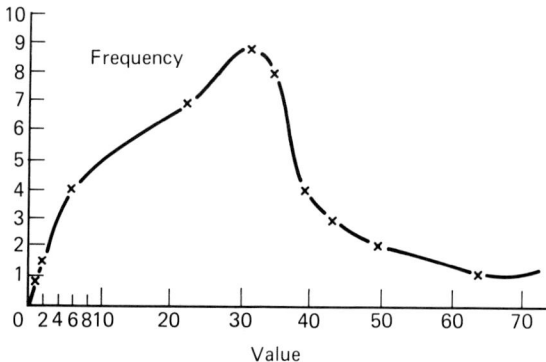

Figure 14.3 Typical failure-density distribution curve

The large number of tests required to give failure rates with reasonable CLs should be noted. Thus for electronic equipment to have a failure rate of 1% per 1000 hours, at least 92 000 unit test hours would be needed to give a CL of 60%, and 230 000 unit test hours would be needed to give a CL of 90%. The fewer the failures there are and the higher the reliability of the item, the larger the number of tests that are needed. Thus when one sees very high reliabilities (say, 0.9999) quoted for certain items, although one may not doubt that they are very reliable, the CL of such a figure may be quite low.

There are various possible theoretical distribution curves used to fit test results. The simplest is the normal or 'Gaussian' distribution which is

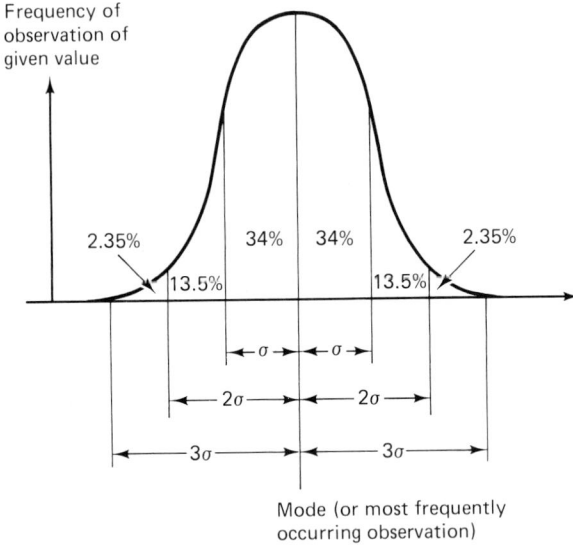

Figure 14.4 The Gaussian distribution

shown in Figure 14.4. This is symmetrical about a mean with tails stretching away on either side. The shape of the curve whether tall and narrow or short and broad is characterised by its 'standard deviation', expressed by the Greek letter σ. This is calculated as the square root of the sum of the squares of the individual deviations (both positive and negative) from the mean value. For a normal distribution, the percentage of the total area under the curve between $+\sigma$ and $-\sigma$ is 68%, and 68% of the failures per batch fall within these limits. Similarly, 95% of the failures per batch fall within the limits $+2\sigma$ and -2σ and 99.7% of the failures lie within the limits $+3\sigma$ and -3σ.

Other theoretical distribution curves determined by mathematical expressions with one or more constants sometimes fit the data better. These include the log-normal, the exponential, the 'Gamma',the 'Weibull' and the modified Weibull distributions. The Weibull distribution is particularly useful for the correlation and interpretation of failure data from field tests, and is facilitated by the use of special Weibull graph paper (Figure 14.5). The Weibull distribution has three constants, the values of which are readily determined as those giving the best fit to any set of data. The first constant t_0 is the period before failures start and generally represents the intrinsically reliable life of the item studied. The second is a scaling constant which also represents time and determines how the distribution is stretched along the time axis. It represents the time measured from t_0 by which 63.8% of the population can be expected to fail. The third, β, is a shaping constant which controls the shape of the curve. For $\beta < 1$, the failure rate decreases with age. For $\beta = 1$, the failure rate is constant and for $\beta > 1$, the failure rate increases with age. The value found for β is often characteristic of the mechanism of failure (e.g.

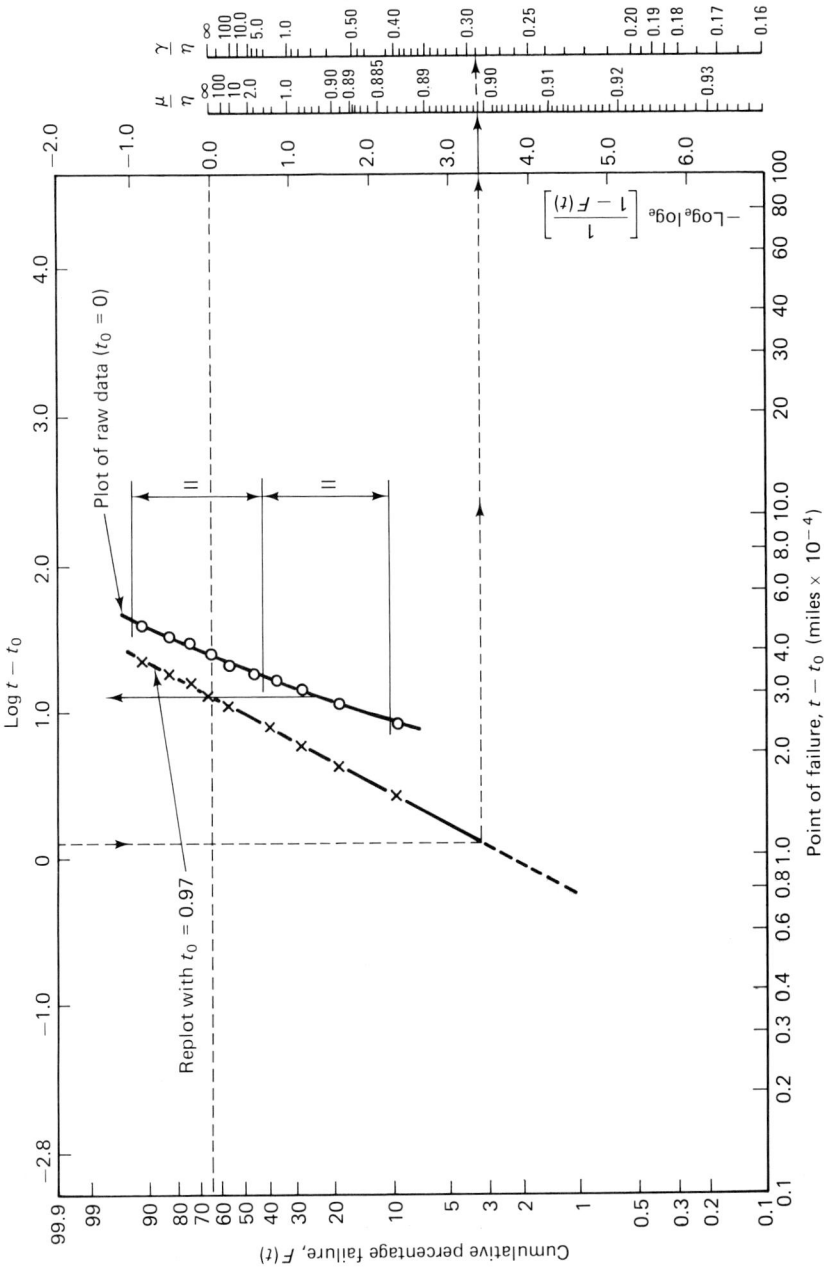

Figure 14.5 Presentation of failure data on Weibull paper

stress-rupture, fatigue, corrosion), about which it can provide valuable clues.

For items which exhibit an early failure rate (i.e. the first part of the bath-tub curve), a modified Weibull distribution can be used, e.g. by using a time-inverted Weibull distribution for this period followed by a normal Weibull distribution for the wear-out period[2].

14.1.5 Factors of safety and loading roughness[2]

On looking behind failure rates of engineering components one finds that both the strength of the components and the loads to which they are subjected vary about mean values. These variations, like those of failure rates, may be fitted to distribution curves which have standard deviations. When the applied load and the strength of the component are expressed in the same units, both curves can be plotted on the same chart (Figure 14.6).

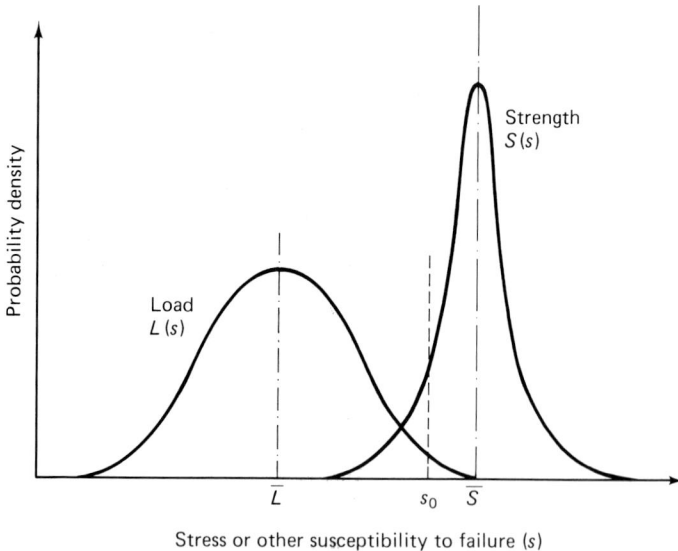

Figure 14.6 Distribution of load and strength

Failures then occur when the tails of both curves show a certain overlap, i.e. when a low-strength item is subject to a load at the top of the load range, shown by the small area which lies under both curves. This area corresponds to the finite probability that a weak item may encounter a load in excess of its strength. If it never encounters such a load, it will not fail. Thus although reliability is a probability, it is a joint probability that a random load is less than a random strength.

Loads generally show a greater variation than the strengths of components. Expressing strength as S and load as L, and their averages as S_m and L_m, the factor of safety used in the older engineering technology was expressed as S_m/L_m, with an average margin of safety of $S_m - L_m$.

This is not an adequate criterion since it takes no account of the variations of load or strength. A non-dimensional safety margin which takes variations both of load and strength into account is thus preferable. It is given by:

$$SM = \frac{S_m - L_m}{(\sigma_S^2 - \sigma_L^2)^{1/2}}$$

where σ_S and σ_L are the standard deviations of S and L.

Carter introduces a further dimensionless parameter to represent the relative shapes of the load distributions to which the item is subjected and the strength distributions of the population of 'identical' items themselves. This parameter is termed the 'loading roughness' and is defined by the expression:

$$\frac{\sigma_L}{(\sigma_S^2 + \sigma_L^2)^{1/2}}$$

The distribution of load and strength curves for various loading roughnesses at a constant (dimensionless!) safety margin is shown in Figure 14.7. For low values of loading roughness (smooth loading) the load distribution is confined to a small range and the strength distribution is much wider. But as the loading roughness increases, the load distribution becomes wider than that of strength distribution. The terms 'load' and 'strength' may also apply in a broad sense, using appropriate units, to electronic and electrical equipment as well as mechanical items.

Practical examples cover a wide range of loading roughness. Electronic equipment is generally subject to low loading roughness. Most fixed-speed mechanical equipment in the process industries operates in a narrow range of throughputs. Its loading is therefore generally smooth, provided it is operated carefully. Variable-speed machinery such as turbine-driven pumps and compressors, and centrifuges which have to pass through critical speeds and are sometimes out of balance, are subject to rougher loading. Safety and protective equipment used in the process industries such as pressure relief systems and flares, sprinklers, deluge systems, fire appliances and monitors are also subject to rough loading. Motor vehicles are another example of rough loading. They are driven in different climatic conditions over a range of speeds, surfaces and gradients by people of widely differing skills.

Loading roughness has important implications for reliability. For both series and parallel systems (such as the electric light bulbs discussed in 14.1.2 and 14.1.3), Carter demonstrates that the Product Rule can only apply to smooth loading (loading roughness <0.2). For rough loading, (loading roughness >0.8), the Product Rule does not apply in either case. The reliability of a series or parallel system approaches the mean reliability of its components, as the ratio σ_L/σ_S become infinitely large. For most series systems the reliability lies somewhere between these two extremes.

14.1.6 Intrinsic reliability, safety limits and deterioration

The four sets of plots in Figure 14.7 show probability distributions for load and strength plotted on a common scale for particular items used under

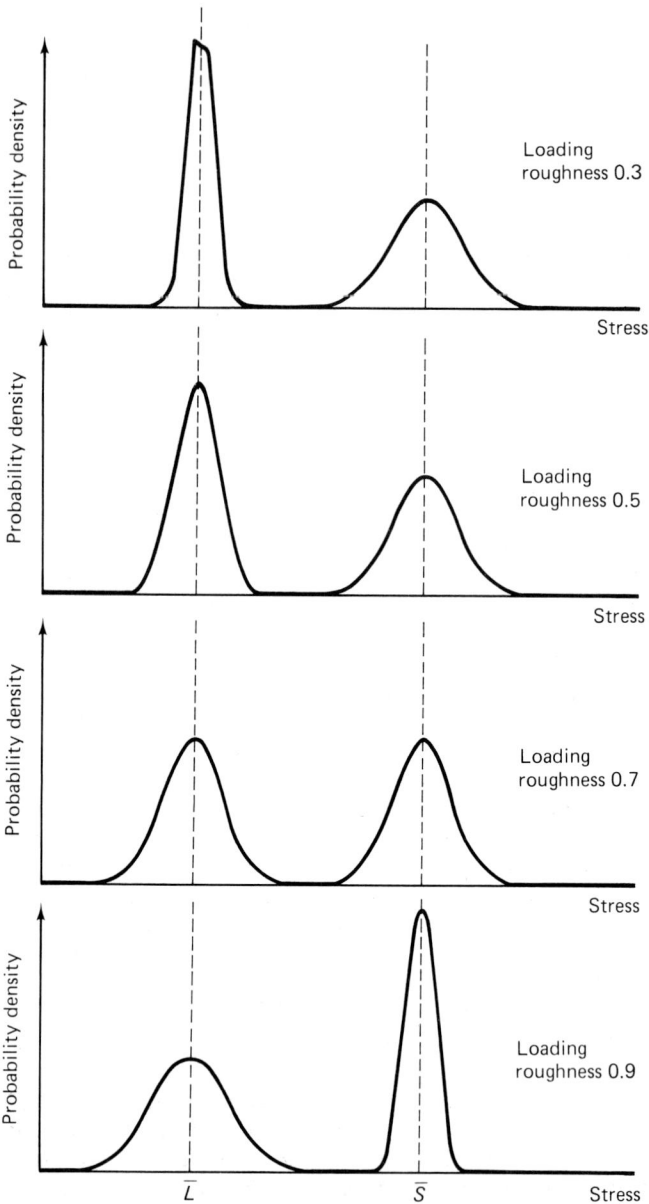

Figure 14.7 Distribution of load and strength for various loading roughnesses at safety margin of 4.5

defined conditions. The fact that the tails of each pair of curves do not overlap implies that there should be no failures. Such items when used under the conditions stated are therefore described as 'intrinsically reliable', with a reliability of 1.0. Although the dimensionless safety margin between the mean strength and load is the same in each case, the gaps between the tails of each pair differ and depend on the loading roughness. The minimum (dimensionless) safety margin required for intrinsic reliability with substantially normal distributions of load and strength increases as the loading roughness increases from 0 to 1.0 (Figure 14.8).

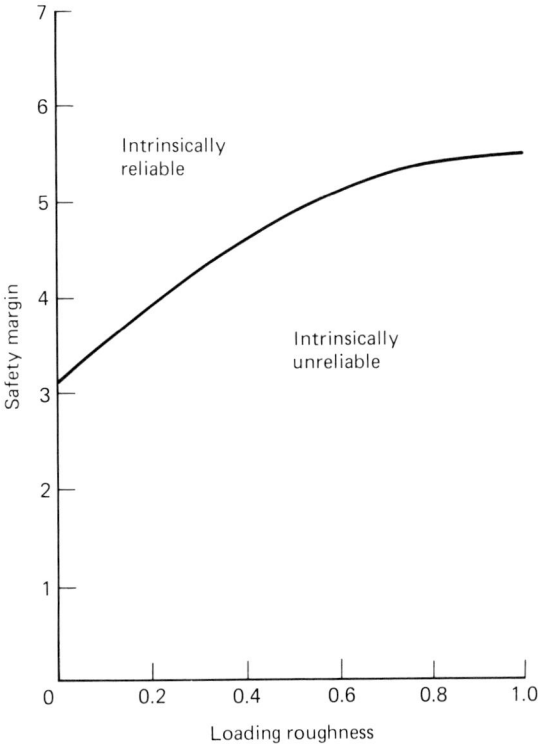

Figure 14.8 Minimum safety margin for intrinsic safety for normal distributions of load and strength

The different processes of deterioration discussed in Chapters 11 and 13, e.g. corrosion and mechanical wear in the case of metals, loss of antioxidant and light stabilisers with chemical degradation in the case of plastics and synthetic fibres, come into play from the time a component is placed in service, if not earlier. The result is a gradual loss of strength and a spread in its distribution. To produce intrinsically reliable components with useful periods of failure-free life, there must be an adequate margin in the initial separation between the tails of the load and strength distributions. This separation progressively decreases during the intrinsically reliable life of the component until the tails of the two distributions meet. This marks the end of the intrinsically reliable life of the component, after which an overlap appears in the two tails, failures begin to appear, and the wear-out period starts. This process is shown in Figure 14.9.

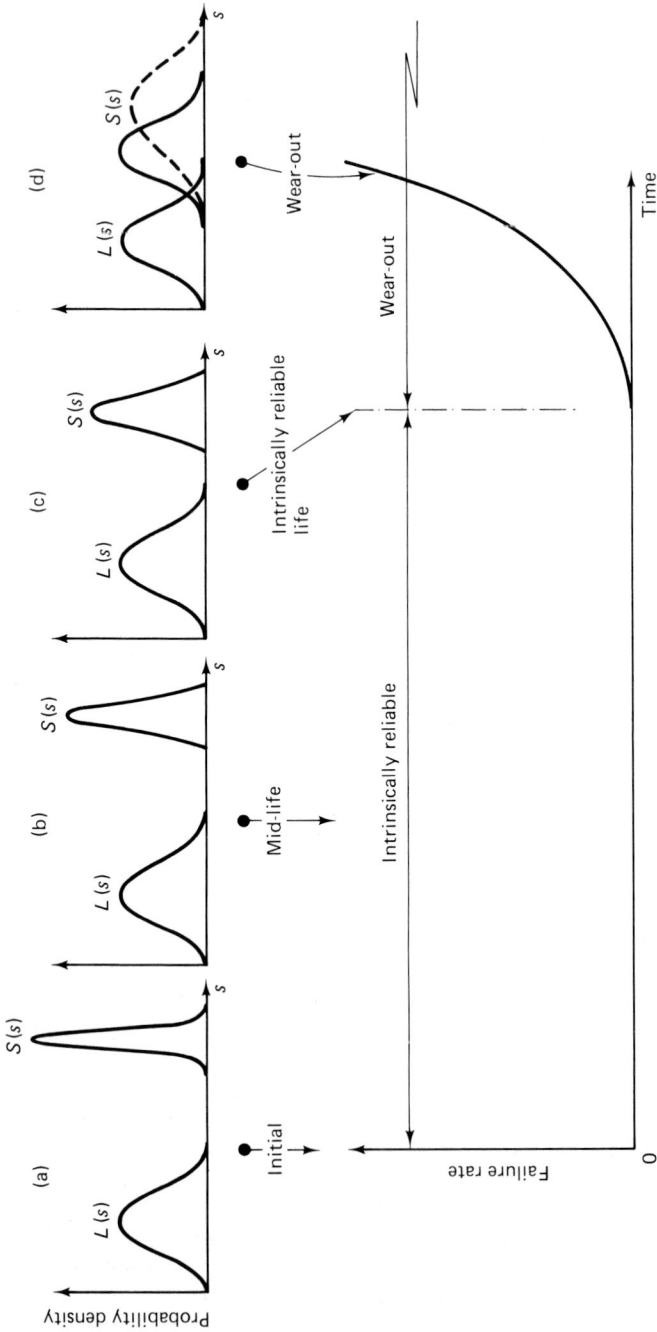

Figure 14.9 Change of strength and strength distribution with age and consequent effect on failure rate

The shapes of the failure-rate/age curves during this period are reflected in the scaling factor β of the Weibull distribution, and can provide useful clues about the predominant mode of failure[2]. Thus fatigue is characterised by a sharp jump in failure rate after the end of the intrinsically reliable period, followed by a long period of falling failure rate, which can last considerably longer than the intrinsically reliable period. Erosive wear, by contrast, is generally associated with a continuously increasing failure rate after the end of the intrinsically reliable period.

The fact that an item may be intrinsically reliable for its intended function for an assured period when it leaves the maker's works is, of course, no guarantee that it will not encounter loads or conditions which are well outside the normal range before its intrinsically safe period in use has expired. Exposure to such abnormal loads and conditions is here broadly described as 'misuse'. Examples of such misuse include accidents in transport and installation, excessive stresses in the installed items caused, for example, by force used in jointing badly fitting pipework, improper site welding and heat treatment, steam and water hammer, cavitation in operation, fires, internal explosions and unintended contact with corrosive fluids outside the normal limits of the process. Even where the item does not immediately fail as a result, its strength may be so impaired that it is liable to fail prematurely.

Misuse can thus introduce doubts about intrinsic reliability and make equipment suppliers reluctant to give service guarantees. Every effort must be made to prevent such misuse in the first place, to report it when it does occur, and to properly inspect and test items subjected to or suspected of damage through misuse, rejecting, repairing or derating them as appropriate.

14.1.7 Implications for maintenance

All process equipment is made up of a large number of components, and even when they may all be considered to be intrinsically reliable for a certain period, there may be large differences in the intrinsically reliable periods for different types of component. The failure rate, reliability and MTBF of the equipment will then depend largely on the maintenance policy [17.1] adopted for the equipment and its components, and whether the equipment in question can continue to operate while the component is being repaired or replaced. If it cannot, and a policy of repair maintenance is adopted for the component(s), the MTBF of the equipment can be no higher than the MTTF for the component. If there are several such components in series, the equipment is liable to fail at random. The product rule of reliability may then apply to the set of components in series, and the MTBF of the equipment will be correspondingly low. If, however, all components are intrinsically reliable for a limited period, and a preventative maintenance policy is adopted whereby they are all replaced together after shorter intervals, the MTBF of the equipment will be equal to the average of these intervals. Since the maintenance policy for the plant as a whole is often a mix of policies for different equipment, many combinations with different reliabilities and MTBFs are possible. These need to be considered by the designer when deciding, for example, whether to install spare stand-by pumps for particular duties, whether to fit common spares, and whether to employ pumps with packed glands, mechanical seals or pumps without shaft seals.

Mathematical models and simulation techniques are sometimes used to solve or throw light on reliability and maintenance problems. These include Reliability Block Diagrams[2], Markov models, and Monte Carlo simulation techniques[3]. Their main application is to parallel systems, rather than the series systems of which process plant is mainly composed. The information needed for their use, particularly the failure rates of specific equipment in specific circumstances, is seldom known with any accuracy at the design stage. Their use then can lead to a make-believe world, which is rudely shattered when the plant starts to operate. They are more useful in helping to plan maintenance on plants which have been operating for several years, and for which a good deal of reliability data has been collected.

In conclusion, too much importance should not be attached to system reliability estimation, particularly for mechanical equipment. The real problem lies in achieving acceptable component reliabilities over the range of loading and environmental conditions to which they will be subjected in real use, and in planning maintenance to maintain that reliability. The important question is whether the design team have thought of all eventualities. For this their track record is probably the best guide.

14.1.8 Types of failure

It is often difficult to decide what is failure and whether an item has failed. Dummer and Winton[1] classify failures by cause, time, degree, and combinations of these three:

1. *Cause*:
   (a) Misuse, e.g. due to the application of stresses beyond the stated capacity of the item;
   (b) Inherent weakness, i.e. in the item itself when subject to stresses within the stated capacities of that item.
2. *Time*:
   (a) Sudden failure, i.e. that could not be anticipated by prior examination;
   (b) Gradual failure, i.e. that could be anticipated by prior examination;
3. *Degree*:
   (a) Partial failure, i.e. resulting in deviations in the characteristics of the item beyond specified limits which nevertheless do not cause complete lack of the required function;
   (b) Complete failure, i.e. resulting in deviations in the characteristics of the item beyond specified limits which cause complete lack of the required function;
4. *Combinations*:
   (a) Catastrophic, i.e. failures which are both sudden and complete;
   (b) Degradation, i.e. failures which are both gradual and partial.

Lees gives another classification which is useful in the process industries[3]:

1. Failure in operation;
2. Failure to operate on demand. (This sometimes depends on human knack or touch, or, some would say, on the sex or temperament of the individual!)
3. Operation before demand;
4. Continued operation after demand to cease.

There is a whole hierarchy in types of failure ranging from those which affect safety to those which hardly matter. Our appreciation of failure is also often highly subjective. It is therefore important to define the types of failure covered by guarantees or included in statistics. Those recognised by the supplier are not always those that matter to the user!

A homely example of this difficulty is a leaking washer in the float-operated valve of a domestic cold water storage tank, which causes water to trickle down the overflow. On complaining to the plumber who had recently installed the system that it had failed, he might, with logic, reply that it was functioning as intended, and that the overflow pipe has the dual function of safety, and of warning that action is needed before failure occurs. While you might not accept this argument, you would have to admit that the system had 'failed safe'. But if the situation arose in winter so that the overflow froze and the tank flooded, your logical plumber would have to admit that the system had failed. This illustrates the fact that environment plays a part in the functioning of any item, and that success or failure is not an absolute property of the item alone.

An example of different degrees of failure (and reliability) in the process industries concerns valves on process pipes. Few valves can be guaranteed to give a 100% positive shut-off of a process fluid under pressure on one side when the other side is open to the atmosphere, yet despite this they may function adequately for normal operation. But when equipment has to be opened and entered for inspection and/or maintenance, a much higher degree of reliability is demanded, so that an elaborate system of 'block and bleeder' valves, blank flanges, spades, etc. is required for isolation [18.6].

Calculations of the reliability of systems, based on statistical data for the reliability of their components, are sometimes made during design to check that equipment made up of such components has the necessary expectation of reliable service.

Such expectations or probabilities should not be confused with guarantees. A probability of, say, one failure in 40 years' operation cannot indicate when such a failure may occur. Hammer[5] cites the case of a solid propulsion unit for a new missile, whose carefully researched reliability indicated that only two out of every 100 000 would be expected to fail. The first to be tested blew up!

14.1.9 Redundancy and its problems

Redundancy (in this connection) generally means providing more than one way of doing something. It may take many forms, e.g. partial, active and stand-by, of which the following are examples.

Partial redundancy
- A lift suspension with two or more ropes which can still operate even if one fails;
- A four-engined aeroplane which can take off on three.

Active redundancy
- Duplication or triplication of working parts as in instrument landing systems for aircraft;
- Fire detectors which operate automatic sprinkler, deluge and shutdown systems to protect process plant or storage installations.

Stand-by redundancy
- Duplicate installed spare pumps, filters, relief valves one of which remains idle until the other fails or is taken out of service. Here the MTBF is less than the maximum interval between planned shutdowns.

When instruments are used to detect an emergency condition such as fire and initiate action of some sort, two or more detectors which operate on different principles are sometimes installed and connected to an electronic voting system. This only initiates the action required (e.g. a warning siren and an automatic water deluge), when the majority of the detectors agree that the fire, etc. really exists. This is often done when there is difficulty in finding a single reliable instrument which always detects the condition in question, and does not give false alarms. (These can have costly consequences, such as flooding a warehouse full of tobacco because someone was surreptitiously smoking in it.) Redundant detectors of the same type are also used with voting systems in 'high integrity protective systems' [14.2.1].

The installation of redundant items other than instruments and electronic systems to improve the safety or reliability of process plant should only be considered when no intrinsically reliable items guaranteed for periods in excess of the maximum interval between planned shutdowns are available..

There may, of course, be other reasons for installing two or more items in parallel, e.g.:

1. Where the duty required exceeds the capacity of the largest suitable item available;
2. Where the required range of throughputs exceeds the turndown ratio of the intended item to be installed.

The simultaneous operation of two or more items installed in parallel increases the instrumentation required and can cause control problems [15.7.3]. Any reader who has tried to run two steam turbine-driven centrifugal pumps in parallel will appreciate this.

The installation of redundant equipment or components [14.1.9], whatever benefit it may have on reliability, also adds to the things that can go wrong and require inspection and maintenance. Failures may occur in changing over from a running item to an installed spare. Some types of deterioration, such as brinelling of ball-bearings and dezincification of brass alloys, may proceed faster on the idle installed spare than on the item in service. Problems of thermal stresses increase when a pump handling a hot liquid operates alongside a cold installed spare. Sediments may cause problems during the start-up of spare pumps. Where spare components or equipment are installed to improve reliability, it is important to rotate their use periodically, to ensure that the spare item is always ready for instant service.

Redundant instrumentation is rather a different matter. We as humans are blessed with five, some would say six, senses, and if we were nocturnal creatures, God or evolution would doubtless have provided us with a sonar system and antennae like those of the bat. The same or similar information (such as the outbreak of fire) is often provided by several of our senses, yet unlike Hindu deities, we do not have redundant arms and legs with which to deal with the situation. Thus direct additional information provided through independent channels about what is going wrong is often invaluable in assessing a potentially hazardous situation. It is furthermore unfortunate that instrument reliability [14.2.1] is not generally as high as that of the equipment which it controls, thus adding to the need for redundant instrumentation.

14.1.10 Reliability and safety

Reliability and its assessment are important to design safety. Yet this importance varies widely. High reliability, while important to the economic success of a project, is not always essential to human safety. In some cases the requirements of operational reliability and safety may actually conflict. An example is the provision of intermediate storage for hazardous process materials between plants or plant units which form parts of a manufacturing chain. Intermediate storage 'buys time' to cope with minor upsets in individual units, thus reducing the need to shut down the entire plant or process unit when something goes wrong. It also provides greater operating flexibility, facilitating the movement of intermediate materials to or from external suppliers or users. Thus it contributes to operational reliability. Yet it greatly multiplies the inventory of hazardous material on the site and the possible consequences of a release, so making the site a more dangerous one. While the provision of storage for 60 tonnes of methyl isocyanate at Bhopal [5.4] may have increased the reliability of insecticide manufacture, it had disastrous consequences for safety. Similarly, the doubling of the number of operators on a hazardous plant may increase its reliability, but it also increases the numbers liable to be killed in an explosion.

In air transport, the reliability of the airframe, propulsive system and controls are vital to safety. In process plant, on the other hand, many failures (such as power supply or of particular motors or valves) usually have more effect on production and profits than on safety, provided the plant has reliable shutdown and protective systems to safeguard against such failures. These systems must, of course, be kept in a state of readiness, and the operating team must be capable of taking any action required of them promptly. When control equipment does fail on process plant, it is arranged as far as possible to 'fail safe' [15.7.2], and often to provide a warning that it has failed.

14.1.11 Reliability of safety equipment

Plant safety demands a specially high degree of reliability for the protective and automatic shutdown systems, and for those pipes and pressure vessels whose rupture could lead to large escapes of hazardous materials. This requirement increases with the degree of hazard of the plant (e.g. as indicated from its Dow or Mond Indices [12]). The consequences of such escapes, whether they be fires, explosions or personal injuries, may be as serious as an aircraft failure over mid-Atlantic. Unfortunately, the very systems for which high reliability is most required for safety usually play no other part in production, are sometimes difficult to test adequately when installed, and are only called into use in rare emergencies. Unlike production equipment which may fail through wear or fatigue after many hours of operation, one of the problems of protective equipment is that it may fail to operate properly when required to do so. This may be caused by friction, deposits, corrosion, birds' nests or human negligence. Even when properly specified, designed and installed in the first place, the designer has no control over how such equipment is used or maintained. The Bhopal disaster [5.4] is an object lesson of this failure to ensure that critical protective systems are both reliable and available when needed. Some difficulties in ensuring and checking a high degree of reliability for complete pressure relief systems (not just the relief devices) are discussed in 15.2.

14.1.12 Key features of reliable systems

The first essential for a reliable system is to ensure that all its components are themselves reliable and appropriate for their duties in the environments in which they operate. Some of the failures that occur are the result of manufacturing flaws; most result from some type of corrosion, wear or other form of deterioration during service as discussed in Chapters 11 and 13. Strict quality control during manufacture of engineering components is essential to their reliability. This should include testing a proportion of the components to destruction, and proof testing the others to stresses higher than those to which they will be subjected in service. These aspects of reliability are more important than price when items critical to safety are being selected.

A second but less often appreciated need is simplicity. Some process flow diagrams and the piping and instrumentation diagrams (P&IDs) which are developed from them are unnecessarily complicated. Many plants contain a bewildering proliferation of pipes and valves, some of which are never used. These only cause confusion and present obstructions, and should be carefully pruned during design [16.4.5]. Designers of electronic circuits face the same problems of eliminating unnecessary items and connections. Simplication of these is helped by the application of Boolean algebra[6], named after its inventor George Boole. Boolean algebra systematises logical reasoning and is one of the foundation stones of computer science. It can also be helpful in simplifying P&IDs and is essential to making 'truth tables' and other forms of risk analysis [14.3].

14.2 Process equipment reliability

A comprehensive bank of reliability data is maintained by the Safety and Reliability Directorate of the UK Atomic Energy Authority. This is available to users on payment of a small charge.

Despite the reservations voiced in 14.1.2 over the general validity of constant failure rates for process equipment, most reliability data are quoted on this basis. For reasons explained later, this assumption appears to be most valid for instruments. Here failure rate data are often used to assess the need for redundant instruments to increase safety.

For other types of process equipment, such data are used as a guide to selection in design, and may be employed in some kinds of risk analysis 14.3.

Published data on the failure rates of different types of equipment show general and broadly repeatable patterns, rather like those of accident rates for different industries and occupations. The fact that the latter are nearly always capable of improvement can be readily seen by comparing rates for the same occupations in different countries. Similarly, the failure rates of equipment can usually be improved by better standards of design, construction, inspection and maintenance, and by choosing materials of construction which are more resistant to corrosion, fatigue, etc. in the working environment. Where the failure rate for particular equipment is thought to be low, this should not be used to justify lower standards of safety.

Some failure rates for instruments are given and discussed first in 14.2.1. This is followed in 14.2.2 by some typical failure rates for other equipment, and by some figures used for the Canvey study of 1978[7].

14.2.1 Instrument reliability

The operation of most process plant is critically dependent on instruments, which may account for 3–7% of the total plant cost[8]. While improvements are constantly being made, instrument reliability generally falls short of that of the equipment which it monitors or controls. This places special demands on the skills of human operators, who often have to operate plant with some instruments temporarily out of order, and automatic control valves by-passed or on remote manual control.

Simple logic tells us that the more links there are in a single chain, the greater the risk of failure, while the use of several independent chains to do the same job reduces the risk. Moving parts of instruments are more likely to go wrong than stationary ones, and solid state circuits and transistors are more reliable than wired circuits and valves. An instrument operating in a clean, cool, vibration-free, non-fouling and non-corrosive environment is far less likely to fail or stick than one operating in a hot, dirty, corrosive and/or vibrating environment.

Although control room instruments operate in an almost ideal environment with smooth loading, the detecting, measuring and transmitting instruments on the plant outside, as well as their connecting cables and impulse lines, are subject to worse conditions and are more likely to get rough treatment, e.g. during plant maintenance.

The failure of instruments which form parts of control loops, emergency trips and alarms usually have more immediate and serious consequences than those which merely indicate or record particular variables. When one of the last types develops an error, an alert operator will often discover that the instrument reading is wrong before acting on it. Although there may be no other instruments to provide a direct check on the faulty reading, other instruments should and generally do provide sufficient related information for the operator to suspect or diagnose a fault in a particular one.

For control loops whose failure could lead to only minor damage or lost production it is common to incorporate redundant instruments with active functions into the same loop. As an example, a level-indicating controller will often incorporate a high- or low-level alarm which depends on the same transmitted signal. But where instrument failure could cause human injury, and/or serious consequences to plant or products, a complete back-up system, including its own power supply and valves to function in specific emergency situations, may be justified.

A review paper with 21 references by Lees provides extensive information on failure rates of several different types of instruments used on UK process plants in the 1970s[9]. Most of the data resulted from surveys of maintenance records. Three surveys in which Lees was involved came from works manufacturing heavy organic chemicals, inorganic chemicals and glass. Others originated from an oil refinery and the UK and US atomic energy authorities.

Much of the data in this review are presented as the average number of faults per year for a particular type of instrument in a particular type of environment. These faults included partial as well as complete failure, and failure in safe as well as dangerous modes. Constant failure rate with time was assumed for all data presented. According to Lees (personal letter), little work has been published on the dependence of instrument failure rates on time. It may be that instrument failure rates in a large works resemble the imaginary example of an underground lighting system given in 14.1.2, which showed a pseudo-constant failure rate. A repair maintenance policy is generally employed for process instruments since it is difficult to justify a replacement policy if an instrument is just as likely to fail after replacement as before.

Many of the instruments had some of their working parts exposed to process fluids which varied in dirtiness and corrosivity. Instruments in contact with process fluids, e.g. for measurement of pressure, level and flow, and flame failure devices had between 1 and 1.5 failures/year. This was substantially higher than for instruments not in contact with process fluids, such as valve positioners, solenoid valves, current-pressure transducers, controllers and pressure switches. These had between 0.25 and 0.55 failures/year. Average failure rates quoted for control valves and differential pressure transmitters for service with dirty fluids were five times higher than in the case of clean fluids.

To allow for such effects and to correlate data for the same type of instrument in different environments, an environmental factor ranging from 1 to 4 was introduced. This was taken as the actual failure rate divided by the base failure rate. An environmental factor of 1.0 (the base rate) applied to instruments which had no working parts exposed to process fluids, and to those whose exposure was limited to a clean and non-corrosive process fluid at low to moderate temperatures and pressures. For instruments exposed to process fluids, the highest environmental factor of 4 was found in inorganic chemical and glass manufacture, while an environmental actor of 1 to 2 appeared to apply to similar instruments in an oil refinery.

Base failure rates for individual instruments (which might form parts of a loop) generally lay in the range 0.02–1.0 failures/year, but failure rates for complete loops generally lay between 1.0 and 2.5 failures/year.

Some data on failure rates of control loops quoted in Lees's review, which have particular significance for safety, are given in Tables 14.1 and 14.2.

Table 14.1 Instrument loop failure rates

Type of loop		Faults per year
PIC	Pressure-indicating controller	1.15
PRC	Pressure-recording controller	1.29
FIC	Flow-indicating controller	1.51
FRC	Flow-recording controller	2.14
LIC	Level-indicating controller	2.37
LRC	Level-recording controller	2.25
TIC	Temperature-indicating controller	0.94
TRC	Temperature-recording controller	1.99

Loop element	Percentage of faults
Sensing/sampling	21
Transmitter	20
Transmission	10
Indicator, recorder	18
Controller	18
Control valve	7
Other	17

Table 14.2 Failure rates for an FIC loop

UKAEA data	
DP transmitter	0.76
Controller	0.38
Control valve	0.25
Total	1.39
Anyakora, Engel and Lees's data	
Impulse lines	0.26
DP transmitter	0.58
Controller	0.29
Control valve	0.30
Valve positioner (for 1 in 5 loops)	0.09
Total	1.52
Skala's data	
FIC loop	1.51

It seems that most instrument failures do not have immediately dangerous consequences. Thus a study of the failure rate of a pressure switch showed a total failure rate of 0.13 faults/year, of which only 0.025 were dangerous. More detailed data for trip systems are given in Table 14.3. A breakdown of faults in control valves showed leaks 29%, failures to move freely 24%, blockages 14%, failure to stop flow 7% and diaphragm faults only 3%.

Table 14.3 Failure rates for components of protective systems

Components	Faults/year		
	Fail hazardous	Fail safe	Total
Trip initiator			
Impulse lines (blocked or leaking)	0.09	–	0.09
Pressure switch	0.10	0.03	0.13
Cable (fractured or severed)	–	0.03	0.03
Loss of electric power	–	0.05	0.05
Total	0.19	0.11	0.30
Steam shut-off system			
Relay (complete with wire)	0.01	0.07	0.08
Solenoid valve	0.10	0.20	0.30
Loss of electric power	–	0.05	0.05
Trip valve	0.10	0.15	0.25
Air supply line (blocked, broken)	0.01	0.01	0.02
Loss of air supply	–	0.05	0.05
Total	0.22	0.53	0.75
Pump shut-off system			
Relay, etc. as above	0.01	0.07	0.08
Pressure relief valve			0.02
Flame-failure detector			1.69

The highest failure rates were found among automatic on-line analysers, with an average of 8.5 faults/year, although analysers with no moving parts such as hydrogen analysers and infra-red liquid analysers had only 1.0 and 1.4 faults/year. GLC equipment had the highest rate with 31 faults/year, followed by electrical conductivity meters for liquids with 17 faults/year.

Despite the general belief that electronic instruments were more reliable than pneumatic ones, the survey showed little difference in their failure rates.

High reliability is specially important for plant protective systems such as those listed in Table 14.3. On oil refineries and organic chemical plants there is often a strong economic incentive to employ a system to shut off steam to the reboiler of a large distillation column if the column pressure rises [15.6] (e.g. in the case of reflux or condenser coolant failure). This is far cheaper than the time-honoured solution of installing a large pressure relief valve (on the top of the column or reflux drum) which discharges through a long and large-diameter pipe to a large flare stack. Although the relief valve is far more reliable than the steam shut-off system shown in Table 14.3, the relief valve is only part of a relief system which may have other faults [15.2]. On the other hand, there may be other causes for excess pressure in the distillation column which do not respond to shutting off the steam supply (e.g. fire exposure, exothermic reaction or the presence of water in the column).

High-integrity protective systems for very hazardous plants which have considerably lower failure rates than the simple steam shut-off system shown in Table 14.3 are described by Wells[10]. These depend on the use of redundant instruments in parallel, sometimes incorporating a voting system. A 'triple modular redundancy system' in which three processors carry out 'software voting' and produce a value based on the majority vote is advocated by Brammer[11]. Such systems also enable component failures to be detected very rapidly and 'hot repairs' to be carried out while the system continues to operate. High-integrity systems are used for the

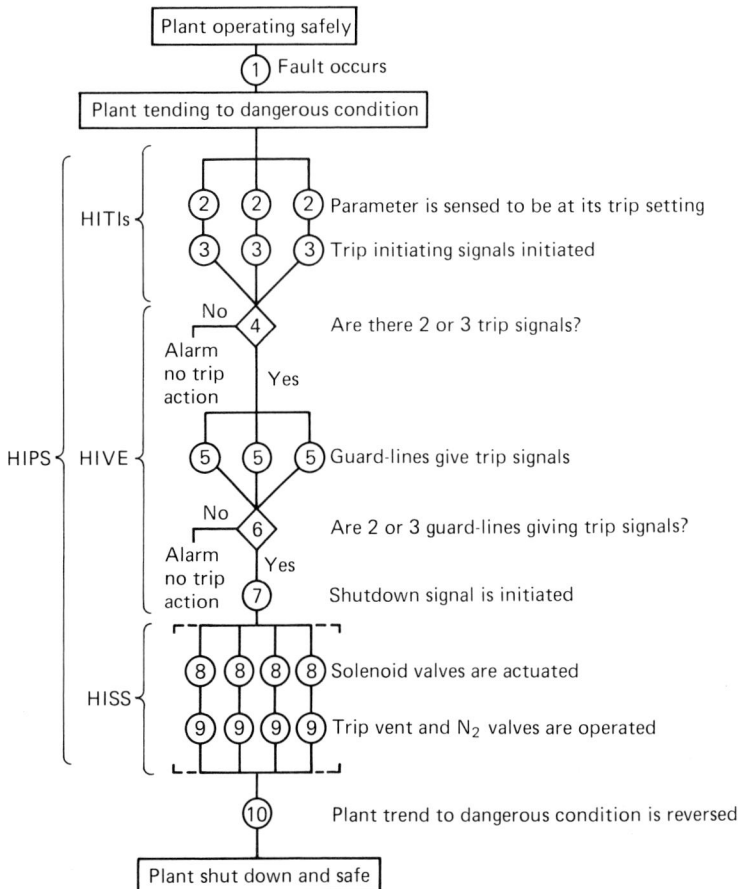

Figure 14.10 High Integrity Protective System for emergency trips or automatic plant shut-down (from G. L. Wells, *Safety in Process Design*, courtesy George Godwin, London)

complete automatic shutdown of a plant in an emergency, and also for relieving particular hazardous conditions such as over-pressure in a distillation column. Their principle is shown in Figure 14.10.

14.2.2 Some frequencies of equipment failure

Lees gives a brief survey and a number of selected references on frequencies of process equipment failures, and events such as leaks and spillages (including their ignition), furnace explosions, pump and tank fires, and vapour cloud explosions (including probability of ignition)[3].

Selected data on equipment failure rates published by the UKAEA are given in Table 14.4. Some failure data used in the Canvey study are given in Table 14.5.

Table 14.4 Selected equipment failure rates

Equipment		Failures/10^6 h
Pressure vessels	(general)	3.0
Pressure vessels	(high standard)	0.3
Pipes		0.2
Pipe joints		0.5
Gaskets		0.5
Bellows		5.0
Diaphragms	(metal)	5.0
Diaphragms	(rubber)	8.0
Unions		0.4
Hoses	(heavily stressed)	40.0
Hoses	(lightly stressed)	4.0
Relief valves	(leakage)	2.0
Relief valves	(blockage)	0.5
Valves	(hand-operated)	15.0
Valves	(ball)	0.5
Seals	(rotating)	7.0
Seals	(sliding)	3.0
Seals	('O' ring)	0.2
Filters	(blockage)	1.0
Filters	(leakage)	1.0
Pins		15.0
Nuts		0.02
Bolts		0.02
Boilers	(all types)	1.1

Table 14.5 Assumed frequencies of catastrophic failures used in the Canvey study[7]

Equipment, etc.	Occurrence	Times/10^6 years
Pressure vessels	Spontaneous failure	10– 100
Pressure circuit (HF)	Spontaneous failure	100
High-speed machine	Disintegration of rotor	100–1000
Pipework (LPG)	Failure (whole refinery)	5000
Pump (LPG)	Catastrophic failure	100
LNG tank (above ground)	Serious fatigue failure	200
Jetty pipework (LNG)	Catastrophic failure	100–1000
Pipeline (butane)	Failure	300/km

14.3 Risk analysis and its scope[3, 12, 13]

The term 'Risk Analysis' as used by the European Federation of Chemical Engineering means answering three questions[12]:

1. What can go wrong?
2. What are the effects and consequences?
3. How often will it happen?

The Institution of Chemical Engineers' booklet on terminology in the hazard and safety field, while not using the term 'Risk Analysis', defines 'Hazard Analysis' as dealing with question (1) and 'Risk Assessment' as dealing with questions (2) and (3)[14]. 'Risk Analysis' should therefore cover the whole field. It came into prominence in the UK in connection with the planning of nuclear power stations, and is now required when planning potentially hazardous process plants and storage installations, particularly those falling within the scope of the NIHHS and CIMAH regulations [2.4]. It is also needed during plant design. The subject has several branches which are well covered in the literature and only an outline is given here.

The methods used fall into four main areas:

1. Those used to identify hazards and how they could materialise;
2. Those used to estimate the probability of such accidents;
3. Those used to estimate their potential consequences;
4. Those which combine the results of (3) and (4) to estimate the risks to individuals, groups of individuals and property.

Some methods cover more than one area. Clifton groups identification techniques into three categories: comparative methods, fundamental methods and methods which use failure logic diagrams[13].

14.3.1 Comparative methods

These rely on experience in the form of checklists, codes of practice and the calculation of hazard indices [12] from plant and process data. Several checklists have been developed covering site selection, layout, process materials, various aspects of design, commissioning, operation and plant shutdown[3, 10, 15]. Their use during design is discussed in 16.1. While a checklist requires definitive answers to specific questions, it should also stimulate thought on open-ended ones. Although checklists are valuable methods of hazard identification, they usually contain many questions which are irrelevant to the case in hand, and their use is time-consuming. It is essential to use them before and not after decisions affected by their questions have been taken. Reliance on checklists alone is not considered sufficient to uncover all hazards and their modes of materialisation on major hazard installations.

The Dow and Mond hazard indices [12] can be applied from an early stage in project planning and focus attention on high-hazard areas. They employ a number of empirical factors which are related to the probable frequency and magnitude of potential accidents without specifying precisely how these could arise. The Dow and Mond indices represent combinations of probable frequency and magnitude of process accidents.

14.3.2 Fundamental methods

These are structured ways of stimulating a group of people to identify hazards based on their own knowledge. The two main techniques available are Hazard and Operability studies (HAZOP) and Failure Modes and Effect Analysis (FMEA). HAZOP studies, which are discussed in 16.3, are particularly valuable for discovering hazards from P&IDs and reducing them, and in writing plant-operating procedures. FMEA has more application to assessing reliability than risk, but it may identify faults which have more serious consequences. Individual components and systems (particularly instrument loops) are examined and all possible failure modes are listed, if possible with their probabilities. The failure modes and rates for a steam shut-off system shown in Table 14.3 is an example. The response of the system to each failure mode would then be analysed.

14.3.3 Fault logic diagrams

These represent fault logic pictorially. Their use encourages speculation about the events which could lead up to or result from a particular situation, e.g. a leak of some process fluid. They provide a structure to the failure logic and show the ways in which events and system states depend on each other. Two principal techniques are Fault Trees and Event Trees. Fault tree analysis was developed by Bell Telephone Laboratories in the early 1960s to improve the safety of control systems used for launching rockets. (Despite this they still pose problems!)

Fault trees are mainly used to define all possible routes leading to a particular type of failure. Event trees are used to discover the ways in which a particular failure could cause a serious accident. Both techniques use binary logic where events and states are connected on the diagram by lines which must be either 'on' or 'off'.

14.4 Fault trees

This section can only serve as an introduction to a complex subject which holds many pitfalls for the unwary novice. Fault tree analysis needs undivided attention and often expert guidance.

Before attempting a fault tree analysis, we should be thoroughly familiar with the system analysed, and have a suitable model in front of us. This should show all items and their interconnections whose faults (basic events) could affect the top event under study. For this a P&ID is a good starting point, though we may require more detail as we move down the tree.

Starting with a particular type of failure termed the 'top event', we work back to a first level of faults which, acting singly or in combination, could have led to it. Next we look for all faults which could cause the first-level faults and continue down the tree until we reach failures for which data are available (or can be invented!). These failures are called basic events.

Both failure rates (dimensions $time^{-1}$) and probabilities (dimensionless) that an item is in a failed state (sometimes referred to as fractional dead-time) are used in building fault trees. Confusion can readily arise between them. The failure rate is usually expressed as so many times per year (often less than one) that the item is expected to fail. This is usually assumed to be constant (corresponding to the constant failure rate period in the life of the item as represented on the 'bath-tub' curve [14.1.2]). The probability that the item is in a failed state is a function both of its failure rate and the period during which the unit of which the item is a part continues to operate while the item is in a failed state.

If the unit does not operate continuously and has to be shut down to repair (or replace) the item, as well as for general inspection and maintenance, the period is simply that between the occurrence of the failure and the shutdown of the unit. If the unit operates for 2000 hours per year and the average time lapse between discovery of the failure and shutdown of the unit is 10 hours, and the item fails, on average, three times per year, then the probability of the item being in a failed state while the unit is operating is $3 \times 10/2000 = 0.015$. Provided the failure rate of the item remains the same after repair or replacement as before it, and the failure and repair do not affect the number of hours per year when the unit operates, the probability of an item being in a failed state is given by:

$$\frac{(\text{Failure rate, years}^{-1}) \times (\text{hours between failure and shutdown})}{\text{Hours of operation per year}}$$

If the item is tested, and if necessary, repaired or replaced while the unit is in operation, the calculation of the probability of the item being in a failed state or out of commission while the unit is running becomes more complicated.

Lines from lower-level faults shown on a fault tree can join at one of two kinds of 'gates' (similar to those used in Boolean algebra) from which a single line passes to a higher fault.

Figure 14.11 shows the principal symbols used in drawing fault trees. At an OR gate, any of the lower events is sufficient to propagate the fault, and their probabilities or failure rates are added. At an AND gate, all the lower events leading to it must have happened in order to propagate the fault. While probabilities can be multiplied at an AND gate, two or more failure rates cannot be multiplied, since this would make the figures for the higher fault dimensionally inconsistent with the failure rates for the lower ones, and the result would be rubbish. *At an AND gate a rate can only be multiplied by one or more probabilities. Great care must be taken to keep the numbers used in fault tree analysis dimensionally consistent.* If the top event of a fault tree analysis is to be given as a rate, we need to know the failure rates of most items and the probabilities of others being in a failed state or out of commission. If the top event of a fault tree is to be given as a probability, we need to know the probabilities of all items corresponding to low-level events being in a failed state.

14.4.1 Example of fault tree

The method is best illustrated by an example, for which I use a simplified model of my 30-year-old domestic gas central heating system which also provides me with hot water. A P&I diagram of the system is shown in Figure 14.12 (which should on no account be taken as a model for a safe

 Intermediate event, resulting from a combination of other events

 Basic fault event that requires no further development (spontaneous event)

 AND gate — the output event (above) occurs when *all* the input events occur

 OR gate — the output event (above) occurs when *any* of the input events occurs

 Failure event. Not a primary failure event, but one for which secondary failure causes are not expressed on the fault tree

 A transfer to another page

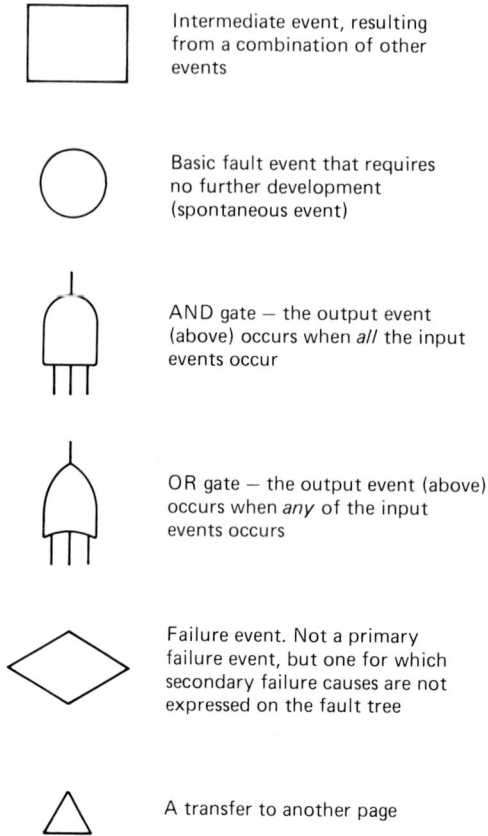

Figure 14.11 Some symbols used in fault trees

central heating system!). The pilot burner is lit and the mains gas valve open for 340 days per year (i.e. except when the system is being serviced or the house is unoccupied for several days).

The top event considered for this tree is the flow of gas from the main burner into the combustion chamber of the so-called boiler for at least 10 minutes without igniting. We want to know how often this is liable to happen. (Although technically incorrect, the term boiler is used here following general usage.) This stipulation that the failure lasts at least 10 minutes excludes some shorter temporary failures which would rectify themselves automatically within that period. It has been made because it is one of the conditions of an explosion in the boiler-room which are discussed later in 14.5.1. The basic events which could lead up to this top event are considered to be failures of the gas-control system, human error and interruption of the gas supply. This analysis does not include:

- The entry of unburnt gas from the pilot burner, which is considered a minor hazard;
- Leakage of gas into the boiler-room which could result from a leak in the piping system, or from unignited gas in the boiler.

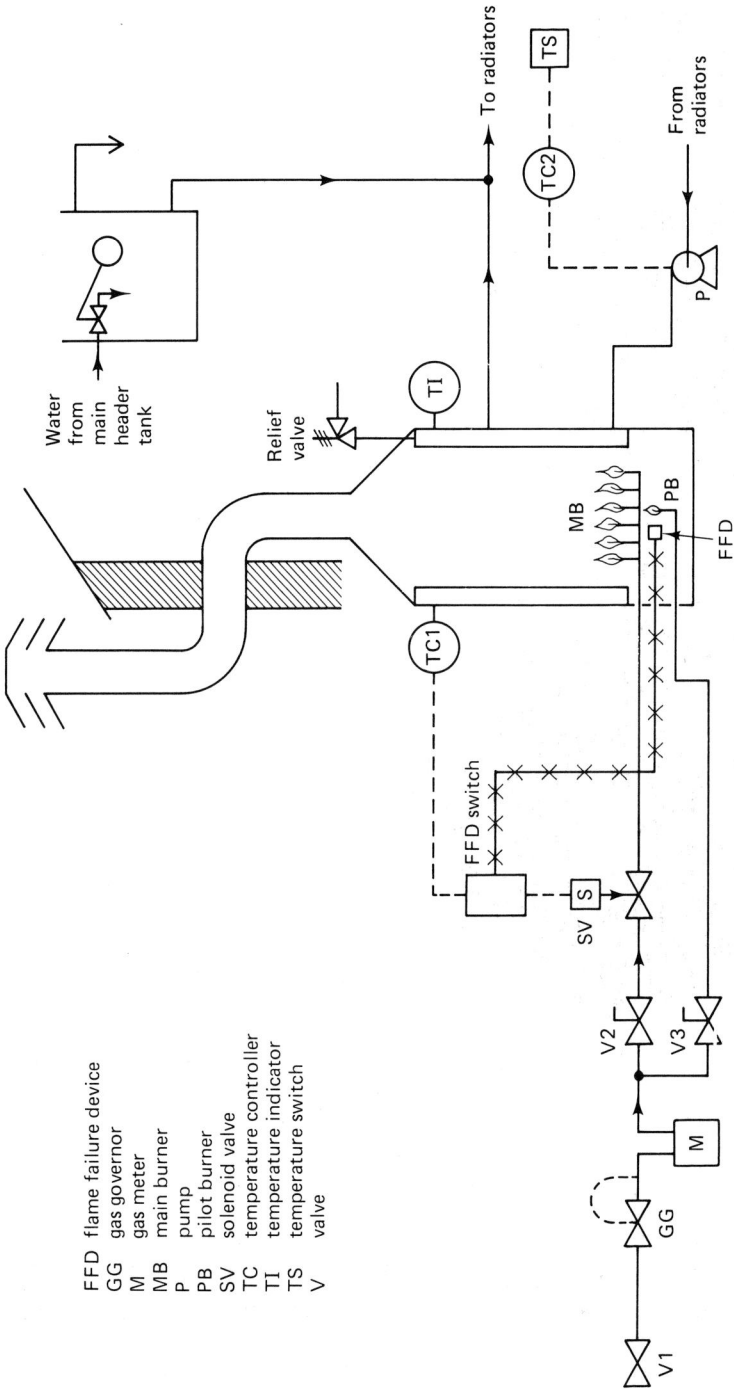

FFD   flame failure device
GG   gas governor
M   gas meter
MB   main burner
P   pump
PB   pilot burner
SV   solenoid valve
TC   temperature controller
TI   temperature indicator
TS   temperature switch
V   valve

Figure 14.12 P&I diagram of domestic central heating system

Gas from the mains passes through isolation valve V1, gas governor GG and meter M to the boiler, where it splits. A small flow passes through valve V3 to the pilot burner PB. This is lit manually and V1 is never closed while the boiler is in service. The main flow passes through valve V2 and solenoid-operated valve SV to the main burner MB. SV closes when the solenoid is de-energised, and is operated by current from a thermostat TC1 on the the boiler. This current is cut off by switch S which is actuated by the flame-failure device FFD when PB is unlit. FFD consists of a fluid-filled bulb connected to S by a copper capillary.

Water is circulated through the boiler, pipework and radiators by pump P whose switch is actuated by a time-switch and room thermostat TC2. On starting the pump the water temperature in the jacket falls, causing TC1 to switch on current to open SV, provided S allows it. MB is lit from PB and continues burning until the water temperature rises and TC1 closes SV again. So the cycle is repeated. Any loss of water from the system is automatically made up from an elevated head tank, and the boiler is protected against overpressure by a safety valve which discharges into the space below the roof of the house.

On starting to draw the fault tree (Figure 14.13) from the top event, two alternate cases or intermediate events are considered. An OR gate is drawn below the top event, with lines leading to it from the two intermediate events:

1. Pilot burner unlit and solenoid valve fails to close on demand;
2. Pilot burner lit, solenoid valve open, but main burner fails to light.

As (1) requires two conditions to be met simultaneously, an AND gate is drawn below it, under which the two conditions are shown side by side and connected by lines to the AND gate. As a rate is wanted for the top event of the fault tree, we need one of the conditions below the AND gate to be shown as a rate or frequency and the other as a probability. Frequency will be used for the first condition, the PB being unlit when V3 is open and probability will be used for the second condition, failure of SV to close on demand.

As we identify four possible alternate causes (A to D) for the first condition, 'PB unlit', we draw an OR gate under it. These causes are basic events which are shown on the fault tree under this OR gate and connected to it by lines. The frequencies assigned to them are given in Table 14.6. Both A and B result from human faults and their frequencies depend on the occupants of the house. C depends on wind and other variables and D on the gas supply.

We also identify three possible alternate causes E, F and G for the second condition, 'SV fails to close on demand', again drawing an OR gate under it. They are mechanical faults to which we can assign failure rates. We have to convert these to probabilities by the equation given in 14.4 in order to use them in the fault tree. If SV is open while PB is unlit, the boiler temperature and pressure will rise, causing the relief valve to blow, discharging steam and hot water. We assume that this will be noticed within a day or two when someone will investigate, ring up the 24-hour emergency service of the gas supply company and be advised to shut gas valves V2 and V3 pending the arrival of an engineer. We estimate the

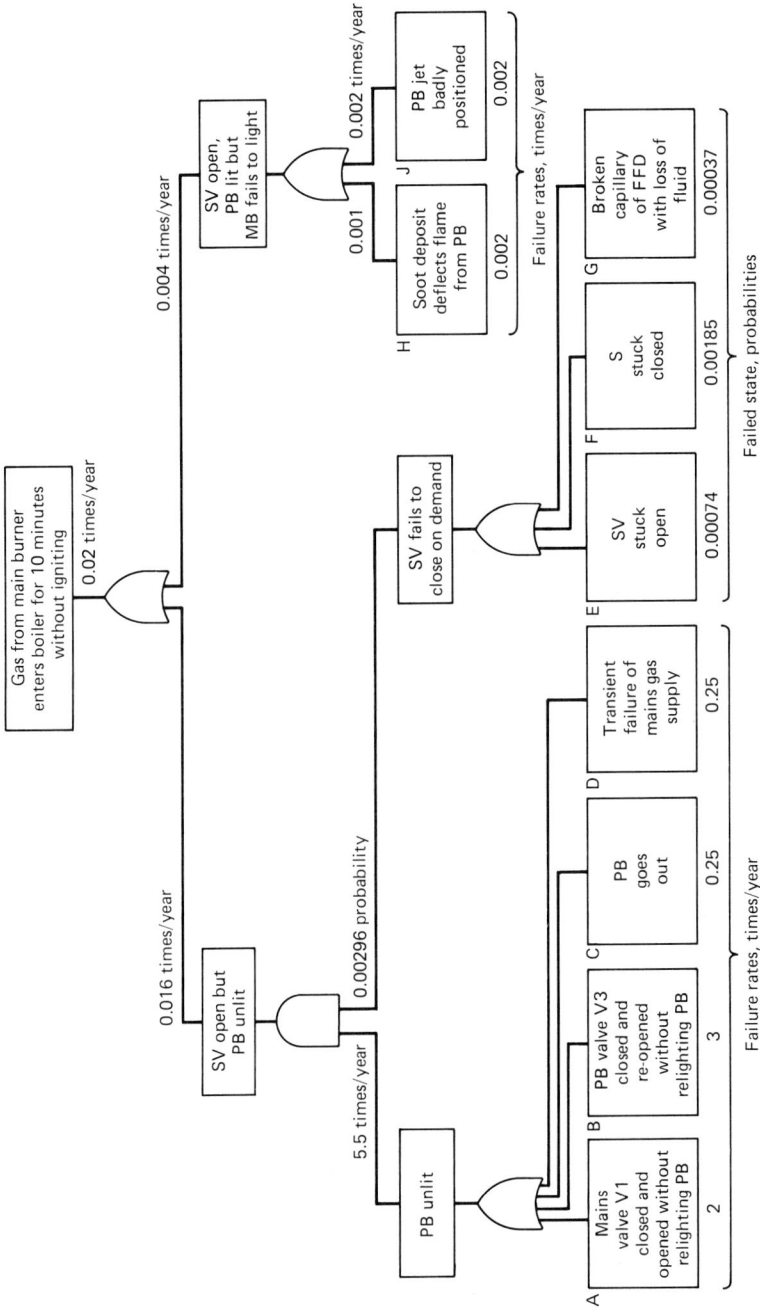

Figure 14.13 A fault tree for the central heating system shown in Figure 14.12

Table 14.6 Assumed probabilities of basic events in fault tree

Basic event or condition	Comment	Rate times/year	Probability $\times 10^{-4}$
A V1 closed and reopened without relighting PB	Gross human error	2	
B V3 opened without lighting PB	Same, more frequent than A	3	
C PB light goes out	Draughts	0.25	
D Transient failure of gas supply	Householder unaware of event	0.25	
E SV stuck open		0.2	7.4
F S stuck closed		0.5	18.5
G Fault in FFD	Broken capillary	0.1	3.7
H MB does not ignite because of soot	Boiler badly needs service	0.002	
J MB does not ignite because PB out of place	Poor design or installation	0.002	

average period between the occurrence of the failure and the closing of V2 as 15 hours. Dividing this by (340×24) gives the factor 3.7×10^{-3}, by which the failure rates must be multiplied to give failed state probabilities. These three causes with their rates and probabilities are also given in Table 14.6.

For (2), we identify two possible alternate causes H and J which we are prepared to treat as basic events. They are given with the rates assigned to them in Table 14.6 and marked on the fault tree under an OR gate.

The resulting frequency of unburnt gas flowing through the main burner into the combustion chamber for at least 10 minutes without igniting is thus 0.02 times per year.

If we consider this is excessive, the analysis can suggest ways of improving the situation. Thus the gas company might reduce the frequencies of the human errors A and B by posting warning notices to the householder with the gas bill, and of events E to J by providing a free annual service of the installation.

14.4.2 Comment on fault trees

It should be apparent that the construction of a fault tree and the assignment of frequencies to various base events and probabilities to various conditions is as much an art as a science, and much depends on the experience, judgement and mood of the person doing it. One person might immediately recognise an important branch of a fault tree which another had missed entirely. There is nothing absolute about the probabilities assigned to the various base events. The main value of fault tree analysis may be that it stimulates our curiosity about the various ways in which a system can fail, so that we learn more about it and understand it better. It also puts these failure modes into some perspective, and concentrates our attention on the more serious ones. The educational effect of the exercise can be more important than its numerical results.

Many fault tree analyses include recommendations that one or more safety gadgets be added to the system. The writer has yet to see one which showed how the risk could be reduced by simplifying the system and throwing out superfluous items!

The main drawback of fault tree analysis is the large amount of time which it takes to investigate a single top event of even a simple system. Here the writer has a suggestion. Shift operators of process plants usually have time on their hands (especially at night) when the plant is running smoothly. They could be encouraged (say, through a works safety committee) to use some of this time in making their own fault tree analyses on selected sections of the plant, for particular top events. Fault tree competitions could be organised between different shifts on the same plant. Provided care was taken that operators did not become so engrossed in their fault trees that they neglected their normal duties, they could learn a great deal about their plants in this way, and thus make a real contribution to their safety. Judging of the fault tree analyses submitted by the different teams could also set in motion a useful dialogue between the teams and the judges, during which misconceptions would be exposed and discarded, and new and useful information would come to light.

Finally, while fault trees are generally used as an aid in studying how a potentially disastrous event might arise in an existing plant or one under design, more use might perhaps be made of fault trees in investigating causes of disasters. Accident inquiry reports, after weighing up the merits of different theories, often base their conclusions about the causes on 'a balance of probabilities'. Why, then, not present the various theories in fault tree form? After identifying and allocating probabilities to the different possible base events on which each theory is based, a fault tree could be drawn for each theory to show its intermediate events and provide an estimate of its probability. This could help to rationalise accident investigation even though it does not solve the more difficult problem of recognising base events. Nevertheless, if objective fault tree analysis shows that all the theories hitherto put forward have low probabilities, it is a fair bet that some base event or vital link in the chain of causation has been missed!

14.5 Truth tables and event trees

Truth tables are a way of expressing relationships between events such as those shown in fault trees. One of their uses is in helping process operators to diagnose faults in complex plant from instrument readings, warning lights on a display panel, or other perceived signals such as particular sounds or smells. Three of the most likely faults on a plant may, for example, cause particular combinations of signals to emerge. Table 14.7 shows such an extract from a truth table. The whole truth table would be much longer and show $2^6 - 1 = 63$ signal combinations as well probably as additional faults.

A microcomputer linked to electric signals from alarm switches can be programmed to recognise combinations of signals, diagnose the fault and inform the operator. In difficult cases, the computer program can give

Table 14.7 Example of truth table used in fault diagnosis

	Perceived signal					Fault diagnosed
	Process alarms			Sound	Smell	
A	B	C	D	Hissing	Acrid	
Yes	No	No	Yes	No	No	Pump X failed
No	No	Yes	Yes	No	Yes	Fire in dryer
No	Yes	Yes	No	Yes	No	Excess MP steam

weightings to different signals which reflect their importance in diagnosing particular faults.

The construction and verification of truth tables is fairly simple to anyone who has mastered Boolean algebra. They are described in several references quoted here[3, 10, 12].

Event trees are diagrammatic representations of truth tables. They provide a means of systematically working from an initiating event, through secondary events or conditions, to its end effects. The forward thinking inherent in an event tree distinguishes it from the 'thinking backward' of a fault tree. The secondary events should be independent of each other. Each has a path for success or failure, to which probabilities are assigned. Although the events and their probabilities may appear to be independent of each other, closer study may often show them to depend (partly at least) on other (often disguised) factors, such as the time of day or year, the financial solvency of the company, or even the winner of the Grand National! This complicates their use in that one needs to know and quantify the dependence of the secondary events and their probabilities on each other, and on other factors which may not be apparent.

Although we usually have some serious outcome such as an explosion in mind when we start an event tree, it can also show multiple outcomes. Event trees are useful for portraying sequences of events, such as those shown in Figure 14.14. The thought processes depicted by an event tree are as old as mankind, and are used by chess players and armchair strategists of all kinds.

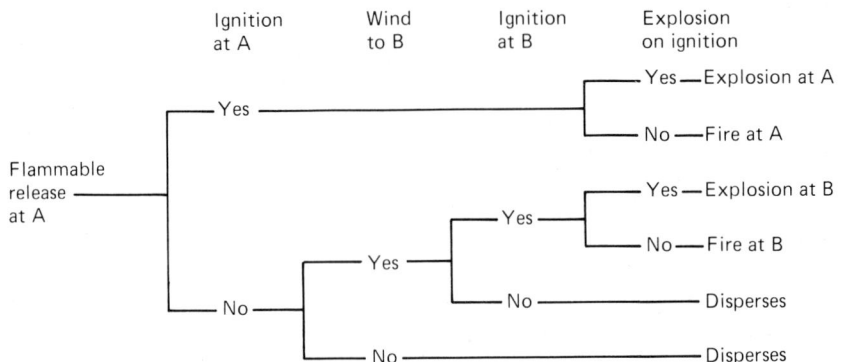

Figure 14.14 Example of event tree portraying sequence of events

14.5.1 Example of an event tree

As an example of an event tree, let's work out what might happen to the fuel gas which entered the boiler without burning, whose probability we considered in the fault tree analysis [14.4.1]. The outcome we fear most is explosion damage. We soon conclude that there is little danger from an explosion in the boiler. This is because of the small volume and solid construction of its combustion chamber and the large port and flue areas through which explosion products would escape. (This boiler is quite different from an industrial furnace with brick walls and thin metal casing.) But what is the risk of an explosion in the boiler room?

The boiler burns $0.057\,\text{m}^3$ of natural gas (density 0.6) per minute and is located in a ground-floor room with a volume of $6\,\text{m}^3$ on a corner of the house. The time-clock, pump and all electric switches and relays for the central heating system are located in this room. It has no windows, but has a single door with ventilation holes in it from the kitchen, and a perforated ventilation panel (area $0.04\,\text{m}^2$) in an outside wall. A flue ($0.15\,\text{m}$ diameter) from the top of the boiler passes horizontally through a sealed hole in an outside wall, then turns upwards to an elevation of $5\,\text{m}$ where it is capped with a cowl. The kitchen has one exterior and one interior door, a covered concrete floor, two windows, no flue or chimney, but it has an electric extraction fan mounted in one wall near the cooker.

The minimum time required to build up an explosive atmosphere in the room if all gas flowing through the main burner entered the room is easily estimated. The lower flammability limit of methane in air is 5% by volume, so that $6 \times 0.05 = 0.3\,\text{m}^3$ is about the minimum volume of gas needed. Allowing for the entry of some air into the room with the gas, and the escape of some gas from the room, we can roughly double this figure and say that about $0.6\,\text{m}^3$ of gas would have to enter the room to cause an explosion. This corresponds closely to the amount of gas flowing in 10 minutes through the main burner, which was a condition set for the top event of the fault tree.

We next consider how unburnt gas in the boiler, which is now assumed to be cool, could enter the room. We conclude that since the gas is lighter than air, and since the main burner jets which are above the air ports direct the gas upwards, all but a very small proportion of it will normally pass up the combustion chamber and out through the flue, even when these are cold. This flow would only be reversed if the pressure in the boiler-room were lower than that of the air outside. After considering the possible effects of wind on various open doors and windows, we conclude that these could only cause gas to enter the room in exceptional circumstances when there would be so much air movement that the gas concentration in the room would not reach the lower flammable limit. We make a note to check this conclusion later (and add any necessary branches to the event tree). But we do notice one combination of circumstances in which an explosion might occur in the boiler-room:

- The boiler has been off for several hours, e.g. in the early hours of the morning, but valves V1 and V2 are open.
- The pilot burner is also unlit (condition A, B or C of the fault tree), and the boiler and flue are cool.

In addition, the following situation could exist:

1. There have been drifting snow and freezing temperatures for several hours which have covered the ventilation panel and sealed gaps in and around the outside kitchen door and window.
2. All doors and windows in the kitchen are closed.
3. The extraction fan in the kitchen wall has been left running.
3A. Following (3) a slight negative pressure has been created in the kitchen and boiler-room causing a small flow of air (less than $10\,m^3$/min) down the flue and into the room. This corresponds to an air velocity down the flue of about 10 m/s. A velocity higher than this would dilute the gas to below its lower explosive range as it entered the room.
4. There is nobody in the kitchen who might smell gas, turn off the main gas tap, or open doors or windows.
5. There is an ignition source (electric) in the boiler-room.

The first two of the conditions listed arise out of the fault tree analysis. They are thus not numbered or included in the event tree which is shown in Figure 14.15. Before assigning probabilities to the secondary events, we note the selective effect of two of them on the probabilities of the others. The first arises from the fact that the ventilation panel could hardly have frozen while the boiler was operating. It must therefore have frozen at night when the gas was shut off by the time-clock to save fuel. This was probably set to turn the gas on again at 06.30 (as assumed though not stated in the fault tree analysis). The second was that it could only have frozen in a cold spell in mid-winter. At these times and dates, the probabilities applying to events (2), (3) and (4) are not the average probabilities which would apply all the year round. The actual probabilities which apply to these limited occasions are marked on the event tree. The yearly average probabilities are shown inside brackets.

Granted a minimum amount of gas entering the boiler without ignition, the chance of it reaching the boiler room and exploding there is calculated by multiplying the sequence of 'Yes' probabilities together. What does the answer mean? Multiplying the rate of the top event given by the fault tree in 14.4.1 (0.02 times/year) by the probability given by the event tree gives a frequency of a gas explosion in the boiler room as 2×10^{-6} per year. If there were 500 000 similar installations in Britain, a serious gas explosion could be expected in one of them every year.

The analyst may be tempted to juggle with the probability data which he used, in order to produce a result which agrees with the statistics of gas explosions. There is no harm in this as the object of the exercise is to harmonise theory and reality. It may help to provide better data which may be used for making predictions in a new situation for which there are no statistics. (Fortunately for the analyst it generally takes many years to confirm or refute these predictions!)

14.6 Consequences of accidental releases

Since this book is mainly about the prevention of process-related accidents, their consequences are only discussed briefly, despite the publicity they

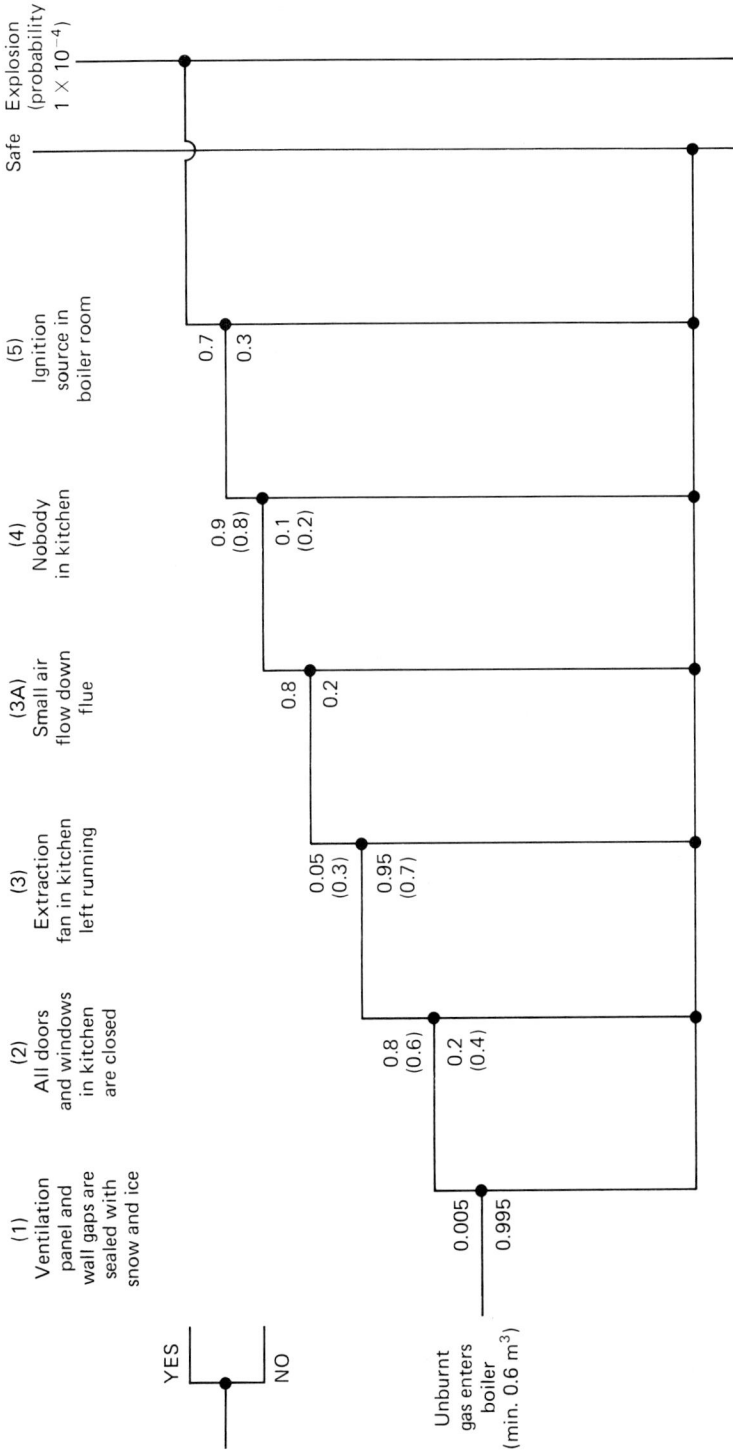

Figure 14.15 Event tree for boiler room explosion

receive. They do, however, affect the siting and layout of process plant, the protection provided for company personnel and property, safety training and its orientation, emergency planning and emergency services, including those of local authorities.

Although the consequences of a serious process accident are largely unpredictable, their general pattern is governed by the following considerations:

1. The nature of the hazards of the process materials present;
2. Their quantities, conditions of temperature and pressure, and their physical state in the various storage and process units;
3. Their release mechanisms and subsequent fate including explosion, combustion, dispersion in the atmosphere, deposition, their accumulation and concentration on land and water, in vegetation and in the human and animal populations;
4. Their effects on exposed individuals and groups of people, and how these are affected by distance from the incident.

Of these, (1) has been discussed in some detail in Chapters 7 to 11, (2) and to a lesser extent (3) in Chapter 12, and the toxic aspects of (4) in Chapter 7. Emergency planning is discussed in 20.3. Lees discusses damage from blast and thermal radiation and the atmospheric dispersion of gases, vapours and particulate clouds and gives exhaustive references[3]. Other recent discussions of blast and radiation effects include those by Lewis[16], Marshall[17], Roberts[19] and Vervalin[19]. The Safety and Reliability Directorate of UKAEA has developed two computer programs on behalf of HSE for the atmospheric dispersion of dense gases, DENZ[20] is for instantaneous releases and CRUNCH[21] for continuous releases such as can follow a pipeline break. There are many unsolved problems in our understanding of gas dispersion which are discussed by Griffiths[22].

14.7 Risks to life – quantification and levels of acceptability

Some risks to life are inherent in all human activities. It is now generally recognised that if we are to control risks in a rational and balanced manner, we need to quantify them. Otherwise we may devote disproportionate effort to reducing hazards whose risk is already very small, while ignoring others which pose far more serious risks to life. The risk to life of any activity is usually expressed as the probability of an individual being killed as a result of one year's exposure to it. The level of risk to which it is considered acceptable to expose employees and members of the public has been widely discussed[3, 23–26]. A distinction is usually made between risks which are imposed on people involuntarily, e.g. as members of the public, and those which people accept voluntarily, including those inherent in their work. In the case of the public a further distinction is made between the risk of single and multiple deaths from any incident. It has been suggested that the seriousness with which a disaster is regarded varies with the square of the number of people killed[26]. (Their race and status also affect the issue.) Three categories of risks to life are recognised: occupational, individual and societal.

14.7.1 Occupational risks

Current risk levels of different industries and occupations are apparent from the fatal accident rates (FARs) reported annually by HSE[27]. The average FAR for the whole of the UK manufacturing industry generally lies in the range of 3 to 4 per 100 000 at risk per year, i.e. an average occupational risk of between 3 and 4×10^{-5}. Some injury statistics for the process industries were discussed in 1.5.1. The occupational risks can readily be calculated from them. If we were to treat these figures as acceptable, we could easily reach the absurd conclusion that each manufacturing company should be allotted a quota of workers who may be killed at work in any year before HSE are required to investigate, and perhaps prosecute. On the other hand, if it were possible during the design and planning stage to forecast the number of employee deaths which this or that design alternative would probably lead to, we would have to set risk targets to assist in decision making, just as a general tries to evaluate the number of casualties his forces will suffer through following this or that strategy on a battlefield.

An alternate way of looking at the matter is to apply cost-benefit analysis and consider how much extra money a company should spend to reduce the probability of killing one man in a year or a lifetime. It seems from some of the references cited that this is exactly what some major companies now try to do. While this may seem a very cold-blooded exercise, it is surely better to face the issue than to shirk it, although there are several dangers in this. One is that the real calculations on which decisions are made are treated as commercial secrets, while other figures are given to employees and the public.

14.7.2 Individual risk (to members of the public)

Individual risk is simply the frequency at which an individual is expected to suffer a specified type injury of injury, which in this instance is death. While debate continues on what is an acceptable level, the most widely accepted criterion is based on a Royal Society report[28]:

> The expected frequency of death for any individual in the population due to the plant is not to exceed a value of 1×10^6 per calendar year.

Perhaps the chief thing to note about this is that it is a *maximum figure*. When a risk assessment is made for the area around a major hazard plant, contour lines of constant risk are drawn round the plant, and if this criterion is to be applied strictly, no member of the public who is not employed on or engaged in some business connected with the plant should be allowed inside the contour line which corresponds to this level of risk.

Unresolved problems arise if we try to apply this criterion in Third World countries. Although the ILO are against applying different standards of safety in different countries [23.2], the writer's own experience suggests that most Third World countries place greater emphasis on development than on safety. A perfectly logical argument in support of this, together with a 'formula' for assessing a level of acceptable risk for members of the public, was advanced by Bowen[29]. This is to balance the loss of life expectancy due to the industrial activity against the gain in life expectancy from this activity. Thus in countries where the life expectancy is low and stands most to gain from an industrial activity, a higher level of risk would be tolerated. (One wonders how this would be applied to a cigarette factory or a brewery!)

14.7.3 Societal risk

The Societal Risk from an activity, sometimes known as its communal risk, is the relationship between the frequency (of a serious accident) and the number of people harmed (or in this case killed). This is often plotted as a cumulative $F(N)$ curve where F is the frequency of accidents which exceed a certain severity and N is the number of persons harmed to some specified degree[12].

The frequency naturally decreases as the number of persons killed increases. It is customarily plotted on log–log graph paper. For a particular hazardous installation which is surrounded by a populated area, a risk analysis produces a series of points on such an $F(N)$ diagram, and through which a continuous line may be drawn. The limits of acceptability and unacceptability are shown as two parallel straight lines on the same diagram, with an area between them in which reduction is desirable (Figure 14.16).

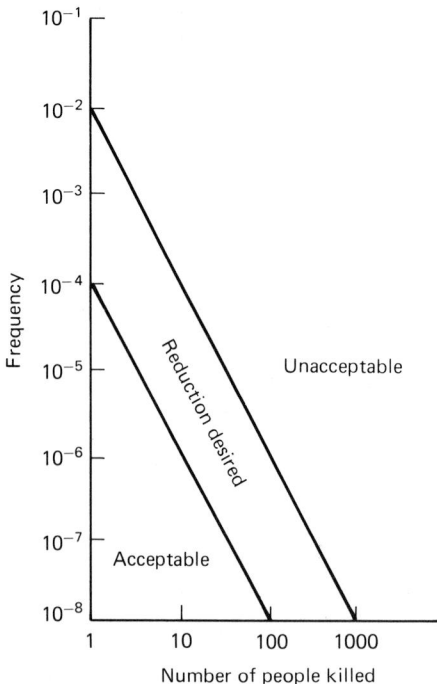

Figure 14.16 Provisional risk criteria for the policy on external safety in the Netherlands (courtesy Dr David Slater, Technica Inc.)

Quantified risk criteria are currently used in three areas:

1. Planning, and Regulatory submissions;
2. Design criteria and operating targets;
3. Insurance and liability planning.

The report of the Canvey study made by SRD on behalf of HSE provides a detailed example of the methods used in risk analysis in a complex planning case involving several hazardous installations[7].

Risk quantification is of particular concern to the Major Hazards Assessment Unit (MHAU) of HSE when advising local authorities on planning permission for developments which involve Notifiable Installations [2.4.1]. Such applications are more often for innocuous developments near an existing Notifiable Installation than for extensions to the installations themselves. In several cases the MHAU has advised the local authority against granting planning permission for innocuous activities which would increase the population at risk. These included a new housing estate, general community development and extensions to an old peoples' home. Following the last mentioned the applicant appealed against planning refusal by the local authority, and a public inquiry was held at which the MHAU was represented[30]. The Secretary of State dismissed the applicant's appeal.

In conclusion, risk analysis is a growing activity involving a number of professional disciplines. Judging from past trends, the perceived man-made hazards confronting the world will surely increase. At the same time, our options for satisfying most of our needs seem doomed to decrease as the human population expands to its limit and fossil fuel reserves dwindle. Perhaps the 1990s will be known as the 'Decade of Risk Analysis'!

References

1. Dummer, G. W. A. and Winton, R. C., *An Elementary Guide to Reliability*, 3rd edn, Pergamon, Oxford (1986)
2. Carter, A. D. S., *Mechanical Reliability*, 2nd edn, Macmillan, London (1986)
3. Lees, F. P., *Loss Prevention in the Process Industries*, Butterworths, London (1980)
4. Caplen, R. H., *A Practical Approach to Reliability*, Business Books, London (1972)
5. Hammer, W., *Handbook of System and Product Safety*, Prentice-Hall, Englewood Cliffs, N.J. (1972)
6. South, G. F., *Boolean Algebra and its Uses*, Van Nostrand, New York (1974)
7. HSE, *Canvey: an investigation of potential hazards from operations in the Canvey Island/Thurrock area*, HMSO, London (1978)
8. The Institution of Chemical Engineers, *A New Guide to Capital Cost Estimating*, Rugby (1977)
9. Lees, F. P., 'A review of instrument failure data', paper in I.Chem. E. Symposium series 47, *Process Industry Hazards – Accidental Release, Assessment, Containment and Control*, Rugby (1976)
10. Wells, G. L., *Safety in Process Design*, George Goodwin, London (1980)
11. Brammer, S., 'Is "safe enough" really safe?' *The Chemical Engineer, Loss Prevention Supplement* (August 1987, p.17)
12. The International Study Group on Risk Analysis set up by the Loss Prevention Working Party of the European Federation of Chemical Engineering, *Risk Analysis in the Process Industries*, The Institution of Chemical Engineers, Rugby (1985)

13. Clifton, J., 'Risk analysis and predictive techniques', paper given at Major Hazards Summer School, Cambridge, organised by IBC Technical Services Ltd, London (1986)
14. The Institution of Chemical Engineers, *Nomenclature for Hazard and Risk Assessment in the Process Industries*, Rugby (1985)
15. Balemans, A. W. M. and others, 'Check-list', in *Loss Prevention and Safety Promotion in the Process Industries*, edited by Buschmann, C. H., Elsevier, Amsterdam (1974)
16. Lewis, D, 'The blast, radiation and interactions with people, control rooms and surroundings', paper given at Major Hazards Summer School, Cambridge, organised by IBC Technical Services Ltd, London (1986)
17. Marshall, V. C., 'Historical experience' and 'Dust explosions and fireballs', *ibid.*
18. Roberts, A. F., 'Vapour cloud explosions and BLEVEs', *ibid*
19. Vervalin, C. H., *Fire Protection Manual for Hydrocarbon Processing Plants*, Gulf Publishing, Houston, Texas (Volume 1, 1984, Volume 2, 1982)
20. Fryer, L. S. and Kaiser, G. D., *DENZ – A computer programme for the calculation of the dispersion of dense toxic or explosive gases in the atmosphere*, SRD Report R 152, UKAEA, Culcheth, Lancashire (1979)
21. Jagger, S. F., *Development of CRUNCH – A dispersion model for continuous releases of a denser-than-air vapour into the atmosphere*, SRD Report R 229, UKAEA, Culcheth, Lancashire (1983)
22. Griffiths, R., 'Unsolved problems – the need for research', paper given at Major Hazards Summer School, Cambridge, organised by IBC Technical Services Ltd, London (1986)
23. The Council for Science and Society, *The Acceptability of Risks*, Barry Rose (Publishers) Ltd, London (1977)
24. Kletz, T. A., *HAZOP and HAZAN – Notes on the identification and assessment of hazards*, The Institution of Chemical Engineers, Rugby (1985)
25. Davies, P. C., 'The role of the Major Hazards Assessment Unit', paper given at Major Hazards Summer School, Cambridge, organised by IBC Technical Services Ltd, London (1986)
26. Slater, D. H., 'The use of risk results in planning and design', *ibid.*
27. HSE, *Manufacturing and Service Industries*, HMSO, London (published annually)
28. Royal Society, *Risk assessment – a study group report*, London (1983)
29. Bowen, J. H., 'Individual risk vs. public risk criteria', *Chem. Eng. Prog.* **72**(2), 63 (1977)
30. Purdy, G., 'Risk assessment and the generation of planning advice – three case studies', paper given at Major Hazards Summer School, Cambridge, organised by IBC Technical Services Ltd, London (1986)

Chapter 15

Active protective systems and instrumentation

This chapter discusses both active protective systems [12.1.6] used to protect operating plant against over-pressurisation and release of flammable and toxic materials, and process control instruments which are closely linked with safety. Active protective systems often provide a cheaper and more practical solution than is possible by incorporating equivalent passive protection into the plant design. Thus it is generally cheaper to provide a pressure relief system than to design the plant to withstand the maximum pressure which could be reached under any conditions.

The choice and specification of any protective system requires a careful study both of the events it is intended to mitigate or avert and of the extent to which such protection is provided in the basic design. General codification is thus difficult. For fire and explosion hazards both the Dow (3rd edn) and the Mond (2nd edn) guides[1,2] [12.1] and [12.2] provide lists of basic and special preventative features, with guidance on their application. The American Petroleum Institute's recommendations on the choice, design and installation of over-pressure relief[3,4] systems for oil refineries, which are summarised in this chapter, have wide application throughout the process industries. For most complex and hazardous processes, the specification of protective systems is best done in conjunction with a hazard and operability study [16.5.5].

15.1 Overpressure relief – general

The risk of rupturing process equipment and releasing its contents through accidental overpressurising beyond design limits has led to the extensive use of pressure-relief systems which release the contained process fluids when a certain pressure is exceeded and dispose of them safely. It is ironic that the blind reliance on a pressure-relief device on equipment which would better have been left vented should have led to the Seveso disaster [5.3.8]. While codes and guides have their uses, particularly as *aide-mémoires* to many of the things that may go wrong, they are no substitute for knowing one's plant and its hazards!

Pressure-relief valves were first used to avert mechanical explosion of boilers, for which they have long been required by law [2.5.1]. The requirements for such systems vary considerably throughout the process industries. A range of spring-loaded valves which open and discs which rupture when a certain pressure is exceeded are available. Methods of disposal which depend on the fluid released and local factors include recycling to a lower pressure part of the process, venting, burning, chemical reaction and adsorption. Compressed air and steam may be vented directly provided the hazards of noise and fog are kept within bounds. The disposal of flammable and toxic gases and vapours presents more serious problems which are not always properly recognised. The failure of such a disposal systems was at the root of the Bhopal disaster [5.4.8].

Other means of pressure-relief using trip systems incorporated in the process instrumentation are discussed in [15.6].

15.2 Pressure-relief devices and definitions

The definitions given here are mainly abbreviated from the API codes[3,4] (and not from the less comprehensive British Standard[5]).

All pressurised equipment is assumed to operate at some pressure below (usually 90% of) its 'maximum allowable working pressure' (MAWP). (This is referred to in UK legislation as 'maximum permissible working pressure' when referring to boilers, and 'safe working pressure' when referring to steam and air receivers[6].) The MAWP is the maximum allowable gauge pressure at the top of the vessel in its operating position for a designated temperature. It is often identical to the 'design pressure' which incorporates a factor of safety [14.1.6].

The 'set pressure' at which a pressure-relief device starts to open is usually the same as or within ±3% of the MAWP.

'Accumulation' is the percentage pressure increase over the MAWP during discharge through a pressure-relief device. The maximum allowable accumulated pressure other than for fire exposure is normally 10% above the MAWP for single-valve installation and 16% above the MAWP for multiple-valve installation (when the settings are usually staggered). For fire exposure the maximum allowable accumulated pressure is normally 21% above the MAWP.

'Back pressure' is the gauge pressure at the outlet of a pressure-relief device caused by pressure in the discharge system. It is the sum of the 'built-up back pressure' which develops from flow through the pressure-relief device and the 'superimposed back pressure' in the discharge manifold which results from other sources.

'Overpressure' is the pressure increase over the set pressure of the primary relief device and is the same as the accumulation when the set pressure equals the MAWP.

'Blowdown' is the difference between the set pressure and the reseating pressure of a pressure-relief valve and may be expressed as a percentage of the set pressure.

These various pressure levels are shown diagrammatically in Figure 15.1. They play an important part in the choice and sizing of pressure-relief devices and in the design of pressure-relief systems.

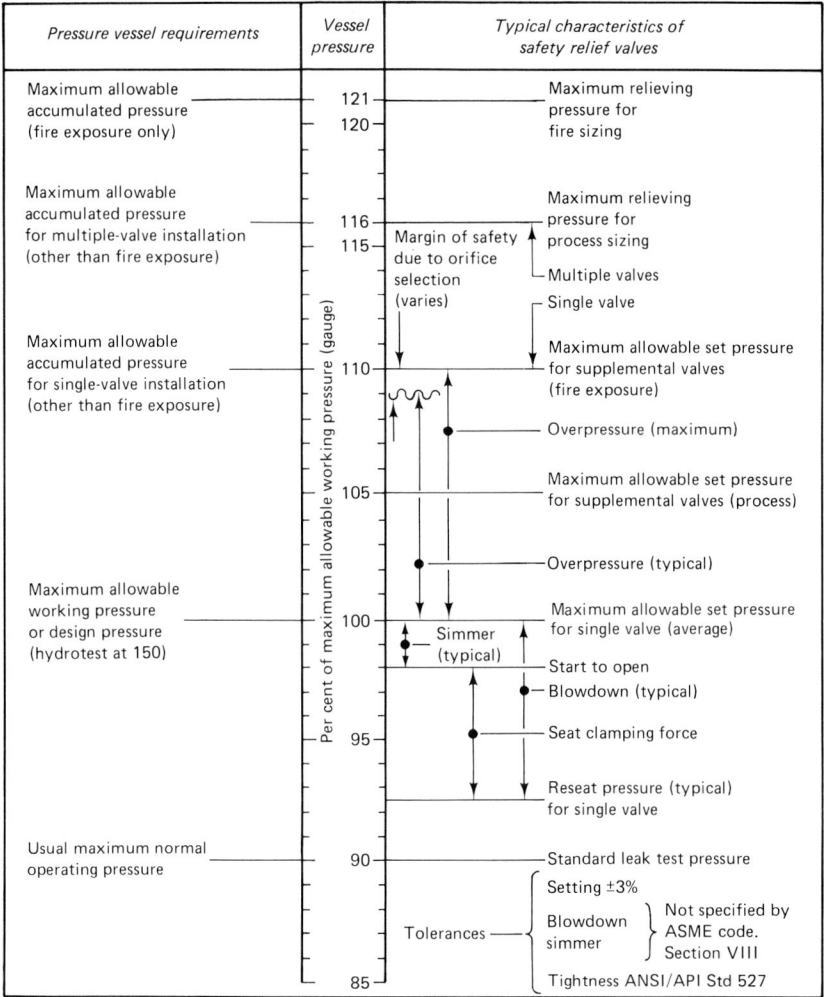

Pressure vessel requirements	Vessel pressure	Typical characteristics of safety relief valves
Maximum allowable accumulated pressure (fire exposure only)	121 — 120 —	Maximum relieving pressure for fire sizing
Maximum allowable accumulated pressure for multiple-valve installation (other than fire exposure)	116 — 115 —	Maximum relieving pressure for process sizing Margin of safety due to orifice selection (varies) Multiple valves Single valve
Maximum allowable accumulated pressure for single-valve installation (other than fire exposure)	110 —	Maximum allowable set pressure for supplemental valves (fire exposure) Overpressure (maximum)
	105 —	Maximum allowable set pressure for supplemental valves (process) Overpressure (typical)
Maximum allowable working pressure or design pressure (hydrotest at 150)	100 —	Maximum allowable set pressure for single valve (average) Simmer (typical) Start to open Blowdown (typical)
	95 —	Seat clamping force Reseat pressure (typical) for single valve
Usual maximum normal operating pressure	90 —	Standard leak test pressure Setting ±3% Tolerances — Blowdown simmer } Not specified by ASME code. Section VIII
	85 —	Tightness ANSI/API Std 527

Per cent of maximum allowable working pressure (gauge)

Notes:
1. The operating pressure may be any lower pressure required.
2. The set pressure and all other values related to it may be moved downward if the operating pressure permits.
3. This figure conforms with the requirements of the *ASME Boiler and Pressure Vessel Code*, Section VIII, 'Pressure vessels', Division 1.
4. The pressure conditions shown are for safety relief valves installed on a pressure vessel (vapour phase).

Figure 15.1 Pressure levels referred to in connection with pressure relief devices (courtesy API)

A 'pressure-relief valve' is any type of automatic pressure-relieving valve operated by its upstream pressure. It generally has a spring (enclosed in a bonnet) which loads a disc closing the inlet port against the pressure. This generic term includes the three following types:

- A 'relief valve' is one which opens in proportion to the pressure increase over the opening pressure and is used primarily for liquid service.

- A 'safety valve' is one which 'pops' wide open when the opening pressure is reached, and is used for gas or vapour service (generally for steam and air).
- A 'safety-relief valve' can be used either as a safety or relief valve depending on application. There are two main types:
  - A 'conventional safety-relief valve' is a closed bonnet pressure-relief valve whose bonnet is vented either to the atmosphere or to the discharge side of the valve (Figure 15.2). Its performance is affected by the back pressure and it is generally used only when the back pressure will not be more than 10% above the set pressure.
  - A 'balanced safety-relief valve' incorporates means of minimising the effects of back pressure on the performance characteristics. There are two types, the piston type and the bellows type, both shown diagramatically in Figure 15.3.
  - There are also 'pilot-operated pressure-relief valves' whose major flow is controlled by a self-actuating auxiliary pressure-relief valve.

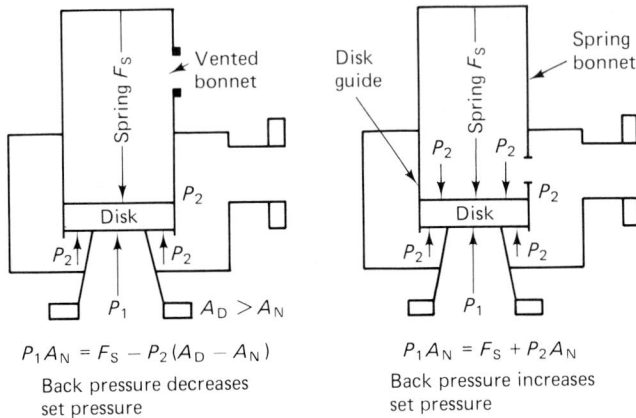

$$P_1 A_N = F_S - P_2 (A_D - A_N)$$

Back pressure decreases set pressure

$$P_1 A_N = F_S + P_2 A_N$$

Back pressure increases set pressure

Figure 15.2 Conventional safety relief valves, schematic. Left: bonnet vented to atmosphere. Right: non-vented bonnet (courtesy API)

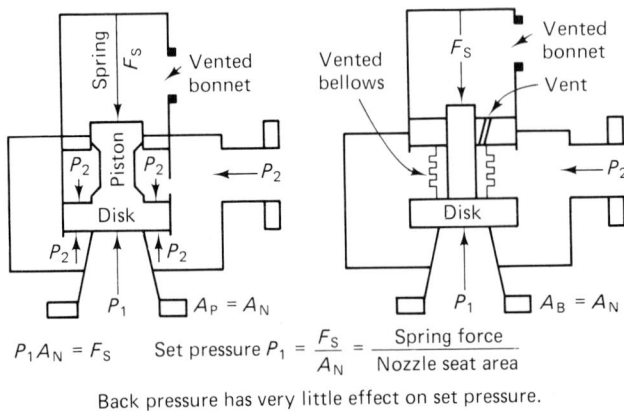

$$P_1 A_N = F_S \qquad \text{Set pressure } P_1 = \frac{F_S}{A_N} = \frac{\text{Spring force}}{\text{Nozzle seat area}}$$

Back pressure has very little effect on set pressure.

Figure 15.3 Balanced safety relief valves, schematic. Left: balanced disc and vented piston type. Right: bellows type (courtesy API)

The internal parts of pressure-relief valves must be suited to the temperature range and corrosivity of the process fluid. The temperature at which a pressure-relief valve can be expected to open (which may differ considerably from its temperature when it is relieving) must be considered in relation to its set pressure and the pressure at which it reseats. Types of pressure-relief valves which allow a small leak of process fluid to the atmosphere when the valve opens should not be used for highly toxic fluids, or flammable fluids when the relief valve is in a building.

'Bursting discs' which are held between pipe flanges are defined by API as 'rupture disc devices'. They are used instead of pressure-relief valves where the pressure rise may be so rapid as virtually to constitute an explosion, and upstream of pressure-relief valves in cases where minute leakage cannot be tolerated, or where blockage or corrosion might render a relief valve ineffective. Bursting discs are available in various metals, plastics, combinations of both, and carbon. Metal bursting discs are usually very thin, and domed. They are secured between special flanges and supported on the high-pressure side by a perforated concave metal plate to prevent flexing of the disc if the pressure difference is reversed, e.g. when the equipment is evacuated[7]. When fitted in series with a pressure-relief valve it is essential to prevent debris from the burst disc from interfering with the operation of the valve, and to provide a device to indicate leakage into the space between the discs and the valve (e.g. a pressure gauge). When used alone, bursting discs have the obvious disadvantage of allowing most of the contents of the equipment to escape. When used in conjunction with a pressure-relief valve, the process should be shut down when a bursting disc has ruptured. The cause of the rupture should be satisfactorily accounted for before replacing the disc and restarting the process.

If it is necessary to fit a valve on either side of a pressure-relief device, e.g. to isolate either the equipment protected by the device or the device itself for maintenance, it must be secured (e.g. by lock or interlock) in the open position while the pressure-relief device is in active service. On difficult duties two pressure-relief devices are sometimes installed in parallel, with isolation valves upstream and, where necessary, downstream, so that one device is always functional while the other is available for maintenance or cleaning. Three way isolation valves with locks or interlocks [15.7.1] should then be chosen which ensure that one pressure-relief device is always on duty.

15.3 Causes of overpressure

Overpressure in process equipment, pipelines and boilers can result from a variety of causes. Those given here are mainly based on the API guide[4] for oil refineries. While these include most causes to be found in the process industries, the lists are for guidance only and are not claimed to be complete.

Two groups of independent causes are given:

1. Those designated 'normal' where the rise in pressure is not too fast for pressure-relief valves to handle;

2. Those designated 'abnormal' where the rise in pressure may be too fast for pressure-relief valves to provide proper protection. For these, additional or alternative means of protection should be provided.

15.3.1 Normal causes of overpressure

1. Closed outlets on vessels, resulting from inadvertent closure of a valve;
2. Inadvertent opening of a valve from a higher pressure source, such as high-pressure steam or a process fluid;
3. Failure of an automatic valve in the open position from a higher pressure source. There is a special danger of this with self-actuating pressure regulators which are not designed to 'fail safe' [15.7.2]. It can also happen with signal failure to pneumatic controllers which fail safe only on air failure.
4. Electricity failure, affecting pumps, fans, compressors, instruments and motor-operated valves;
5. Cooling-water failure, affecting condensers, coolers and cooling jackets on machines;
6. Instrument air failure, affecting transmitters, controllers, process-regulating valves, alarm and shutdown systems;
7. Steam failure, affecting steam-driven machines (pumps, compressors, generators, etc.), steam-heated equipment, steam injection to processes and steam ejectors;
8. Fuel failure (gas, oil, etc.), affecting boilers, fired heaters, engine- and turbine-driven machines;
9. Inert gas failure, affecting seals and purges;
10. Mechanical breakdown of fans, pumps, compressors and other machines;
11. Instrument failure, particularly of temperature, pressure and flow controllers;
12. Heat exchanger tube failure, leading to flow of fluid at higher pressure into equipment operating at lower pressure;
13. Failure of tube in fired heater for process stream, causing increased combustion in furnace and excessive heat input to waste heat boilers, etc.;
14. Thermal expansion of liquids in liquid-filled pipes and equipment (especially heat exchangers), on which valves have been closed ('boxed-in equipment'). This can be a serious problem with long pipes carrying liquids which are exposed to the sun.
15. Plant fires, causing expansion and vaporisation of fluids in equipment and pipes. Pressure-relief systems for plant fires are usually designed on the assumption that external thermal insulation on equipment remains intact, but if it is destroyed or removed, e.g. by an explosion, the pressure-relief system may be unable to cope. Metals are also weakened when heated by a fire, so that equipment may fail at normal operating pressures with the pressure-relief valves closed. To protect against fire, pressure-relief systems need to be backed up by other forms of protection, including depressurisation, to reduce the danger of a vessel bursting.

16. Chemical reactions, particularly exothermic ones [8.6]. These are normally controlled by other means to prevent overpressure, but where they fail, the pressure-relief device should be able to save the situation, provided the reaction is not extremely fast. The danger of very fast run-away reactions can sometimes be averted by carrying out the reaction in a boiling solvent under a reflux condenser, which returns condensed solvent to the reactor. This removes the heat and controls the temperature.

17. Presence of unexpectedly volatile material in feed to distillation column caused, for example, by failure of reboiler heat in immediately preceding column.

15.3.2 Abnormal causes of overpressure

1. 'Water hammer' is frequently caused by quick-closing valves in pipes through which liquid is flowing, but it can also be caused by pulsating liquid flow, e.g. from reciprocating pumps. The response time of most pressure-relief devices is too slow to prevent damage from water hammer, which can best be avoided by other means, e.g. preventing valves in liquid lines being closed too fast, and fitting pulsation dampeners [13.6.3].

2. 'Steam hammer' produces effects similar to water hammer and occurs in wet steam lines where slugs of condensate are propelled against bends in the pipework. This is usually cured by fitting steam traps to remove condensate and ensuring that they are functioning.

3. Cavitation [11.4.2] in pipes and vessels, which happens when steam or another vapour is injected into a colder liquid, produces similar shock waves. It is caused by the collapse of vapour bubbles within the liquid. Its effects are minimised by distributing the steam or other vapour through a number of small holes, and by gas injection into the vapour condensing.

4. Physical explosions [6.4.3] are caused by 'latent superheat', which can arise where there are two unmixed liquid phases in contact (especially water and hydrocarbons) and the liquid present in larger quantity is heated to a temperature higher than the boiling point of the mixture[8,9]. Since the resulting pressure rise is practically instantaneous, few pressure-relief devices (except perhaps large bursting discs) can be of much value. This situation can also arise in tanks and vessels which are vented to the atmosphere without pressure-relief devices.

5. Chemical explosions and some very fast reactions lead to the sudden formation of large amounts of gas or vapour. Here again pressure-relief valves cannot usually give much protection. Large bursting discs which discharge directly to the atmosphere are often used to protect vessels and closed tanks against deflagrative explosions [9.1], although they are of little use against detonations. A relief area of $0.065\,m^2$ per cubic metre of vapour volume is probably adequate to protect tanks containing liquid hydrocarbons at temperatures above their flash-points against air/hydrocarbon-vapour explosions[4].

15.4 Calculation of individual relieving rates

Pressure-relief devices are usually fitted to the top rather than the sides or bottom of equipment for which they are required, and except for liquid-filled equipment the fluid released is predominantly gas or vapour rather than liquid. It is, however, dangerous to assume that this will always be so, and besides determining relieving rates, the state and properties of the fluid released must be realistically assessed. While an open pressure-relief device will allow more weight of liquid than vapour to be released in a given time, the volume of liquid released will be considerably less, and the same applies to two-phase flow of liquid and vapour. A pressure-relief device sized for vapour flow only may thus be choked if liquid is also relieved, even though its mass flow rate is greater.

Individual relieving rates are part of the information needed to determine the size and type of relief valves and to design the pressure-relief systems. They should be determined by energy and material balances over the equipment concerned when one or more causes of overpressure come into play. The peak relieving rate is generally taken as the maximum rate that must be relieved to protect equipment from overpressure due to any single unrelated cause. The calculations for which computer programs are available should be made by experienced process engineers. The API guide[4] and Fitt[10] provide useful guidance to these calculations.

The most common causes of overpressure are heat input resulting in vaporisation or thermal expansion, and direct pressure input from higher pressure sources. Pressure-relief is generally required over a period of time. While some allowance can generally be made for the responses of operators in relieving overpressure, these responses may be delayed for several minutes. Pressure-relief is required for most equipment and pipes in which fluids, particularly liquids, may be shut in by isolation valves. In the case of heat exchangers in which a cold liquid could be locked in when the exchanger is shut down, relief valves may be omitted if operators are trained in proper draining and venting procedures on shutdown. As the required capacity of most pressure-relief devices for hydraulic expansion is very small, it is usually possible to standardise on one of the smallest sizes available. Their set pressures should not exceed the MAWP of the weakest component in the system yet should be high enough to ensure that they open only under hydraulic expansion conditions.

The possibility of blocking condensing surfaces by accumulation of non-condensables should be considered among causes of loss of cooling. Credit may be given for control valves which can be relied on to fail open or 'safe' [15.7.2], but not for those which fail in their last position on power failure since this position is unpredictable.

In the case of possible exothermic runaway reactions [8.2], normal overpressure-relief is often insufficient and it may be necessary to provide rapid relief before pressure and temperature rise to exponentially accelerating levels. This hazard can often be minimised by incorporating a volatile non-reacting solvent into the reaction mixture so that its evaporation absorbs the excess reaction heat.

When sizing pressure-relief devices for fire conditions, the use and effect of other forms of protection, e.g. insulation, automatic water spray and deluge systems, and of vapour depressurising should also be considered. If a vessel partly filled with liquid is engulfed in a fire, the upper part which is not cooled by the liquid is particularly vulnerable, due to weakening of the metal, and it may fail below the set pressure of the relief device. Any protection offered by insulation against fire can only be counted on if its insulating properties are retained up to about 1000°C, and where it is protected against dislodgement by fire water jets or possible external explosions, e.g. by an adequate steel covering.

The vapour relieving rates of air-cooled heat exchangers, particularly liquid coolers, when exposed to fire can be very large and become the dominant loading factor for the relief system.

Often two or more pressure-relief valves of different sizes with settings staggered so that the smallest one relieves first are preferable to a single pressure-relief valve. One reason is that pressure-relief valves which are leak-tight after installation or maintenance often fail to reseat properly after opening and leak slightly. A small valve is then likely to leak less than a large one. There is also likely to be less chatter with a small valve than a large one. For pressure storage of LPG in spheres, a relatively small pressure-relief device set at the MAWP and discharging via a liquid catchpot to a remote flare may provide the best protection against liquid overfilling and thermal expansion of the contents, and against LPG with an abnormally high vapour pressure. For fire exposure, larger pressure-relief valves set at 105–110% of the MAWP and discharging directly to the atmosphere at a safe distance from the spheres might be used.

15.5 Disposal of released fluids

Three principal methods are used for the disposal of fluids released by pressure release devices:

1. By discharge to the atmosphere;
2. By burning;
3. By disposal to a lower-pressure system. This includes the use of separators to collect released liquids and special chemical scrubbers to prevent the release of highly toxic gases and vapours to the atmosphere.

The API guide[4] deals with (1) and (2) in detail. The design of systems for dealing with specific toxic gases follows normal chemical engineering practice[11].

Factors to be considered when deciding on the method of disposal include toxicity, odour, smoke, particulate matter, noise, heat, reliability and ease of maintenance

Refinery gases consisting of combustibles with small amounts of hydrogen sulphide are an interesting case. Hydrogen sulphide has an OEL of 10 ppm, which is higher than that of sulphur dioxide (OEL 5 ppm), its main combustion product. So far as the toxic hazard is concerned, there is at first sight little to chose between discharging the gas at high level and at high velocity or burning it in a flare. In the latter case the sulphur dioxide will be carried to a higher level by thermal convection, and its concentration at ground-level will be low. A quantitative assessment of all the factors involved needs to be made when deciding on the disposal method.

Released-fluid disposal systems, for which a high degree of reliability is required, unfortunately tend to be the 'Cinderellas' of plant design. A serious incident which occurred a few years ago when a pressure-relief valve blew on a large batch reactor on a plant which the writer was visiting is described in 16.6, together with the (not widely appreciated) lessons to be learned from it. Serious deficiencies in the discharge arrangements of the pressure-relief system of a new LPG installation which the writer encountered in a Third World country are described in 23.5.1.

The pipework of disposal systems must be designed and supported to withstand the thermal strains resulting from the entry of hot and cold fluids, and the shock loading resulting from the sudden release of compressible fluids and slugs of liquid. If compressed liquefied gases are relieved, flash evaporation will occur at the release device subjecting the pipework to very low temperatures. It is then essential to choose an alloy which does not suffer embrittlement at the lowest temperature which may be reached, and to provide adequate bends in the pipework to cater for contraction. If solid materials (such as rubbers and polymers) are liable to be released into or formed in the pipework, adequate flanged joints must be provided to allow it to be cleaned and inspected internally. Means must also be provided (e.g. locks on valves or line blinds [18.6.1]) for leak testing the pipework after installation and maintenance. (This is often overlooked.)

If liquid is liable to be present in the discharge, the piping should be self-draining into an adequately sized knockout pot before the vent stack, flare or scrubber, and effective means of removing liquid collected must be provided.

15.5.1 Discharge to the atmosphere

This is usually the cheapest method and is commonly used not only for air and steam and other gases of low toxicity but also for flammable gases and vapours. Individual vent pipes near the equipment relieved are generally used. The levels of noise and atmospheric toxicity to which nearby workers and members of the public are liable to be subjected by the discharge should be evaluated and within accepted limits for the chosen location. Mist is generally only a problem with steam discharges since most saturated vapours superheat on passing through a relief valve.

In the case of flammable gases, these must leave the vent at a sufficiently high velocity to entrain enough air to reduce its concentration to below the lower explosive limit before the energy of the jet has dissipated. Hydrocarbon gases and vapours are then generally diluted to below their lower flammability limit [10.2.1] at a distance from the vent exit equal to 120 times its diameter. When the cross-sectional area of the vent exit is equal to that of the full valve opening, this condition is usually met so long as the valve is passing gas only and is at least 25% open. This applies to 'pop-open' relief valves which close positively when the pressure in the equipment relieved falls below the set pressure. The vent should be located on the top of a tall structure and so directed that the jet does not impinge on any solid object which would cause a stable flame if ignition occurred. The vent should preferably point upwards, and be adequately supported against reaction from the jet discharge. A small drain hole is needed at the bottom of the vent pipe to ensure drainage of rain-water. This should be arranged so that a little air is sucked into the vent pipe by the momentum of the gas rather than allowing gas to escape through it at low level.

There is always some risk of ignition when flammable gases are discharged to the atmosphere, especially if the relief device opens as a result of fire. This happened in the Fézin disaster of 1966[12], when the gas released ignited, causing a huge flame which radiated heat back onto the LPG spheres and was a major factor in causing their disastrous rupture. The consequences of radiation from burning gas at the vent exit must therefore be considered when deciding its position. Possible causes of ignition include lightning, static electricity and iron oxide particles. The risk of ignition increases if the gas contains molecular hydrogen [10.2.4]. The exit of the vent can can be designed to reduce the chance of static ignition when hydrogen-containing gases are vented, and also to ensure that any flame which tends to form is blown out by the jet. There should be no risk of forming an explosive vapour cloud with a properly designed vent discharge since the concentration of flammable gas is only momentarily within explosive limits in the jet itself. Considerations of jet dispersion favour separate elevated vents for each pressure-relief valve rather than manifolding several discharge lines together into a common stack. For similar reasons, process vents of flammable gases should be kept separate from pressure-relief vents.

Wells[13] recommends that the following criteria should be met before venting to atmosphere under emergency conditions is permitted:

1. Molecular weight of vented material <61;
2. Liquids and solids are absent (including liquid condensate in lines and liquid droplets in the discharge resulting from condensation);
3. The vapour release velocity is adequate to ensure that in the case of flammable gases and vapours the expanding jet entrains sufficient air for the mixture to be non-flammable by the time it has ceased to move under the influence of the jet;
4. Concentrations of toxic and noxious materials in the air meet industrial hygiene specifications;
5. Harmful airborne materials cannot enter buildings, enclosed areas or other workplaces to a dangerous extent;
6. The discharge tip is >25 m above grade and >3 m above the highest working level or roof within a radius of 12 m;
7. The discharge of any flammable material should be more than 30 m horizontally from heaters, air intakes and sources of ignition;
8. Should the material ignite, the radiant heat flux should not exceed 4700 J/m^2 s on any neighbouring access floor.

15.5.2 Disposal by burning

Disposal of combustible gases by burning is generally accomplished in elevated flares which can be a nuisance to local communities, especially at night-time. Flare stacks and burners are best designed by, obtained from and installed by specialist firms. This is because of the difficult problems of ensuring positive pilot ignition, flame stability and acceptable levels of noise, thermal radiation and luminosity. Smokeless operation generally requires considerable quantities of steam, high-pressure waste gas or forced-draught air. Carbon deposited on the lip of the burner presents an ignition hazard, since red-hot particles may be detached and travel considerable distances in the air, with the risk of igniting pockets of flammable vapour, e.g. above floating roof tanks. Another potential hazard of flare stacks is that of explosion within the stack resulting from air entering it when no gas is flowing through it. Care should be taken to avoid holes in the piping, knockout vessel and stack through which air could enter, and a small continuous flow of inert gas through the system is recommended to prevent back diffusion of air through the burner.

If space permits, a ground-level flare has the advantage of facilitating maintenance and allowing a light shield to be constructed which obscures visible flames under all conditions other than major releases. When continuous flaring is necessary, smokeless combustion can be achieved through the use of a number of small burners at ground-level without the need for steam or forced-draught air. Combustible liquids released by pressure-relief valves can where space is available be discharged to a burning pit in which a fire is maintained continuously.

15.5.3 Disposal to a lower pressure system

For this method to be safe the lower-pressure system must have adequate capacity to take the maximum amount of fluid liable to be discharged from the higher-pressure system without itself becoming overpressurised. This method might be used in an LPG storage terminal with spheres and a considerable amount of pipework which requires thermal expansion relief for liquid LPG trapped between closed valves. Here the small amount of liquid LPG relieved could generally be routed to a storage sphere provided all possible hazards are checked.

15.5.4 Treatment of toxic gas and hot fluid discharges

Properly engineered systems for dealing with releases of highly toxic and hot fluids are essential. Highly toxic gases must be treated in adequately sized towers with liquid absorbent or solid adsorbent, or catalytically to form non-toxic compounds. Companies and industries which have these problems usually develop their own methods and codes for dealing with them, such as BCISC's chlorine code[14].

Even where an adequate system is installed, the fact that it may not be needed for a long time may induce managements to question its need and allow it to lapse into a state of disrepair, as appeared to happen at Bhopal [5.4]. The rate of discharge which such systems may have to deal with in an emergency must also be properly appreciated when they are designed. The high cost of an adequate system is a temptation to provide a 'cosmetic' but inadequate solution.

15.6 Other means of pressure-relief

The size and cost of conventional pressure-relief systems which are sometimes needed has led to the partial acceptance of other solutions involving instrument 'trips' with microswitches and solenoid valves. If, for example, a pressure-relief system including relief valves, disposal line and flare had to be designed to guard against reflux or coolant failure on a large distillation column separating iso- from normal pentane (normally carried out at 2 to 3 bar g.), the size of the system could be considerable, because of the large amount of vapour generated in the steam-heated reboiler of the column. By installing a pressure-sensitive switch at the top of the column, which actuates a three-way solenoid valve on the compressed air to the diaphragm of the steam control valve to the reboiler, the steam flow can be automatically stopped when a given pressure is reached in the column. Although this does not entirely obviate the need for a pressure-relief system for the column, since there are other possible reasons for a pressure rise, it enables a much smaller one to be used. In such cases, special steps, such as the installation of a duplicate trip system, are needed to ensure that it is quite as reliable as the larger pressure-relief valve which it is replacing. The consequences of suddenly closing a large steam flow to the reboiler must also be considered and guarded against.

Means of dumping liquids and of depressurising equipment to pressures well below the normal working pressure are also frequently needed for protection against fire exposure, which may result in serious weakening of the metal. Manual operation of valves which may be near the fire cannot be relied on. Systems commonly used involve the remote manual operation of pneumatically controlled valves from two different locations, one in the control room and another at some distance from it, with minimal risk of fire exposure. 'Dump tanks' are needed to retain liquid suddenly released in this way. Sometimes it is possible to retain vapour released on depressuring. but more often it has to be vented or flared. Depressurising systems should contain some temperature-actuated safety feature of last resort which will automatically release the gas or vapour in the event of fire if operators are unable to do this before evacuating the plant.

15.7 Instrumentation for control and safety[11, 15-17]

The importance of instrumentation, which typically accounts for 5% of the capital cost of process plant, can hardly be exaggerated. It has considerably reduced the number of operators required and increased the capital investment per employee. It has enabled more complex and efficient plants with lower material inventories to be built. Such plants may incorporate several reaction and separation stages, internal recycle streams, and integrated fuel and power supply combined with waste-heat recovery. If the same process had to be designed for manual operation without control instruments, it would have to be split into several much simpler units, with considerable intermediate storage between them and a large team of operators. Today many processes are controlled by programmable electronic systems (PESs), with only a handful of highly trained operators in supervisory and monitoring roles. While the term PES includes a single monolithic computer such as those installed in control rooms in the 1960s and 1970s, it also covers more modern systems consisting of networks of individual electronic devices linked together with coaxial cables or fibre optics. These, like single computers, can select the optimum conditions to give the required product output and specifications at minimum cost.

The design of plant instrumentation and the selection of instruments is, even more than pressure relief systems, a specialised engineering field. It cannot, however, be left entirely to instrument engineers, but requires close cooperation with process engineers who design the process, operating staff who will run it, and safety and reliability specialists. While these people cannot all be instrument experts, they need a general appreciation of instrument and control theory and of the instrument symbols used on P&I diagrams. These generally follow the Instrument Society of America's standard S5.1[18]. A selection of symbols is shown in Figure 15.4.

The reliability of process instruments is discussed in 14.2 and again under hazard and operability studies in 16.5.5. While complexity is inevitable in modern plants, every effort should be made to reduce the instrumentation to its essentials, and keep it as simple as possible. The measuring elements of control instruments should be as close as possible to the point where control is most needed, and response times should be minimal.

Instruments are used both for normal plant control and for special safety functions. Many of those used for plant control incorporate such safety features as interlocks [15.7.1], 'fail-safe' systems [15.7.2], and microswitch-operated alarms and automatic plant shutdown systems. Special and separate safety instrumentation includes automatic 'dump' systems to dispose of reactions which have got out of control, water sprinkler and 'deluge' systems for fire protection and explosion-suppression systems [10.4.4]. These only come into play when a critical situation has been reached which demands action which is beyond the scope of the normal control instrumentation.

15.7.1 Interlocks and other aids to safe operation

An interlock is something which connects two or more adjustable devices that prevents one or more of them being set in a particular mode (e.g. open or shut) which would be hazardous while others are also in particular modes. The first interlocks were mechanical and applied to the hand-operated levers used to switch railway signals and points. Interlocks today may be electrical, electronic, pneumatic and hydraulic as well as mechanical, or may involve combinations of these signalling or actuating media. They are very widely used and a few examples follow:

- In starting plant or machinery, interlocks are used to ensure that all prestart conditions (e.g. adequate lubricating oil pressure) are met and that the correct sequence is followed;
- Interlocks are used to prevent unauthorised entry to electrical switchrooms, process vessels, and cubicles where explosives are tested;
- Isolating valves fitted in series with pressure relief devices have interlocks to prevent all of them being closed at the same time;

Connection to process, or mechanical link, or instrument supply

Pneumatic signal, or undefined signal for process flow line

Capillary tubing (filled system)

Electrical signal

Hydraulic signal

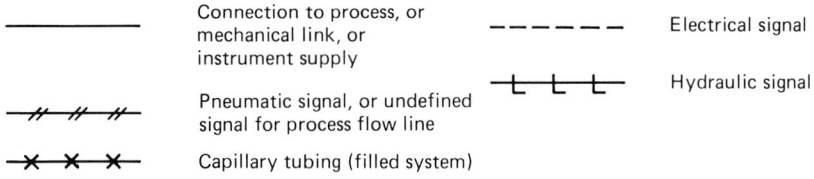

(a) Instrument lines

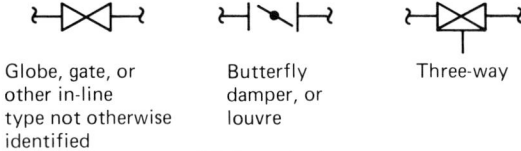

Globe, gate, or other in-line type not otherwise identified

Butterfly damper, or louvre

Three-way

(b) Control valves

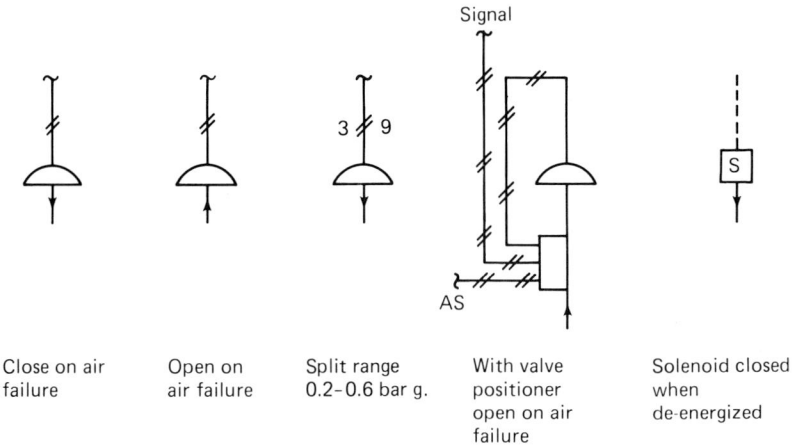

Signal

3 9

AS

Close on air failure

Open on air failure

Split range 0.2–0.6 bar g.

With valve positioner open on air failure

Solenoid closed when de-energized

(c) Control valve actuators

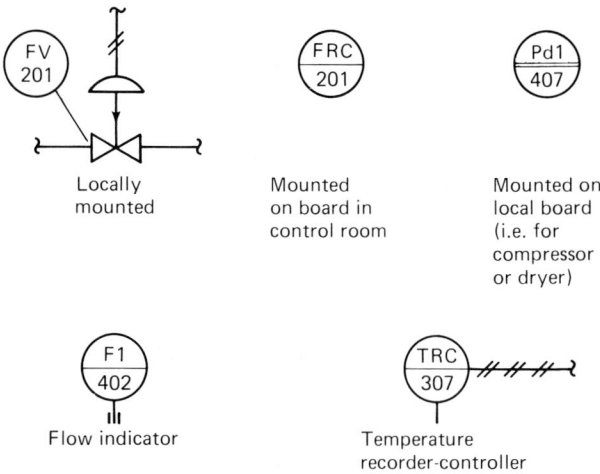

FV 201

FRC 201

Pd1 407

Locally mounted

Mounted on board in control room

Mounted on local board (i.e. for compressor or dryer)

F1 402

TRC 307

Flow indicator

Temperature recorder-controller

(d) Instrument and valve number balloons

Figure 15.4 Instrument symbols used on P&I diagrams

- Interlocks are used to prevent control instruments being decommissioned for calibration or maintenance unless certain safe conditions are met.

Many hazards revealed by hazard and operability studies [16.5.5] can be eliminated by interlocks.

Features of a good interlock system as stated by Lees[12] are that it

1. Controls operations positively;
2. Cannot be defeated;
3. Is simple, robust and inexpensive:
4. Can be readily and securely attached to the devices associated with it;
5. Can be regularly tested and maintained.

Microswitches and solenoid valves are widely used to actuate alarms and shutdown devices, and also as components of electro-pneumatic interlocks to prevent accidents arising from plant malfunctions or human errors. Richmond describes several examples, including the fitting of microswitches to the levers of cocks on the oleum charge lines to three sulphonation reactors, and a method of protecting a chemical reactor from the consequences of agitator failure[19].

15.7.2 'Fail-safe' design

This philosophy which is found in all branches of engineering applies particularly to automatic control systems which contain several sub-systems, any one of which may fail. Claims that a particular system or instrument can only 'fail safe' need critical appraisal. The term 'fail-safe' is something of an euphemism, and cannot disguise the fact that failure has occurred, thereby causing loss and sometimes hazard. Thus it is important to concentrate on choosing reliable instruments and instrument systems with little likelihood of failure.

In opting for one mode of failure as 'safer' than another, the choice may be between the devil and the sea. With control systems the main criterion for fail-safe is that the valve or other control element should assume one of three positions if failure occurs: open, closed and the last position of the valve before failure. Possible failures affecting the valve include:

1. The power supply to the valve;
2. Measuring and sensing elements;
3. The control system itself, which may depend both on electric and pneumatic power sources.

While a valve can be arranged to fail safe in case (1), this may be difficult or impossible for the other two since the valve has no way of knowing whether such failure has occurred. The extent to which 'fail-safe' applies is therefore limited.

With powered control valves with spring return mechanisms, particularly those with pneumatic actuators, the designer can choose between an actuator which opens or closes the valve on power failure. With a pneumatic actuator he can also arrange for 'last position failure' by the addition of a solenoid valve in the air line to the actuator. Thus it is important for the process engineer to study which is the least dangerous position of each control valve on loss of control and specify this when ordering and installing the valve.

Richmond[19] has shown through the analysis of several types of self-actuating pressure and temperature regulators that all types of pressure regulator failed in the unsafe condition and that only one type of self-acting temperature regulator would 'fail-safe'. This type, however, is operated by a fluid under vacuum which has limited power to move the valve stem.

The consequences of all possible modes of failure of critical instruments need to be studied. For modes in which it is not possible to secure fail-safe operation, other means of protection should be provided, e.g. a pressure relief valve or a high-temperature trip switch.

15.7.3 Process stability

Some processes are inherently stable or self-adjusting, so that any deviation from the required value of an important variable produces some effect in the process which tries to restore the variable to its original value. Others are inherently unstable, in that a change from the desired value of the variable produces some effect which increases and accelerates the change. Sometimes the same process has both a stable and an unstable regime. A pilot cracking plant which the writer designed and which was fired by its own cracked gases, a mixture of hydrocarbons and hydrogen, was an example of a very stable process. If (due to some upset) the cracking temperature increased, the calorific value of the cracked gas (and its Wobbe Index [15.8.4]) fell, so that with other conditions unchanged the heat input to the furnace fell, thus restoring the original cracking temperature. This resulted in a very constant cracking temperature without any temperature control instrument. An example of a process with both stable and unstable regimes is a water-cooled exothermic reactor for which there is critical temperature below which the temperature naturally falls to a stable value, and above which it accelerates upwards [8.2.1].

When planning a new process and how to control it, the stability of different process variants and control methods should be explored. An inherently stable natural control system is likely to be safe and simple. The characteristics of the system adopted should be also be studied to discover any unstable control regimes and delineate their boundaries. Precautions are needed to prevent crossing such boundaries while special protective instrumentation may be needed in case an unstable control regime is reached.

15.8 Component features of instrumentation

Some or all of the following features appear in the instrumentation of all process plant.

1. A control room;
2. One or more power sources;
3. Basic control systems;

4. Measurement and sensing of process variables;
5. Receivers, i.e. indicators, recorders, controllers and alarms, etc.;
6. Final control elements;
7. Signal and power transmission systems;
8. PES control.

Within each of these there are various alternatives to choose from. In doing this, reliability and safety are key issues.

15.8.1 Control rooms

Most process plants have a control room as their nerve centre. This typically houses a control desk or console with visual display units (VDUs), one or more control panels, sometimes a computer, and operating personnel. On the panel is mounted a network of receiving instruments (known as receivers) which receive information from the plant. These receivers increasingly tend to incorporate PES devices which extend their functions and allow plant data to be displayed on VDUs built into the console instead of on panel-mounted instruments. Receivers continuously receive pneumatic or electronic signals which relate to variables such as temperature and pressure and are transmitted by measuring instruments in the plant or elsewhere. From the control room pneumatic and/or electric signals are automatically sent to valves, motors, etc. to control the variable as required. No hazardous materials should be allowed to enter control rooms via pipes or tubes (e.g. to reach meters or gauges).

Control rooms are usually located close enough to their plants for operators to be able to move quickly from the control room to inspect plant items or open and close valves, etc. Good communications (e.g. by radio-telephone) are also essential between those working in the control room and those on the plant. Factors which affect operator morale and instrument performance such as the temperature, humidity, air movement, noise, vibration, illumination, interior decoration and furniture in a control room should be considered.

On oil and chemical plants where flammable gases and vapours are present, the control room may have to be located in an area classed as 'hazardous' and in which special precautions against electrical ignition are required [6.2.3]. It is usual here to 'pressurise' the control room with air supplied with a fan from a 'non-hazardous' area. The only admittance to the control room is then through an air lock with self-closing doors which allow a slight positive air pressure to be maintained inside it. This allows standard industrial electrical equipment to be used in the control room.

Since the Flixborough disaster [4] in which 18 men were killed in a control room, most control rooms built near plant where there is an explosion risk have been designed as separate blast-resistant single-storey buildings[12, 20] [16.6.5]. Windows, if fitted, contain only small panels of toughened glass.

15.8.2 Power sources

Most instrument systems rely on pneumatic or electrical power for signal transmission and control and electricity for chart drives and illumination.

Pneumatic systems should create no ignition risks in electrically 'hazardous' areas, while pneumatic valve actuators are generally cheaper and simpler than electric ones. Pneumatic signal transmission is, however, relatively slow, thus limiting transmission distances to about 100 m. Low d.c. intrinsically safe [6.2.3] electric current signal transmission and electronic instruments in the control room are often combined with pneumatic valve actuation.

All instrument power sources must be completely reliable. While failures of plant electrical power, steam, fuel gas and water usually cause an emergency shutdown, sudden and complete loss of control instrumentation is even more serious, since without it, it is far more difficult to shut the plant down safely, particularly if one of the other services fails at the same time.

On a large or medium-sized works, a reliable supply of clean, dry, compressed air is conveniently provided by an electrically driven air compressor, with a diesel-driven stand-by compressor, and a large compressed air reservoir capable of meeting all likely demands until the stand-by unit is operating in the event of electrical failure. Instrument air should be distributed by a ring main, to give two alternative supply routes to every plant. Branches from the main should be provided with excess flow valves or other means of protection against escape of air and loss of pressure if a branch pipe is broken. The instrument air supply should be kept entirely separate from compressed air used for pneumatic tools and other purposes. Local compressed air reservoirs and sometimes even separate air compressors may be justified for large or hazardous process units. The compressed air is reduced in control rooms to the pressures required by the instruments.

Fitt[10] made the point that some instrument rooms are fed from a single leg of the air main. Failure of this leg alone without failure of the air supply to the control valves could lead to valves assuming 'unsafe' positions. The consequences of partial failure of instrument air can thus be more serious than total failure.

A reliable supply of high-quality electrical-power, with emergency supply to critical instruments from batteries and/or a standby generator, is needed for electronic instrumentation. The instrument circuits should be kept entirely separate from the normal power and lighting circuits, and must meet the relevant flameproofing or intrinsically safe requirements.

15.8.3 Basic control systems[11, 15, 17]

The main control systems considered here are of analogue type employing continuous feedback, and are based either on pneumatic signals in the pressure range 0.2–1.0 bar g. or d.c. electronic signals in the current range 4–20 mA. The various control modes and other features considered here apply to both pneumatic and electronic systems despite their other differences. Pneumatic systems employ small mechanical devices with moving parts, e.g. baffle-nozzle amplifiers and metal bellows. Electronic systems use electronic amplifiers, switches and relays as well as electro-mechanically balanced potentiometers and Wheatstone bridges. Pneumatic signals interface directly with mechanical control devices.

Electronic signals generally interface via pneumatics with mechanical control devices.

Also considered briefly here is 'Power Fluidics'[21], a control system with no moving parts, and self-actuating controllers which depend for their power on the moving process fluid. PES control which today is mostly of the digital type is considered later [15.8.8].

Instrument loops and degrees of freedom

The feedback control loop is basic to process control. It includes the process, the measuring element, the controller which receives its signal from it and the final control element (generally a valve) which manipulates a process variable and receives its signal from the controller. It is a 'closed loop' when controlling automatically (Figure 15.5) but an 'open loop' when the control element is actuated manually or not at all. (*Warning*: Some

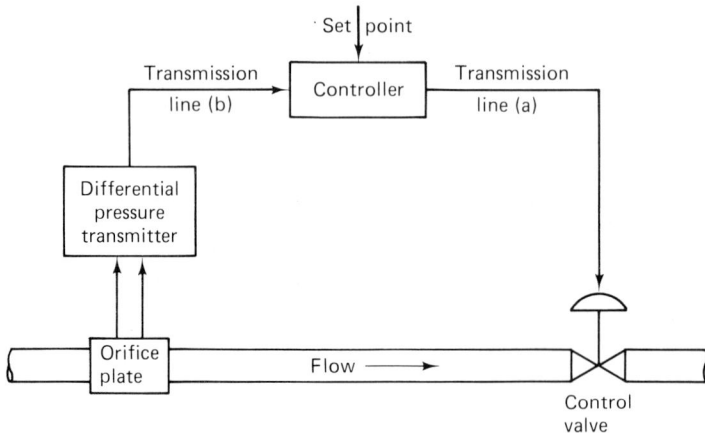

Figure 15.5 Process-flow control loop (from Perry and Chilton[11], courtesy McGraw-Hill)

writers refer to a loop as closed when a human operates the control valve!) In flow control loops the variable manipulated (by a valve) is generally the same (flowing fluid) as the one being controlled, but in most other control loops, one variable is manipulated in order to control another, e.g. the flow of steam may be manipulated to control the temperature of a continuous liquid mixer (Figure 15.6).

Most controllers have an adjustable 'set point' which can be set at the required value of the variable under control, and they measure the deviation between the set point and the measured value of the variable. Their signal to the final control element aims to correct deviations which result from process disturbances (known as 'upsets'). When steady conditions are re-established, and depending on the control mode, the deviation will have been eliminated, or a residual deviation known as the 'offset' will remain (see 'Proportional control' later).

With feedback systems and their components, there is a time lag referred to as 'dead time' between any stimulus and the response of the system or component to it. This sometimes causes serious problems.

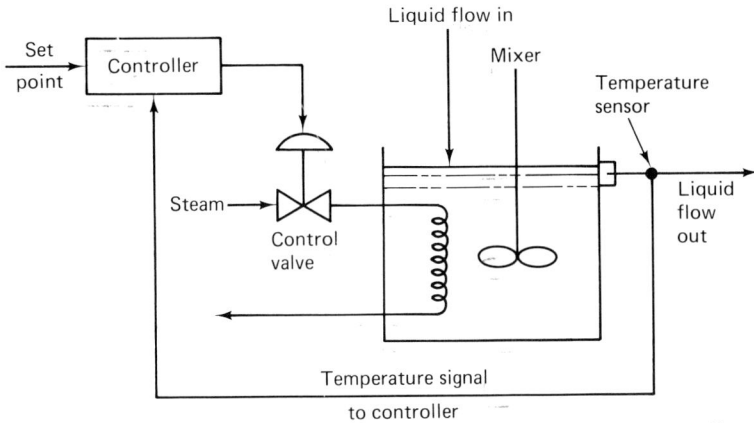

Figure 15.6 Temperature control loop for liquid mixer (from Perry and Chilton[11], courtesy McGraw-Hill)

The number of closed control loops needed for any process equals the number of independent variables or degrees of freedom (e.g. temperature, pressure, flow rate, level, concentration) which have to be controlled. But if too many control loops are provided, they will fight each other for control and the system will not work. This can happen because some variable was mistakenly thought to be independent whereas it was in fact dependent on and uniquely determined by other variables. As a simple example, consider two liquids which flow into a vessel, mix there and flow out as one stream. This gives three degrees of freedom, the flows of each stream into the vessel and the liquid level in it. We can control these three independent variables but we cannot at the same time control the flow of mixed liquid from the vessel.

Most cases are, of course, more complicated than this, but the same principles apply. In complex cases Boolean algebra [14.1.12] can help to determine the irreducible minimum number of independent variables which need to be controlled, thereby simplifying the plant design and avoiding superfluous instrumentation.

Control modes
Analogue controllers may incorporate one or more of the following characteristic modes, the choice of which depends, *inter alia*, on the inertia of the process, the various time lags and any hysteresis in the instrument loop. The different modes of any controller have to be tuned for its particular process application (see later).

1. On–off or 'bang-bang' control
This is a simple, accurate but rather drastic method which is widely used (e.g. for refrigerator thermostats) where there are no process constraints (e.g. water hammer) which prohibit sudden control action. It is used where the controlled variable changes slowly in relation to changes in the manipulated variable. Where a constant value of the controlled variable is needed a single set point is used, but when control is not critical, e.g. the level of liquid in a supply tank, two set points may be used, between which the level oscillates slowly.

On–off control is used more in batch processes than in continuous ones. Microswitches used in safety instruments to protect plants by flaring, venting or dumping process materials, when the temperature, etc. exceeds a critical value are a form of on–off control.

2. Proportional control, single mode (P)

In proportional control the difference in output of the controller from its mid-point is proportional to the deviation of the input signal from the set point. The relationship between the ranges of the output and input signals is called the proportional band and is expressed as a percentage. Thus a 20% band is narrow and gives sensitive control because 100% output change is produced by a 20% change of the input signal (i.e. the measurement scale).

The controlled variable only steadies out at the set point when the output signal is at 50% of its range. For other outputs the controlled variable will steady out at a value which differs from the set point by the offset. The maximum and minimum values of the offset for any proportional band would be plus or minus half the proportional band. If the proportional band is too wide, the offset will be large and the control insensitive. If it is too narrow, although the offset will be small, the loop will cycle and the controlled variable may never reach a steady value. Responses of a purely proportional controller to a process upset with too high, too low and correct proportional band settings are shown in Figure 15.7. The best setting of the proportional band for the range of process conditions likely to be encountered is generally found by trial and error.

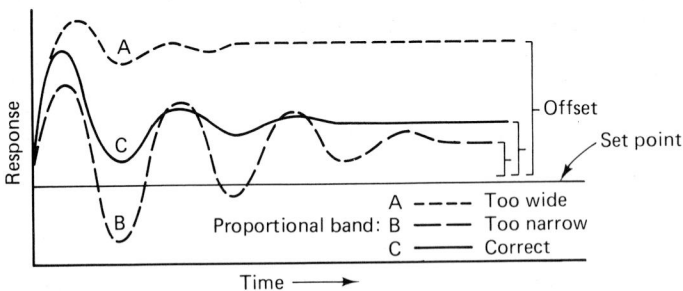

Figure 15.7 Responses of proportional controller to a process upset (courtesy The Foxboro Company)

Proportional control works best when the controlled variable responds rapidly to changes in the manipulated variable, and where the value of the controlled variable is not critical, e.g. in some temperature and level-control loops.

3. Proportional plus integral (reset) control, two-mode (PI)

Integral action, formerly known as 'reset', is a means of correcting the offset when proportional control is used. It produces a change in the output which is proportional to the integral of the deviation with time. It is combined with proportional control on critical loops where no offset can be tolerated, and on loops requiring such a wide proportional band that the amount of offset would be unacceptable. Integral action is expressed as 'minutes per repeat', the time needed for the integral mode to repeat the response of the proportional mode for a sudden upset. If the integral time is too long, the offset will persist for a considerable time, and if it is too short, the system will tend to cycle continuously. Figure 15.8 shows the PI system response to a process upset with different integral times. The addition of integral to proportional control, while eliminating the offset, increases the initial response of the loop to an upset and causes some tendency to cycle.

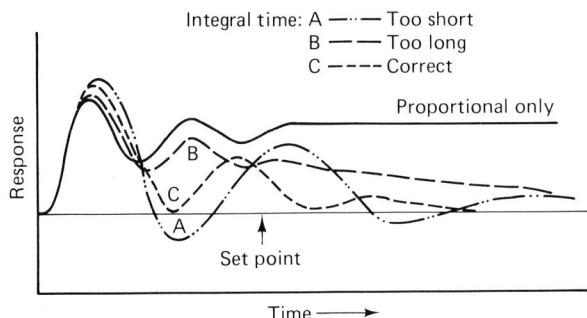

Figure 15.8 PI system response to process upset with different integral times (courtesy The Foxboro Company)

A problem of integral action which is specially found with batch processes is known as 'integral wind-up'. Here the controller (say, of a reactor temperature by a valve on the steam supply to the jacket) may start to work while the controlled variable is well below the set-point (and perhaps below the temperature scale of the controller). So much integral action may then have been stored up that the steam valve opens fully and remains so until the reactor temperature passes the set point. The temperature continues rising for some time before the output of the controller falls and the steam valve starts to close.

To overcome this problem, a 'batch switch' may be fitted to a controller with integral action. This cancels the integral action when the offset exceeds a preselected limit. Beyond this limit only proportional control operates, but the integral action restarts once the offset has been reduced to within this limit. Batch switches are used mainly in discontinuous processes, and in continuous ones where speed of response is vital to equipment protection, e.g. on anti-surge flow controllers for compressors. Tuning of a PI controller with a batch switch may be very critical.

4. Proportional plus integral plus derivative control, three-mode (PID)
Derivative action is a means of speeding up the control action when both proportional and integral modes are used. (It can also be used with proportional control only.) This measures the rate of deviation of the controlled variable from the set point and applies a control action proportional to this rate. It is usually expressed in minutes which represent the reduction in response time which the derivative action gives when added to a proportional controller. The action adds stability to the loop

and is useful in cases where the overall lag is rather large or where upsets tend to be large and rapid. It is seldom used in 'noisy' control loops where random fast-moving signals are superimposed on the measured variable. Derivative and integral action are in many ways complementary. Figure 15.9 compares the responses of loops with PI and PID control to a process upset.

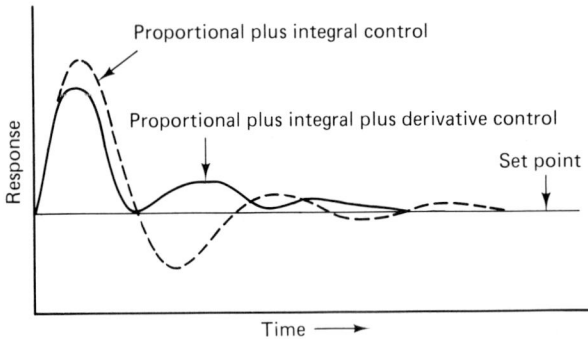

Figure 15.9 Comparison of system response to a process upset with PI control and PID control (courtesy The Foxboro Company)

Feedforward

Another control system, known as feedforward, reduces the time lag. Here a process variable which would affect the variable to be controlled is itself measured, and changes in it are compensated for without waiting for changes in the controlled variable. If the temperature of a continuous mixer is the main one to be controlled, feedforward control might be applied to compensate for changes in the flow and temperature of its main feed stream (Figure 15.10). The temperature of the main feed stream but

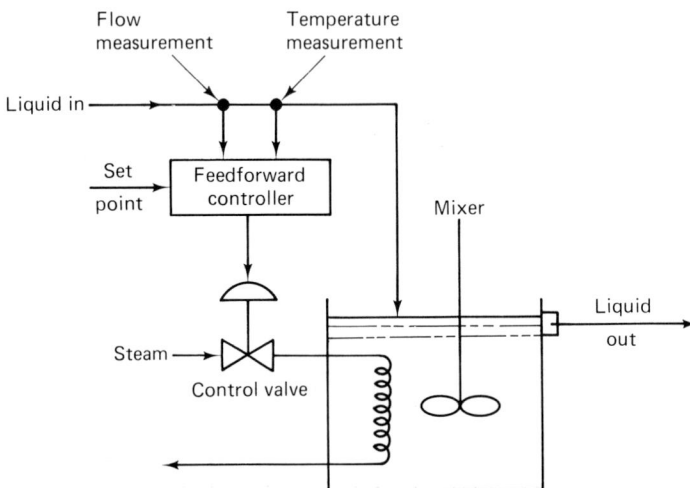

Figure 15.10 Feedforward control of a continuous mixer (from Perry and Chilton[11], courtesy McGraw-Hill)

not that of the reactor would then be part of a closed loop. This system requires a computational model of the process which can be pneumatic or electronic and which calculates and applies the correct compensation. Feedforward control works faster than feedback control, but is less accurate if there are other process variables whose changes are not allowed for by the model.

Cascade control
In cascade control a master controller which has to keep a particular variable constant operates by resetting the set point of a slave controller which controls another, faster, variable which affects the first one. The feedforward control system just discussed could, for instance, be converted to cascade control by installing a master controller to control the mixer temperature by adjusting the set point of the feedforward controller (Figure 15.11). Cascade control might, for example, be used to control the (slowly changing) composition at the mid-point of a large distillation column operating with a constant reflux rate and top pressure by a master controller which adjusts the set point of a steam pressure (slave) controller for the column reboiler (Figure 15.12). Cascade control can produce significant improvements where upsets in the supply variable which affects the main controlled one are large and frequent, where the main variable responds slowly and where the supply variable can be controlled in a rapidly responding loop. Where integral action is included (as is usual) in the primary controller, or in the slave controller, special precautions have to be taken in the link-up of the two controllers to counter the effects of 'integral wind-up', which can otherwise cause serious trouble.

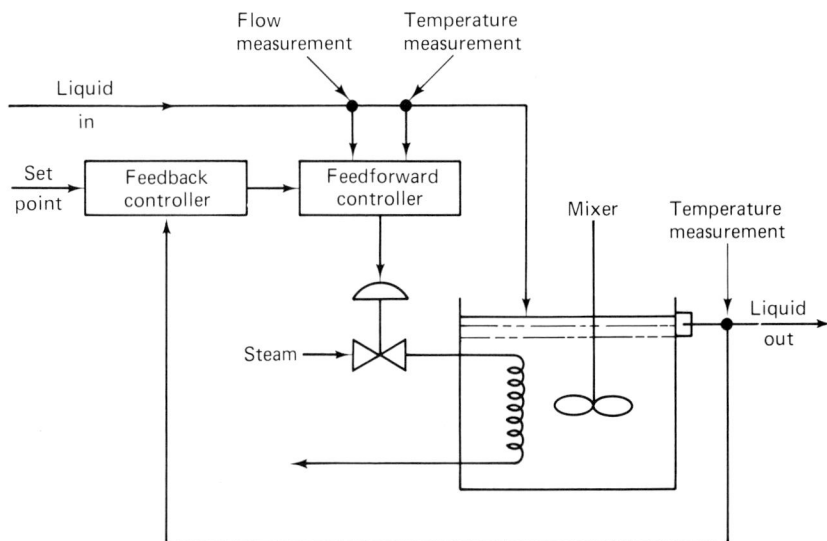

Figure 15.11 Cascade control of a continuous mixer (from Perry and Chilton[11], courtesy McGraw-Hill)

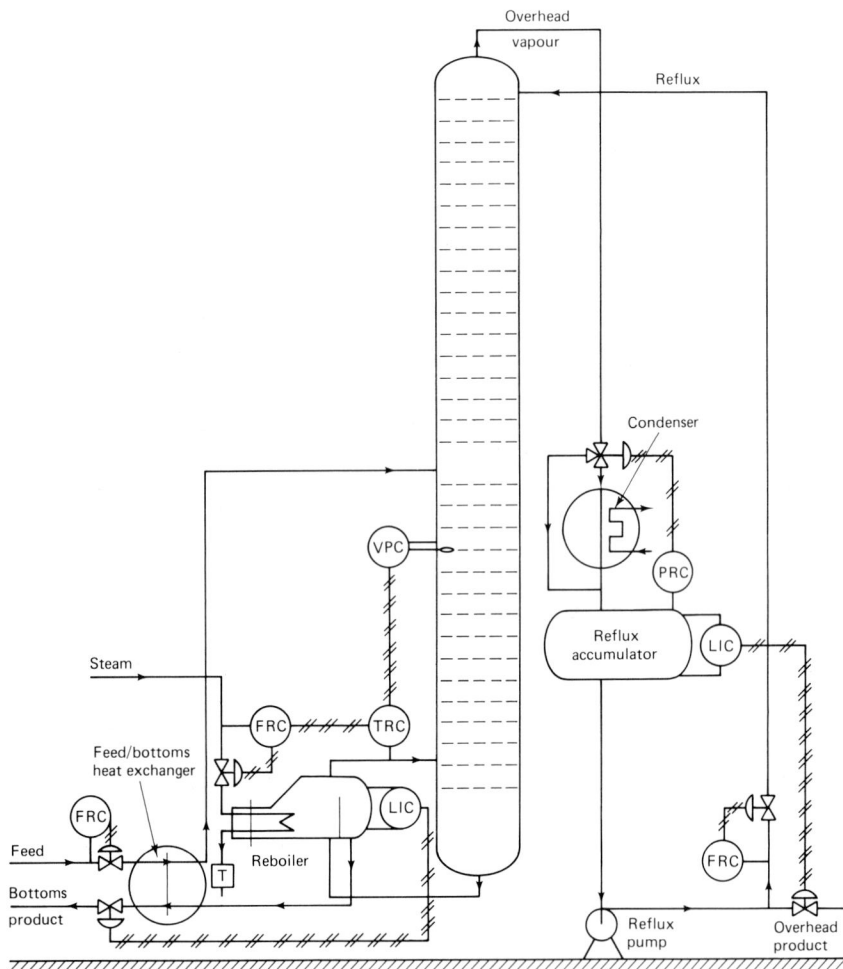

Figure 15.12 Cascade control of a large distillation column. Vapour pressure controller (VPC) measures difference between vapour pressure of reference substance and column pressure at a sensitive point in the column, and resets bottom vapour temperature controller, which resets steam flow control

Other compatible control systems[17]

The 'auto-selector' control system is used to keep several variables which are affected by one control valve within safe operating limits. Each variable has an associated controller whose output goes to a device which selects the controller with the highest or the lowest output, provided it is above or below a certain value. This controller then takes over control of the valve. It might, for example, be used on a pipeline pumping station to throttle a normally open valve if the pump suction pressure falls too far, if the motor load gets too high or if the pipeline pressure rises too high. Although somewhat complicated, the system can form a useful first line of defence and prevent a complete shutdown if the trouble is a minor one. If, in spite of it, one variable goes further into an unsafe operating area, the use of microswitches may be needed to actuate complete automatic shutdown.

Ratio control is used for blending two flowing fluids where the flow of one fluid is measured by adjusting the set point of a flow controller for the second fluid.

Tuning control loops

The width of the proportional band and the speeds of integral and derivative action need to be tuned finely enough for each control loop to respond adequately over all likely combinations of plant conditions without 'cycling'. The resulting oscillations in the controlled and manipulated variables may lead in complex plant to worse cycling of other variables. This can be a serious hazard, causing pumps to lose suction and pressure relief valves to blow. Tuning should be done only by trained and authorised persons, and a record kept of the adjustments made. Tuning a proportional controller is largely a matter of trial and error. An empirical method of tuning three- and two-mode controllers is given by Foxboro[17] and summarised as follows:

1. *Tuning three-mode controllers.*
   Step 1: Set the integral time of the controller at maximum and the derivative time at minimum (giving only proportional control). Reduce the proportional band until cycling starts. Measure the natural period (between two successive crests or troughs, Figure 15.7).
   Step 2: Set the derivative time at 0.15 times the natural period and the integral time at 0.4 times the natural period and observe the new period. If it is less than 75% of the natural period, reduce the derivative time, but if it is longer, increase it.
   Step 3: Finally, adjust the proportional band to give the required degree of damping.
2. *Adjusting two-mode controllers (proportional-plus-integral).*
   Step 1: Same as step (1) for three-mode controller.
   Step 2: Set the integral time to the natural period and observe the new period which should be 140% of the natural one. If it is longer, increase the integral time.
   Step 3: Same as step (1) for three-mode controller. (Adding integral always increases the proportional band required for stable control.)

More sophisticated methods of tuning are sometimes used[11, 15].

When correctly tuned, a disturbance in the process causes the value of the controlled variable to somewhat overshoot its final value at which it should settle after three or four damped oscillations (Figure 15.9).

A loop which has been in satisfactory operation for some time may start to cycle because a control valve is sticking, or because backlash has developed in an instrument mechanism. Regular lubrication of valve spindles, adjustment of glands and maintenance and checking of all instruments are essential. If despite proper tuning the control is still unsatisfactory, the use of cascade control and/or a valve positioner [15.8.6] should be considered.

In plants with conventional instrumentation, the tuning should be checked under various process conditions so as to give stable operation under the worst combination of flows, temperatures, etc. which may occur. This may result in rather poor responses under normal conditions. In plant where the controllers are incorporated into a PES, it is possible to arrange for the tuning constants to be changed automatically to correspond to the current process conditions.

Remote manual control

Most control instruments can be switched by the operator from automatic to (remote) manual control. Limits should be imposed on the number of control instruments which may be put on manual control at any one time, and on the length of time during which a controller may remain on manual control. The compatibility of various combinations of automatic and remote manual control should be checked. The switching of one control instrument from automatic to manual control may upset the operation of other control instruments. This can cause special problems in PES-controlled plant.

For critical variables (especially flows of dangerous fluids) which may have to be controlled in an emergency, duplicate and well-separated remote manual control stations in well-chosen locations are often needed so that one can function if the other is knocked out.

Self-actuating controllers

Most self-actuating controllers are small, self-contained mechanical devices which incorporate a measuring element, a simple adjustable control system and a valve whose stem is actuated by the force of the process fluid balanced against a spring or weight. They do not usually show the value of the controlled variable. These controllers are often used on simple plants for control of pressure, temperature, flow and level, and on more complex plant in auxiliary control functions (e.g. P&V valves for nitrogen blanketing of vessels). They are mostly of on–off or proportional type. It is usually difficult to incorporate 'fail-safe' features into them [15.7.2].

Power fluidics[21]

One common feature of the 'power fluidic' devices which have been developed for control and other applications (which include pumping,

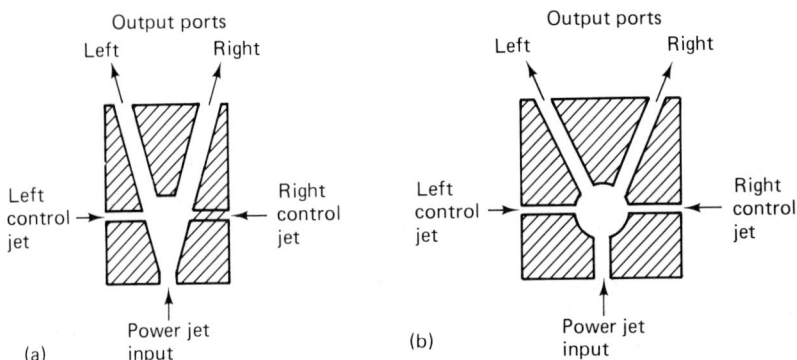

Figure 15.13 (a) 'On-off' and (b) proportional fluidic controllers (from Perry and Chilton[11], courtesy McGraw-Hill)

ventilation, precipitation and phase contacting) is that they have no moving parts and should hence be more reliable and require less maintenance than conventional devices. Their main applications to date appear to be in the processing of nuclear fuels and other very hazardous materials. Their control applications include fluid diversion, flow restriction and pulsing. The principle of a power fluidic diverter is shown in Figure 15.13. The main limitation of power fluidics as applied to control valves is that it is seldom possible to achieve a positive shut-off without employing some moving part.

15.8.4 Measuring and sensing process variables

Process variables include the physical variables, temperature, pressure, flow and level, physical properties such as density, viscosity, thermal conductivity, refractive index, calorific value and Wobbe index of fuel gases, vapour pressure and boiling point of liquids, and chemical composition. Each of these can be measured by several different methods, the choice of which needs careful study. Important considerations include simplicity, speed of response, accuracy, reliability, the consequences of failure (e.g. safe or unsafe), cost, standardisation and familiarity by the operating organisation, ease of installation and servicing. In the writer's experience, more problems were caused by faulty installation and servicing and by inappropriate choice of instrument than by failure of instruments themselves.

Clear thinking is needed about what should be measured and how this relates to other measurements. One should consider the principle of the meter and whether it measures the required variable directly, or whether it measures some related variable and converts it. Thus some thermometers measure the vapour pressure of a liquid and convert this to a linear temperature scale, whereas the vapour pressure itself may be a better control parameter than the temperature. Some flow meters measure volume flow (L^3T^{-1}), some measure mass flow (MT^{-1}), while variable head and variable area meters measure volume flow divided by the square root of the fluid density ($L^{1\frac{1}{2}}M^{-\frac{1}{2}}T^{-1}$). If the fluid density is liable to change and the figure required is either mass flow or volume flow, a continuous density meter and a computing device would be needed as well if a variable head or variable area meter were used. The flow of a fuel gas measured by such a meter is, however, directly compatible with its Wobbe index and not its calorific value (J/m^3), since

Wobbe index = (calorific value)/(gas density)$^{\frac{1}{2}}$

Continuous measurement of calorific value of a gas involves measuring its Wobbe index and continuously compensating this for changes in its density. Thus to control the heat output of a large burner by controlling the volumetric flow of the fuel gas by an orifice meter, as well as its calorific value, means putting opposing compensators for gas density into each meter so that their corrections balance out.

Measurement of the physical variables is next discussed followed by a very brief discussion of the measurement of physical properties and chemical composition.

Temperature measurement[22]

Thermocouples and electrical resistance thermometers are commonly used for direct intrinsically safe cable transmission to receivers. Instruments with fluid-filled bulbs and bimetallic strips used for local temperature indication are simple and generally reliable. Fluid-filled bulbs with long flexible metal capillary tubes (which are easily damaged) are used for remote control and recording, usually on small plants with little instrumentation. Radiation pyrometers of various types are used for measuring temperatures of hot visible objects. Temperature-sensitive pigments are available as paints and self-adhesive labels which change colour reversibly or irreversibly at known temperatures up to 350°C in steps of 5°C. They are useful for monitoring the temperature of bearings, etc. and showing the temperature at various levels of a burning oil tank.

Thermocouples give a small electromotive force (e.m.f.) between the working junction in the plant and the cold junction in the receiver. The receiver includes a device which compensates for variations in the cold junction temperature. Several combinations of thermocouple wires are available for temperatures from −250 to 2600°C. The combination must be suited to the corrosivity and temperature range of its working environment. Only couples whose e.m.f. increases continuously with the temperature difference between the cold and hot junction over its working temperature range can be used. With strip recorders the e.m.f. is measured by a potentiometer, but for temperature indicators a galvanometer or an electronic amplifier is often used. A failure in the thermocouple circuit (e.g. caused by breakage of the welded hot junction) causes a drop in e.m.f. which results in a lower apparent temperature than the real one. Thus it fails in the unsafe mode.

Since most thermocouples are fragile and liable to chemical attack, their working junction is generally enclosed in a metal sheath, or inserted in a thermowell on the plant, or both. These reduce the speed and accuracy of the measurement, particularly of gas temperatures. A bare hot junction may reach thermal equilibrium in less than 30 seconds, but may take several minutes to do so in a thermowell. If the closed end of the thermowell is its lowest point and the temperature is not too high, the time lag may be reduced by the use of a heat transfer liquid in the well.

Platinum **resistance thermometers** are available for temperatures from 15 K to 800°C and nickel ones for the range −200 to 350°C. Their resistance increases with temperature, so that any failure of the resistance circuit gives too high a temperature reading and their failure mode is safe.

Thermistors with a negative temperature coefficient are available for measuring temperatures over five ranges between −100 and 300°C. A break in the resistance gives an unsafe failure mode. Thermistors with a positive temperature coefficient are also available and used to protect the electrical windings of motors and transformers.

Resistance thermometers are protected in the same way as thermocouples, and the same problems apply.

Pressure measurement[23]

Pressures are measured as absolute, relative to a perfect vacuum, or as gauge, relative to the surrounding atmosphere. While devices measuring gauge pressure are simpler than those measuring absolute pressure, gauge pressure are often converted to and quoted as absolute pressures [3.1.1].

There are three main types of pressure-measuring devices: those based on the distortion of an elastic element, electrical sensing devices and manometers. Measurement of low absolute pressures is not discussed here.

All pressure-measuring devices except manometers need regular checking and calibration, generally against a 'dead-weight' tester. Their performance varies widely, not only as a result of their basic design and materials of construction but also because of their service conditions. Care should be taken to ensure that the element can withstand the maximum possible process pressure. Errors can arise from hysteresis, corrosion, changes in temperature, friction, backlash and pulsations. An isolating valve is generally needed between the pressure-measuring device and the process. Elements should not be mounted where they are subject to vibration or extremes of temperature. Pulsations in the process pressure should be damped by a restriction between the process and the measuring device. Any pressure element has its own natural frequencies of vibration, and if one of these coincides with the frequency of a process pulsation, it is likely to be quickly damaged.

Elastic elements include Bourdon tubes, spiral and helical elements (Figure 15.14), bellows and diaphragms. The effects of temperature changes are minimised by using elements made from an alloy with a low coefficient of thermal expansion. Elements from which the process fluid is excluded by a thin diaphragm are available for corrosive and fouling process fluids.

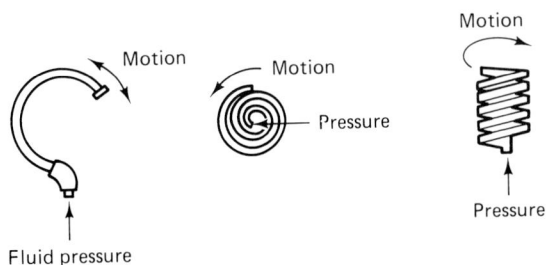

Figure 15.14 Bourdon tube, spiral and helical pressure elements (Foxboro, Figure 24)

Locally mounted Bourdon tube gauges are widely used for pressure and vacuum. Bellows are widely used for transmitters and receivers of pneumatic signals. Diaphragm elements are of two main types, those which depend on the elastic properties of the diaphragm and those which are opposed by a spring.

Electrical sensing elements include strain gauges and piezoelectric transducers, all of which lend themselves to electrical transmission of the signal by intrinsically safe circuits to potentiometer-type receiving instruments.

An electrical strain gauge uses a fine electrical resistance wire or a semiconductive wafer which is attached to an elastic element and stretches with it, its resistance increasing with the strain. If the electrical circuit of a strain gauge is damaged or broken it would indicate a higher pressure than the true one and fail safe.

Piezoelectric transducers which generate a potential difference proportional to a pressure-generated stress are used for measuring rapidly changing pressures but are not usually suitable for measuring fairly static ones.

Manometers are based on the height of a liquid column, usually mercury, water or an organic liquid of low volatility. They are simple and reliable and mainly used for measuring small differences in pressure, up to a maximum of about 2 bars. There is a heath risk with mercury manometers but although once common, they are now little used in industrial instruments.

Flow measurement[24]

The term 'flow' usually refers to the volume of fluid passing in unit time, the term 'mass flow' being used for mass of material passing in unit time. Here we are mainly concerned with the flow of a single liquid phase or a gas.

The main classes of flow-measuring devices used in the process industries are variable-head, variable-area, positive displacement, turbine, electronic flow and mass flow meters, and weirs and flumes for flow in channels. Variable-head and variable-area devices are commonly used for clean liquids of low viscosity and gases flowing in pipes. Corrections have to be applied for changes in fluid density and sometimes viscosity. For this reason, the gas density needs to be controlled at the constriction producing the differential head, while only small temperature changes can be tolerated at this point.

A **variable-head** device has a fixed restriction with upstream and downstream pressure tappings which allow the differential head to be measured with a differential pressure (d.p.) cell and transmitted to a receiver. As the flow is proportional to the square root of the differential pressure, these instruments show a maximum sensitivity at their full-scale reading while their sensitivity at 10% of maximum flow is low. 'Square root extractors' are often used to convert the differential pressure into a signal proportional to the flow before transmission. Orifice plates are the most commonly used form of restriction while Venturi and Dall tubes have lower energy losses. Errors arise through deposits of solids on either side of the restriction and through the presence of gas or a second liquid phase in the lines between the pressure tappings and the d.p. cell.

The most common form of **variable-area** meter is the rotameter, a tapered vertical tube with a linear scale and a plumb-bob spinner (with a higher density than the fluid) which moves up and down the tube as the flow increases or decreases (Figure 15.15). The area between the spinner and the wall of the tube is proportional to the height of the spinner in the tube, and the flow is nearly proportional to this height, giving an almost linear scale. Transparent glass tubes are used for local metering of small flows, and metal tubes with electromagnetic sensing of the position of the spinner are used for larger flows and dangerous fluids, and allow easy signal transmission.

Several types of **positive displacement** meter are used to give the total

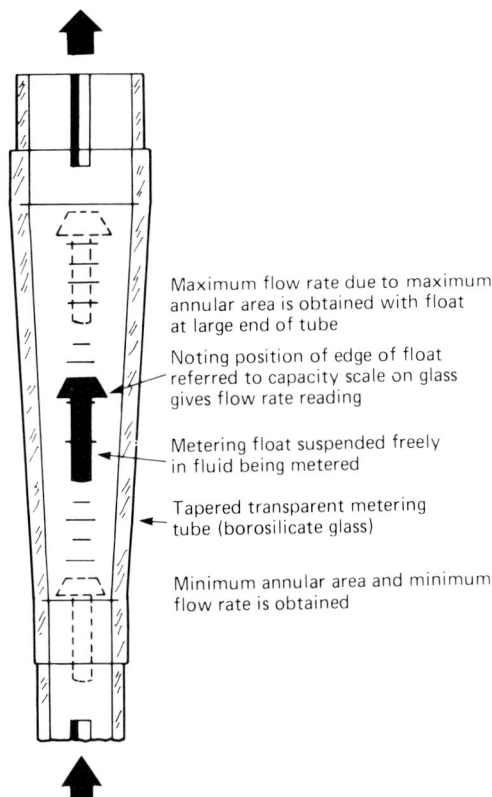

Maximum flow rate due to maximum annular area is obtained with float at large end of tube

Noting position of edge of float referred to capacity scale on glass gives flow rate reading

Metering float suspended freely in fluid being metered

Tapered transparent metering tube (borosilicate glass)

Minimum annular area and minimum flow rate is obtained

Fluid passes through this annular opening between periphery of float head and I.D. of tapered tube. Of course, flow rate varies directly as area of annular opening varies

Figure 15.15 Variable area meter (Rotameter) (courtesy Fischer and Porter Ltd)

volume of fluid which has flowed since the last reading. They are mostly of rotary type, of high accuracy, and are little affected by pulsations in flow and by changes in fluid density or viscosity. They are mainly used for accounting purposes and dispensing known quantities of liquids and not for flow control.

A **turbine** meter consists of a tube with an axially mounted turbine wheel rotating between almost frictionless bearings. The rotational speed of the wheel is proportional to the volumetric flow rate and is sensed by an electric pick-up coil outside the tube. It can be used as the measuring element in a flow control loop as well as for accounting purposes.

Several types of **electronic flow** meters have been developed over the past two decades. Of these, electromagnetic flowmeters have special uses, being able to measure dirty liquids, slurries and pastes provided these have an electrical resistivity greater than 1 microhm/cm. Their accuracy is largely unaffected by changes in temperature, viscosity, density or conductivity. Ultrasonic flowmeters based on the Doppler effect and on the transmission of an ultrasonic pulse through the flowstream also have considerable promise, as do those based on the frequency of vortex shedding from a bluff body in the flowstream.

Mass flow meters are of two types: true mass flow meters which measure the mass flow directly, and inferential mass flow meters which measure both a combined function of the mass flow and the density and the density itself and compute the mass flow from the two measurements. Several rather sophisticated types of true mass flow meter have been developed, including ones based on angular momentum in the fluid and on the Coriolis effect.

Weirs depend on the flow of liquid through a shaped notch to a lower level. The shape of the notch may be be triangular, rectangular, trapezoidal, parabolic, etc. **Flumes** are used where there is not enough liquid head for a weir or where the stream carries much suspended matter.

Level measurement[25]

There are two circumstances with rather different requirements where levels are measured or sensed:

- Storage of liquids and particulate solids including the feed and product tanks of processes. Here accurate measurement without automatic level control is needed for accounting purposes.
- For the control of liquid levels (including liquid/liquid interfaces) in vessels of continuous processes. Here the inventory is often as low as possible, the difference between the maximum and minimum level is usually small and the flow of liquid through the vessel is high, giving a residence time of only a few minutes. The level, which in this case may not be critical, is measured or sensed in order to control it.

In both cases high- and low-level alarms and sometimes trips are often needed to prevent accidents. In the case of level instruments one can rarely speak of fail-safe or unsafe since any failure is often unsafe.

Methods of liquid level measurement include dipsticks and tapes, gauge glasses, float-actuated devices, displacer devices, head devices and electrical methods based on electrical conductivity and dielectric constant. Radioactive level gauges are used to some extent for measuring the depth of solids in silos. While these are adequately shielded in use, these may augment fire risks since nobody usually knows what has happened to the radioactive source if the silo is involved in a fire, and firefighting is thereby inhibited.

Some useful guidelines on the selection of level measuring devices are given by Sydenham[25]. The floats of float-operated devices must be protected against the forces of moving liquids. Backlash and leakage must be avoided with the glands of lever and shaft mechanisms. Head devices have the advantages of no moving parts in contact with the liquid and of giving the weight of liquid in a tank without correcting for its density. Specially designed head device systems are needed for boiling liquids.

For critical applications it is sound practice to use a device based on one principle for level measurement and/or control and one based on a different principle for safety alarms and/or trips.

For measurement of liquids in storage, the traditional **dipstick or tape** is still widely used, since it is simple, accurate and usually reliable. For light petroleum fractions there is here a risk of discharging static electricity and igniting a vapour/air mixture in the tank if appropriate precautions [6.3.6] are not taken. There may also be other hazards to the dipper such as toxic vapours, slipping on a tank roof on a windy night and even falling into the tank (as happened to a man whom the writer knew).

Although special **gauge glasses** are used for quite high pressures and temperatures, they should never be improvised and only complete assemblies from reputable suppliers should be used, within their design specifications for temperature, pressure and process liquid. All gauge glasses should have isolation valves or cocks and drain valves, and those used under pressure or for dangerous liquids should have safety devices which will cut off the flow if the glass breaks. It is good safety practice to avoid gauge glasses entirely on vessels containing flammable and other hazardous liquids.

Physical properties and chemical composition
It is impossible to discuss here the many different methods and automatic measuring instruments available. It should be stressed that for most control purposes, and especially for safety, simplicity, speed of response and reliability are more important than absolute accuracy. Measurement of a physical property is usually simpler than chemical analysis and for process streams consisting essentially of two components, the composition can be inferred from an appropriate physical property such as density, viscosity, refractive index or thermal conductivity.

Continuous sampling of process streams can present serious problems, e.g. in securing a representative sample, in reducing time lag, in removing interfering materials such as water from the sample and in disposing of toxic or flammable sample streams. According to Giles[26], 'Analytical instruments are out of commission more frequently due to troubles in the sampling system than to any other cause'. Whereas the instrument itself may have been developed over many years, the sampling system is often designed on a 'one-off' basis by someone who only understands part of the problem. A complete system contains the following five elements: the sample probe, the sample transport system, sample conditioning equipment, the analyser itself and sample disposal. The subject is treated in detail by Cornish[27].

Monitoring for safety
Continuous analytical control is sometimes vital to safety. An example is the continuous monitoring of the vinyl acetylene content in a butadiene-purification column [9.8.2]. Another is the monitoring of the oxygen content of the mixed feed gases to a direct hydrocarbon oxidation process [8.5.2]. Many process analysers whose main function is to monitor the purity or quality of the product also provide a warning when something goes dangerously wrong with the process.

Other continuous instruments detect fires and flammable and toxic [7.7] gases in the atmosphere. Several principles are used for automatic fire detection, i.e. detection of flame (by IR and UV methods), of heat and of smoke. The choice of method and location of detectors require case-by-case study, not only to ensure rapid response but also to avoid false alarms. Most flammable gas detectors are based on detecting the rise in temperature of an electrically heated capsule of a combustion catalyst (a 'pellistor') in contact with air containing the flammable gas. Designers of such detection equipment have to overcome two problems with conflicting requirements, (1) to ensure that the detector does not become a source of ignition and (2) to make certain that it responds fast enough for effective action to be taken.

Errors in measurement

Some measuring instruments are prone to errors which may develop gradually while the plant is operating and which cannot be corrected until it is shut down. These may be due, for example, to deposition of carbon or polymer on a thermocouple pocket or on the moving vanes of a turbine-flowmeter. Much ingenuity has been spent in finding ways of living with and allowing for this problem. Computer programs have been developed for reconciling the simultaneous outputs of all relevant plant instruments, e.g. by means of mass balances, showing which ones are most likely to be in error, and suggesting correction factors to be used. This 'solution' is most applicable on plants under PES control.

Instrument errors also result from condensation and deposition of solids in connections and short lines between plant and transmitting instruments, and in the instruments themselves, and in compressed air signal lines. To avoid the former, inert gas purges, diaphragms and steam or electrical tracing are sometimes needed, and to avoid the latter, precautions are required to ensure that the compressed air remains clean and dry.

15.8.5 Receivers

Controllers and their tuning were discussed in 15.8.3.

Marking of instrument scales

Normal (safe) and unsafe operating ranges should, where possible, be marked prominently on instrument scales. There is a distinct danger with modern recorders which simply have a linear scale with graduations of 0–100, of operators, supervisors and even plant managers thinking of these as mere numbers, without being clear what temperatures, pressures, etc. they represent, or knowing, for example, whether the figure for a pressure being controlled is in gauge or absolute units. Conversion factors with unambiguous units should be clearly marked on all receivers.

Ergonomic considerations[28]

Instruments on a control panel should be arranged logically for the operator and the most critical instruments should be well within his or her field of view. 'Mimic diagrams' (basically enlarged flowsheets [16.4.2]) on the control panel, with colour-coded lines and lights to show the condition of the process, can greatly assist operation. (Similar but more versatile and interchangeable diagrams are shown on VDU screens at the console control stations used with PES control.)

The accuracy of instrument reading should be appropriate to the need. Too fine subdivisions increase reading errors. These are less likely with digital displays than with pointers and dials, but the latter are better for rough checks and for assessing rate of change. The best arrangement for reading accuracy is a moving dial with a fixed pointer. The next best is a circular or semi-circular scale with a moving pointer. Horizontal and vertical scales are most prone to reading errors.

Instruments should react quickly enough to show expected movements and be sensitive enough to show the smallest meaningful change in the variable measured.

Where the readings of several dials have to be checked, e.g. the temperatures or flows of several parallel streams, all pointers should point in the same direction for the same plant condition. Identical instruments and scales should be used for identical duties on parallel processes.

Dials and indicators should be logically laid out and related to the control involved. If the dials and their controls cannot be placed side by side, they should be arranged in a pattern where they relate to each other.

Controls should be grouped according to their function and to the part of the plant which they affect. Those which have to be operated in sequence should be placed near each other and, if possible, in their order of operation, even though this leads to an asymmetrical layout. Different types of control knob should, if possible, be used for different functions.

The fact that most of an operator's bodily movements result from habit, and from subconscious rather than from conscious thought, must be appreciated by the designer. Most manual skills depend on the development of reflexes. Thus the arrangements of manually operated valves, switches, and gauges for similar equipment should follow similar patterns. Errors readily creep in when there is 'an odd one out'. Habits die hard, and most of us become aware of the extent to which we are creatures of habit when a small change in our domestic arrangements (such as the position of a light switch) interferes with a long-established routine.

Alarms and trips

All critical points of operation in a plant are normally protected by alarms and/or shutdown devices which are actuated by microswitches triggered by high or low pressures, flows, levels, temperatures, etc. An audible/visual alarm is actuated when the variable deviates from normal and reaches a certain figure, to allow the operator to take corrective action. If this is unsuccessful and the variable deviates further to reach another figure, a shutdown device may be actuated, which, by means of solenoid valves in the appropriate instrument air lines, shuts down one or more sections of the plant. The alarm consists of a horn or other audible device and lights which appear at labelled positions on an annunciator board mounted on the instrument panel.

The operator can usually stop the audible device, but cannot switch off the light completely, although he or she may be able to alter it in some way (e.g. change from flashing to continuous) so that it is not confused with a new warning signal from another point in the plant which appears on the board. All warning lights continue to show until the variables which set them off have returned to their normal range. More sophisticated alarm display and cancelling arrangements are available with PESs.

Trouble shooting and checking

When some malfunction, perhaps of an instrument, develops suddenly on a highly interactive plant, it generally actuates an alarm for a particular part of the plant. It is then often surprisingly difficult to pinpoint the source of the malfunction and distinguish between cause and effect. Misconceptions readily arise[14]. The plant has the same propensity to disguise its ailments as the human body, where even doctors can be misled, and such dictums have arisen as 'for pain in the knee, treat the spine'. Confusion is compounded by the simultaneous blowing of a pressure relief valve, with an ear-splitting roar. The near-disaster on the Three Mile Island atomic power plant in the USA was a classic example of this kind of difficulty in fault diagnosis[30].

On PES-controlled plant it is possible to provide software programs to help operators in rapid fault diagnosis. These programs are designed by experts who have carefully studied the various faults which might develop and how to diagnose and handle them[31]. Even here the operators generally need special training. For this special process simulators with similar VDU- and keyboard-equipped control stations are being increasingly used[32] [21.3.5].

To avert complete plant shutdowns because of minor malfunctions, the panel instruments should allow operators to take some holding action, such as putting the plant on total recycle or columns on total reflux, while the problem is being investigated.

Procedures must be established between operating and instrument department personnel for the calibration, zeroing and other necessary testing of instruments carried out while the plant is running. This generally involves temporarily putting each control instrument in turn on remote manual control and similarly de-activating alarms and trips. Clearly marked tags or notices should be displayed on any panel instruments which are out of service. Verbal and written warnings of this must be given to incoming shifts.

15.8.6 Final control devices[11, 15, 17]

The final control device consists of two parts, an actuator and a valve or other mechanism (such as a variable-output pump) which adjusts a flow or other manipulated variable. The actuator translates a signal, usually pneumatic, electric or hydraulic, into a force which operates the valve, etc. Pneumatic signals usually lie in the range 0.2 to 1.0 bar g. and are transmitted through 6 mm tube. With pneumatic actuators the air pressure acts on a diaphragm attached to a stem and is opposed by a spring. Two versions are available, one which opens and the other which closes the valve with increasing air pressure (Figure 15.16). Ideally, the actuator should respond quickly and assume a position proportional to the signal pressure. In practice the action may be sluggish and the position reached may deviate from linearity. The delay is caused by the time taken for sufficient air to travel through the transmission tube to pressurise and enlarge the space behind the diaphragm in order to move it. This delay increases with the volume of the diaphragm chamber and the length of the transmission line, and typically amounts to 5 seconds for a line length of

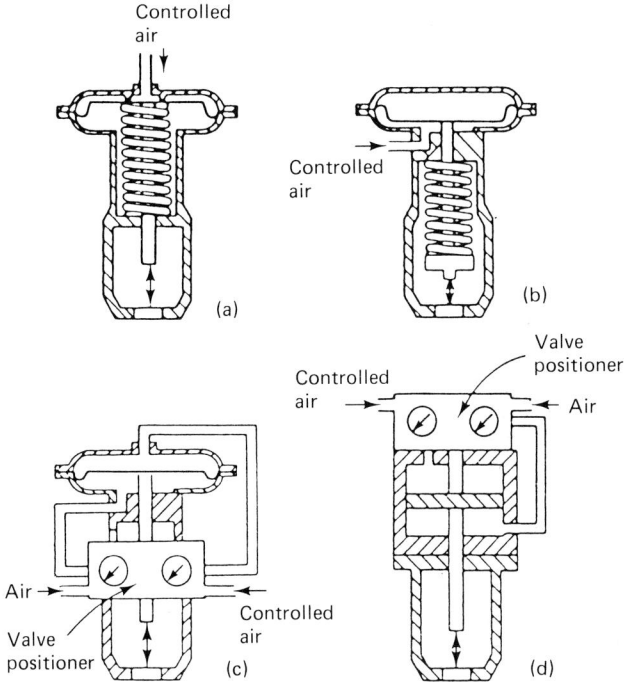

Figure 15.16 Pneumatic linear valve actuators: (a) simple actuator, air to close; (b) simple actuator, air to open; (c) actuator with positioner, air to close; (d) actuator with positioner, air to open (from Perry and Chilton[11], courtesy McGraw-Hill)

100 m and a terminal volume of 3 litres. Lack of linearity may be caused by the force of the process fluid on the valve plug and friction in the valve gland.

To overcome these problems valve positioners and signal boosters are often used. The pneumatic valve positioner uses a separate air supply to operate the valve and adjust its position to correspond exactly to the signal pressure.

Direct-current electrical signals are generally converted to variable air pressures by electro-pneumatic transducers mounted on the control device. A typical electro-pneumatic actuator is a combination of a current-to-pressure transducer, a feedback positioner and a pneumatic spring-diaphragm actuator.

Hydraulic actuators employ a fluid acting on one or both sides of a piston. In the first case the force of the fluid is opposed by a spring and the operation is similar to a pneumatic actuator.

Purely electrical actuators are also used in which an electric motor or solenoid provides the driving force. These can easily be used to open or close a valve completely, but those which adjust their position to correspond to an electrical signal are more complicated.

A wide choice of valves and valve plugs, some of which are illustrated in Figure 15.17, is available for operation by pneumatic and other types of actuator. These give a choice in the flow characteristics of the valve which are shown in Figure 15.18 as a plot of the percentage of maximum flow

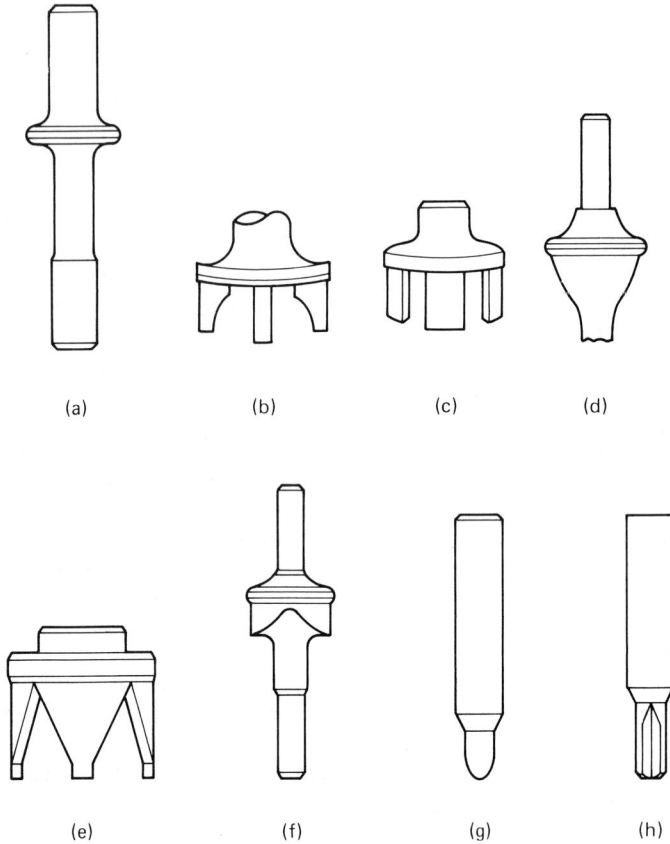

Figure 15.17 Shapes of typical valve plugs. (a) Top and bottom guided single-port quick-opening; (b) port-guided quick-opening; (c) rectangular (linear) port; (d) throttle plug (modified linear); (e) V-port (modified linear); (f) equal-percentage V-port; (g) miniature throttle plug (equal percentage); (h) miniature fluted plug (equal percentage) (from Perry and Chilton[11], courtesy McGraw-Hill)

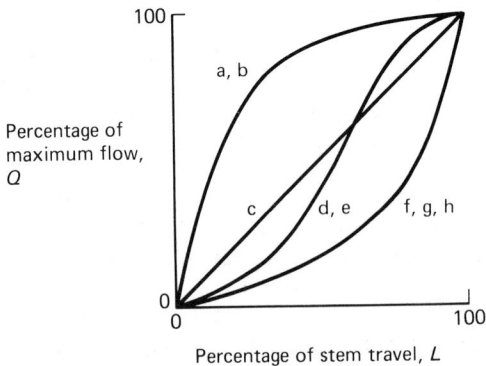

Figure 15.18 Flow characteristics of valve plugs shown in Figure 15.16 (from Perry and Chilton[11], courtesy McGraw-Hill)

against the percentage of stem travel. Some control valves do not give a positive shut-off of the process fluid unless this is specified, when they can usually be provided with soft seats. For good control the size of a control valve is usually considerably smaller than the pipe where it is installed. Valve manufacturers will advise on the best type and size of control valve for different applications.

Actuators can also be used to control the speed of variable-speed motors which are used, for instance, with solids-metering valves and for variable-output pumps used for slurries and metering small flows (e.g. chemical additives and catalysts). Most of these are either constant-volume pumps operated by variable-speed drives or variable-volume pumps operated by constant-speed motors.

Formerly a hand-operated by-pass valve was usually installed in parallel with an an automatic control valve for use while the controller was being adjusted or the control valve was being serviced. This requires a local indicator of the controlled variable within view of the human by-pass operator. A manual by-pass valve might also be installed to assist in start-up or to reduce the sensitivity of the control valve.

A by-pass valve can, however, cause hazards, particularly if the control valve 'fails-safe' in the closed position. Such by-pass valves can generally avoided, e.g. by a better choice of control valve. The installation of a manual by-pass valve round any control valve thus requires special justification, e.g. by a simple hazard analysis [16.5.5].

Isolation (block) valves on one or both sides of an automatic control valve are more often needed, particularly where the control valve does not give a positive shut-off. For very critical duties the duplication of the entire control loop, valve included, with block valves on either side of each control valve, may be justified on grounds of reliability [14.2].

15.8.7 Signal-transmission systems

Most telemetering systems used in process control involve fixed electrical or pneumatic conductors. Apart from thermocouples whose millivolt output is transmitted directly by wires to the receiver, most instrument outputs are transformed into electrical or pneumatic signals, which in the case of receiver instruments, are reconverted to a numerical value representative of the measured variable.

Analogue signals are used exclusively with pneumatic and often with electrical transmission, while digital signals are also used with electrical transmission. Analogue electrical signals which are now usually of d.c. current type with the range 4–20 mA are carried by wires, as are voltage signals.

Digital signal transmission within a PES can be by frequency change or pulses, and sent by coaxial cable, optical fibres or short-wave radio. Complex information can be sent rapidly in these ways. Binary codes are generally used, with electronic coders and decoders.

Electrical 'noise', that is, unwanted signal, can be a serious problem with electrical signal transmission, particularly with digital signals. A false or distorted signal in a PES-controlled plant can result in the wrong opening or closing of valves, with resultant hazards. The main sources of electrical noise are electrostatic and electromagnetic fields, instruments and sensors, and grounding problems. The subject lies outside the scope of this book and requires expert attention during plant and system design.

15.8.8 PES control

PESs may be used as an aid to the operator, e.g. by logging data and performing calculations, or for automatic process control. Here there are two main alternatives, both supervised by the operator:

1. Using the PES to adjust the set points of conventional controllers;
2. By direct PE control, where all controllers are part of a PES.

With PESs, the panel indicators, recorders and controllers used with conventional instrumentation are replaced by VDUs and keyboards at the control station. Instead of scanning a row of instruments on a panel the operator sees a more logical representation of present and past values of process variables on a VDU. By touching selected spots on the screen he or she can call up other 'windows' onto the screen which provide more detailed and appropriate information which is needed about the state of the process. Instead of tweaking a knob on a conventional controller to change its set point, the operator types abbreviated commands on the keyboard, with the assistance of 'help' menus on the VDU screen in case he has forgotten the procedure laid down in the operating manual [20.2.2]. At a simple command the PES will alter the set point for him gradually over a period of an hour or so, saving him the tedium of altering the set point a little at a time every few minutes. The PES will also will arrange the 'bumpless' introduction of cascade or manual control. But if something goes wrong with a screen and his 'window' onto the plant is lost, he may be in for serious trouble. This is when he is in most need of traditional operating skills.

The advantages claimed for PES control are mainly economic and apply particularly to large-scale process plant and complexes.

A hierarchy of control systems exist when PESs are used, those at a lower level receiving instructions from the one above. Systems at the lowest level control unit operations involving individual plant items such as heat exchangers and dryers. Next comes the control of individual process units each containing several plant items whose operations must be coordinated to meet specific objectives such as product purity, yields on raw materials and production rate. The third level, plant control, involves the coordination of all process units in the plant to achieve the required performance at minimum cost. The PES will adjust the control parameters of other units to cater for upsets in the operation of any particular one. On a fourth level is the control at a single location of several interdependent plants (such as a petrochemical complex) to optimise the yields and qualities of several products within the constraints imposed by the plants, processes, raw materials and utilities available. Above these levels come department and company control. Although computers are used here to assist decision making, such decisions fall outside automatic process control and are made by humans.

One problem with PES control is that the variables controlled at higher levels include chemical compositions of process streams as measured by on-line analytical instruments which have higher failure rates than most other types of instrument [14.2.1]. In normal operation the analytical controller may reset the set points of other controllers in cascade fashion, but if it develops a fault the operator must take the analytical controller 'off-line' and reset the set points of the slave controllers manually in the light of the results of laboratory analyses.

PES control improves the performance of unit operations hitherto controlled by conventional instrumentation in cases where uncontrollable disturbances in the process occur without compensation, and where the variables controlled are of only indirect interest. With PES control systems it is possible to calculate, specify and control performance variables which previously could be neither measured nor controlled. A further advantage with direct PES control is that it can automatically adjust the tuning of individual controllers [15.8.3]. The PES generally allows the number of uncontrolled variables or degrees of freedom to be considerably reduced.

PESs used for control are now almost exclusively of digital rather than analogue type. They are used for on-line control of both batch and continuous processes, for process simulation, as a diagnostic aid to human operation, for logging, and monitoring operating variables and for actuating special protective features to safeguard the plant in particular emergencies. They range from single microprocessors to mainframe computers. Although they can greatly increase control efficiency and safety, they pose additional and different safety problems compared with non-electronic control and protection equipment. By distributing programmable electronic devices and personal control stations and incorporating a degree of redundancy among them, the consequences of any single failure can be reduced and fault location can be made easier. A high standard of safety and reliability, which should be at least as high as for non-PES systems, is needed when equipment is controlled or protected by a PES.

HSE has published guidelines on PES in safety-related applications[33] in two parts, a general introduction for the non-specialist and technical guidelines which are mainly intended for specialists. The Engineering Equipment Material Users Association (EEMUA) has also published 'companion guidelines'[34].

15.9 Features of PES systems used for control and safety[33]

The first feature which strikes the novice is the strange, new vocabulary used by the initiated. Most readers will be familiar with the terms 'hardware' and 'software'. Hardware refers to all the physical components of the PES system, the chips, bits of wire, keyboards, VDU screens, printers, etc. Software refers to the instructions which the system follows. Some software is incorporated into the PES, while other software consists of programs written in electronic language on hard and 'floppy' discs, magnetic tape, etc. The software can be introduced into the 'programmable electronics' (PE) of the machine which is, in effect, its brain. The PE communicates with the installation which is being controlled, and with its

Input units   Output units

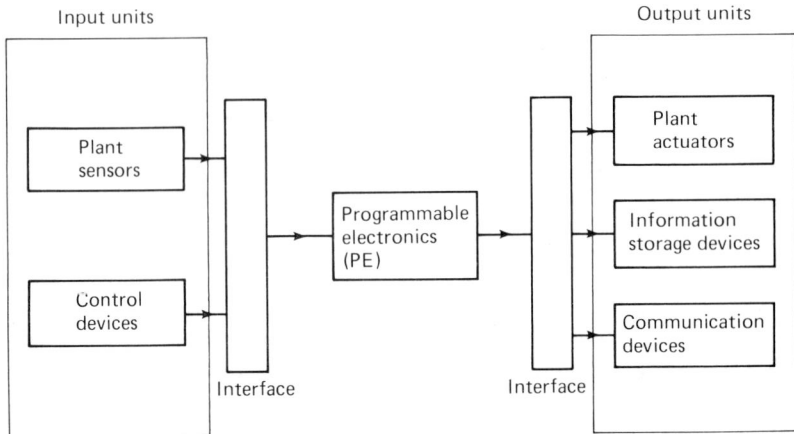

Figure 15.19 Structure of a programmable electronic system (PES) (courtesy HSE)

human operators, through input and output units (Figure 15.19). Input units include plant sensors, e.g. conventional temperature- and pressure-measuring instruments and transmitters which send coded messages to the PE unit, and control devices which enable operators to give instructions to it in the same way. The PE analyses the data thus received and transmits instructions and information to the output units. These include plant actuators (e.g. pneumatic control valves), information-storage devices and communication devices (e.g. VDU screens, printers). The input and output units are linked to the PE by electronic interfaces. Information is stored in two types of electronic memory, Read-Only Memories (ROMs) which are mostly built into the machine and Random Access Memories (RAMs) which are used for the temporary storage of information which can be read and used later by the machine. PESs may control process plant continuously through sensors and actuators, or act only in emergencies as revealed by similar sensors, to protect the plant by opening and shutting valves, stopping motors, etc. Possible failures of such PESs are of two types, *random hardware failures* and *systematic failures*.

15.9.1 Random hardware failures

These are caused by the wear-out of one of the large numbers of components in the PES which are assumed to be unpredictable and to occur at random [14.1.2]. Redundant back-up components or systems which continue working if one breaks down and control of the physical environment in which the PES operates are used to increase reliability. The PES should then have the means of identifying the failed part so that it can be replaced or repaired promptly to restore the reliability of the system protected by a redundant item.

15.9.2 Systematic failures

These happen every time a particular set of conditions occurs and are due to mistakes in the specification, design, construction or operation of a system. Such failures may lie dormant until the particular circumstances arise and then lead to accidents. Three types of error which can lead to systematic failures are:

1. *Specification errors* made when the system was planned;
2. *Equipment errors* in design, construction, installation, etc.;
3. *Software errors* left in the original program or introduced into it during modification.

To ensure that the safety of the plant or equipment controlled is adequate, five steps should be taken:

1. A hazard analysis to identify the hazards and how they could arise [14.3 and 16.5.5];
2. Identification of the systems on which the safety of the installation depends;
3. Determination of the required safety level;
4. Design of the systems identified in (2);
5. Analysis (e.g. by a check questionnaire) to ensure that the installation meets the safety requirements.

There are three fundamental aspects of an installation which require expert examination and may need improvement: *reliability* [14.2], *configuration* and *overall quality*.

Configuration means the way in which the hardware and software components of a PES are organised and the ways in which PESs and non-programmable equipment are linked to make up a complete installation. Overall quality depends both on the quality of manufacture (e.g. whether the PE was manufactured using an established quality-assurance system) and on the competence of the designers.

Control and protection systems should be completely separated, with their own sensors and actuators (Figure 15.20). Although some control systems include protective features, these may fail if the control system fails.

PESs must be protected from accidental damage (by position), from environmental hazards (temperature, humidity, dust and corrosive vapours), from fire, electrical interference and electrostatic effects.

15.9.3 Concerns about the use of PESs for process control and protection

Many of HSE's concerns about the application of PESs to control and protect process plant were voiced by Dr Jones[35]. Failures can occur in both the hardware and the software.

Large companies usually have specialists who can match the process parameters to the control and/or protective system. Medium- and smaller-sized companies who lack such specialists tend to buy control and protective packages 'off the shelf'. Such companies need to call in specialist help to match the process chemistry to the electronics.

Sound systems for the validation of software are not yet generally available. Even in high-quality software, errors can exist which may lie dormant until triggered by an unusual combination of process parameters.

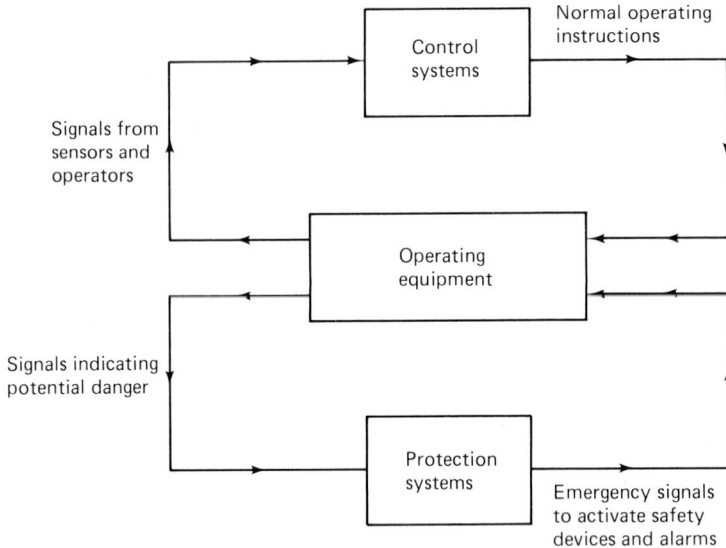

Figure 15.20 Separation of control and protection systems (courtesy HSE)

The sensing elements and actuators (valves, etc.) on the plant form an essential link in any system. Their state and that of instrument housing, wiring runs and installation workmanship are often below standard, particularly on plants where the atmosphere is dusty or contains acid fumes.

P&IDs do not always correspond to what actually exists on the plant [16.5.6]. Sometimes vital protective items have not been fitted and sometimes they have been removed or by-passed during plant maintenance or modification. Plant managers therefore need to check process instrumentation thoroughly and ensure that it corresponds to design drawings during plant audits.

The protection of the plant control system prompts four questions:

- Can the equipment be easily damaged by accident or by tampering?
- Are the instruments and controls appropriate to the atmospheric environment?
- Could electrical interference or electrostatic discharges affect the operation of PESs?
- Could the safety back-up system(s) be knocked out by a fire or flood?

The security of PESs is also critical and access to software should be restricted to key personnel. (Visits by school and college parties can spell danger to unprotected software.)

Appropriate staff training is needed when PESs are introduced into a process plant to ensure that they are used and maintained safely.

Exact procedures for operating PES-controlled plant need to be spelt out (e.g. in the plant operating manual [20.2.2]) and must be followed religiously. Critical points in these procedures can be shown as 'menus' on one of the VDU screens.

The importance of electrical and process isolation before maintenance must be stressed [18.6], and a check should be made after maintenance or modifications to ensure that the software has not been corrupted.

HSE are also concerned about the use of 'expert systems' [19.4] for on-line process control, although agreeing that they may be a valuable diagnostic aid in tracing the causes of faults arising during the operation of complex plants. There is still truth in the statement that 'the best expert system for process control is the well-trained, experienced and dedicated plant operator'.

15.10 Hazards of instrument maintenance and modifications

Instruments, particularly control valves, require scheduled preventative maintenance and lubrication in the same way as mechanical and process equipment [17.1], although prompt repair maintenance is also essential to deal with faults in operation.

The ranges of transmitters and receiving instruments are mostly readily changed. If this is done without operating staff being aware of the change and its implications, the results can lead to serious accidents. An operator may think he had a certain flow through a line or level in a tank, whereas it was double that figure. Such modifications to instruments, despite their apparent minor nature and the ease with which they can be made, require proper scrutiny and authorisation before they are carried out, positive notification of all concerned when they are undertaken, and old conversion factors replaced by new ones. Similar precautions are needed when instrument loops themselves are modified, which should only be done after thorough discussion with operating and safety personnel.

References

1. Dow Chemical Company, *Fire and Explosion – Safety and Loss Prevention Guide – Hazard Classification and Protection*, 3rd edn, Dow Chemical Company, Midland, Mich. (1972)
2. ICI Mond Division, *The Mond Index*, 2nd edn, Imperial Chemical Industries plc, Explosion Hazards Section, Technical Department, Winnington, Northwich, Cheshire (1985)
3. API, RP 520 *Recommended practice for the design and installation of pressure-relieving systems in refineries*
   Pt 1, Design, 4th edn, (1976),
   Pt 2, Installation, 2nd edn, (1973), The American Petroleum Institute, New York
4. API, *RP.521 Guide for pressure-relieving and depressurising systems*, 2nd edn, *ibid.* (1982)
5. BS 1123: 1972, *Specification for safety valves, gauges and other safety fittings for air receivers and compressed air installations*
6. The Factories Act 1961, Sections 32 to 36, HMSO, London
7. BS 2915: 1984, *Specification for bursting discs and bursting disc devices*
8. King, R. W., 'Latent superheat – a hazard of two phase liquid systems' in *Chemical Process Hazards*, vol. 6, The Institution of Chemical Engineers, Rugby, (1977)
9. AMOCO, Booklet No.1, *Hazard of water*, 5th edn, AMOCO, Chicago (1971)

10. Fitt, J. S., 'The process engineering of pressure relief and blowdown systems' in *Loss Prevention and Safety Promotion in the Process Industries*, edited by Buschmann, C. H., Elsevier, Amsterdam (1974)
11. Perry, R. H. and Chilton, C. H., *Chemical Engineers' Handbook*, 5th edn, McGraw-Hill, New York (1973)
12. Lees, F. P., *Loss Prevention in the Process Industries*, Butterworths, London (1980)
13. Wells, G. L., *Safety in Process Design*, George Godwin, London (1980)
14. BCISC, *Codes of Practice for Chemicals with Major Hazards: Chlorine*, British Chemical Industry Safety Council, London (1975)
15. Anderson, N. A., *Instrumentation for Process Measurement and Control*, 3rd edn, Chilton, London (1980)
16. Noltingk, B. E., *Instrumentation Reference Book*, Butterworths, London (1988)
17. Foxboro, *Introduction to Process Control,* The Foxboro Company, Foxboro, Mass., USA (frequently revised)
18. ISA Standard S5.1, 'Instrument symbols and identification', The Instrument Society of America (1985)
19. Richmond, D., 'Instrumentation for Safe Operation', in *Safety and Accident Prevention in Chemical Operations*, edited by Fawcett, H.H. and Wood, W.S., 2nd edn, Wiley-Interscience, New York (1982)
20. Forbes, D. J., 'Design of blast-resistant buildings in petroleum and chemical plants', *ibid.*
21. UKAEA, 'Power Fluidics – Control without movement', *Loss Prevention Bulletin 040*, The Institution of Chemical Engineers, Rugby (1981)
22. Hagart-Alexander, G., 'Temperature measurement' in ref. 16
23. Higham, E. H., 'Measurement of pressure', *ibid.*
24. Fowler, G., 'Measurement of flow', *ibid.*
25. Sydenham, P. H., 'Measurement of level and volume', *ibid.*
26. Giles, J. G., 'Sampling', *ibid.*
27. Cornish, D. C. *et al. Sampling Systems for Process Analysers*, Butterworths, London (1981)
28. Grandjean, E., *Fitting the Task to the Man: an ergonomic approach*, 2nd edn, Taylor and Francis, London (1980)
29. Turner, B. A., *Man-made Disasters*, Taylor and Francis, London (1978)
30. Marshall, V. C., 'What happened at Harrisburgh?', The Chemical Engineer, *346*, 479 (1979)
31. Forsyth, R. (ed.), *Expert systems*, Chapman and Hall, London (1984)
32. Pathe, D. G., 'Simulator a key to successful plant start-up', *Oil and Gas Journal*, 7 April (1986)
33. HSE, *PES – Programmable electronic systems in safety related applications: 1 An introductory guide, 2 General technical guidelines, HMSO, London (1987)*
34. EEMUA *Guidance notes on programmable electronic systems*, Engineering Equipment Material Users Association, 14 Belgrave Square, London SW1 (1988)
35. Jones, P. G., 'Some areas of HSE concern about safety of computer control systems', in symposium *The Safety and Reliability of Computerised Process Control Systems*, NW Branch of the Institution of Chemical Engineers, 24 March (1988)

Chapter 16

Designing for safety

The importance of sound design for the safety, as well as the reliability and performance, of process plant can hardly be overstated. One book sponsored by the Institution of Chemical Engineers and published in the UK is devoted entirely to design safety[1]. Several chapters of the books by Lees[2] and Fawcett and Wood[3] deal with design safety, while several I. Chem. E. symposia have been devoted to it. Here I concentrate on those aspects with which I am most familiar from personal experience.

Hazards which have to be considered during design are discussed in most chapters of this book. Their combined effects are reflected in the process hazard indices discussed in Chapter 12. They also have important consequences for plant layout, engineering standards and protective features. Reliability and the methods of risk analysis used to assist planning and design were mentioned in Chapter 14. Automatic protective devices and control instrumentation were discussed in Chapter 15. A balanced appreciation of the interrelationship between hazards of different types is vital to a safe design.

Design shortcomings which have contributed to major accidents and aggravated their consequences include:

- Poor siting;
- Poor layout and inadequate separation of hazardous units, especially from inhabited buildings and busy areas;
- The use of incorrect data, and errors in process engineering calculations;
- Flow diagrams which allow reactive materials to mix and react violently in the wrong places;
- Unsuitable instruments and illogical instrumentation;
- Poorly engineered protective systems, particularly for pressure relief;
- Neglect of the potential for runaway reactions;
- Overlooking mechanical stresses, particularly those to which pressure vessels and pipework may be subjected;
- Unsuitable choice of materials of construction for some process conditions;
- Valves wrongly chosen and/or located;
- Non-existent or inadequate drainage, containment and disposal arrangements for leaks of process materials;
- Control rooms for high-risk plants unsuitably located and constructed;

- Inadequate fire protection, e.g. of structures, pipework, process plant and storage areas;
- Failure to consider the future maintenance of the plant, and how this is to be achieved safely. A questionnaire for designers on this point is given in Appendix J.

Many hazards inherent in a design result from a failure to consider them and budget for their control in an earlier planning stage. Unless a realistic sum is budgeted for safety before design starts, the organisation's commitment to it is suspect. Although Kletz has written a booklet with the enticing title *Cheaper, Safer Plants*[4], he would be the last person to advocate economising in design at the expense of safety. Kletz's slogan applies to innovations which allow plants to be designed with lower inventories of hazardous materials. Its scope for application seems to be rather limited and generally a 'once only' benefit. When the fruits of any technical innovation are reaped, we often find that a price has to be paid for health and safety. The export of hazardous industries to Third World countries with low health and safety standards [23.3] provides ample evidence of this.

One obstinate problem in achieving a safe design lies in recognising all the hazards. Hazard studies and checklists are used to try to overcome this [14.3]. A hazard study (The Safety Case) is legally required for every plant falling within the scope of the CIMAH Regulations [2.4.2]. An integrated programme of six separate studies at different stages of the project was described by Gibson for new continuous chemical plants in ICI's Mond Division[5]. The phasing of these studies which should be integrated with the appropriate project activities is shown in Figure 16.1.

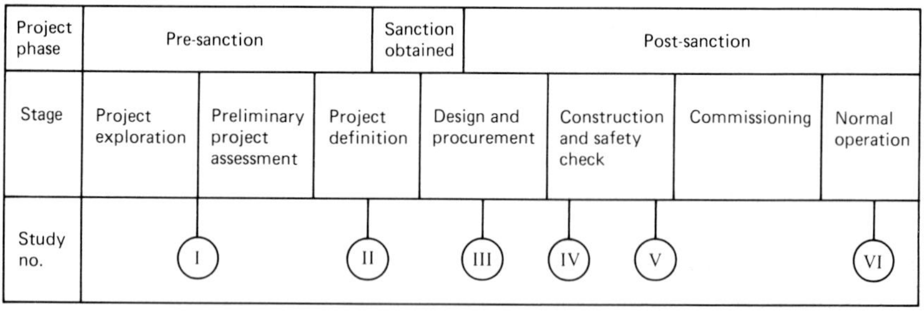

Project phase	Pre-sanction		Sanction obtained	Post-sanction			
Stage	Project exploration	Preliminary project assessment	Project definition	Design and procurement	Construction and safety check	Commissioning	Normal operation
Study no.	I		II	III	IV	V	VI

Figure 16.1 Phasing of project hazard studies (from S. B. Gibson, 'The design of new chemical plants hazard analysis', in IChemE Symposium Series No. 47, courtesy Institution of Chemical Engineers)

16.1 Checklists

Comprehensive checklists contain points which should be examined at all stages in a project from its first conception. Such lists have been published by Balemans and others[6], and by Wells[1], and a bibliography of checklists is given by Lees[2]. The principal headings of Balemans's comprehensive 25-page checklist (1974) are given in Table 16.1. The main headings of

Table 16.1 Main headings of Balemans's checklist

No.   Main heading and first sub-heading

---

1   *Choice, situation and layout of site*
      1.1 Choice and situation
      1.2 Site layout

2   *Process materials*
      2.1 Physical properties
      2.2 Chemical properties
      2.3 Toxicological properties

3   *Reactions, process conditions and disturbance analysis*
      3.1 Reactions
      3.2 Process conditions
      3.3 Disturbance analysis
      3.4 Causes of abnormal conditions
      3.5 Abnormal conditions
      3.6 Critical situations

4   *Equipment*
      4.1 Introduction
      4.2 Design
      4.3 Choice of material
      4.4 Construction
      4.5 Location of equipment
      4.6 Special provisions

5   *The storage and handling of dangerous substances*
      5.1 The storage of dangerous substances
      5.2 The handling of dangerous substances

6   *Handling and removal of hazardous waste products*
      6.1 Introduction
      6.2 Aspects of disposal
      6.3 Reduction of disposal

7   *Civil engineering aspects*
      7.1 The ground
      7.2 Foundations
      7.3 Drainage systems
      7.4 Roads
      7.5 Buildings (see also section 9)
      7.6 Additional points related to installations

8   *Division of site into areas* (for hazards of igniting flammable vapours, etc.)

9   *Fire protection*
      9.1 Introduction
      9.2 Fire protection of buildings and plants
      9.3 Fire-fighting organisation
      9.4 Fire detection and alarm
      9.5 Classification of fires according to European Standard EN2 of 1973

10 *General emergency planning*
      10.1 Introduction
      10.2 Operational emergency situations
      10.3 Escape of liquids and gases
      10.4 Fire and explosion
      10.5 Personal protection
      10.6 Training
      10.7 Communication systems
      10.8 Briefing and information services

Table 16.2 Main headings of Wells's checklist

A	Basic process considerations
C	Some overall considerations
D	Operating limits
E	Modes of plant start-up, shutdown, construction, inspection and maintenance, trigger events and deviations of system
F	Hazardous conditions
G	Ways of changing hazardous events or the frequency of their occurrence
H	Corrective and contingency action
I	Controls, safeguards and analysis
J	Fire, layout and further precautions
K	Documentation and responsibilities

Wells's 18-page checklist which are given in Table 16.2 follow a different pattern which is based on the structure of his book.

Where process hazards are above average and where time permits, it is recommended that Balemans's checklist (whose structure makes it easier to use) be applied first, followed by that of Wells in case it reveals hazards missed in the first check. The period during which such checklists can be applied is usually short and sandwiched between the time a provisional decision has been taken to build and when it is endorsed and a budget approved. The discovery of a serious hazard after design decisions have been taken and money has been committed to them can be traumatic.

16.2 Pre-sanction planning and preliminary hazard studies

Various other studies usually precede budget approval and the detailed design of any major process plant. They include the market for the product, and pre-feasibility and feasibility studies for its production[7]. It is important to consider possible inherent hazards carefully during these studies to ensure that their implications are reflected in the project's budget and that detailed design starts on a safe basis. The first two of Gibson's recommended hazard studies[5] are usually done at the pre-sanction stage.

16.2.1 Pre-feasibility studies and Hazard Study I

Pre-feasibility studies are carried out to examine the viability of different raw materials, process routes, sites, sources of technology and capital, and plant capacity. Preliminary process designs [16.4] are needed in these studies so that capital and manufacturing costs can be estimated. Of the process alternatives to be selected, one may already be in use by the organisation, another may be on offer from a licensor or contractor and a third may be a new one which has just been developed.

Hazard Study I should be made at the same time as the pre-feasibility study and its findings should be included in it. Its main purpose is to discover the potential hazards of the raw, intermediate and final materials in the process. This may be done by comparing their properties with checklists, and querying any uncommon materials whose hazards may not have been thoroughly investigated. Wells[1] gives an example of how easily a hazardous material may be selected for a process in place of a safer alternative, simply because the designer has more data on the hazardous one. Thus chemical engineering students engaged in a design exercise on hydrogen purification chose an arsenical liquid to remove carbon dioxide rather than a less toxic one on which fewer data were available.

A 'process package' including requirements of materials, utilities and labour and an equipment list and cost is usually available for each process route considered. Due to time constraints, other items which contribute to the plant cost (including design, erection, civils, instrumentation, piping and fire protection) tend to be estimated at the pre-feasibility stage as percentages (based on past records) of the equipment cost.

A pre-feasibility study should state which of the various possible combinations of processes, raw materials and other factors satisfy the economic and safety criteria adopted, and it usually recommends one of them. Since the cost of meeting safety requirements may tip the scales in favour of one or other alternative, it is important that an evaluation of this cost be included in the terms of reference of the pre-feasibility study. In the absence of more detailed information, cost factors for special safety features may be estimated from the Dow F&E index of the process [12.1]. Two vital questions for plants with significant inventories of hazardous materials are whether these would bring the installation within the scope of the NIHHS or CIMAH regulations, and whether these design inventories could be safely brought below these limits [14.7]. In considering sites in the UK, it is necessary to ensure that the possible impact of accidents on neighbouring communities is sufficiently low to meet the criteria of the Major Hazards Assessment Unit (MHAU) of HSE.

As an example of the effect of material hazards on the choice of process route, a company contemplating the manufacture of chloroprene (the monomer of the synthetic rubber Neoprene) could base it on either acetylene or butadiene. Although both routes have inherent hazards, the acetylene route involves the production of vinyl acetylene, an explosive compound which has destroyed at least one chloroprene plant[8] [9.8.2]. The butadiene route might therefore be chosen even if it was the more expensive of the two.

16.2.2 Feasibility study and Hazard Study II

Before final project approval is given a full-blown feasibility study for the process, site and raw materials selected in the pre-feasibility study[7] is usually undertaken. Typically it requires several months' work by a team which comprises an economist, a process engineer and draughtsmen, with part-time inputs from a cost estimator and several other engineers and specialists. This study considers a range of issues including soil, climate, transport, fuel, power, water supply, by-products, effluents, manpower, finance, housing and social infrastructure, on all of which data must be acquired. Process flow diagrams and material balances are prepared at this stage and many decisions which will affect the design and safety of the project are taken. The purpose of the study is to provide closer estimates of capital and manufacturing costs, forecast the financial viability of the project, and establish a framework of tasks and specifications required for design, construction and subsequent manufacture.

Hazard Study II is recommended at this stage. This is a fault tree type analysis as discussed in 14.4, and is referred to as HAZAN by Kletz, who gives detailed guidance on its application.to process plant[9]. In it the sections which make up the plant are examined in turn. By taking particular hazards such as release of toxic gas as the top event, the fault tree is continued downward until every branch ends in some quantifiable primary event. The conclusions of Hazard Study II should be included in the feasibility study.

Once the study has been completed and a project budget has been prepared and approved, negotiations are concluded with all parties involved in the project, and design proper begins.

Not all projects start or evolve in this orderly way. In 16.5.6 we discuss what can happen when no proper planning or design is done at all, and the plant can be said to have just 'growed like Topsy'.

16.3 Design organisation and parties involved

A safe design depends on a well-managed and balanced design team whose members understand the process in question and its hazards, as well as being qualified and experienced in their own professional fields. Furthermore, they must be provided with sufficient motivation and time to identify all serious potential hazards, and with adequate financial resources to eliminate them, or provide adequate protection against their possible consequences.

Effective communication between members of the team is vital. Regular and well-prepared meetings are needed to discuss and resolve problems before they become inbuilt hazards of the design. A technical problem which is not solved in the design stage may become a safety problem later, when the cost of solving it is much higher.

The design of a process plant is a highly creative activity. It has to proceed in a series of logical steps, each generating information needed for the following ones. Appropriate hazard studies and safety checks should, in the first place, be made by members of the design team for those aspects of the design for which they are responsible, as the work proceeds. There should be a safety specialist to whom they can turn for advice, and they should have ready access to data banks. Such studies and checks should be essentially practical. They should not be allowed to dominate design activities, like the novice gardener who digs up his potatoes each week to see how they are growing.

16.3.1 Organisational complexity in design

Several organisations may be involved in the design of a new project, thus preventing the work from being done 'all under one roof'. Some of it may be carried abroad using different languages and codes.

The organisations involved may include:

A *The transnational parent company* – the principal owner of,
B *The operating subsidiary company*,
C *The process licensor* (if different from A),
D *Contractors* for design, procurement, construction and commissioning, etc.,
E *Other owners of B* (e.g. banks, other companies),
F *Consulting engineers*, used mainly where the owner(s) lacks particular experience and expertise.

Where the plant is owned by a transnational company with experience in the particular manufacture, this company usually negotiates and places contracts with contractors, and makes deals with licensors. The controlling transnational company usually also supplies and/or trains the top management of the operating company, and provides most of the technical know-how and back-up in case of difficulties.

The operating company, which is owned by A and E, provides the operating, maintenance and local sales staff and may place and/or control some or all of the construction contracts.

The licensor who has developed the process in question usually owns a plant where it is used. He often sells a 'process package' which includes the process design, patent rights, technical assistance and certain guarantees.

Contractors undertake all aspects of design including process, mechanical, civil, electrical, instrumentation and fire services. They also procure equipment and deliver to site, construct, erect and commission. Usually there is one main contractor to whom other contractors are responsible. Contractors range from large transnational companies to small local ones for painting and insulation.

Other main owners and sources of finance include the World Bank and various regional development banks, government departments and nationalised industries, insurance companies and cooperatives. Thus a farmers' cooperative with no manufacturing experience might decide to make its own fertilisers, and engage a contractor to design, build and even operate the plant for it.

Consulting engineers are employed by the owners to fill gaps in their expertise or staff capacities, and in areas where they have been traditionally used, e.g. civil work and power plant. They carry out pre-feasibility and feasibility studies, risk analysis and environmental impact studies and negotiate with licensors and contractors on behalf of the owner. Much of the work of implementing a project may be done either by contractors or by consulting engineers. The main difference between them lies in the terms of the contract. Consulting engineers tend to work on a time-plus-expenses basis, whereas contractors often have to tender and work on a lump sum basis, thus standing to make profits or losses on their contracts.

With such a mixture of organisations involved, careful delineation of responsibilities at the times contracts are signed is essential. Each part of the design has to be frozen at an early stage, even when it is later discovered that the overall result is far from optimal. The abilities and performances of the various participants in the design can be very uneven.

Sometimes one of them is painfully aware of the weaknesses of another upon whom the safety and performance of his own work will depend, and yet be powerless to intervene. The more fragmented are the responsibilities for various aspects of design, the greater become the problems of communication and control. This easily leads to organisational misconceptions as discussed in 3.3, thereby introducing hazards into the design which are not discovered until the plant becomes operational.

16.3.2 The role of the main contractor in safe design

Two things most influence a client when choosing a contractor – the price tendered and the contractor's track record, including the safety of his designs. The fear of losing his reputation is thus a powerful incentive to a contractor to produce a safe design. With fixed price or negotiated bid contracts the contractor is under pressure to keep design and equipment costs as low as possible. Fitt, then a contractor's senior manager, wrote in 1976[10]:

> It would be much more satisfactory if costs relating to safety features always received special treatment contractually, in order to remove all incentive to trim expenditure on them in order to produce a superficially 'keener' proposal.

In the same paper he noted that:

> An increasing number of clients are demanding that a summary of the safety problems of a process and the means proposed for dealing with them be included in the bid document.

(Contractors would probably like to make similar demands of their clients, who are often in a better position to appreciate these problems!)

Anderson, the safety adviser to another large contractor, produced extensive evidence in support of the same point in an article published in 1980, 'The effect of safety costs on the competitiveness of tenders'[11].

Safety problems with contracts in the petroleum industry are discussed by HSC's Oil Industry Advisory Committee, in a 'Health and Safety Guideline', although its emphasis is on the contractor's work on site and not on design.

Figure 16.2, reproduced from Fitt's paper (on which much of this section is based), shows the responsibility for safe design in a contractor's office. The ultimate responsibility usually rests with the project manager as the authority responsible for job specifications. For process safety he is mainly dependent on the senior process engineer (process supervisor) and specialists on various topics (e.g. corrosion, instruments). These specialists (lead engineers) provide guidance to more junior process and other engineers and draughtsmen.

The basis for a safe design must be laid down in the process engineering department, beginning with flowsheets. This generally applies even when the contractor's design is based on a process package (supplied by the client or a licensor) which includes process flow diagrams, P&I diagrams and material and energy balances. These generally have to be developed further by the contractor to meet particular local conditions or other

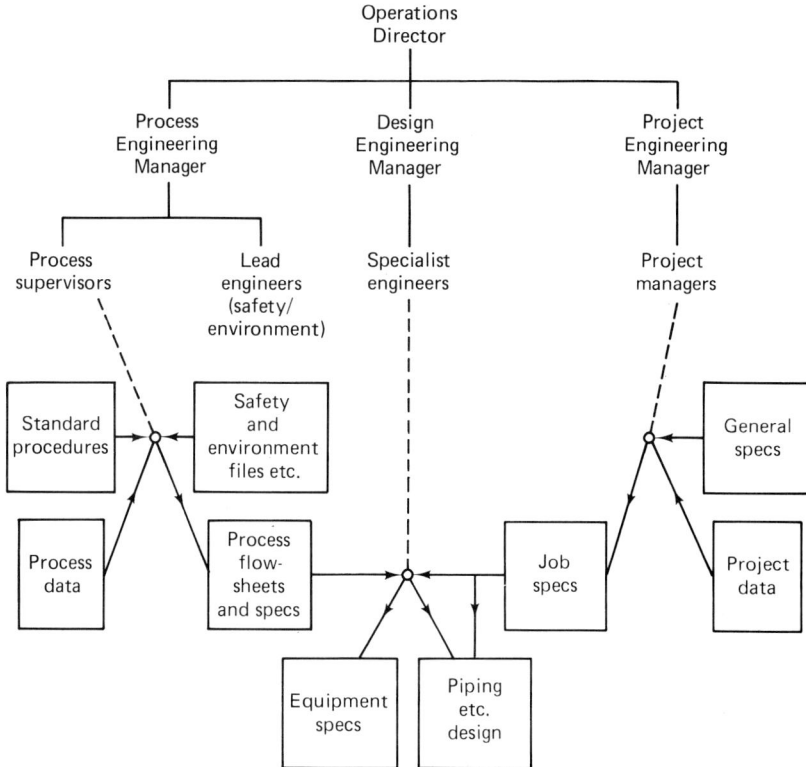

Figure 16.2 Responsibility for safe design in a contractor's office (from Fitt[2], courtesy Institution of Chemical Engineers)

constraints. (In the case of the plant described in Chapter 4, the entire process design was apparently 'frozen' by the transnational owner and licensor before being issued to the contractor, who was not officially aware of the compositions of the process streams!)

This safe basis must be consolidated throughout the stages of equipment specifications, unit operations, and design of safety facilities such as relief and blowdown systems, trips and interlocks.

Most contractors keep standard compilations of design data and procedures which include recommendations on relief valve sizing, vent header design, etc. and relevant codes of practice, routines and checklists.

Job specifications which incorporate safety standards must be agreed with the client at the beginning of a project. These cover all equipment, piping, civil work, etc. Some of these are determined by legal requirements in the country where the plant is built. Some clients have extensive safety standards of their own and loan copies of these to the contractor under an undertaking of confidentiality. The contractor should always provide those of his staff who are working on a particular project with copies of all relevant safety standards and codes. Some contractors now have safety training programmes for their staff at all levels. Safety reviews which involve the contractor are recommended by Fitt[10] at the following stages of a design:

1. Issue of preliminary flowsheets and mass and energy balances;
2. Issue of preliminary equipment specifications;
3. Design of utilities systems;
4. Review of layout and civil design;
5. Line-by-line review of flowsheets (line diagrams) and piping model (in which we would include P&I diagrams, general arrangement drawings and piping isometrics);
6. Final design of pressure relief and other safety systems;
7. Writing of operation instructions;
8. Detailed design review with client;
9. Review of vendor designs.

These reviews should, where possible, be made jointly by the contractor and the client, although the consulting engineer may have to act here for the client if this is his first manufacturing venture and he lacks experienced staff.

Besides economic constraints, two other (often related) factors can severely prejudice the best endeavours of a contractor in designing and building a safe plant:

- Time limitations with penalty clauses, and
- The extensive use of sub-contractors who may have little appreciation of safety issues and the operating hazards that may arise.

16.4 Process engineering

Process engineering is a skilled profession which operates at the nerve centre of process development and plant design. While vital to safety, it is often misunderstood. Most process engineers are chemical engineers, but only a minority of chemical engineers work as process engineers.

There is constant competition between processes for making the same end-product. This is particularly felt in the drive to reduce consumptions of raw materials, power, steam, fuel and cooling water and to utilise cheaper raw materials. It has led to the development of new catalysts which give higher yields and reduce operating temperatures and pressures, and to the extensive use of power and heat recovery by the use of waste-heat boilers, gas turbines, low-temperature expansion turbines (which also provide refrigeration) and heat exchange between outgoing and incoming streams. Many cracking, reforming, dehydrogenation, ammonia, gas separation, hydrocarbon oxidation and even polymerisation plants now have many resemblances to power stations. This is not always an unmixed blessing for safety, particularly when the process design becomes so 'tight' that operators have little room for manoeuvre if things go wrong. This situation can usually be improved by the provision of recycle loops, particularly round distillation columns, which allow production to be halted while vital equipment is kept 'ticking over' ready to come on-stream again as soon as the problem is sorted out.

The wastes and effluents from a process are also determined by its process design [16.4.2 and 16.6.7] and process engineering is involved in the design of their treatment facilities.

16.4.1 The role of the process engineer

The highest skills are needed by process engineers who design new processes from the results of small-scale experiments. They require a good knowledge of chemistry (particularly of the process in question), chemical engineering and process hazard evaluation and a working knowledge of economics, metallurgy, corrosion engineering, computer programming and mechanical and other branches of engineering, a varied operating experience and sometimes a foreign language or two.

Some companies distinguish between *process design* and *process engineering design*. For them, *process design* is oriented towards generating the process information given in a licensor's process package or a feasibility study. This includes process flow-schemes, material and heat balances and many other items. They regard *process engineering design* as part of the contractor's function and oriented more towards the preparation of piping and instrument diagrams, equipment data sheets and specifications, and information required by mechanical, civil, electrical and other engineering departments. This division of the subject into two functions needs to be borne in mind, but is not adhered to here.

Hazards such as those discussed earlier in this book creep readily into design in these early stages, and are often difficult to recognise and eradicate later before they cause mischief. Examples are given in 16.5.

Ludwig[12] summarises the responsibilities of the average process engineer thus:

1. Preparing studies of process cycles and systems for various product production or improvements or changes in existing production units;
2. Preparing economic studies associated with process performance;
3. Designing and/or specifying items of equipment required to define the process flowsheet or flow system;
4. Evaluating competitive bids for equipment;
5. Evaluating operating data for existing or test equipment;
6. Guiding flowsheet draughtsmen in detailed flowsheet preparation.

The process engineer must understand the interrelationship between research, engineering, purchasing, construction, operation and safety, and appreciate how any of these may affect process engineering decisions.

The scope of process design includes:

1. Process material and heat balances;
2. Correlation of physical data, and data from research, pilot plants and test runs on other plant;
3. Material, heat and energy balances for power, water, steam and other auxiliary services;
4. Development, detailing and completion of flowsheets;
5. Specifying conditions and performance of equipment shown on flowsheet consistent with mechanical practicality;

6. Specifying instrumentation required for process requirements and safety and interpreting this to instrument specialists;
7. Interpretation of process needs to all other engineering departments involved in the project.

Process engineers are generally expected to record all their calculations and sources of information in a design book so that these can be checked. Most process engineers, if asked to check the calculations of a particular design, prefer to start with a clean sheet and see what conclusions they reach, before checking someone else's calculations. This may take longer but it provides a better check on the data and methods used and on the assumptions made.

Process engineers are consulted on plot plans and plant layout (both in plan and elevation), process drains and effluents and usually on most other matters which form part of a design. To advise on these they need, *inter alia*, to evaluate the fire, explosion and toxic hazards of the process, e.g. as determined by the Dow or Mond hazard indices [12]. They may also need to modify their designs to reduce their hazard potential.

16.4.2 Flowsheets

Flowsheets are maps of a process and define the responsibilities of other engineers (mechanical, electrical, etc.) who are not expected to understand the process itself. Most flowsheets show the process steps in sequence pictorially, using standard symbols for different types of equipment, valves, pipes and instruments. Austin gives sets of drawing symbols according to British and American standards[13], but most contractors appear to follow American practice and use the symbols given by Sherwood and Whistance[14].

Many flowsheets are drawn and printed from left to right on long strips of paper, of the same height (generally A4) as used for office documents, into which they can be bound and folded, to be extended concertina-wise when required. These are more convenient to use than large drawings, although they cannot provide as much detail. Separate flow diagrams are used for each process and utility unit of which the plant is composed. Flowsheets include block diagrams, process flow diagrams (PFDs) and

Ammonia process

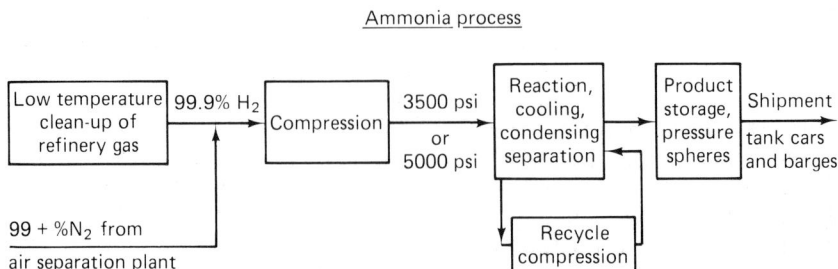

Figure 16.3 Process block diagram (from Ludwig[12], courtesy Gulf Publishing Co., Houston)

piping and instrument diagrams (P&IDs). Material balances are prepared and used in conjunction with process flow diagrams and sometimes printed on the same strip, above or below them.

Block diagrams (Figure 16.3) are the simplest form of flowsheet and set out the basic process concept. They show only what each step has to achieve but not how it is done.

16.4.3 Process flow diagrams (PFDs) and material balance sheets

Process flow diagrams (Figure 16.4) show all major equipment items with descriptions and interconnecting flow lines for process materials. Most show no valves or instruments. Some indicate only control valves (automatic or manual). Important flowlines are usually 'flagged' or otherwise marked with the stream number, and the temperature, pressure and design flow quantity. Design flow rates, temperatures, pressures and stream compositions are also shown.

The stream number refers to a material balance sheet which accompanies or forms part of the diagram. This sheet contains descriptions, quantities and components of each numbered stream, with its composition. It is useful if it also shows the physical state, enthalpy and important physical properties such as specific gravity and viscosity of the flowing material. Stream data given on material balance sheets are, in the case of a new process, calculated theoretically from basic data and principles, and from experimental results. The preparation of material balances is an important part of the process engineer's job, and being often very time-consuming does not always receive the care and attention needed. Thus whenever a process stream containing several components enters a separation step from which two or more streams leave, one needs to know what proportion of each component is present in each of the outgoing streams. The problem becomes more difficult in process steps in which reactions occur. Material balance data are, where possible, checked against information from similar plants or from pilot plants or laboratory data. They often have to be checked by measurement and analysis when the plant which is being designed comes on-stream. Computer programs are developed for many of these calculations. These may have to be repeated several times with different inputs to select the optimum design and operating conditions.

The material balance sheet should also show the nature and quantity of all solid wastes and liquid and gaseous effluents from the process. This information is vitally important in order to comply with anti-pollution requirements [2.2.3] and to design the treatment plant required [16.6.7]. This may be an important cost element in the process and has to be considered when choosing between different processes and raw materials.

The 'information' (hopefully, not misinformation!) developed in the course of preparing PFDs and material balance sheets is used to specify design duties of equipment, pipes and valves, as well as in calculating how much steam, electricity and other ancillaries are needed. These calculations are important to the technical and economic success of the project as well as to its safety.

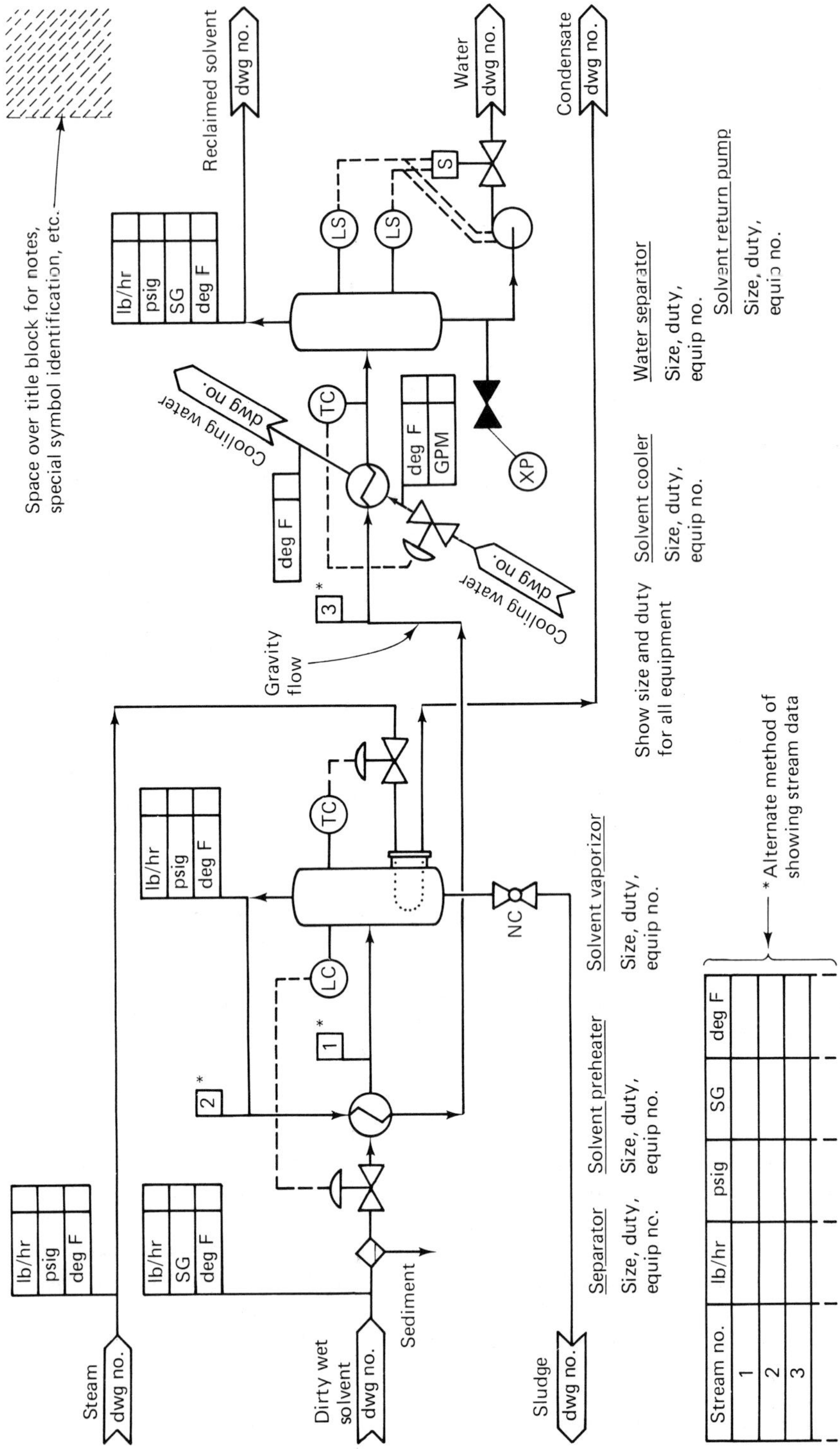

Figure 16.4 Process flow diagram (PFD) (from Sherwood and Whistance[14], courtesy Chapman and Hall)

16.4.4 Piping and instrumentation diagrams[14] (P&IDs)

The object of a P&ID (Figure 16.5) is to indicate all service lines, instruments and controls and data necessary for the design groups. The PFD is the principal source of information for developing the P&ID. This should define piping, equipment and instrumentation well enough for cost estimation and for subsequent design, construction, operation and modification of the process. Material balance data, flow rates, temperatures, pressures, etc. are not shown, nor are mechanical piping details except for permanently installed 'spectacle plates', pipe flanges which may require to be broken to insert line blinds, and removable spools needed for isolation. P&IDs show all process pipes and valves (including drains and vents) with numbers and sizes which refer to pipe and valve lists, all pressure relief valves and bursting discs. (Not all P&IDs show valve numbers, although these are particularly needed on P&IDs used to illustrate operating manuals and procedures [20.2.1].) P&IDs are laid out in the same way as PFDs, but with greater horizontal spacing between equipment, and the process relationship of equipment should correspond exactly. P&IDs are not to scale, but, where possible, equipment is drawn in the correct vertical proportions, and critical elevations are noted. P&IDs show symbolically all instruments and instrument loops (including trips and alarms) and control valves, with conventions to indicate whether they are located in the control room or on the plant. They show impulse and power supply lines (pneumatic, electric or hydraulic) to control instruments and valves, and the mode of valves if power fails (open, shut or same position). Separate PFDs and P&IDs indicate the pressure relief systems with headers, catchpots and flares, etc.

From these P&IDs (or sometimes replacing them) more detailed drawings (engineering line and general arrangement drawings) on which equipment is shown to scale in correct elevation are prepared. On these, piping, valves and instrumentation are developed in greater detail.

Utility flow and piping and instrument diagrams (UFDs and UP&IDs), similar to PFDs and P&IDs, are prepared for steam, water, compressed air, vent and relief systems, effluents and fire water. Electrical 'single-line' diagrams are prepared for power and lighting. Both PFDs and P&IDs are used in the preparation of operating manuals, and also in HAZOP studies which are best done at the same time. These are discussed later [16.5.5].

16.4.5 Simplification of PFDs and P&IDs

Once PFDs and P&IDs have been drafted two or three times, process engineers should try to simplify them (an exercise in which Boolean algebra can be helpful). For this they need to reconsider all the possible needs for pipes and valves during start-up (including the use of recycle circuits), normal operation, turndown, normal shutdown, emergency shutdowns in various circumstances, emptying and flushing. This done, they should systematically work out the simplest network of pipes and valves which allows each of these operations to be carried out safely, and make notes on how these are to be done. This generally results in improved and simplified PFDs and P&IDs. The notes written at this time form a

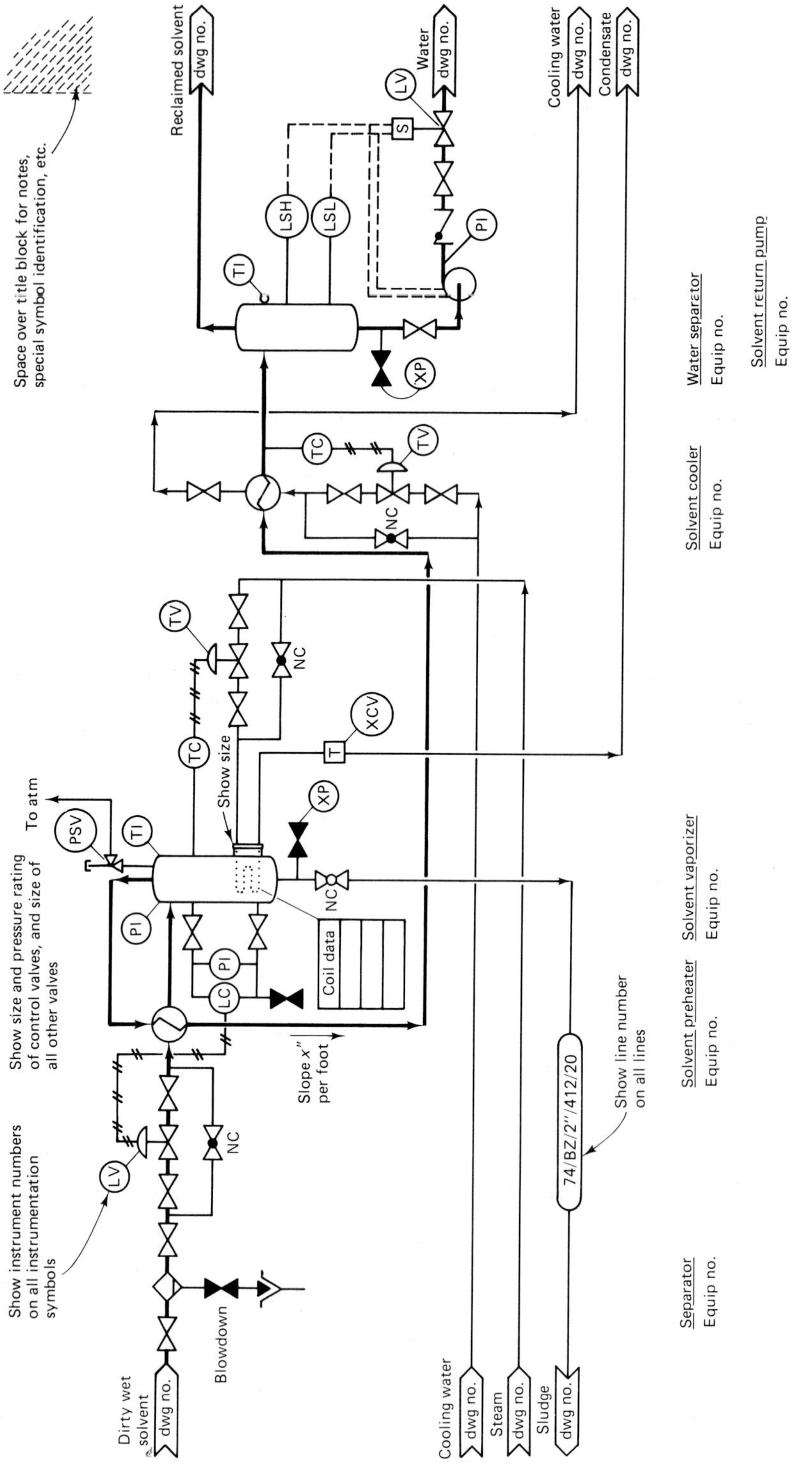

Figure 16.5 Piping and instrument diagram (P&ID) (from Sherwood and Whistance[14], courtesy Chapman and Hall)

basis for the plant-operating manual, when the PFDs and P&IDs are finalised. This manual is best written by a small working party which includes start-up and operating personnel as well as designers. It is best combined with the hazard and operability (HAZOP) study discussed in 16.5.5. There is not, of course, always time for such exhaustive reviews of the PFDs and P&IDs, but where they can be done they generally result in significant reductions in the plant costs which more than pay for the additional design hours, as well as improving operability and reliability.

16.5 Process engineering hazards and Hazard Study III

Process engineering covers such a wide field and presents so many opportunities for mistakes that general checklists are of limited value. Some typical problems in process engineering are next noted.

16.5.1 Scale-up and decrease in surface:volume ratios

Heat removal from exothermic reactions is particularly affected by scale-up. With liquid-phase reactions it is sometimes possible to carry them out in a boiling solvent, remove the heat by condensing the vapour and return the condensed liquid to the reactor. A solid-catalysed gas reaction can often be carried out in a fluidised bed which may be cooled in various ways, including the recycle of gas from the bed through a cooler. The scale-up problem becomes acute if it is not faced squarely, as when a tubular reactor consisting of a large number of parallel tubes filled with catalyst and surrounded by coolant is used for a strongly exothermoc reaction. It is then virtually impossible to control the reaction in individual tubes (parallel flow instability), so that hot spots and preferential flow result.

The danger of explosions in equipment containing flammable mixtures of gas, vapour and dusts with air increases as the volume increases, e.g. in furnaces and storage silos.

16.5.2 Errors in data used and in calculations

Published basic physico-chemical data from different sources may vary considerably, and errors are sometimes found even in reputable publications. There are often ways of checking such data for self-consistency, but most process engineers do not have time to do this. In a major design office there should be someone responsible for collecting, checking and issuing all relevant physico-chemical and other data used in the design.

Process engineers, being human, are prone to make calculational errors. These include misplaced decimal points, those arising when converting from one system of units to another, from using approximations, and from the use of formulae and calculational methods which do not apply to the case in hand. Results range from the installation of undersized pipes and valves which are discovered and replaced when the plant is commissioned to the creation of unsafe conditions which were thought and asserted by the designer to be 'perfectly safe' [4.5].

16.5.3 Unexpected formation of additional phases in process streams leading to chokes and blockages, etc.

This is generally due to inadequate information on the part of the process engineer. The formation of liquids in gas streams causes two-phase flow, with reduction in pipe and valve capacity, and can cause severe damage to compressors and turbine blades. The formation of solids in liquids or gases leads to blockages of valves and equipment, the clearing of which has caused many accidents. The formation of gases in liquids causes pumps and level instruments to fail, sometimes with serious consequences, while the formation of gases and liquids in solids leads to nasty, sticky messes which nobody wants.

16.5.4 Gaps in knowledge of process chemistry and trace components

Such gaps sometimes arise with a newly developed process and lead to the the unsuspected formation and accumulation of a hazardous compound at some point in the plant. The process engineer may have considered all the known by-products of a process which has been developed on a small scale, and designed the plant to take care of them. He may be unaware of an unidentified pimple on the chart of the GLC analysis of the crude reaction product. This was not identified during experimental development because it was not found in any other analyses, and it disappeared when attempts were made to concentrate it. On building and starting up a continuous commercial plant with a capacity of 40 000 t/a, what was a small pimple on the GLC was now found to be an unidentified by-product formed at the rate of several kilograms per hour, although there may still have been no sign of it in any of the product streams. It may in fact have been trapped inside the plant like a *jinee* in a bottle from which it cannot escape. This happens in distillation columns if the compound is more volatile than the 'bottoms' product and less volatile than the 'top' one, and is quite common when the compounds present are of different types. So it simply accumulates somewhere inside the column until it reaches a high concentration. The material may be harmless, but it may be explosive, highly toxic or corrosive.

One example was an air-separation plant built in the 1950s in Manchester (UK) where there was a trace of acetylene in the air intake. One day the plant suddenly exploded without warning. Something similar happened with vinyl acetylene in the purification train of a butadiene plant, which also exploded[15] [9.8.2].

Other examples discussed in this book where unrecognised by-products caused trouble were:

- Dioxin in trichlorophenol production [5.3];
- Elemental sulphur in the condensers of a crude oil distillation unit [11.7.4].

The HAZOP type of study discussed in 16.5.5 helps the study team to face up to such a problem provided one of its members was dimly aware if it in the first place. It is hard to see how the study can help here if none of the team has the slightest inkling of the problem. One would need to run courses for process engineers on 'how to develop a sixth sense'.

Checks and double-checks of their calculations and design assumptions may be needed on hazardous plants. Sources of help for such checks include recently retired technical staff with appropriate backgrounds.

16.5.5 HAZOP – Hazard Study III

This is considered the most important of the seven hazard studies recommended by Gibson, and since it is also described in some detail by Kletz[9] and Lawley[16], only a brief account is given here. It is based on a team study of the P&I diagram as this reaches its final stage. The need for and the thoroughness required of a HAZOP study depend much on the degree of hazard of the process (e.g. as given by the Dow Index [12.1]. Since it is also an operability study, it may be combined with the writing of the operating manual [20.2.3]. Although a HAZOP study is undoubtedly expensive, it frequently saves much greater expenditure later when the plant may have to be modified because of some problem which would have come to light in a HAZOP study.

A team of four or five people each provided with copies of the PFD and the P&ID sit round a table in sessions of two to three hours and study the P&ID in a formal and systematic way two or three times per week. Because of the degree of attention required, and the need to incubate ideas which the study may release, longer and more frequent sessions are not recommended. Gibson in 1976 estimated about 200 man-hours per million pounds of capital for these studies, i.e. about eight sessions. Typical members of the team would be:

- The process engineer responsible for the P&ID;
- The project engineer responsible for the mechanical design;
- The commissioning engineer responsible for start-up;
- The plant manager who will be responsible for operating the plant;
- The instrument engineer responsible for plant instrumentation;
- A research chemist (specially needed if new chemistry is involved);
- The computer programmer (for computer-controlled plant);
- A hazard analyst who acts as independent chairman. He records the ground covered at each meeting and issues reports of each meeting with agreed action items and responsibilities before the next meeting.

If the plant has been designed by a contractor, a mixed team of contractor's and client's staff is needed, and some functions may have to be duplicated.

Not all of these team members listed above need to be present at every meeting, but the process engineer, the commissioning engineer and/or the plant manager as well as the chairman should be present throughout.

The team studies each pipe and vessel in turn using a series of expressions to stimulate thought as to what would happen if the fluid in the pipe were to deviate from the design intention. The expressions recommended for a continuous process are[9]:

NONE	MORE OF	LESS OF
PART OF	MORE THAN	OTHER THAN

These are then applied to each possible variable in the conditions in the pipe or vessel, e.g. flow, temperature and pressure in all cases, with level, reaction, mixing, etc. added for vessels. The causes and consequences of all these variations are then considered and, where necessary, rough estimates of their probability are made.

Start-up, turndown, shutdown and other special operations such as catalyst regeneration should be considered as well as normal operation, and the study may be extended to utility units.

Batch operations are studied in a similar way by listing the sequence of operations and applying these expressions to each step. On computer-controlled plants the instructions to the computer should be studied as well as the P&IDs.

While a P&ID provides a good general plant model for HAZOP studies, its limitations must be recognised. It is purely a diagram, is not to scale, and does not always show equipment items in their proper vertical relationship to each other. It does not show all mechanical and instrument details, nor whether drains for removing water from low points in a distillation column are all correctly situated. What may appear to be a short-straight pipe in a P&ID may in fact be long and tortuous, and vice versa. The chairman should have the ability to draw out useful information half-buried in the minds of team members. Drawings, manuals and a three-dimensional plant model (if one exists) should be available to the study team as needed.

The usual outcome of HAZOP studies is a number of (hopefully minor) design modifications, and for example, the provision of special communication channels and lighting, changes to the program of a computer-controlled plant and to the analytical programme for quality and hazard control. The design modifications might include changing some materials of construction, adding, subtracting or altering the type or position of valves (particularly drains and vents) and vessels, modifying pipework, altering the instrumentation, particularly emergency trips and alarms, steam tracing and/or insulating lines, and improving access to particular valves or equipment. Since these modifications sometimes bring fresh and unsuspected hazards, the exercise must be repeated to the modified design.

While HAZOP studies may extend over weeks or months, they have to be fitted into the relatively short interval between the advanced drafting of P&IDs and other drawings and their freezing and approval for construction. Many of the points that emerge from a HAZOP study could have been avoided if the engineers had been more experienced, better informed, or had checked their own work more thoroughly.

A HAZOP study may also reveal that it would have been difficult or impossible to have carried out certain steps necessary during operation or maintenance on the plant as originally designed. This could have been anticipated if those writing the plant-operating manual had done their homework more thoroughly, and had 'gone through all the motions' on paper of opening and closing valves and starting and stopping motors, etc. as needed for all operations. (The importance of valve identification and numbering [20.2.1] might again be mentioned.)

One principal reason for Hazard studies I and II at an earlier stage in the project is to reduce the expense involved if the HAZOP study shows that major design modifications are needed. The time required for the HAZOP study will also be much reduced if all information given on the P&IDs has been properly worked over by the process engineer and others concerned. Correct timing is therefore very important to HAZOP studies.

16.5.6 Unplanned plant modifications

While most process plants were designed either for unique functions or as general-purpose ones, there are a number of older plants whose function has been changed more than once and for which no up-to-date P&I diagrams or drawings exist. The changes have often been initiated by the plant manager who has made rough sketches of what he wants and had them executed without drawings by the maintenance department. The company probably has no process engineering department. The mechanical engineer responsible for maintenance will, hopefully, have checked the mechanical soundness of the modification, but no drawings will usually have been made.

Such modifications usually work after a fashion, although there may be no record of them. The plant manager has apparently saved time and money, and probably been complimented on his action which may have raised production or had some other desirable effect. Such modifications then tend to become the norm ('We don't need to put *this* through the drawing office, *do we?*'). The company is, after, all in business to make various materials in bulk, and it earns no dividends on the output of its tiny drawing office.

Whether this is acceptable from a safety viewpoint depends on possible hazards which the modification has introduced. While this may be difficult to appreciate, the Dow Index which is generally easy to estimate provides an approximate indication of the process hazard. While it may be acceptable to by-pass the drawing office for plants of low hazard potential, this is not acceptable on highly hazardous plants.

Nevertheless the writer does not accept the frequently quoted verdict of the Flixborough inquiry that the disaster was due entirely to an improvised modification for which no mechanical study had been made. As pointed out in Chapter 4, the same error (failure to appreciate the compressive thrust of bellows) had been made in the design of the reactors for the conditions to which they were subjected. Thus had the design of the Flixborough by-pass pipe gone through the 'proper channels', there is no reason to think that the original mistake would not have been repeated. Had it then been discovered, the plant should have been shut down for the necessary modifications (which might have taken several months).

One thing, however, is clear. A proper record, at least in the form of reproducible P&IDs, should be made of all plants with their modifications, however and whenever they were made. In the absence of process engineers, the plant manager should be responsible for these P&IDs, either making them himself or getting a suitably qualified member of his staff to make them, if necessary with the assistance of a consultant. These P&IDs can then serve as 'maps' to the operators and form the basis for operator training and instruction. If the Dow hazard index of the plant is in or above the moderate range, a HAZOP study can then be undertaken, if necessary with a consultant experienced in hazard analysis acting as independent chairman. No future modifications should then be allowed without first modifying the P&ID and then making a formal HAZOP study to check what new hazards the modification would introduce. The recommendations of the study must then be acted on.

Without at least an up-to-date P&ID or equivalent drawing it is virtually impossible to carry out a useful process hazard study.

The writer knows one firm which had converted some old batch plants to continuous ones by modifying and adding to them in the way described here, with no P&IDs and no hazard studies. Perhaps because there was no proper check on modifications, they became rather common. Eventually there was a serious escape of a toxic vapour from one of these plants which caused several casualties. This was followed by an HSE investigation and prosecution, when the firm was blamed and fined and received adverse publicity.

16.6 Other design activities

Most other design activities for a process plant follow from the process design, although this itself can seldom be finalised until they are well advanced. The operator and the plant he controls are both parts of a closed loop which interact on each other. The design of a plant and its controls affects the performance of its operators. There is a constant need in all design activities to pay adequate attention to safety, ease of operation and maintenance [J]. Not only must the hazards to maintenance workers be considered by the designer. He needs to ask himself if others may arise from physical difficulties in maintaining the plant properly and manifest themselves in its subsequent operation.

An example which made a lasting impression on me happened while I was visiting a plant containing several tons of a C_4 hydrocarbon under pressure. A relief valve on a reactor suddenly 'blew' (as a result of a faulty thermocouple which formed part of the control system). The relief valve discharged into a common relief line several hundred metres long, leading to a liquid catchpot and elevated flare. Unfortunately a flanged joint immediately downstream of the relief valve leaked badly as soon as the relief valve blew. This caused the best part of a ton of boiling C_4 hydrocarbons to escape into a busy process area. Fortunately the staff were on their toes, and the escape was eventually dispersed harmlessly. On considering the causes of the leak which might have caused a serious fire or explosion, the following emerged.

1. Some time before the incident the vessel to which the relief valve was fitted was taken out of service, isolated, purged and entered for inspection. In isolating the vessel the flanged joint which leaked during the incident was opened and a line blind was fitted to prevent gas passing into the relief line via other relief valves from entering the vessel while it was out of service.
2. Before putting the vessel back into service, the blind was removed and the joint was remade, although it was both difficult and dangerous to

clean the flange faces and remake the joint properly. The difficulty lay in the fact that the flanges had to be sprung and could only be separated by a few millimetres to work on the joint. The danger lay in the fact that several other relief valves were connected into the same relief line, and had any of them blown or been leaking at that moment, there would have been an escape of heavy flammable gas from the open joint.

3. It was impossible to pressure or leak test the relief line after the joint had been remade.

This hazard seems to be a common one to which there is no universal remedy. The main points about it which should be considered by designers are:

1. There should be no risk of a significant escape of a dangerous fluid when a joint is broken;
2. It should not be physically difficult for fitters to remake pipe joints;
3. The design should allow any joint that has been remade to be pressure and leak tested before it is put back into service.

The other design activities discussed in this section are:

1. Plant siting;
2. Site and plant layout;
3. Mechanical and piping design;
4. Equipment specification and ordering;
5. Civil and structural design;
6. Utility systems;
7. Waste and effluent treatment and disposal systems;
8. Road and rail tanker loading and off-loading stations;
9. Fire-fighting services and fire protection.

Many of these are mentioned elsewhere in this book. Instrumentation, pressure relief and emergency devices dependent on special instruments have been discussed in Chapter 15.

16.6.1 Plant siting

The choice of site is usually made only once in the lifetime of a process plant, but it is a vital decision, and it is all too easy for economic considerations to take precedence over those of safety. Economic considerations include markets, raw material, fuel, power and water availability, transport, skilled labour and industrial and community infrastructure.

The consequences of several major process plant disasters in different countries in recent years were much magnified by unsatisfactory plant siting and failure to control housing and other activities around such plants. The Bhopal, Seveso [5] and Mexico City disasters illustrate this point. The health and safety implications of plant siting are now considered by local UK planning authorities, advised by the Major Hazards Assessment Unit of HSE, before permission to use a site for a new process plant is granted. Considerable effort may be required by the applicant in order to comply with the NIHHS [2.4.1], CIMAH [2.4.2] and various pre-HSWA Regulations discussed in 2.3.

The most comprehensive safety study for a proposed industrial site made in the UK was the Canvey study carried out for HSE by the UKAEA Safety and Reliability Directorate in 1978[17]. This arose from a public inquiry over planning permission for an oil refinery on Canvey Island. This already had three tanker unloading and storage installations for LNG, petroleum products and other hazardous liquids. On the mainland nearby there were two oil refineries, a nitric acid and an ammonium nitrate plant with storage for ammonia and ammonium nitrate. Explosives were trans-shipped at a neighbouring anchorage. The island measured about 8 × 4 km and had about 33 000 permanent residents, as well as caravan sites and holiday camps. Many inhabitants had been drowned in floods in 1953, but the sea wall had since been strengthened. It was linked to the nearby mainland by two road bridges.

The investigation involved some 30 engineers and cost about £400 000. The work included:

1. Identification, location and quantification of potentially hazardous materials in storage and in process;
2. Reviewing the flammability, toxicity and other relevant properties of these materials;
3. Identifying ways in which plants could fail thereby endangering the community, and identifying possible routes to such failures; these included operator error, metal fatigue, corrosion, loss of process control, overfilling, impurities, fire, explosion, missiles and flooding.
4. Quantification of the probability of selected failures and of their consequences.

The last of these tasks was the most difficult and hypothetical. The report recommended:

• That certain improvements should be made to reduce the risks of existing installations;
• That three proposed new refineries in the area would cause an unacceptable increase in the risks as they were planned, but that the plans could be modified to reduce the risks to acceptable levels in every case.

The study used statistical data collected by the Directorate on the probabilities of various events, in particular of equipment failure. It presented societal risks [14.7.3] as probabilities in units of 10^{-6} per year for numbers of casualties from 10 to 18 000 for 33 selected events. These included ships' collisions, process explosions, major fires, hydrogen fluoride release, pipeline failure and sea flooding.

Restrictions have long applied to sites used for munitions manufacture, military establishments and exercises, public water supply and sewage treatment. They now cover many other land uses, e.g. power stations (particularly nuclear ones), pipelines, storage of many industrial materials and transport systems. Controls also apply to the building of new plant on existing industrial sites and to the use and adaption of existing plants for purposes not covered by the original licence or permission. The complex problems involved have led to the widespread use of risk analysis [14.3].

The obvious safety requirements of a site for a hazardous process plant are that it should be remote, windy, flat, and large enough for the project and all foreseeable extensions to it, but not subject to flooding. These requirements depend largely on the nature of the hazards and the inventory and type of hazardous materials. The main concerns are usually fourfold:

1. Escapes of dense flammable or toxic gases, vapours and dust clouds should disperse and become diluted to harmless levels before reaching populated target areas;
2. The effects (blast and radiation) of fires and explosions at the site on surrounding property and populated target areas should be below the threshold at which they cause harm or damage;
3. Escapes of hazardous liquids and solids should be contained and not enter public sewers, streams and waterways where they may cause widespread damage to the environment and to public water supplies;
4. An unbuilt and unpopulated zone surrounding the site which is large enough to satisfy (1) and (2) should not be built on or populated during the life of the plant.

The radii of the zones referred to in (4) may be considerable, i.e. five or more kilometres for zones surrounding installations with potential for vapour cloud explosions.

One reason for the preference for flat land is that such sites are generally windy, but this must be checked against meteorological data. Ideally one would like a site with a steady wind blowing in the same direction day and night throughout the year with an unpopulated area downwind of it. Such sites are very rare.

Valleys surrounded by hills and mountains are mostly unsuitable for projects which give rise to concern (1), although they may sometimes be advantageous for those leading to concern (2). Similarly, although flat ground is generally preferred, a site which slopes down to a marsh or uninhabited waste land below it can be advantageous for the dispersal of dense gases. In this case one needs to know the extent of the waste ground, and what lies beyond it. If several thousand tonnes of butadiene were stored in pressure spheres on the side of a hill sloping down to a marsh-fringed estuary within a few miles of a busy seaport, the seaport could be at greater risk from a massive escape than the installation where the butadiene was stored. A hilly site may have safety advantages for the manufacture and storage of explosives.

The ideal site should not be subject to earthquakes, hurricanes, flooding, subsidence or extremes of temperature. Unfortunately the ideal requirements for a safe site usually conflict with commercial requirements which place a premium on very different qualities. In the UK the MHAU of HSE recognise three situations when considering planning permission:

- Situation 1 – negligible risk;
- Situation 2 – marginal risk. Although there are identifiable risks, they do not constitute clear reasons for refusing planning permission;
- Situation 3 – substantial risk. Safety is a major issue. This is the situation where quantitative risk analysis as discussed in 14.7 may have to be applied.

Balemans's checklist[6] contains a number of useful points about siting, which is also discussed at some length by Lees[2].

16.6.2 Site and plant layout

These are different but interrelated subjects which are commonly treated together. They are discussed by Mecklenburgh[18], Lees[2], Liston[19], and Wells[1], and also by Lewis[20], who suggests a means of relating plant and equipment spacing to the its degree of hazard as indicated by the Mond hazard index [12.3].

The layout of the plant determines what an operator can see from any position, how quickly and easily he can move around the plant, operate valves, inspect, make adjustments and do any work required. The position of pipes, valves, pipebridges, stairs, platforms, plinths, electrical and instrument cable trays needs close attention to avoid tripping and bumping hazards, and to ensure ease of access and operation. Two-dimensional drawings and isometrics are seldom sufficient for engineers and draughtsmen to visualise how the plant will appear to its operators. As an example, during investigations into the Flixborough disaster, even after visiting the site and inspecting it and several relevant drawings, I failed to appreciate what laboratory assistants could see of the 'by-pass assembly' [4.3] from their laboratory window until I laid out a scale model of the plant and the laboratory on a basement floor at the NCB's head office.

A three-dimensional scale model of the plant is often needed to assist in solving layout problems. Computer graphics which are now playing a growing role in design may in time supplant solid models[21].

16.6.3 Mechanical and piping design

Faulty mechanical and piping design can lead to breakages and leaks. Causes of metal failure and other hazards of mechanical equipment were discussed in Chapters 11 and 13. Much of the training and professional knowledge of mechanical engineers and piping draughtsmen is vital to safety, but it is impossible here either to separate the safety content of this knowledge from the rest or to cover the whole field of mechanical engineering, including pipework design. Mechanical engineering is covered in textbooks, codes, standards and engineering courses. Pipework design for process plant generally follows US practice as described by Sherwood and Whistance[14]. The best guarantee that the mechanical design, including pipework, is safe is to ensure that the work is done by well-trained, competent and experienced engineers and draughtsmen, who are provided with the proper design basis in the first place. An I. Chem. E. training module [M.2.7] deals with safer piping.

Most of the equipment employed in process plant is purchased from specialist equipment manufacturers, whose mechanical design engineers may not be familiar with the processes in which they are used. The mechanical engineers of the operating company and contractor play more of a monitoring and inspection role. They first help to draw up data sheets and specifications for the equipment, are then involved in the choice of equipment and vendor selection, and later inspect and test the equipment.

They do, however, sometimes become more involved in equipment design. Thus a large international company participating in a process industry may find there is no suitable equipment on the market for one of its needs and may decide to develop the equipment, e.g. a special extruder for a new plastic. On other occasions the operating company or contractor may modify available equipment for special purposes. Reactors and fluidised-bed dryers are examples.

Pipework typically accounts for a third of the capital investment in the plant and its design often requires more man-hours of engineering time than any other design activity. The work is done mainly by contractors who employ specialist piping draughtsmen. Mistakes in piping design can lead to breakages (through overstressing, vibration and corrosion, etc.) and to operating difficulties, e.g. through badly positioned valves. The piping department usually starts with P&IDs prepared by the process engineers, who will have calculated pipe diameters from flow and pressure-drop calculations, specified materials of construction, maximum pressures, maximum and minimum temperatures, valves, instruments and all flanged joints which may have to be broken when equipment has to be isolated [18.6.1] for maintenance or inspection. The P&IDs show the pipework diagramatically, with each section numbered and with the above information, and any other special requirements such as:

- The need to slope a particular pipe from A to B, with no low points in it where liquid could collect;
- The need to steam trace a pipe;
- The need to keep a particular pipe well clear of electrical cables, instrument lines or certain other pipes.

The piping draughtsman has to design these pipes and their supports and arrange them in three dimensions, with exact lengths, bends, Tees, connections and fittings, and provide in a satisfactory way for thermal expansion and other stresses. The piping design is done when all details, sizes and position of the equipment are provisionally known, while the structures to support it are being designed and its exact position and orientation are being decided. These in turn depend to some extent on the pipework arrangement. The final result is a compromise between these various requirements. The piping draughtsman produces plan and elevation drawings of the pipework, and isometric drawings of each section of pipe which requires to be individually fabricated. The preparation of isometric pipe drawings is today largely done with the aid of a computer. A three-dimensional plant model is specially useful to the piping draughts-man, without which it is all too easy to get two pipes crossing each other in the same element of space. The piping draughtsman must ensure that all valves are readily accessible, that pipes do not obstruct operators and other workers, that pipebridges are installed where they are essential and that the pipework provides for easy maintenance [J] and does not impede cranes and other maintenance equipment. Care must also be taken to avoid the danger of pipework being struck by moving vehicles, especially at road and rail car loading and unloading stations, and at road crossings. The piping draughtsman also prepares lists of all materials required, pipe, valves, fittings, supports, etc.

He has to be thoroughly familiar with all the codes[22] and standards used[23, 24], pipe classes and materials, methods of fabricating and joining, and stresses arising from differential thermal expansion and pressure (particularly from pipe bellows). Welds rather than flanged or other mechanical joints are used as far as possible. Where flanged joints and gaskets have to be used their types should be selected with special care in relation to the process fluids and their conditions. Spirally wound metallic gaskets are considered safer for high-pressure and critical duties than compressed asbestos fibre gaskets. The piping draughtsman is usually responsible to a mechanical engineer to whom he refers any special problems. The work of a piping draughtsman should be checked line by line by another piping draughtsman, a stress specialist and a member of the commissioning team or operating staff before it is approved for fabrication. Pipework is usually fabricated in a pipeshop adjacent to the site, and joined as far as possible by welding as it is erected.

On some jobs, in attempts to save time and money some of the pipework is 'site-run' by a sub-contractor, often for a fixed sum without piping drawings, after most other plant items are in position. The results in the writer's experience were far from satisfactory and the economies largely illusory. Apart from the dubious quality of the pipework, it resulted in a number of tripping and scalping hazards, and poor access to vital valves, etc.

16.6.4 Equipment specification and selection

Purchasing specifications were discussed in 2.8.5 and reliability of plant equipment in 14.2. When several offers have been received by the purchasing officer for particular equipment, these should be carefully reviewed by the engineers who prepared the data sheets and specifications for it in the first place. If a particular manufacturer offers equipment which appears to have technical merit but is otherwise unfamiliar, it is worth making enquiries from other users and even to visit the site and see the equipment in use. When comparing the prices of equipment from different suppliers, one must make sure that they have the same or comparable safety features (e.g. guards, electrical flameproofing, fire protection). It is also worth making enquiries from appropriate data banks about the reliability of the equipment offered, although the information may be misleading if the equipment is to be used in conditions different from those under which its reliability was examined. Customer service, availability of spares and guarantees are other factors to be considered as well as price and delivery.

One point about ordering valves is worth making. One should make sure that an operator can tell at a glance whether any valve is in the open or closed position. Most but not all valves meet this requirement.

16.6.5 Civil and structural design

Here we are concerned with the design of foundations, drains, roads and buildings as well as structures to support plant equipment, pipes and services. Blast-resistant control rooms, tank bunds and firewalls also come under civil design, which is well covered by codes and standards and by Balemans's checklist[6].

Three common problems are ground settlement, vibration and corrosion from both the atmosphere and the ground. Settlement can lead to excessive stresses in pipework, causing leaking joints and even rupture. In soft ground even the foundations for pipe supports may have to be piled. Foundations and structures may have to be designed to enable large vessels to be filled with water and hydraulically tested after erection.

Special corrosion protection of foundations, floors and structures is needed when corrosive process fluids are present. Many companies have their own standards for control room construction, fireproofing of structures, ground slope and process drains. Recommendations on these and other safety features and standards which relate to the hazards present are given in the Dow guide [12.1]. Wells[1] gives several formulae for estimating the likely effects of different types of flame and fire on plant structures, which may be used to check the fireproofing recommendations of the code followed.

Plant structures should include alternative means of escape for an operator who may be trapped on a high platform by a fire. The gap (fully 3 m) between the platforms of the depropaniser and debutaniser towers of a cracking plant at Abadan refinery became picturesquely known as 'Dai Jones's leap' after an operator saved his life by jumping from one to the other, a feat he never believed he was capable of performing.

The questions of when a blast-proof control room is needed, to what standards it should be designed, and whether any windows should be allowed are difficult to answer in general. Figure 16.6 shows Nypro's control room (in which all 18 men present were killed) after the 1974

Figure 16.6 Nypro's control room after the explosion

explosion [4]. Recommendations have been made by Wells[1], Langeveld[25] and Marshall,[26] whose suggestions as summarised by Lees[2] are as follows:

1. The control room should contain only the essential process control functions;
2. There should be only one storey above ground;
3. There should be only the roof above the operator's head;
4. The building should have cellars built to withstand earthshock and to exclude process leaks and should have ventilation from an uncontaminated intake;
5. The building should be oriented to present minimum area to probable centres of explosion;
6. There should be no structures which could fall on the building;
7. Windows should be minimal or non-existent and glass in internal doors should be avoided;
8. Construction should be strong enough to avoid spalling of the concrete, but it is acceptable that, if necessary, the building be written off after a major explosion.

The only one of these suggestions about which the writer has reservations is (4), since it is difficult to ensure that dense gases and vapours do not enter cellars. The easiest way in which such gases an enter a control room is through drains and pipe and cable trenches, which should be filled and sealed.

The provision of ventilation from an uncontaminated source is more easily said than done, since the very escape of flammable gas or vapour which could cause a major explosion is liable to cover a large area which includes most air intakes that would normally be considered safe. Most control rooms, although located in zones classed as hazardous under electrical safety codes, contain non-flameproof electrical equipment and therefore have to be kept 'pressurised' with ducted air supplied by a fan, with air-locks for personnel entry and exit. As a life-protection measure, the provision of air from a source which would be uncontaminated in the event of a massive escape of flammable or toxic materials seems as important as providing blast-proof construction for the control room.

As Lees[2] points out, the foundations of refrigerated storage tanks must be designed to prevent heaving caused by the earth freezing. The bottom of the tank is usually insulated and supported above the ground on a structure under which air is free to circulate.

The bunding of storage tanks has been partly covered in 10.5.1, and is also discussed by Lees[2]. Full bunding is usually provided for atmospheric storage tanks containing Class A or B flammable liquids, whether at ambient temperature or refrigerated, but not for pressurised storage of liquefied flammable gases. Pressurised storage vessels rarely fail, and such leaks that do occur are mainly in the form of vapour spray, the dispersion of which is hindered by full bunding. Low walls are usually provided in these cases to protect the vessels and pipework from vehicle collisions, and the spread of liquid from other sources. The ground under pressurised storage vessels should slope away from them, preferably to a large shallow depression in open ground to which any liquefied gas released can escape and where it can burn with least risk to the storage vessels, personnel, pipes and other equipment.

The Oil Insurance Association[27] stresses the need for individual bunds capable of containing, at the very least, 100% of the tank capacity round large oil storage tanks of $50\,000\,m^3$ and more.

16.6.6 Design of utility systems

Process plants depend on reliable supplies of utilities which include:

1. Electricity;
2. Instrument air;
3. Steam;
4. Fuel, usually natural gas, LPG or fuel oil, or a by-product of the process;
5. Cooling water;
6. Process water;
7. Process air;
8. Inert gas;
9. Compressed air for breathing apparatus;
10. Special heating and cooling media.

Reliability in the supply and quality of any utility is important both to performance and safety of process plant. Lees reviews this very large subject[2] and gives many references. Useful booklets on the hazards of water, air, electricity and steam are available from AMOCO[28]. The design of utility systems and pipework starts with UPFDs, UP&IDs and utility balances in the same way as that of the process equipment and pipework.

Plants should be designed so that the sudden failure of one or more utility supplies does not prove disastrous. In cases where such failures could have serious consequences, back-up supply is needed for emergency use, e.g. electricity for lighting, communications and critical instruments and compressed air for instruments. Risk analysis [14.3] is used to assess both the need for and the reliability of back-up utility supplies. An event tree for the loss of grid power supply given by Andow[29] is quoted by Lees[2].

The sudden failure of a utility is more dangerous than a gradual one, and for this reason steam turbines may be chosen for the drives of critical pumps rather than electric motors, despite their higher cost. Experience had shown that steam failures are generally gradual, allowing about 10 minutes in which to shut down the plant, whereas electricity failures are sudden.

The routing of service supplies should be as secure as possible. The position of underground supplies should be clearly marked on drawings and, where possible, by ground posts. Underground supplies must be adequately protected from heavy vehicles, etc. Service supplies are usually run as ring mains with the necessary valves, to ensure continuity of supply if an accident puts part of the main out of action. The consequences of such accidents to the flow in the main should be analysed.

When critical control functions are duplicated to increase reliability and safety, the duplicate cables and impulse lines should go by different routes, which, in any case, should be as safe as possible.

Special precautions are sometimes needed to prevent process fluids from entering utility mains, and vice versa. A catastrophe caused by the entry of hot hydrocarbon vapour into a steam main was described in 10.5.2. Such precautions include double block and bleed valves, the use of special detection and warning instruments and of an intermediate heat transfer medium (e.g. hot or cold oil or glycol solution) for heating or cooling. The heat transfer medium should not itself introduce a significant hazard.

1. Electricity

Electricity is used as a source of power, for lighting, for instrumentation and sometimes for heating. It is supplied from the national grid and often from a works power station, with back-up arrangements between them. The use of flameproof and intrinsically safe electrical equipment and conduit to reduce the risk of flammable gases, vapours and dusts was discussed in 6.2.3. A 'hazardous zone drawing' is made when all sources of leaks of flammable materials have been identified. This shows the categories of electrical equipment allowed in each zone.

Loss of electrical power is specially serious if it causes materials to solidify in a process or pipeline, air-cooled and refrigerated heat exchangers to cease functioning and cooling water and process pumps to fail. One has to decide for each motor whether, after a temporary power interruption, it should be arranged to restart automatically or be manually restarted when the power supply is resumed. Reliable back-up systems known as 'uninterruptible power supply' (UPS) which are ultimately reliant on local battery power packs are specially important for emergency lighting and instrumentation.

2. Instrument air

This has been discussed in 15.8.2.

3. Steam

Steam is used for heating, driving machinery, direct introduction into a process, purging and snuffing fires. Supplies are seldom duplicated since a steam failure is usually gradual and allows plant to be shut down safely. Particular hazards of steam are steam hammer [15.3.2], erosion [11.4.1] (especially by wet steam), static electricity from steam leaks [6.3.5], scalds, harmful noise from relief valves, local fog from steam traps and the creation of vacuum in closed systems heated by steam if the steam fails. Steam pipework, expansion bends and supports must be designed to high standards to resist the stresses caused by high velocities. Careful thought must be given to the collection and disposal of steam condensate from reboilers, steam heaters and traps. Whether this is returned to the boiler or not, the system should be designed to prevent the escape of steam near the plant which causes local fog, condensation on nearby surfaces, corrosion and accidents.

Some processes generate sufficient steam in waste-heat boilers to meet their own requirements and even produce a surplus. These, however, nearly always require an adequate supply of steam from an external source for start-up and purging.

Steam quality, particularly from waste-heat boilers, is sometimes poor and it may be contaminated by entrained solids and gases from the boiler feed water. These aggravate corrosion and can cause scaling and heat transfer problems.

4. Fuel

Fuel, usually natural gas, LPG or fuel oil, or a by-product of the process, is used for heating, internal combustion engines, and gas turbines which are used typically to drive centrifugal compressors and generators. Fuel gases always present fire and explosion hazards, both within the equipment (e.g. furnaces) where they are used and through leaks into the atmosphere. Bought fuel gases are normally 'stenched' by the addition of strong-smelling compounds which enable leaks to be detected. Fuel gases produced as by-products of a process and used locally are sometimes highly toxic (e.g. containing carbon monoxide, hydrogen sulphide) as well as flammable. In such cases suitable gas detectors and alarms should be provided, taking special care over the choice and location of the detector heads.

Firebox explosions in boilers are not uncommon, while there have been a number of highly damaging fires and explosions in fired heaters. Guidelines for their safe design and operation are available from the Oil Insurance Association[30] and given in API codes.

As with steam, some processes (e.g. ethylene cracking) are self-sufficient in fuel, but generally require an auxiliary source of fuel for start-up. When two sources of fuel gas are used, care is needed that the burners can use both, with little or no adjustment, or that two sets of burners are provided. If the Wobbe index of the fuel gas is liable to vary, it may need to be controlled automatically by the injection of vaporised LPG.

5. Cooling water

The use of water for cooling in process plants has been in decline for some time as air-coolers with fans have taken its place. The reasons for this are both economic and environmental. The discharge of large quantities of warm water from plant-cooling systems into waterways can destroy some aquatic life. The greatly increased concentration of dissolved salts in the water discarded from circulating cooling systems creates similar problems. The discharge of process materials in the return water caused by leaking heat exchangers can lead to even greater damage.

Cooling water is provided both from closed-circuit systems with cooling towers and by once-through systems using river or sea-water. Problems found with cooling water include corrosion, scaling, the growth of molluscs and the presence of fish and other solid objects which block heat exchanger tubes, small valves and lines. The chemistry of the cooling water determines the choice of metals used in pipes and coolers, and any water treatment used. This requires expert study and may include flocculation, sedimentation, chlorine addition to kill algae and molluscs, deaeration, and the addition of chemicals to control pH, scaling, precipitation and corrosion. Some CW additives are highly toxic. An economic balance has to be struck between the costs of treatment and the consequences of no treatment. A sometimes serious problem of circulating cooling water systems is the day-to-night temperature fluctuations which cause process disturbances unless special means of preventing them are provided. The same also applies to air-cooled heat exchangers. An example of a complex corrosion problem to which untreated cooling water contributed was given in 11.7.4.

In cases where the consequences of cooling water failure are serious, stand-by water pumps provided with an alternative power source (e.g. diesel) are often installed. In cold climates cooling water mains are run below ground (under the frost line). Pipes and manifolds above ground which are liable to freeze should be provided with drains at low points so that they can be emptied when the plant is shut down.

6. Process water

Water used directly in processes and for boiler feed is usually taken from the public water supply and/or from steam condensate returned from process heaters. In the case of public water supply, the water should first enter a vented break-tank (via a float-operated valve) to prevent any possibility of process fluids entering the public main. This water often requires further treatment (hardness removal or complete de-ionisation, deaeration and/or the addition of treating chemicals) before it is used in a process, and always before it is employed as boiler feed.

The use of hot steam condensate is economically attractive, especially for boiler feed, but provision should be made for detecting and, where possible, removing impurities which may be introduced into it from leaking reboilers and process heaters. Such contamination can result in deposits inside boiler tubes which have led to overheating and explosions.

Where such impurities cannot be removed from the condensate, provision must be made for changing over to treated water from another source, and for treating the contaminated condensate in a way which satisfies local pollution requirements before it enters a sewer or waterway. Usually a leaking heat-exchanger requires the plant to be shut down for maintenance.

7. Process air

Process air is normally taken from a local air intake provided with a filter, and compressed to slightly above the pressure at which it is to be used by a suitable blower or compressor, which in many cases has to be of an oil-free type. Failure of the compressor usually results in an automatic emergency plant shutdown, but as the consequences are seldom disastrous, a stand-by air compressor is not usually provided. The quality required of process air depends entirely on the process in which it is used. It may have to be dried, sometimes to a low dewpoint, and sometimes sterilised. Air entering air-separation plants usually requires special treatment to remove traces of hydrocarbons and other impurities to prevent plant explosions [16.5.4]. The need to limit the outlet temperature of oil-lubricated air compressors is discussed in 17.1.5.

There are some processes in which liquid air is used. Here there is the danger of the air becoming concentrated in oxygen, which has a higher boiling point than nitrogen, if it has been stored so long that much of it has evaporated.

8. Inert gas

An inert gas, usually nitrogen but sometimes carbon dioxide, argon, or de-oxygenated air from an inert gas generator, is used for a variety of purposes which include inerting (blanketing), purging, pressurising and as a feed material for the process. Nitrogen may be purchased as a compressed gas in cylinders, as a liquefied gas (in special insulated containers) which is evaporated before use in specially designed evaporators, or it may be produced on-site by cracking ammonia or from a small gas-separation plant using a pressure swing adsorber or a membrane. The use of liquid nitrogen poses certain hazards which are common to most cryogenic fluids. These include splashes onto the skin or into the eyes, causing serious frostbite or worse, and the entry of liquid nitrogen from an evaporator into a process line causing contraction, metal embrittlement and often fracture. This may happen if the instrumentation on the evaporator is faulty. Only purpose-built nitrogen evaporators should be used and they must be regularly inspected and maintained. These and similar hazards are discussed in the *Cryogenic Safety Manual*[31].

When an inert gas is used to blanket vessels and tanks containing volatile flammable or toxic liquids, P&V valves [3.1.2] are commonly used to ensure that inert gas only enters the tank, etc. when the pressure is below a preset level, and that gas (loaded with vapour) escapes when the pressure rises above another preset level. (Such valves are better described here as 'pressure and vent valves' rather than 'pressure and vacuum valves'.) Only valves specially designed for this purpose should be used. Their preset pressures are usually controlled by weights. Where two or more tanks are used for the same liquid (one filling while another is emptying) it is usually possible to manifold their vent lines and use a single P&V valve. This saves inert gas and reduces the escape of vapour. The vent manifold must then be designed so that the gas space in any tank can be isolated from that of the others and taken out of service (for maintenance, etc.) while the others are still connected via the P&V valve to the inert gas supply [J].

When an inert gas is used for purging process gases and vapours from tanks and equipment before admitting air for entry and inspection, etc. it is vital that the design should allow adequate air purging and testing before entry [18.6.2]. Several fatal accidents have been caused by men entering vessels filled with inert gas, although not always because of design faults.

A shortage of inert gas at a critical point in process operation can be dangerous, and it is necessary to ensure that supplies are not interrupted. A tank blanketed with inert gas using a P&V valve will usually collapse if liquid is still being withdrawn when the inert gas supply is exhausted. An appropriate alarm to show when the inert gas availability has fallen below a critical level is then essential.

9. Compressed air for air-line respirators

Neither process air nor instrument air are generally quite free from oil and decomposition products formed from it during compression. They are also often saturated with water vapour and liable to form slugs of rusty water in pipes which may cause valves to stick. Their suitability for human breathing is thus suspect. Since the quantities of air needed for air-line respirators are usually small, it is generally safest and most economical to purchase compressed contaminant-free and dry air in cylinders for air-line respirators. If there are processes such as paint spraying where air-line respirators are in regular use, a special mobile oil-free compressor may be needed, with an adequate compressed air cooler, dryer and reservoir. Arrangements are needed to warn when the compressed air available is approaching exhaustion.

10. Special heating and cooling media

Special heating media are used for temperatures above those attainable with the steam supply. Refrigerants, sometimes utilised in conjunction with brines, are utilised for cooling below ambient temperatures.

Heating media include pressurised hot water, a range of high-boiling liquids, and molten solids. The liquids include the Dowtherms, mineral oils, various silicones, chlorinated diphenyls and mercury. Molten solids include heat transfer salts (usually mixtures of sodium and potassium nitrites and nitrates), sodium–potassium alloys and other low-melting alloys. All require to be pumped from a reservoir through a pressurised heater (fired or electrical) where in some cases they are vaporised, the vapour or hot liquid being piped to the equipment heated, from which the condensed or cooled liquid is returned to the reservoir.

Apart from hot pressurised water which (because of the high pressures needed) is not much used for process heating at temperatures above 250°C, all have significant hazards. These include fire, release of toxic mists and vapours, and internal explosions on start-up (if water is present in the system). They are also usually quite expensive. If at all possible it is better to design the plant in such a way that they are avoided. Systems using heat-transfer media are generally used for heating (and cooling) reactors through their jackets and internal coils, to achieve process temperatures in the range of 200–400°C. Alternate means of heating include direct heating by fuel gas, by the use of electricity including microwaves, and by heat exchange with hot process streams. Fuel gas heating is generally the cheapest option. To avoid direct flame impingement on the vessel, an oven can be built round it, or gas-fired radiant panel heaters may be used. For external electrical resistance heating, a shaped block of aluminium in which the heating element is encased may be cast to fit snugly against the base of the vessel being heated, any gaps being filled with a heat-transfer cement. Electrical heating coils may be used inside equipment in place of coils filled with heat-transfer media.

The main reason for using heat-transfer media rather than a direct fired or electrical heater is that parts of the inner vessel walls which are not covered by liquid or on which deposits of solids have settled are not exposed to temperatures higher than those of the heat-transfer medium itself, whereas with fired or electrical resistance heating, these parts are more likely to become overheated. Nevertheless, this disadvantage can usually be overcome by careful design.

If there is really no alternative to the use of a heat-transfer medium, the hazards of the various possible systems need careful assessment before one is selected, and a contract for its design and supply should be placed with a company specialising in this field.

Many of the refrigerants formerly in common use are flammable (e.g.

propylene) or toxic (ammonia, sulphur dioxide). A range of safer chlorofluorocarbon (CFC) refrigerants allows temperatures down to −80°C to be achieved with single-stage refrigeration, and down to −120°C with two-stage refrigeration. These include R-14 (carbon tetrafluoride), R-13 (monochlorotrifluoromethane), R-12 (dichlorodifluoromethane), R-11 (trichloromonofluoromethane) and R-22 (chlorodifluoromethane). Unfortunately, CFC refrigerants are now implicated in the depletion of the ozone layer in the stratosphere, which removes harmful UV rays from sunlight. In 1987 an international protocol to protect the ozone layer was signed in Montreal. This will restrict the future availability of CFC refrigerants, particularly the two most widely used ones, R-12 and R-11. A conference on 'Refrigerants and the Environment' was held in London by the Institute of Refrigeration and the National Economic Development Council to discuss the effects of the protocol on future plant design and alternatives to CFCs.

When a process requiring refrigeration produces one or more materials such as ethylene, propylene or ammonia, which are themselves good refrigerants, there is always an urge to employ them as such. Here one should first examine how far the use of the material as refrigerant (as compared with an inert one) increases the inventory of flammable/toxic materials in the plant, and hence its inherent hazards (e.g. from its Dow or Mond F&E Index). By careful design such increases can usually be minimised. The option of designing the plant for the use of the safer CFC refrigerants may not always be available. Refrigeration plants using refrigerants other than water should be designed so that the refrigerant is always under a positive pressure when the plant is running. The use of chilled water evaporating under vacuum is, however, a safe and useful system when temperatures below 5°C are not required.

16.6.7 Wastes, effluents and their disposal

Wastes and effluents and their treatment and disposal affect both pollution and safety. Although there is no clear distinction between the two, pollution and its control is too large a subject to be dealt with here. *Industrial Pollution*, edited by Sax[32], is perhaps the best-known book on the subject. Other information sources are given in Appendix M.

Sometimes the methods used for hazard and pollution control seem to be in conflict. Thus the use of local exhaust ventilation to reduce toxic hazards inside a building will increase pollution outside it. The flaring or atmospheric discharge of vapours from a pressure relief system, while preferred methods for hazard control, cause atmospheric pollution. This partly accounts for the growing use of trip systems to suppress causes of plant overpressure in place of relief valves. Considerations of pollution may thus reduce the options open for safety.

The treatment and disposal of wastes and effluents from most process plants require a substantial engineering effort. Much of this has to be done early in the project. These requirements should emerge from the mass balances which accompany the PFDs [16.4.3]. Having created the problems, the process engineer must help to solve them. The environmental impact of the proposed plant must be studied. The design of aqueous drains and many treatment plants is done by civil engineers.

Wastes from a plant are considered as solids and semi-solids.

Effluents from a plant may are considered as gases, vapours, liquids and water-borne materials. Liquid effluents may be classed as domestic, cooling water return, process, storm and soil drainage, and fire water. Effluent gases may come from the process, from utility generation and waste incineration.

Non-combustible wastes are generally removed by a contractor for disposal in approved 'holes in the ground', and sometimes dumped at sea. Combustible wastes are generally incinerated and the ashes similarly disposed of. If the flue gases from the incinerator contain harmful compounds such as hydrogen chloride or sulphur dioxide, they may have to be removed in a special scrubber attached to the incinerator. The design and operation of incinerators are often quite critical, requiring narrow ranges of temperature and excess air ratio in the combustion chamber. They should also be as simple, reliable and easy to maintain as possible [J]. They should be built close to where the waste is produced so that its manual handling is minimised. Some pretreatment such as compacting may be needed.

Process effluents are of two kinds: relatively steady emissions of liquids and gases with low contents of flammable or toxic constituents, and large and sudden emissions of short duration during emergencies. The treatment and disposal of steady emissions is usually straightforward. Designing for large and sudden emergency emissions is much more difficult.

Steady emissions of gases containing harmful constituents such as oxides of nitrogen, hydrogen fluoride and odorous impurities should be treated by scrubbing, catalytic conversion, adsorption or combustion before entering the atmosphere. Means of sampling and monitoring the final emissions should be provided. Care must be taken that different vents which may release traces of contaminants which react with each other at very low concentrations to produce visible clouds (ammonia and hydrogen chloride) or unpleasant smells (ammonia and chlorine) are well separated.

Large and sudden emissions of process gases and vapours in emergencies can only be effectively dealt with if the discharge takes place through a vent or pressure-relief device. The treatment which must be provided is discussed in 15.5. Water and steam curtains are of limited help for massive escapes.

Leaks of non-aqueous liquids from pump glands, valves and sample points should be collected in special containers rather than allowed to enter process drains where they may become greatly diluted and difficult to recover or treat. In some cases it is possible to provide bunds and excavations for the collection and segregation of large spillages of process liquids from process plants and road and rail tanker stations to prevent them from entering drains.

The design of drains, sewers and liquid effluent treatment for process plant is seldom easy. Where flammable liquids may be released, the first requirement is to slope the ground (usually concreted) under the process equipment and storage vessels towards a shallow open drain which has no structures or equipment above it [12.1.6]. This drain will link up with other similar ones and be routed by sewers, usually covered, to liquid separators and/or other treatment needed to satisfy the pollution requirements of the public sewage system or other destinations into which the effluent will be discharged. These requirements vary widely, from those of a salmon river to a those of a pipe discharging some distance out to sea. The corrugated plate interceptor for oil removal (Figure 16.7) is compact and efficient. Treatment methods used include most of those listed for cooling water treatment in 16.6.6.

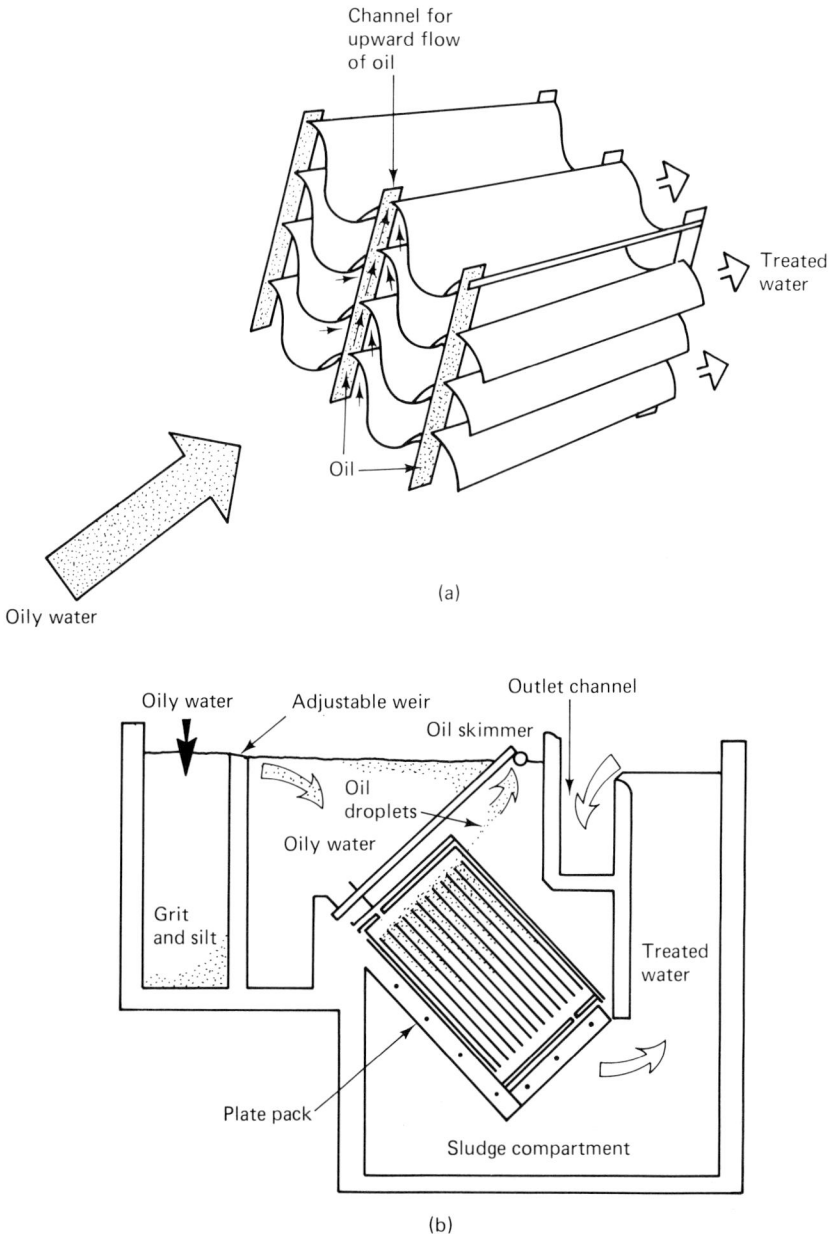

(a)

(b)

Figure 16.7 Corrugated plate oil interceptor: (a) corrugated plate pack, (b) unit in sump

The principal safety consideration in the design of process sewers is to prevent fire spreading through them. Here much depends on the flammability, water solubility and density of the liquids. There are usually several separate drainage systems:

1. **For drains in curbed and paved process areas and road and rail tanker loading and offloading stations.** Process spillages of all kinds and aqueous effluents from the process itself (e.g. washings, water from process separators) enter this system. Care must be taken in handling solutions which would react in the drain, e.g. dilute acids and alkalis or sulphides. If this is liable to cause a hazard, some pretreatment of one or more of the liquids is needed before they enter the system. The sewers are usually trenches lined with suitably resistant tiles and covered with concrete slabs which can be lifted for cleaning. The trenches are often built in short sections with syphon-type liquid seals between them to act as flame traps, and they may kept be purged with inert gas. This system leads into appropriate separation and treatment plant located on the periphery of the site at a point of natural drainage.
2. **For storm water from roads, roofs and open areas of the works.** These follow normal civil engineering practice.
3. **Hot cooling-water discharge.** This is usually to a natural drainage channel.
4. **Domestic drains which discharge into public sewers or septic tanks.**

The problem in segregating (1) from (2) is that a considerable amount of storm water usually enters (1), while there is always a risk of process liquids entering (2). If (1) and (2) are combined, substantially larger treatment plant may be needed. The drainage area within tank bunds sometimes has valved outlets leading to either (1) or (2). The outlet leading to (1) is provided with a liquid-sealed lute if the tank contains a liquid lighter than and immiscible with water.

A further problem lies in the drainage and treatment of the large volume of fire water used in emergencies. This will normally be contaminated with process fluids and enter (1) but even if the drains and sewers are designed to handle it, it may be difficult and expensive to design the separator and treatment plant to do so.

In cases where the presence of raw process effluents in the final watercourse could be very damaging and costly, the provision of large impounding basins should be considered. These should be provided with pumps with alternative discharges, to the separator and to the final destination. The impounding basins would normally be kept pumped empty of rain water but would be used to impound raw process effluent in emergencies when the flow is too great for the separators. This raw effluent can then be pumped through the separator and treatment plant when the emergency is over.

Where flammable hydrocarbons may be present, the effluent-separation and treatment area and the impounding basin and its pumps should be treated as a hazardous zone from the point of view of electrical ignition, and access to personnel and vehicles should be restricted. The impounding basin should then include some simple means (e.g., baffles and weirs) for removing hydrocarbons from the water and returning them to the separator.

16.6.8 Road and rail tanker loading and unloading stations

Many accidental releases of dangerous liquids in process plants have occurred at road and rail tanker loading and unloading stations. This is hardly surprising considering the different hazards which concentrate and interact at these points. They include:

1. Vehicle hazards – collisions, errors in parking position, unauthorised starting, and damage to and poor maintenance of vehicles outside the control of the project;
2. Hazards of temporary hose and solid pipe connections;
3. Hazards of ignition both from vehicle engines and from static electricity;
4. Hazards of overfilling and errors in sequence of valve opening and closing.

As practically every material handled as a liquid in process plant is liable to be transported in a road or rail tanker, the spectrum of possible material hazards is very wide. Liquids as diverse as liquid oxygen, petrol, molasses, aqueous caustic soda, milk, beer, molten sulphur, liquefied chlorine, liquefied propane, oleum and hydrazine have to be loaded, transported and unloaded.

While the design of the facilities and vehicles needed and the precautions necessary are highly specialised, certain common (and possibly obvious) design precautions are needed in every case.

Loading and offloading should only be allowed at carefully selected sites where the necessary facilities have been installed. An example of a completely unplanned process plant which had simply 'grown like Topsy' was described in 16.5.6. This included an unplanned offloading site where a volatile liquid which was both toxic and flammable was unloaded from an internal works road by a makeshift arrangement into a works tank.

Loading and offloading sites should be within but near the perimeter fence, in paved or concreted areas which are used for no other purposes. The areas and rail tracks at loading and unloading bays should be perfectly level, and, if possible, the whole site should be level. The bays must be clearly marked and generally provided with light roofs and upper side coverings to protect against sun and driving rain, but without end walls, so as to allow through-transit of vehicles and good natural ventilation.

There should be a weighbridge at the entrance to the site over which all vehicles should pass on entering and leaving. Close to the weighbridge there is usually a despatch office where written loading or unloading instructions are given or confirmed. Only vehicles approved for the liquid in question may be used. These should have the maximum contents and in the case of trailers and rail tankers, the tare and maximum full weights stamped on them. They should also be provided with appropriate means of gauging the contents, and of indicating or warning when the approved filling ratio is nearly reached. Vessels carrying liquefied gases under positive pressure have pressure and temperature gauges, a valved vent and in many cases a pressure-relief device.

Only the vehicles to be loaded and unloaded, and authorised fire and emergency service vehicles, should be allowed within the site, and the number of vehicles permitted within it at any one time should be strictly restricted. Separate sites should be used for different products or types of materials, and the loading and unloading of different materials such as ethylene oxide, chlorine, ammonia and molten phenol on the same site should not be permitted. If necessary, a special vehicle park for road tankers may be arranged outside the works perimeter with its own security fence and gatehouse. Road tankers should follow a planned route through the site, proceeding from the weighbridge first to a small adjacent waiting area where the vehicle can be inspected. It is then driven forwards into the transfer bay, and forwards again after transfer through the far end of the bay and round by a different route to the weighbridge.

Wheel chocks and adequate vehicle earthing leads and connections, with any necessary interlocks, must be available at the loading bay. Special permanent markings on the ground and/or structures should be made to assist drivers of vehicles of different makes and models to position them easily in the correct location for coupling-up at the first approach. If the driver does not succeed the first time, he should not have to reverse and manoeuvre in the bay itself, but go through and out of the bay, round it and approach it again from the front.

The pipework, loading arms, couplings, hoses, valves, pumps, compressors, meters, instrumentation, line heating or refrigeration (if required), pumps and permanent storage tanks or vessels should be designed for the liquids in question. Non-flammable liquids in vented tankers are usually pumped through a single filling line, but liquefied gases have a liquid filling line with a return line for uncondensed gas leading back to the appropriate vessel on the installation. In some cases the liquefied gas is pumped. In others a compressor is installed on the gas return line to raise the pressure in the vessel being emptied and cause liquid to flow through the liquid connection. The installation should be designed by a specialist firm with experience in this field. The loading arms or hoses should be provided with special valves or valved couplings which reduce spillage when disconnecting to the barest minimum, and with emergency valves (usually excess flow type), alarms and interlocks to stop the flow and provide warning of line rupture. The installation must be carefully protected against vehicle collisions. The installation should allow the contents of a tanker which is being loaded to be unloaded back to storage without disconnecting any couplings or flanges.

Means must be provided to catch liquid spillages and render them harmless. Special ventilation and pressure-relief systems with appropriate means of disposal or neutralisation may have to be provided. Inert gas may have to be suppled for vented tankers carrying certain flammable or reactive liquids, or for displacing air from new or recently maintained tankers. Monitoring equipment for hazardous gases and flames may also have to be provided, as well as emergency protection such as water sprinklers, monitors and curtains, foam and/or dry powder systems. Sufficient emergency showers, protective clothing and firefighting equipment should be available at strategic points. Adequate lighting and means of communication must be provided, with emergency back-up.

Solid lines for tanker connection are preferable to hoses, but where there is no alternative to the latter, they must be appropriate to the duty and frequently inspected. For measuring and checking on the quantity filled, two independent methods should be used while filling is taking place.

Wells[1] stresses the dangers of a vehicle being driven away while a hose or loading arm is still attached to it, and suggests some safeguards. Lees[2] emphasises the dangers of ignition of a flammable vapour/air mixture by static electricity [6.3] when a vented tanker is being filled. This danger is greatest when the tanker is filled from the top with a high liquid velocity. Lees gives detailed advice on how to prevent static ignition when filling tankers.

16.6.9 Industrial fire-fighting services

The types of service included here are:

- Emergency 'first-aid' equipment to enable operatives to fight small fires when they start, e.g. water hose reels, snuffing steam lances, portable extinguishers, sand buckets and blankets made of safe non-combustible fibres;
- Permanently installed wet and dry sprinklers, water deluge systems, foam pourers in tanks of flammable liquids, water and steam curtains, water and foam monitors and gas-inerting and explosion-suppression systems;
- Fire-water systems including water supply, tanks and reservoirs, fixed pumps, ground and rising mains, hydrants and stand pipes;
- Mobile appliances, pumps, hoses, branch pipes, nozzles.

Unlike most design activities discussed in this chapter, the fire-fighting services provided within a works or factory make no direct contribution to production. Historically, fire-fighting services grew from the needs of the insurance industry, and today the services provided on any installation are decided largely by insurance considerations.

The main subject of this book, as explained in the Preface, is the prevention of process accidents rather than dealing with their consequences. Industrial fire-fighting services have become such a large subject, including several specialised fields, that it is quite impossible to cover it in this short sub-section. Underdown's informative and practical guide[33] to the whole subject, including Fire Offices' Committee (FOC) rules, public and private water supplies, automatic sprinklers and portable and fixed extinguishing systems, is recommended reading. Several other sources of help are suggested in Appendix M.

Apart from the technical, insurance and economic aspects, the selection of services for any installation depends on knowledge of the services available from public water companies, fire authorities, police and insurance companies, and the legal and contractual conditions attached to such services.

An important factor to be considered carefully in the civil and structural and electrical design of process plant is the damage that may be caused by fire water and foam. There are accounts of material stores protected by sprinkler systems where the damage caused by the accidental triggering of the sprinklers was comparable to that of a major fire. Drains designed to cope with large amounts of fire water readily become choked with debris in a real fire, and the weight of fire water on floors in buildings has caused them to collapse. The quantity of fire water used has often overloaded the effluent-treatment facilities, causing severe pollution downstream. It is thus essential for civil engineers to know the maximum flows of fire water to be expected in any part of the plant. In the same way the fire authorities and fire protection engineers may have to be advised on any limitations imposed by the overall design on the maximum quantities of fire water and foam that may be used.

Two mistakes which the writer has found in the design of ring mains and fire pumps for LPG and similar installations are mentioned here:

1. A large fire-water ring main is often kept pressurised by a 'jockey' pump with high–low pressure switches, set to cut out at say, 14 bar g., and to cut in again at, say, 12 bar g. Unless there is an adequately sized air-filled cushion chamber connected to the pump discharge, a small water leak from the ring main or through a non-return valve on the pump discharge will cause the jockey pump to start and stop every few minutes, causing frequent pressure cycling in the main, with the risk of fatigue, and premature wear of the pump, motor and starter.

2. The main fire-water pumps usually have a 'kick-back' line with a back-pressure control valve from the pump discharge leading back to the fire-water tank to protect the pump when demand is very low. The diameter and length of this line should be such that the minimum flow needed to protect the pump occurs when the back-pressure control valve is fully open. If the diameter of the line is greater than this, there will be a large pressure and energy drop across the half-open valve, causing cavitation and rapid wear of the valve.

Finally the point must be made that any form of fire protection provides protection only for the situation for which it was envisaged, and in the case of explosions it may be just as vulnerable as the plant it is designed to protect. Thus it is sobering to re-discover how:

The best laid schemes o' mice an' men
    Gang aft a-gley,
An' lea'e us nought but grief an' pain
    For promised joy.

Tucker of the Fire Research Station described the fire protection equipment at Flixborough before the explosion thus[34]:

There were many fire protection systems and some are listed below. In addition the plant was fitted with the usual safety devices, e.g. overpressure safety valves, flammable gas detectors.

1. An electrical alarm system, using break-glass units, connected to an annunciator board and linked to Fire Brigade Control.

2. Vessels containing flammable liquids were fitted with deluge systems operated both manually and by quartzoid bulbs linked to the fire alarm.

3. Tanks storing flammable liquids were contained within bunded areas fitted with foam pourers. The cyclohexane tanks in the tank farm were fitted with foam injection as well.
4. Tanks and other vessels were inerted with nitrogen.
5. There were 44 hydrants fed by a main supplied by two 5700 l/m (1250 gal/min) pumps from a 2.3 M litre (500 000 gallon) tank.
6. Two additional hydrants near the site perimeter were fed by towns mains.
7. The river Trent is close to the site but is tidal at this point (Figure 16.8). However emergency water was obtained from the six 4.5 M litre (1 M gallon) lagoons near the wharf used to supply the steelworks.

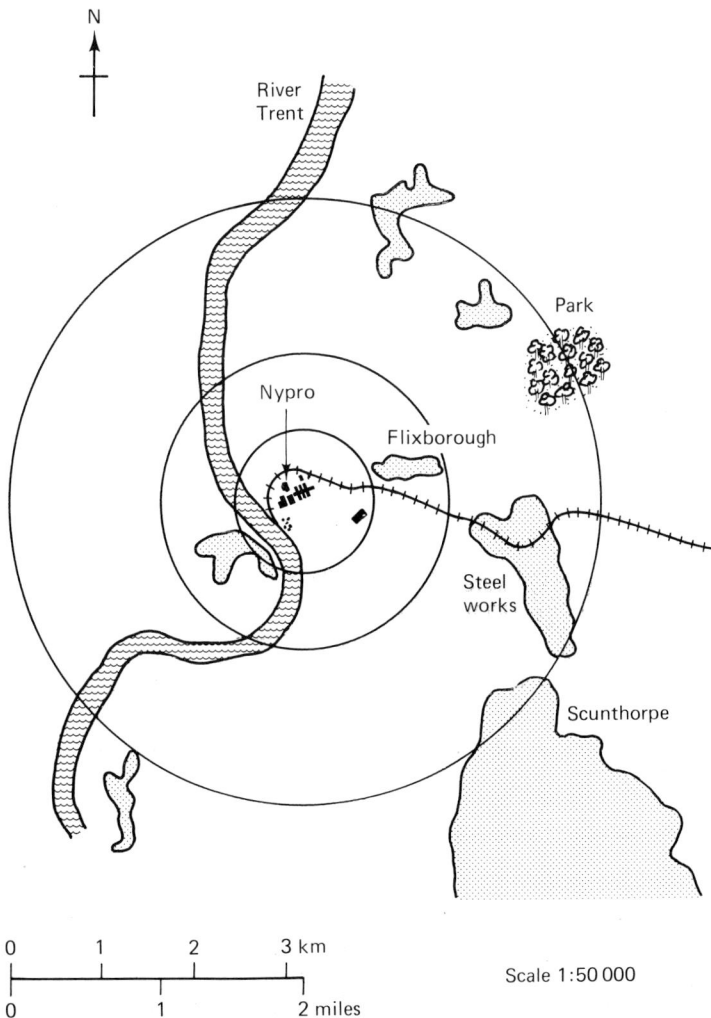

Figure 16.8 Area around the Nypro works (courtesy the Controller, HMSO)

The part played by these systems during the disaster is further described by Tucker:

> The fire alarm was actuated before the main explosion and the initial fire brigade assistance was despatched [from Scunthorpe Fire Station].
> The explosion wrecked most of the fire protection equipment on the site. The fire pumps were put out of action by the collapse of the pump house and destruction of the electric power lines. The fire-water tank did not rupture since it was full and the inertia of its contents protected it. The fire main was broken in several places but the two town's mains survived although the hydrant on the large main was inaccessible due to the fire in that area. The dry sprinkler system in the caprolactam warehouse was completely destroyed along with the building itself. The bunds around the tanks in the tank farm remained fairly intact.

Thus apart from the half-minute advance warning of the explosion given by the alarm, and the limited protection against fire spread given by the tank bunds, the many fire protection systems installed at Flixborough played no useful part in fire protection or fire fighting. Fire fighting was done entirely by outside fire services with improvised water supplies.

16.7 Hazard Studies IV to VI

The three remaining hazard studies recommended by Gibson[5] are now briefly described.

Hazard Study IV is a check made by the senior manager responsible for commissioning to confirm that all the actions called for in the previous hazard studies have been fully implemented before commissioning starts.

Hazard Study V is first, a safety inspection of the new plant, made before commissioning by representatives of the works or factory and of the design organisation, to check, among other things, that all statutory requirements have been met, e.g. provision of adequate access and escape routes, machine guards, showers and respirators. If significant modifications are made to the plant during commissioning, these must be properly recorded and flow diagrams, drawings, etc. amended. The study should then be extended to consider the effects on such modifications on plant hazards, and their relationship to decisions taken in earlier hazard studies.

Hazard Study VI is an audit made as soon as possible after the plant is in production, to check that it satisfies the criteria adopted for major hazards and that the original design assumptions have not been invalidated by changes made during commissioning and early operation. The report of this audit should be combined with reports on Hazard Studies I to V for future reference. The record should draw attention to deviations from design concepts, to design shortcomings, equipment failures and operating difficulties which affect plant hazards.

While this approach cannot claim to identify every hazard, the systematic and formally structured studies recommended by Gibson and described here are appropriate techniques for identifying problem areas at the design stage.

16.8 Computer-aided design (CAD)

Computers are widely used both for design calculations, drawings, e.g. for material balance calculations [16.4.3], in place of three-dimensional models [16.6.2] and for piping isometric drawings [16.6.3]. Process engineers use computers in conjunction with theoretical models of reactors and entire plant to run mass and energy balance calculations needed for equipment sizing and capital and manufacturing cost calculations. Programs are available for an increasingly wide range of design calculations. Those used for pipework isometric drawings are claimed to saving drafting time and reduce the number of errors. Important design decisions are taken on the bases of electronic calculations.

Concerned with the legal and safety implications if things go wrong, the Institute of Chemical Engineers has published a booklet on computer-based design decisions[35]. The result of a wrong design decision is, of course, the same whether the data search and/or calculation was made manually or by a computer. It may only be that the plant does not perform satisfactorily, but it could be a serious accident.

One safeguard provided by most computer software is that it will not accept inconsistent data, which leaves the user tearing his hair and wondering what he has done wrong. A good computer system usually reduces the level of errors. 'The danger lies in them not being spotted, or even looked for, because of the extra credibility that data printed on continuous stationery offers.'

The following guidelines are important for engineering management to ensure that they discharge their obligations under HSWA 1974:

- The use of a computer with design software cannot be regarded as a substitute for design expertise. Engineers using computer aids must fully understand the engineering problems and have the ability to check the computer outputs.
- Records of errors or failures during the initial testing and use of software, and the circumstances in which they occurred, should be kept and their lessons learnt.
- The mere filing of computer print-outs is not acceptable. They must be accompanied by a clear statement of the problem to be solved and an interpretation of the results.
- Engineers must be able to justify the use of a particular computer-aid and data for a specific problem.
- Untried programs written by design staff to solve particular problems should be treated as highly suspect. Proven programs are available for a wide range of applications. Only after a search has shown that no appropriate and satisfactory program is available should a new one be written. This must be thoroughly tested before being used for design purposes.
- Programs used should be designed to minimise opportunities for misinterpretation or error by the user and should make failures obvious.
- The solution methods used and their applicability, accuracy and range of validity should be provided by the vendor with any program. The engineer needs sufficient information to select a suitable method, to

judge the reliability and accuracy of the results, and to check them by hand calculations.

The booklet contains much other useful information, including appendices on common sources of error, typical manual checks and legal issues for the engineer.

References

1. Wells, G. L., *Safety in Process Design*, George Godwin, London (1980)
2. Lees, F. P., *Loss Prevention in the Process Industries*, Butterworths, London (1980)
3. Fawcett, H. H. and Wood, W. S., *Safety and Accident Prevention in Chemical Operations*, 2nd edn, Wiley-Interscience, New York (1982)
4. Kletz, T. A., *Cheaper, Safer Plants or Wealth and Safety at Work*, The Institution of Chemical Engineers, Rugby (1984)
5. Gibson, F. B., 'The design of new chemical plants using hazard analysis', in I. Chem. E. Symposium Series no. 47, *Process Industry Hazards, Accidental Release, Assessment, Containment and Control*, Rugby (1976)
6. Balemans, A. W. M. *et al.*, 'Checklist – Guide lines for safe design of process plant' in *Loss Prevention and Safety Promotion in the Process Industries*, Elsevier, Amsterdam (1974)
7. United Nations Industrial Development Organisation, *Manual for the Preparation of Industrial Feasibility Studies*, United Nations, New York (1978)
8. Vervalin, C. H., *Fire Protection Manual for Hydrocarbon Processing Plants*, Gulf Publishing, Houston, Texas (Volume 1, 1984, Volume 2, 1982)
9. Kletz, T. A., *HAZOP and HAZAN – Notes on the identification and assessment of hazards*, The Institution of Chemical Engineers, Rugby (1985)
10. Fitt, J. S., 'Design for safety in the contractor's office', in I. Chem. E. Symposium Series no. 47, *Process Industry Hazards, Accidental Release, Assessment, Containment and Control*, Rugby (1976)
11. Anderson, P. W. P., 'The effect of safety costs on the competitiveness of tenders', in *Protection*, London (March 1980)
12. Ludwig, E. E., *Applied Process Design for Chemical and Petrochemical Plants, Volume 1*, 2nd edn, Gulf Publishing Company, Houston, Texas (1977)
13. Austin, D. G., *Chemical Engineering Drawing Symbols*, George Godwin, London (1979)
14. Sherwood, D. S. and Whistance, D. J., *The 'Piping Guide'*, Chapman and Hall, London (1979)
15. Jarvis, H. C., *Chemical Engineering Progress*, **67**(6), 41–44 (1971)
16. Lawley, H. G,. 'Operability studies and hazard analysis', in A. I. Chem. E. Symposium, *Loss Prevention in the Chemical Industry*,, New York, (November 1973)
17. HSE, *Canvey: an investigation of potential hazards from operations in the Canvey Island/Thurrock area*, HMSO London (1978)
18. Mecklenburgh, J. C., *Process Plant Layout*, George Godwin, London (1985)
19. Liston, D. M., 'Safety aspects of site selection, plant layout and unit plot planning' in *Safety and Accident Prevention in Chemical Operations*, edited by Fawcett, H.H. and Wood, W. S., 2nd edn, Wiley-Interscience, New York (1982)
20. Lewis D. J., 'Application of the Mond Fire, Explosion & Toxicity Index to Plant Layout and Spacing Distances', *Loss Prevention*, **13**, 20, (1980)
21. Barton, J. C., Finley, C. R. and Faulkner, J. R.,'CAD system used for major offshore project', *Oil and Gas Journal* (3 March 1986)
22. ANSI B31, *Pressure Piping*, American National Standards Institute, New York
23. BS 3351: 1971, *Piping systems for petroleum refineries and petrochemical plants*
24. BS 3974, *Pipe supports*, Part 1 (1974), Part 2 (1978), Part 3 (1981)

25. Langeveld, J. M., 'Structural design of control buildings', in I. Chem. E. Symposium Series no. 47, *Process Industry Hazards, Accidental Release, Assessment, Containment and Control*, Rugby (1976)

26. Marshall, V. C., 'The siting and construction of control buildings – a strategic approach', *ibid.*

27. Oil Insurance Association, 'Jumbo oil storage tanks', *Loss Control Bulletin no. 400-2*, Chicago (March 1975)

28. American Oil Company, *Refinery Process Safety Booklets, 1. Hazard of water, 2. Hazard of air, 5. Hazard of electricity, 6. Hazard of steam*, AMOCO, Chicago

29. Oil Insurance Association, *Boiler Safety*, Chicago (1971)

30. Oil Insurance Association, *Fired Heaters*, Chicago (1971)

31. British Cryogenics Council, *Cryogenic Safety Manual*, London (1970)

32. Sax, N. I. (ed.) *Industrial Pollution*, Van Nostrand, New York (1974)

33. Underdown, G. W., *Practical Fire Precautions*, 2nd edn, Gower Press, Aldershot (1979)

34. Tucker, D. M., 'The explosion and fire at NYPRO (UK) Ltd Flixborough on 1 June 1974', I. Chem. E. Symposium, *Technical Lessons of Flixborough*, Rugby (1975)

35. The Institution of Chemical Engineers, *The Engineer's Responsibility for Computer-based Decisions*, Rugby (1987)

Chapter 17

Maintenance and inspection

Without regular inspection and prompt maintenance, plant and machinery sooner or later lapse into a dangerous state as a result of wear, fatigue and corrosion. Regular inspection is needed to determine in detail how far such deterioration has proceeded, and since its message often runs counter to immediate production pressures, those responsible for inspection should, as far as possible, be insulated from them. The person responsible for plant inspection in any organisation should be given the authority to over-rule production pressures, especially when the safety of personnel and the integrity of the plant are clearly at stake.

Prompt maintenance is needed to keep plant in sound running order. In many small and medium-sized organisations the functions of maintenance and inspection are combined in a single department. In larger organisations and those with hazardous processes, a separate and independent inspection department is common. This is particularly needed in organisations where much of the maintenance is carried out by production personnel.

Although virtually all manufacturing organisations recognise the need for sufficient maintenance to support production, there are wide variations in the way inspection and maintenance are organised and performed, in the competence and authority of those performing these functions, and in the backing given by management to inspectors and engineers who recommend urgent action which conflicts with current production targets. The Flixborough disaster [4] provided an object lesson in these points.

Inspection starts with that of the materials and equipment ordered for the construction of the plant and continues throughout its life. Maintenance starts later, after a plant is commissioned, but it should be planned for during design. Some inspection has to be done by maintenance workers even where there is a separate inspection department.

Most maintenance and much inspection require the plant or equipment to be shut down and specially prepared [18.6]. A limited and clearly defined number of maintenance jobs including lubrication are done while the plant is running. 'Condition monitoring' [17.5] is the name given to various inspection techniques carried out while the plant is running. Mandatory inspections which are usually carried by specialist companies are required for cranes, boilers and some other items. Similar special inspections may also be required by the plant insurers. Inspections may be of various kinds (mechanical, electrical, civil, safety, etc.) and it is essential that their limitations are clearly stated and recognised.

A detailed equipment register and a good record system [17.1.1] are vital to maintenance and inspection to ensure that items are not overlooked during inspections and that the proper lessons for maintenance and replacements are learnt. The register should include the following details of each item:

1. Identification number, order number and drawing number;
2. Specification, design parameters and process materials in contact with the item;
3. Inspection/test reports during manufacture. For pressure vessels and other critical items, the reports should include tests on materials of construction, radiographs and other weld tests and pressure and leak tests;
4. Inspection/maintenance categories, intervals, methods;
5. Features relevant to corrosion, wear, failure;
6. Materials/parts list;
7. Date of entry into service.

Reports on maintenance, modifications, inspections, tests and changes of duty of the equipment should be filed in the register. Mechanical inspection reports on operating plant should state the condition of the equipment and how much wear, corrosion, etc. has occurred since the last inspection. The inspector may require certain repairs or replacements to be carried out, and a further inspection before it enters service again (as with motor vehicle tests). He may certify the plant or equipment as fit to resume normal service for a stated maximum period before the next inspection, or he may 'de-rate' it for less onerous conditions of temperature, pressure, load, etc., or he may condemn it to the scrap heap.

17.1 Maintenance

Hazards connected with maintenance are of three kinds:

1. Those caused by lack of it, thus allowing dangerous conditions to arise;
2. Those occurring during it, e.g. gassing of a maintenance worker inside a vessel;
3. Those caused by faulty maintenance, e.g. nuts not properly tightened.

Hazards of type (1) and the maintenance organisation needed to prevent them are discussed first. Maintenance workers and their safety are discussed next. Hazards of types (2) and (3) are dealt with in Chapter 18.

Most plant and machinery is subject to wear, fatigue and corrosion, and unless its working life is very short, maintenance will become necessary at some stage. Anticipated annual maintenance costs regularly appear on manufacturing cost estimates for new projects as percentages on capital invested (3–6% on plant costs and 2–3% on civil work and offsites).

Many companies' maintenance costs exceed their profits, to which they make no direct contribution. Cost cutting in maintenance is therefore common. Yet maintenance is critical to safety, and maintenance workers generally require a high degree of skill and safety awareness.

Table 17.1 Maintenance costs as percentage of sales in various industries[1]

Industry	Maintenance costs as percentage of sales
Chemical and allied	6
Food, drink and tobacco	2
Paper, printing and publishing	4
Textiles	3
Vehicle manufacture	2
Electrical engineering	2
Shipbuilding and marine engineering	1
Clothing and footwear	2
Total, all manufacturing industries	3

In the process industries, particularly chemicals, maintenance costs (in relation to sales revenue) are higher than in other manufacturing industries, as Table 17.1 shows[1].

Most maintenance requires the plant to be shut down. A minor job is usually known as a repair and a major job as an overhaul. A limited and clearly defined number of maintenance jobs are done as 'running maintenance'. Servicing (which includes lubrication) often comes under this category, and is defined as maintenance in which only consumables are replaced[2].

Maintenance may be 'planned' or 'unplanned'. Unplanned or 'breakdown' maintenance means operating the plant until something breaks down. It may be classed as 'emergency' or 'corrective', depending on its urgency, although the work done may be the same in both cases. Although economical and appropriate in some industries, there is less scope for breakdown maintenance in the process industries because of the hazards caused by breakdowns, and of the high costs of interruptions to complex continuous processes which depend on long production runs for profitability.

Planned maintenance involves both preventative maintenance and corrective maintenance. The latter is still necessary to some extent, because it is impossible to guarantee that all items covered by the preventative maintenance scheme will never break down, and because it is not practical to include everything in the scheme. In corrective maintenance where the work to be done is seldom known precisely in advance, the item to be maintained must first be inspected.

Preventative maintenance is based on servicing and overhauling key plant items, hopefully before they break down or their performance deteriorates. This is generally done at pre-selected intervals, when it is known as 'scheduled preventative maintenance'.

The performance of heat exchangers, furnaces, catalysts or pumps in many continuous processes deteriorates with run-time, leading to loss of temperature, pressure, yield or throughput. A point is finally reached where it is more economical to shut the plant down and overhaul it than continue running at reduced efficiency, even though there has as yet been no breakdown. The economic run-length is usually known from experience, and the scheduled maintenance is planned accordingly. But if conditions require that the run has to be terminated earlier for maintenance, this may be described as 'condition-based maintenance'. In other cases preventative maintenance may be based entirely on the condition of the plant, machinery or equipment, without any fixed time schedule. This is not the same as breakdown maintenance, but depends on some form of condition monitoring to indicate when maintenance is needed.

In the simplest cases the human sense organs, particularly the ears, eyes, touch and sometimes the smell of an experienced operator or mechanic, are often the best indicators of when something is going wrong, and needs to be shut down for inspection and maintenance. For costly and complex machinery subject to wear or fatigue, and for pipes and equipment liable to corrosion, a range of condition-monitoring equipment [17.5] is available to complement the human senses, and indicate when maintenance is needed.

The amount and cost of maintenance can be reduced in the design stage by the use of better and more corrosion-resistant materials and components. It is no less important to ensure during design that all plant items can be properly prepared for maintenance as well as maintained [J]. An economic balance has to be struck between the probable cost of breakdowns and the capital cost of the plant. In some industries such as nuclear energy and underwater oil production, maintenance is so difficult and expensive that extreme measures have to be taken to design out the need for it in the first place.

17.1.1 Preventative maintenance

The legal requirements in the UK for the periodic inspection and maintenance of steam boilers, cranes, power presses, fire-warning devices and other items form a basis on which a more thoroughgoing system of preventative maintenance can be built up. Preventative maintenance has been practised for many years in the process industries, particularly oil refining and petrochemicals. Its principles are thus described here only broadly.

It would generally be uneconomic to apply preventative maintenance to every plant item, to the extent that complete freedom from breakdowns can be guaranteed, although the need for this is greatest when many lives are at stake, as with passenger aircraft. It is generally accepted that the most cost-effective system of maintenance is a mix of preventative and corrective maintenance[3].

The introduction for the first time of a system of (scheduled) preventative maintenance into a works takes time and preparation. It is apt to be resisted by maintenance workers, who complain that it downgrades their skills and turns them into robots. Unplanned maintenance, on the other hand, uses human time inefficiently, since tradesmen may be idle for long periods just waiting for something to go wrong. Thus as a young and eager plant supervisory chemist in a wartime munitions factory, I suggested to a group of maintenance tradesmen who were playing cards that there were several useful jobs they might do. 'You should not talk to us like that', one replied, 'You are only the effing chemist but *we* are the effing fitters'.

Scheduled preventative maintenance is best introduced in stages, starting with one department where the cost of frequent interruptions to production caused by breakdowns has demonstrated a case for it. First a complete inventory of all the facilities (machinery, equipment, buildings, etc.) to be maintained has to be prepared. Every item should be shown on a scale plan of the plant and given a specific code number. The numbering system adopted must be chosen with care to ensure that it describes all relevant present and future aspects of each item. The numbering system shown, for instance, on a P&I diagram will not be sufficient if there are identical pumps, motors, valves, etc. with different plant numbers performing different duties. Once a pump or motor has been removed and maintained, it may be returned to the stores before being later re-installed, perhaps in a different location where it performs a different duty. Each pump or motor, etc. therefore needs an identification number which will stay with it throughout its life, so that a record can be kept of its complete history with its present and former duties (which are identified by plant numbers, e.g. depropaniser overhead pump P.204B). Few people, however, will want to know the history of every nut and bolt.

Two important digits should be included in the identification number for every item. One is a priority rating, e.g. from 1 to 5, to indicate how important the proper functioning of the item is to safety and production. Priority rating 1 implies that breakdown would create a serious hazard, or an interruption to production affecting a large area of the plant (e.g. a centrifugal compressor with no installed spare). On hazardous processes, care is needed to include in priority rating (1) those items and parts which have been identified from hazard studies [14.3 and 16.5 5] as ones whose deterioration or failure could cause a serious hazard.

Priority rating (5) signifies equipment which would not affect output or safety in any significant way if it broke down (e.g. one of several identical vacuum cleaners). Priorities (2), (3) and (4) are used for items whose breakdown would have intermediate consequences (such as a pump for which there is an installed spare, ready to be started immediately). Criteria in which safety plays a large part have to be established for these priority ratings.

The other digit should reflect the maximum interval allowed between successive preventative maintenance operations. This may depend on likely wear, corrosion or reliability, on which information must be gleaned from various sources. These include:

- The designers of the plant, who should have considered the question carefully in the first place;
- Maintenance records of similar plants;
- Data banks such as that kept by the Systems Reliability Service of the UK Atomic Energy Authority, Culcheth, Lancashire.

The interval between maintenance operations should be reviewed periodically in the light of the plant's own maintenance and inspection records.

From the inventory, a list of items to be included in the preventative maintenance programme is drawn up based on their priority ratings. A maintenance schedule is prepared for each item on this list, specifying both the maintenance work required and the frequency. Items such as pumps which have installed spares, and jobs such as lubrication which can be safely carried out without shutting the plant down, impose the least limitations on the frequency of (preventative) maintenance. There are also no serious limitations for plant operated for five days per week which can be maintained at weekends.

For large-scale continuous processes in the oil and petrochemical industries, which may aim to achieve long runs of two years between planned shutdowns, the situation is different. Most maintenance can then only be done when the whole plant has been shut down, isolated and freed of process materials. This requires careful study in the design stage of all items which cannot be changed or maintained while the plant is running. They must all be so reliable and free from corrosion and wear that there is little chance of their failure during operation.

Preventative maintenance requires a great deal of organisation. All jobs need to be clearly defined and times allotted to each, so that a balanced team of tradesmen in the different specialities required can be recruited and employed with least interruption. Careful stock control and timely ordering of all spares and engineering materials which may be needed are essential. Standard job report cards, which may be in the form of a checklist, should be prepared, issued to, and completed by tradesmen for all equipment being maintained. These list the components, materials and spare parts used when doing the job, with the worker's impressions on the state of the equipment before maintenance, and his opinion on the need for further action later. Sometimes it may be preferable to use separate cards for repair and inspection work.

From the completed job report cards, a 'history card' for each equipment item, which collates all reports and operating experience on it, is prepared and kept up to date. This records the maintenance work done on the item in the past, the defects found and the corrective action taken, as well as all labour and material incurred in maintenance.

From these two cards, management can modify the future maintenance programme, extending the intervals between inspections of equipment showing little or no wear, and reducing it for equipment which has broken down between inspections.

The next stage in planned maintenance is to use a computer. Once a suitable program has been written and debugged, and staff have been trained, the computer can print out the maintenance programme week by week, with all necessary job cards, stating the class of tradesmen and all tools, machines, spare parts and materials which may be needed. It can be extended to analyse trends in maintenance, to control spares and to order materials and components.

While the merits of centralised versus decentralised maintenance organisation are not discussed here, the growing trend to use outside contractors for maintenance work can pose serious hazards. Contractors' men are seldom as familiar with the plant, personnel, hazards and safety procedures as the manufacturing company's own employees.

17.1.2 Preparing for maintenance on new plant

Preparations for maintenance should start in the design stage [J]. The future maintenance engineer for the plant should, if possible, be a member of the design team, or at least be in close touch with it. Three aspects are important:

1. The reliability and maintainability of machinery and equipment under the proposed working conditions should be considered when they are selected. This is especially important for plants which have to run continuously for long periods between maintenance shutdowns.
2. The corrosion, fatigue and wear resistance of the materials of construction are vital points in their selection. They should be considered both in relation to their normal operating conditions and to any expected though less common conditions which could be far more corrosive, e.g. during start-up, shutdown and the removal of process materials (e.g. acid tars). Any special welding, heat treatment or other requirements of the materials also need to be considered such as argon arc welding of titanium.
3. The way in which the equipment will be maintained needs to be thought out in some detail to ensure that there is adequate access for personnel, cranes and mobile plant. Davits and permanent beams for travelling cranes may be required for removing compressor rotors and the tube bundles of heat exchangers.

The makers' maintenance manuals, guarantees and obligations in case of failure for all items of plant and equipment should be reviewed. An overall maintenance manual for the whole plant should be prepared which incorporates this information. The manual should include detailed schedules for preventative maintenance, lists of all spares and materials needed, with the names and addresses of suppliers, the quantities required and available as initial inventory, and the anticipated annual consumptions. Arrangements for segregated storage and identification of all maintenance materials and spares must be made, and a system of stock control, authorised withdrawals, ordering and records must be instituted. Suitable tools, machines, equipment, workshop facilities and personnel must be made available, and personnel organised and trained.

Training must include instruction and practice in the use of permit systems as discussed in Chapter 18 and other relevant safety procedures for the works, including duties in emergencies [20.3] (evacuation and re-assembly and special duties as first-aiders and firemen). Arrangements should be made with outside contractors for work which will not be done by company personnel and for safety training of contractors' personnel.

Inspection requirements, schedules and information exchange should be reviewed between with those responsible for maintenance and inspection.

17.1.3 Maintenance personnel and their work

Maintenance of a large plant may, in the course of time, require as many different trades, skills, tools and equipment as it took to construct it. Steel rigger-erectors, scaffolders, crane and mobile plant operators, ground workers, bricklayers, carpenters, concrete pourers, painters and roof workers may all be needed as well as the more obvious tradesmen such as of welders, electricians, mechanics and instrument engineers. The special hazards of these many trades were discussed in the writer's earlier book[5]. Some of the work is done on site and some of it requires equipment to be moved to a workshop for repair or overhaul.

The importance of maintenance work to safety can hardly be overemphasised. Maintenance workers should be selected for their experience, ability and alertness, and specially trained in accident prevention. They are faced with a complex and changing pattern of hazards and engineering standards. In the morning a mechanic may be dealing with a part with coarse dimensional tolerances and in the afternoon with a shaft with very fine ones.

Whatever their basic trades, they should be trained in the safe use of ladders, slings, ropes, safety lines and harnesses, and other protective equipment which they may need. Their training includes recognition of excessive wear and other defects in the equipment and its removal from service. Maintenance workers who are not continuously engaged in their work often form the backbone of fire-fighting and first-aid teams.

Clothing should fit closely with the minimum of pockets, with breast pockets removed to prevent items falling into machinery, etc. when the wearer bends over. Neckties and loose clothing should not be worn. For some work this also applies to rings, jewellery and wrist watches. Each worker should have gloves and eye protection appropriate to the job in hand. Those working inside tanks, vessels and silos and in high places should wear a harness attached to a lifeline and be trained in its use.

Maintenance workers should make a habit of checking all tools used for wear and defects, and should be trained how to do this. The use of non-ferrous spark-proof tools (generally of beryllium bronze) in areas where flammable gases and vapours are liable to be present has been much debated, since they are weaker and more liable to splay than steel tools. Sparks from the latter are only powerful enough to ignite a few gases and vapours (principally carbon disulphide, acetylene, hydrogen and ethylene oxide). It is widely agreed that steel spanners may be used so long as these and similar substances are absent, but that spark-proof hammers should be used in areas where any flammable vapours could be present.

Where the number of tools carried does not warrant a tool box or bag, a belt may be used with tool carriers at the side of the body rather than behind it, to minimise injuries in case of a fall. Maintenance workers should make a habit of checking their tools before and after a job to make sure nothing has been left inside the equipment on which they were working.

Maintenance workers should be advised of the properties of any hazardous materials which they may encounter in their work and trained how to deal with them.

Before starting a non-routine maintenance job, the maintenance crew should meet to discuss the hazards and plan safe working methods. For specially hazardous jobs, the safety manager should attend the meeting. After a safe procedure has been discussed and agreed, a record of the various steps should be made and copies distributed to the workers involved.

The inspection duties of maintenance workers should be clearly thought out by management and defined in writing.

Permits and isolation procedures discussed in Chapter 18 must be established for work on plant and equipment handling hazardous or hot and very cold materials, or which might become energised during maintenance.

Maintenance engineers should keep up to date with new methods, products and equipment and arrange for workers to attend courses as required. Maintenance procedures should be periodically reviewed for safety and a suggestion scheme set up.

17.1.4 Special on-site maintenance services

Several special maintenance services developed for reasons of economy and safety are now available. They include:

- On-stream leak sealing;
- On-site machining;
- Bolt and stud tensioning;
- On-site testing of pressure relief valves.

The last of these is as much a part of inspection as of maintenance but is included here for convenience.

On-stream leak sealing

The high costs of shutting down a large and complex plant when a leak develops at a flanged joint or a valve has led to the development of methods to seal such leaks while the plant is operating and under pressure. One specialist contractor offering this service on an international basis is Furmanite International Ltd, based in Kendal, UK. Its technicians are specially trained to work safely in potentially hazardous conditions and provided with high-quality protective clothing, respiratory, eye and hearing protection appropriate to the fluid escaping (Figure 17.1). The customer is responsible for providing safe working conditions including access and working platforms.

Fluid leaks sealed in this way include steam, hydrocarbons, a wide range or organic compounds of varying toxicity, flammability and corrosivity, and a range of inorganic fluids including fluorine, hydrogen sulphide, nitric and sulphuric acids. Fluid temperatures and pressures can vary from sub-zero to 600°C, and from vacuum to 350 bar g.

In the commonest case of a flange leak, a path for the leak is first provided through a special adapter fitted to the stud or bolt nearest to the leak, and an injection device fitted in the same way on the opposite side of the joint. The gap between the flanges is closed by a peripheral sealing ring. An injection-moulding compound compatible both with the fluid to be sealed and the temperature of the joint is injected to fill the annulus between the failed gasket and the peripheral seal to form a tough, resilient mass. This forms a new secondary gasket to surround the defective original one. It is important to prevent excess compound from entering the pipework and causing downstream blockages in small-diameter instrument connections, etc. Some 30 000 flange leaks are sealed annually world-wide by this process.

Figure 17.1 In-service flange leak sealing by Furmanite technician (courtesy Furmanite International Ltd)

Maushagen[6] has published a summary of a study made for a West German electricity company on the safety of on-stream leak sealing of flanged joints. Some of the methods he describes differ from those used by Furmanite. He quotes five case histories of accidents, which occurred between 1977 and 1981, while leak-sealing processes were being used. Two accidents resulted from severe internal corrosion of pipework, but two involved the bolting of flanged joints. He raises the important point that bolt stresses inevitably increase when a new annular joint of diameter larger than that of the original failed gasket is formed around it. The greatest stress increase is at a bolt immediately adjacent to the final injection (Figure 17.2). Its magnitude depends on the characteristics of the compound and the skill of the operator. He proposes three precautions for the use of compound sealing processes:

1. The initial bolt stress laid down in the appropriate standard should have been complied with during assembly of the flange.
2. The flange to be sealed and its bolts and nuts should not have been weakened by corrosion or erosion. This usually means that the leaking joint must be sealed promptly after it starts to leak.
3. The increased stresses which will occur in the bolts when the sealing process is employed must be checked individually before each application to ensure that there will still be sufficient safety margin in their yield strength.

Somewhat similar methods are used to seal leaks on valve bonnets and joints on heat exchangers. Seals are also effected on pressure vessel casings such as turbines and condensers.

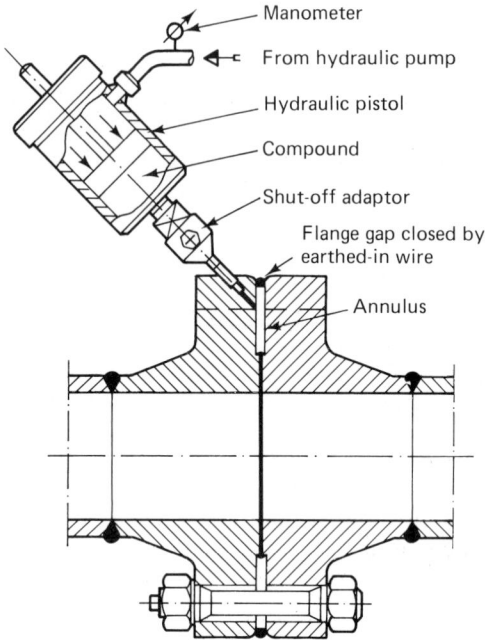

Figure 17.2 Injection of sealing compound around flange joint ring (courtesy H. K. Maushagen of TUV Rheinland, and Institution of Chemical Engineers)

Close cooperation between the competent engineer of the operating company and that of the specialist contractor is necessary for safety.

On-site machining
Various on-site machining services which avoid the need to remove heavy equipment to engineering workshops are also available from Furmanite and other specialised maintenance contractors. These are particularly useful for refacing the flanges of pipes, heat exchangers, cover joints and boiler manholes. Other specialist services include milling of bedplates, cutting of thick-walled pipe, weld preparation, line boring and drilling out corroded studs.

Bolt tensioning
A common cause of flange leaks is uneven and incorrect tensioning of bolts and studs. For flanges operating at high or very low temperatures, the problem is often accentuated by differential thermal expansion between the threaded fasteners and the flanges, which may be of different materials. Even the use of torque wrenches at operating temperature for the final tightening does not entirely solve the problem, for the tension in the threaded fasteners depends on both the torque applied to the nut and the friction between the nut and both the fastener and the flange or washer, which are hard to allow for and can vary considerably. The tightened fastener is then left under a combination of tensile and torsional

stresses, and if the torsional stress is relieved by slippage between the nut and the flange, the tensile stress also changes.

The proper solution is to apply a known and calculated tension to all the bolts simultaneously, and then tighten the nuts to lock in this tension. Special hydraulic devices as well as specialist services are available for this. They are very valuable for tightening flange joints on high-pressure plant where they should be employed during construction as well as maintenance.

Pressure relief valve testing

Another useful specialist site service is the testing of pressure relief valves while the plant is operating under normal pressure. The 'Trevitest' service, also provided by Furmanite, makes use of special equipment for connection to the valve spindle, a hydraulics unit and a recorder. A known force is applied to overcome the spring loading and open the valve for a second or two, while recording the applied force and either the system pressure or the spindle displacement. This can replace normal testing under operating conditions when the plant pressure has to be raised until the valve opens, with the attendant loss of steam or process fluid. The Trevitest technique can also be used when the plant is shut down and depressurised, or for the cold testing of pressure relief valves in a workshop. The test is accepted, with qualifications, by HSE for the statutory examination of safety valves, and has been approved by many insurance companies.

17.1.5 Lubrication

Lubrication is primarily needed to reduce friction between solid surfaces in relative motion, but at the same time it can remove heat, protect against corrosion, seal gaps and scavenge contaminants. Examination of the lubricant in use can provide much useful information on the condition of machinery and is a form of condition monitoring [17.5.4]. Since lubricants are used over a wide range of temperatures and other conditions, many different ones have been developed, which often causes problems.

The lubrication requirements of a works or factory are generally served by a section of the maintenance department on a scheduled basis, although some lubrication is often done by operating personnel. A full survey should be made showing all parts of every machine, etc. which requires lubrication, the type of lubricant used, how it is to be applied, and how often. Machine and equipment suppliers provide information on the correct lubricants to be used. The survey should include a physical inspection of the lubrication points. Missing fittings and oil cups should be replaced and oil holes that have become blocked should be cleaned out.

Lubricants for machine tools are now classified according to BS 5063: 1982[7] and lubrication instructions are presented according to BS 5739: 1979[8]. While intended for machine tools, these standards cover most needs of the process industries. The corresponding ISO standards are ISO 5169 and ISO 3498. Lubricant suppliers will assist works and factories with advice on lubricants and in carrying out lubrication surveys.

Lubrication should follow a carefully prepared programme in which each worker has a supply of all necessary oils, greases, pumps and guns, and lubrication diagrams for all machinery. These should show which parts require lubrication, the type of lubricant required and how often it should be applied. Motors and machines should have been chosen so that they can be lubricated safely without stopping them or removing guards. In cases where this has not been done the programme must draw special attention to any motors and machines which must be stopped and any guards which must be removed for lubrication. Clear responsibilities must then be established for stopping, isolating and restarting machinery, and for removing and replacing guards. A permit system [18] should then apply. Lubrication workers must always have safe access to points requiring lubrication and should never have to straddle running machinery with their legs or ladders. Pump oil cans with extra-long spouts are sometimes required.

Lubricants must be kept and handled in closed or covered containers and kept free of contaminants, especially gritty material and water. Excessive lubrication can in some cases cause as much damage as lack of it. Over-oiling of motor bearings may cause oil to enter the insulation of motor windings, causing deterioration and accumulation of dirt and hiding defects. It must not be forgotten that all lubricants used must end up somewhere, usually either in the process materials or in the local environment. There they cause greasy patches on floors and machine parts which eventually spread through handling to switches, instrument knobs, clothing, papers and parts of the body.

Oil-lubricated air compressors present an explosion hazard when excessive discharge temperatures are reached. A current recommendation is a maximum outlet temperature of 145°C for an outlet pressure of 10 bar g., decreasing to 124°C for an outlet pressure of 100 bar g. with automatic shutdown at 10°C higher[4]. It is particularly important to prevent the presence of lubricating oils in the discharge of the main air compressors of air-separation plants, where they end up in the liquid oxygen and have caused serious explosions.

Lubrication workers must be on the alert and report any symptoms they observe of incipient trouble, bearings running hot, unusual noises and vibration, excessive lubricant consumption, high-pressure build-up across oil filters, rapid deterioration and the presence of metal debris in circulating-oil systems and in oil filters. In this way they are the advance scouts of the inspection organisation. It is recommended that they be provided with standard forms listing points to be checked so that faults can be brought to the prompt attention of inspectors and management.

17.1.6 Maintenance of civil work and buildings

Some common faults which may develop and cause hazards are discussed here briefly. This section is not intended for experts in building maintenance or for structural engineers.

Foundations and column bases

A watch must be kept for settlement, cracks and water seepage. Settlement which affects the stability of buildings and may damage plant and pipework requires prompt investigation by experts. This and the resulting repairs generally need excavation, and the plant to be shut down while it is being done. Cracks in foundation walls which may allow water to enter and corrode steel columns and reinforcement should be repaired promptly. Rust at the base of columns should be removed and an anti-rust coating applied.

Structural members
Steel members should be checked yearly for rusting, concrete members for cracks, spalling and chipping, and wood members for dry rot, shrinkage and slippage. Columns should be checked for distortion and holes cut or drilled in them, and horizontal members for sagging. These faults, where found, should be reported for investigation and action by the structural engineer concerned.

Walls
Exterior and interior walls should be examined for cracks and loose joints which should be raked and pointed. Walls and columns liable to be struck or scraped by vehicles should be guarded by substantial steel railings near floor level which are fastened to the floor or ground.

Ceilings and floors
These should be checked for excessive sag (generally more than 1 in 360 of the span) and the need for cleaning, repair and painting. Floors should be inspected for holes, irregularities and excessive wear. These should be repaired and the cement allowed to set while traffic is kept off them. Waxes which make floors slippery should not be used. Care must be taken not to overload floors with heavy equipment and stored materials. Where there is a danger of this, warning signs showing the maximum quantities of materials which may be stored should be painted or fastened to walls or columns.

Loading platforms
The edges should be protected against vehicle damage, by angle or channel iron. The surfaces should be kept in good repair.

Roofs
Maintenance workers have suffered many serious accidents through falling off the edges of roofs and through fragile ones. Side rails should be provided and secured before work on a roof is commenced. Only experienced roof workers should only be allowed access to fragile roofs, using duck boards or crawling boards as appropriate. Leaking roofs should be repaired without delay. All roofs must be securely anchored. Flashings should be checked to ensure that the metal is tight with the roof. Gutters must be kept clean and leak-free.

Stacks
Stacks which are liable to damage by wind, weather, flue gases, lightning and settlement need frequent examination. The lightning conductors of brick and concrete stacks should be checked to ensure that they are continuous from the top of the stack, and well grounded.

Underground maintenance

The many hazards of ground work are discussed in an earlier book[5]. Air testing and ventilation of sewers, tunnels and pits and other precautions for work in them are covered in 18.7. The walls of all excavations deeper than 1.5 m should be shored or protected by sloping or battering, although exceptions can be made in hard rock. The positions of underground pipes and cables must be carefully checked before excavation is started. Underground pipes should not be cut into or opened until they have been completely isolated, depressurised, vented and purged, unless they are known to have contained only air or water. Waste-disposal channels and trenches should be kept in good repair and checked for settlement which may occur as a result of leakage. Excavations should not be made beneath buried pipelines in use. Open trenches at night must be protected by barricades, signs and lanterns.

Interior decoration

This is more important than is often realised, both for illumination and for the morale of workers in the building. Light-coloured surfaces with a good balance between the cool and the warm colours are recommended, despite the fact that they will require more regular cleaning than darker surfaces. If the surfaces become dirty very quickly, this is an indication that excessive concentrations of particulate matter are present in the atmosphere, and probably a risk to health as well as illumination. Painting itself, especially of ceilings, can usually only be done safely when the plant is shut down and most personnel are absent.

17.1.7 Electrical and lighting systems

Electrical systems need periodic inspection by an electrical engineer. Switchgear, wiring and insulation are susceptible to corrosive attack by acid fumes and spillages of process liquids from which they should be protected.

Replacement of lamps and the cleaning of reflectors, etc. which require mobile platforms or ladders is best done when the plant is shut down and when most personnel are absent, to reduce exposure to broken glass and dust. Lamps of the same type which were installed together are usually best replaced together towards the end of their working lives, even though only a small proportion of them have actually failed. The mobile platforms used should have built-in compartments, trays or fixtures for lamps, cleaning buckets and materials. Provided the lighting scheme has proved satisfactory, the replacement lamps should be similar to the original ones. A record should be kept of the type of lamp in every position. If the lighting was unsatisfactory, an illuminating engineer should be consulted.

Gloves should be worn when replacing lamps. If the spent lamps need to be broken before disposal, this should be done in the open, after wrapping the lamp in newspaper to contain glass and dust which may be toxic.

17.2 Pre-operational inspection

Pre-operational inspection involves many different branches of engineering each with its own range of standards. It is done primarily to ensure that plant is safe and sound when handed over by the contractor to the operating company. Faults found at this stage should be corrected by the contractor and not by the operating company's maintenance department. Yet however thorough the inspection, errors and omissions sometimes slip through. Those which affect the operation of the plant such as leaking pipe joints and glands, faulty thermocouples and switches are usually spotted and corrected during commissioning. More serious ones [4.6] which may have no immediate effect on plant operation are not always recognised.

Plant and equipment makers use metal tubes, plates and sheets whose compositions, mechanical properties, dimensions, surface finish, freedom from flaws and internal strains, heat treatment, etc. are guaranteed by their suppliers to meet agreed specifications. These are first inspected by the supplier, again where necessary by the equipment manufacturer, and in case of disagreement, checked again by an independent inspector as agreed in the supply contract. Various inspection methods are employed ranging from simple visual checks to the use of complex testing and analytical equipment.

The equipment is inspected by the manufacturer before it leaves his premises. Pressure vessels and heat exchangers, particularly non-standard items, require special attention. The welding of pressure vessels needs properly trained and qualified welders. Training, testing and certification of welders, welding engineers, testers and inspectors in the UK in different welding and inspection techniques, on different materials and at different levels is given by the Welding Institute's School of Welding Technology (SWT) at Abington, Cambridge and at Paisley College of Technology. The SWT operates a Certification Scheme for Welding Inspection Personnel (CSWIP)[9] and also gives courses and awards certificates of competence in practical safety in welding.

Welds on vessels must be inspected in accordance with the applicable pressure vessel design code (usually BS 5500[10] in the UK).

This 'quality control' function should be organised independently of the actual welding work, to prevent it coming under commercial pressures. The main methods used in welding inspection are ultrasonic, radiographic, magnetic particle and liquid penetrant. These and other non-destructive testing methods are discussed in 17.4.

The ultimate users of the plant may also insist on inspecting equipment before it leaves the manufacturer's premises. Dimensions are checked to be within the specified tolerances. Materials of construction may be analysed to ensure that they are as specified. Metal surfaces may have to be inspected for flaws which could start cracks. Sharp angles, changes of section, nozzles and lugs which intensify stresses [13.1.1], and parts which may be subject to wear, are checked that they meet the design requirements. Joints between dissimilar metals which might cause local galvanic corrosion unless special precautions are taken [11.2] should be studied. The suitability of joints, gaskets and seals as well as lagging and protective finishes (which must be properly specified in the first place) need to be checked. The adequacy of vents and drains should be ensured.

Sometimes design inadequacies are found at this stage, despite the fact that the drawings, which have been modified several times already, were double-checked and approved. To correct these now may require altering both the design and the equipment. This is expensive, time-consuming and usually contentious, and requires staff of high calibre to see it through, especially as there is usually strong commercial pressure to sweep such faults under the carpet.

Further inspection takes place as the plant is being built. This usually involves a good deal of site welding, especially of pipework, for which inspection is called for under the contract.

Although the onus for such inspection generally falls on the contractor, who has to cut out and remake defective welds, the actual work of inspection is often sub-contracted to specialist firms. The client for whom the plant is built may require copies of all inspection records, including radiographs which his own specialists can interpret.

Besides welding, many other things have to be inspected while the plant is being built. These include concrete, reinforced foundations, underground pipes and pipe coatings, electrical and instrument wiring, and especially items which cannot easily be inspected or rectified when the plant has been completed. The final inspection in which both the contractor and the operating company are involved takes place when the plant is commissioned and handed over.

17.3 In-service inspection

Once in service, the plant is subject to internal wear, corrosion and fouling of surfaces, external corrosion, damage from earthquakes, flooding, hurricanes and accidents. These create needs for descaling, maintenance and cleaning. Where plant failure could lead to serious accidents (e.g. the bursting of a pressure vessel containing a hazardous fluid), there is also a need for periodic inspection, to check how far any critical parts of the plant have deteriorated.

This need for periodic inspection of process plant is legally recognised in the UK for a limited range of items, mainly steam boilers and air receivers [2.5.1]. The law also provides for the periodic inspection of other items used in construction and other industries including hoists, lifting machinery and appliances, chains, ropes and tackle, power presses, dust-extraction equipment and electrical installations. In other fields where equipment failure can have even more serious consequences – transport by air, sea road and rail, atomic energy, and munitions – mandatory inspections are even more exacting.

In cases not covered by specific legislation, the need for in-service inspection is implied by the general duty of the owner to provide safe equipment, which HSWA 1974 reinforces. Insurance companies sometimes also make periodic inspection a condition for a low insurance premium, or even for insuring the plant at all.

These special needs for inspection over and above that normally done in the course of operation and maintenance have led to the formation of companies, often closely linked with insurance, which offer an inspection service to the process and other industries, particularly for pressure vessels. In the UK these include Plant Safety, National Vulcan and British Engine. They recruit staff with experience in their own and allied fields, and train them as inspectors.

This need for regular engineering inspection during the life of the plant is not a uniform requirement throughout the process industries, but depends on the consequences of failures which it should avert. While it is difficult to visualise the worst consequences of failure in every case, a hazard assessment of the plant, using the Dow or Mond methods discussed in Chapter 12, should indicate whether such inspection will be needed for safety.

For each part of a plant which is inspected, there can be several alternative outcomes:

1. The condition may have deteriorated so little that it still lies within the design limits of temperature, pressure, etc. and can be returned to operation without maintenance;
2. The condition may have deteriorated below the design limits but can be restored by appropriate maintenance;
3. The condition may have deteriorated below the design limits, to which it would be impractical to restore it by maintenance; the part can, however (with or without maintenance), be 'de-rated' and used again in some less arduous duty with lower design limits;
4. The condition has deteriorated to the point that the part is no longer safe to use and must be scrapped.

17.3.1 Statutory inspections of pressure vessels

Sections 32–35 and 37–38 of the Factories Act 1961 and supporting regulations govern the periodic examination of steam boilers, including waste-heat boilers, economisers, steam superheaters as well as steam ovens and hot plates. Section 36 governs that of air receivers and Section 39 that of water-sealed gas holders. There are also requirements for the inspection of containers for the transport of compressed and liquefied gases, and of vessels for the underground storage of petrol and other liquids.

Boilers must be examined within 14 months of initial service, and at subsequent intervals not greater than 26 months in most cases, although certain small boilers, as well as large ones after 21 years' service, must be inspected at intervals not greater than 14 months. Steam and air receivers must be inspected after service intervals not greater than 26 months, except for solid-drawn air receivers for which the interval is four years.

The regulations call for boilers, etc. to be shut down, isolated and made safe and thoroughly examined by a 'competent person', who completes printed forms which list the points to be examined. The examination is made in two parts, the first when the boiler is cold, and the second part when steam is first raised after the cold examination. The following types of defects are looked for during the cold examination:

- *External*. Corrosion at joints between fittings and shell, tubes and tubeplates; wastage of manhole, mudhole, and handhole joint seatings; general physical damage;
- *Fire side*. Wastage in combustion chamber plates due to leaky tubes or stays; furnace flame impingement damage; erosion of metal by entrained particles; overheating damage due to scale or sludge build-up, oil or water shortage; bulges in furnace;

- *Water side*. Corrosion including general wastage, pitting and thinning; mechanical damage caused by expansion and contraction, grooving, and cracking at joints of tubes and tube plates.

This inspection is partly visual, assisted by optical aids such as endoscopes and fibre optics (Figure 17.3), and partly by a variety of other techniques including hammer testing, drilling, non-destructive flaw testing, thickness measurement and hydraulic pressure testing.

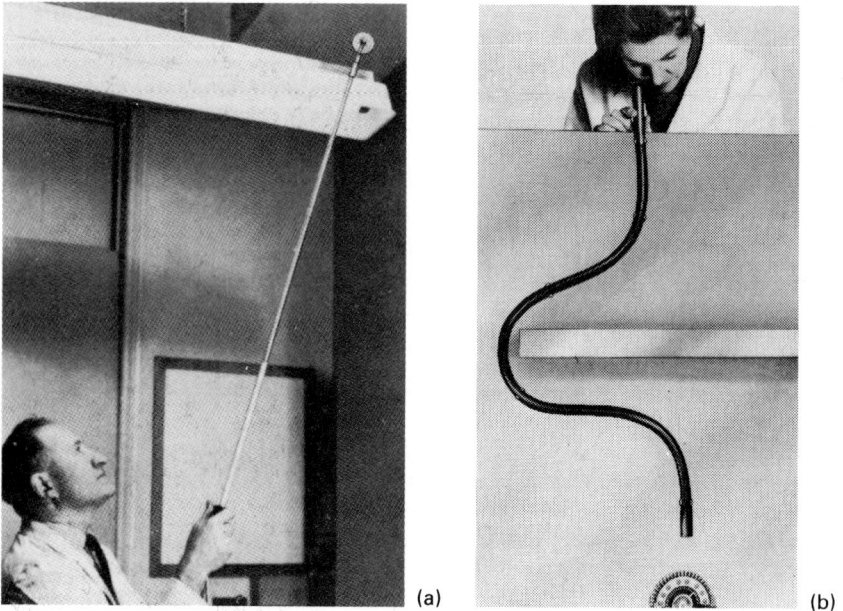

(a)

(b)

Figure 17.3 Optical inspection aids: (a) telescopic inspection mirror, (b) flexible periscope

During the working examination the safety valves are set to lift at the correct pressure and then locked to prevent their being tampered with. Instruments such as pressure gauges, water gauges, level controls, alarms and cut-outs are checked to be working properly and in accordance with their calibrations.

Where the inspection reveals defects which affect safe operation at the current maximum permissible working pressure, the report must define the repairs needed to make the boiler safe for operation at that pressure, or the maximum working pressure for the boiler (calculated from plate thicknesses and other data obtained from the inspection) in its present state. Repairs specified in the report must be carried out before a certain date, and may be followed by a further inspection before the boiler is used again. A copy of the examination report has to be sent to the HSE Inspector of Factories within 28 days by the person making the examination.

Steam and air receivers must be thoroughly examined internally and externally and, in the case of air receivers, cleaned. Where internal corrosion of the shell has led to a reduction in wall thickness, the vessel may be de-rated and a new safe working pressure recalculated. The necessary fittings are checked as present and in good working order, and are adjusted where necessary. The vessel marking is checked for identification number and safe working pressure.

Although these statutory inspections are generally carried out by a specialist inspection firm as discussed earlier, there is no legal reason why they should not be undertaken by a 'competent person' on the staff of the operating company.

The main complaint in industry today about these inspections is that they are more frequent than they need be when modern methods of corrosion control are employed. They also do not allow full scope for the use of modern inspection methods which might be used while the equipment is in operation.

17.3.2 Non-mandatory inspections of pressure vessels

Pressure vessels, or rather pressurised systems containing hazardous and/or corrosive materials, have on any logical basis a greater need for inspection than those containing only compressed air, which the law requires to be inspected at intervals not greater than 26 months. This is recognised by HSE, which issued a consultative document containing proposals for new legislation for pressurised systems in 1977 [2.5.1]. This is in abeyance and HSE now appears to be in favour of a self-regulatory scheme in industry which has the support of the engineering insurance industry[11].

The only (readily) available British code on the inspection of pressure vessels is that published for the petroleum industry by the Institute of Petroleum[12] in 1976. This treats pressure vessels for which inspection is mandatory as Class A and others as Class B. It recognises several inspection grades for Class B pressure vessels, 0, I, II and III, and gives maximum periods between inspections for each grade and various duties, which are shown in Table 17.2.

Most equipment is placed in Grade 0 until it has received its first inspection. Equipment in Grade III is subject to review at the periods stated to check that the conditions which led to that classification still apply. The maximum period between inspections for a pressure storage

Table 17.2 IP inspection periods for Class B pressure vessels

Equipment		Inspection period (months)				Review
	Grade:	0	I	II	III	
Process pressure vessels and process						
vacuum vessels		24	36	72	108	72
Pressure storage vessels		60	60	90	120	90
Heat exchangers		24	36	72	108	72
Protective safety devices		24	36	60	–	–

vessel of Class B of 60 months is in striking contrast to the statutory interval of 26 months for a welded compressed air receiver (Class A).

A comprehensive guide for inspection of all refinery equipment including pressure vessels has been published by the American Petroleum Institute[13]. A detailed code on the subject for the chemical industry was prepared by ICI (who hold its copyright) and published through RoSPA in 1975[14]. This is referred to extensively by Lees[4]. Since then ICI has withdrawn from joint safety publications with RoSPA, apparently for legal reasons, and has revised its pressure vessel inspection code for its own internal use. Other large companies which make extensive use of pressure vessels also have their own pressure vessel inspection codes which explain when and how inspections should be done.

Before a pressure vessel is inspected, it is normally taken out of service, isolated, emptied, cleaned and gas freed following the type of procedure described in 18.6. Lagging may have to be removed and surfaces specially cleaned before inspection, when all fittings which affect its safe operation are also examined and tested.

The types of defects which inspection of all types of process equipment may reveal include internal and external corrosion, fretting, fatigue, erosion, surface, weld and wear defects, deposits, high stresses and inadequate drainage. Causes and types of corrosion are discussed in Chapter 11 and of mechanical damage in 13.1. Non-destructive inspection methods are discussed in 17.4.

Fresh moves are now afoot in the UK for greater standardisation of engineering inspection procedures – not merely for pressure vessels – as well as for the qualification and registration of inspection bodies. In this the Engineering Inspection Authorities Board of the Institution of Mechanical Engineers, and the Council of User Inspectorates which mainly represents insurance interests, are working to promote a scheme under the umbrella of the British Standards Institute for the 'Certification of Bodies Performing In-Service Inspection'[10].

17.4 Non-destructive testing (NDT)

Not long ago, in order to measure the wall thickness of a large vessel the inspector had to drill a hole through it, which would be plugged or welded after the measurement had been made. Today, by the use of ultrasonic methods, the wall thickness can be measured from the outside without interfering with the operation of the vessel. There is now a bewilderingly wide choice of inspection methods and equipment which do not involve destruction of the item. Their use, however, requires a considerable degree of training. A number of British Standards on the use of NDT methods for the inspection of equipment used in the process industries are listed in Table 17.3.

For some 'in-service' NDT inspections, the plant has to be shut down, isolated, emptied, cleaned and opened up. For some it has only to be shut down but not opened up, and for others the measurements are made continuously or intermittently, while the plant is operating. The last group of methods, known as condition monitoring (CM), are discussed in 17.5.

Table 17.3 Selected British Standards on NDT methods

BS	Title
2600	Radiographic examination of fusion welded butt joints in steel. Part 1: 1983, Part 2: 1973
2633: 1973 (1981)	Specification for Class 1 arc welding of ferritic steel pipework for carrying fluids
2737: 1956 (1985)	Terminology of internal defects in castings as revealed by radiography
2910: 1973	Methods for radiographic examination of fusion welded circumferential butt joints in steel pipe
3451: 1973 (1981)	Methods of testing fusion welds in aluminium and aluminium alloys
3683	Glossary of terms used in non-destructive testing (in five parts, from 1965 to 1985)
3889	Methods for non-destructive testing of pipes and tubes (in three parts from 1966 to 1986)
3923	Methods for ultrasonic examination of welds (in three parts from 1972 to 1978)
3971: 1980 (1985)	Specification for image quality indicators for industrial radiography (including guidance on their use)
4069: 1982	Specification for magnetic flaw detection inks and powders
4080: 1966	Methods for non-destructive testing of steel castings
4206: 1967	Methods of testing fusion welds in copper and copper alloys
4331	Methods for assessing the performance characteristics of ultrasonic flaw detection equipment (in three parts from 1972 to 1983)
5044: 1973 (1982)	Specification for contrast and paints used in magnetic particle flaw detection
5289: 1976 (1983)	Code of Practice. Visual inspection of fusion welded joints
6072: 1981	Method for magnetic particle flaw detection (1986)
6443: 1984	Method for penetrant flaw detection

There is, however, no clear dividing line between static NDT methods and dynamic CM ones, since the same method is sometimes used in a static mode and sometimes in a dynamic one.

NDT inspection of equipment for the process industries takes place in the following cases, and the training and competence of NDT workers has to be considered for all of them:

- Pre-service inspection during or after the manufacture of the equipment and the building of the plant;
- Mandatory and insurance in-service inspections done by a specially trained inspection body (either an outside firm or a department of the operating company);
- Inspection by maintenance personnel during the course of their work.

Difficulties may arise, especially in the last case, in choosing the most appropriate method, in deciding who should do the work, and in ensuring that the persons chosen have the necessary competence.

Besides the training and certification of inspectors performed by the Welding Institute and discussed in 17.2[8], the British Institute of Non-destructive Testing has established a Central Certification Board for standards of competence in all methods of NDT under an umbrella scheme known as PCN (Personnel Certification in NDT[15]). This is built round the needs of four industrial sectors: welding, castings, wrought products and aerospace. CSWIP is responsible for the welding sector.

The NDT requirements for new process equipment, particularly for welded pressure vessels, are given in BS 5500[9] and discussed by Lees[4]. The very classification of a pressure vessel depends primarily on the extent of NDT carried out. Category 1 pressure vessels require 100% radiographic or ultrasonic inspection of all welds for internal flaws, and magnetic particle or penetrant examination of all welds other than full-penetration butt welds for surface flaws. Category 2 pressure vessels require only 10% of the above NDT examination while Category 3 pressure vessels require no NDT for internal flaws.

The principal methods of NDT used for equipment in the process industries are next explained briefly.

17.4.1 Visual examination

Many surface cracks, oxide films and weld defects as well as notches and misalignment can be detected visually by experienced inspectors. Mirrors, lenses, microscopes, borescopes, fibre optics and other visual aids increase the scope (Figure 17.3). Examples of visual aids for special purposes quoted by Collacott include[16]:

- Deep-probe endoscopes incorporating fibre optics which allow the inside of boiler and heat exchanger tubes up to 21 m long to be illuminated and examined at various angles and magnifications;
- Examination of the fracture surfaces of failed components by taking a cellulose acetate impression of the surface from which a carbon–chromium copy of the fractured surface is made; examination of this copy by electron microscope often allows the mode of fracture to be determined, so that appropriate remedial action can be taken.

17.4.2 Radiography

The amount of absorption when X-rays and γ-rays pass through optically opaque objects depends on the thickness and nature of the material. The transmitted rays produce images on fluorescent screens or photographic film from which the thickness of the material or the presence and position of voids or refractory inclusions can be determined. The method has been developed and refined over many decades. X-rays, which have a smaller and more concentrated source than γ-rays, give better-quality images and are generally preferred, despite the greater size and cost of the equipment. The use and interpretation of radiography require high degrees of skills and training.

17.4.3 Ultrasonics

Sonic testing has long been used by wheeltappers for detecting flaws in wagon wheels, while the ear of an experienced motor mechanic is a valuable CM tool in fault diagnosis. Ultrasonic testing which relies on frequencies higher than 20 000 Hz, the limit of the human ear, has been developed to a much greater degree than testing with audible frequencies, and in many cases offers a better alternative to X-rays. Most metals readily transmit ultrasonic vibrations, whose short wavelengths are about the same size as many metal flaws which scatter or reflect them.

Any ultrasonic method requires the use of a transmitter and a receiver. Techniques available include transmission, where the transmitter and receiver are on opposite sides of the object being tested, pulse echo (Figure 17.4), where they are both on the same side, resonance, which is a development of the transmission technique, frequency modulation, acoustic imaging, which has now been largely superseded by holography, tube thickness probes, contour scanning and triangulation fault location.

Figure 17.4 Ultrasonic thickness gauge with built-in data logger (courtesy Baugh and Weedon Ltd)

Ultrasonic methods have many possible applications including measurement of the wall thickness of pipes and vessels, detection and location of internal flaws, examination of surface roughness and pitting, and of coatings on metal surfaces. In the pulse echo method, which is probably the most widely used, the transmitter emits an ultrasonic pulse, then stops and receives an echo which is displayed on a cathode ray tube. This is a combination of the return pulses from the opposite side of the object and from any internal defects.

17.4.4 Eddy current testing

Variants on this method are used to detect cracks, pits, and hydrogen embrittlement in ferrous and non-ferrous metals, as well as the thickness of non-metallic coatings on metal surfaces. Coils carrying alternating currents with frequencies from 1 to 5 MHz placed close to the metal surface induce eddy currents in the metal which are measured in a variety of ways. Changes in the eddy current when the coil is moved over a flaw in the metal allow it to be located and in some cases measured. The method can give very useful results, although these are liable to be distorted by extraneous factors which affect the electromagnetic properties of the metal. The work is highly skilled.

17.4.5 Magnetic particle

This method is used to detect defects on or near the surface of ferro-magnetic materials. The item to be tested is magnetised by one of several methods, and fine particles of iron or iron oxide are spread over the surface either as a dry powder or in suspension in a liquid as a coloured or fluorescent 'ink'. The particles concentrate at cracks and other places where there are discontinuities in the magnetic path. The method reveals cracks invisible to the eye (Figure 17.5).

Figure 17.5 MPI inspection of forged gate valve bodies (courtesy Magnaflux)

17.4.6 Liquid (dye) penetrant

This is used to detect surface cracks in non-magnetic metals such as austenitic stainless steel and non-porous non-metals such as glazed ceramics. It is an extension of visual inspection and depends on the capillary attraction of the defect for the liquid which it must retain after surplus material has been removed. The liquid usually contains a dye or a fluorescent compound which is visible in ultra-violet light. Several dye penetrants, washes and developers are available.

17.4.7 Choice of NDT method for in-service inspection

The methods used should be appropriate to the types of mechanical damage and corrosion liable to be encountered. A summary by Collacott of the suitability of the five methods described earlier for different types of corrosion is reproduced as Table 17.4[16]. Ultrasonic methods which are most suitable for uniform corrosion are also suitable for general mechanical wear, while radiographic methods used for corrosion fatigue are also appropriate for mechanical fatigue.

Table 17.4 NDT techniques applied to corrosion evaluation (NA = not applicable)

Corrosion type	Technique				
	Radiography	Ultrasonic	Eddy current	Magnetic particle	Dye penetrant
Hidden wastage of unknown mechanism	Best general technique	Limited	Limited	NA	NA
General wastage	Poor	Best general technique	Limited	NA	NA
Pitting corrosion	Best general technique	Can detect	Good method	NA	NA
Intergranular corrosion	Can detect	Best general technique	Good method	Can detect	Can detect
Dezincification	Can detect	Best general technique	NA	NA	NA
Corrosion fatigue	Best general technique	Good method	Good method	Good method	Can detect
Stress corrosion	Good method	Best general technique	Good method	Good method	Good method
Hydrogen embrittlement cracking	Can detect	Best general technique	Good method	Good method	Good method

17.5 Condition monitoring (CM)[4, 16, 18]

Although a wide range of CM methods are available, CM is less clearly defined than NDT. CM methods are rarely specified in standards for inspection, and only one British Standard has been found which relates to it[17]. In one book on maintenance, CM is classified as of two types, 'on-load' and 'off-load', the latter being NDT under another guise[18].

The only CM methods discussed here are those used *primarily* to monitor the condition of equipment during operation, in order to provide early indication of the need for maintenance or of its impending failure. Such CM may be continuous or intermittent, but once installed it is used while the plant is operating.

There are many border-line cases, e.g. of instruments used for warning and protection [15.7] which serve a secondary CM function.. They include high and low temperature, pressure and flow switches, e.g. on coolant and lubricant systems for engines, compressors and other machines, flame failure detectors on furnaces, flammable and toxic gas detectors.

The condition of process equipment is often reflected in its performance, as calculated from the readings given by the normal plant instrumentation. Thus the slow decline in performance of a continuously running multistage centrifugal pump, which may be judged from its rotor speed, discharge rate, differential pressure and other measured variables, may usually be attributed to corrosion of the impellers. A process computer receiving signals from instruments measuring these variables may compute and monitor performance and thus warn of a deterioration in condition.

CM methods and instruments whose sole purpose is to monitor the condition of expensive equipment have been developed over many years. These are in wide use in the nuclear, aerospace and off-shore oil industries. Their use in the process industries is increasing, mainly to detect and measure wear, cracks, corrosion, deposits, deterioration of protective coatings, electrical and thermal insulation, lubricants, heat transfer fluids and catalysts. Only the main methods are discussed briefly here.

17.5.1 Eyes, ears, nose and fingers

The unaided senses of an alert person can detect symptoms of a very wide range of defects. The senses can be extended by visual aids such as those listed in 17.4.1, aural aids such as stethoscopes, and tactile aids such as brittle lacquers. There is a risk here of people exposing themselves to hazards in order to get a better view, feel or sound of a component that seems to be behaving abnormally. (The tongue has been omitted from the sense organs listed here for obvious reasons!) Careful training and safeguards are needed to prevent people's curiosity leading to their undoing.

17.5.2 Vibration monitoring

This has the widest application of all CM techniques and can detect a wide range of faults in machines such as imbalance, misalignment, damage to bearings, gears and other transmission components, looseness and cavitation. The instrumentation required consists of a transducer (usually a piezoelectric accelerometer), a signal-conditioning amplifier and a readout meter. It may be portable for use in occasional checks, or fixed to the machine or structure. Several different techniques are available.

17.5.3 Corrosion monitoring

There are several methods of monitoring corrosion [11], not all of which comply strictly with our definition of CM. The principal ones are given here[19].

Weight loss specimens
Carefully cleaned and weighed metal coupons are mounted in the required position on special insulating holders which facilitate easy removal and replacement. After exposure for a predetermined period, the specimens are removed, cleaned and weighed, and the corrosion rate is calculated as mils per year (mpyr). Specimens should be carefully examined after exposure since the appearance may indicate the cause of corrosion, e.g. a black coating on steel indicates hydrogen sulphide and a red oxide coating the presence of oxygen.

Electrical resistance probes
The specimen is mounted so that its change in electrical resistance can be measured while it is in position in the corrosive environment. This gives a rapid and continuous measurement of corrosion rate without disturbing the specimen, but it is of limited value for assessing pitting.

Polarisation-resistance probes
This is a rapid method which involves applying a potential difference to electrodes placed in the liquid, and measuring both the potential and the current flowing. The corrosion rate can be calculated from the resistance obtained in this way. The method requires expert knowledge.

Galvanic corrosion probes
Two dissimilar metal electrodes such as copper and brass are placed in the corrosive fluid and the current is measured. This is very sensitive to the dissolved oxygen content of water, and can be interpreted by an expert to give a virtually instantaneous measurement of the corrosion rate.

Water analysis
Analysis of water samples taken at various places in a flowing system can often be useful in monitoring corrosion, but like other methods it requires expert interpretation. Such analyses include dissolved iron and other metals, dissolved oxygen and pH.

Hydrogen probes
This is another specialised method which depends on the rate at which hydrogen is formed. It can only be used in (acid) conditions where hydrogen is a by-product of the corrosion.

Ultrasonics
By attaching ultrasonic thickness measuring heads to the outside of parts of equipment or pipework where rapid corrosion is anticipated, its course can be followed while the plant is running.

17.5.4 Lubricant monitoring

Magnetic drain plugs are commonly used in oil sumps to collect magnetic debris, which also collects in oil filters. The systems should be designed so that both plugs and filters can be periodically removed and cleaned without stopping the machine. Measurement and examination of the debris provides evidence of wear of gears, bearings and sliding surfaces. Monitoring of the pressure drop across lube oil filters provides earlier evidence of wear. Spectrometric oil analysis procedures (SOAP) have been developed to a fine art, particularly for the inspection of aircraft engines. These and other contaminant analysis procedures are described by Collacott[16].

17.5.5 Thermal monitoring

In addition to its use in process control, temperature monitoring is used to detect bearing damage, coolant failure, build-up of unwanted materials (sediment, dust, corrosion products, etc.), damage to thermal insulation and faults in electrical components. Temperature-sensing devices include scanning infra-red cameras, optical pyrometers, thermocouples, liquid-in-metal sensors, bimetallic elements and temperature-sensitive alloys, labels and paints.

17.5.6 Selection and use of CM methods

Before a decision is taken to use any CM method, the frequency and effects of critical plant breakdowns, particularly of costly machinery, should be surveyed to establish whether there is a case for it. This may need the services of a consultant. The survey should be done first during the plant design, so that the CM sensors and monitoring equipment can, if necessary, be incorporated into it. A further survey may be needed after the plant has been in operation for several months. The following examples illustrate the need for CM:

- A costly gas-turbine driven compressor is liable to develop a bearing fault, which, if detected early, can be remedied at little cost, whereas serious damage would be caused if the machine continued to run in that condition.
- A small pinhole in a glass-lined vessel containing a corrosive and toxic mixture, might if detected early, be repaired with a simple tantalum plug, whereas the vessel could be damaged beyond repair, accompanied by a dangerous escape of toxic material, if it continued in service.

Once the need for CM has been identified, the most appropriate method must be selected. Questions to be decided include whether the CM device should be portable or fixed to the equipment, and where the sensors should be located. It is then purchased and one or more engineers should be trained to use it.

One then has to decide how the results given by the monitor will be used. First, the 'signature' of the machine, etc. to be monitored must be determined in its normal healthy condition, like the temperature or pulse rate of a patient. Next, the limits or limiting envelope of monitored responses at which action needs to be taken must be determined. Sometimes two levels are most appropriate. The first level, when breached, may be the signal to prepare for a planned maintenance shutdown within a limited period, whereas the second level may be the signal for an immediate emergency shutdown and replacement of certain items.

17.6 Pressure, leak and acoustic emission testing

Pressure testing of pressure vessels, associated pipework and equipment is carried out at pressures higher than the maximum operating ones for which the equipment was designed as an assurance that it will not fail in service. This is in some cases a statutory obligation and is usually required by insurance authorities. The test pressures used are generally determined by formulae given in the vessel design codes. These formulae take into account corrosion allowances and differences in the strength of the material at operating and test temperatures. Vacuum tests are similarly required for vessels used under vacuum, and leak tests, mostly on joints, are needed on plant before it goes into service. These tests are partly or fully repeated during in-service inspection or maintenance.

17.6.1 Pressure testing

Most hazards associated with pressure testing are of two kinds:

1. The testing may be inadequate or unrealistic and so fail to reproduce all the forces caused by pressure when the plant is operating;
2. The equipment may fail under test, causing injury and damage.

The first type of hazard is the least recognised. An example of this was given in 4.6. Here pressure vessels were designed and individually tested to take into account the normal forces resulting from internal pressure. The tests, however, failed to take account of the thrust forces on the sides of the vessels caused by the pressure inside the large flexible metal bellows which connected them. This would have required submitting the whole train of reactors, including the bellows, to the vessel test pressure. The reactors which had not been designed to withstand these large thrust forces would then almost certainly have failed, and the subsequent disaster would have been averted.

The second type of hazard, which is discussed by Lees[4] and is dealt with in an HSE guide[20], arises mainly when the pressure testing is done pneumatically rather than hydraulically, because of the danger of a mechanical explosion. This, fortunately, is rarely necessary. Hydraulic testing is generally done with water, although for large or very tall vessels, the weight or hydrostatic head of water when the vessel is full may cause problems which must be considered before filling. In such cases it is usually possible to employ a liquid of lower density such as kerosene.

Tests are normally carried out at ambient temperature. Precautions are required to prevent the temperature of steel vessels falling to the point where brittle fracture may occur, and water of specially low dissolved salt content should be used when testing vessels of austenitic stainless steel, to avoid risks of stress corrosion. The vessel is normally first filled with water at low pressure while venting all air from the highest points until water emerges, when all valves are closed. The pressure is then carefully raised to the test pressure by a hydraulic pump, while it is carefully monitored and controlled throughout the test to prevent over-pressurisation. The pressure should be released as soon as the test is complete, and vessels should not be left full of liquid under test pressure when the heat of the sun may cause liquid expansion and overpressure. A recommended procedure is given in BS 5500[10]. Precautions to protect personnel against vessel failure must be specially stringent when pneumatic pressure testing is employed.

17.6.2 Leak testing

Leak testing is usually done pneumatically at pressures up to 10% above the design pressure, but only after a full pressure test has been carried out. Some methods are described in BS 3636[21]. Flanged joints are usually covered with strips of adhesive tape, with a small hole to which a soap or detergent solution can be applied. There are various methods of discovering and locating air leaks, including the addition of small amounts of special gases such as CFCs (Freons) to the pressurised system, and using a probe and sensitive detector to search for the gas on the outside.

Another method employs an ultrasonic probe to detect the high-frequency pressure waves caused by a small leak. This can be employed when the plant is running and pipe joints are hot (Figure 17.6). The use of soap or detergent solution is still widely used and hard to beat. When all leaks appear to have been found and rectified, it is usual to leave the equipment for 24 hours under pressure at ambient temperature to observe any fall in pressure, which should be zero or less than some pre-specified and very low figure.

Leaks of process fluids from pressurised systems and of air into vacuum systems can be very hazardous and a high standard of pressure and vacuum tightness is usually necessary. Hydrogen is more prone to leak than air. After plant which will contain hydrogen under pressure has been leak tested with air before commissioning, a further leak test with hydrogen should be carried out during commissioning.

17.6.3 Acoustic emission testing[22]

This depends on detecting and analysing the noises made when flaws in vessels, tanks, pipes and machines under stress propagate further. It uses a system of sensing transducers, amplifiers and recorders. Although it is hardly a true NDT method, it is often classed as one. It is particularly useful for detecting flaws in fibre-reinforced plastic (FRP) vessels and pipework which cannot easily be detected by other means. It has also been used on liquefied gas storage vessels and large reactors used in oil refining, etc. As generally used, it depends on subjecting a flaw-containing structure to a higher stress than it had previously been exposed to. Hence it is employed in conjunction with the periodic pressure testing (requalification) of vessels which have been in service. It is a useful but not infallible

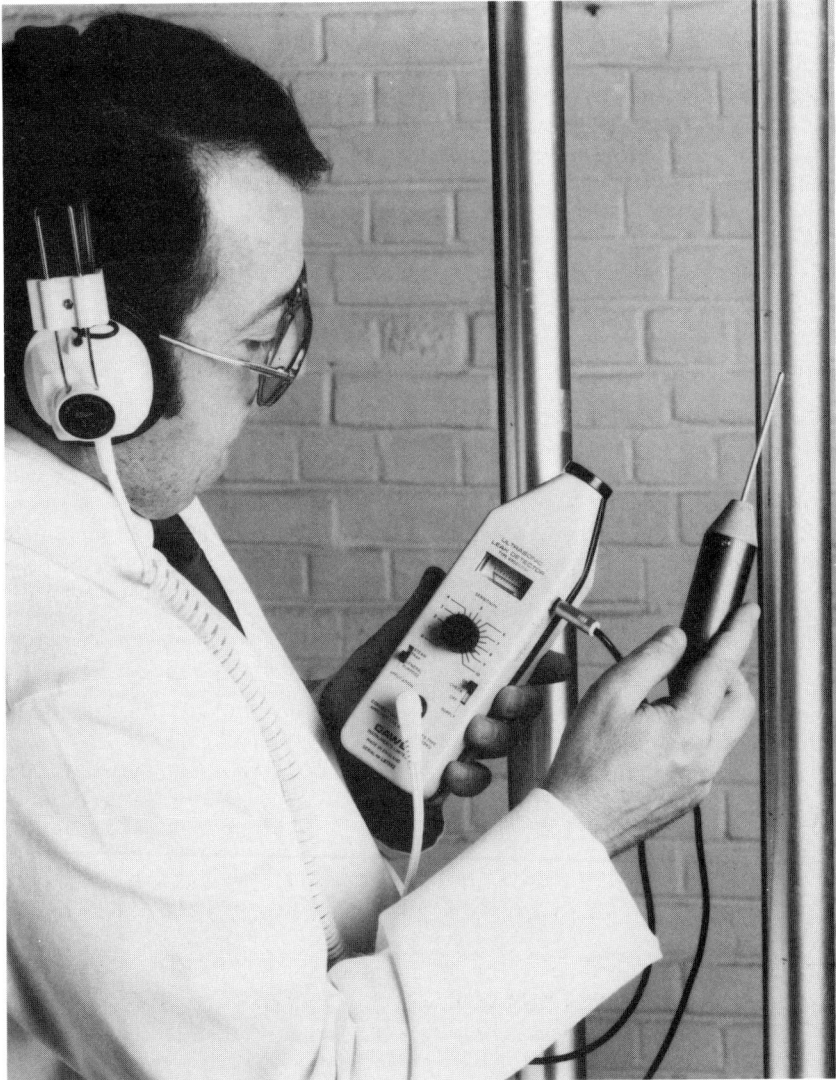

Figure 17.6 Ultrasonic leak detector (courtesy Lucas CEL Instruments Ltd)

method of detecting cracks caused by fatigue and stress corrosion. While requiring special equipment and experts who can use it and interpret the results, the cost of an acoustic emission test on a large vessel while in service may be only a fraction of that of isolating, emptying, purging and opening the vessel and inspecting it internally. It is used among other methods for continuous monitoring of vessels in nuclear installations, and is also employed to detect high-pressure leaks and pump cavitation.

References

1. Ministry of Technology, *Study of Engineering Maintenance in Manufacturing Industry*, London (1969)
2. BS 3811: 1984, *Glossary of maintenance management terms in terotechnology*
3. Husband, T. M., *Maintenance Management and Terotechnology*, Saxon House, Farnborough (1976)
4. Lees, F. P., *Loss Prevention in the Process Industries*, Butterworths, London (1980)
5. King, R. W. and Hudson, R., *Construction Hazard and Safety Handbook*, Butterworths, London (1985)
6. Maushagen, H. K., 'On-stream sealing of flanged joints using a compound sealing process', *Loss Prevention Bulletin 055*, The Institution of Chemical Engineers, Rugby (April 1984)
7. BS 5063: 1982, *Classification for a rational range of lubricants for machine tool applications*
8. BS 5739: 1979, *Method of presentation of lubricating instructions for machine tools*
9. *A Guide to CSWIP Certification Scheme for Welding Inspection Personnel*, 7th edn, The Welding Institute, Abington (1986)
10. BS 5500: 1985, *Specification for unfired fusion welded pressure vessels*
11. Advisory Council for Certification of Inspection Bodies, *National Certification Scheme for Inspection Bodies – Prospectus*, The Institution of Mechanical Engineers, London (24 March 1987)
12. The Institute of Petroleum, *IP Model Code of Safe Practice in the Petroleum Industry, Part 12, Pressure Vessel Inspection*, John Wiley, Chichester (1976)
13. American Petroleum Institute, *Guide for Inspection of Refinery Equipment*, New York (20 chapters published separately and periodically updated)
14. ICI/RoSPA, Code IS/107, *Registration and Periodic Inspection of Pressure Vessels*, RoSPA, Birmingham (1976)
15. *PCN – Personnel Certification in non-destructive testing*, Document no. CP8, The Central Certification Board, The British Institute of Non-Destructive Testing, 1 Spencer Parade, Northampton
16. Collacott, R. A., *Mechanical Fault Diagnosis and Condition Monitoring*, Chapman and Hall, London (1977)
17. BS 4675, *Mechanical vibration in rotating machinery*, (Part 1: 1976 (1986), Part 2: 1978)
18. Kelly, A. and Harris, M. J., *Management of Industrial Maintenance*, Butterworths, London (1978)
19. Uhlig, H. H. and Revie, W. R., *Corrosion and Corrosion Control*, 3rd edn, John Wiley, New York (1985)
20. HSE Guidance Note GS4, *Safety in pressure testing*, HMSO, London (1977)
21. BS 3636: 1963 (1985), *Methods for proving the gas tightness of vacuum or pressurised plant*
22. The International Study Group on Hydrocarbon Oxidation (ISGHO), *Guidance Notes on the Use of Acoustic Emission Testing in Process Plants*, The Institution of Chemical Engineers, Rugby (1985)

Chapter 18

Safe work permits

This chapter owes much to Professor Trevor Kletz[1], who gives details and underlying principles of the permit system used at ICI's Wilton works; to Bill Sampson, safety officer of Dow Chemical Company's King's Lynn works, who kindly provided information about the permit system[2] developed and used there; and to HSE's guidance note *Entry into confined spaces*[3]. Kletz also gives numerous examples of real accidents which could have been avoided had an effective permit system been in force. Probably the one best known to British readers was the fire and explosion at Dudgeon's Wharf in 1969 in which six firemen were killed[4] during the attempted demolition of a tank which had been steamed out but which still contained a layer of gummy residue. However much we detest red tape, the systems and principles described here undoubtedly save many lives.

18.1 Why permits are needed

Maintenance is essential for the safety and integrity of process plant. Yet many plant accidents have occurred during or following it because of misunderstandings and neglect of essential precautions when plant was handed over from production to maintenance workers, and vice versa. The maintenance workers may be company employees in a particular section of the engineering department, e.g. instruments, electrical, general mechanical, rotating machinery, thermal insulation, drains, painting, or they may be employed by an outside contractor. The possibilities of misunderstandings between operating and maintenance personnel are aggravated by shift work and by the use of outside contractors.

Production workers usually play only minor roles in maintenance [17.1], e.g. lubrication and the replacement of filter-elements, while maintenance personnel play no direct part in production. Serious injuries to maintenance workers have occurred because some flammable, corrosive, toxic, hot or asphyxiating material had been left in equipment which they were required to work on, or which had entered the equipment while they were working on or in it, or because an electric motor driving a pump, compressor or stirrer had been accidentally started while they were working.

Typical accidents involving production personnel have occurred because the wrong pipe flanges were disconnected by a maintenance worker, or because a motor was wrongly connected, so that when started it ran backwards, or because a joint which had been disconnected during maintenance had not been tightened and tested before restarting production. The writer still vividly remembers his narrow escape over 40 years ago from a 30-foot fountain of concentrated sulphuric acid, which was caused by a fitter removing a pipe bend from the discharge line of the wrong pump.

Errors in instrument maintenance are particularly insidious since instruments form an extension to the eyes, mind and hands of the operator. Thus the connection of the wrong thermocouple wires to the temperature controller of a reactor may lead to a runaway reaction and an emergency shutdown or a serious accident.

To prevent such errors and the accidents to which they readily give rise, most enterprises in the process industries, particularly oil and chemicals, have found it necessary to use formalised procedures whenever work is done on a plant by personnel other than those who are normally in charge of it. There are also such potentially dangerous activities as entry into pits and confined spaces (where the atmosphere may be unsafe to breathe), digging holes in the ground (with the risk of cutting a buried pipe or cable) or working at a height without guard rails. For these a written permit system which incorporates the necessary safeguards is also needed.

In many countries there are legal requirements for a written permit to be issued by a competent person before certain things are done. In the UK this applies to entry into various types of confined spaces under the following legislation:

- Chemical Works Regulations 1922;
- Kiers Regulations 1938;
- Shipbuilding and Ship Repairing Regulations 1960;
- Construction (General Provisions) Regulations 1960;
- Factories Act 1961 – sections 30 and 34.

Official forms are available for such permits which are dealt with in HSE's Guidance Note, *Entry into confined spaces*[3] [18.7]. Other UK legislation which requires permits-to-work includes:

- The Electricity at Work Regulations 1989
  (safe use of electricity);
- Factories Act 1961 – sections 27(7)
  (precautions needed to prevent persons being struck by overhead travelling cranes);
- Factories Act 1961 – section 31
  (precautions regarding explosive or flammable materials);
- The Ionising Radiations Regulations 1985
  (safe use of radioactive sources);
- Health and Safety at Work etc. Act 1974 – section 2(2)(a)
  (a general duty for safe systems of work).

18.2 Principles of permit systems

A written permit-to-work should identify the precise item on which work is to be done, record the hazards anticipated and detail the precautions needed before work starts. Bamber[5] gives the following principles for operating a permit-to-work system:

1. The information given in the permit must be precise, detailed and accurate.
2. The work to be done – and who will supervise and undertake it – must be clearly stated.
3. It must specify which apparatus or plant has been made safe and should outline the safety precautions already undertaken.
4. It should also specify the precautions still to be taken by the employees prior to commencement of the work (for example, fixing locking-off devices, siting of warning and danger boards, etc.).
5. The permit should specify the time at which it comes into effect and for how long it remains in effect. A re-issue should take place if the work is not completed within the allocated time.
6. The permit should be regarded as the master instruction which – until it is cancelled – overrides all other instructions.
7. Work must not be undertaken in an area not covered by the permit.
8. No work other than that specified should be undertaken. If a change is considered necessary to the work programme, a new permit should be issued by the authorised person who issued the original permit.
9. The authorised person who is to issue any permit must, before signing it, assure himself – and the persons undertaking the work – that all the precautions specified as necessary to make the plant and environment safe for the operation in question have in fact been taken.
10. The person who accepts the permit – i.e. the person who is to supervise or undertake the work – becomes responsible for ensuring that all specified safety precautions continue in being, and that only permitted work is undertaken within the area specified on the permit.
11. A copy of the permit should be clearly displayed in the work area.
12. All persons not involved in the work should be kept well away from the defined area.
13. Where relevant, regular environmental monitoring should be undertaken throughout the time the permit is operative.
14. The precautions to be followed for cancelling the permit should be clearly stated so that a smooth hand-over of plant or machinery occurs.

18.3 Permits for maintenance

In the hand-over of single plant items, or even complete units, from production to maintenance personnel the main points of information to be transmitted by the production personnel are:

- The identity of the item or items on which maintenance is needed.
- A clear description of the work to be done.
- A statement and guarantee that all normal and necessary precautions for the work as specified have been taken, e.g. complete isolation of the item or items from process materials, electricity and service fluids, as well as the elimination of process materials, fuels and heat transfer fluids from the items isolated.

- A statement of the remaining hazards facing the maintenance worker(s), and any special protective clothing, equipment and precautions which he or they should use.

It should be clear at this point that the nature of the work to be done largely determines the precautions which have to be taken. Thus the changing of an external bearing on the shaft of a stirrer in a vessel may require little more than that the process in the vessel and the electrical drive be isolated. But if it is found that an internal bearing has to be changed, far more careful precautions will be needed, e.g. complete isolation of the item concerned, removal of all process materials from it, and purging. If workers have to enter the item to replace the bearing, further precautions, including testing of the atmosphere within the item, will be needed.

A permit system should not require every simple and straightforward maintenance job to be referred to senior management level for authorisation. This would be enormously time-wasting and would blunt priorities. The system should act as a filter which allows straightforward jobs of low hazard potential to be dealt with at senior operator or shift supervisor level, but catches more potentially hazardous tasks and refers them for review and authorisation to appropriate staff of higher seniority.

Where the maintenance job is a routine one, the precautions needed are usually clear at the outset. But where the work appears to be a little unusual, discussion is necessary between the production and maintenance personnel to decide how to tackle it, and what precautions to take. If it is discovered after the work has been started that the job is more complicated than was foreseen, and requires additional preparations by production personnel, then the maintenance work must be stopped. The work to be done and the precautions needed must be redefined, and any further ones necessary must be taken before the maintenance continues. In such cases it often saves time to assume the worst initially and take the more stringent precautions at the outset.

18.3.1 Types of permits and certificates used

Although the principles of permit-to-work systems are well established in the process industries[6], there is no universally agreed system of permits for maintenance and other work and there is even ambiguity about the use of the terms 'permit' and 'certificate' in this connection. Most large companies appear to have developed their own systems for their own internal use. The permit and system as used at ICI's Wilton works have been described at length by Kletz[1]. In this the preparations and precautions needed for most maintenance work are covered by one general 'permit-to-work' (also called a 'clearance certificate'!). Special permits are used for purposes not covered by the general permit, e.g. hand-over of new equipment by contractors, entry into vessels and other confined spaces, excavations, use of portable radiation producing equipment, work on pipelines connecting units under the control of different supervisors, and work on equipment sent outside the plant.

The scope of a general permit-to-work can be extended to cover special cases by the use of certificates. Such certificates are not in themselves permits to do anything and can only be used in conjunction with a permit. For example, a vessel entry permit might allow a person wearing breathing apparatus to enter a process vessel, subject to certain conditions. It would not, however, allow him to enter it without breathing apparatus unless the permit is backed up by a valid gas test certificate, signed by the person who tested the atmosphere inside the vessel, and certifying that it is safe to breathe, again subject to certain conditions.

The Dow Chemical Company uses a two-tier system for its UK operations. This was developed by discussion with the personnel concerned and consensus over a period of 18 months at their King's Lynn works. It was instigated by Dow's safety officer, Bill Sampson, after studying permit systems in use by other UK companies. It attempts to match the hierarchy of hazards inherent in different maintenance and other tasks with a hierarchy of professional experience and expertise among those who authorise and supervise the work. It has been used without serious incident since 1982.

18.4 Outline of the Dow system

This uses a single multi-purpose permit-to-work (Figure 18.1) which can be authorised by a single designated signatory for most maintenance work, but which requires to be backed up by a 'Safety Planning Certificate' (Figure 18.2) authorised by higher-level signatories for several specified jobs of greater hazard.

A permit-to-work is not required for work covered by 'Job safety analyses' [21.2.1] or 'Safe operating procedures', which is done by people within their own departments but is required for any site work involving construction, installation, alteration, dismantling and maintenance.

In addition to a permit-to-work, a safety planning certificate is required for the following types of work:

1. Projects (for which capital has been authorised);
2. Any plant change (affecting its design or integrity);
3. Work in confined spaces (defined in a company standard);
4. Hot work in electrical zones (1) or (2) [6.2.2] capable of igniting a flammable vapour or dust [10.4];
5. Use of an open flame anywhere on site capable of igniting combustible liquids [Table 10.1] and solids;
6. Critical line breaking (where a reasonably sized leak could lead to a major accident). Critical lines are defined as those carrying certain hazardous fluids and are designated on boards close to plant control rooms;
7. Asbestos (refers to a company standard);
8. Excavating or digging (refers to a site standard);
9. Use of a mobile crane (refers to a site standard);
10. Roofwork and work higher than 5 m where permanent access is not provided, and excluding scaffold erection;
11. High-voltage (>1000 V) electrical work.

Dow KING'S LYNN **PERMIT TO WORK**   No: 109608

1. AREA/TANK/VESSEL/EQUIPMENT/PIPELINE
EXACT LOCATION:

2. WORK TO BE DONE:	COMPANY:	No. of Men:
	Man in Charge:	
	Attendant for Line Breaking is:	

3. SAFETY PRECAUTIONS:

		Time									
a)	Gas Tests (when applicable)	Flammable Vapours									
	(Results and initials in boxes)	Toxic Gases									
b)	Protective Equipment to be worn (ring items which apply)	Oxygen									

1. Safety Helmet	2. Safety Spectacles	3. Chemical Goggles	4. Face Shield	5. Updraft Helmet	6. Air Hood	7. Self-contained C.A.B.A.	8. Compressed Air-line B.A.	9. Dust Mask	10. Ear Muffs
11. Gloves	P.V.C. General Special	12. Rubber Boots (Steel Toecaps)	13. Protective Footwear	14. Boilersuit	15. PVC Suit	16. Neoprene Suit	17. Disposable Suit	18. Plant Overalls Rubber Overboots Rubber Gloves	19. Safety Harne

c) Other Precautions

4. STATE OF ISOLATION

	No. of Lines	Depressurised and Drained	Positive Isolation	Tagged Off	Valve only	Initials as applicable	Not Isolated	N/A
						Steam		
a) Although the job may be isolated and depressurised,						Gas/Vapour		
small residual quantities of hazardous chemicals may						Liquid		
still be present so wear protective clothing suitable						Solids		
for the risk.						Air		
b) All Motive Power has been isolated and any logic						Nitrogen		

b) All Motive Power has been isolated and any logic control interrupted. (LOCK/TAG/TRY).   Yes | No | N/A   Signed: .Approved Signator

c) I have placed my lock.   Signed: . . . . .Person undertaking wo

d) Electrical fuses have been withdrawn, all circuits are dead   Signed: .Electrician

e) Electrical circuits are live for 'Troubleshooting' only.   Signed: . . . . .Electrician/Inst.

5. I certify that a Safety Planning Certificate is **not** required because the work does not involve Projects, Plant Changes, Confined spaces, Hot work in Zone 1 or 2, Open flame, Critical line breaking, Asbestos, Excavations, Mobile Cranes, Roofwork or heights $>$5m, H.V. Electricity.

Signature: .Permit Signator

6. a)   CONFINED SPACE ENTRY (Cancelled if Site Alert (pips) sounds)
In accordance with Regulation 7 of the Chemical Works Regulations, 1922, and Section 30 of the Factories Act, 1961, I have inspected the above confined space, it has been tested, is fully isolated, has been safely prepared according to the precautions above and on Safety Planning

Certificate No: . . . . and is, therefore, safe to enter from . . . . to . . . .
on. . . . . Signed: .Approved Signator
Name of competent Attendant outside vessel: . . . .

b)   HOT WORK (Zone 1 or 2, or Open Flame) (cancelled if Site Alert (pips) sounds)
I have inspected the above job which has been safely prepared according to the precautions outlined above and on Safety Planning
Certificate No: . . . . therefore work may start from . . . . to . . . . on . . . .
Signed: .Approved Signator

c)   OTHER HAZARDOUS WORK (See Safety Planning Certificate)
I have inspected the above job which has been safely prepared according to the precautions outlined above and on Safety Planning
Certificate No: . . . . , therefore work may start from . . . . to . . . . on. . . . .
Signed: .Approved Signator

7. APPROVAL OF PERMIT TO WORK
I am satisfied that this permit is properly authorised and that safe access is provided and that no work is taking place above or below this job. Work may proceed
from . . . . to . . . . on . . . . .Date.   Signed: . . . . .Permit Signator

8. ACCEPTANCE OF PERMIT TO WORK
I have read and understood the above precautions and agree that for our/my protection we/I will observe them. I confirm that all our/my Power Tools and Equipment have been registered and inspected as required by Dow Standards and that we/I understand the Site and Area Emergency Plans.
Signature: . . . .

9. COMPLETION OF PERMIT TO WORK
I certify that this job is complete/incomplete (ring appropriate word), that all guards have been replaced and secured in position, that all Tools and Equipment have been removed and the Job Site has been left clean and tidy.
Signature. . . . . .Time:. . . . . .Date:. . . . .

RENEWAL OF PERMIT TO WORK (CONSECUTIVE SHIFTS ONLY)
10. Approved until . . . . Time/Date . . . . Permit Signator. . . . .
Approved until . . . . Time/Date . . . . Permit Signator. . . . .

NOTE: When a job is finished this Permit must be signed off in Section 9 and returned. Should the job not be completed by the time specified this Permit must be renewed. This Permit is cancelled if Area Alarm (warble) sounds.

Figure 18.1 Example of permit to work (courtesy Dow Chemical Co.)

SAFETY PLANNING CERTIFICATE

DOW

(A PERMIT TO WORK IS REQUIRED BEFORE WORK STARTS) CERTIFICATE No: 6953

KING'S LYNN

1. FOR WORK INVOLVING:	Projects	Confined Spaces	Open Flame	Asbestos	Mobile Crane	H.V. Electricity	Other:
	Plant Change	Hot work in Zone 1 or 2	Critical Line Breaking	Excavations	Roofwork and Heights > 5m		

2. CERTIFICATE APPLIED FOR BY: Department/Contractor:

Area/Tank/Vessel/Equipment/Pipeline:

Exact Location:

WORK TO BE DONE:

TOOLS TO BE USED:	Welding Cutting Equipment	Gas	Mobile Crane	Mobile Pump		Compressor	M/Vehicle	Cold Tools only
		Arc	Excavator	Temporary Lights	110V / 24V	Electric Power Tools	Other	

3. USE OF A MOBILE CRANE: I have inspected this job and it may proceed subject to the following precautions: [N/A]

Signed: Approved Crane Supervisor.   Date:

4. EXCAVATIONS: I have inspected this job and it may proceed subject to the following precautions: [N/A]

Signed: Approved Construction Signator.   Date:

Signed: Approved Electrical Signator.   Date:

5. ROOFWORK – WORKING AT HEIGHTS AND ASBESTOS: I have inspected this job and it may proceed subject to the following precautions: [N/A]

Signed: Approved Construction Signator.   Date:

6. H.V. ELECTRICITY: I have inspected this job and it may proceed subject to the following precautions: [N/A]

It will be switched by . . . .

Signed: Approved H.V. Electrical Signator.   Date:

7. I confirm that the Area/Tank/Vessel/Equipment/Pipeline, as described above, will be safe for the proposed work provided the precautions listed above, together with those ringed on the check list opposite, are taken.

Additional precautions: (if none, write none)

A permit-to-Work must be obtained from: before work starts. This Safety Planning Certificate is

valid from   hours on   to   hours on

Section 6(C) on the Permit to Work may be signed by . . . .

Signed: Approved Safety Planning Certificate Signator.   Time: Date:

8. HOT WORK IN ZONE 1 OR ZONE 2 AREAS OR H.V. ELECTRICAL WORK OR ANY PLANT CHANGE: [N/A]

I confirm that the above work may take place provided all the stated conditions are satisfied.

Signed: Authorised Signator;   Time: Date:

9. RENEWAL: Subject to the provisions and precautions stated above and opposite this certificate is further valid.

Renewed from	hours on	to	hours on	Signature (approved SPC):
Renewed from	hours on	to	hours on	Signature (approved SPC):
Renewed from	hours on	to	hours on	Signature (approved SPC):

NOTE: (a)   A separate signature is required for Section 8.   **(b)**   This Certificate is not valid until all necessary signatures have been obtained.

Figure 18.2 Example of safety planning certificate – continued on page 518 (courtesy Dow Chemical Co.)

SAFETY PLANNING CERTIFICATE CHECK LIST (ALL REQUIRED PRECAUTIONS TO BE RINGED)

SIGN

GENERAL PRECAUTIONS

001. All power tools and equipment (including steps and ladders) must be registered with valid label affixed.
002. All power tools must be 110 volts maximum.
003. Ensure that power supply cables to transformers and welding sets above 110 volts are less than six feet long.
004. Suitable steps or ladders to be used.
005. Scaffolding to be erected and inspected by competent persons and notice fixed before use (mobile or fixed).
006. Provide life-line.
007. Use inertia fall arrestor (e.g. Sala Block).
008. Cordon off work area, above and below.
009. Notify adjacent plants/areas.
010. Check that all holes, excavations, work areas where covers or drains are removed are barricaded off and warning notices affixed. At night any such hazards must be adequately lit.
011. Isolate all power driven equipment before work starts — LOCK, TAG and TRY.
012. Check showers and eye bath units before work starts.
013. Instigate safe procedures for materials containing asbestos to comply with King's Lynn Site Standard No. 20 — Asbestos.
014.
015.

PROTECTIVE CLOTHING
100. Protection required:

1. Chemical Goggles	2. Face Shield	3. Updraft Helmet	4. Air Hood
5. Self-contained C.A.B.A.	6. Compressed Air line B.A.	7. Dust Mask	8. Ear Muffs
9. PVC Gloves Gen. Spec.	10. Rubber Boots (Steel Toecaps)	11. Protective Footwear	12. PVC Suit
13. Neoprene Suit	14. Disposable Suit	15. Plant Overalls Rubber Overboots Rubber Gloves	16. Safety Harness

ATMOSPHERE TESTING
200. Test for flammable vapours (explosimeter) BEFORE WORK STARTS/REPEAT EVERY . . . . HOURS/MONITOR CONTINUOUSLY.

201. Test for oxygen BEFORE WORK STARTS/REPEAT EVERY . . . . HOURS/ MONITOR CONTINUOUSLY

202. Test for toxic gas BEFORE WORK STARTS/REPEAT EVERY . . . . HOURS/ MONITOR CONTINUOUSLY.

LINE BREAKING
300. Positively identify by tagging, taping or painting.
301. Before cutting into a pipeline a 'test' hole should be drilled in the pipe.
302. Process operator to 'stand by' (protected to same standard as craftsman).
303. Check pipeline suspension.
304. Drain and isolate line, lock off pump(s).
305. Provide scaffolding — fitter should work at waist height.
306. Blank off open ends of pipelines.
307. Flush area with water after job to ensure no spillage left.
308. Decontaminate tools, protective clothing and boots, gloves, face and eye protection (keep goggles on until last and then remove in safe area wearing clean or disposable gloves).
309.
310.

HOT WORK
400. Guard against falling sparks and slag.
401. Keep work area and below wet with running water.
402. Instigate fire watch.
403. Check area 30 minutes after cessation of work.
404. Check work area every . . . . minutes.
405. Run out fire hose.
406. Provide fire extinguisher, Type:
407. Clear all combustible materials from work area.
408. Remove all full and empty drums from area.
409. Use only approved welding set, see Safety Standard No.17.

SIGN

HOT WORK . . . . . continued

410. Check welding cables are in good condition and where they must cross pipelines a suitable insulating bridging must be used to prevent possible contact. Weld return routing via installed equipment is prohibited.
411. Site gas cylinders so as to be clear of sparks and slag.
412. Check detachable cylinder key in situ.
413. Check compressed gas cylinders are used in metal wheeled trolley (not free standing or fixed to a structure).
414. Test all compressed gas connections using soap solution before work starts.
415. Check that oxygen and fuel gases have flash-back arrestors fitted between regulators and supply hose and that non-return valves are fitted between torch and supply hoses.
416. Check that all hoses are in good condition and located away from traffic. They should not present a tripping hazard to personnel.
417. Erect screens to safeguard personnel from U.V. radiation.
418. Site diesel driven D.C. generating sets in open air to prevent fumes accumulating in work area.
419. Check that smoke detectors are isolated.
420.
421.

ENTRY INTO CONFINED SPACES

500. All pipelines must be isolated, either by removing spool pieces and blanking off live ends or by inserting spade in lines.
501. Isolate agitator by removal of fuses, followed by LOCK, TAG and TRY.
502. Trained attendant to stand by outside vessel (must be named on Permit-to-Work).
503. Use mini-winch with life-line and full hoister-type safety harness.
504. Check vessel is cool enough to enter (< 35°C).
505. Use air mover or fan (must be grounded).
506. Use 24 volt lamp.
507. Check adequacy of means of vessel entry/exit.
508. Provide portable alarm for attendant.
509. Provide two sets of breathing apparatus outside vessel.
510. Compressed gas cylinders must be kept out of confined spaces.
511.
512.

MOBILE CRANES
600. Simple lift — banksman to be named on Work Permit(3c)
601. Qualified Dow representative in control—Name . . . .
602. Critical lift — check list completed —Construction Supervisor or Owner's Representative (mech.) in control.
603.

EXCAVATIONS
700. Over 1.2 metres deep — Construction Department in control.
701. Hand dig only.
702. Sides of excavation made secure.
703. Test ground water for contamination.
704.
705.

ROOF WORK & HEIGHTS GREATER THAN 5 METRES WHERE THERE IS NO PERMANENT ACCESS
800. Crawling boards must be used.
801. Working method and safety devices to be approved and recorded by Construction Signator.
802. Provide working platform with handrail and toe boards.
803.
804.

Figure 18.2 (continued)

18.4.1 Safety planning certificates

Except in an emergency or exceptional circumstances, the safety planning certificate must be raised at least two days before work is planned to start. The person initiating the work must determine if it needs a safety planning certificate and if so, which sections of it will apply. He then fills in sections (1) and (2), thus indicating which other sections should apply.

All persons authorised by management to approve certain sections of the safety planning certificate should have appropriate job and safety training and minimum periods of service with the company (five to ten years) and of experience in the job or area concerned (three months to two years, depending on the job). Two to four signatories have been approved for each function and their names and duties are made known to all personnel. They are as follows:

- *Approved crane supervisor* must inspect the job, check ground and overhead conditions, specify the precautions needed and sign section (3) of the safety planning certificate.
- *Approved construction signatory* is responsible for inspection and precautions for any work involving excavations, roofs, working at heights and asbestos. He liaises with the approved electrical signatory regarding buried power cables and is responsible for signing sections (4) and (5) of the safety planning certificate.
- *Approved electrical signatory* checks for underground power cables prior to any excavations.
- *Approved high-voltage electrical signatory* is responsible for safety on work involving over 1000 volts. He will name the electrician who will do the power switching and sign section (8) of the safety planning certificate.
- *Approved safety planning certificate signatory* is responsible for the overall safety planning aspects of the job, reviewing, liaising and specifying any additional safety precautions needed with particular emphasis on process hazards and for signing section (7) of the safety planning certificate
- *Approved signatory for section (8)* is generally the plant superintendent or senior process engineer and must confirm that the work specified in section (8) of the safety planning certificate may proceed by signing it. While an approved safety planning certificate initiator may initiate a safety planning certificate as well as signing section (7), sections (7) and (8) must always be signed by different signatories.

The original safety planning certificate is given to the person responsible for carrying out the work, the first copy is filed in the plant control room or area where the permit for work is normally issued and the second copy is retained by the initiator. The safety planning certificate may be issued for an initial period of up to two weeks. If more time is needed it may be renewed up to three times by an approved signatory. *Before work can start a permit-to-work is required.*

18.4.2 Permits-to-work

Approved signatories for permits-to-work require training and experience broadly similar to those of safety planning certificate signatories. Their authority to sign should be limited to their normal areas of work. Persons undertaking the work, and electricians, etc. who sign section (4), should have had appropriate training and a minimum of two years' appropriate work experience, at least three months of which should generally have been spent in the present job and in the same plant or area.

To initiate a permit-to-work *the person undertaking it* reports to the plant/area and contacts the permit signatory. If the work requires a safety planning certificate, the person undertaking the work must take this with him whenever he requests a permit-to-work, which can only be issued within the times specified on the safety planning certificate.

The permit signatory completes sections (1) and (2) naming the person in charge of the job, the company (where relevant), the name of the attendant if line breaking is involved and specifying the number of people working. He then arranges for any gas tests to be carried out and the results entered in section 3(a), rings the items of protective clothing to be worn and specifies any other precautions. He also arranges for the required isolation of pipelines and electricity to be carried out.

The person undertaking the work will place his own lock (on the power switch), try the equipment to ensure that it is isolated and sign section 4(c). Where an electrician or instrument has to 'trouble shoot' on live equipment, he will sign section 4(e).

The authorisation of a permit-to-work depends on whether or not a safety planning certificate is involved. If it is, an approved safety planning certificate signatory will inspect the job, sign the appropriate parts of section (6) and stipulate time constraints. The permit signatory then authorises work to start by signing section (7). He will assume responsibility in an emergency.

If no safety planning certificate is involved the permit signatory will sign both sections (5) and (7) of the permit-to-work.

The individual undertaking the work signs section (8) of the permit thereby accepting it before starting work.

The permit-to-work may be renewed by the signatory for up to two shifts, providing the work is continuous and that he is satisfied that all stipulated safety conditions are still in force and that it is safe to continue the work. If a safety planning certificate is involved, any time constraints stipulated on it should still be valid.

On completion of the job or period of work, the person responsible for it signs off his permit to work in section (9), returns it to the designated control area and attaches it to the plant copy.

Permits become void when the works evacuation alarm sounds and can only be renewed or reissued when the incident is over.

18.5 Precautions before issuing a permit

The following recommendations are due to Kletz[1]. Before a permit is issued, the person who issues it should go with the person who will be doing the job, or his supervisor, to the job-site and there discuss with him or her the work to be done and the precautions to be taken, as detailed on the permit. If the item to be maintained has no number, a tag should be attached to it which bears the same number as that used on the permit. The issuer of the permit should point out all valves which have been closed and locked, all electrical isolation switches which have been locked, and all fuses which have been removed as part of the isolation procedure. These should be verified on the spot by the person accepting the permit. Tags should also be attached to all unnumbered joints on pipes and equipment which will have to be disconnected. The working copy of the permit should be placed in a transparent cover and hung in a prominent position near the job while it is in force so that it can be readily referred to and signed as required by those doing the job.

Precautions before entry into confined spaces are discussed later [18.7].

18.5.1 Hot work

Operations classed as 'hot work' include welding, burning, the use of industrial (non-flameproof) electrical equipment, the entry or use of vehicles and plant with internal combustion engines, and the use of pneumatic chippers, hammers and rock drills. Three hazards must be recognised:

- Those due to the presence of flammable materials in the item on which hot work has to be done;
- Those due to the presence of combustible materials in the item which burn or give off flammable vapours when heated as a result of the work being done;
- Those due to the presence of flammable gases and vapours in the surrounding atmosphere.

The first hazard will have been removed if the isolation and purging discussed in 18.6 have been carried out effectively, and confirmed by a negative test for flammable gases [18.7.1].

The second hazard is more difficult to eliminate, and has caused several explosions. Where hot work has to be done on the outside of vessels, etc. which may still contain combustible materials in inaccessible places, the vessel, etc. should either be filled with water, purged with inert gas, or filled with a stable foam containing only inert gas before starting hot work. Where possible, cold methods of work should be used rather than hot ones.

Where welding has to be done on pipes which may contain residues or polymers, there should be at least two openings for fumes to escape, fire extinguishers should be available, and the welder should have a clear escape route.

The third of these hazards arises when work has to be done near running oil and chemical plants. The area within which hot work is to be carried out should be cordoned off and warning notices displayed. The atmosphere in this area should be tested for flammable gases, and no hot work should be allowed if the flammable gas content exceeds 10% of the lower explosive limit. Portable combustible gas detector alarms should be placed nearby, mainly upwind of the hot work operations, and hot work should cease at once if an alarm sounds. Supervisors of nearby plants from which flammable materials might escape should be consulted. They should be required to give loud audible warnings of any escape of flammable materials from their plants, or of any abnormal plant conditions which might herald such escapes.

The hazard of flammable vapours reaching the hot work area from nearby drains should be checked by flammable gas tests near the drains and by warning those liable to discharge flammable materials into the drains.

Many works have internal roads where vehicles, plant and cranes are allowed without a permit and where the risk of flammable vapour is low. Before vehicles, etc. are allowed to leave these roads, a permit may be needed.

18.5.2 Excavations, etc.

Excavations have a wide range of hazards including the striking of buried cables, pipes and other buried objects (even unexploded bombs!), those of persons and vehicles falling into them, the collapse of their sides onto people working, flooding, and the accumulation of dangerous heavier-than-air gases such as propane, carbon dioxide and chlorine. Their constructional hazards are covered in the UK by The Construction (General Provisions) Regulations 1961. These, among other things, require them to be inspected at least once every day by a competent person while people are working in them (Regulation 9).

Besides excavations, the operations of levelling ground and driving piles, poles and stakes often carry the risk of striking a buried cable or pipe, and should also require a permit. Before any such work is started, a responsible person should study the site plans and check whether buried cables or pipelines are present within 1 m of the place of work or not. Problems arise when records are incomplete or inaccurate, and in such cases cable- and pipe-locating devices should be used[7].

Where there is a risk of hitting a buried cable or pipe, the person issuing the permit should clearly mark the limits of the excavation, etc. on the ground with paint or pegs. The permit should state whether the area is clear of buried cables and pipes, whether machinery may be used, and within what boundaries. If cables or pipes are present within 1 m of the work, only hand tools should be allowed, taking care not to disturb the cable or pipe which should, if possible, be isolated before the permit is issued. The permit should also state whether the cable is alive or the pipe is in service.

18.5.3 Ionising radiation

Before portable equipment producing ionising radiation (as used for radiographic weld inspection) is introduced into a plant, all necessary precautions (e.g. barrier fences round radiation areas) should be in force before the permit is signed.

When instruments (such as level and density gauges) which contain sealed radioactive sources are present, these may have to be removed by a 'competent person' before a permit-to-work is issued. Plants which contain unsealed radioactive substances (e.g. catalysts containing uranium) also require a competent person to be satisfied that all necessary precautions and legal requirements have been met before a maintenance permit is issued. (See The (UK) Ionizing Radiations Regulations 1985.)

18.5.4 Work on live electrical circuits

Occasionally work has to be done on live electrical circuits to prevent a plant or factory coming to a complete standstill. The permit should be authorised by the chief electrical engineer, and only certain named electricians should be allowed to do it, under close supervision. Special training, techniques and equipment are needed. (See The (UK) Electricity at Work Regulations 1989 [2.8.7].)

18.6 Practical preparations for maintenance

As people belatedly discover to their cost, maintenance starts in the design stage. Wherever possible, the future plant manager and maintenance engineer should be members of the design team [16.3 and 17.1]. The team should ensure that all likely maintenance work can be done expeditiously and that every item needing maintenance, etc. while the plant is running can be isolated, and made safe for the work [J]. It is, however, as important to avoid unnecessary valves and fittings as it is to install enough of them. Every one represents a potential leak and will itself require inspection, maintenance and access.

18.6.1 Equipment isolation

Few valves can be guaranteed leak-free after a period in service. For all critical isolation duties, particularly when equipment has to be entered, spectacle plates should be turned or line blinds fitted, or a spool pipe section removed between the closed valve and the item being isolated and a blank flange fitted after the closed valve. The methods of isolation used for various process types of duty are given and illustrated in Appendix J.

Valves used for isolation should be capable of being locked shut, whether a blank or blind is fitted or not. The commonest procedure is to close and lock all isolating valves first, then remove residual material from the item, and only then break joints between the closed valve and the item to fit blinds or disconnect pipe sections.

Blank flanges, spectacle plates and line blinds should be of as high engineering standards as the flanges to or between which they are fitted. Permanently installed spectacle plates should be used on rigid lines. These only have to be turned after loosening the bolts and renewing the gaskets. Line blinds should only be used on flexible lines. Blank flanges should be used when a pipe spool section adjacent to a closed valve has been removed if a leak through the closed valve could present a hazard.

Approved line blind valves (Figure 18.3) past which leaks are a proven impossibility offer an alternative to the above methods. Although dearer, they save much time in use and are the cheapest solution when frequent isolation is needed.

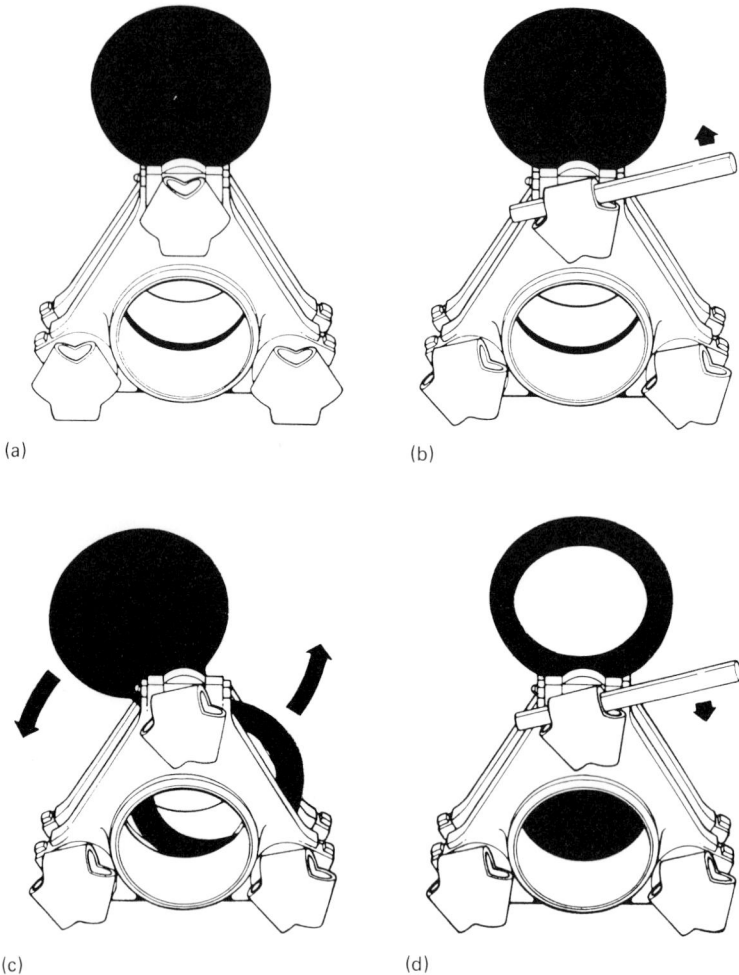

Figure 18.3 Operation of line blind: (a) line open; (b) bolts slackened, line spread; (c) spectacle plate reversed; (d) bolts re-tightened (courtesy Hindle)

Entire plant or plant units are usually isolated at their battery limits by spectacle plates on all incoming and outgoing lines, thus avoiding the need to isolate individual plant items.

Special care is needed when isolating plant items from relief and vent headers. The relief or vent valve from an item being isolated may discharge into an valve-free common relief header, since some codes and organisations do not allow valves to be placed downstream of a relief or vent valve or bursting disc. It is then dangerous to disconnect a joint on the discharge side of the vent valve or relief device to fit a blind when there may be pressure in the header (Figure 18.4). Unless a blind can be fitted between the item and the closed relief device, or unless the relief device can be relied on completely for isolation, the item should not be isolated

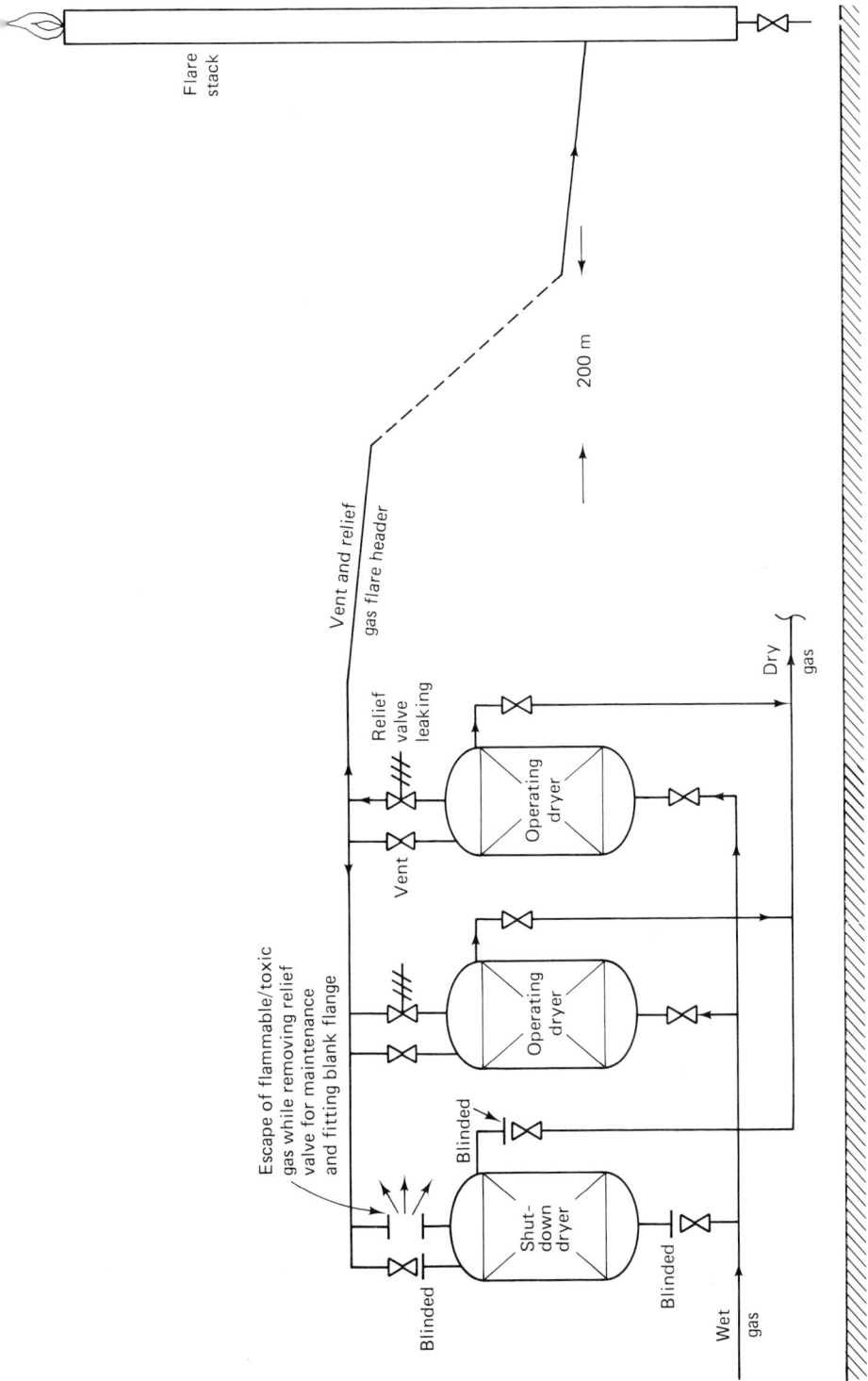

Figure 18.4 Danger of blanking discharge flange of relief valve of shut-down process vessel

alone. Maintenance on it should only be done when all items connected to the common relief header have been shut down and depressurised.

Vent valves and lines used for emptying, cleaning and purging should not be isolated or disconnected until this work is complete.

Fuses should be removed as part of the electrical isolation procedure. (Locked switches do not always prevent live circuits.)

18.6.2 Removing residual hazards

Equipment should not be opened up for maintenance, etc. until risks to those doing the work have been removed. The worker is least exposed when all the work can be done from outside the equipment without applying heat. Exposure is greater if he has to enter the equipment, and greater still he has to do 'hot work' while inside it.

It is impossible to lay down universal procedures because of:

1. The wide range of hazards which may be encountered;
2. The ease or difficulty of removing hazardous materials, and the many different methods which may have to be used;
3. The wide differences in the degree of worker exposure, depending on the job to be done and the method used.

Hazardous materials may be biologically active, carcinogenic, super-toxic, explosive, pyrophoric, highly reactive, flammable and/or corrosive. They may be gases, liquids and/or solids, including tars, resins, rubbers and fibres. The production department is generally responsible for removing them but the problems involved in this should have been studied and provided for during design [J].

For totally enclosed processes, the item or unit is first depressurised by venting gas or vapour through the normal disposal system (e.g. flare, scrubber, vent stack). If volatile liquids are present, they are generally next removed by draining to a safe place, displacing by or dilution with water, followed by blowing through with steam, air or inert gas, usually for several hours, until no trace of the liquid can be found in the steam or gas leaving the item. A special sampling point and condenser and some simple test may be needed to check this. Where steam is used the flow should be enough to ensure that the entire equipment is swept through. The item steamed out must be left vented until cool to prevent formation of a vacuum with risk of collapse. Most solids can be removed by high-pressure water jets. Their use requires training and personal protection including shields.

Difficulties will arise if there are no drain valves or connections at the lowest points of lines and plant items. These may have been omitted because of the risk of leakage of some very hazardous material such as carbon disulphide through a joint or valve.

The use of special solvents and chemicals to remove difficult materials is sometimes necessary, but should only be considered as a last resort. Most strong solvents are flammable, toxic or corrosive, and their final removal from the equipment and disposal may cause as many problems as that of the material they are intended to remove.

Chemical reactions with the surfaces of equipment which is being prepared for maintenance and entry can pose problems. When steel tanks which have contained concentrated sulphuric acid are washed out with water, the dilute acid formed attacks the steel and generates toxic gases derived from impurities in the metal, as well as hydrogen.

The inside surfaces of steel tanks, vessels, pipes, etc., even after cleaning, are sometimes still coated with iron oxide or sulphide, which react when air is admitted and reduce the oxygen content of the air inside. Such reactions with air also generate heat and the material may burn unless kept wet.

Water containing dissolved chlorides can damage stainless steel. It is thus best to use demineralised water or steam condensate for washing out stainless steel equipment.

The removal of residual hazards from process equipment before maintenance, and especially before entry or hot work, can give rise to unexpected problems (Figure 18.5). It is not a task to be undertaken lightly or hurriedly and may require expert advice.

The vessel looked clean so the welder
was allowed to go inside

A deposit behind
the baffle caught
fire

If you cannot see the whole of a vessel, assume it contains
hazardous materials

Figure 18.5 Hidden hazard to welder inside process vessel (courtesy T. A. Kletz and John Wiley & Sons)

18.6.3 Staged maintenance jobs

Many maintenance jobs proceed in three stages:

1. Disconnection of joints and fitting blinds/blank flanges, etc. as part of the isolation procedure;
2. The maintenance work proper, after residual hazards have been removed from the job;
3. Removal of blinds, etc. and reconnection of joints. Before this is done, the production team must have satisfied themselves that the job has been completed satisfactorily, and that the item is clean and in a fit state to bring back into production. (There are endless anecdotes of fitters' cleaning rags and even sandwiches being left inside equipment when production was resumed.) In most cases a leak test [17.6.2] is required, and in some processes it is critically important to remove all water from equipment before process materials are admitted. A suitable inspection

procedure and checklist for handing equipment back to production is often necessary.

In this three-stage process, the maintenance worker is more exposed to process materials in the first and last stages than in the second, when the real maintenance work is being done. It is therefore recommended that separate permits be made out for each stage. The personal protection (gloves, goggles, etc.) specified for stages (1) and (3) will be of a standard higher than that required for stage (2). The first permit should then have been signed off, and all work called for under it completed, before the next permit is issued.

Some straightforward jobs such as renewing a gland seal in a valve or pump can often be done in less time and with less hazard exposure of maintenance workers than is involved in disconnecting and reconnecting joints, etc. in stages (1) and (3). In these cases it is safer and more practical for the work to be done under a single permit, using shut and locked valves and electrical isolation switches as the only form of isolation. Here the person doing the job must wear appropriate protection against leaks of the process material. To assist persons issuing permits to decide when this simpler procedure may be used, a list of such jobs should be discussed and agreed between senior members of the production, engineering and safety departments, and given for guidance to those authorised to issue permits.

18.7 Entry into confined spaces

Section 30 of the UK Factories Act (1961) refers to 'any chamber, tank, vat, pipe, flue or similar confined space', and Chemical Works Regulation 7 (1922) refers to 'any absorber, boiler, culvert, drain, flue, gas purifier, sewer, still, tank, tower, vitriol chamber or other place where there is reason to apprehend the presence of dangerous gas or fume'. HSE's Guidance Note GS 5[3] refers to 'reaction vessels, closed tanks, large ducts, sewers and enclosed drains', as well as 'open topped tanks and vats, closed and unventilated rooms, and medium-sized and large furnaces and ovens'. Kletz[1] recommends that excavations more than 1 m deep be treated as confined spaces and that entry permits be required for them unless their width at their widest point is more than twice their depth.

Before deciding to enter a confined space, all reasonable possibilities of doing the work from the outside should have been explored.

The requirements of the Factories Act are summarised thus in HSE's GS 5[3]:

Atmospheres in which dangerous fumes are liable to be present
No-one may enter or remain for any purpose in a confined space which has at any time contained or is likely to contain fumes liable to cause a person to be overcome, unless:
(1) He is wearing approved breathing apparatus;
(2) He has been authorised to enter by a responsible person;
(3) Where practicable, he is wearing a belt with a rope securely attached;

(4) A person keeping watch outside and capable of pulling him out is holding the free end of the rope.

Alternatively, a person may enter or work in a confined space without breathing apparatus provided that:

(1) Effective steps have been taken to avoid ingress of fumes;
(2) Sludge or other deposits liable to give off dangerous fumes have been removed;
(3) The space contains no other material liable to give off such fumes;
(4) The space has been adequately ventilated and tested for fumes;
(5) The space has been certified by a responsible person as being safe for entry for a specified period without breathing apparatus.

No-one should be allowed to enter a vessel or confined space in any circumstances unless the size of manhole or other opening is large enough to allow a person to enter and be rescued while wearing breathing apparatus. The Factories Act 1961 stipulates the minimum internal diameter of circular manholes as 18 inches for stationary vessels. Kletz recommends a suitable manhole diameter of 24 inches[1].

The person issuing a permit to enter a confined apace must be satisfied that:

- Ventilation (forced where necessary) is adequate not only for breathing and comfort but also to cope with possible changes in the atmosphere inside a vessel as a result of disturbing scale, burning, welding and painting;
- Rescue facilities and persons trained in their use are available in case a person is injured or becomes ill while working inside a vessel;
- Adequate illumination complying with electrical safety standards is provided inside the vessel, etc.;
- There is safe access to all parts within the vessel, etc. which may have to be reached;

The permit should only be issued after a gas test has been carried out and attested by a gas test certificate, or the signature of the tester on the permit itself (depending on the permit system in use). Special care is needed in removing hazards and making rescue plans from vessels containing baffles or other obstructions. A group of vessels with large interconnecting lines and no valves or obstructions between them which have been isolated together may be treated as a single vessel. Where parts of the inside of the vessel, etc. cannot be seen from the outside, the permit should first be issued on a provisional basis for inspection only. Only after this has been done should the permit be re-issued or endorsed as a working permit.

Kletz recommends three types or grades of vessel entry permits with a maximum validity of 24 hours from their time of issue[1].

Type A permits are issued where a gas test certificate shows that the atmosphere within a vessel, etc. is fit to breathe.

Type B permits are issued when the atmosphere in the vessel, etc. is hazardous or unpleasant, but would not prove immediately fatal to a person breathing it. These require the following further precautions:

- Appropriate, well-fitting and adjusted respiratory protective equipment [22.8.4] and a harness and life-line should be worn by anyone entering.

The harness and line need to be adjusted and worn so that the wearer can be drawn up head-first through any manhole or opening. An armlet attached to the life-line and fastened to the wrist or forearm of the wearer will facilitate this. Any lifting gear needed for rescue should be ready in position.

- Another person qualified in the use of the breathing apparatus and resuscitation equipment should remain at the entrance to the vessel, etc. as long as anyone is inside, and be in constant communication with him. He should have another set of breathing apparatus for himself, and resuscitation equipment with him, and should have the means of summoning a rescue team without leaving his post.
- A rescue team within easy reach should be available at short notice.
- Rescue plans should have been practised so that all know what to do in an emergency.

Type C permits are issued when the atmosphere in the vessel is so deficient in oxygen or contains so much toxic material as to present an immediate danger to life. This situation should be avoided if at all possible, but where it cannot, in addition to the precautions listed for type B, two trained rescue workers should be on duty at the entrance to the vessel, keeping the one inside continuously in view and in radio contact with rescue and medical services. Only breathing apparatus of a type approved for use in atmospheres immediately hazardous to life should be allowed. Further precautions may be needed depending on the hazards involved.

18.7.1 Gas tests

These are required as a condition for issuing permits for vessel entry and for hot work. They may be attested by special certificates and/or on the permit forms. Only trained and authorised gas testers should sign. Where the test is made by a laboratory technician, a separate certificate is usual. Most tests are made on the spot with portable apparatus. This is connected via a flexible tube to a long sample probe which can be inserted from outside the space to be tested to any point within it. The outlet of the apparatus is connected to an aspirator bulb or suction pump. At least two of the three following tests are usually required:

1. *Oxygen content.* For a normal vessel entry permit, this should lie between 20% and 22% by volume. An oxygen content below 15% presents an immediate danger to life.
2. *Flammable gas content.* For a hot work permit, this should give a maximum reading at any point in the space of less than 10% of the lower explosive limit in air. The apparatus should first have been calibrated with the same flammable vapour as that present in the space.
3. *Toxic gas or vapour.* Since tests for toxic gases and vapours are specific to particular compounds, one can only test for those likely to be present. Tests which depend on the extent of a colour change in an adsorbent-filled glass tube through which a known volume of the gas sample has been drawn[8] are available for many gases and vapours. For those for which no portable test apparatus is available, air samples

should be taken and analysed in a laboratory. The analyses together with HSE's 'Occupational Exposure Limits'[9] and the toxicity data discussed in 7.4 provide guidance on the type of vessel entry permit which should be issued.

Gas test certificates should state:

- The time and exact place where the air was sampled;
- The oxygen content;
- The flammable gas or vapour content as a percentage of its lower explosive limit;
- The names of any toxic gases and vapours liable to be present and tested for, the concentrations found by testing and the degree of hazard which these represent;
- The signature of the tester;
- The period of validity of the test (usually not more than 24 hours from the time of sampling).

If the work for which the test was required is not complete by the time of expiry of the certificate, a further test is needed before it proceeds.

For work in an area where dangerous gases and vapours may intrude from sources other than the item which has been isolated, continuously monitoring gas detectors should, where possible, be placed round the work being done. Otherwise air samples should be taken at frequent intervals and tested promptly.

18.7.2 Hot work inside confined spaces

The hazards which can arise when hot work, especially oxy-fuel gas cutting and welding, has to be done inside vessels, etc. must be specially appreciated and the following precautions observed:

- Fuel and oxygen cylinders and their pressure regulators should not be brought into the vessel, etc.
- These gases should only be introduced into the vessel, etc. at reduced pressures via flexible hoses which are in good condition with secure connections.
- Forced ventilation should be used when hot work is done inside vessels.
- Workers must be warned of the dangers of gas escapes from damaged hoses, unlit burners, and of carbon monoxide poisoning from incomplete combustion. Carbon monoxide is liable to be produced when a large, cold, metal object is heated directly by a flame.
- Another worker should always be present outside the entrance to any vessel, etc. in which hot work is being done. He or she should be able to shut off fuel gas and oxygen instantly if needed.
- Continuous monitoring of the atmosphere in the vessel may be needed.

18.8 Other permits and certificates used

In addition to those used for maintenance, Kletz[1] discusses some other types used with process plant.

18.8.1 Permits to work on inter-plant pipelines

The normal responsibility for pipelines between plants may lie with two or more departments or plant managers, e.g. those at each end of the pipeline, and of any areas through which the pipeline passes. Close liaison between all of them is required. The general permit-to-work discussed in 18.4.2 might be used provided it has sufficient space for authorised persons in all plants and areas concerned with the pipeline to sign and write the precautions they have taken, and those which the maintenance team should take. The pipeline and those parts of it on which work is to be done should be clearly identified.

18.8.2 Certificates for used equipment sent outside plant

Before process equipment leaves a plant for repair, maintenance or scrapping, all hazardous material should have been removed from it. This is not always possible, and since accidents have resulted from this cause, a certificate should be used. The certificate should be issued by the head of the process department which is sending the equipment and should accompany it. The certificate should state that the equipment:

- Is free of all hazardous materials, or
- Contains certain named and potentially hazardous materials.

In the second case, the certificate should state what hazards may arise from the materials, under what circumstances this may happen, and what precautions should be taken. If necessary, someone from the department despatching the equipment should visit the organisation receiving it and explain the hazards, or even supervise the work done on it until all hazardous material has been removed.

18.8.3 Certificates of plant hand-over by contractors

Before new plant and equipment is handed over by contractors to an operating company and/or connected to operating plant, a formal hand-over procedure which includes inspections, test-runs and the issue of a certificate [20.1.2] should be followed. A normal permit procedure for all subsequent engineering work would then be applied. Connections from new to existing plant are preferably made by the engineering department of the operating company, under an appropriate permit, rather than by the contractor.

18.9 Pitfalls that must be avoided

Of the many pitfalls in maintenance work which the use of a good permit system helps to prevent, several common ones are mentioned.

18.9.1 Inadequate isolation

Several accidents have occurred through failure to isolate equipment from all lines connected to it, from reliance on leaking valves for isolation, and from failure to lock closed isolation valves to prevent them from being opened.

18.9.2 Faulty identification

Faulty identification by the maintenance person of the item to be maintained or the joint to be disconnected can have fatal consequences. The procedure of handing over and explaining the job to be done on the spot can be rendered ineffective by shift changes and other interruptions. It is essential for the permit issuer to write clearly and legibly, and to double-check against slips of the pen. Mistakes can also occur if the equipment numbering does not follow a consistent and logical pattern.

18.9.3 Changes in intent

Changes in the scope of work to be done once the permit has been issued and the work has started can also have serious consequences. It is important that the work requested be clearly detailed on the permit. If it is found that the job is more complex than appeared at the outset, this must be discussed with the issuer, and a fresh permit issued which states all precautions needed for the redefined job.

18.9.4 Inadequate communication

Lees[10] quotes a case of a maintenance fitter who left a job unfinished over-night intending to complete it the next day. The job was, however, finished and signed off by a night fitter. When the original fitter began work again the next day, the plant was no longer safe for work.

18.9.5 Unauthorised and unrecognised modifications

There are many temptations during maintenance to improve, simplify or streamline pipework, always with good intent and often with justification. Sometimes, however, some important reason for apparently untidy or complicated pipework is totally missed by the maintenance worker. Kletz[1] quotes a case of the air supply used for breathing apparatus which came from a branch-pipe originally 'teed' in to the top of a horizontal compressed air main. During maintenance a fitter thought the pipework would be neater if the branch-pipe were teed into the main from below. When the next person used breathing apparatus fed from the branch-pipe for work inside a vessel, he received a faceful of water, fortunately without serious consequences.

References

1. Kletz, T. A., 'Hazards in chemical system maintenance permits', in *Safety and Accident Prevention in Chemical Operations*, edited by Fawcett, H. H. and Wood, W. S., 2nd edn, Wiley-Interscience, New York (1982)
2. Dow Chemical (internal standard), *Permits to work and safety planning certificates*, Dow Chemical Company Ltd, King's Lynn, Norfolk, PE30 2JD

3. HSE, Guidance Note GS 5, *Entry into confined spaces*, HMSO, London (1980)
4. *Public enquiry into a fire at Dudgeon's Wharf on 17 July 1969*, HMSO, London (1970)
5. Bamber, L., 'Techniques of accident prevention' in *Safety at Work*, edited by Ridley, J., 3rd edn, Butterworths (1990)
6. Chemical and Allied Products Industry Training Board, *Permit-to-work systems*, CAPITB, London (1977)
7. National Joint Utilities Group, *Cable locating devices* (available from any Area Electricity Board) (1980)
8. Lee, G. L., 'Sampling: principles, methods, apparatus, surveys', in *Occupational Hygiene*, edited by Waldon, H.A. and Harrington, J.M., Blackwell Scientific, London (1980)
9. HSE, Guidance Note EH 40, *Occupational Exposure Limits*, 1985, HMSO London (1985)
10. Lees, F. P., *Loss Prevention in the Process Industries*, Butterworths, London (1980)

Chapter 19

Management for Health and Safety (HS)

By health and safety, abbreviated HS, we include here many matters and activities which are often classified under other headings such as 'loss prevention', 'reliability' and 'risk analysis', so long as they relate to or contain an element of unintended risk to human life or health. However, we exclude matters and activities which relate mainly to security and protection against deliberate acts of theft and destruction.

In discussing industrial management, it is hard to escape its economic role as put by Professor Drucker[1]:

> Management must always, in every decision and action, put economic performance first. It can only justify its existence and its authority by the economic results it produces.

The cost of major accidents such as those discussed in Chapters 4 and 5 can, however, be crippling to a company. The legal penalties in the UK for managerial failure to implement proper HS policies are also increasing, as in 1989 one large oil company was fined £500 000 for a repeated safety lapse, and manslaughter charges were brought against another company and some of its senior managers. Accidents also make for poor and costly relations between a company and both its employees and the general public. Thus even by Drucker's yardstick, managements need to include HS in the economic equation and pay special attention to protecting members of the public.

This leads to cost-benefit analysis, which attempts to balance the probabilities and costs of various types of accidents against the costs of preventing them (Figure 19.1[2]). The total cost of accidents in any year is taken as the sum of two parts, A the direct costs of the accidents plus P the costs incurred in preventing them. As more money P is spent on preventing accidents, their number and direct cost A fall. When both A and P and their total are plotted against some index representing the degree of risk reduction achieved, the total cost passes through a minimum. This approach has stimulated reliability studies and risk analysis [14] and the development of hazard indices (e.g. Dow and Mond) [12]. We are back to the analogy made earlier between company managements and skippers of racing yachts [11].

Cost-benefit analysis of HS is still, however, at an early stage and has two main weaknesses. One is the difficulty of costing the benefits arising

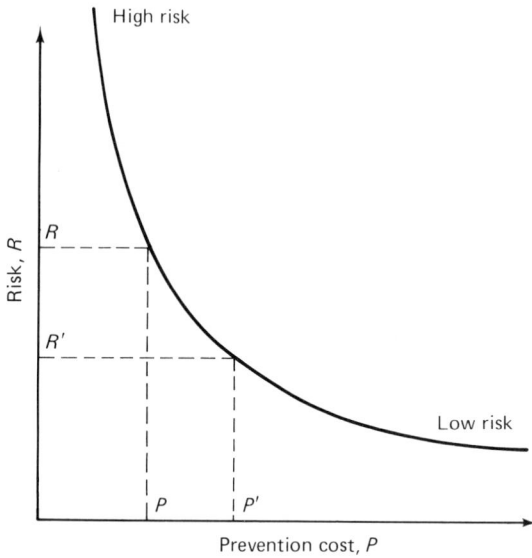

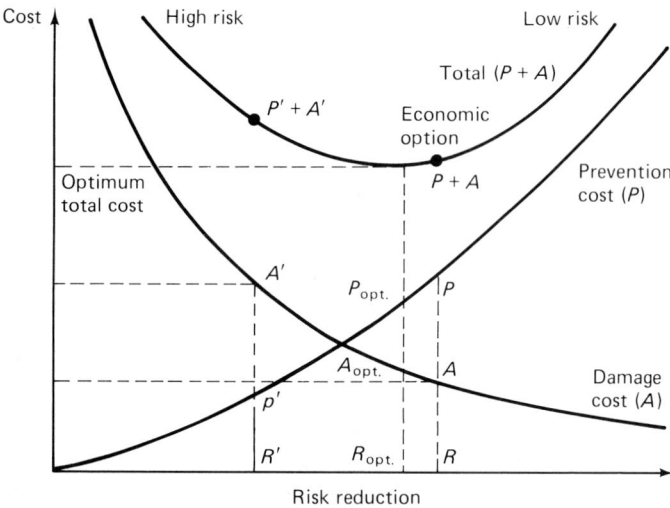

Figure 19.1 Accident costs versus risk reduction

from any level of expenditure on HS, and the other is the reluctance felt by many people at equating human life in money terms.

The safety of modern process plant is largely determined by the competence of those who design, manage, operate and maintain it.

A pipe carrying a hot, highly flammable liquid under pressure may break because there was no stress engineer in the organisation which designed the plant, and the piping draughtsman had not provided sufficient flexibility to allow for thermal expansion. On another plant an austenitic stainless steel pipe may break because it was carrying an aqueous process fluid containing chloride ions, and there was no metallurgist or corrosion engineer to advise the piping design group that this would result in rapid corrosion. On a third plant a pipe may burst as a result of the explosive decomposition of a peroxide inside it, which could have been avoided if the process engineer had a better knowledge of peroxide chemistry.

Someone capable of recognising all these hazards before they caused damage would need a very broad technical education, including mechanical engineering, chemistry, metallurgy and corrosion engineering. To have been able to avert them he would have needed the authority to shut down an operating plant for inspection when he felt that something was amiss. A safety professional who combined all these qualities would be a rarity. The responsibility for such accidents can therefore lie only with the technical executives and their staff who caused the hazards in the first place. Safety in the process industries is thus everybody's business, especially management's. It requires that all employees, particularly those doing technical jobs, have the necessary training, experience and *hazard awareness* to avoid accidents arising from their work.

19.1 Management's responsibilities for health and safety (HS)

Management's responsibility for the safe performance of its workers was expressed by Heinrich in 1931[3]:

Management's responsibility for controlling the unsafe acts of employees exists chiefly because these unsafe acts occur in the course of employment that management creates and then directs. Management selects the persons upon whom it depends to carry on industrial work. It may, if it so elects, choose persons who are experienced, capable and willing to do this work, not only well, but also safely. Management must also train and instruct its employees, acquaint them with safe methods, and provide competent supervision. In following the principles of delegated authority, management, through its representatives in the supervisory staff, may set a safe example, establish standards for safe performance, and issue and enforce safety rules.

In the UK, after the recommendations of the Robens Report of 1972[4] had been embodied in HSWA 1974 [2.3], management's key role in the safety of operations under its control was given legal emphasis. HS should now be treated as integral parts of the work processes. Managerial competence in HS must match the risks inherent in the undertaking and be no less than that needed to run the business successfully. HS at work requires methods, time and money similar to those devoted to other business objectives.

Management responsibility for HS does not apply only to top management or to shop-floor supervisors, but it must be recognised and accepted at all levels of line management, particularly by middle managers who may find themselves torn by conflicting requirements of production and HS. To resolve these they need help and support from senior managers. Senior managers, while motivating those at lower level in HS, should not, however, become too involved in detailed decisions which subordinates can take, given proper training and guidance.

The directors of companies involved in the process industries are responsible for ensuring that all staff participating in the design, operation and maintenance of the plant have an appropriately high degree of technical competence, hazard appreciation and personal responsibility. They also must ensure that adequate checks are made by hazard studies such as those discussed in Chapter 16.

The appointment of safety professionals whose roles are discussed later [19.1.7] cannot relieve line management of its safety responsibilities. The safety professional has no direct responsibilities for production or maintenance and the role is mainly advisory. The Accident Prevention Advisory Unit (APAU) of HM Factory Inspectorate (HMFI) provide the following key questions[5] which managers should ask themselves to assess their effectiveness in their HS performance:

1. Do we have a safety policy?
2. Is it up to date?
3. Do the subsidiary parts of our organisation have a policy?
4. Who is in charge of health and safety?
5. Are the technical problems of safety handled by competent persons?
6. Do we have a system to measure safety performance?
7. What is the worst disaster that could happen?
8. If the worst happened could we cope?
9. Would our workforce know how to react in an emergency?
10. What do our employees think of our safety standards?
11. What are we trying to achieve?
12. How much effort are we putting into safety?
13. Is the effort directed to the right place?
14. Is there an efficient system of checking that the duties are being carried out efficiently?
15. What are our long-term objectives?

19.1.1 Accident causes – technical and organisational

Industrial accidents have technical causes which need to be identified in order to prevent their repetition. Although the law has in the past been concerned to identify single 'proximate' causes for major accidents, careful analysis has shown that a combination of two or more technical causes was usually involved. But behind nearly all technical causes lie organisational weaknesses which allowed them to be present [3.3]. This can be illustrated by the simple example of a man falling from a ladder. The technical cause was identified as a defect in the ladder. Getting rid of the defective ladder will prevent it causing more accidents, but more questions need to be asked to discover the faults in the organisation, e.g.

1. Was a system of regular ladder inspection in force and who was responsible for it?
2. When was the defective ladder last inspected and why was the defect not then found and the ladder removed from service?

3. Did the injured employee's supervisor examine the job and the ladder before the accident? Why did the supervisor allow the ladder to be used?
4. Did the injured employee know that the ladder was defective and that he should not use it?
5. Was the injured employee properly trained in the use of ladders?

The answers to such questions might lead to the following organisational improvements:

1. A better ladder-inspection procedure
2. Better training
3. Clearer responsibilities
4. Better job planning by supervisors.

The same types of question need to be asked and the same types of organisational corrections made time and again following the more complex accidents typical of the process industries. It is even more important for management to anticipate the hazards and create an organisation in which the accidents do not occur.

19.1.2 Policies and degrees of hazard

Since 1974 all undertakings employing five or more persons have been required under HSWA to have a written policy for HS. The policy, which is a statement of intent, will have limited value unless it is backed up by an effective organisation, adequate resources and motivated personnel. The statement and the degree to which it is implemented provide acid tests of managements' attitudes and commitment to HS.

Each policy should be unique to the special needs of the organisation for whom it is written. 'It cannot be bought or borrowed nor can it be written by outside inspectors or consultants[6].' In large enterprises typical of the process industries, the most senior management specifies the overall objectives and top-level organisation for HS while each section of the enterprise amplifies its organisation and arrangements needed to meet these overall objectives. The policy should give a clear, unequivocal commitment to HS, be agreed by the board and be signed and dated by a director. It should be regularly reviewed, agreed with trade union representatives, brought to the attention of employees and should state that its operation will be monitored at workplace, divisional and group level. A 1980 review of the effectiveness of company policies for HS by the APAU of HMFI[6] contains a checklist of questions (which appears here as Appendix K) to probe the applicability, strengths and weaknesses of HS policy documents.

The effort and organisation which need to be devoted to HS is determined very largely by the magnitude and nature of the inherent hazards of the operations. In the process industries, which commonly employ flammable, reactive, toxic and corrosive substances from which to make useful products, the hazards may range from the mundane to the major. The HS resources needed in different cases have parallels with traffic control. Pedestrian precincts and shopping arcades from which motorised traffic is excluded are akin to safe processes such as solar evaporation of sea-water and salt crystallisation, which have only minor hazards. The accidental bodily contact of shoppers is quite common, but unless there is a sudden panic causing a stampede, injuries are rare. Normally there is little need for control.

On roads and motorways, which are akin to moderately hazardous processes, special protection has to be devised and procedures with formal safety rules have to be laid down by experts, enforced by police or inspectors.

In busy air-lanes near major airports, which correspond to highly hazardous processes, more sophisticated safety equipment and much stricter controls are needed. These must be continuously monitored by highly trained people.

19.1.3 HS goals

Several things have to be considered in setting HS goals. They should be practical, comprehensive, within the capabilities of the management and relevant to the conditions in the undertaking. It is even more important that every manager should understand his role in meeting them and that his progress in this can be and is monitored, preferably by his immediate superior. It will then be the latter's responsibility to check how far the junior manager is able to fulfil his HS goals, (particularly if he has other urgent production tasks), to compliment him if he succeeds and to work out with him how to improve his performance if he fails.

All broad HS objectives of the organisation, both short term and long term, should first be listed by the most senior managers (assisted by HS professionals), together with the names and functions of those next in line who will be responsible for implementing them. After the senior manager has discussed and agreed with each of his junior managers which of the broad goals apply to him, the latter should list his own goals in greater detail with the names and functions of those of his own subordinates responsible to him. In this way a complete, detailed and often unexpectedly long list of HS goals for the enterprise is built up. (The length of the list is less surprising when one considers that most chapters of this book include several different HS goals.). The list should then serve as the basis for HS planning, budgeting and costing. Some typical and rather broad goals follow by way of example:

- Provision of adequate resources both financial and in terms of man-hours of competent persons to meet the goals set. These includes not only full-time HS professionals but a reasonable percentage of the time of line managers to meet their HS responsibilities.
- Maintenance of a sound organisational structure, with accepted job descriptions which include HS responsibilities;
- An adequate level of competent staff which leaves no gaps, especially in positions considered critical to safety and plant integrity;
- Achievement of lower-than-average accidental injury rates for the type of industry. This implies the reporting and analysis of all accidental injuries: better still, all accidents, whether causing injury or not, within the organisation. Here one needs to consider accident severity as well as frequency. Some authorities use a combined rate, the product of

frequency and severity rate, as the best overall criterion [19.2.5]. One difficulty in judging progress toward this goal is that the figures for any one year may have little statistical significance unless large numbers of workers are involved. Changes in the type of manufacturing operation and in methods of accident reporting can be further difficulties.

- Compliance with relevant codes of practice, standards and regulations (such as COSHH), and the setting and monitoring of company HS systems (such as permits-to-work);
- Identification and elimination or reduction of specific hazards arising from the work. This includes careful investigation of accident causes;
- Adequate or improved HS training and commitment at all levels;
- Regular inspection, testing and and maintenance of protective systems for emergency use as well as the plant and machinery itself;
- Adequate protection of visitors, the public and the environment;
- Adequate monitoring of programmes in support of the HS goals.

Table 19.1, based on headings given in Dow Chemical's publication *Minimum Requirements*[7], provides a list of items to be included in most lists of safety goals.

Table 19.1 Items to be covered in safety and loss-prevention goals

Safety	Loss prevention
Accident/incident investigation and reporting	Buildings and structure design
Audits	Capital project review
Confined space entry	Combustible dusts
Contractor safety	Electrical
De-energising and tag procedures	Emergency planning
Employee training and job-operating instructions	Equipment and piping
Government regulations	Fired equipment
Guarding and interlocking	Fire protection systems
Hot work and smoking	Firefighting capability
Job and process-operating procedures	Flammable liquids and gases
Ladders, scaffolding, work surfaces, etc.	Flexible joints in hazardous service
Line and equipment opening	Fragile devices in hazardous service
Personal protective clothing and equipment	Instrumentation
Safe operation of motor vehicles and motorised handling equipment	Leak and spill control/containment
Testing of emergency alarms and protective devices	Means of exit
	Pressure vessels
Related requirements	Process computers and data-handling equipment
Distribution emergency response	Reactive chemicals
Industrial hygiene and medical programme	Risk analysis
Material hazard identification	Rotating equipment
Product stewardship	Storage
	Technology centres

19.1.4 Management systems and accountability for HS

There are as many views on organisation and management systems as there are systems, whose structures vary from the nearly vertical to the nearly horizontal. From experience of several organisations, the writer prefers one which contains the following features:

1. A minimum of hierarchy;
2. Structure oriented towards the project or activity;
3. Decisions made at the lowest possible level with guidelines provided by higher levels;
4. Supporting staff and techniques available where they are most needed;
5. Communications across vertical lines of authority;
6. Effective feedback of information.

Special problems of management including stress are discussed in 19.3. The position of safety professionals, particularly in organisations with hierarchical structures, can cause problems if their advice is ignored by their seniors [19.1.8].

Performance in HS should be included in staff assessments. Managers with successful HS records need encouragement. Those who fail must be made aware of where they have failed and the appropriate lessons discussed with them. Provision should be made to protect the 'whistle blower' (who calls attention to particular hazards or dangerous practices) from being victimised[8]. Managers at all levels need to be convinced of the importance of HS goals, that the organisation intends to achieve them, and that they will be personally accountable for their part in it. The cue will be taken from the top.

19.1.5 Job descriptions and their HS content

The drawing up of job descriptions should go hand in hand with that of organisation charts. It is of immense importance in selecting applicants and determining the remuneration of various posts. No job description is complete unless it includes an agreed list of HS duties. The APAU of HMFI, which has seen many efforts at HS organisation within companies, gives the following advice about the issue of job descriptions for safety[6]:

1. The construction of a job description, a defined list of tasks for each manager, is a valuable exercise, so long as it is personally relevant to each person.
2. Since the person who will monitor the degree of success in meeting the tasks listed in the job description is the employee's superior officer, it is that officer who should first of all offer the suggested list of duties. He should then have an interview to discuss the job holder's view on his own work.
3. At the interview any uncertainties about the duties, on the part of the person for whom the job description is written, should be discussed and resolved.
4. Agreement between the two employees should then be followed by the issue of a final personal job description.
5. Monitoring by the senior employee should be against the job description, and reference should be made to it in assessing performance against the allotted and agreed tasks.
6. When either person changes, the exercise should be repeated, in order to establish the clearest possible common attitude to HS between the new parties.

7. The procedures listed above should be carried out progressively through the company, starting at senior management and finishing with the most junior manager in the organisation.

19.1.6 Resources for HS – time and money

A good deal has been written on the economics of loss prevention (see Lees[9]), but less has been said on what resources should be provided to meet companies' HS goals, particularly to support line managers. The full-time HS professional knows what he is being paid to do but the line manager, who among his other duties is directly responsible for HS performance, often has little idea of how much of his time is, should or may be devoted to HS matters. If he finds himself overstretched with other (e.g. production-related) duties which are given priority over HS, he may see all the paperwork about his HS goals and duties as merely a trap to make him a scapegoat if something goes wrong. If he is required by company accountants to record how he spends his working time under various cost centres, one or more of these should clearly relate to HS activities. This is a matter to be decided between the director who signed the HS policy statement and his chief accountant.

It is hardly enough to say that the time that the manager spends on HS matters should be charged to production activities to which they may relate. In this case nobody is any the wiser about how much time he actually spends or should reasonably spend on HS matters, and figures like 5% or 15% are simply plucked from the blue. Thus there generally seems to be a need for research on how much of their time line managers need to spend on discharging their HS duties. Top managers need to consider these findings and discuss them with their line managers. Individual records of time spent on particular HS matters should be kept, and managerial staffing levels should be adjusted to ensure that no line manager can plead that pressing production problems left him no time for his HS duties.

Likewise, when budgeting for capital expenditure, a special allowance should be made for safety items and special protective equipment needed over and above the normal protection incorporated into plant and equipment to meet the requirements of codes and standards. This is closely related to money spent on buying insurance. The application of the Dow and Mond hazard indices [12] sheds useful light on the subject even if it does not provide exact answers. The idea of financing a company's HS expenditure from a special fund rather than from its current account is raised in 19.3.6.

It is very important when choosing between different projects or process alternatives on which to invest capital that the differences in loss potential and safety/insurance expenditure are fully considered. Any project with a high accidental loss potential needs to show a correspondingly high potential for profit.

19.1.7 Motivation for HS

Motivation for HS is needed, particularly to change deeply ingrained habits and attitudes and to indoctrinate new employees. The key element is knowledge of work hazards and the effects they are likely to have if certain precautions are not taken. The dilemma for managers is how to put these facts over without creating undue anxiety and stress. As one worker interviewed by an industrial psychologist put it, 'If I knew all the hazards I faced at work, I'd never sleep at night'[10]. The relevant information needs to be presented in a way which assures the individual that the hazards have been assessed, provides him with the means of coping with them, and eliminates uncertainty. Sound knowledge is not, however, always enough to change bad habits, while the act of changing one bad habit can easily lead to another. A genuine desire to change must be there. If it does not or is only half-hearted, it must be fostered, preferably by objective discussion of the facts. But will alone is seldom enough to overcome subconscious impulses and something more is generally needed.

Management more than committees has the tools for the job. If it can organise people to achieve results in other fields, it should be able to do the same for HS. Its main tools are *communication*, *assignment of responsibility*, *granting of authority* and *fixing accountability*. Yet each individual will still make his or her own decisions. Management has to recognise those influences over which it has little or no control and extend its own influence in areas where it can. These include:

- Group attitudes to safety,
- Selection and placement,
- Training,
- Supervision,
- Special emphasis programmes,
- The media.

Studies on industrial motivation[11] suggest that the factors which motivate people are separate and distinct from those that cause dissatisfaction. Examples of both are:

Motivators	*Dissatisfiers*
Achievement	Company policies
Recognition	Supervision
Quality of work	Working conditions
Responsibility	Salary
Advancement	Relationships
	Status

Factors labelled as potential dissatisfiers need first to be brought up to the level of contentment. The next step is to work on the motivating factors.

19.1.8 HS professionals and safety organisation

Full-time HS professionals employed in industry include inspectors, doctors, nurses, industrial hygienists and fire officers as well as safety specialists. The last named act mainly as advisers. Their number and need depend both on the size of the undertaking and the nature of the hazards. The APAU of HMFI quote the following advantages in having a safety adviser[5]:

1. He can keep abreast of HS developments and changes in legislation and provide line managers with such information as is relevant to their needs.
2. Training effort can be concentrated and specialist experience widened by seeing a range of problems.
3. He can advise whether the safety policy is being consistently implemented throughout the organisation's premises – particularly important in large undertakings such as local authorities or conglomerates having multiple premises.
4. Coordination of safety effort is simplified. He can avoid the duplication of effort that inevitably results from each location or department trying to resolve its own problems in isolation.

There is a clear role in the process industries for professionals who can advise and assist management in controlling the hazards common to most industries, and who are familiar with the law on safety. While they require some technical education, this need not always be to degree level. Otherwise they should be full members of the appropriate professional organisation, i.e. the Institution of Occupational Safety and Health. One prime requirement is that they should be good communicators. Hearn[12] has given a good description of their work and duties, of which an extract follows:

> In small or medium-sized units a safety officer may be appointed for duties which cover industrial accident and fire prevention, road safety, security, welfare, personnel, etc., but in large units a safety officer specialising in safety and hygiene is essential.
>
> He should advise on the formulation of a company's safety policy and guide all employees on the implementation of this policy, and he must be given the status necessary for him to carry out his duties *vis-à-vis* all levels of line management.

Duties of safety officer
> The duties of the safety officer will depend to a great extent on the size and nature of the works. In general they will consist of advice on safety measures to all members of management, and to specialist employees such as architects, designers and purchasing agents, etc.
>
> He should ensure that basic safety principles are incorporated in the design stage of buildings, plant, processes, storage and distribution, not only to provide a safe working environment, but also to facilitate production. He should advise on the incorporation of safety measures in all operational procedures, machine usage, use of hoists and other lifting equipment; on the provision and usage of personal protective equipment; on the preparation of safety and emergency instructions; on reporting and investigating accidents and the preparation and analysis of records; on training, safety propaganda, safety incentive schemes, etc. He must at all times work in close harmony and collaboration with line management and the trade unions to ensure that no aspect of safety is neglected.

Such safety officers cannot, however, be expected to assume responsibility for a variety of special hazards in the process industries which they do not have the necessary technical training to recognise.

This still leaves a need for a safety professional at a higher technical level, who, while he may not be *au fait* with all the possible hazards which may arise in a process plant, can nevertheless discuss technical matters and hazards on equal terms with technical executives, audit their work, assess their own attitudes to safety, and strive to improve these where necessary.

Petersen[11] gives the following criteria on where safety professionals should be located in an organisation:

1. Report to a boss with influence,
2. Report to a boss who wants safety,
3. Have a channel to the top,
4. Perhaps – install safety under the executive in charge of the major activity.

In large organisations where several HS personnel in different fields are employed, the question often arises whether they would not be better organised in a separate department, responsible to a director or top manager. Arguments in favour are that it ensures a channel to the top, that all members of the department have a common objective and are not liable to find themselves reporting to hostile bosses. Against this it might be argued that a separate HS department can become so isolated from other professional staff that it defeats its own objectives.

An organisation chart for a large chemical company (making both toxic and flammable products) in which the HS functions are slotted in where there is most demand for their services is shown in Figure 19.2. This contains four functions exclusively concerned with HS, i.e. health, safety, inspection and fire brigade.

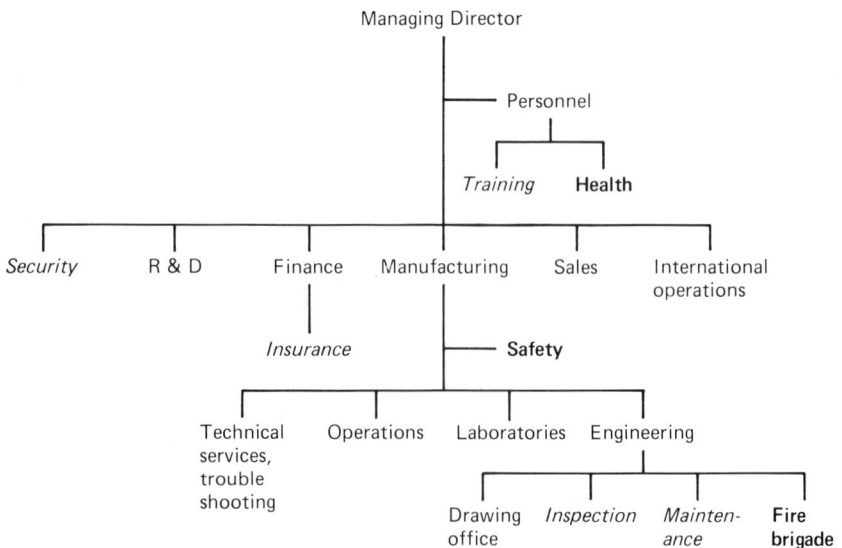

Exclusive HS function
HS shared with other functions

Figure 19.2 Organisational chart for a chemical company making both toxic and flammable products

The HS organisation at Dow Chemical Company's King's Lynn site is shown in Figure 19.3[13]. This includes both full-time HS professionals (the safety officer and the industrial hygienist) as well as departmental managers and other senior staff and workers' safety representatives. The works manager heads the organisation.

The safety council is the policy-making authority for site safety. It consists of the works manager (chairman) and the safety officer (secretary) as permanent members, with one member from each department having an accident-prevention committee. Meetings are held monthly and their minutes are circulated to departmental heads and displayed on the main notice boards.

The safety officer who is directly responsible to the works manager is responsible for the incident controller, one of several senior managers appointed in rotation to coordinate during an incident on or off the site.

The health and ecology council is the policy-making body for health and ecology and meets on a quarterly basis. The works manager is again chairman, the industrial hygienist is secretary and the other members are plant managers and appropriate staff. The council deals with industrial hygiene, environmental health and waste disposal and the occupational physician reports to it.

There are also several part-time safety directors, usually senior staff members, who monitor specific areas of the safety programmes of which they have special knowledge. This work is very important since it would not be practical to employ a full-time safety specialist in each of these areas.

19.1.9 Safety committees

Safety committees may be set up either as a result of a management decision or at the request of two or more workers' safety representatives. In either case workers should be represented as well as managers and supervisors since they are the best people to ensure that hazards which may affect them are removed or contained. Small departmental committees with no more than twelve members each are more effective than a large organisation-wide committee. The departmental accident-prevention committees (APCs) at Dow Chemical's King's Lynn works shown in Figure 19.3 provide a good example of this. Each departmental head is encouraged to form his or her own accident-prevention committee and may act as chairman. Small departments may amalgamate to form a single APC and rotate the chairmanship. Each APC is required to draw up and work to an annual safety programme which includes the following elements:

1. *Safety meetings* involving all employees for a minimum of 10 hours per year for process, craft and service personnel and 3 hours per year for office staff;
2. *Safety inspections* (at least one per month);
3. *Fire drills and simulated emergency trials* (two per employee per year);
4. *Training* including job-related, firefighting, breathing apparatus, permit-to-work systems, first-aid, etc.;

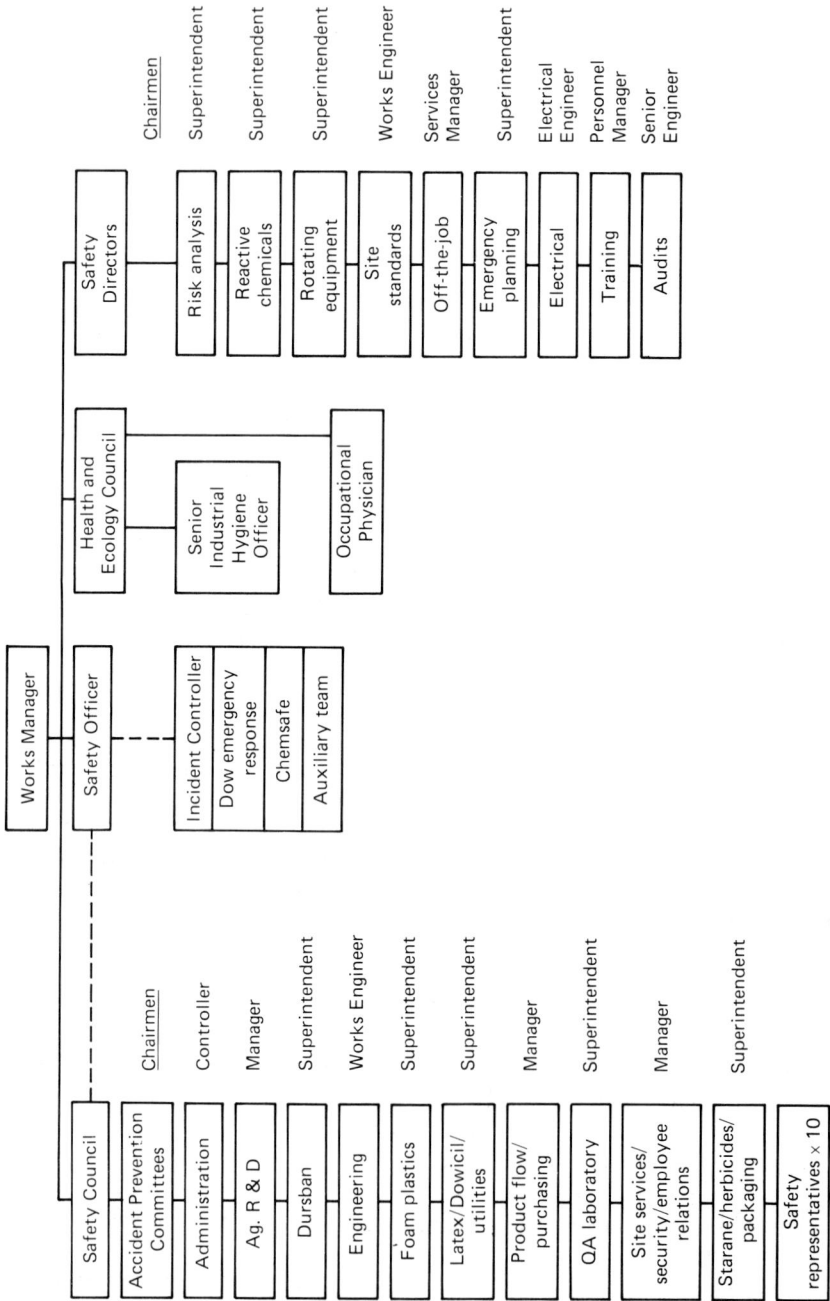

Figure 19.3 Dow Chemical Company's site safety organisation at King's Lynn

5. *Monitoring and enforcing safety standards and practices;*
6. *Job safety analysis* with regular reviews and use in training.

Each APC should send a report or minutes to the safety council every month.

Each departmental head appoints one or more safety representatives who receive special training and are encouraged to take an active role in the safety programme.

19.1.10 Documentation

Most managements are well aware of the large amount of paperwork which must be kept and filed. Lees[9] lists some of the principal subjects on which documentation is needed for a chemical plant, with details of the documents themselves. The subject areas only are listed in Table 19.2.

Table 19.2 HS-related subject areas on which documentation must be kept

Company systems	Fire protection
Standards, codes and legal requirements	Plant operation
Organisation	Training
Process design	Safety equipment
Plant layout	Hazard identification and assessment
Mechanical design	Security
Services design	Plant maintenance
Electrical, civil, structural design	Plant inspection
Plant buildings	Emergency planning
Control and instrumentation	Environmental control
Effluents, waste disposal, noise	Medical

A secure and efficient central filing system is required, with arrangements to ensure that working copies of documents are available to those needing them. Duplicate copies should be kept on microfiche or computer discs at a different, secure and fireproof location in case the only existing copies in the site office are destroyed by fire or explosion (as happened at Flixborough).

19.2 HS programmes and their elements

The following matters, most of which are discussed elsewhere in this book, should be considered when preparing a complete HS programme:

1. The involvement of the safety department and of relevant line management at the planning stage and the analysis of health and safety factors in new projects [16];
2. The HS performance criteria required of new plant, machinery and equipment [13];
3. Evaluation of toxic properties of process materials and the precautions required in their use [7];
4. Instructions for the use of machines, for maintaining safety systems and for controlling health hazards [20 and 7];

5. Specific training for operatives, particularly those whose activities affect the safety of other workers [21];
6. Arrangements for medical examination and biological monitoring [7.7];
7. The provision of personal protective clothing and equipment, in consultation with those who have to use them [22];
8. Permit-to-work systems [18];
9. Emergency [20.3] and first-aid procedures [7.9];
10. Procedures for visitors and contractors;
11. Relevant instructions at all levels.

The main elements in the programme fall under the following headings, each of which is then discussed briefly:

1. Information and communications
2. Training
3. Hardware and protective systems
4. Inspections and audits
5. Accident reporting
6. Measuring safety performance.

19.2.1 Information, communications and consultation

Personnel at all levels should know precisely where to get the information and decisions needed for their work. Management should ensure that all such information is correct, adequate and relevant for those who receive it. Most enterprises have developed their own jargon and abbreviations. To ensure that these are generally understood, it is recommended that a glossary of jargon, technical terms and abbreviations used within an enterprise be included in a handbook issued to all new entrants.

The problem of communications within an organisation varies with its size. In a small company where everyone knows everyone else and the boss is on first-name terms with the staff, communication is seldom a real problem. Person-to-person communication is the accepted norm.

In a large concern the directors are only names to most employees and are often separated from them by many layers of management. Internal communications then require special care to ensure that they are effective. Procedures and channels of communications between different departments should be established as close as possible to the working level. It is then important to know who is responsible for providing the various items of information which a particular group or individual will need. Permits for maintenance personnel to work on process plant provide examples of such procedures [18.3].

Communication has many pitfalls, including:

- Language and translation problems [19.3 and 23.4.1];
- Inundation with too much information, much of it irrelevant to the needs of the person and the job [3.3.3];
- Problems of communication between different shifts working on the same plant [20.2];
- Ambiguity and poor self-expression;

- Lack of explanation, and essential details;
- Inadequate consultation;
- Information lost or distorted through being relayed through too many intermediaries between its originators and those who need it.

A common fault of management is to instruct employees to achieve a certain result without explaining the reasons for it or how it is to be achieved. In many cases the only way to ensure that unusual instructions apply and are properly understood is to present them first as a draft to the recipients and then discuss it with them. Only when management is sure that the instructions apply and are understood are they presented in their final form. This, of course, takes rather longer than the simpler course of issuing a straight edict. The results, however, are far more satisfactory and pay for the extra trouble taken.

Consultation plays a major role in HS matters and requires a type of manager who is ready and willing to understand other people's point of view. Although consultation is easiest on an informal basis, formal channels of consultation between management and employees often have to be established. There is, however, a danger of consultation being carried too far and placing too great a strain on managers.

Information is liable to be lost or distorted during transfer when it is passed by word of mouth through several intermediaries. If the originators of the information do not have the time to discuss it with all its recipients personally, they should take pains to brief thoroughly those whose task it will be to explain it.

19.2.2 Training, codes and standards

Training methods are discussed in Chapter 21. Management is responsible for organising the training of its staff and workforce, particularly new employees. This must be appropriate to their work and its hazards and to the safety content of their job descriptions. Management should keep individual training records, review training annually, and ensure that every individual receives sufficient training, instruction and supervision to carry out his or her work without risk to health and safety. Special attention should be given to 'one-off' jobs which should be reviewed in advance jointly by supervisory staff and those who will do the work. Specialist training should be provided for safety representatives, for those joining fire-fighting and first-aid teams and for those who will have special duties in an emergency. Management needs to monitor the quality of the training given.

Codes and standards are discussed in 2.7/8. Management must ensure that it is familiar with and has copies of all relevant codes and standards and that these are incorporated into the appropriate training programmes and manuals. It must be alert to the issue of new and revised standards, particularly those affecting health and health monitoring, and ensure that steps are taken to comply with them. Management must ensure that records are kept of all inspections made to check compliance with standards. It should maintain a system of internal company standards for areas not covered by national or other published standards or where existing standards are inadequate.

19.2.3 Safety audits and measuring safety performance

Safety audits, surveys and inspections are valuable management tools so long as their recommendations are followed. They are used for assessing the strengths and vulnerability of management systems and technical features in HS programmes. The following definitions are given in the Chemical Industry Safety and Health Council (CISHEC) publication *Safety Audits*[14]:

- A *safety audit* examines and assesses in detail the standards of all facets of a particular activity. It extends from complex technical operations and emergency procedures to clearance certificates, job descriptions, housekeeping and attitudes . . . An audit might cover a company-wide problem or a total works situation (say, its emergency procedures or effluent systems) or simply a single plant activity.
- A *safety survey* is a detailed examination of a narrower field such as a specific procedure or a particular plant.
- A *safety inspection* is a scheduled inspection of a unit carried out by its own personnel.
- A *safety tour* is an unscheduled tour of a unit carried out by an outsider such as the works manager or a safety representative.

The type of audit discussed here covers all aspects of a process site, plant or unit in regular production. It should be carried out by a team whose members are not involved in the plant or activity being audited. The expertise of the team should be compatible with the type of audit. Audits which delve into technical matters such as pressure vessel inspection require appropriate specialists. It is beneficial to include line managers of other plants or units in an audit team as well as one previous auditor of the same unit. Self-auditing, e.g. by the management or safety committee of the same plant, is, however, a valuable complementary exercise.

Audits are carried out in a formal way using a carefully drawn up checklist of items and descriptive standards for each item. Usually there are four standards for each item examined: poor, fair, good and excellent. Suitable checklists and standards for the chemical industry are available from BCISC[14] and Dow Chemical Company[15], while Lees[9] gives extensive references on safety audits and hazard identification.

Dow[15] recommends four features in a safety audit:

1. *Pre-audit survey questionnaires* which are handled by line management;
2. *Establishment of standards of performance* through interviews with management and examination of documents and records;
3. *Employee perception and implementation* through interviews with employees to assess their knowledge, involvement in and perception of the HS programme. These are supplemented by observations of employees at work by knowledgeable observers to assess compliance with site rules and standards;
4. *Inspection/observation of work environment* to identify work hazards from unsafe design, lack of protective features and exposure to materials in or evolving from the process.

Dow also recommends that advance warning of an audit be given. The object of the audit is to correct faults rather than find them, and the announcement that an audit will be made results in the elimination of some obvious faults and hazards. Auditors need to develop their powers of observation and perception in spotting disguised hazards. Dupont has published a programme[16] which helps to develop such skills.

A pre-audit meeting with the management of the plant, etc. being audited is advisable to ensure that the audit activities go smoothly and that personnel are available to be interviewed at mutually convenient times. Auditors should not hesitate to ask to see documents (e.g. operating instructions) to verify verbal statements. Both collective and individual interviews should be used, the former encouraging group dynamics while the latter often reveal information which would not be given in front of colleagues. The anonymity of persons interviewed must be respected.

A line manager or supervisor of the plant, etc. under audit should be asked to accompany the auditors inspecting it. He should be informed of all corrections and improvements required by the auditors so that he can start taking the necessary steps before the audit report is submitted to management. The main object of the inspection should be to determine whether the layout, design and condition of equipment and protective features are up to standard and to ensure that the protective features will work in an emergency.

The audit team should give a verbal report to management on completion of the audit followed by a clear and concise written report within two weeks. This should highlight features in urgent need of improvement.

Dow's outline of a site or plant audit falls under six main headings:

I   PROGRAM FUNDAMENTALS
II   FUNCTIONAL ASSISTANCE TO LINE
III   PHYSICAL FACILITIES
IV   MAINTENANCE
V   GENERAL COMMENT
VI   SPECIAL SYSTEMS

There are an average of about eight sub-headings for each main heading with an average of about six items to be audited for each sub-heading. As examples, under the sub-heading *Accident Investigation* which comes under main heading I, there are five audit items – Injuries, Losses, Potentially severe incidents, Cause determination and follow-up, and Analysis and trends. Under the sub-heading *Building construction* which comes under main heading III there are also five audit items – Type (open/closed structure), Materials (combustible walls, floors, roof), Structural members (fireproofing), Explosion relief, Smoke and heat venting.

As an example of the standards for activities audited, the following are given for *Supervisor safety training*:

Poor   All supervisors have not received basic safety training
Fair   All supervisors have received basic and some specialised training
Good   Annual training required on some phase of safety and loss-prevention program with documentation

Excellent   In addition, specialised sessions conducted on specific operational problems.

It is clear that a thorough safety audit is a demanding exercise which requires meticulous planning. The need for such audits depends very much on the hazards of the processes used, on the worst consequences which could result, and on the actual HS record of the site, plant or unit. An independent safety audit is the most thorough method of measuring safety performance.

A list by Williams[17] of subjects for which audits are recommended and their frequency is given in Table 19.3. Most of these, however, come under the definitions given here of safety surveys and inspections rather than full audits and not all relate exclusively to safety.

Table 19.3 Some safety audit activities

Activity	Description	Interval (months)
Plant safety review	Adequacy of operations, equipment and building safety	12
Job safety analysis	Standard operating procedures, to be updated where necessary	12
Operator review	Check for deviation from standard operating procedures and on work habits	6
Supervisors' safety meetings	Education, training, drills, follow-up	3
Management development seminar	Development of management competence	1
Supervisory training	Training of foremen for supervisory role	1
Safety committee	Motivational safety suggestions	1
Plant managers' meeting	Communication, education, training, innovation, follow-up	1
Foremen's meeting	Communication (vertical, horizontal), motivation, education, training	
Critical incident technique	Observation of unsafe acts, conditions. Reports of near-misses	1[a]
Central plant safety committee	Safety policy	As needed
Safety review committee	Review of safety of new processes and/or equipment	As needed
Works safety procedures review	Review of works safety procedures	As needed

[a] Plus continuous observation as needed

19.2.4 Accident reporting and investigation

Employers in the UK process industries are legally bound under RIDDOR[18] [2.5.5] to report many types of accident and most industrial diseases which occur on their premises to the area office of HSE. They are also obliged to keep written records of these events. Items which have to be reported under RIDDOR are summarised in Appendix L. The investigation of accident causes is an important part of any HS programme and is frequently required to meet claims under common law.

Accidental injuries and industrial diseases in the UK which were sustained in the course of work and which keep an employee off work for more than three days should be reported by or on behalf of the employee to the Department of Social Security (DSS). This serves several purposes:

1. It enables the victim to claim industrial injury benefit while off work. This is significantly more than unemployment benefit and is tax free.
2. It may form the basis for a claim for a disability pension under National Insurance.
3. It may form the basis for a lawsuit against the employer.

On receipt of the victim's report, DSS will send the employer two copies of a report form which gives some details of the victim's report of the accident. The employer is required to supply further details by completing both copies of the report form and returning them to DSS.

Managements must establish procedures to:

- Ensure that they are informed promptly when an accident happens, and whether anyone was injured;
- Provide first-aid and call for an ambulance, medical assistance, the fire brigade and the police where necessary;
- Investigate and report on the causes of the accident, and delegate appropriate persons to carry this out;
- Keep an adequate record of all accidents which are reported to HSE under RIDDOR and/or form the basis of a claim by an accident victim on DSS.

Managers' first action on hearing of an accident should be to ensure that these things have been done. In the event of a serious injury, management should inform HSE, the victim's family and the employer's insurers immediately, and should obtain details of the injuries as soon as possible. In the case of a fatality, the police and coroner's officers should also be informed.

Once any victims have been removed for treatment and everything necessary has been done to prevent further injury and loss, the site of the accident should be isolated and nothing should be disturbed. The supervisor of the area where the accident occurred should make an immediate examination. In the case of serious accidents, further examination by appropriate specialists such as safety advisers, chemists and engineers should be made. Photographs and samples should be taken and tests made which might shed light on the accident cause or prove useful in future training, taking care not to destroy evidence in the process and to preserve samples of materials involved for further tests.

Most serious accidents in the UK are investigated by HSE and many of their accident reports are published. This should not prevent the enterprise which had the accident from making its own investigation first, providing it does not thereby destroy evidence or otherwise prejudice the official investigation.

The investigation should start as soon as possible while witnesses' memories and evidence are fresh. Where a workers' safety representative requests it, a joint investigation should be carried out. Injured persons should, whenever possible, be interviewed to obtain their versions of events. The immediate investigation should be concerned with assembling information which enables the causes of the accident to be established rather than who was to blame. Since most accidents, especially major ones, have more than one physical cause, the investigator should not be content with finding a single cause, but look also for contributory and sub-causes. Once the technical causes are clear, management needs to discover the personal and organisational factors which allowed the accident to happen. Prompt steps should be taken to remove physical causes and correct organisational weaknesses to prevent a repetition of the accident, and to ensure that all its lessons are properly learnt. The scope of accident investigation is greatly widened when all accidents, whether causing injury or not, are investigated as part of a damage-control[19] programme.

Accident reports should contain the following features[20]:

- Title
- Contents list (unless the accident was minor and the report short)
- Summary
- Findings (the information gathered during the investigation)
- Conclusions (based on the findings)
- Recommendations
- Appendices (where necessary, with tables, photographs, etc.).

The investigation of accident causes without apportioning blame is easier said than done, since most people will know who or which department was responsible for different possible causes. Thus in the case of a burst pressure vessel, the cause might lie with its design, its construction, its maintenance, the way it was operated, or with something very unusual for which nobody could be held responsible. It is then hardly surprising that these causes tend to be regarded as implications of responsibility by the persons or departments concerned. This makes the task of the investigators more difficult, especially when the accident was a serious one. Some of the pitfalls in investigating major accidents were discussed in 2.6. These are clearer from the conclusions of a paper[21] which this writer gave on the subject:

1. Accidents do not just 'happen', they are *caused*. The investigator's duty is to find the cause.
2. Frequently a major accident has been preceded by a number of minor incidents with a common cause. It is worth checking for these, both by interviewing personnel and by studying log books and instrument records.
3. Beware of bias in 'data' supplied by the parties concerned. Keep an open mind and investigate every source of data thoroughly. Expect some contradiction in statements.
4. Account for all the evidence. The theory with the least assumptions is the one most likely to be correct. Where alternative theories are plausible, estimate and quote their relative probabilities (e.g. from an event-tree analysis [14.5]).
5. Most hypotheses put forward will be seen by some persons as threats to their reputations and by others as self-justification. They may react in ways which reduce the chances of reaching an objective conclusion. So

keep your favourite hypothesis to yourself until you have enough evidence to nail it conclusively.
6. Remember that people are sensitive to criticism, especially after an accident. Their confidence in your discretion will often improve the chance of the truth being discovered. (This is not to suggest that you should assist in covering it up, although you may have the difficult task of persuading them that it should be revealed!)

Lees[9] gives several references on accident investigation and shows a useful procedure used by the Safety in Mines Research Establishment for examining faulty equipment.

19.2.5 Accident records, statistics and analysis

When considering any record, management should ask a few practical questions about it such as:

1. Where is it kept?
2. Who is responsible for keeping it?
3. Is it catalogued, with cross references?
4. Is it kept in its original paper form or on computer discs or transferred onto microfilm or fiche?
5. Is it kept because the law says it should be kept? If so, for how long? Are there other copies and who keeps them?
6. Who has access to it?
7. What is its value and why is it being kept?
8. For how long will it be kept before being discarded?
9. What does it cost to keep it?

Of these questions, (7) is often the most difficult to answer. This writer finds there is a sort of 'Peter Principle'[22] about records which says 'You never need them until you have thrown them away'. It is truly remarkable how often following a major accident, records which might have shed valuable light on it are missing.

Most managers keep records which are important to them in their own offices, but there is a limit to how much paper can be kept there. Decisions are constantly having to be taken whether to discard a record entirely, make a microfiche before discarding it, or consign it to a central archive (if there is one). These decisions should be based on a common system within the organisation and not left to the whim of the individual or the moment.

Apart from legal obligations to keep them, most accident reports with details of investigations and the causes found are worth keeping in their entirety for possible future study. New facts may come to light which put a totally different complexion on the causation. When a new accident occurs under similar circumstances, it is important to know whether the causes were the same and if so, why the lessons of the first one were not applied. The same applies to medical, occupational and accident records of employees and to records of monitored health hazards such as noise and harmful substances in the working environment. These may contain useful data for epidemiologists in correlating diseases with particular substances or conditions. They may also provide data for assessing the need and effectiveness of protective measures, e.g. ventilation equipment. Fire records provide similar evidence about fire protection. Records of instrument and equipment failures and plant engineering inspections are essential to maintenance and safe operation. They also provide useful data on which to judge the reliability of instruments and equipment in particular service conditions. This may influence future designs and decisions on equipment purchasing.

Petersen[11] recommends that the system of injury records set up should enable the line manager's safety performance to be judged. For this he recommends:

- That accident records should be kept by the supervisor (by department);
- That they should give some insight into how the accidents seem to be happening;
- They are expressed eventually in terms of dollars by department (by supervisor);
- They conform to any legal and insurance requirements.

It is certainly useful to classify accidents and accidental injuries, not only by cause but also by severity, and in the case of injury, by the parts of the body affected, the occupation of the injured person and where he or she was working. It is best here to follow some recognised method of classification, either the one recommended by the International Labour Office (ILO)[23], or that used by the HS authority of the country concerned. There is no need to use complicated classification systems in their entirety. A few selected headings appropriate to the industry or enterprise are usually enough, sometimes with one or two special additions.

The ILO classify accidental injuries in five different ways[23]:

1. *According to the degree of injury.* Fatal, permanent disablement, temporary disablement and minor injuries.
2. *According to the type of event causing the injury.* Here there are nine main categories (e.g. falls of persons, being struck by flying objects), each with several sub-categories.
3. *According to the agency,* of which there are seven main agencies (e.g. machines, means of transport), each with several sub-divisions.
4. *According to the nature of the injury,* of which there are 16 headings such as fractures, dislocations, contusions and sprains.
5. *According to the bodily location of the injury,* of which there are seven main headings (e.g. head, neck and trunk) and a number of sub-headings.

Classified injury figures are far more use to management than total injuries, which give little guidance to prevention or protection.

When accidents which include losses and/or near-misses as well as injuries are recorded, it is important to classify them by cause as well as by effect and cost. These are essential features of a total-loss-control programme[24] which should result in the elimination of many hazards before they cause serious injuries. Heinrich's famous dictum[3] should be remembered:

For a group of similar accidents, for every one causing a major injury there will be 29 producing minor injuries and 300 near misses.

This is not, of course, a hard-and-fast ratio. The actual ratio depends mainly on the occupation. That of minor to major injuries is lower for steel erectors than for carpenters.

Accident statistics are more useful when making comparisons or looking for trends than mere lists of accidents, but one should beware of making comparisons unless the statistical significance of the figures is known. Boyle[25] gives a salutary diagram, shown here as Figure 19.4, of steps needed when designing a project using statistical analysis. He concludes by advising a trial check of the proposed analyses:

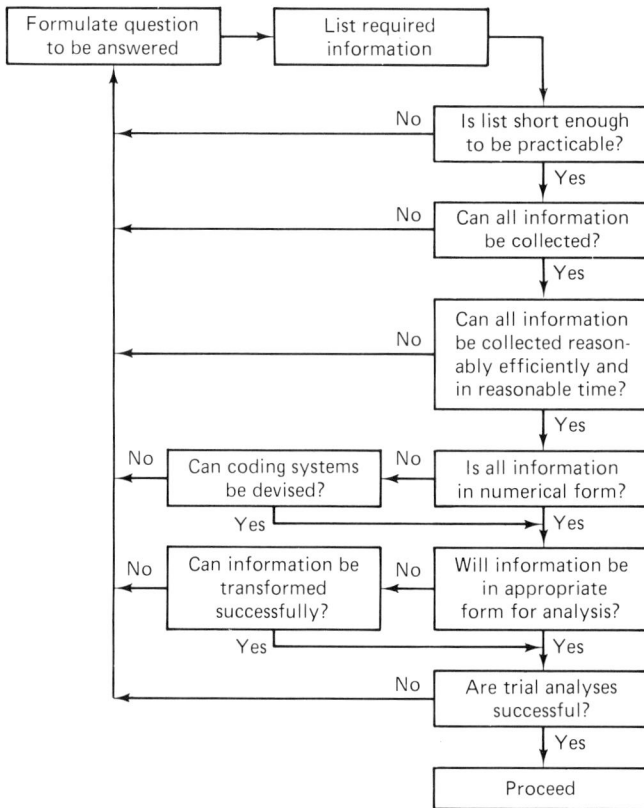

Figure 19.4 Steps in designing a project using statistical analysis

The main reason for doing this is that it enables an estimate to be made of the time the complete analyses will take, there being little point in collecting so much information that the analysis takes too long to be useful. An additional benefit of a trial run is that it often shows problems which were not identified earlier, and enables final refinements to be made to the formulation of the questions before the full analysis is commenced.

Two comparative methods of presenting injury statistics are as frequency rates (FR) and as incidence rates (IR). Unfortunately there are differences in the bases of frequency rates recommended by the ILO and used by the UK authorities. In the USA the ILO recommendations are followed. These are:

$$FR \text{ (frequency rate)} = \frac{\text{Total number of accidental injuries (of stated severity)} \times 1 \text{ million}}{\text{Total number of man-hours worked}}$$

$$IR \text{ (incidence rate)} = \frac{\text{Total number accidental injuries per year} \times 1000}{\text{Average number of workers at risk}}$$

In the UK accidental injury frequency rates are more usually quoted per 100 000 hours worked except for fatal injury frequency rates, which are quoted per 100 million hours worked.

Another important indicator is the accident severity rate (ASR). In the USA this is given as:

$$ASR = \frac{\text{Total days charged} \times 1 \text{ million}}{\text{Employee-hours of exposure}}$$

In the UK it is given as:

$$ASR = \frac{\text{Man-hours lost} \times 100\,000}{\text{Man-hours worked}}$$

Some American companies like to combine their frequency and severity rates into a single measure, the frequency severity indicator, which is given as the square root of the product of their frequency and severity rates divided by 1000.

Several schemes have been proposed for presenting statistics of damage to plant but none, according to HSE's APAU, have widespread support[5].

The analysis of accident statistics requires the same statistical tools, especially analysis of variance, as those used for quality control and many other industrial activities. It is not proposed to discuss the technicalities of the subject. It is best for a statistician to set up a system in the first place, choose appropriate computer programs for the calculations, instruct those who will apply the system, and be available to assist in interpreting results where necessary.

There are several reasons why, in the process industries, injury statistics tend to supply less useful information than in other manufacturing industries.

- Because of the high capital : employee ratio of most process industries, the number of employees per enterprise of given size is lower than in most other manufacturing industries. As a result, the number of injuries per year of given type and circumstances tends to be too small for differences from one year to the next to have much statistical significance. Of course, the situation may be different in very large enterprises which carry out similar operations at several centres.
- The process industries are more highly mechanised than most others so that there are fewer accidents arising from manual work.

- The process industries are characterised by a few large accidents each of which may involve a number of employees, but which occur only rarely, perhaps once only or never in the lifetime of a plant. This leads to a high 'standard deviation' and a low level of significance in the annual injury statistics. It also leads to a situation where the emphasis is concentrated on preventing the last major accident rather than the next one.

Another factor which increases apparent accident rates in all industries is an improvement in accident reporting. Managements should be prepared for this and be careful not to blame supervisors for a rise in accident rates when they should be commending them for better reporting.

One benefit from the relatively small number of employees and hence accidental injuries in the process industries is that it allows more time and effort to be spent on investigating near-misses and non-injury accidents. Special attention should also be given to accidents during maintenance work [17.1] and the operation of permit systems [18] and those caused by health hazards in the working environment [7.7].

19.3 Special management problems

Health and safety pose some difficult and often unexpected problems to management. Some which the writer has met are discussed here briefly.

19.3.1 Organisational blind spots

Whatever the technical cause of an accident, there is usually a blind spot in the organisation which allowed it to happen. The disasters discussed in Chapters 4 and 5 provide clear examples of such blind spots. As the subject has already been discussed in 3.3 under the heading of 'Organisational misconceptions', little further needs be said on it here. In high-technology fields where major hazards exist, the need for truly independent monitoring of of management decisions is recognised as perhaps the only way of avoiding blind spots. Unfortunately, political factors often intervene which undermine the independence, integrity and objectivity of the monitor. A former colleague, when introduced to a new consulting assignment of this nature, addressed those of us already working on it thus: 'Spare me the facts. Just tell me what are the politics of this assignment.' He showed a certain worldly wisdom in this question and was soon in a leading role. This does not only happen in consultancy but is found in all walks of life, including the management of large companies. By checking at the outset what the client or boss wants to hear, the resourceful consultant or junior can ensure that he does not upset him. This is fine so long as the ideas of the client or manager are sound. But when the consultant or junior merely reinforces firmly held misconceptions, the last state is worse than the first. Unfortunately, the management which most needs independent advice is often the one most wedded to its own misconceptions, which are difficult to remove without a damaging confrontation.

Of the factors which contribute to blind spots in the minds of otherwise competent people, mental stress is perhaps the most important and is discussed next.

19.3.2 Stress – a contributory factor

As Atherley has pointed out[26], individuals and even entire organisations are sometimes unable to cope with the mental stresses generated within industry. This leads us to consider briefly the following psychological danger points which Lord Ennals in a foreword to Kearns's readable book[27] considered were 'totally ignored' by the Robens Report:

1. *Overpromotion* – stress, anxiety and often breakdown can follow from responsibilities beyond a person's capacities.
2. *Underwork* – sometimes agreeable for a time but leading to dissatisfaction, doubts about the worker's capacities, demoralisation and frequent spells of absence for minor complaints.
3. *Job definition* – it is essential that the employee should know the requirements of the job and to whom he or she is responsible. Uncertainties over these issues can become a major strain. The resulting stress can be taken by seniors as implying inadequacy for the job, when the real cause of trouble is lack of clarity by higher management.
4. *Lack of effective consultation and communication* – an all too common fault of management. Sudden unexplained changes in policy, take-overs and fears of redundancy can cause stress disorders. There is a correlation between the level of morale of employees and the quality of concern of senior management for the people they employ.
5. *Lack of financial security* – is especially felt by manual workers who have no job security, sick pay or pension schemes, which clerical and professional groups largely take for granted. This has obvious deleterious effects upon the psychological and physical health of workers.

Further consideration of these factors lies outside the scope of this book.

19.3.3 Changes in hazard awareness

Many serious hazards, especially to health, have only been recognised belatedly after they have been present in industry for many years. Examples are alpha naphthylamine, asbestos, vinyl chloride monomer, benzene and beryllium. In some cases the increased awareness of the hazard has led to a total ban on the use of the offending substance. Here managements are faced with a clear-cut choice: either to change to the use of a safer substance, or, where this is impractical, to shut down the operation altogether. In other cases the increased hazard awareness has led to very much tighter physical controls. In the case of vinyl chloride monomer (VCM) the controls needed are so stringent that the normal decision would probably have been to ban its use. The material is, however, so widely used and the investment in its production and use is so high that industry responded to the challenge by devising control methods to cope with the situation. Major companies such as ICI and Shell, which are normally in intense competition and take pains to conceal their research from each other, cooperated to develop the required analytical and control techniques.

While large companies generally have the financial and technical resources to improve control techniques in line with greater awareness of existing hazards, the problem can be a very serious one for smaller companies. It is also very much more difficult and costly to apply tighter controls to an existing plant which was designed for lower control standards than it is to apply them during the design of a new plant.

19.3.4 Problems in growth and decline

Technical management tends to be at its weakest both during the rapid expansion of a company or its use of a particular process and also during periods of falling sales and declining production.

In the first case the reason is that the technical expertise is very thinly stretched. This appears to have been the case at Flixborough [4], where the plant was only one of eight or nine similar ones which were designed and built within a few years for operation by DSM and its subsidiary companies. The top management in most cases were staff seconded from the parent company. This needs to be borne in mind when considering the organisational weaknesses which allowed the disaster to happen.

In the second case the organisation or one branch of it may be running down its staffing levels and economising as far as possible in preparation for final closure. All process plants are profitable only above certain production levels. Below a certain throughput the operation can only run at a loss. Sometimes a conscious decision is then taken to shut down the plant for good, sometimes to put it into 'mothballs'. Often a conscious decision is deferred in the Micawberish hope that something will turn up to rescue the situation. Something of this sort appears to have been the case with Union Carbide India at the time of the Bhopal tragedy [5.4]. Perhaps managements of large companies ought to consider creating and training special shutdown teams for controlling the closure of very hazardous plants.

Top management needs to be specially alert to the risks involved in the growth and declining phases of a major hazard project. In the first case the main need is for more experienced staff. In the second case both men and and money are needed.

19.3.5 Problems of subsidiary companies

Of the five major accidents discussed in Chapters 4 and 5, four occurred on the site of an operating subsidiary of a foreign transnational company whose expertise had been developed and whose technical headquarters lay in the country of the parent company. In one case (Seveso), the accident occurred in a subsidiary of a subsidiary. In such cases there is a tendency for the parent company to supply only sufficient information to its subsidiary as it considers necessary to operate and maintain the plant, and the staff of the subsidiary company are chosen with these objects in mind. If the foreign subsidiary is acquired as a result of a take-over, there is often a brain drain from it to the parent company. It is obviously cheaper to concentrate higher technical effort and expertise at one centre than at several, and modern communications should allow top technical experts in the parent company to deal quickly with problems beyond the skills of the subsidiary company. There is often a gap between the competence of the staffs of parent and the subsidiary company. This is sometimes exacerbated by hierarchical structures, chauvinism and differences in language and standards. These problems are accentuated when the subsidiary company is in a Third World country [23]. Staff of the parent company also seldom feel so involved with problems of the foreign subsidiary as they do with those closer at hand. There is no simple solution to this problem although decentralisation of technical expertise and a less hierarchical organisation should help.

19.3.6 Funding difficulties

A company may experience real problems in providing adequate funds to meet HS objectives when it is operating at a loss and all expenditure is being cut to the bone. There is therefore much to recommend the creation of a special central trust fund for HS expenditure (including loss prevention) for the various projects of a major company and its subsidiaries. This would be quite separate from their individual operating budgets and be a charge on company reserves and current profits. Its purpose would be to ensure that money needed for essential HS expenditure is available when a project or company is in economic difficulties and the integrity of the plant and safety of its workers and the public are most at risk. Items covered by the fund should include the inspection and maintenance of protective systems (such as the vent gas scrubber at Bhopal), general maintenance of plant with significant hazard potential, and the salaries and wages of personnel with essential HS functions. There is evidence that some large companies have adopted similar measures as part of their internal insurance schemes. Governments should give practical encouragement.

19.3.7 Contractors

The safety role of contractors has been discussed in design [16.3.2], during commissioning [20.4] and during work on site [21.4]. Accidents during construction[28] lie outside the scope of this book. Management should ensure that the purchasing department of their company issues appropriate safety rules, terms and conditions as part of any contract involving work on site. Every contractor should attend a safety induction course before starting work unless the work is of such short duration that this is impracticable. In that case he or she should be accompanied while on site by an experienced employee of the operating company. All contractors' powered equipment and accessories should have a valid safety certificate issued by an independent authority. Contractor's ladders and steps should be inspected and registered by the appropriate department (e.g. the main stores) before being used on site.

19.3.8 Informing the public

Management should take the initiative in issuing warnings of special hazards and news of emergencies [20.3.2]. If it fails to do this, rumours will grow and the chances of erroneous reports will increase. It is recommended that a senior manager be appointed as the company's sole authoritative source of information in an emergency, assisted where possible by an experienced press officer. Other employees should be instructed not to comment but to refer inquiries to the company spokesman.

19.4 Computers and safety

The present phenomenal growth in the use of computers and robots is comparable to the Industrial Revolution which started over 200 years ago and seems unstoppable. A robot is seen here as a computer with the equivalent of arms, legs, hands, fingers, eyes and ears. The robot may not actually move but sit, as it were, permanently at a desk and control a large process plant.

There have been many failures in the use of computers, partly because those who had bought them had failed to analyse their own objectives, and partly because employees who had to work with them wanted to prove that they would not work.

Properties in which computer-controlled machines and robots outclass human workers are given in Table 19.4[2].

Table 19.4 Properties for which robots are preferred to human workers

Property	Human	Robot
1. Obedience	Variable, often sensitive to abuse	Good, only does what it is told
2. Dependence on oxygen supply	Complete	Seldom required
3. Resistance to radiation	Poor, needs special protection	Much improved
4. Temperature range	Narrow	Wider
5. Resistance to toxic hazards	Poor, needs special protection	Improved
6. Mechanical power	Low and limited for long periods	Limited only by design
7. Performance of repetitive work	Fair and easily impaired	Good if properly maintained
8. Information searching	Fair	Excellent
9. Disposability	Expensive	Easy

Robots are chosen for jobs which are dull, dirty and dangerous (the three Ds), and hot, heavy and hazardous (the three Hs). There is more concern, however, over the employment of 'smart robots' equipped with 'expert systems' which are capable of doing the work of highly trained specialist staff. An expert system is defined as[29]

A computer system which reflects the decision making processes of a human specialist. It embodies organised knowledge concerning a defined area of experience and is intended to operate as a skilful, cost-effective consultant. An expert system comprises a knowledge base, inference engine, explanation programme, knowledge refining programme and natural language processor (Figure 19.5).

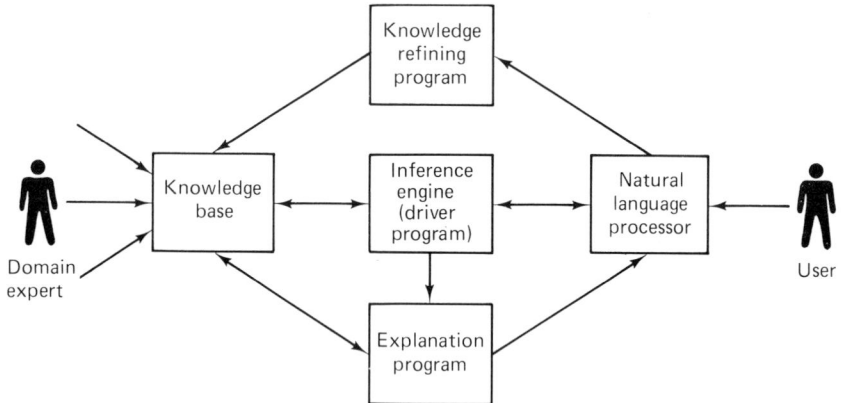

Figure 19.5 Components of an expert system

Barnwell and Ertl[30] describe an expert system computer program as a 'shell' into which knowledge is placed by a human expert, when it becomes a 'loaded shell'. They believe that the use of expert systems in the process industries will be commonplace by 1992. Typical design applications include pipe sizing, control valve selection, materials of construction selection and steam/condensate system design. Plant management applications include capital cost estimating, reactor simulation and optimisation and utility usage strategy. (Shells could probably be loaded for most subjects discussed in this book.) The power and effectiveness of the plant manager or design engineer is greatly enhanced when he has several such loaded shells at his disposal, while fewer human professionals of his kind will be needed. One danger which many fear in this is that decisions may be taken on the basis of knowledge encapsulated in chips and about which the engineer or manager responsible for the decision is largely ignorant. The expert system is a marketable commodity, and while it may solve problems, the basis on which it does so is valuable information which its owner is probably anxious to protect.

Because of their ability to improve the flow and speed the processing of information of all kinds, computers are widely used for all types of office duties in the process industries, as well as for computer-aided design (CAD) [16.8] and computer-aided manufacture (CAM). The latter generally involve the use of 'programmable electronic systems' which may be used for the on-line control of process plant in 'real time' [15.9]. (That is, the computer functions simultaneously with the plant in response to signals from it.

Management must appreciate that computer technology is highly specialised, and its vocabulary often difficult for non-experts to understand. Even words such as *reliability* and *hazard* do not mean quite the same to computer specialists as to other professionals, and communication problems thus easily arise. Managers and safety personnel may therefore need training in the use of computers, particularly in the operation and maintenance of PES systems used to control and protect process plant.

Computers do not prevent the 'pollution of information' (such as leads to the organisational misconceptions discussed in 3.3). Unfortunately they can greatly magnify the effects of human errors while making them harder to trace. Computerised crime such as the theft of valuable information from electronic data banks, and the interference with computer software is also growing and very difficult to trace. According to Miller[31]:

> A programmer with less than a month's training can break the more elaborate procedures currently being used in large data banks within five hours,

and

> The programmer, for instance, could insert a secret 'door' in the monitor programme that would enable unauthorised people to bypass protective devices, or could 'bug' a machine in such a sophisticated manner that it might remain unnoticed for an extensive period.

Computer criminals include 'crackers' and 'crashers'[29]:

> A cracker is a hacker who specialises in overcoming software protection systems.
> A crasher is a hacker who deliberately attempts to cause a serious interference with the operation of the computer system.

19.4.1 An expert system and a young cracker

Unwittingly I bought an expert system for £13. It is a chess-playing program against which I can play, using the word-processing computer with which I wrote this book. Its skill can even be downgraded to match its human adversary. I was once a fair chess player, having played in a local team, yet now I can hardly beat the program at its second-lowest skill level. So I have no doubt it is an expert system!

Chatting one day with a seven-year-old friend Adam, whose mother is a computer consultant, and has a computer and chess-playing program similar to mine, I told him how hard I found it to beat the program. 'Why, it's simple,' said Adam, 'I can beat my mum's computer any time I want to'. Intrigued, I coaxed from him the secret of his success. 'I simply get the computer to change all my white pawns to queens, and then surround the black king with them.' I did not think the program allowed me to do that, and on checking found that it did not. Adam had 'fixed it'. He has a brilliant future open to him as a software cracker!

References

1. Drucker, P. F., *The Practice of Management*, Heinemann, London (1955)
2. King, R. W. and Majid, J., *Industrial Hazard and Safety Handbook* Butterworths, London (1980)

3. Heinrich, W. R., *Industrial Accident Prevention*, 4th edn, McGraw-Hill, New York (1968)
4. Lord Robens's Committee, *Health and Safety at Work*, HMSO, London (1972)
5. The Accident Prevention Advisory Unit of HM Factory Inspectorate, *Managing Safety*, HMSO, London (1981)
6. HSE review of work of The Accident Prevention Advisory Unit of HM Factory Inspectorate, *Effective policies for health and safety*, HMSO, London (1980)
7. Dow Chemical Company Corporate Safety and Loss Prevention, *Minimum Requirements*, 3rd edn, Dow Chemical Company, Midland, Mich. (1984)
8. The Council for Science and Society, *Superstar Technologies*, Barry Rose (Publishers) Ltd, London (1976)
9. Lees, F. P., *Loss Prevention in the Process Industries*, Butterworths, London (1980)
10. Powell, P. I., Hale, M., Martin, J. and Simon, M., *2000 Accidents*, National Institute of Industrial Psychology, London (1971)
11. Petersen, D. C., *Techniques of Safety Management*, McGraw-Hill, New York (1971)
12. Hearn, R. W., 'Management responsibilities for safety and health' in *Industrial Safety Handbook*, edited by Handley, W., 2nd edn, McGraw-Hill, Maidenhead (1977)
13. 'Health & Safety Policy and Principles of the Dow Chemical Company King's Lynn, 1st July 1986'
14. Chemical Industry Safety and Health Council, *Safety Audits*, London (1973)
15. Dow Chemical Company, *Guidelines for safety and loss prevention audits*, Dow Chemical Company, Midland, Mich. (1980)
16. E. I. DuPont, 'Safety training observation program' in *Safety Training Course for Supervisors*, DuPont, Wilmington, Delaware
17. Williams, D., 'Safety audits', in *Major Loss Prevention in the Process Industries*, The Institution of Chemical Engineers, Rugby (1971)
18. S.I. 1985, No. 2023, *The Reporting of Injuries, Diseases and Dangerous Occurrences Regulations (RIDDOR)*, HMSO, London
19. Bird, F. E. and Germain, G. L., *Damage Control*, American Management Association, New York (1966)
20. Adrian, E. W., 'Accident investigation and reporting' in *Safety at Work*, edited by Ridley, J., 3rd edn, Butterworths (1990)
21. King, R. W. and Taylor, M., 'Post accident investigations – in the aftermath of a catastrophe', in *Eurochem Conference – Chemical Engineering in a Hostile World*, Clapp & Poliak Europe Ltd, London (1977)
22. Laurence, J. and Hull, R., *The Peter Principle*, Souvenir Press, London (1969)
23. The International Labour Office, *International Recommendations on Labour Statistics*, Geneva (1976)
24. Fletcher, J. A. and Douglas, H. M., *Total Loss Control*, Associated Business Programmes, London (1971)
25. Boyle, A. J., 'Records and statistics' in *Safety at Work*, edited by Ridley, J., 3rd edn, Butterworths (1990)
26. Atherley, G. R. C., 'People and safety – stress and its role in serious accidents', in *Chemical Engineering in a Hostile World*, Clapp & Poliak Europe Ltd, London (1977)
27. Ennals, D., 'Foreword' to Kearns, J. L., *Stress in Industry*, Priory Press, London (1973)
28. King, R. W. and Hudson, R., *Construction Hazard and Safety Handbook*, Butterworths, London (1985)
29. Longley, D. and Shain, M., *Data and Computer Security – dictionary of standards, concepts and terms*, Macmillan, London, (1987)
30. Barnwell, J. and Ertl, B., 'Expert systems and the chemical engineer', *The Chemical Engineer*, No.440, p. 41 (September 1987)
31. Miller, A., 'Personal privacy in the computer age', *Michigan Law Reporter*, **67** (6) 1090–1246 (1969)

Chapter 20

Plant commissioning, operation and emergency planning

The term 'commissioning' as used here includes all activities needed to bring a newly constructed plant into regular production. These activities, which require a great deal of planning, organisation and training, are usually carried out in four stages. These correspond to specific points in the hand-over of the plant by the contractor to the owner.

- Inspections, cold trials and preparations made before mechanical completion;
- Pre-commissioning, in which steam and other utilities are introduced, further preparations and hot-running trials are carried out, and any necessary engineering corrections are made;
- First start-up. This begins with immediate preparations for the introduction of process materials, and continues until the plant is operating with all systems working at or near design throughput. It may typically last for a month. It usually includes test runs to prove licensors' and contractors' guarantees of yield, throughput, product quality, utility consumptions, etc.;
- Post-commissioning. This term refers to the period which starts when the plant first comes on-stream and ends when it has settled down to regular production. It may take the best part of a year during which adjustments and modifications are made, faults are corrected, performance is brought up to scratch and test runs are carried out.

Operation [20.2] includes normal start-up, normal and emergency shutdown and most activities carried out by the operating or production department. Emergency planning [20.3] covers both controllable emergencies which lead to a safe plant shutdown (e.g. failure of electricity, cooling water or an internal plant function) and uncontrollable ones caused or accompanied by fire, explosion or toxic release. There is no sharp dividing line between these two types of emergency. What starts as an apparently controllable emergency sometimes escalates into a more serious one requiring speedy evacuation of plant personnel as well as plant shutdown. In planning for emergencies, local authorities and their fire and medical services will be involved as well as HSE.

The subjects of this chapter are so extensive that it is possible to deal with them only generally, with special emphasis on safety. The importance of starting to plan and train for commissioning, operation and emergencies *while the plant is still under design* cannot be overstressed. These exercises also usually reveal design shortcomings which, if not corrected, would hinder first start-up or cause hazards in operation.

20.1 Commissioning

The following critical dates correspond to points in the four stages of commissioning listed earlier:

1. The date of mechanical completion when the plant or unit [12.1] is provisionally accepted by the owner;
2. The date when process materials are first introduced into the plant or unit;
3. The date when the plant or unit is first brought on-stream;
4. The date on which test runs which meet performance guarantees are completed to the everyone's satisfaction and the plant or unit is fully accepted by the owner.

Up to (1) the plant is normally under the control of the contractor, and from then it is under the owner's control.

The utilities (power, steam, etc.) and other offsites (storage, effluent treatment, etc.) on which the plant or unit depends should, if possible, have been commissioned before (2), although sometimes they have to be started up at the same time as the plant. Regular supplies of raw and auxiliary process materials must be available before (2) as well as proper means of handling and disposing of the products, which initially may be unsaleable.

The hazards and safety precautions needed undergo an abrupt change when point (2) is reached. The constructional hazards which previously predominated now give way to the hazards of the process materials, which may be flammable, toxic or highly reactive.

On large and complex plants some units may be still under construction while others are already in operation. The above critical dates are then staggered, to allow effort and personnel to be deployed most effectively. Where different units of a plant, or different plants in an new integrated complex, can be operated independently by introducing or withdrawing semi-processed material, it is generally best to get the last unit of the plant completed and working first, using bought-in materials. This enables markets for the product to be built up and problems (including product quality) solved before the front end of the plant is working. As an example, on a polymerisation plant, the powder-handling unit comprising drying, blending, pelletising and bagging operations might be commissioned first with imported wet polymer powder, and the equipment run-in before starting up the polymerisation unit. This is not then held back by difficulties with the powder-handling section.

Temporary hook-ups involving pumps, pipes, hoses, drums, tanks and hoppers, etc. are often needed for circulation of cleaning chemicals and other fluids during pre-commissioning, and for the transfer of process materials when the commissioning of a plant or unit is staggered. Such once-only operations, especially the use of hoses, tend to introduce hazards and need careful checking for safety.

Staged commissioning makes it necessary to erect temporary fencing to segregate areas where construction and welding are still proceeding from those in which process materials are being handled. Flammable and toxic process materials must be prevented from entering pipes and drains in areas where construction is still being carried out.

20.1.1 Roles, organisation and planning

Commissioning involves the personnel of several organisations with different interests, e.g. the operating company (and often its transnational parent), the main contractor, the process licensor, suppliers of packaged units and specialised equipment, sub-contractors and independent inspectors working on behalf of government, insurers and the plant owner. Many different but overlapping activities have to be are crammed into a short time-span. Unless the roles and responsibilities of these organisations and their personnel have been carefully worked out, understood and agreed in advance, conflicts are likely to arise. From the date of mechanical completion and provisional acceptance by the owner, commissioning is best controlled by a single person, usually the owner's commissioning manager. In cases where the owner has no suitably experienced staff, he may arrange to have the contractor start up the plant, hand it over in a fully operational state, and train his personnel. Few contractors, however, have teams of experienced operators capable of starting up a complex plant. Owners thus generally need to recruit and train their operating and maintenance personnel well before start-up.

The owner should set up a task force at the start of design to ensure that its basis is safe and realistic, vet design work, carry out HAZOP and other safety studies, prepare operating and maintenance manuals, permit forms, checklists and other documentation and advise management of progress and problems. The task force should be led by the commissioning manager, preferably an experienced production specialist who will later be the plant's production manager. The task force should include the plant's project manager, process design manager and maintenance manager. Other experts including safety specialists should be available to assist the task force.

After completion of design, the task force leader moves to site to head the commissioning team where its members familiarise themselves with the plant, check that the contractor's work complies with design drawings and standards [20.1.2], complete the extensive documentation needed for commissioning, and arrange and negotiate any modifications found necessary for safe operation and maintenance. They will be responsible for recruitment, training, liaison with site contractors, plant acceptance testing, pre-commissioning and first start-up. The commissioning team has three main sections, each with its own leader, as well as experienced shift leaders for start-up. See Figure 20.1.

The commissioning team should include the key personnel who will later operate and maintain the plant, together with extra experienced personnel. If these cannot all be released temporarily from other plants, some may have to be recruited on a short-term basis. Panic recruiting can result in square pegs in round holes. The writer recalls one fitter on a

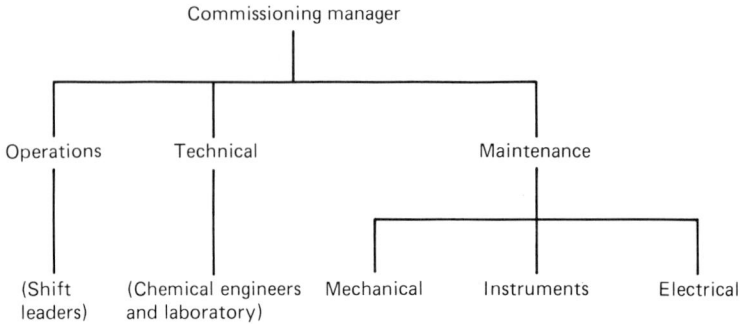

Figure 20.1 Structure of a commissioning team

commissioning team who seemed strangely ill at ease with spanners. Enquiry revealed that he had previously been a tailor's fitter!

The human problems can be quite as serious as the technical ones and confrontation is often necessary to resolve them. There has to be clear agreement between the members of a commissioning team about who will do what and when. Group discussion is needed to allot tasks and ensure that each member is able and willing to play his or her part, especially when this involves working long hours on shift and coping with the unexpected. Good training [21] of personnel at all levels is also vital to successful commissioning. The commissioning manager should visit and become familiar with other plants which use the same process. For a large modern plant controlled from a central control station, where process information is shown on video display units, there is a growing trend to use a specially designed simulator to train operators [21.3.5].

20.1.2 Mechanical completion and provisional acceptance

The site contractor is responsible for building the plant in accordance with the terms and specifications of the contract. The owner's construction organisation, reinforced by maintenance personnel from the start-up team, checks the contractor's work as it proceeds. Good communications are essential and prompt and effective feed-back from site to the design office is necessary.

Electrical and compressed air systems are usually commissioned before mechanical completion of the plant or unit.

On mechanical completion the contractor carries out tests, inspections and checks such as those outlined below in conjunction with the owner's commissioning team, which carries out a complete inspection following a planned procedure. Each item is checked against a systematic checklist and the completed checksheets are incorporated into a dossier. Typical instructions include:

- Check that there is safe and unrestricted access for operators, lubrication and other workers to valves, switches, gauges, pump glands, sample points and other items where they may need to work during operation. Look out for tripping hazards and other booby traps and valves with inaccessible handwheels. Check that guard rails are securely fitted to the sides and floor openings of platforms and stairs, that vertical ladders are enclosed, and that non-slip flooring is used.

- Check that no process fluids can enter the unit via feed and product lines, drains, vent and blowdown lines. Check that spectacle plates, line blinds and blank flanges are in position at the battery limits and that these can readily be turned or removed when first start-up begins.
- Check that pipework conforms to P&ID [16.4.4] specifications and is not unduly strained at joints[1].
- Check that all pipes and equipment connected with unrestrained expansion bellows are designed and tested to withstand the thrust forces generated by the bellows at their maximum operating pressure [4.6].
- Check that screwed plugs are used only for air, water and nitrogen at pressures below 7 bar g. and on pipes not larger than 1½ inches NB.
- Check that welds have been inspected as per contract and found acceptable [17.2].
- Check that all gaskets, bolts, studs and other fastenings used in joints on pipes and equipment are suitable for the service conditions and that the fastenings have been correctly tensioned [17.1.4].
- Check that pipes are properly supported for both ambient and operating temperatures and pressures and for any likely vibration[1].
- Check that spring-hangers are in sound condition and correctly set with retaining pins in position.
- Check that there are adequate valved vents and drains on high and low points on pipework, including temporary ones installed solely for pre-commissioning, and that they are in working order.
- Check that there are no drain holes on relief valve discharge lines inside buildings, nor on those handling flammable and toxic fluids when there is risk of a dangerous discharge to the atmosphere [15.5.1].
- Check that there are no small-bore pipe projections or equipment that can be accidentally broken off.
- Check that all small pipe branches, especially drains, are sufficiently clear of pipe supports.
- Check that drain lines lead to the correct drainage system [16.6.7] and do not discharge over paved areas.
- Check completion of insulation and cladding except for flanged joints, for which properly fitting insulating covers should be ready to install after pre-commissioning.
- Check that painting, colour coding and numbering of pipes and equipment are complete.
- Check that valves can be easily identified, preferably by numbers marked on them or on adjacent pipework [20.2.1].
- Inspect internal parts of vessels, drums and heaters.
- Check that pipework is clean internally and clean it (chemically) where necessary.
- Pressure test process and utility equipment and pipes [17.6.1].
- Test, tag and check installation of relief valves.
- Test and commission electrical systems, check motors for rotation and run them uncoupled.
- Exercise motor control centres adequately.

- Flush, clean and fill lubricating systems and bearings with correct lubricants.
- Install pump seals and packings and suction strainers.
- Align and couple shafts of rotary machines and test-run them cold.
- Remove transport restrictions from instruments, and clean, pressure test, lubricate, adjust and calibrate them.
- Check proper interconnection and functioning of instrument loops including alarms and trips.
- Pressure test, check and commission pneumatic and hydraulic instrument systems and and stroke and check control valves. (It is now possible to pre-commission, check and adjust the control system of a complex plant before any process materials are introduced with the aid of a process simulator[2].)

All outstanding tests and faults brought to light are recorded onto 'punch cards' which are divided into two types:

1. Faults and omissions,
2. Site alterations (which are generally to facilitate operation and maintenance and frequently for safety).

On completion of this work the plant unit is provisionally accepted by the owner.

20.1.3 Pre-commissioning

Pre-commissioning follows mechanical completion on which it is dependent and is carried out by the commissioning team with assistance from the contractor. All utility, effluent and other auxiliary systems which do not involve toxic or flammable materials are now commissioned, and all process systems and equipment are prepared for start-up. Typical tasks during pre-commissioning are:

- Blow out steam pipework with steam, commission steam and condensate systems, check steam traps.
- Commission all non-hazardous utility systems systems which are not yet working (e.g. cooling and process water, inert gas).
- Check, test and commission effluent and effluent-treatment systems.
- Check and test vent and pressure relief disposal systems.
- Dry out furnace linings.
- Water-wash and/or gas-purge piping systems, and drain water from low points.
- Dry piping and equipment as required.
- Install flow elements and orifice plates.
- Run steam turbines uncoupled and later coupled, where possible, to compressors and other driven equipment.
- Where possible, circulate suitable safe fluids through process systems with drivers coupled to pumps and other machines; bring these up to normal operating temperature and align them correctly.
- Remove temporary drain and vent valves and close openings as specified on piping drawings.
- Install non-hazardous solid catalysts, drying agents, adsorbents in process vessels, with supports and retaining grids etc.

Detailed reports are made on all work done during pre-commissioning including repairs, modifications and improvisations. A copy of these should be kept by the works maintenance department. Table 20.1, based on Pearson[3], lists questions to be asked before first start-up.

20.1.4 First start-up and post-commissioning

These two phases are here treated as one. The dividing line between them is the final acceptance of the plant by the owner at the end of start-up. By then most of the temporary and visiting personnel present during commissioning will have left. There may, however, still be many outstanding problems. It is usually left to the regular operating, technical and maintenance personnel to cope with them, although there is often a need for additional experienced staff for 'trouble shooting' during post-commissioning.

Table 20.1 Questions to be asked before process fluids are admitted to a new plant

1. Have the following been removed from the new plant area?
   (a) Contractors' huts, tarpaulins, etc.;
   (b) Non-flameproof equipment;
   (c) Rags, paper, wood, rubbish, dry grass, weeds.

2. Are the following in place and ready for use?
   (a) Perimeter fence and gate(s);
   (b) Security gateman and cabin;
   (c) Hazard and safety notices (no smoking, matches, general vehicles, etc.);
   (d) Nitrogen and/or other inert gas purge systems;
   (e) Oil/water separators and effluent-treatment plant;
   (f) Fire alarm system;
   (g) Fire main, hydrants, hoses, monitors and foam system and supplies;
   (h) Fire extinguishers;
   (i) Eye-wash bottles, first-aid boxes and kit;
   (k) Emergency personal showers;
   (l) Water sprinklers and deluge systems;
   (m) Steam hoses;
   (n) Gas detectors;
   (o) Plant lighting (normal and emergency);
   (p) Pressure relief, flare and/or blowdown system;
   (q) Instrument air and electrical supplies and back-up;
   (r) General utilities, water, power, steam, fuel, etc.

3. Have the following been informed of the plant start-up and its consequences?
   (a) Construction personnel (including restrictions on smoking, welding, etc.);
   (b) Fire services;
   (c) Local authorities;
   (d) Neighbouring plants;
   (e) Records section.

4. Other questions
   (a) Are operating and maintenance personnel properly trained and organised?
   (b) Have drains been flushed and checked free of obstructions?
   (c) Can isolation blinds, etc. at battery limits be readily turned or removed?
   (d) Have shift fire and first-aid teams been nominated and trained?
   (e) Has all welding been done?
   (f) Is there an effective permit system [18] for further engineering work, particularly welding, which may be required?

First start-up is usually the owner's responsibility. It is the acid test when both aspirations are realised and errors and omissions are relentlessly exposed.

First start-up should not begin until pre-commissioning is complete and all questions listed in Table 20.1 can be answered 'Yes'. The first steps in first start-up are usually to:

- Introduce any hazardous solid catalysts, adsorbents, etc. into the plant;
- Remove all line blinds and turn all spectacle plates at the battery limits on pipelines carrying hazardous materials into and out of the plant or unit;
- Purge the plant or unit with inert gas and leak test any joints broken and remade since the last leak test;
- Line up valves in the plant or unit as detailed in the operating manual for the initial introduction of process fluids.

The start-up now depends entirely on the process and should follow the procedures laid down in the operating manual. First start-up (like regular operation) consists very largely of opening and closing valves in the right order and when certain pre-conditions are met. The special hazard of mistaken valve identity is discussed later [20.2.1].

Batch and semi-batch processes can usually be started up in a more relaxed manner than continuous ones, since the former usually allow the process to be interrupted to take stock of the situation. Most continuous units consist of a number of stages. It is important to ensure that the products from any stage have reached their design composition, temperature and pressure before being admitted to the next one. Where the stage is one of physical separation only, such as fractional distillation, the process piping should allow its products to be mixed and recycled to the inlet of the stage. This enables the stage to be brought on-stream on total recycle until its products reach the required composition, before entering the next stage. Where the stage involves chemical reaction, such a recycle may be impossible. It is thus sometimes necessary to provide for withdrawal and disposal of off-specification reactor products until these have reached the required composition. Once the start-up of a continuous process gets under way, its operation becomes a very dynamic affair which often requires both quick and correct reactions from the operators.

At the same time, start-up cannot be rushed. Sufficient time must be allowed in heating up equipment or allowing operations such as distillation, crystallisation and reaction to settle down to a steady state. Catalytic reactors can be specially temperamental and sometimes seem to be governed more by black magic than by science. Two examples from personal experience are given. In one, several weeks were spent trying unsuccessfully to start a stirred semi-batch reactor in which gases were to react with solid particles immersed in some of the liquid product. Only then was it discovered that the stirrer was too high in the reactor and not reaching the liquid at the start of the reaction. In another a fixed-bed gas-phase catalytic reactor failed to show any reaction at all for several weeks. After an expert on the process had visited the site, tried several remedies without success, and left, the reaction suddenly started when everyone had lost hope and the process was about to be written off. From that time the reaction never failed to start and the reason why it failed earlier was never discovered.

One hazard of starting up exothermic reactors, both semi-batch and continuous, is that the reaction may fail to start as intended when the concentration of one of the reactants is low. Unless this failure to react is recognised and the flow of one of the reactants is halted until it is cured, the reaction may start suddenly when the concentrations of reactants is high, and become uncontrollable. This is a common hazard of autocatalytic reactions. Sometimes it can be cured by adding reaction products to the reactor before introducing feed materials.

Eventually the whole unit is, hopefully, operating at reduced throughput under instrument control. Controllers are tuned, plant samples are taken and analysed, readings are recorded, and, where appropriate, extrapolated to design conditions. From instrument readings, mass balances are made over sections of the plant and discrepancies investigated. It is advisable to bring the whole plant or unit up to its design throughput as soon as practicable in order to check performance of individual items and of the plant or unit as a whole against design expectations.

Careful monitoring of signs of future trouble is needed during first start-up. These include leaks, vibration, unusual noises, corrosion, differential thermal expansion, cavitation, steam and water hammer. First start-up is a time when damage readily occurs. Furnace tubes may be overheated and warp, compressor and turbine rotors may be overstressed and their shafts bent, and vessel linings and mechanical seals may be damaged by hard particles left inside equipment.

First start-up can proceed smoothly like clockwork, or it can be long drawn-out and beset by many problems. Some plants (such as the two discussed in 11.7) could not be started up at all and had to be written off and scrapped. Others such as the Nypro works rebuilt after the 1974 explosion were scrapped soon after commissioning because they were not economically viable in the changed economic climate. Well-established processes usually give fewest problems. Plants employing brand-new technology and which have been designed on the basis of laboratory and pilot plant experiments tend to give most trouble, particularly when the process fluids pose severe and unusual corrosion or fouling problems.

Troubles in first start-up are frequently attributed to equipment failure, when in fact the wrong equipment had been chosen in the first place. As an example, a high-speed centrifugal pump was installed to handle a polyethylene slurry from a polymerisation reactor. The high shear forces shredded the polymer particles into a mass of fibres resembling asbestos which quickly blocked the pump casings and discharge lines. A similar problem was found when the flow of slurry was throttled by a control valve. Both problems were solved by changing to a slow-running 'Mono' single-rotor screw pump with a synthetic rubber lining, which was fitted with a variable-speed drive. This both pumped the slurry gently and controlled the flow, thus eliminating the control valve.

Similar troubles arise when the equipment is suitable for the normal process fluid in contact with it but cannot cope when the properties of the fluid change significantly. Thus a centrifugal compressor may surge or overheat if the properties of the gas change. A settler separating a light hydrocarbon from water will not work if traces of a strong emulsifying agent are present.

The short residence times in much of the equipment of modern plant often demands a very rapid response which only a finely tuned control instrument can give. Thus first start-up can reveal many a 'chicken and egg' situation which the process designer had not foreseen, although experienced operators often develop special knacks to deal with them. The writer once designed a pilot cracking plant which only a certain green-fingered Norman could start up. When asked what his secret was, he would only say 'It's all quite simple and I just don't understand what you are doing wrong'. Eventually the plant had to be modified so that less-skilled operators could start it up.

There is an increasing tendency to use computers to assist in commissioning, even when it is not intended to employ them during normal operation. Simulators are used to tune control instruments [15.8.3] and to train operators [21.3.5]. Computers are sometimes linked to the plant instruments during commissioning and used to log data, for simple calculations such as mass, heat and energy balances, and for reconciling discrepancies between instrument readings and suggesting where the fault lies. They may also be used for additional alarms to alert operators of possibly dangerous changes in process variables at an early stage. A further use for computers is to forecast changes in the consumptions of steam and other utilities during commissioning. This can be of considerable help in planning.

It not infrequently happens that a plant being started up has to be shut down, isolated, made safe for welding and modified and the whole process of commissioning begun again. Unfortunately such modifications may nullify HAZOP studies on the original design on which much time and money had been spent, and a further HAZOP study has to be done hurriedly on the modified design.

Generally a process plant reaches its highest state of integrity at the end of the post-commissioning phase, after which it slowly deteriorates. Successive overhauls partly restore its integrity but seldom to the original level.

20.2 Plant operation

The plant manager is responsible for the safe and successful operation of process plant around the clock. For this he must ensure that there are proper job descriptions for the operators and supervisors working under him, that their numbers are adequate and that they are well selected and trained. Many of these will be working on shift, which causes special problems of stress, rest and home-life to which he must respond. He also needs to ensure that fairly standardised procedures are adopted, even when there appears to be more than one safe way of carrying out a particular operation. The main reason for this is that if different shift teams develop their individual operating methods, with different orders in which things are done, continuity is lost at shift change. Jobs which should have been done are left undone and the chances of accidents increase.

These procedures should be incorporated into specific operator training for the plant (including simulator programs [21.3.5]). They also form an essential part of the plant-operating manual which should be available in draft form when the plant is commissioned, even though it may have to be modified in the light of commissioning experience. Many instructions in the operating manual fall into the category of 'Safe systems of work'. These should be highlighted in the text according to their importance to safety.

The plant manager needs to motivate his supervisors and operators and monitor their performance, both during and outside his own normal working day. Above all, he needs to set a good example in giving clear instructions, listening to problems and complaints, following agreed safety guidelines and promptly investigating dangerous incidents whether or not they caused loss or injury. He and his senior day supervisors should occasionally visit the plant at night to audit the standards of safe operation and compliance with laid-down procedures, and he should encourage or reprimand where this is called for.

Before discussing the plant-operating manual, it is appropriate to say something about valve identification and operation, since a high proportion of the operations performed on process plants consists of opening, closing and adjusting valves, which must be done at the right time and often in a particular order.

20.2.1 Identification and operation of hand-operated valves

Mistakes in valve identification, in the order in which they are operated, and failures to pass correct information about valve settings from shift to shift are a potent cause of accidents.

Over 40 years ago while working as a shift supervisor in an oil refinery, this writer mistakenly caused the wrong valve to be opened on a high-octane gasoline plant so that over 100 000 gallons went to the wrong tank. The loss cost the company several times the writer's annual salary. The mistake would probably not have occurred if the valves had been clearly identified. Without valve numbers it is also difficult for shift operators to leave clear reports of the settings of their valves to the following shift.

Clear identification of valves, both on the plant itself and in operating manuals, instructions and P&IDs should be a cardinal point of good operating procedure. Unfortunately this is not always appreciated, and contractors do not usually number hand-operated valves unless the contract requires them to do so. It is thus often left to the operating company to provide valve numbering. This may involve redrawing the P&IDs since there is often not room on the original ones to add valve numbers.

Valve numbers can most simply follow as an extension to the pipe numbers as shown on the P&ID. The valve number is then a composite of the pipe number plus a serial-plus-type number which is unique for that particular valve. Since tags attached to valves have an unfortunate tendency to disappear, and valves are sometimes changed during maintenance, it is recommended that the composite valve number (with an arrow pointing to the valve) be stencilled on the pipe (or pipe insulation cladding) next to the valve connected to it.

The correct sequence of valve operation is sometimes vital, e.g. on batch processes, on start-up and shutdown of continuous ones, when taking samples and when draining water from vessels containing a gas or liquefied gas under pressure. The expansion or evaporation of gas passing through the valve may cause ice to form in it which prevents it being closed. At least one major disaster started in this way (Fézin, 1966). There should be two valves of different sizes in series on such drain lines, the larger one closest to the vessel. To drain water, the larger valve is opened first and the smaller one then 'cracked open' to allow the water to drain under supervision. Should gas escape and the second valve freeze, there should then be no difficulty in closing the first one across which there is little pressure drop.

On plants whose operations are controlled or assisted by PES programs [15.9], the correct sequence of valve operations should be written into the program so that if valves have to be opened or closed manually, the operator receives a 'prompt' on the VDU to remind him of the correct operating order.

Clear information about the settings of all hand-valves must be available and passed on at shift change. This can be given on a log sheet with a list of valve numbers in the first vertical column, and subsequent vertical columns for each shift to mark whether they left the valve open closed or partly open. Another method which gives a better visual impression is to mount a P&ID showing all valves on a panel, with a small captive screw with a distinctively coloured slot inserted at the centre of each valve symbol to show the state of the valve. Every time a valve is opened or closed the corresponding screw is turned. (The writer first saw this done many years ago but has rarely met it again.) A permanent record of the state of the valves at each shift change can then be made by a Polaroid camera mounted in front of the panel.

20.2.2 The plant-operating manual: general

The manual is probably best drafted during plant design at the same time that the HAZOP study is made, by a small team which includes the commissioning manager (destined later to become the plant manager), the process designer or (design manager) and a safety specialist. A technical writer and an illustrator may also be needed. The question of who writes the manual is, however, less important than that it be written. The team should go through all the motions of testing and prestart-up activities, start-up, operation at normal and reduced throughput, and shutdown, as an imaginative round-table paper exercise with the aid of suitable 'models' (e.g. PFDs, material and energy balances and P&IDs). It should choose between alternative methods of operation before finalising the draft. The order in which operations (including valve opening and closing) are carried should be stated, all valves being referred to by their number. The times required for various operations should be given. Attention should be drawn to operations which have to be completed or brought to a particular state before others are started.

The manual should contain not only detailed operating instructions for all likely situations but also essential information about the processes, plant, equipment, hazards and protective features, as well as P&IDs and other illustrations. The manual may be of loose-leaf type to enable revisions to particular sections to be incorporated. It is suggested that personnel of shift leader status and higher should be given their personal copies of the complete manual, and that a copy be available (e.g. in the control room) for all operating personnel to refer to. Those below shift leader status should be provided with personal copies of those parts of the manual needed for their work, i.e. the operating instructions for the units on which they are working.

So far as safety is concerned, the depth and detail needed in the manual depend largely on the degree of hazard of the plant and the possible consequences of failing to follow the procedures laid down. (Sufficient detail is always needed for any plant to be operated economically.) The manual should be an important part of the training for personnel new to the plant. They should be examined on their understanding of it before they play an active operating role.

The operating manual should always be kept up to date, but revisions should only be made with the signed approval of the plant manager, after considering their implications for safety. If modifications to the plant become necessary during commissioning, the manual should be revised accordingly.

The operating procedures drafted before a plant is commissioned can never be perfect. Thus they should be carefully assessed and revised after first start-up.

Operators often find improvements and short-cuts. These should be recorded and discussed with the plant manager, and, if possible, with others involved in writing the manual, before any revisions are made. There may be sound (perhaps safety) reasons against a suggested short-cut which an operator with limited technical education fails to appreciate. (In Iran the writer encountered uneducated operators who believed that the performance of a distillation column depended on the behaviour of a benevolent *jinnee* inside it!) Warnings with explanations about the dangers of short-cuts should be written into the manual.

The normal operation of a plant may be continuous, batch or a blend of the two. Thus a plant consisting of two continuous distillation columns and a continuous reactor may contain a cyclic adsorption dryer with two drying vessels which operate alternately, one always being on continuous drying service while the other is being purged and regenerated. Although batch processes are usually simpler and easier to control than continuous ones, they involve more operational steps and valve changes, which need to be detailed in the manual.

20.2.3 Suggested contents for a plant-operating manual

The following chapter headings are suggested:

1. Introduction, safety and general
2. Units, symbols and nomenclature
3. Plant description

4. Start-up on commissioning or after major overhaul
5. Normal plant start-up
6. Normal plant operation
7. Normal plant shutdown
8. Emergency shutdown
9. Emptying and purging plant.

The following additional information should be included (e.g. in appendices):

- Requirements, properties and hazards of raw materials, chemicals, catalysts and other ancillary materials;
- Detailed equipment, instrument and valve lists and performance details such as characteristic curves of centrifugal pumps and compressors;
- Operational details of special packaged units;
- Lists and details of special safety items and protective systems such as showers, first-aid kit, personal protective equipment, monitoring equipment for harmful substances, pressure relief devices and disposal systems, fire-fighting equipment, sprinklers, fire-water systems;
- Sampling methods and programme, on-stream analysers and tests (e.g. S.G.) made in plant.

The scope of each chapter is described briefly below.

Chapter 1 should explain the purpose and scope of the manual, the purpose and general principles of the plant, with specifications or typical analyses of the main feed materials and products. It should state the hazards of the process and materials processed, their possible consequences, and describe the protective features used. It should explain and stress the need to follow the procedures given in the manual, especially those categorised as 'Safe systems of work'. For these, checklists should be prepared and issued to operators against which to tick off each step as they take it. It should cover communications with other plants, departments, managers and emergency services, giving their telephone numbers. It should discuss plant staffing, job functions and responsibilities, log books and plant hand-over by outgoing to incoming shifts and arrangements for overlap at shift change (which is especially important for shift supervisors).

Chapter 2 should explain the units of measurement and technical terms used and also the drawing symbols employed for equipment and instrument items. In the absence of company standards, the drawing symbols and abbreviations given in *The Piping Guide*[1] are recommended. These include those of the Instrument Society of America[4] which are widely used internationally.

Chapter 3 should describe the various process, storage and utility units of which the plant is composed, referring to drawings and lists included in the manual and drawing attention to critical and safety features.

Chapter 4 should list and describe the steps required in the initial start-up [20.1.4] in a logical sequence so that each operation follows smoothly from the preceding one. It should list the various problems and hazards which may be encountered, with their symptoms, and explain how to diagnose and resolve them. It should give instructions for the disposal of solid wastes, liquid effluents, vent gases and processed materials leaving the unit, and the steps to be taken to ensure that downstream units are ready to receive the latter. Special attention should be given to the start-up of reactors and how to recognise whether a reaction is proceeding properly or not.

Chapter 5 should repeat those parts of Chapter 4 which apply to a normal start-up (i.e. after a short shutdown). This and the following chapters should be included in the standard operating instructions issued to all operators.

Chapter 6 should include guidelines on:
• Pump changes and gland adjustments;
• Record keeping;
• Checking relief valves, alarms, controls, the temperature of pump glands and bearings, the current consumption of electric motors, and other items essential to safety;
• Lubrication;
• Checking and reconciling instrument readings and recognising faults in instrument control loops;
• Recognising and dealing with fault conditions, especially when several alarms are triggered in sequence[5];
• The scope of 'running' maintenance and the conditions, under which it may be done;
• Relief of excess pressure;
• Maintenance of throughput and product quality when throughput, feed composition, product grade, catalyst activity, heat transfer, weather and other process conditions change;
• Action required in the event of leaks, instrument failures, excessive vibration and other danger signals.
The ranges of 'safe' operating conditions of equipment and process units and their limits should be discussed. While maximum and minimum operating temperatures, pressures and shaft speeds [13.3.1] are easily appreciated, regions of unstable conditions often exist within these limits. Certain concentrations of process materials (e.g. caused by mixing them in the wrong order or proportions) may lead to runaway chemical reactions [8.2] or rapid corrosion [11]. While the plant and its controls should have been designed to give a wide range of safe operating conditions, deterioration in performance and reduced integrity which occur with age can narrow this range considerably. Guidance should be given on the steps to be taken in upset conditions and how to decide when an unscheduled plant shutdown is necessary[5]). The use of condition-monitoring equipment for corrosion, vibration, lubricant contamination, etc. should be explained [17.5].

Chapter 7 should give the procedure for normal plant shutdown as a result of shortage of feedstock, full product storage or plant inspection and maintenance. It is assumed here that the shutdown is brief and that equipment will not have to be opened. The final state of the plant including services and the approximate quantities of process materials to be left in each part of it after shutdown should be stated, together with all the steps needed to reach this state. Special requirements on shutdown, such as leaving parts of the plant filled with nitrogen under slight positive pressure, or at a temperature higher than ambient with some means of heating, should be stated.

Chapter 8 should give the procedure for rapid shutdown in emergencies such as fire or failure of electrical power, cooling water, steam or instrument air. Besides total shutdown, it should deal with partial emergency shutdowns which may follow the failure of an important plant item such as a compressor or turbine.

If the plant presents major hazards, the actions required of operators in an emergency shutdown should be as few and simple as possible, so that they can escape almost immediately. There is usually an interval (only 45 seconds at Flixborough) between the first warning of an explosion or other event likely to kill or seriously injure personnel still remaining on the plant, and the event itself. This suggests that the plant and its instrumentation should be designed so that by actuating a special switch, it shuts itself down automatically without creating further hazards. This is most readily achieved when PES control is used. The switch might, for example, do the following:

- Shut off the air and/or electric power supply to all control instruments and manual valve loading stations in the control room, thus causing all control valves to take up their 'fail-safe' positions.
- Shut off the normal electric power supply thus stopping electrically powered pumps and other motors, and shutting off electric heaters.
- Close remotely operated isolation valves on feed, product and service lines at the plant battery limits.

Such drastic action may itself lead to other hazards which should have been considered and provided for when the plant was designed. This means that the design team (and the HAZOP study team) should have systematically considered the consequences of automatic emergency action for each of the possible events leading to it. For example, the loss of power to the reflux pump of a distillation column with a steam-heated reboiler could lead to a temporary pressure rise in the column and the opening of pressure relief valves, even if the steam supply was shut off when the pump stopped. Another example is the sudden loss of stirring and cooling on a reactor in which an exothermic reaction is taking place. Valves cannot be closed too fast on liquid pipelines without risk of liquid hammer and rupturing the pipe or valve.

It is therefore often necessary to provide emergency supplies of electricity (e.g. from a stand-by diesel-driven electricity generator) and compressed air (e.g. from an air reservoir located on the plant) which come into operation automatically when the normal supply fails, in order to operate certain valves and motors for a limited time at the start of an emergency. Such operations should again be as automatic as possible and should ideally require no action on the part of an operator. Emergency lighting (often from batteries) is also essential and this should switch itself on automatically when needed.

A tricky problem can arise over the starting of an emergency electrical generator which takes over the power supply automatically if the normal supply fails. In this case there may be an interval of a minute or more with no power between the failure of the main supply and the availability of emergency power. There will then be a peak power demand as some motors start on full load. Provided this temporary loss of power can be tolerated by the process, the instrumentation which controls automatic plant shutdown could contain a time-delay which applies only to power failure, so that the shutdown is only triggered if the emergency supply fails to start and continue functioning within a short given period. But if at the time of the power failure the steam and fuel supplies remained normal, the dangers mentioned earlier of excess pressure and run-away reactions would be accentuated. This problem can sometimes be reduced by the use of steam turbines instead of electricity for critical pumps such as column reflux [16.6.6].

Most plant emergencies which result from a service failure are handled quite safely without any need for plant operators to abandon their posts. For these 'safe' emergency shutdowns it is usually possible to devise a modified shutdown procedure which falls between a normal and a major emergency shutdown.

The manual should explain the consequences of failure of each service in turn, unit by unit, and advise on the most appropriate action to take. It should also deal with the consequences of failure of critical equipment such as compressors and turbines. Thus in the case of electrical power and/or compressed air failure, operators may be able to use emergency power and air supplies to empty excess process materials (especially flammable ones) from the plant to a storage unit as well as shutting manual isolation valves at critical locations. They may even feel compelled to do this in the case of a major emergency, so as not to endanger other lives. (One brave operator at the time of the Flixborough disaster went through the burning wreckage to close manual isolation valves on a liquid ammonia storage sphere when the lines to the sphere had been damaged and the sphere itself had been displaced by the explosion.) Positive action may be required on the part of operators in many fire situations, including small or incipient plant fires which may be dealt with by portable extinguishers, hose reels and steam snuffing hoses, fires on storage tanks which may require the contents of a burning tank to be pumped out, and fires under LPG storage spheres. The last may require the sphere to be depressurised and sometimes water to be pumped into its base, e.g. to prevent the escape of LPG from a leaking joint on the pipework below it. Planning for such fire emergencies should be done jointly with those responsible for fire fighting. It must be clearly understood who will have overall control and what the operators' responsibilities will be.

Chapter 9 should describe the steps (discussed in 18.6) needed to empty, clean, and purge each unit and section of the plant after a normal or emergency shutdown in preparation for inspection and/or maintenance. It should state the operators' responsibilities when plant and equipment is handed over before and after inspection/maintenance, and it should explain the permit system used.

20.3 Planning for major emergencies

Under the CIMAH Regulations [2.4.2], all manufacturers who operate Major Hazard Installations must prepare, test and and rehearse emergency plans to the satisfaction of HSE and their local authorities, who are also obliged to prepare emergency plans. Though separate, these plans should be prepared in conjunction and dovetail together. The subject is dealt with by Lees[5], who gives many references to pre-1980 literature, and in CIA and HSE booklets[6,7]. More recent papers about on-site[8] and off-site arrangements[9] and sites with toxic hazards[10] include further references. Training for emergencies is discussed in 20.3.9 and 21.3.4. Although the main emphasis is on Major Hazard Installations[11], a major emergency (defined as one which involves several departments, serious injuries, loss of life and/or extensive damage[6]) can happen on many other installations.

Emergency plans by their very nature deal with incidents which have already started or taken place, and do not prevent them from happening in the first place. Only the main points of such plans are discussed here. The first step in preparing one is to check the safety of the installation as regards its design, operation and maintenance.

The main objectives of an emergency plan are to:

- Rescue victims and treat them;
- Safeguard others, arranging for their escape or evacuation where necessary;
- Contain the incident and control it with minimum damage;
- Identify the dead;
- Inform relatives of casualties;
- Provide reliable information to the news media;
- Preserve relevant records, equipment and samples which may be needed as evidence for subsequent investigations.

To achieve these objectives the plan should make the best possible use of works and outside services and personnel. Unlike most plans which start from a clearly defined situation, an emergency plan may have to start from any one of several abnormal ones. Emergency plans should therefore be simple and flexible. They should be tested as they are made, and updated later as necessary. Simplicity increases the chances that the plan can be followed and flexibility allows for adaptation to a changing situation.

While simple, the plan must be appropriate to the emergency which arises. Unfortunately this has not always been the case. A careful objective examination of the main risks is needed to ensure that it does. The most unlikely scenarios such as an aircraft crash or a severe earthquake (in the UK) on the site of the installation can usually be ignored. Those involving the explosion or release of hazardous substances present on the installation must, however, be considered, as must the effects of toxic and ecotoxic substances which may be formed as a result of fire or abnormal plant conditions (e.g. dioxin and the combustion products of polyurethane foams). The 'domino' effects of incidents on neighbouring installations should also be considered. Thus the disruption of cooling water supply to a neighbouring steel plant by the Flixborough explosion in 1974 would have produced a further major emergency had not the fire brigades been able to quickly improvise an emergency water supply.

Nypro's emergency plan at the time which called for all plant and laboratory personnel to assemble in the control room in the event of fire is an example of an inappropriate plan, since it completely failed to cater for the consequences of a large escape of hot flammable liquid under pressure in the plant. A simplified plan of the works is shown in Figure 20.2. All those already in the control room were killed, and one man who had followed the plan and run from a safe place about 200 metres north of the control room was killed just outside it. Those in the adjacent laboratory who saw the initial release of vapour from the cyclohexane oxidation unit ignored the plan, escaped by running to the north west and survived, despite being caught by the blast of the explosion in the open.

Two lessons might be learnt from this organisational misconception [3.3]:

1. A bad emergency plan is worse than none at all because it inhibits independent thought and gives people an illusory impression that they will be safe by following the plan.
2. Persons whose lives and safety may depend on the plan should be consulted on how the plan will affect them, if possible while it is being prepared. They should be encouraged to think out for themselves and make notes on the various possible emergency situations, and how they would react to them. Those responsible for preparing or revising the plan should study these notes and discuss them with their writers before the plan is finalised. By taking such ideas into account when the plan being is made, mistakes and shortcomings may be avoided and greater confidence created. There is a clear role here for safety committees and representatives [19.1.9].

20.3.1 Identification of major hazard situations [2.4] and assessment of risk [14.3–14.7]

Both subjects are dealt with elsewhere in this book, and in the references quoted. Potentially major hazard installations include:

1. Those where more than minimum quantities (in most cases a few tons) of toxic and flammable materials, particularly gases and vapours, may be released on loss of containment;
2. Those containing more than minimum quantities of the following:
   (a) Any gas contained at pressures of 100 bar g. or higher;
   (b) Liquid oxygen;
   (c) Explosive, unstable and highly reactive materials and mixtures of them.

The minimum quantities (not always identical) are given in the NIHHS and CIMAH Regulations and in the reports of HSC's advisory Committee on Major Hazards[11].

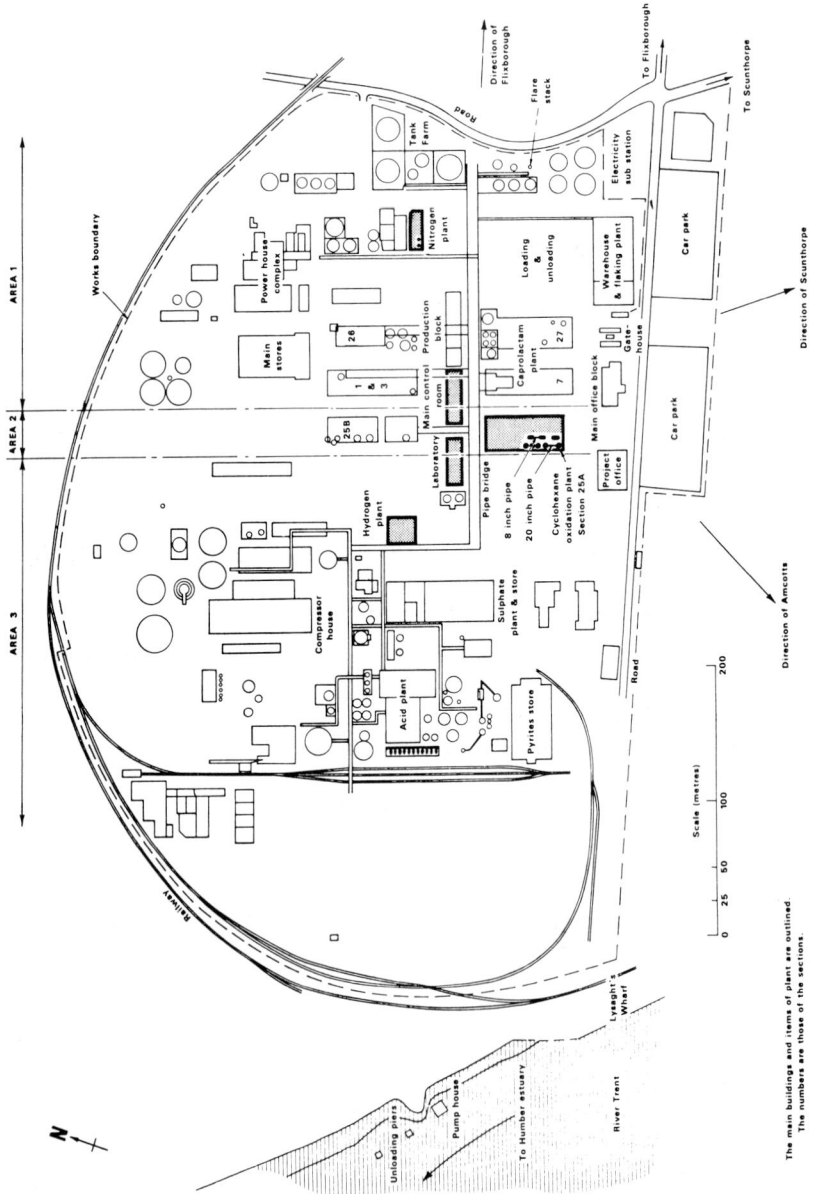

Figure 20.2 Simplified plan of Nypro's former works at Flixborough (courtesy the Controller, HMSO)

In assessing the risk of major hazards to people and the spread of damage, the following must be considered:

1. The type of incident expected (fire, explosion, toxic release) and its probable duration;
2. The location of the incident in relation to neighbouring plants, storage areas and built-up areas;
3. Prevailing winds;
4. Areas most likely to be affected;
5. Population densities in the areas possibly affected;
6. Possible damage or contamination of drains, crops, water supplies;
7. Possible 'domino effects', i.e. an explosion in one area causing the release of flammable or toxic materials elsewhere;
8. Possible effects of collapse of buildings and structures;
9. Presence of radioactive sources;
10. Topography and physical features of site.

Models may be used to assess the spread and dispersion of emissions of flammable and toxic materials in various circumstances [14.6]. Since airborne toxic materials are harmful at much lower concentrations than merely flammable ones, their effects extend to much greater distances.

Having assessed the risk, one should examine the adequacy of existing resources (first, from the works and second, from the local fire, ambulance and hospital services) to handle the most serious foreseeable emergency, and then in collaboration with outside services decide what further provision or action is needed.

Typical questions raised at this point relate to:

- Adequacy of works fire-fighting resources before arrival of outside fire service;
- Adequacy of drains for fire water;
- Ease of isolation of plant and equipment; should more isolation valves be motorised for remote operation?
- Adequacy of fire protection of instrument and electrical cables, structural steel and tanks;
- Adequacy of relief valves for fire conditions;
- Adequacy of plant alarm, evacuation assembly and roll-call arrangements for the scenarios postulated, including the presence of contractors, drivers and visitors;
- Alternative methods of communication if the usual one fails or proves inadequate.

20.3.2 Liaison with outside authorities

Outside authorities must be prepared to deal with the effects of major accidents both inside and outside the installations in which they originate. Resources include police, ambulance, and hospital and other local authority services and district inspectors of the HSE. They may also include the services of neighbouring firms under mutual-aid schemes.

In a major emergency the police will coordinate the activities of the various emergency services, and a senior police officer will be designated as the Incident Controller.

The fire service, while maintaining a brigade adequate for normal requirements, may not be able to cope with serious incidents in isolated locations. This should be discussed between the works management and the chief fire officer. On being called to an incident, the senior fire officer present will take charge of fire fighting, rescue and salvage, and the brigade will also assist in dealing with the escape of toxic materials.

The HSE need to see and approve in advance the plans for dealing with emergencies and satisfy themselves that they are practical, sufficiently detailed and rehearsed. The local HSE inspector should be advised promptly of a major emergency once outside services and key personnel have been informed.

Liaison with outside services should ensure that:

- There is a properly coordinated plan which will be effectively controlled;
- Works procedures are in harmony with plans developed by outside authorities;
- Outside services understand the nature of the risks and have appropriate knowledge, equipment and materials to deal with them;
- The equipment of the works and outside services is compatible;
- The personnel of the works and outside services who are likely to have to cooperate in an emergency already know each other;
- The appropriate type and number of outside services reach the scene promptly.

20.3.3 Works organisation for major emergencies

No universal organisation suitable for all circumstances can be given but the following basic features generally apply. Nominated persons should be trained to fill and deputise for two key roles in a major emergency – *works incident controller* and *works main controller*. Their duties are spelt out in more detail in the references quoted earlier. They will normally be day staff (e.g. plant or works managers) but senior shift staff should be nominated and trained to deputise for them to ensure that trained persons are available for these roles at any time. If a major emergency arises when the day staff nominee is away, his shift deputy will act for him until he arrives and can take over.

Other personnel who may be needed in addition to the works fire, first-aid, engineering and security personnel include:

- Persons with special knowledge of the plants, processes and their hazards;
- Checkers for assembly points;
- A mobile analytical team for monitoring the environment for harmful gases and vapours, etc.;
- A public relations officer or team;
- Additional incident controllers, telephonists and assistants for the main key staff.

All personnel liable to be needed in an emergency should carry means of quick identification to avoid delays at police check points, etc.

The works incident controller will proceed to the scene of the incident, assess the scale of the emergency, take responsibility and activate the major emergency procedure if this has not already been done. He should wear a distinctive hat and/or jacket, have a portable two-way radio-telephone and an assistant with him. In the absence of the works main controller or pending his arrival, the incident controller will normally direct shutting down and evacuation of affected plant and ensure that outside services and key personnel have been called in.

The works main controller will go to the emergency control centre as soon as he is aware of the emergency and take over from whoever is deputising for him. His duties include:

- Calling in outside services and key personnel;
- Informing other organisations in the neighbourhood;
- Establishing communications and liaising with the works incident controller;
- Exercising operational control over those parts of the works outside the affected area, directing the shutdown of plants, and the evacuation of personnel as necessary;
- Ensuring that casualties are attended to;
- Ensuring that relatives are informed;
- Liaising with police and fire services and HSE and advising them as necessary, particularly on the possible effects of the emergency on areas outside the works;
- Controlling traffic movement in the works;
- Recording or arranging for a chronological record of the emergency to be made (e.g. on a long-playing tape recorder kept in the emergency control centre);
- Arranging for personnel to be relieved and provided with food and drinks;
- Contacting the local meteorological office for early warning of weather changes;
- Issuing statements as required to news media;
- Liaising with his company's head office;
- Controlling the clean-up and rehabilitation of affected areas after the emergency.

Two 'musts' for the works main controller are good communications and reliable up-to-the-minute information.

Checkers on each shift should be nominated to proceed immediately to each assembly point once a major emergency is declared with a list of names of all persons known to be on the works at the time. Their duties are to record the names of all reporting to the check point and relay these to the emergency control centre. If time clerks are present they are an obvious choice, but otherwise shift workers who have no specific emergency duties should be nominated and trained as checkers.

A **mobile analytical team** is recommended by Essery[8] of ICI. Its main purpose is to provide rapid analyses of the atmosphere for harmful substances at various locations and relay them by radio-telephone to the emergency control centre, so that the works main controller has reliable information on the extent and spread of toxic and flammable materials.

The team would normally be laboratory personnel but their organisation, equipment and means of transport depend on the size and layout of the works and its hazards.

20.3.4 Emergency control centre

At least one pre-arranged emergency control centre should be established and provided with adequate means of communication with areas inside and outside the works, as well as maps, site plans and relevant data and equipment to assist those manning the centre, not forgetting toilet facilities. The centre should be reasonably close to the scenes of possible incidents, yet sufficiently far from them and well enough protected to be able to function in a major emergency. It should be close to a road to allow for ready access by persons and equipment, etc. needed. It will be manned by the works main controller, other designated key personnel, and senior officers of the outside services. To cater for the possibility that a single emergency control centre becomes inoperable as a result of the incident (e.g. through being downwind of a toxic or flammable gas release), an alternative emergency control centre at a different location (usually on the opposite side of the works) should also be established. The police will, if necessary, assist in the establishment of an emergency control centre. A list of suggested equipment for an emergency control centre is given in Appendix 5 of the CIA booklet[6].

20.3.5 Assembly points

Two or more clearly marked assembly points should be chosen in safe places on different sides of and well away from the areas at risk. Their purpose is twofold:

1. To enable the names of those known to have been present on the plant to be checked as safe and not in need of rescue;
2. To enable those assembled there to be recalled to duty if and when it is safe to do so.

Employees should proceed to the assembly point if escape is imperative or if evacuation proves necessary. They should know their location and have simple instructions on which one to make for, depending on the wind direction and the apparent site of the incident. Generally they should be in a building whose risk of collapse resulting from an explosion is realistically low and where there is adequate protection from the hazard of flying glass fragments. If toxic gases or vapours are liable to be released in an emergency, the assembly room should have tight-fitting doors and windows and no other openings to the outside atmosphere. If the latter is heavily contaminated, those in the room should be able to survive in it until the danger passes, or until they are rescued or able to leave, wearing escape breathing apparatus. Unless the risks of a heavier-than-air gas or vapour emission or of flooding can be ignored, the assembly point should not be below ground-level. Where possible, the assembly point should be at or near a manned gate in the works perimeter fence. Access to the assembly point should be possible at all times, if necessary by breaking a special seal on a door lock. Every assembly point should be provided with emergency lighting (torches or batteries) and means of communication with the emergency centre. Every assembly point should be manned by a checker [20.3.3] as soon as possible after a major emergency starts.

20.3.6 Raising the alarm and declaring a major emergency

Every works should be provided with a sounding alarm, duplicated where necessary, so that it can be heard everywhere in the works. Very large works may be divided into zones, each provided with an independent alarm. It is best to have an alarm which gives at least two distinctive sounds, one for 'alert', the other for the real emergency. Electronic devices which produce a pulsed tone which varies between two frequencies are recommended by Underdown[12]. The alarm should be actuated by an electrical signalling system with enough call points spread over the works (or zone) for the 'alert' to be raised by anyone without going far. The alert may also be triggered by suitable fire and/or gas detectors. Only a limited number of senior personnel (including those on shift) should be authorised to sound the major emergency alarm. Once this is raised, the police/fire services, key works personnel and neighbouring firms must be informed immediately.

If more than one type of major emergency (e.g. release of flammable or toxic gases) is possible on the same site, consideration should be given to providing distinctive alarm signals for each, although their numbers should be strictly limited and all employees carefully instructed in their meaning in order to avoid confusion.

The Major Emergency may announce itself to most of those present before warning can be given. Even so, the alarm should, if possible, be sounded.

20.3.7 Evacuation, searching and accounting for personnel

The possible hazards to personnel while escaping from their workplaces to assembly points should be considered, and special readily accessible protective equipment as well as advice and training may have to be provided for this. In some circumstances it is safest for personnel not to attempt to escape at once, but to remain where they are until the danger has passed or until they can be rescued. The safest place much depends on the type of incident. Provided there is no nearby explosion, people are less at risk from toxic gases and thermal radiation in a closed building than in the open. Buildings, however, are more vulnerable to explosion damage than the human body[13], so it seems that people may be safer in the open if there is an explosion. The draught of a large fire and the blast of an explosion speed the dispersion of airborne toxic materials.

Escape-type breathing apparatus [22.8.5] may be needed for toxic hazards, and skin and eye protection for thermal radiation. Hand torches and mobile spotlights will be required at night if normal lighting fails, while alternative escape routes will be needed if the normal one is impassable.

The importance of accounting for personnel following a major incident, and the consequent need for them to report at once to the checker at the assembly point on making good their escape or evacuating, require emphasis. If because of injuries or other reasons a person escaping cannot report at the assembly point, he or those helping him should ensure that his escape is reported to the main controller in some other way, e.g. via the security guard at the works gate or the police.

Escape may be spontaneous in the area where a severe incident arises but evacuation will be more controlled in areas further from it. The emergency plan should include measures which ensure that personnel in these areas are quickly warned if and when they should shut down operations and evacuate.

A good security system whereby the names of all visitors, contractors' men, drivers, etc., with their times of entering and leaving and their locations on site are recorded at the main gate-house is essential. Despite such measures, there may still be uncertainty, particularly at shift change, in knowing precisely who was present on the site at the time of an emergency.

The following measures are required to account for personnel:

1. The incident controller should arrange as soon as possible, subject to his discretion that he is not thereby seriously endangering other lives, for a search to be made to locate and rescue casualties. A further search should be undertaken by the local authority fire brigade on arrival, advised as necessary on particular risks by the incident controller.
2. Nominated personnel should record the names and works numbers of casualties taken to hospitals, mortuaries, etc. and the addresses of these places. The names and addresses of fatal casualties should be reported to the police.
3. Nominated works personnel (checkers) should record the names and departments of those reporting at assembly points, and advise the emergency control centre, with special emphasis on persons feared to be missing.
4. A responsible person at the emergency control centre should collate the lists, check them against the nominal roll of those believed to be on site, and inform the police of any thought to be missing. Where missing persons might reasonably have been in the affected area, the incident controller and senior fire officer should be informed and a further search made.

Those responsible for drawing up emergency plans should consider the use of electronic aids in searching for and locating casualties on a disaster site. One is the Breitling Emergency wrist-watch (made by Breitling Montres SA, PO Box 1132, CH-2540, Grenchen, Switzerland). This contains a miniature radio transmitter which emits an uninterrupted signal for 20 days or more which can be detected at a radius of up to 20 km by a receiver tuned in to its frequency. The signal may be switched on manually or automatically. By issuing all persons entering a major hazard site with such devices and equipping the incident controller and/or the search team with locating radio receivers, the task of searching for survivors should be simplified. Such devices must, however, comply with electrical safety requirements [6.2.4] so that they do not introduce ignition hazards into zones where flammable gases may be present. Another device, now being used by fire brigades to locate casualties after a disaster, is the thermal image camera[14].

20.3.8 Post-emergency duties

Many other special duties which mostly lie outside the scope of this book follow any major emergency. These include comforting and attending to the needs of relatives, ministering to the injured and dying, investigating and publicising the causes of the incident, public relations and rehabilitation of affected areas and property.

20.3.9 Training and rehearsals

Training for emergencies has a threefold aspect: that of works personnel, that of outside service personnel, and joint training exercises involving both.

Works personnel are generally not used to the types of incident discussed here and all require training and rehearsal. A Millbank film *Rescue Team Alert* [M.2.1], an I. Chem. E. video training module and a computer-based training program which simulates process plant emergencies are available [M.2.7]. Rediffusion Simulation Ltd offer tailor-made 'emergency response trainers' for oil and gas plant[2]. Special training is needed both individually and as team members for key personnel, including part-time works teams for fire fighting, first-aid, atmospheric monitoring and emergency engineering operations [21.4]. All on-site emergency procedures (mostly shutdown, isolation and evacuation) for every process and storage unit should be rehearsed regularly, where possible by doing the real thing but where more appropriate by simulation. All employees liable to be involved in emergencies should receive initial and refresher training.

Professional emergency service personnel may be assumed to have adequate general training, but they need to become familiar with the hazards, geography, facilities and special problems of the works, and with key works personnel.

Joint exercises involving works and emergency service personnel are essential and probably form the most important part of emergency training. Due to the cost and interruptions to production caused by full-blown exercises, these should be complemented by 'table top' and 'control post' exercises. In the former it is possible to 'go through the motions' of the parts played by personnel of different organisations following a particular hypothetical event, several of which may be treated in a single session, with the various personnel sitting round a table with a tape recorder running. The more realistic 'control post' exercises are designed to test communications, with key personnel working in the locations they would use in an emergency. These should allow the different professional emergency services to test their own roles and their coordination with other organisations. They should also prove the accuracy of telephone numbers and other plan details, and the availability of special equipment and materials needed in various emergencies. Periodic full-scale practices are also needed to increase confidence and ensure that nothing important has been ignored or forgotten. All exercises need careful preparation. Results should be studied and the lessons learned should be circulated and discussed.

References

1. Sherwood, D. S. and Whistance, D. J., *The Piping Guide*, Chapman and Hall, London (1979)
2. Rediffusion Simulation Ltd, *Simulation in the Oil and Gas Industry* (brochure), Manor Royal, Crawley
3. Pearson, L., 'When it's time for startup', *Hydrocarbon Processing*, **58**(8), 116 (1977)
4. ANSI/ISA S5.1-1984, *Instrumentation Symbols and Identification* (available from London Information (Rowse Muir) Ltd, Index House, Ascot, Berks SL5 7EU)
5. Lees, F. P., *Loss Prevention in the Process Industries*, Butterworths, London (1980)
6. Chemical Industry Safety and Health Council of the Chemical Industries Association Ltd, *Major Emergencies*, 2nd edn, London (1976)
7. HSE, *Control of Industrial Major Accident Hazard Regulations – Further Advice on Emergency Planning*, Booklet HS(G)25, HMSO, London
8. Essery, G. L., 'Planning for the worst – on-site arrangements', paper given at Major Hazards Summer School, Cambridge, organised by IBC Technical Services Ltd, London (1986)
9. Cooney, W. D. C., 'The role of public services – off-site emergency arrangements', *ibid*
10. Lynskey, P., *The Development of an Effective Emergency Procedure for a Major Hazard Site*, European Federation of Chemical Engineering, Publication Series No.42, available from the Institution of Chemical Engineers, Rugby
11. HSC Advisory Committee on Major Hazards, *First, Second and Third Reports*, HMSO, London (1976, 1979 and 1984)
12. Underdown, G. W., *Practical Fire Precautions*, 2nd edn, Gower Press, Aldershot (1979)
13. Roberts, A. F., 'Vapour Cloud Explosions and BLEVES', paper given at Major Hazards Summer School, Cambridge, organised by IBC Technical Services Ltd, London (1986)
14. Treliving, L. 'Thermal image cameras: the new lifesavers', *Fire International*, **92**, 81 (April-May 1985)

Chapter 21

Safety training for process workers

The need for safety training of all ranks in the process industries is clear from most chapters of this book. That of supervisors is specially important because of their many responsibilities for safety.

Safety training requires:

- A properly prepared programme and allocation of adequate time and resources;
- Careful analysis of the jobs to be trained for and their hazards;
- Good instructors who are thoroughly familiar with their subjects;
- Appropriate training media and methods. A good programme usually employs a combination of several different ones.

Training films, aids and sources of information on process safety are listed in Appendix M. Safety training should form part of normal job training. Psychological factors are important in training, which is most effective when it satisfies three important human motivators in the trainee[1] [19.1.7]:

- Recognition of his effort and achievement;
- Acceptance by members of his group;
- Maintenance of his self-respect.

Adult trainees should not be treated as children. Three important laws of learning which apply in training are *primacy*, *recency* and *frequency*[1]. Primacy means learning the right way of doing something from the start. This is more difficult if the worker has been doing it the wrong way and has first to break a bad habit. Recency means that we tend to remember what we have just learnt better than what we learnt, say, last year or the year before, and frequency means that we need regular reminding or revision of what we have learnt to prevent it being forgotten.

Studies in the USA[2] have shown that most people forget 90% of what they were told, 80% of what they were shown but only 35% of what they were both told and shown three days earlier. They are most likely to remember instruction when they perform a task and describe the task while doing it.

21.1 Training aims and framework

The (UK) Chemical and Allied Industries Training Review Council (CAITREC) states some essential features of training in the process industries[3]:

1. All new employees should receive Induction Training which includes relevant Health and Safety information.
2. All new employees, and all people changing job in-company, should receive job-specific training based on an analysis of training needs, including relevant Health and Safety aspects.
3. All employees should receive health and safety information and training.
4. All training programmes should specify:
   ○ The name of the person responsible for ensuring delivery of the programme,
   ○ The objectives, i.e. what the trainee will be able to do on completion of the programme,
   ○ The trainers, i.e. who will carry out each part of the instruction,
   ○ The time and place of training.
5. All training programmes should:
   ○ Build upon the trainee's existing skills, knowledge and experience,
   ○ Specify the standard of performance required after training.
6. Trainee job-performance should be monitored during and after completion of training.
7. After training, the trainee's manager/supervisor should assess his performance against standards set for his job.
8. All training should be recorded to show:
   ○ The date of training,
   ○ The training that has yet to be completed,
   ○ The time taken to reach acceptable performance.
9. All training programmes should be monitored and revised as necessary.

21.1.1 Training needs

Training is most needed:

1. For new employees,
2. When new equipment and processes are introduced,
3. When procedures have been revised,
4. When new information has to be imparted,
5. Where performance and morale need to be improved.

Indicators of the need for training are:

- Above-average accident and injury rates for the type of work,
- High labour turnover,
- Excessive waste, poor yields and poor-quality products,
- Works, factory or plant expansion.

21.1.2 Training objectives and levels

Training objectives must be set, based on:

1. Detailed descriptions of the jobs being trained for, including possible hazards,
2. The initial levels of knowledge and skills of trainees,
3. The levels to be reached during training.

Training courses for skilled jobs (such as plant operation and maintenance) should be graded, e.g. as basic, intermediate, advanced and supervisory. New and inexperienced employees should be required to take the basic course and pass its test before being assigned to work involving risk to themselves or others. They should have a stipulated amount of appropriate work experience before starting more advanced training. Similar considerations apply at all levels. Awards such as a badges or ties showing the level reached by successful trainees (e.g. bronze, silver and gold like those for life-saving and dancing) offer useful incentives as do cash bonuses or steps in the wage scale. It is best to engage new trainees for skilled jobs on a probationary basis. They should be required to pass an appropriate test before being taken on the permanent pay-roll.

Jobs to be trained for should be analysed and broken down into steps. The time and method required to teach each step and test the trainee must be assessed when planning a training programme. While training should not be so narrowly based that its logic remains unclear, it should keep to its subject.

Management should ensure that jobs which are critical to safety are performed only by workers who have successfully taken the relevant training course and test, or preferably one a stage more advanced. The latter is of great advantage in ensuring a pool of trained personnel when production facilities are expanded.

21.1.3 Induction of new workers

It is important to include a positive health and safety (HS) message in the information imparted to every new employee before he starts work. He may, however, absorb only part of it because of the newness of his surroundings and the many other things he has to take in at the same time. The message should therefore be reviewed and amplified soon afterwards, i.e. about two weeks later. It should include the following[4]:

1. Management's interest in preventing accidents. This should be illustrated by the company's safety policy, programme and record.
2. To prevent accidents, certain rules and procedures must be followed. A company safety-rule booklet should be given to the new worker and explained. All its rules should be logical and have been discussed and agreed with workers' representatives. They should be enforceable, e.g. by suspension of offenders. The rule booklet should include the following items where applicable:
   - Smoking rules,
   - Permits-to-work,
   - First-aid and its organisation,

○ Personal protective clothing and equipment (including its issue, safe-keeping, inspection and maintenance),
○ Work clothing (including the above provisions),
○ Raising a fire alarm, fire fighting and its organisation,
○ Electrical equipment,
○ 'Housekeeping',
○ Emergency and evacuation procedures,
○ Procedures for reporting accidents and injuries and getting medical attention.

3. Every employee should report to his supervisor unsafe conditions which he encounters in his work.
4. No employee is expected to undertake a job until he has learned how to do it and is authorised to do it by his supervisor.
5. No employee should undertake a job which appears to him to be unsafe.
6. If an employee suffers an injury, even a slight one, he is required to report it at once.

Training should be discussed, including essential training provided by the company. The new entrant should also be encouraged to take any good and appropriate training courses run externally.

This message should be given by someone with proven ability (and, where necessary, training) as an instructor. He or she may be a member of the personnel department or preferably, the manager of the worker's future department, or a safety professional. The message should be carefully prepared and presented and may be supplemented by a brief safety video or film which reinforces the points discussed. New workers should be encouraged to ask questions which should be answered in a friendly way. Any safety rules which are a condition of employment should be enforced from the start.

A checklist of safety topics should be discussed with the new entrant and each should be ticked off as it is discussed. On completion the form should be signed by both the person giving and the person receiving the indoctrination. It should then be attached to the employee's record to confirm that the safety indoctrination has taken place.

If the new employee is examined medically, the doctor and/or nurse should explain the work of the medical department, encourage him or her to use its services and stress the importance of reporting all injuries, sicknesses, skin complaints, dizziness or irritability.

21.1.4 Initial safety training

On reporting for work to his supervisor, the new employee is usually entrusted to an experienced worker who will first familiarise him with the welfare facilities, the plant, its safety features and controls, the work done and any special terms and jargon used. For about the first two weeks the new employee's work is usually limited to assisting experienced workers in non-hazardous routine tasks. When he is familiar with his new surroundings, his supervisor, who should know the HS message given to him on his first day, should meet him and review it in detail. At the same time, the supervisor should get to know him personally, assess his experience, and discuss with him specific safety aspects and potential hazards of the department and the job for which he will be trained. He should discuss any departmental safety rules, fire protection, the use of any personal protective equipment needed for the job, location of emergency showers and eye-wash units and the department's safety programme. The supervisor should record the points covered on the employee's safety orientation checklist. He should discuss what training the new entrant will next receive and what work he will be doing during training, and make proper arrangements for both.

21.2 On-the-job training

Most on-the-job training should be to a definite programme and given by the supervisor or by an experienced worker with proven training ability nominated by the supervisor. The programme organiser is responsible for reviewing and, where necessary, developing the training methods used.

The first source to be considered is the plant documentation. This should include operating and maintenance manuals [20.2.2], process material data sheets [C], permit systems [18], operating and maintenance reports, emergency procedures [20.3] and sometimes departmental safety rules (as distinct from those covering the entire works). The documentation should contain information on potential hazards, how to recognise them and how to control them mechanically or procedurally. The training organiser has to select what is relevant to the trainee's educational level and needs. For the average new worker who may not understand the meaning of the terms 'flash-point' and 'occupational exposure limit', a set of data sheets giving detailed properties of all process materials in a plant would be confusing. The training organiser might therefore short-list the more hazardous process materials using the criteria shown in Table 21.1[1], and provide abridged safety data sheets for their safe handling as shown in Table 21.2[1]. Trainees should, however, have the right to see and obtain copies of more detailed material data sheets.

Table 21.1 Criteria for selection of hazardous process materials for chemical safety data sheets used for training new process workers[1]

Health	1. Materials absorbed by the skin
	2. Dusts with OELs $\leqslant 1\,mg/m^3$
	3. Vapours and gases with OELs $\leqslant 1000\,ppm$
	4. Corrosive and irritating materials
	5. Others recommended by the industrial hygienist
Flammability	6. Materials with flash-point $<38°C$
Reactivity	7. Materials readily capable of detonation, explosive decomposition or reaction at ambient temperature and pressure
	8. Materials capable of detonation or explosive decomposition under a strong initiating source or when heated while confined
	9. Materials which react explosively with water
	10. Other unstable materials capable of reacting violently, especially with water and materials which may form explosive mixtures with water

Table 21.2 Contents of chemical safety data sheet for training new process workers[1]

1. Name and description of material (including chemical name)
2. Nature of hazard (toxic, flammable, reactive, corrosive, etc.)
3. Prompt first-aid procedure for exposed personnel
4. Protective equipment needed
5. Handling precautions
6. Prompt spill-control method

Other information where appropriate on:
7. Engineering control method
8. Fire-extinguishing agent
9. Incompatibility with other chemicals
10. Unusual fire and explosion hazards
11. Simple description of symptoms expected in exposed persons

A trainee operator for a particular unit of a plant must know and understand the operating and emergency procedures for that unit, but needs only a general understanding of others which affect it. The trainer should present the information in small doses, checking that each is learnt and understood before proceeding.

Three job training tools which are applied systematically in the USA are Job Safety Analysis, Job Instruction Training and 'Over the Shoulder Coaching'.

21.2.1 Job safety analysis (JSA)[1, 4]

This is a technique for identifying potential hazards in each step of a job and eliminating them by specifying a safe procedure or changing an existing one, or by the use of particular equipment or tools. Jobs usually selected for JSA are those in which:

- Accidents have frequently occurred,
- There have been disabling injuries,
- There is a high potential for severe injury or damage,
- The work is new, resulting from a change in equipment, process or procedure.

Typical jobs for which JSA may be used in the process industries are breaking a flanged pipe joint, clearing a blockage in a pipe, taking a sample of a process fluid under pressure, starting a compressor or adding chemicals to a reactor. Broadly defined jobs such as building a plant and narrowly defined ones such as pressing a button or tightening a screw are not suitable for JSA.

The trainee should participate in and study safety analyses of jobs which will form part of his or her work. A JSA is best done by the line supervisor for that job with its work crew. The job is first broken down into a sequence of steps each starting with an action word such as 'remove' or 'open' with a description of what is being done. An experienced and cooperative worker is briefed on the purpose of the exercise and asked to do the job while the other participants watch him and record each step, where possible with the help of a video camera. The participants then attempt by a brainstorming type of approach to identify all hazards and potential accidents associated with each step, regardless of their probability and without at that time attempting to devise means of controlling them. It helps here to consider various types of injury and look for exposures which could cause them.

The participants then review the hazards and develop corrective action to control them, and define a safe procedure for the job. This may involve a change from the previous one used. Changes in tools or equipment may also be required, or improved guarding or interlocks may be necessary. The analysis is recorded on a form (Figure 21.1) with three columns, the first showing the basic steps of the job, the second the hazards of potential accidents associated with each step and the third the recommended safe job procedure and any other corrective action which needs to be taken.

An incidental benefit of JSA is that it often develops new ideas for saving material and labour which more than pay for the time and effort required to do it.

21.2.2 Job instruction training (JIT)[1]

Where a simulator is available for the job this may be used until the trainee has mastered it. Otherwise most JIT is done on or at the controls of real plant or equipment, often under normal operating conditions. JIT can be applied to a sequence of operations which, when started, often dictate their own pace.

The instructor may be the trainee's supervisor or a worker with experience of the job and the method of instruction, or a special instructor. He first has to decide on the speed of teaching and what skills the learner should acquire in the time available. He needs to analyse the job and check the following:

- Every step in the standard procedure being taught;
- Any health risks and adequate provision for them;
- Any personal protective equipment needed;
- The safety of methods used for handling materials;
- Opportunities for trainee error and their consequences;
- Any equipment hazards present and provisions against them;
- Protection against fire and explosion hazards;
- Potential emergency situations and arrangements to control them;
- Adequate emergency shutdown procedure.

He must make sure that all equipment and materials are ready and that each trainee can be placed in such a position while observing the instructor's demonstration that he will see the task as he would if he were doing it himself. (He should not be placed opposite the instructor but could sit or stand beside him and view his actions in a mirror facing them both.)

JIT is broken down into four parts:

1. Preparation,
2. Presentation,
3. Application,
4. Testing.

JOB SAFETY ANALYSIS TRAINING GUIDE	JOB:		DATE:
	TITLE OF MAN WHO DOES JOB:	FOREMAN/SUPR:	ANALYSIS BY:
DEPARTMENT:	SECTION:		REVIEWED BY:
REQUIRED AND/OR RECOMMENDED PERSONAL PROTECTIVE EQUIPMENT:			APPROVED BY:

SEQUENCE OF BASIC JOB STEPS	POTENTIAL ACCIDENTS OR HAZARDS	RECOMMENDED SAFE JOB PROCEDURE
Break the job down into its basic steps, e.g., what is done first, what is done next, and so on. You can do this by 1) observing the job, 2) discussing it with the operator, 3) drawing on your knowledge of the job, or 4) a combination of the three. Record the job steps in their normal order of occurrence. Describe what is done, not the details of how it is done. Usually three or four words are sufficient to describe each basic job step. For example, the first basic job step in using a pressurized water fire extinguisher would be: 1) Remove the extinguisher from the wall bracket.	For each job step, ask yourself what accidents could happen to the man doing the job step. You can get the answers by 1) observing the job, 2) discussing it with the operator, 3) recalling past accidents, or 4) a combination of the three. Ask yourself: can he be struck by or contacted by anything; can he strike against or come in contact with anything; can he be caught in, on, or between anything; can he fall; can he overexert; is he exposed to anything injurious such as gas, radiation, welding rays, etc.? for example, acid burns, fumes.	For each potential accident or hazard, ask yourself how should the man do the job step to avoid the potential accident, or what should he do or not do to avoid the accident. You can get your answers by 1) observing the job for leads, 2) discussing precautions with experienced job operators, 3) drawing on your experience, or 4) a combination of the three. Be sure to describe specifically the precautions a man must take. Don't leave out important details. Number each separate recommended precaution with the same number you gave the potential accident (see center column) that the precaution seeks to avoid. Use simple do or don't statements to explain recommended precautions as if you were talking to the man. For example: "Lift with your legs, not your back." Avoid such generalities as "Be careful," "Be alert," "Take caution," etc.

Figure 21.1 Job safety analysis form (courtesy National Safety Council, Chicago)

During preparation the instructor describes the job and the way it should be done, discusses the more important points about it and tries to find out what each trainee knows about it already. The instructor should follow a format such as that used by the American National Safety Council (Figure 21.2). In stating what must not be done, he should explain why.

HOW TO GET READY TO INSTRUCT

Have a Timetable—
how much skill you expect him to have, by what date

Break Down the Job—
list important steps. pick out the key points. (Safety is always a key point.)

Have Everything Ready—
the right equipment, materials and supplies.

Have the Workplace Properly Arranged—
just as the trainee will be expected to keep it.

*Use JSA, Job Safety Analysis breakdown to locate hazards.

JOB INSTRUCTION TRAINING (JIT)

HOW TO INSTRUCT

1. Prepare
Put trainee at ease.
Define the job and find out what he already knows about it.
Get him interested in learning job.
Place in correct position.

2. Present
Tell, show, and illustrate one IMPORTANT STEP at a time.
Stress each KEY POINT.*

3. Try Out Performance
Have him do the job—coach him.
Have him explain each key point to you as he does the job again.
Make sure he understands.
Continue until YOU know HE knows.

4. Follow-Up
Put him on his own.
Designate to whom he goes for help.
Check frequently. Encourage questions.
Taper off extra coaching and close follow-up.

*Safety is always a key point.

**SAFETY TRAINING INSTITUTE
NATIONAL SAFETY COUNCIL**

Figure 21.2 Format for Job Instruction Training (courtesy National Safety Council, Chicago)

The instructor then illustrates and, if possible, demonstrates each step of the operation, referring frequently to the standard procedure of which the trainee should have a copy. He discusses it with the trainee to find out how much he has absorbed and, if necessary, repeats all or part of the demonstration.

He then checks what the trainee has learnt by getting him to explain how the process is operated and describe what goes on in each step. He next asks the trainee to operate the process and describe the key points of each step as he does it. The instructor watches, corrects and explains as needed. The trainee then repeats the cycle with the instructor watching him until both are confident that he can do it safely on his own.

The trainee is then invited to carry out the operation on his own after being told from whom to get help if needed. He should be impressed with the dictum 'if you are not sure – don't do it'. The supervisor must check frequently until certain that the trainee has mastered the operation, and then occasionally to ensure that he is able to cope on his own in unusual situations.

21.2.3 Over-the-shoulder coaching[4]

This is a flexible and direct method of training which allows the trainee to develop and apply his skills under the guidance of a skilled and safe operator who has the time, patience and desire to help him. As with JIT, it can be applied in conjunction with simulators of complex plant [21.3.5]. The coach now sits in a room separated from the trainee by a window with 'one-way vision' so that he can observe the trainee's actions. They should be able to communicate with each other by telephone. The coach feeds the trainee with various simulated plant situations on VDU screens or verbally and the trainee has to respond to them.

The coach should keep a careful record, e.g. on a chart, of the progress of each of his trainees. Because of its personal nature the method can be very effective.

21.3 Training media and methods

A well-designed training programme may use several different media and methods.

21.3.1 Printed media

Books and notes are easy to produce, survive minor mishaps such as spills of hot coffee and are invaluable for reference and records. Each trainee can be given a personal copy which he or she can study almost anywhere. Whatever other methods are used, the booklet or manual is usually an essential accompaniment.

The written word is, however, quickly forgotten. Books, moreover, are passive aids to learning. While the student may respond to the book, the book cannot detect any difficulties which the student is having.

Written training material may include exercises to which answers are available, but there is nothing to stop the student from cheating or losing interest unless the book is used in conjunction with personal instruction. Printed instructions are most useful for tasks which do not require rapid hand and eye coordination, although even for these an illustrated pocket book can be a helpful supplement to practice.

21.3.2 Programmed instruction

This method was first used with special printed texts (developed for correspondence courses), and later as computer programs [21.3.5]. It depends on breaking the subject down into many small parts and then concentrating on each in turn. The student is given encouragement whenever he demonstrates that he has mastered a task, and is then allowed to progress to the next one. If he cannot master a task the first time, he will have the option of further study in which the subject matter is presented differently, usually in even smaller pieces. It provides a more thorough training than the mere reading of a book.

21.3.3 Personal instruction

This can take a variety of forms ranging from individual coaching and counselling to a lecture to a large audience. Instructors are more versatile than books and can sense and probe students' learning problems. The spoken word alone is not, however, a good means of communication because it is easily misheard, misunderstood and forgotten. To overcome these drawbacks the speaker may articulate slowly to allow students to take notes, which is a distraction.

The teacher's voice is best reinforced by demonstrations or visual aids, such as a chalkboard, flip-chart, overhead slide projector, slide and film strip projector or three-dimensional models, as well as printed notes and illustrations. Personal instruction is generally used in conjunction with a printed text. Conferences, discussions, case studies, role-playing exercises, demonstrations, drills and panels are forms of personal instruction which are discussed later [21.3.7/8].

21.3.4 Films and video cassettes

Film and video programmes can be powerful training tools. Several excellent films on specific safety topics are available for hire or purchase [M.2]. They require considerable skill and planning to produce, with a carefully prepared script, often professional actors, proper lighting and subsequent editing. Although a film can be stopped for discussion, most projectors do not allow single frames to be viewed as 'stills' and it takes time to wind a film forwards or backwards if one wants to project only part of it.

Low-budget films and videos produced in-house with works personnel as actors can be effective if the training programme is well designed and there are enthusiastic and skilful amateur actors and a film maker on the staff who are willing to cooperate. Otherwise it may be possible to involve a small commercial film unit or a film/video teacher at a local art college or evening class.

Video cassettes are largely replacing film as an instructional medium. They can be used with conventional TV sets and video players for small audiences and with video projection screens for larger ones. Video training cassettes produced in-house with amateur video cameras can be indexed to allow a particular 'rush' to be selected and viewed with far less delay than in the case of a film. Most video projectors can be stopped to allow a single shot to be examined, although the resolution may not be good.

Video discs are a newer development which allow very rapid transfer from one part of the disc to another during viewing. At present all video sequences have to be shot first on tape and then transferred to discs, which adds to their cost.

21.3.5 Computers[5]

Computers were introduced as a training aid to extend the availability of training when teachers were in short supply. They are used in three principal ways: computer-assisted learning (CAL), computer-managed learning (CML) and for keeping student records. CAL developed as a means of programmed instruction, using a computer terminal or desk-top computer with a screen, keyboard and sometimes a printer and a 'mouse', at which one or two students study from a program 'written' on 'floppy' disc. It is not difficult to make a program which can carry out a restricted silent dialogue with a student on a selected subject. A text similar to a page in a programmed instruction book is displayed on the screen. This is followed by a display of questions, usually with a multiple choice of answers from which the student selects by pressing a key. In other cases the answers are not displayed and the student has to type his own. The CAL program checks and evaluates the answers, records the student's mark, and proceeds to the next step if the answer was correct, or returns to remedial work if it was wrong. At the end of a lesson the student's marks are added, and if a printer is available a record of the lesson can be made for the student to retain. Once the program has been developed and polished, the computer can act as an individual tutor of unlimited patience to a number of students, although it cannot entirely replace a teacher. It can also be used for drill and practice programs where the student has to answer a number of questions drawn at random from an 'item bank'.

In another form of CAL known as 'revelatory', the program contains a hidden model of some real-life situation and allows the student to develop a feeling for its behaviour under various circumstances. This form of CAL has been developed into sophisticated and expensive multi-media training simulators. Special tailor-made simulators are now used to train operators of complex PES-controlled [15.8.8] oil and petrochemical plant[6]. In these the operator sits at a control station provided with VDU screens which give him detailed information about the state of the plant, and a keyboard on which he issues instructions to the control system.

The simulator has a nearly identical control station at which the operator sits observed by a trainer who can feed him with various plant scenarios to which he has to respond. Such simulators are proving themselves in reducing start-up times, improving operator performance, and contributing to safer operation[6]. Simulators have also been installed inside plant control rooms as an aid to operation, where they can be used to try out the effect of a particular operating strategy before applying it to the real plant [20.2.2].

CML is used to assist the examination and assessment of students and for the general administration of a teaching establishment. Computers are also used to keep students' records and report on their work.

21.3.6 Interactive video

Portable equipment which combines an audio-video system with a computer, keyboard, mouse, graphics facility and a colour monitor is now proving popular for training operators for complex chemical plant, as well as for communications, exhibition display and other purposes. An example is Ivan Berg's 'Take Two' system[7] (marketed in the UK by Quadrant Network) which uses video tape or discs controlled by a computer program. It allows computer text and graphics to be superimposed on the video picture, or the video sequence to be interrupted while the computer program is run.

Such systems lend themselves well to courses of programmed instruction. These can be made from bought or in-house-shot video discs which are combined with computer programs authored in-house. The effect of seeing and hearing a scene on the video screen while assimilating its lesson, on which the student is then interrogated by the computer, makes it a more powerful training tool than either computer or video alone. The complete program naturally requires more effort to produce. Examples of its use are:

- Instruction and examination of trainee operators on the operating manual for a hazardous plant;
- Training operators and maintenance fitters on permit systems and plant isolation, hand-over and maintenance procedures;
- Warning and instructing operators on changes in procedures such as those brought about by plant modifications;
- Instructing operators on the diagnosis and treatment of plant fault conditions as revealed by alarm signals;
- Training personnel on procedures and special duties in major emergencies.

21.3.7 Conferences

Conferences are widely used for teaching management subjects and for solving problems common to a special groups such as plant supervisors whose contacts with each other are normally limited. Good leaders who can draw out information and opinions from participants and sum up conclusions are vital. A conference called to discuss a particular problem can yield good returns in educating its participants. It is important that its leader and members know its scope and limitations well in advance. The leader, while keeping speakers within the terms of reference of the conference, must not succumb to the temptation of trying to steer the discussion in the way he wants it to go. A closely controlled 'conference' is one in name only.

The safety professional is often faced with the need to call a conference of production supervisors and others to discuss problems such as putting a policy or directive into practice, on-the-job training, plant hand-over at shift change, the high incidence of a certain type of accident, the company's permit-to-work system and first-aid. If a purpose of the conference is to recommend action and it does this, its members should know what becomes of their recommendations. The best conferences are usually on matters which only affect those present, when the conclusions drawn are mainly for their own guidance.

Safety professionals and managers at all levels should develop their skills at leading conferences. This involves the following sequence:

1. The leader states the problem;
2. He tries to break this down into segments to keep the discussion orderly;
3. He encourages free discussion;
4. He makes sure that all significant points have been properly understood and that members have given sufficient thought to them;
5. He notes any conclusions reached;
6. He states the final conclusions which truly represent the group's findings, any agreed action items and the names of those responsible for implementing them.
7. He ensures that a concise report of the conference with its conclusions and action items is made and copies distributed to all conference members and others concerned.

21.3.8 Other methods

Other methods of group training include discussions, case studies, role playing, drills, demonstrations and panels. All are useful provided they encourage participation by the trainees.

Discussions
These should be held formally as mini-conferences, to exchange ideas and standardise procedures and techniques. They should allow students to participate and pool their knowledge.

Role playing
Incidents based on real situations are re-enacted by selected members of the group, playing roles and making their own decisions. These are then discussed by the group and its instructor to highlight behaviour patterns.

Drills
These consist of repetition of the task and its various components under guidance to develop important and fundamental skills. They are important to ensure the safe performance of tasks critical to safety such as firefighting and first-aid, starting pumps, taking samples and lubricating moving machinery. The safe method must be well established before a drill is carried out and the limits of its applicability should be made clear and stressed.

Demonstrations
The operation is demonstrated by the instructor as in job instruction training [21.2.2].

Panels
These are planned sessions in which two or more 'experts' in turn answer questions from a selected audience (such as trainees) on various aspects of an assigned subject. A panel benefits from having a good leader. It sometimes reveals differences in the views of experts which would not otherwise come to light.

21.4 Training for special safety responsibilities

Groups of personnel with special responsibilities include:

- Professional managers, engineers, chemists, etc.,
- Health, safety and security professionals,
- Supervisors,
- Process plant operators,
- Skilled engineering tradesmen – electricians, crane, truck and mobile plant drivers, welders and others,
- Trained volunteer workers with special duties in emergencies – part-time firemen, first-aiders, personnel 'checkers' and telephonists,
- Contractors,
- Safety representatives.

Training within industry in the UK is now largely in the hands of non-statutory training organisations (NSTOs) which are run by the industries themselves. The names and addresses of NSTOs involved with the process industries are given in Appendix M. Those for the chemical and petroleum industries which have high hazard potential are:

1. The Chemical and Allied Industries Training Review Council (CAITREC)[3]. This is closely linked with the Chemical Industries Association[8].
2. The Petroleum Training Federation (PTF), which runs several courses by arrangement on its members' sites[9].

While all personnel need safety training which is often specialised, only that of supervisors [21.4.1] and process plant operators [21.4.2] is discussed in any detail here. In the UK the Chemical Industries Association has established eleven self-help regional training organisations within the chemical industry. Its training department runs short residential HS courses in the UK for managers, supervisors, safety representatives and other groups. It cooperates with the City and Guilds of London Institute [M] in the training, assessment and certification of skilled craftsmen, process workers and laboratory technicians and has developed computer-based training packages in several of these fields. Other sources of HS training and information are given in Appendix M.

Much of managers' safety training consists of self-study in which this book can play a part. Lees's list[10] of topics in the safety training of managers is given in Table 21.3.

While all employees should have the most basic instruction in first-aid and fire duties, most companies in addition need volunteer workers trained to higher levels for special emergency duties. This need depends on the extent of both the company's and the local authority's full-time fire, medical and ambulance services and the speed with which they will respond. Thus a petrochemical works might have a medical centre with a trained nurse available during office hours only, and a fire appliance with a professional skeleton crew available for 24 hours a day. The works would need trained voluntary first-aiders able to give artificial respiration and to cope with fractures and other injuries outside office hours, and trained voluntary firemen to make up fire crews for the fire appliance at any time.

Table 21.3 Some topics in safety training of managers (by kind permission of Professor Lees)

Managerial responsibility for safety and loss prevention
Legal requirements
Principles of safety and loss prevention
Company safety policy, organisation and arrangements
Hazards of the particular chemicals and processes
Accidents and accident prevention, statistics and case studies
Pressure systems
Trip systems
Principle of independent assessment
Plant maintenance and modification procedures, including permits-to-work and authorisation
    of modifications
Fire prevention and protection
Emergency planning arrangements
Training of personnel
Information feedback
Good housekeeping
Sources of information on safety and loss prevention including both people and literature
Case histories

The training of these volunteers is usually arranged through the company's or the local authority's fire and medical services.

It is important that adequate numbers of trained volunteers for these special tasks are present while work is being done. To ensure this, managements should provide proper recognition and incentives.

Before contractors are employed in works under the control of the operating company, it is essential that they and their personnel should have received appropriate HS training and be familiar with the relevant works' safety rules and procedures. These conditions should be stated in the contract. Dow Chemical Company, for instance, insists that every contractor must attend a safety induction course before starting work unless the job is of such short duration that it would not be practicable to do so. In this case the contractor would be treated as a visitor and accompanied at all times by a Dow employee while on the site. Table 21.4 gives the headings of Dow's safety rules and procedures[11] for contractors working on their sites.

Table 21.4 Headings of Dow's safety rules and procedures for contractors

Accident procedure	Cranes, hoists, lifting machines and lifting tackle
Emergency plan	Power tools and equipment
Assembly points	Welding, burning and cutting
Working on site	Cutting into drums, tanks and vessels
Chemicals	Ladders
Asbestos	Scaffolds
Housekeeping	Roofwork
Machinery safety	Piling operations
Tags	Right of search
Excavations	

To ensure that its own safety guidelines for contractors are enforced on its numerous sites, Dow lists the following questions to be asked by its executives:

- Does the site have a safety and loss-prevention manual for contractors?
- Is safety referred to in the purchase contract?
- Are site safety rules, emergency procedures and special hazard procedures a part of each job specification?
- Are pre-job meetings held? What is their content?
- Does Dow receive a written outline from each contractor of an accident-prevention programme prior to bid acceptance?
- How are 'service' contractors handled?
- Are field audits done to check for violations?
- Who performs these field audits?
- How is contractor's performance recorded?
- Are site safety rules translated into local languages used by construction personnel?
- Is there a policy of restricting access of non-essential people to plant while it is being started up?
- Are contractors permitted to use plant air as a source of breathing air for respirators, or under hoods of sandblasters?

21.4.1 Supervisors

Supervisors have many duties which are closely linked with safety such as:

1. Establishing methods of work;
2. Instructing people how to do jobs;
3. Assigning people to jobs;
4. Supervising people at work;
5. Maintaining equipment and the workplace.

The company should recognise the key role of supervisors in safety and ensure that they understand and accept it. This often requires patient discussion with the supervisor, e.g. by the safety professional. There is also a special need for supervisors' safety training, the details of which depend on the manufacturing operation and the work of the supervisor's department or team. The American National Safety Council's (NSC) twelve-hour 'Key man development course'[12] for which instructors' course notes and visual aids are available forms a useful foundation for developing a safety training course for supervisors and for more specialised training later.

The instructor should be a company safety professional, a division manager or a general supervisor. He should know the supervisors' work from first-hand experience and have proven talent for training. He needs the time to study the course material available and adapt it to his company's situation, using examples and problems drawn from it. The twelve training sessions of the NSC course have the following headings:

1. Safety and the supervisor;
2. Know your accident problems;
3. Human relations;

4. Maintaining interest in safety;
5. Instructing for safety;
6. Industrial hygiene;
7. Personal protective equipment;
8. Industrial housekeeping;
9. Materials handling and storage;
10. Guarding machines and mechanisms;
11. Hand and portable power tools;
12. Fire protection.

The NSC recommends that the course be formally organised, where possible in company time, in weekly one-hour sessions. Ideally, it should be opened by a senior manager or director of the company. Records of student attendance and performance should be kept and some form of certificate or award presented to candidates who complete the course satisfactorily, if possible at a dinner to mark its successful completion.

Any of the training methods and media discussed earlier may be used for more specialised safety training of supervisors and for updating those who have taken the basic course previously.

21.4.2 Process plant operators

Mistakes made by process plant operators can have disastrous consequences, and while every effort should be made to design plants which reduce the consequences of human error, high levels of skill are needed for many operating jobs. Crossman[13] suggested that operators should be responsible, conscientious, reliable and trustworthy. Lees[10] classifies the work of process operators as (1) simple tasks, (2) vigilance tasks, (3) emergency behaviour, (4) complex tasks and (5) control tasks, and discusses the problems of operator error and fault diagnosis. He gives lists (reproduced here with his permission as Tables 21.5 and 21.6) of topics which should feature in the general and safety training of operators and all workers.

The question is often raised as to how much theory a process plant operator needs to know. A higher technical education is no substitute for specific training in process tasks. While it should enable the operator to adapt readily to different types of plant, he may become bored with the routine and constant shift work. A few years spent as a process plant operator is, however, a useful background for future managers and designers.

In the UK the revised City and Guilds training scheme[14] in which CAITREC are closely involved provides a substantial technical education for process plant operators, particularly in the chemical, petroleum, pharmaceutical, food and other related industries. It is offered at several technical colleges for part-time and block-release training and is suitable for workers of all ages, being intended to supplement the training and experience gained in their employment. The scheme consists of three courses:

Part I   Basic knowledge, including science and communication
Part II   Industrial science and process calculations
Part III Physical chemistry of processes

Table 21.5 Topics in the training of process operators

Process goals, economics, constraints and priorities
Process flow diagram
Unit operations
Process reactions, thermal effects
Control systems
Process materials, quality, yield
Process effluents and wastes
Plant equipment
Instrumentation
Equipment identification
Equipment manipulation
Operating procedures
Equipment maintenance and cleaning
Use of tools
Permit systems
Failure of equipment and services
Fault administration   Alarm monitoring
                        Fault diagnosis
                        Malfunction detection
Emergency procedures
Fire fighting
Malpractices
Communications, record-keeping, reporting

Table 21.6 Some topics in safety training of workers

Workers' responsibility for safety
Legal background, particularly in the UK's Health and Safety at Work etc. Act 1974 and the
    Factories Act 1961
Company safety policy, organisation and arrangements, in particular general safety rules,
    safety personnel, safety representatives and safety committees
Hazards of the particular chemicals and processes
Fire/explosion haard (flammable mixture, ignition source). Ignition sources and precautions,
    including electrical area classification, static electricity, welding, smoking. Fire spread, fire
    doors. Action on discovering fire or unignited leakage
Toxic hazard. Action on discovering toxic release
Emergency arrangements, including alarm raising, alarm signals, escape routes, assembly
    points
Protective clothing, equipment use and location
Fire-fighting methods, equipment use and location
First-aid methods, equipment use and location
Lifting and handling
Security, restricted areas
Accident reporting
Case histories
Permit systems
Good housekeeping
Health, medical aspects

Safety aspects are considered at all stages and form an integral part of
the scheme. The competence of process operators is assessed based on
standards set for the common elements of ten operating tasks – start-up,
running and shutdown of continuous plant, batch plant operation,

materials handling, filling and packing, preparing plant for maintenance, hand-over, emergency procedures and other routines. Special emphasis is placed on fault-finding ability. Assessment is by a combination of internal assessors nominated by the site, and external assessors appointed by CIA and the City and Guilds of London Institute. 'Certificates of Process Operations Competence' with three levels are awarded.

- Level 1 applies to simple operations with basic instrumentation, simple equipment, a limited number of operator/equipment interactions and a small number of parameters under the operator's control.
- Level 2 applies to operations of intermediate complexity.
- Level 3 applies to complex operations with a high degree of instrumentation, many operator/equipment interactions and parameters within the operator's control and a significant frequency of problems.

The assessments include the candidate's knowledge of process technology as given in the revised City and Guilds course 060 on process plant operation[14].

To conclude this chapter, special attention should be drawn to the excellent safety training films, videos and other visual aids listed in Appendix M. While some cater for all workers in the process industries, others cater for special groups – operators, maintenance personnel, contractors and others. Films such as *Nobody's fault* and *Is there anything I've forgotten?* leave a lasting impression. To gain maximum benefit from such films they should be shown twice to the same audience, as many of the finer points tend to be missed on first viewing. After the first showing, a discussion session led by an instructor should be held to highlight the lessons to be learnt and to answer questions. Some time after the second showing it is useful to subject viewers to a written examination in which they answer questions related to the film. Their answers provide a good indication not only of what they have learnt and understood but also of their ability to communicate in writing.

There is no excuse today for lack of safety training in the process industries when there is such an abundance of good training material available.

References

1. Kubias, F. O., 'Tools and techniques for chemical safety training' in *Safety and Accident Prevention in Chemical Operations*, edited by Fawcett, H. H. and Wood, W. S., 2nd edn, Wiley-Interscience, New York (1982)
2. Bird, F. E. and O'Shell, H. E., 'Incident recall' in *National Safety News*, **100** No. 4, 58–62 (1969)
3. Chemical and Allied Industries Training Review Council, *Guidelines for good training*, Chemical Industries Association Ltd, London
4. National Safety Council, *Accident prevention manual for industrial operations*, Chicago (frequently revised)
5. Rushby, N. J., *An Introduction to Educational Computing*, Croom Helm, London (1979)
6. Pathe, D. G., 'Simulator a key to successful plant start-up', *Oil and Gas Journal*, 7 April (1986)
7. 'Take five mark two', *Audio Visual* (April 1987)

8. *CIA Training bulletin*, No. 5, November 1987, Chemical Industries Association Ltd, London
9. *PTF Course information brochure*, Petroleum Training Federation, London
10. Lees, F. P., *Loss Prevention in the Process Industries*, Butterworths, London (1980)
11. Dow Chemical Company Ltd, *Rules and procedures for contractors' personnel*, King's Lynn, Norfolk (1987)
12. National Safety Council, *Key man development course*, Chicago
13. Crossman, E. R. F. W. and Cooke, J. E., 'Manual control of slow-response systems', in *International Congress on Human Factors in Electronics*, Long Beach, California (1962)
14. City and Guilds scheme pamphlet, *060 Process plant operation*, City and Guilds of London Institute, 76 Portland Place, London (1988)

Chapter 22

Personal protection in the working environment

Personal protective clothing and/or equipment (abbreviated PPC/E) is needed against particular hazards of the working environment. Adverse conditions of temperature, humidity, air speed and lighting impose severe restrictions on man's performance, particularly when he has to wear special protection. Thus the use of any form of PPC/E should be considered in the context of the prevailing range of environmental conditions. Some types of PPC/E cause little inconvenience. Others such as respirators and ear-plugs are always inconvenient, require special precautions to be effective, and bring added hazards of their own.

Impervious clothing for protection against chemical splashes becomes almost impossible to wear in very hot climates. The writer recalls a European supervisor in a Middle East oil refinery who suffered extensive acid burns from which he later died through trying to stop the leak of a sulphuric acid/butane spray from a pump gland, wearing only shorts and shirt. With the ambient temperature in the 40s he would probably have collapsed through heatstroke had he worn suitable protection.

Current industrial HS strategy aims to create safe places of work from which hazards of injury and disease are excluded, thus eliminating the need to wear special clothing or equipment. The COSHH Regulations[1] [2.5.6] in the UK and similar ones throughout Europe place special emphasis on monitoring and engineering controls to prevent personal exposure to harmful substances at work, and should thus reduce the need for PPC/E. There are still, however, many jobs for which some form of special protection has to be worn regularly to protect the worker, and others (e.g. in the food industry) where it is needed to protect the product from contamination by the worker. Special clothing and/or equipment are also needed occasionally for tasks such as cleaning large tanks and silos and disconnecting flanges on process pipes, as well as in emergencies caused by leaks, spillages, fire, etc. As such situations may develop rapidly and unexpectedly, it is important that the right PPC/E should be at hand and that employees know when and how to use it and have full confidence in it.

Countless examples could be given of people whose lives, limbs or senses have been lost through failing to use some special PPC/E when it was needed. A few years ago five men, one a fire officer with self-contained breathing apparatus, collapsed and died in a water cistern containing hydrogen sulphide[2]. After one had collapsed in the cistern, two others entered in an attempt to rescue him, and in turn collapsed. As the fire brigade arrived and a fire officer was putting on his breathing apparatus, a bystander entered the cistern, followed by the fire officer *who removed his breathing apparatus*, apparently to issue an instruction. Both collapsed and died.

On the other hand, the successful use of PPC/E is seldom news except where there is visible evidence of a life, eye or limb saved. Figure 22.3a shows the chemical-resistant face-shield which protected the wearer from a splash of molten zinc[3].

The need for some form of PPC/E is more obvious in some cases than in others. It may be a statutory requirement. If it is not, but there is a latent hazard against which some PPC/E could protect, one may need to evaluate the the probability of it being realised (e.g. by some form of fault tree analysis [14.4]). The decision usually becomes one of company policy, which is not always logical, at least on safety grounds (as in the almost universal provision of safety helmets in some industries, regardless of the degree of risk).

PPC/E should be selected to meet the following general requirements:

- Adequate protection against the specific hazard(s) to which the worker will be exposed;
- Minimum weight and discomfort compatible with efficient protection;
- Flexible but effective attachment to the body;
- Weight carried by a part of the body which is able to support it;
- No restrictions on those body movements and sensory perceptions which are essential to the job;
- Durability;
- Clothing should be attractive;
- Accessibility of parts which should be maintainable on the premises;
- No additional hazards introduced by the PPC/Es even when they are misused;
- Compliance with relevant standards.

Selection may require a complete review of what is available and discussion with workers' and manufacturers' representatives, and, if possible, with other users. The possibilities of standardisation and component interchangeability and the provision of training in use and maintenance should be considered before choosing a manufacturer. The PPC/E needed for any job should be included in the written job description and in company standards. HSWA section 2 obliges the employer to provide it and section 7 compels the worker to wear it. This should be made a condition of employment. Managers and visitors who may be exposed to the hazard, even for a short time, should wear the same or equivalent protection as an example.

Personnel who need to wear PPC/E must be trained until they can do so confidently. Precise instructions are needed as to when and in what circumstances it should be worn. Has it to be worn continuously on the job, or only when a particular task is being performed? If the latter applies, this should be made crystal clear by a written job instruction reinforced by an appropriate poster or two at the place where the task is performed.

In most cases the PPC/E should be provided free to each worker who may need it, allocated for his personal use and marked with his name or works number. Generally, the working clothing or protective equipment should not be worn away from work, although the worker should be responsible for its safe care on the premises and may be required to contribute to the cost of replacing it if he loses it. Management is responsible for providing secure changing, washing, toilet and locker facilities on the premises where the worker can change, wash, keep his PPC/E while he is away from work, and his normal clothes, hats, raincoats, etc. when he is working. If there is any risk of damage or contamination of the worker's normal clothing from substances on his PPC/E, and *vice versa*, separate lockers for the worker's normal clothing and PPC/E should be provided. Management is also responsible for arranging for the regular cleaning, disinfection, testing, maintenance and replacement of all PPC/E issued. For some types and uses of breathing apparatus there is a legal requirement for a competent person to examine it every month and keep a record of the results[4].

Visitors' needs for protection are generally confined to safety helmets and sometimes plastic eye shields. The security guard at the main gate should be responsible for issuing and fitting them, collecting them when the visitor leaves, and cleaning and maintaining them.

22.1 Standards for PPC/E

British and Commonwealth readers are now most familiar with British Standards or others based on them. A list of those dealing with the selection, use, construction and testing of PPC/E is given in Appendix N with the numbers of any international (ISO) equivalents. There are as yet few ISO standards and even fewer European ones on personal protective equipment and clothing, although most EC member states have standards of their own.

This position is expected to changed substantially by 1992 as a result of two (draft) EC directives:

100A. Council directive on the approximation of the laws of the member states relating to personal protective equipment,
118A. Council directive on the minimum health and safety requirements for the use by workers of personal protective equipment at the workplace.

National standards of EC member states (including the UK) on PPC/E will in time be replaced by common European ones issued by the European Committee for Standardisation (CEN). These will probably not differ much from present British standards. In the USA there are many (mostly ANSI and NIOSH) standards and codes on PPC/E.

Standards of special interest to users are those on the selection and use of equipment for the same general purpose but in different applications. Examples of such a British Standard and its US equivalent are:

BS 4275: 1974   Recommendations for the selection, use and mainten-
                  ance of respiratory protective equipment

ANSI Z88.2-1980   American national standard practices for respiratory
                   protection

The first of these, although still referred to in current HSE publications[5], is
in fact obsolescent and gives no information on the newer types of RPE.
Another British Standard, 'Recommendations for the selection, use and
maintenance of chemical protective clothing' is being drafted (reference
89/401051).

One area where differences between various national standards can
cause trouble is in colour coding, e.g. for respirator canisters to protect
against different gases. Thus before using any imported and colour-coded
equipment, the maker's code should be checked.

22.2 Comfort and body protection

The UK has a climate in which workers can be comfortable for most of the
year even when wearing protective work clothing. Workers in hot climates
are less fortunate, especially when they are exposed to heat from
high-temperature processes. The writer recalls a plate glass plant built in
Nigeria similar to one built in West Germany. It could not be operated
because no human could work in the heat where certain manual jobs had to
be done.

Unfavourable micro-climates in the working environment can pose a
threefold hazard.

- In severe cases they are a health hazard in themselves;
- By causing discomfort to the worker and reducing his alertness, they are
  a contributory though seldom-recognised cause of accidents;
- High ambient temperatures place severe restrictions on the use of
  PPC/E to protect against chemical hazards.

Designers of process plants thus have to take special account of climatic
factors and the environmental conditions in which its operatives will have
to work. A wide range of protective work clothing is now available. Its
manufacturers have the expertise to survey the working conditions and
hazards where required and advise on the best solutions.

22.2.1 Thermal comfort in the working environment

In normal clothing people can only work safely and efficiently in a limited
temperature range. This amounts to 18–23°C for light work, sitting or
standing, and extends down to 10°C for heavy standing work. Heating,
ventilation and air-conditioning requirements are governed by the fact that
the human body, which generates heat at rates varying from about
8400 kJ/24 hours for light sedentary work to nearly twice this figure for
heavy standing work, must lose it to the surroundings in order to maintain
a constant body core temperature of 37°C. This heat is lost by convection
(25–30%), water evaporation (about 25% in temperate climates) and the
rest by radiation.

While extra clothing can be worn to reduce heat loss and extend the human working range to considerably lower temperatures, there is an obvious limit to how much can be discarded at higher ones.

A better indication of thermal comfort than the dry-bulb temperature is given by the wet-bulb-globe temperature (WBGT). In the absence of direct sunlight, this is the weighted average of 70% of the globe temperature which is sensitive to radiation and air movement, and 30% of the wet bulb temperature which is sensitive to humidity. According to a NIOSH standard[3], male workers should not be exposed continuously for more than one hour to a WBGT above 26°C unless special precautions are taken to ensure that their core body temperature does not exceed 38°C. The corresponding WBGT for women is 24.5°C.

Air speeds in offices and control rooms should be limited to 0.2 m/s for comfort, and a relative humidity of 40–50% is desirable. Relative humidities below 30% encourage the formation of static electricity and dehydrate the mucous membranes of the respiratory tracts, making them more liable to infection.

22.2.2 Hot-work clothing

The appropriate clothing depends on the degree of hazard from heat, fumes, splashes of molten metal, etc. A compromise has often to be struck between allowing a man to work in thin clothing for comfort and chancing the risk of being burned, and encumbering him with thick or impervious clothing so that he collapses from heat exhaustion or endangers himself by discarding some essential item. Flame-retardant cotton fabrics are available which allow the skin to breathe and do not burn easily or form molten droplets. These generally require special treatment after laundering. Suits for protection against heat and flame are made from aluminised wool and carbon fibre which reflects radiant heat, wool/glass fibre composites, Nomex and other materials. One-piece suits are available with internal cooling provided by the expansion of compressed from an air-line connected to an expansion valve attached to the suit. The air temperature falls by several degrees in passing through the expansion valve.

For work near molten metal, no part of the body should be exposed, and clothing should provide no traps or pockets such as open collars or shirt fronts or the tops of shoes or boots where hot metal splashes could lodge and inflict serious local burns on the wearer. Suits should be fastened up to the neck. Safety helmets should have rear flaps to cover the back of the neck. Gauntlet gloves of heat-resistant leather should be worn for hand and wrist protection. The feet and legs are specially vulnerable to hot metal splashes. Boots are available with quick-release fastenings to allow them to be removed quickly in an emergency. Trouser bottoms should be wide and long enough to overlap the top of the footwear[6].

22.2.3 Clean-work clothing

Clothing in the food industry and in the production of some electronic components is needed to protect the product from contamination by the worker. In the production of pharmaceuticals it is required to protect both the worker and the product, and in the production of pesticides, herbicides and other fine chemicals it is needed mainly to protect the worker. Dust elimination is a key objective. The degree of protection needed depends again on the degree of hazard. The following features are generally necessary[6].

1. Clean-work clothing should completely cover the worker's ordinary clothes (or underclothes);
2. It should not be made of a material which is liable to shed fibres;
3. The hair should be confined by a well-fitting head covering since the head is a prolific source of dust and dandruff;
4. There should be no uncovered external pockets (particularly breast pockets), belts or buckles where dust can collect and from which objects may drop when the worker bends;
5. Buttons or string fastenings (which might come loose or be detached) should be avoided. Zips, studs or Velcro fastening should be used;
6. Clothing should be easily washable and of light colour to show any dirt.

Clean-room garments vary widely depending on the degree of cleanliness required. Sometimes only a simple overall and cap are needed. For greater protection, 'coveralls' which leave nothing but the face (sometimes only the eyes) exposed and one-piece suits complemented by a close-fitting head and neck hood and gloves which cover the sleeve cuffs are available. For extreme cases, gastight suits with hoods and transparent plastic faceshields are available. These have a regulator for compressed air which is supplied via a small-diameter flexible hose (Figure 22.1).

Figure 22.1 Air-fed gas-tight suit (courtesy Respirex)

Blends of cotton and synthetic fibres are generally most suitable for washable clothing. A range of disposable clothing and underclothing can be obtained. Disposable aprons of plastic-impregnated fabrics are available for jobs where the front of the work suit is liable to rapid contamination. Handkerchiefs should be forbidden and only paper tissues which are used once and then disposed of should be allowed. Impermeable synthetic or rubber gloves are usually required. No badges, brooches, jewellery or ornaments which might come loose should be allowed on the outside of clean-work clothing.

22.2.5 Clothing for oil and chemical workers

Chemical protective clothing has been divided into three groups, light, medium and heavy duty[7]. Light-duty chemical protective clothing generally consists of uncoated cotton or synthetic fabrics with a water-resistant finish, in the form of overalls, laboratory coats and smocks. It is used by workers who are at slight risk from relatively harmless chemicals (e.g. dilute acids and alkalis).

Medium-duty chemical protective clothing consists of an apron of one of several impervious and resistant materials such as natural or synthetic rubber, PVC-coated fabrics or leather. These are useful for workers exposed to quite a wide range of chemicals, oils and hydrocarbons.

Heavy-duty chemical-resistant clothing is based mainly on PVC or PVC-coated fabrics in the form of boiler suits, long surgical coats, bib and brace overalls, leggings and three-quarter-length suits.

With an ever-increasing need for body protection against a widening variety of chemicals, manufacturers of protective clothing now offer a wider choice of clothing materials and types, and will make special clothing for specific needs. Figure 22.2 gives a general guide to the selection of fabrics for different hazards[8]. Most manufacturers of protective clothing, gloves and footwear maintain lists showing the chemical resistance of the fabrics, etc. which they offer against a wide range of chemicals. Where necessary, tests of the most promising fabrics, etc. should be carried out under appropriate conditions. ISPEMA's current reference book[9] gives a number of useful points which should be considered by purchasers of protective clothing.

The hazardous materials must be considered in relation to their state (gas, liquid or solid), their temperature, the effects they will have on the body of the wearer and on the fabric, its coating and any other materials used in the construction of the garment.

An important point in garment construction is how their seams are joined. It is useless to stitch the seams with a thread which is attacked by a liquid which the clothing is intended to resist. Some years ago the garments worn by workers in the nitration rooms of government explosive factories (where the writer was working) were made from a woollen fabric with a sateen weave which had tolerable resistance to acid splashes. Unfortunately the garments quickly came apart at the seams sewn by cotton thread, which was attacked far more rapidly than wool.

If the fabric has a coating which can be welded, such garments are best constructed entirely by high-frequency welding, with double seams, or by

```
General workwear          General workwear              Airfed          Emergency
requiring no specific     requiring further             workwear*       clothing*
further protection        hazard protection
                                                    ┌───────┴───────┐
                                                   Hoods          Suits
       ├─ Consideration   Nature of hazard
         Style ────────┐                         ┌──────────────────────────────┴──────┐
         Climatic ─────┤         ├─ Mechanical   Gas       Splash                    Fire
         Fabric ───────┤                         tight     contamination
         Cost ─────────┘           ├───────┐              ╲       ╱
                                Leather   Rubber        ┌───┴───┐
  Fabric                                               Single  Multiple
 ┌───────┬───────┐                 ├─ Heat            hazard  hazard
Cotton Nylon Polyester     ┌────┬──────┬──────┐
                          Wet   Dry   Fire    Fire
                                    approach  entry
                          └ Fabrics
                            (usually coated with aluminium or steel)
                            ┌──────┬──────┬──────┬──────┬───────────────┐
                           Asbestos Glass Kevlar Nomex Carbonised fibres

                            ├─ Chemical
                           ┌──────────┬─────┬──────┬──────────┐
                          Classification Gas Liquid Degree of
                          of chemical                exposure, i.e.
                                                     splash or
                                                     immersion
                          └ Fabrics — coated with
                           ┌─────┬──────┬─────────┬────────┬──────┬───────┐
                          PVC  Butyl Neoprene Hypalon Viton Nitrile

                            ├─ Radiation
                           ┌──────┬──────────┬──────────┐
                          Dust  High level  Low level

                            └─ Fabrics
                           ┌───────────┬───────────┬──────────┐
                          Cotton drill Impervious Absorbing
                                        fabrics    fabrics

                            └─ Biological
                           ┌───────────┬──────────┬──────┬──────┐
                          Respiratory Protection Skin  Eyes
                          fabrics
                           ┌──────────┬───────────┐
                          Disposable Sterilisable
```

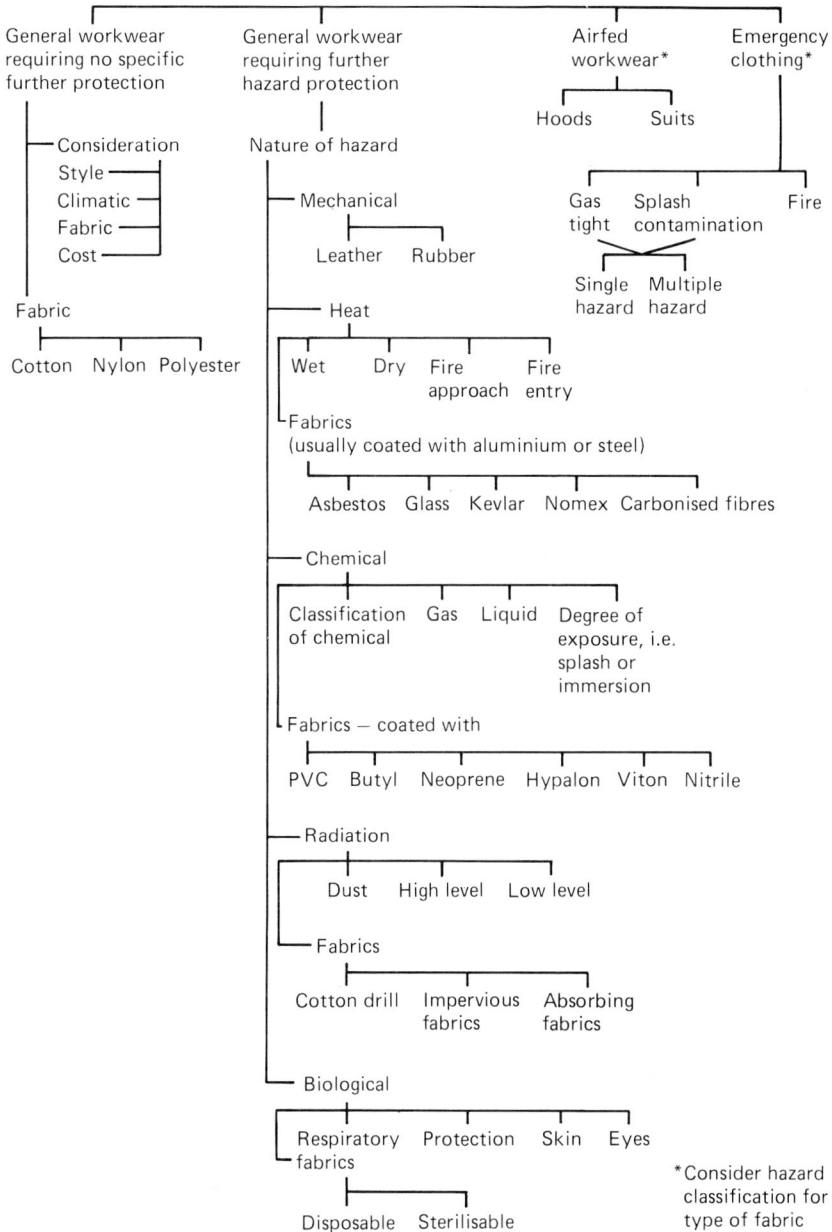

*Consider hazard classification for type of fabric

Figure 22.2 Types of workwear

welding a strip over the seams. If the fabric has to be stitched, the thread must be carefully chosen to resist the chemicals. Special adhesives and chemical methods of joining have been developed for some fabrics.

Fabrics and coatings are available for use in a wide variety of chemical environments. Nylon and polyester scrims are the most commonly used synthetic fabrics. Nomex or carbonised fibres are used to give high fire retardancy or fireproof properties. Two examples of the choice of garment for different situations follow:

1. *Transfer of acetone between storage and mixing vessels.* Virtually all possible synthetic coatings and fabrics are attacked in some way. PVC dissolves, rubbers swell, nylons and polyesters soften. But although acetone vapour is narcotic in high concentrations and liquid acetone causes defatting of the skin, the effect of occasional splashes is not serious. The clothing suggested is cotton overalls, with good ventilation.
2. *A continuous process plant where hydrofluoric acid, hydrocarbons and fluorinated hydrocarbons are present.* Of the available coatings, nylon is destroyed by the acid and plasticised PVC and butyl rubber are severely affected by the hydrocarbons, leaving the choice between Viton (very expensive), Neoprene and Hypalon. Either Neoprene or Hypalon coated on a polyester fabric would be suitable. Looking at the hazards, the eyes, skin and respiratory tract all need protection. A garment which prevents entry completely (i.e. at the wrist, ankle, face and flies) is needed. In emergency and severe contamination conditions, a gas-tight suit would be required. Two types are possible. In the first the breathing apparatus is worn outside the suit, with which the face mask forms an integral part. In the second the breathing apparatus is worn inside the suit. Other possibilities are an air-fed hood worn over jacket and trousers or one-piece overall, or a one-piece air-fed suit. The hood has the advantage of being removable without undressing, while the air-fed suit has the advantage of being cooler and freer.

There are many possible variations from which choices have to be made, e.g. the position of zips on suits and whether ankle protection is needed. ISPEMA suggests the following guidelines[9].

1. *Zip position.* The zip should normally be at the front except where the wearer is not required to dress himself or work alone, and where decontamination by a high-pressure water shower is required.
2. *Hood type.* If the overall will be worn in conjunction with breathing apparatus, the seal between the hood and the BA face mask should be flat and without drawstrings. A rubber or similar grommet is required. If breathing apparatus is not worn, a drawstring or elasticated hood is needed to keep it close to the face.
3. *Wrist seals.* The choice lies between open or elasticated wrist seals which allow the passage of contaminant and tight wrist seals which prevent it. It depends on the properties of the materials handled and the probability of their coming into contact with the hands of the wearer.
4. *Ankle protection.* The choice lies between boots which are integral with the suit, long or thigh-length boots, and boots or shoes and trousers with elasticated or straight trouser ends. If the wearer is liable to have to stand in the contaminant, integral or long boots should be used.

Contaminated chemical-resistant clothing should always be washed or hosed down before the wearer removes it and proper provision must be made for storing, inspecting and maintaining it.

22.2.6 Foul-weather clothing for outside workers

As much process plant is built in the open, operators and maintenance workers need suitable protection from bad weather. Polyurethane-proofed nylon is a reasonably rain-resistant fabric but for work in heavy rain a fabric-supported PVC which retains its flexibility at low temperatures is recommended. Suits may have a removable foam-backed lining for cold weather. Adequate and loosely covered ventilation holes are needed. A nylon zip down the front with a flyover cover secured by press studs provides a good storm fastening. The following further features should be considered[6]:

- Nothing should be worn which can get entangled with moving machinery;
- There should be no gap between jacket and trousers when bending;
- Jackets should be highly visible;
- Pockets should be secured by Velcro fastenings or covered by flaps longer than the pocket opening to prevent their being tucked in;
- The jacket sleeves should allow unrestricted arm movement;
- The collar should not interfere with the wearing of a safety helmet but should otherwise cover as much of the neck as is practicable.

22.2.7 Fall protection and rescue harnesses, etc.

Harnesses, belts, lanyards, fall-arrest devices, shock absorbers and fixed anchorages are available both for fall protection and rescue. The hazards or needs for which these are designed include:

1. Falls from heights;
2. Falls in vats, silos, tanks, etc.;
3. Rescue of persons trapped or incapacitated in tanks, silos, sewers, lift shafts and other confined and inaccessible places.

The first and, to some extent, the second of these are mainly construction hazards which lie outside the scope of this book. The third, however, is a common need in the process industries, particularly during maintenance [18.7].

There are five main types of belts and harnesses:

A   Pole belts, for the use of linesmen. These incorporate an adjustable pole strap and are better described as 'working belts';
B   General-purpose safety belts for use in conjunction with safety lanyards and anchorage points in cases where mobility can be limited;
C   Chest harnesses for use similar to (B), but giving somewhat greater fall protection;
D   General-purpose safety harnesses which incorporate thigh and shoulder straps and may be built into lightweight suits. Their use is similar to (B) and (C) and they offer the best protection;
E   Safety rescue harnesses to be worn by persons working in confined spaces, particularly those wearing breathing apparatus, and where there is a risk of being suffocated or overcome by toxic gases, etc. These should be used in conjunction with approved rescue lines. Such

harnesses, although intended primarily for withdrawal in the event of an accident, also offer some protection against falls.

Special attention is needed in the use of safety rescue harnesses and lines to ensure that the (often unconscious) victim trapped inside the confined space of a vessel, with only manhole access, can be hauled out safely by the person standing guard outside the manhole, with the equipment available. The harness itself should ensure that the victim is in a relatively upright position, and in the case of a narrow manhole, the rescue line may have to be attached to one of the wearer's wrists as well as to the harness to prevent his shoulders jamming in the manhole. Requirements may include a davit over the manhole, a block and tackle to provide mechanical advantage, and a means of securing the line to prevent the victim being accidentally dropped back into the confined space. Such rescue operations need careful rehearsal and training. The harness manufacturers will advise and usually assist with this.

Only approved types of fall-arrest devices and anchorages should be used. ISPEMA[9] gives helpful advice about fall protection and rescue equipment. These need regular inspection and servicing, and careful storage in conditions where they will not deteriorate.

22.3 Hand protection

We use our hands so much that they are injured more often than other parts of the body. Harmful substances often enter the body through the skin of the hands, particularly through cuts and abrasions. While gloves are frequently required for hand protection, they are not universal panaceas and their use when operating drills and other moving machinery is very dangerous, since gloved hands are easily snatched by projections on rotating parts and in-running nips, causing frightful injuries. The first line of defence should therefore be to provide safe systems of work where workers' hands are not exposed to sources of injury. These include guards for machinery, belt and chain drives, and the use of tongs, grips and push-sticks for manipulating hot, cold, dirty and sharp-edged objects, and ladles, spouted containers, syphons and pumps for transferring liquids.

22.3.1 Some hand hazards[10]

Serious hand injuries have been caused during lubrication of unguarded machinery, while cleaning and maintaining machinery and equipment which had not been isolated from power sources and process materials, and by failure to use an effective permit system. Precautions are discussed in 17.1 and Chapter 18.

Many injuries are caused by the use of hand-held tools such as spanners, hammers, screwdrivers and chisels, particularly when they are the wrong size for the job and when their handles are loose and defective. Portable power tools are also a prolific source of hand injuries. Safety training is essential in both cases and care is needed to ensure that defective tools are promptly removed from service and that the right ones are always available for the job in hand. Burns resulting from grasping a hot, uninsulated pipe or object are common hand injuries, from which the writer has suffered.

Rings are a menace to hand safety as they often catch on projections. Synthetic resins cause many minor injuries through getting fingers stuck together. Resin removers should be kept handy for immediate use as the resins rapidly set hard.

There are many risks of infection and poisoning through the hands. Burns carry this risk and should be treated by a nurse or doctor. Tetanus and other germs can enter the bloodstream from open hand-wounds. In all jobs where there is a risk of infection and contact with hazardous substances, all hand-wounds, however trivial, should be treated promptly and kept clean with sterile dressings.

Special hand hygiene measures are needed for workers in food and pharmaceutical industries. They should be examined regularly by a doctor for signs of skin disorders and they should wash their hands regularly with mild soap and hot water. Fingernails should be kept short and well scrubbed and nail varnish should not be allowed. Wearing of rings should be discouraged. Any cuts or abrasions on the hands should be kept covered by waterproof dressings.

There are a number of prescribed occupational diseases involving the hands and forearms. These include *acne* caused by contact with mineral oil, *beat hand* caused by repeated jarring or pressure, *skin cancer* caused by long exposure to arsenic, soot, pitch and mineral oils, etc., *chrome ulceration* caused by exposure to chromic acid, *cramp*, *dermatitis*, *mercury poisoning*, *osteolysis* caused as a result of exposure to vinyl chloride, *vibration white finger*, *vitiligo*, a patchy depigmentation of the skin caused by contact with rubber, plastics, oils and adhesives containing certain antioxidants. Of these, dermatitis is probably the most widespread. Many processes employ one or more substances which can damage the skin or cause dermatitis over a period. Workers must be warned to avoid contact with them and should wear gloves where necessary. Abrasives and solvents should not be used to clean the skin. Barrier creams applied before work make it easier to remove most skin contaminants with soap and water, while skin creams are useful after dirty work to replace natural oils and fat. UV-light-absorbent barrier creams protect the skin against sunlight and radiation from arc welding.

22.3.2 Gloves, etc.

Several types of gloves, gauntlets and mittens in a variety of materials are available to protect the hands from corrosive chemicals and other harmful substances, as well as abrasion, cuts, extremes of temperature and high voltages. In some cases they are required by regulations, in others as a result of discussions between management and workers. Even when gloves fit well and are light and flexible, a certain manual dexterity and delicacy of touch are lost so that they only tend to be worn where the need is clearly felt. While cut and sewn gloves are often preferred for general handling, welding and high-temperature applications, dipped seamless rubber and PVC gloves give better protection against liquid penetration. These, however, do not allow the skin to breathe, so that the skin becomes clammy when wearing them. Chemical resistance charts published by the manufacturers should be consulted to find the best material for different exposures, and BS 1651 on the types of glove for different duties.

PVC gloves are available in various types. Smooth-surfaced gloves protect against acids, alkalis, chemicals, solvents, oils and fats. Gloves with a matt surface coating provide a better grip and abrasion resistance. Cotton linings increase comfort and absorb sweat. Anti-bacteriostatic PVC gloves are available which reduce dermatitis risks. The PVC used for gloves generally contains a plasticising liquid to make them flexible. This is leached out by many organic liquids, which makes the gloves hard and brittle.

Natural and synthetic rubber gloves are available in a range of thicknesses, both lined and unlined. Ultra-lightweight rubber gloves are used for delicate operations and food processing while reinforced and abrasion-resistant rubber gloves are worn for heavy duties. Neoprene, nitrile, butyl and Viton rubber gloves have good though selective resistance to solvents, oils and chemicals. Special natural rubber gloves which are made for electrical work (BS 679) need retesting at regular intervals.

Traditional gloves made from leather and cotton are hardwearing, supple and strong and have good heat resistance. Loop-pile fabric gloves give good protection against sharp edges, rough surfaces and extremes of temperature. Gloves and mitts made of glass and other inorganic fibres have very good heat resistance. Newly developed composite fabric gloves offer protection in a wide range of environments. Metal chain-mail gloves as used by butchers give good protection against injury from sharp tools. Disposable PVC, polyethylene and rubber gloves are also available.

Some rubber and leather gloves contain compounds (chrome used in tanning leather and accelerators and antioxidants used in rubber manufacture) which may cause dermatitis in sensitive persons.

22.4 Head protection

Special headwear is commonly provided for industrial workers, both to protect against particular hazards and, by its colour and special markings, to provide a means of easy identification of the wearer or his trade, status or group. In the oil and petrochemical industries, where most plants are built in the open, safety helmets are widely used. For work inside buildings, other types of headwear are often required, especially to contain long hair and prevent contamination of process materials with human hair, dandruff, etc.

22.4.1 Safety helmets

Safety helmets are provided to protect against:

1. Impact from falling objects;
2. Exposure to abnormal heat (e.g. in metal smelting);
3. Chemical splashes (e.g. from fractured overhead pipes);

4. Contact with high-voltage sources (a risk largely confined to electrical workers);
5. Risk of sideways crushing.

A safety helmet also contains long hair and prevents its entanglement with machinery, and like normal headwear, it protects the wearer from rain and extremes of temperature and also shades the eyes. A safety helmet consists of a strong outer shell which is supported clear of the head by a harness and adjustable headband to ensure a sound fit. Two main types are available:

- 'Hats' with an all-round brim (preferred for risks of chemical splashes);
- 'Caps' with a peak at the front only (preferred in confined spaces);

Three main types of material are used:

- Thermoplastic – ABS, HD polyethylene and polycarbonate;
- Fibreglass reinforced plastic (GRP);
- Aluminium alloy (now less common).

Harnesses are available in plastic and a combination of leather, plastic and textile tapes. Some of these require the use of a chin-strap.

The helmet type and material should be chosen to suit the main hazards likely to be encountered by the wearer. Laminated and fibreglass helmets are most satisfactory against extreme heat, which may deform some plastic ones. A brimmed hat is obviously preferable against chemical splashes, but its material depends on the nature of the chemicals liable to drip on the wearer. Where risks of contact with high voltages exist, metal helmets must be ruled out and the helmets chosen should pass an electrical insulation test. Where there is a risk of sideways crushing, a helmet of high lateral rigidity should be worn.

Some safety helmets provide support for other protective items such as eye and face shields, breathing apparatus and ear-muffs. When purchasing helmets one should ensure that any such items liable to be needed can be fitted to them and be bought from the same supplier. Warm linings, neck, ear and forehead protectors which are supplied as optional extras may be needed in cold weather.

22.4.2 Other safety headwear

This includes:

- Welders' helmets to which visors or face shields are attached and which are generally lighter than industrial safety helmets (Figure 22.3c);
- Acoustic helmets for those exposed to very high noise levels, i.e. above 115 dB'A';
- Bump caps and scalp protectors for protection against bruising and abrasion in confined spaces;
- Caps and hairnets to prevent long hair catching in moving machinery;
- Cape hoods and 'sou'westers' as part of 'foul-weather clothing';
- Shot-blasting helmets;
- 'Air-stream helmets'.

The last type has a hinged face shield and an electric blower powered by a battery worn on the belt. This forces air through a filter in the helmet and blows it over the face of the wearer, giving some respiratory protection against dusts.

22.5 Standing work and foot protection

Since practically all foot injuries occur while we are on our feet, one obvious way of reducing them is to minimise the amount of time we spend standing. Those obliged to stand for many hours are also prone to foot discomfort which, when acute, distracts their attention and makes them more accident prone. The degree of mechanisation in most process industries has reduced the number of jobs which require continuous standing. Skilled workers such as control room operators who spend much of their time sitting at control stations are then in little danger of foot injury. There remain many jobs where people have to stand and sometimes lift and carry hard and heavy objects which could, if dropped, injure unprotected feet. There are other jobs where workers may have to walk in oil and chemical spillages. Many aches and injuries are caused by the use of unsuitable and faulty footwear and flooring, and the adoption of unnatural body postures such as bending continuously over a low workbench while standing with the feet together.

22.5.1 Perils of standing work[11]

We were never designed to stand on our hind legs, while the human foot is constructed for mobility and not for maintaining the upright stance with muscles contracted. Women suffer the further disadvantage that fashion footwear with high heels and pointed toes throws all of the body weight on the forefeet, gives little support when standing and forces the big toe towards the centre. Hence many women's feet are deformed by middle age.

A healthy person with sound feet and footwear merely feels tired after spending a day standing on resilient flooring. Many reasons such as poor general health, advancing years, corpulence, lack of exercise and years of wearing badly fitting shoes cause fallen arches and deterioration in foot health which turn mere tiredness into agony.

Shoes or boots should be fitted by a qualified shoe fitter and should have thick resilient soles or cushion insoles to mitigate the impact of walking on hard floors and reduce heat loss from the feet. They should be kept in good repair and both heels and soles replaced before they are badly worn, particularly if the wear is uneven and mainly on one side. Flooring on which people have to stand for long periods should have good resilience and insulation as well as other qualities demanded by the process conditions. Anti-slip matting which is resistant to wear, water and many chemicals is useful in mitigating the hardness of concrete floors.

22.5.2 Safe footwear

When considering foot safety and the provision of safety footwear, one should think not only about the foot hazards of the job but also about the footwear which workers would otherwise use at work. The latter is often badly worn and sometimes quite unsuitable. If persuasion fails, it may be necessary to ban certain types of footwear at work.

Many firms now use a scheme such as that run by the Golden Shoe Club which enables some or all employees to purchase from a wide range of economical and well-designed safety footwear at a discount by monthly deductions from their wages. When selecting a supplier one needs, however, to check whether the range of widths (as well as lengths) offered is adequate.

Some shoe manufacturers send mobile vans to factory sites with a range of footwear which employees can try on and keep if satisfactory. This overcomes the problem of having to order safety footwear from a supplier, with no assurance that it will fit. When safety footwear is made available through employers, they should arrange for it to be inspected, serviced and repaired or replaced when worn or damaged.

Most safety footwear is designed to protect the toes from impact or crushing. Other hazards against which safety footwear is available include penetration through the mid-sole and upper, oil and chemical penetration, slipping, heat, and metal turnings. Antistatic rubber footwear has sufficiently high electrical resistance to protect against electrocution hazards while allowing electrostatic charges to dissipate harmlessly. Electrically conducting rubber footwear is available for work with sensitive explosive and unstable compounds. Special footwear and gaiters are available for foundry work where there is a legal obligation to provide and use them.

Several government establishments and industrial and commercial undertakings have, with union backing, made the wearing of safety footwear a condition of employment for certain jobs.

22.6 Vision and eye protection

Defective vision and inadequate lighting are fertile causes of mistakes and accidents, while the eye is such an important and sensitive organ that it is vital to protect it.

22.6.1 Vision and lighting

For jobs requiring acute visual perception, e.g. the reading of instruments and the inspection of plant and equipment for wear and corrosion, it is important to ensure that both the worker's eyesight and the illumination provided are more than adequate. The first step is to classify the lighting standards required for the various jobs and check that the existing or proposed lighting is satisfactory. The survey needed for this often reveals other factors such as poor colour contrast or persistent steam leaks which impair visual perception and require correction. The second step is to submit employees whose work requires high visual standards to a sight test by a qualified optician. Only workers whose sight meets the standards required for critical jobs should be employed on them.

Recommended illumination levels and minimum glare indices for different working areas and jobs are given by the Illuminating Engineering Society[12]. General lighting throughout an area should be combined with supplementary lighting in critical areas. Emergency lighting powered by batteries or a generator which automatically comes into operation when the power fails is also essential. An adequate electrical supply of constant voltage is vital for good lighting. The luminaires, reflectors, windows and walls must be kept clean and well maintained. Light-coloured matt surface finishes are recommended. Contrasting colours aid easy recognition of pipework, gas cylinders, danger spots, etc., but the colours chosen must be easily distinguishable in the lighting used. Sudden changes in lighting intensity between different areas and buildings should be avoided. The effects of beams of strong sunlight on illumination levels should be minimised. For those wearing spectacles and other forms of eye protection, steps are needed to prevent accidents resulting from misting of lenses when entering a warm and humid zone from a colder one. Industrial lighting often requires specialist advice and should not be left entirely to architects and electricians.

22.6.2 Eye protection

The eyes are vital organs which must be protected against various hazards. Adequate facilities must also be available at strategic points for treating eye injuries. The most basic of these is the eye-wash fountain which uses plain water. Specially prepared solutions prescribed by the works doctor for dealing with any anticipated eye contaminants may also be provided (e.g. in first-aid boxes).

In the UK the Protection of Eyes Regulations 1974[13] make employers responsible both for providing any necessary eye protection for their workers for a number of tasks and for replacing it when it is lost, damaged or destroyed. Such eye protection must conform to the appropriate specification and be marked to show the hazards against which it protects (Table 22.1).

Table 22.1 Obligatory marking of approved industrial eye protectors in the UK

For protection against	*Marking*
General hazards	BS 2092
Chemicals	BS 2092 + 'C'
Dust	BS 2092 + 'D'
Gases	BS 2092 + 'G'
Impact – Grade 1	BS 2092 + '1'
Impact – Grade 2	BS 2092 + '2'
Molten metal	BS 2092 + 'M'
Glare	BS 1542/647

The main forms of protection are safety spectacles, safety goggles and shields (Figure 22.4). Before making a selection it is essential to survey the eye hazards to which each worker or group of workers will be exposed. Invisible radiation, i.e. infra-red, ultra-violet, X-rays and laser beams, poses special dangers since the eye damage caused is not apparent until some time after the exposure. Specialist advice (which large suppliers of eye protectors can provide) is essential to ensure that the protection matches the hazard.

Safety spectacles, while generally dearer than goggles, are more comfortable and readily accepted, but because they only rest on the bridge of the nose, they offer less protection against impact. They also offer far less protection against gases, vapours, fine dusts and mists which affect the eyes.

Safety spectacles have toughened glass or plastic lenses in various tints, with plastic or metal frames, and may have ventilated side shields for protection against low-energy droplets and particles. Each person requiring safety spectacles should be fitted and provided with one or more pairs for his or her personal use only. The lenses of safety spectacles tend to be thicker and heavier than those of normal ones. Plastic lenses which are lighter than glass ones are thus generally preferred, although they are softer and more easily scratched.

For those who normally wear spectacles with prescription lenses for vision correction, safety spectacles with lenses to similar prescriptions should be ordered, and fitted professionally.

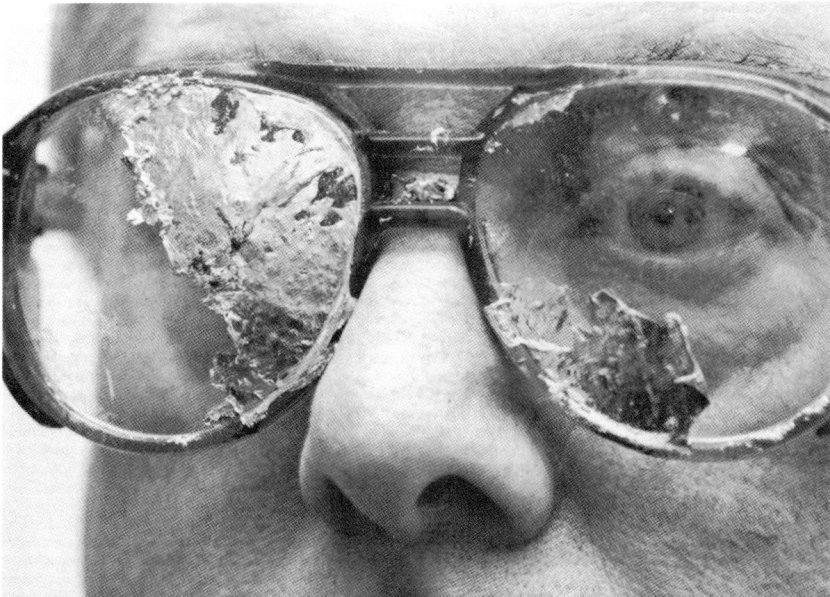

Figure 22.3 (a) Safety spectacles saved employee's eyes from splash of molten zinc

Figure 22.3 (b) High-impact goggles attached to safety spectacles (courtesy Pulsafe)

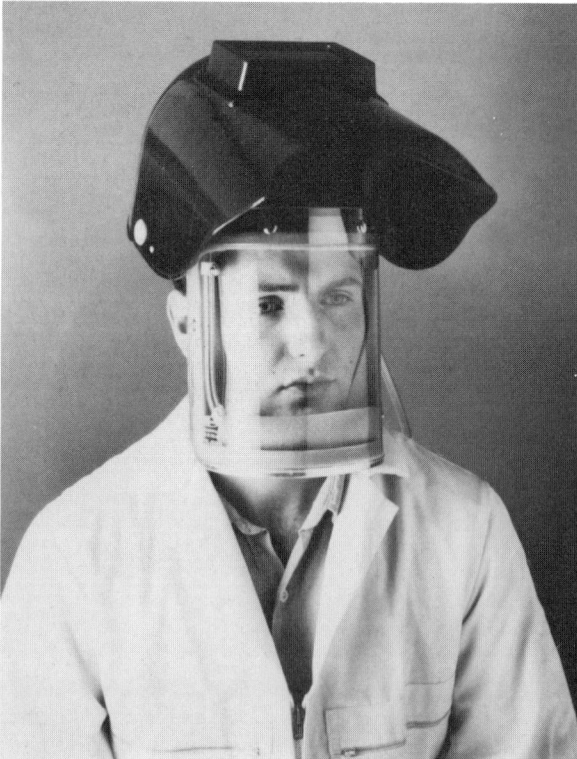

Figure 22.3 (c) Air-supplied welding visor (courtesy Pulsafe)

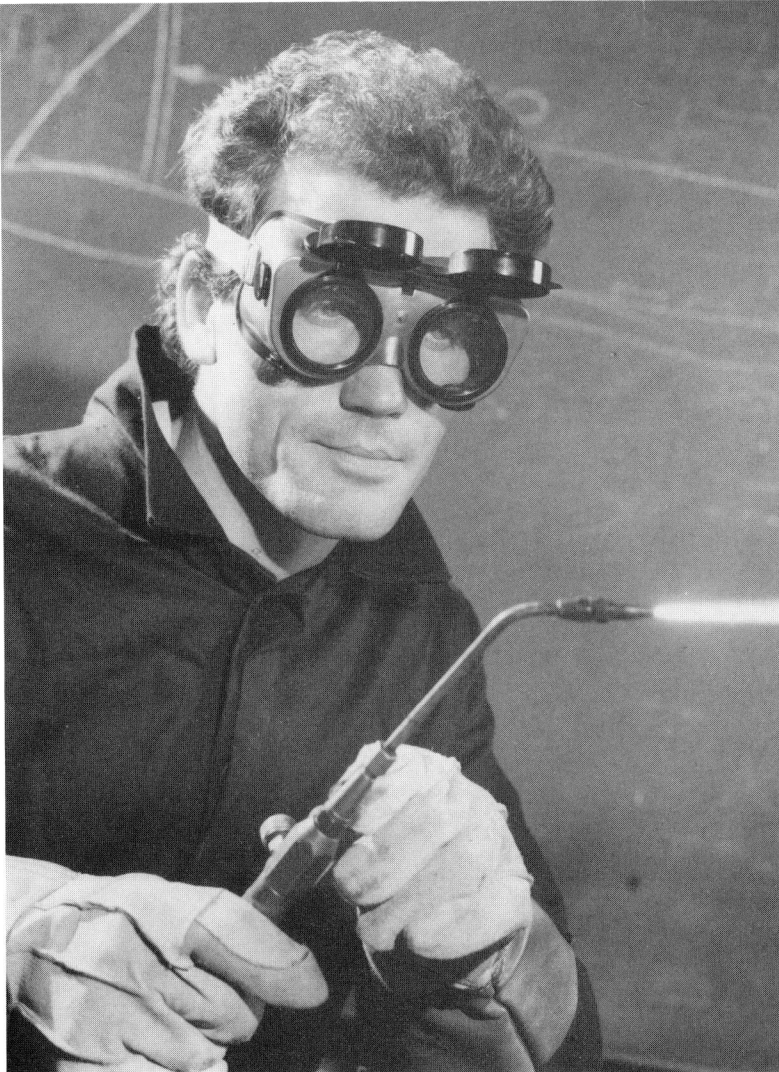

Figure 22.3 (d) Flip-up goggles with gas-welding filter (courtesy Amigo)

A problem arises when safety spectacles have to be worn as well as circumaural ear-muffs, or respiratory protection with full face masks. With normal spectacle side pieces which extend over the ears, an effective seal between the ear-muff or face mask and the side of the face is virtually impossible. Spectacles can be made with special frames to overcome this problem.

Safety goggles which enclose the eyes completely, with or without ventilation, and which are kept in position by an elastic headband, provide more complete protection than spectacles but are less comfortable to wear.

They are not usually supplied with prescription lenses, although some goggles can be worn over spectacles. Conditions for which goggles may be considered but where spectacles are generally inadequate include:

- Gas welding and cutting,
- Work in dusty atmospheres,
- Work with injurious chemicals, particularly when these may be present as fine sprays or mists,
- Work with toxic and irritating gases,
- Work where impact from large particles is possible.

Goggles are classified either as 'box type' with one-piece protective lenses or 'cup type' which enclose each eye with individual protective lenses. Goggles are also classified as 'gas tight', 'dust' and 'chemical'. The inner surfaces of most goggle lenses tend to mist up. Easily applied hydrophilic coatings are available to alleviate this problem.

While some types of goggles and gas masks or respirators can be worn together, a wider range of hazards can be covered by the use of breathing apparatus with a full face mask or a complete protective suit.

Transparent shields of various types are available. Lightweight disposable plastic eye shields (which are sometimes called goggles) offer a fair degree of all-round protection and can be worn over normal spectacles. They are often issued to visitors to factories. Rigid face shields which protect the face and eyes are widely used, especially by welders. They can be hand-held, attached to the helmet, strapped to the head or fixed between the operator's head and his work.

22.7 Noise and hearing protection

Noise can be hazardous by impairing human hearing, by interfering with communication and by reducing morale and general awareness. It is discussed in detail in a UK code of practice[14] and in several books[15] including my earlier one on construction hazards[23]. Apart from construction and maintenance activities, excessive noise in the process industries is mostly found in places such as boiler houses, compressor rooms and where fluids under pressure are discharged to the atmosphere or flared.

Most jobs in the process industries require good hearing which can be tested speedily (Figure 22.5). Hearing records should be kept. Periodic retesting of workers' hearing may reveal excessive exposure to noise at work.

22.7.1 The (UK) Noise at Work Regulations 1989

These regulations, which result from an EC directive, aim at protecting the hearing of workers from damage by noise. Their first requirement on employers is to assess the problem in noisy areas, e.g. where people have to shout or have difficulty in being understood by someone about 2 m away. The assessment should be made and recorded by a competent person. Noise levels should be reduced as far as practicable to below 90 dB(A), e.g. by fitting silencers and enclosing noisy machinery in

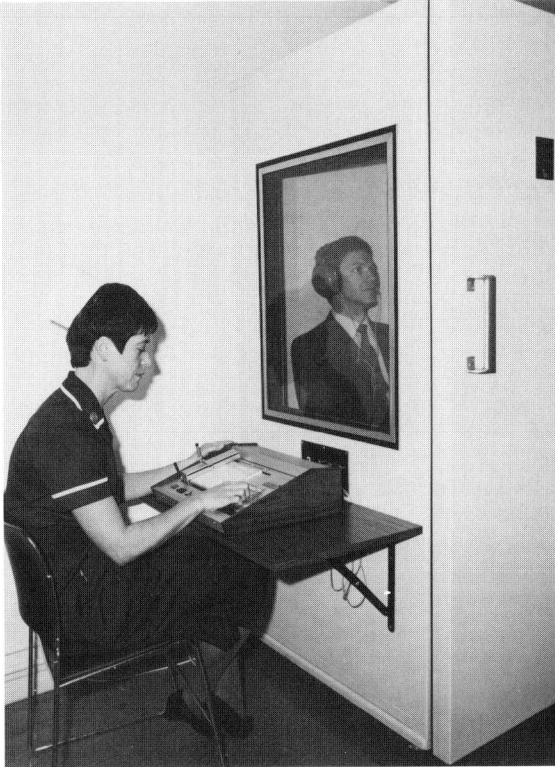

Figure 22.4 Audiometer for measuring employee's hearing (courtesy P. C. Werth Ltd)

acoustic hoods. Noise emissions of new machinery should be checked on purchase.

Managements should restrict the number of workers in zones with noise levels above 85 dB(A) to a minimum, inform them of the risk, and provide workers in zones with noise levels between 85 and 90 dB(A) with ear protectors if they ask for them.

In zones where noise levels exceed 90 dB(A), warning notices should be displayed and all working in them should be provided with ear protectors and trained in their use. Employees are obliged to wear ear protectors when in such zones. Employers are obliged to maintain machines, etc., in these zones to prevent noise from increasing and to maintain the ear protectors provided. Actions indicated at 90 dB(A) also apply where peak sound pressures may exceed 200 Pa.

Makers and suppliers of machines are obliged to provide information on the noise they are likely to generate.

22.7.2 Hearing protection

Ear-plugs (inserts) are mainly for workers continuously exposed to noise levels between 90 and 100 dB(A). Ear-muffs give protection to noise levels up to 120 dB(A). For still higher noise levels acoustic helmets are available. Since ear protectors cause a communication problem, care is needed when selecting them not to accentuate it by overprotecting the wearer.

Ear-plugs are worn in the ear canal, sealing the entrance to the ear. Some are conformable and allow the plug to be compressed before insertion in the ear, where it expands to give a comfortable fit. Others which are premoulded to a predetermined shape are available in a range of sizes. Some have a valve system which, it is claimed, absorbs high noise levels while transmitting speech sounds. There are also semi-inserts attached to a headband which keeps them in the right position.

Disadvantages of ear-plugs are:

- It is difficult for a supervisor to check whether they are being used;
- There is a hygiene problem when a worker decides to fit ear-plugs when his hands are dirty;
- The conformable type tend to be displaced if the wearer moves his jaws sharply.

Ear muffs consist of two hard cups which fit over the ears, foamed plastic or rubber cushions which fill any gaps between the cup and the head, and a semi-rigid headband which keeps the cups and cushions in contact with the head. This can be worn over the head, behind the neck or under the chin. The cups may also be attached to some safety helmets by adjustable side-arms. The cushions are liable to degrade from mechanical abuse or sweat from the wearer and therefore need regular inspection and replacement.

Ear muffs are of two types, circumaural and superaural. The former, which enclose the ears, are commoner and more effective except where spectacles with normal side-arms are worn. The latter, which are lighter, seal against the ears themselves and are less affected by spectacle frames.

22.8 Breathing and respiratory protection

The air which has surrounded the earth for millions of years and made life possible has a remarkably uniform composition, consisting of approximately 79% nitrogen and other inert gases, 21% oxygen and 0.03% carbon dioxide by volume on a dry basis, with varying amounts of water vapour. It also contains many different trace contaminants such as oxides of sulphur and nitrogen, carbon monoxide and hydrocarbons which are mainly man-made and whose concentrations vary widely locally, depending on the closeness to their source.

22.8.1 Respiration and fresh-air requirements

Respiration converts oxygen in the air to carbon dioxide, the maximum concentration of which should not exceed the OEL of 0.5% in the breathing zone of the worker. This sets a minimum fresh-air requirement per person of about 2 litre/s for respiration, depending on the person's rate of activity. The minimum air space per employee required under the Factories Act 1961 is $11.5 \, m^3$, when a complete air change every two hours would be needed for respiration only. However, an air velocity inside buildings of 0.1 to 0.15 m/s at normal temperatures is recommended for comfort. Taking this and the need to remove body odours and water vapour into account, HS guidance note EH 22[16] recommends a minimum *fresh* air rate per person of 8 litre/s in factories, although much higher rates may be required if harmful substances, tobacco smoke, heat and/or combustion products are released into the working atmosphere.

22.8.2 Ventilation[16]

Whenever possible, fumes and other air contaminants released from industrial processes should be contained or controlled at source by local exhaust ventilation. This should be so arranged that contaminants are swept away from the breathing zone of the worker. When local exhaust ventilation is used, care must be taken to discharge the contaminated air so that, whatever the wind direction and atmospheric condition, it does not re-enter the fresh-air intake of the building or that of other buildings, or cause a hazard or nuisance in the outdoor environment. Dust, fume and toxic gases and vapours may have to be removed from the exhausted air by filtration, scrubbing, adsorption or catalytic treatment before it is released to the atmosphere. Fawcett quotes an example where charcoal bed adsorbers had to be installed to reduce the ethylene dibromide content of air exhausted from a building to 1 ppm to meet environmental requirements[17].

There are two principles of general ventilation – dilution and displacement, although most ventilation uses a mixture of both. In dilution ventilation the incoming air, generally moved mechanically, is introduced at a high enough velocity into the room or building to create turbulence and mix completely with the air inside it before leaving. Mixing and dilution also occur in mechanical ventilation systems which recycle part of the air back into the building to conserve heat. General dilution ventilation should be so designed that contaminants released into the atmosphere are continuously diluted to below the OEL at points where operatives are stationed.

In displacement ventilation cool incoming air enters the building at low level at a low velocity, displacing warm contaminated air which is discharged at ceiling or roof level (Figure 22.5). This may require an independent source of heat if there is no process or other source of heat in the building. It appears to be more efficient than dilution ventilation in removing gaseous contaminants of low density and others which mix rapidly with air at their point of entry, but it is probably less effective in removing suspended particles and droplets, and pockets of heavy gases and vapours. Ventilation of buildings where there are processes liable to discharge harmful contaminants into the atmosphere requires careful study, using smoke generators, tracer gases and direct measurements of air flow[18].

Chapter 7 discusses how toxic air contaminants escaping from industrial processes can enter the human body through the respiratory system. It also emphasises the importance of preventing their entry into workplace atmospheres, of monitoring the latter for their presence, and the legal obligations on employers to ensure that this is done. While respiratory protective equipment is less used than most forms of PPC/E, when needed it can make the difference between life and death.

Hot, fume laden air rises to
roof where it accumulates or
escapes through roof vent

Fume which does
not escape may hang
in stratified layer

Hot
process

Cool air enters
at low level

Warm
contaminated
air discharged
at ceiling or
roof level

Warm air is
displaced
and rises

Cool air enters
at low velocity
and flows through
building

Inlet air
diffuser

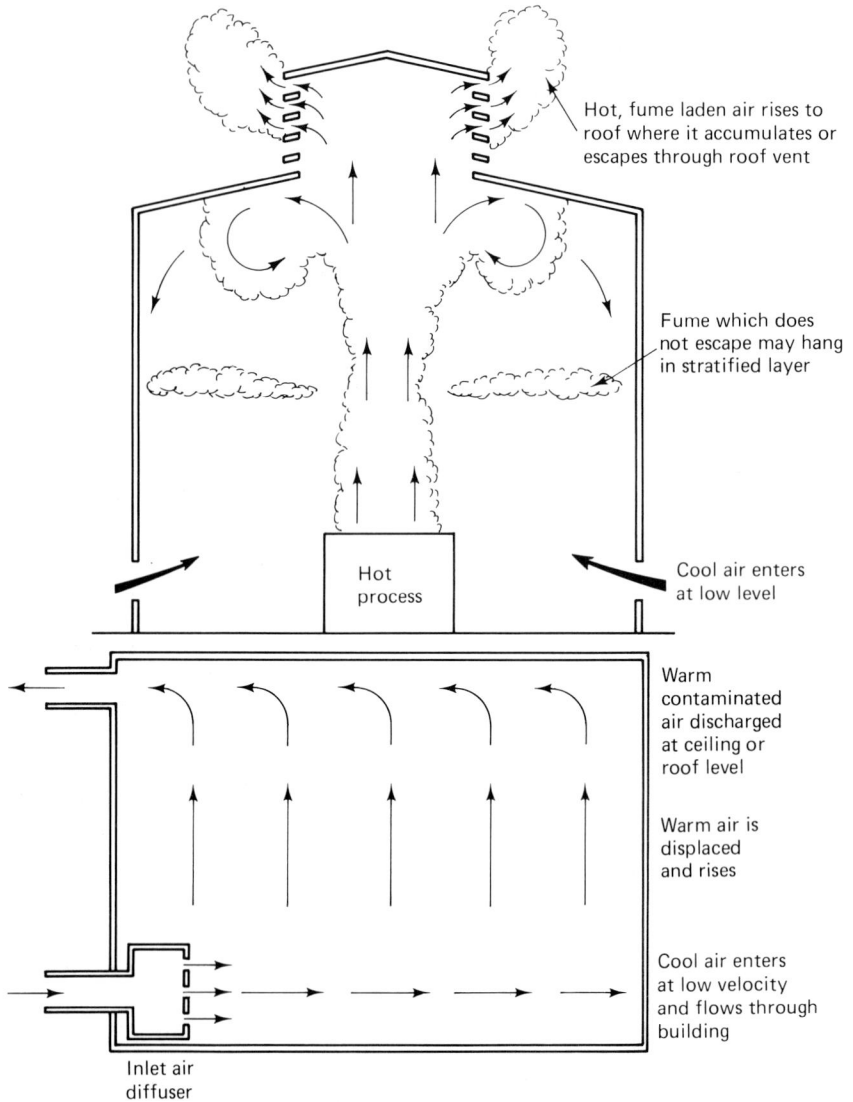

Figure 22.5 Displacement ventilation

22.8.3 Hazardous atmospheres

Atmospheres may be hazardous for one or more of three reasons:

1. Oxygen deficiency;
2. Contamination by hazardous dust, mist or fume;
3. Contamination by toxic gas or vapour.

The degree of hazard, which can vary widely in each case, can be classified crudely as to whether or not it would present an immediate danger to the life of someone without respiratory protection. In the case of oxygen-deficient atmospheres, this applies when the oxygen content is below 16% by volume. Oxygen-deficient atmospheres arise mainly in pits and confined spaces (tanks, vessels, silos, vats, cellars, etc.) either because oxygen in the original air has reacted (e.g. by combustion or with iron sulphide scale in a tank) or because the air has been displaced by some other gas such as carbon dioxide or nitrogen. Personal entry into oxygen-deficient atmospheres should be undertaken only by trained persons such as firemen in extreme emergencies. Otherwise the zone where the oxygen deficiency occurred should be cleared of harmful contaminants, swept through with clean air and its oxygen content tested and checked as normal before anyone is allowed to enter [18.6/7].

22.8.4 Respiratory protective equipment (RPE)

The great variety of respiratory hazards and of RPE available can make the choice of appropriate RPE quite difficult. UK users should, where applicable, follow the recommendations given in BS 4275: 1974[19], and refer also to the current Certificate of Approval for RPE issued annually by HSE[5]. This gives a list, broken down into classes (now 12), of commercially available RPE, with the names and addresses of their suppliers, and conditions of use for which they are approved. Not all classes in this list are quoted here. American users of RPE are referred to standard ANSI Z88.2[20] and to a chapter by Fawcett[17] which includes a list of the types available, the protection provided by each, and the conditions for which they are approved in the USA.

RPE are of two main types: respirators and breathing apparatus.

Respirators depend on the oxygen in the air surrounding the wearer for respiration and provide means for removing the hazardous contaminants from the air before it is breathed. They may not be used in oxygen-deficient atmospheres or in ones which present an immediate danger to life, and the type to be used depends on the contaminants found. If dusts are present a filtering type of respirator is needed, but if there are hazardous gases or vapours the respirator must contain an appropriate adsorbent or chemical which will remove them. Some respirators contain both a filter and an adsorbent. Some types depend on the lungs of the wearer to draw in the air through the purifying device while others include a portable powered fan or blower to do this.

Breathing apparatus provides an independent supply of air or oxygen for the wearer and does not depend for oxygen on the surrounding atmosphere. It can thus be used for a wider range of hazards than respirators and some types have been approved for use in atmospheres which pose an immediate danger to life. There are two main types:

- Self-contained breathing apparatus which has its own portable supply of air, oxygen or an oxygen-generating chemical, and which allows unrestricted movement to the wearer for as long as the supply of oxygen lasts. It is principally used by firemen and in certain types of escape apparatus.
- Air hose and compressed air-line breathing apparatus in which clean air is continuously supplied to the wearer by a flexible hose from an outside source (Figure 22.1). These limit the movements of the wearer and are used for more routine jobs such as paint spraying and tank cleaning.

Both respirators and breathing apparatus require a means of connection to the nose and mouth of the wearer which prevents ingress of the contaminated atmosphere. This may be done in five ways:

1. Ori-nasal (half-mask) facepieces which cover the nose and mouth only, and whose protection is limited by the difficulty in making a good seal on the varying face contours;
2. Full-face masks which cover the nose, mouth and eyes and make a much better face-seal possible, besides providing eye protection;
3. Hoods which cover the whole head;
4. Combined hoods and blouses which cover the upper part of the body;
5. Full protective suits which cover the entire body (Figure 22.1).

Most facepieces have a simple one-way valve for discharging the expired air to the atmosphere. There is a danger in very cold weather of moisture condensing and freezing in the valve, causing it to fail. All respirators and breathing apparatus with facepieces require careful fitting to the wearer by a skilled person. The headstraps should be adjusted until the facepiece fits closely and comfortably. With cartridge and canister types of respirator, the seal should be temporarily removed to check that the wearer can breathe naturally without excessive resistance to inhalation. To check for leakage of air, the air supply should be closed and the wearer should inhale deeply until the facepiece collapses, when it should remain collapsed until the air intake is reopened. The wearing of spectacles, beards and side whiskers can make an effective face-seal impossible, while face-masks for every shape of face are not always available.

The transparent windows of many full-face masks are prone to fogging through condensation of moisture in the breath. This tendency is avoided with some full-face masks which have a double face-seal. This is so contrived that the inside of the window is exposed only to the incoming air.

Speech is impossible with most facepieces so that other forms of signalling have to be arranged between the wearers and their colleagues. Some facepieces incorporate a speech diaphragm, which transmits a distorted message. The act of speaking is liable to disturb the face-seal and cause inward air leakage. Special facepieces for self-contained breathing apparatus used by firemen, etc. incorporate a microphone which is connected to a portable transmitter.

The degree of protection given by any form of RPE to any contaminant is defined by the nominal protection factor (NPF). This is the ratio of the concentration of the contaminant in the atmosphere to that inside the facepiece. It is affected both by the leakage of contaminated air past the facepiece to the nose and mouth and in the case of respirators by the efficiency with which the contaminant is removed. Crockford[21] discusses NPF air leakage and how to assess this in practice. Half-mask facepieces give variable and generally low NPFs (sometimes as low as 5) but full-face masks can readily give NPFs of 100 or more, although leakage is still possible. Breathing apparatus and powered respirators which maintain a slight positive pressure all the time inside the RPE give higher values of the NPF than those where there is a negative pressure inside the RPE while the wearer is inhaling. Quite high air flows are, however, needed to prevent the development of negative pressures when the wearer breathes in if at the same time he is doing hard physical work.

22.8.5 Respirators

Where there is doubt about the effectiveness of a respirator, breathing apparatus should be used. The simplest types of respirator have disposable filtering half-mask facepieces made entirely of porous material through which air is both drawn in and forced out by the lungs. While providing suitable protection against most low-toxicity 'nuisance dusts', their efficiency is low both for very small particles and also generally because of face-seal leakage.

Many respirators with full and half-mask facepieces for harmful dusts, gases and vapours are available in which incoming air is drawn by the lungs through a disposable cartridge attached to the facepiece, or a larger canister attached to the body and connected to the facepiece by a large-diameter rubber tube. The facepiece also has a non-return exhalation valve. The cartridge or canister may contain a layer of chemical and/or of a material of large internal surface area such as charcoal to remove harmful gases and vapours. Some also contain a filter and some only a filter. Colour-coded cartridges and canisters containing different chemicals and adsorbents are available for different gases and vapours such as chlorine, ammonia and chlorinated hydrocarbons. Canister types are generally more effective than cartridge types and can be used in higher concentrations of contaminant. All canisters and cartridges have a limited life and should be kept sealed when not in use. Unless there is some means of testing cartridges and canisters which have become partly exhausted, they should be discarded and replaced after use.

High-efficiency respirators which depend on the lungs of the wearer, with full-face masks and filters for protection against toxic dusts such as asbestos and radioactive dusts, are also available.

Several types of positive-pressure powered respirators in which filtered air is supplied from a power-pack carried by the wearer are available. The power-pack reduces face-seal leakage and cools the wearer in hot conditions.

Positive-pressure powered hoods and blouses which employ a similar power pack and air filter are also available. Exhaled and surplus air is vented through apertures in the hood or blouse. Hoods are less efficient than facepieces since wind and body movements can cause contaminants to enter them. Blouses are somewhat better in this respect.

Emergency escape respirators for once-only use are available which protect against specific air contaminants. Their main use is in mines but they may offer an alternative in some plants to escape breathing apparatus [22.8.6].

22.8.6 Breathing apparatus

There are five main types of breathing apparatus:

1. *Self-contained closed-circuit breathing apparatus* in which the exhaled air is passed through a chemical which replaces its carbon dioxide with oxygen. It is mainly used by professional firemen and rescue workers.

Figure 22.6 Positive-pressure escape set (courtesy Draeger Ltd)

2. *Self-contained open-circuit breathing apparatus* which supplies air on demand (through a lung-governed valve) or at constant flow (through a pressure reducer and flow controller) to a full-face mask with an exhalation valve. With this type, which is mainly used professionally, a limited supply of compressed air is contained in a cylinder attached to the wearer's body.

3. *Compressed air-line breathing apparatus* which is routinely used in industry for tasks such as tank cleaning and paint spraying where it is impossible to provide a safe respirable atmosphere. Compressed air-line breathing apparatus is similar to open-circuit apparatus, with a small-diameter air hose from a safe compressed air source replacing the compressed air cylinder (Figure 22.1). It may also be used with hoods or blouses, but is not recommended in circumstances where the worker's life would be in jeopardy should the air supply fail.

4. *Fresh air hose breathing apparatus* in which air suitable for respiration is drawn from an adjacent uncontaminated area through a larger-diameter hose by the breathing of the wearer, with or without a blower. The latter is needed when the hose length exceeds 9 m or when the atmosphere is immediately hazardous to life. In this case the wearer should also wear a rescue harness attached to a rescue line held by a second person located at the other end of the hose who is able to haul out the wearer in an emergency [22.2.7].

5. *Escape breathing apparatus*, a simple form of self-contained breathing apparatus for short-duration use which is usually of open-circuit type, with its own small compressed air cylinder (Figure 22.6). This is needed in works where there is a possibility of a large escape of toxic gas such as chlorine or ammonia which could trap workers in an emergency. It may also be required where there is a risk of their being trapped by smoke and fumes from a fire. Arrangements for the provision, storage and training in the use of escape breathing apparatus need to be carefully considered in conjunction with emergency planning [20.3.7].

22.9 Other personal hazards

The process industries, while having some hazards which are largely peculiar to them, are not exempt from common ones which affect workers throughout industry. Some of those not treated elsewhere in this book are discussed here briefly. Ropes, cranes, lifting tackle, forklift trucks and other wheeled transport are dealt with elsewhere[22].

22.9.1 Poor housekeeping

This can be especially dangerous where chemicals and volatile flammable liquids are used. These materials should only be stored and handled in authorised and clearly marked containers which are kept closed. Methods to be used for dealing with spillages of particular hazardous liquids and accumulations of dangerous dust must be clearly stated in the standing instructions and followed promptly. Operations involving hazardous dusts should only be allowed where there is proper local exhaust ventilation and other facilities needed to protect the worker. Regular inspection is needed to ensure that good housekeeping and regular cleaning are enforced. This should include inspection of concealed nooks and places such as high ledges which are outside the normal field of vision (Figure 17.3). Clear responsibilities for the receipt, storage and issue of materials used in processes must be established. Proper arrangements for inventory control, records and security must be made.

22.9.2 Manual handling and the slipped disc[6]

There is less manual handling in the process industries than in most others, but the unusual load being carried by someone for the first time often causes injury. Training in lifting and carrying, in good body postures, and the hazards of the slipped disc should be given to all workers. Suitable posters to illustrate these points (Figure 22.7) should be displayed where all will see them.

Figure 22.7 Good and bad body postures

22.9.3 Falls and falling objects

Falls of persons are usually classified in five ways: falls at the same level, through floor openings and into pits, etc., from ladders, through roofs and from high working places.

Common underlying causes of falls at the same level are:

- Poor or badly maintained floor surface;
- Obstacles on floor;
- Poor lighting or sudden changes in lighting;
- Worn or unsuitable footwear;

- Poor eyesight or health;
- Running or jumping;
- Influence of drugs, alcohol or fatigue.

Most falls through a floor opening occur because the person is not aware of it. All floor openings must be adequately guarded or covered.

Falls from ladders and roofs and through fragile roofs are dealt with elsewhere[23].

Many falls occur from unfenced platforms of scaffolding and down lift shafts and stairwells. The most frequent falls from heights are those of steel erectors, whose work of assembly usually proceeds faster than it takes to erect adequate working platforms. Effective and easily used safety harnesses, lanyards and fall-arrest devices [22.2.7] should be used by steel erectors and others working at heights and their use should be properly planned before work starts.

The wearing of (properly adjusted) hard hats and safety shoes and the observation of a few simple safety rules have prevented many serious injuries from falling objects.

22.9.4 Vibration

Vibration can be both a safety and health problem, causing:

- Fatigue and failure in metal parts;
- Fixtures to work loose;
- Objects resting on high surfaces to move and fall;
- Inability to read instruments and instructions,
- Ill-health such as 'white fingers' among persons whose hands or bodies are exposed to excessive vibration.

Like noise, much unwanted vibration can be avoided by careful design and initial choice of machines, etc. and by improved mounting. Again like noise, it can be surveyed by specialists with suitable instruments. Draft UK standards for human exposure to vibration are available[24].

22.9.5 Electromagnetic radiation

Lighting which involves radiation in the visible range of the spectrum has been discussed [22.6.1]. Other forms of radiation sometimes used in the process industries present hazards if not properly controlled. The main ones are listed in Table 22.2, with typical operations in which they are occur and the risks to which they give rise.

Table 22.2 Hazards of electromagnetic radiation

Form of radiation	Use	Health risk
Microwaves	Heat treatment	Deep burns
Infra-red waves	Drying and heat treatment	Eyes
Ultra-violet light	Welding	Skin and eyes
Lasers	Measurement, cutting	Eyes
X-rays	Metal inspection	Whole body
Gamma-rays	Level measurement	Whole body

References

1. S.I. 1988, No. 1657, *The Control of Substances Hazardous to Health Regulations 1988*, HMSO, London
2. Lees, F. P., *Loss Prevention in the Process Industries*, Butterworths, London (1980)
3. National Safety Council, *Accident Prevention Manual for Industrial Operations*, 7th edn, Chicago (1974)
4. S.I. 1961, No. 1345, *The Breathing Apparatus, etc. (Report on Examination) Order 1961*, HMSO, London
5. HSE, *Certificate of Approval (Respiratory Protective Equipment)* HMSO, London (issued annually)
6. Hamilton, M., 'Special clothing for special jobs', in *Industrial Safety Data File* (October 1987)
7. *Reference Book of Protective Equipment*, Industrial Safety (Protective Equipment) Manufacturers Association, 6th edn, London (1982)
8. Simpson, K, 'Choosing work wear, getting it right', in *The Safety Practitioner*, London (September 1986)
9. *Reference Book of Protective Equipment*, Industrial Safety (Protective Equipment) Manufacturers Association, 7th edn, London (1987)
10. Hamilton, M., 'Hands at work' in *Industrial Safety Data File* (April 1987)
11. Hamilton, M., 'Feet at work' *ibid.* (June 1987)
12. Illuminating Engineering Society (London), *Interior Lighting*, London (1977)
13. S.I.1974, No. 1681 as amended by S.I. 1975, No. 303, *The Protection of Eyes regulations 1975*, HMSO, London
14. Department of Employment, *Code of Practice for Reducing the Exposure of Employed Persons to Noise*, HMSO, London (1972)
15. King, I. J., 'Noise and vibration', in *Occupational Hygiene*, edited by Waldron, H.A. and Harrington, J. M., Blackwell, London (1980)
16. HSE, Guidance Note EH 22, *Ventilation of the Workplace*, HMSO, London (1988)
17. Fawcett, H. H., 'Respiratory hazards and protection', in *Safety and Accident Prevention in Chemical Operations*, edited by Fawcett, H.H. and Wood, W.S., 2nd edn, Wiley-Interscience, New York (1982)
18. Ower, E., *The Measurement of Air Flow*, 3rd edn, Chapman and Hall, London (1949)
19. BS 4275: 1974, *Recommendations for the selection, use and maintenance of respiratory protective equipment*
20. ANSI Z88.2-1980, *American National Standard Pratices for Respiratory Protection*, American National Standards Institute, New York (1980)
21. Crockford, G. W., 'Protective clothing and respiratory protection', in *Occupational Hygiene*, edited by Waldron, H.A. and Harrington, J.M., Blackwell, London (1980)
22. King, R. W. and Majid, J., *Industrial Hazard and Safety Handbook* Butterworths, London (1980)
23. King, R. W. and Hudson, R., *Construction Hazard and Safety Handbook*, Butterworths, London (1985)
24. BS DD 32: 1974, *Guide to the evaluation of human exposure to whole body vibration* BS DD 43: 1975, *Guide to the evaluation of exposure of the human hand–arm system to vibration*

Chapter 23

Hazards in the transfer of technology (TT)

The risks of accidents and ill-health generally increase when technologies cross national frontiers, particularly when the recipient is a 'Third World' or 'developing' country (DC). The world's worst industrial disasters have occurred in DCs although the installations where they happened were only of medium size for their type. The release of toxic methyl isocyanate vapour in Bhopal, India, in 1984 caused over 2000 deaths and 200 000 injuries, while two weeks earlier the LPG fire and explosions at Mexico City caused 650 deaths and several thousand injuries. Four out of five of the disasters discussed in Chapters 4 and 5 happened on plants operated by a subsidiary of a foreign company which had developed the technology and had some responsibility for its use by the subsidiary.

There is no simple panacea. Even adjacent countries such as Burma and Thailand, Israel and Jordan, India and Pakistan have very different legal and political systems and cultures. For the industrialist who wishes or intends to introduce a certain new technology into a particular DC, one of the first steps should be to visit the country and find out the ground rules. However, the International Labour Office (ILO) code of practice[1] on HS in TT to DCs, abstracted here in 23.2, covers many common denominators to the problems.

23.1 Definitions and historical introduction

Most of us are used to thinking of technology in terms of complete processes and artefacts, including both hardware and software. But when we consider the hazards of TT, it is often useful to think of technology in the limited sense of 'know-how' which is created and retained in people's minds. This corresponds to the definition given in the *Concise Oxford Dictionary* as the 'science of the industrial arts'. We all use a great deal of (technologically based) hardware in our daily lives which few of us understand and fewer still could service, let alone create. The hardware is assembled from a multiplicity of components, the construction of each of which depends on the specialised know-how of a small group of people.

The importance of this 'know-how' element struck the writer forcibly when working in Uganda in 1976 for a UN agency on the rehabilitation of industries which had collapsed after the exodus of expatriate technical personnel [23.6].

Technology, however, unlike mathematics and pure science, has little significance without the hardware associated with it and the goods it helps to create. In the limited sense technology consists of patents, written instructions, drawings, computer programs, models, manuals, reports and training. To translate it into hardware and use it effectively requires capital, workers, training and a range of supporting infrastructure and services. Research and development are needed to create new technology. Organisation, discipline, standards and quality control are required to apply it.

The development of technologies and their transfer from one region of the world to another have gone on for thousands of years. Countries of the Western World can trace an almost continuous lineage of their technologies from the ancient Egyptians, 6000 years ago, through Greece, Rome, the Arab world and Spain, with implants from China (including gunpowder and porcelain) and other countries. People in other parts of the world, particularly in Africa, the Arctic, Australasia, and the Americas, lived until recently cut off from this mainstream of technological development. This may have led to the present uneven spread of industrial development and the concepts of 'developed' and 'developing' countries, although the latter term is largely a euphemism.

The USA, Canada, Western and most of Eastern Europe, the USSR, Japan and Australia have well-developed industries and broad infrastructures. The DCs have poorer populations, predominantly agricultural societies, and relatively undeveloped industries and infrastructures. They include areas with extremes of climate which hinder most human activities.

At first a country lacking a new technology may import its products until its market for them becomes sufficient to justify importing the technology and setting up indigenous production facilities. With the growth of international trade and multinational enterprises, many new technology-based plants have been built in DCs. Sometimes the products are primarily made for export. Reasons for this include cheaper labour, raw materials or energy, or freedom from controls on HS and environmental pollution which the same industries would face in their country of origin.

In Third World countries with expanding populations, low life expectancies, widespread poverty and unemployment, fatal accident frequency rates in industry are higher than in industrialised countries, although they are only one of many reasons for premature death. Such statistics for 22 countries (chosen solely because they had common data bases) are given in Table 23.1.

Regulations in DCs are generally less stringent than in more industrialised ones and standards such as occupational exposure limits for airborne toxic substances are often lacking. Regulatory agencies in DCs are often thought of as impotent and ineffective and seldom have the equipment and trained staff needed to monitor the observance of even those national limits that exist. Where prosecutions of manufacturing firms for the death or incapacity of workers from occupational exposure are successful, the penalties are far lower than in a developed country. In short, life is cheaper in DCs than in industrialised ones.

Workers in poorer countries are rarely worse off as a result of the import of hazardous industries nor do their representatives always oppose it. If the

Table 23.1 Fertility, mortality and fatal accident frequency rates for 22 countries at different stages of development

A = Annual rate of natural increase per 1000, 1980–1985[2]
B = Male life expectancy at birth, years 1980–1985[2]
C = Fatal accident frequency rate in manufacturing industry per 100 000 worker-years, 5-year average from 1981 to 1986[3]

Country	A	B	C
Developing countries			
Egypt	28	57	17
Togo	29	49	60
Zimbabwe	35	54	13
Panama	23	69	15
Peru	26	57	23
Bahrein	28	66	25
Intermediate development			
Cyprus	12	72	5
Hungary	−2	66	12
Ireland	8	70	27
Korea (South)	13	63	18
Spain	5	73	12
Yugoslavia	6	68	6
Developed countries			
Austria	0	70	7
Canada	7	72	8
Czechoslovakia	2	67	6
Finland	3	70	4
France	4	71	6
German Federal Republic	−1	71	9
Hong Kong	9	74	4
Netherlands	4	73	2
Switzerland	2	74	9
United Kingdom	1	71	2

choice is between starvation and early death or a job with a low but certain wage, accompanied by a one-in-ten risk of cancer after 20 years, a rational worker would choose the latter, as was pointed out by delegates at a 1981 International Labour Office (ILO) symposium on safe TT to DCs[4]. (In some DCs, as Table 23.1 shows, most workers would not expect to live for another 20 years in any case). Strongest opposition to the uncontrolled transfer of hazardous technologies came from workers' representatives from industrialised countries, where tighter health controls were already in force.

23.2 The ILO code of practice[1]

Of the various UN agencies and other international bodies [23.2.12], the ILO [M.4] is specially concerned with the HS problems (which in this context include working conditions) of TT. In 1988, it published a code of practice on the subject[1]. This is specially addressed to designers, importers, exporters and users of technology, to national authorities responsible for HS aspects of imported technology and to contractors involved in installing and operating new technological hardware. Selected parts are abstracted here and passages quoted directly from the code are shown in parentheses.

Much of the code stems from bitter experience and its full implementation will be an uphill task. Like other UN agencies, the ILO has no powers of compulsion over its member states, and depends on voluntary ratification of its conventions, recommendations and codes. This limits their effects unless their provisions are incorporated into strictly enforced national legislation. Important subjects which are not dealt with directly in the code are:

- Economic factors which underlie many of the hazards of TT,
- The need for adequate funding to control these hazards.

Several examples of the problems which led to the code are given later in this chapter.

23.2.1 General provisions

Methods of TT covered by the code include:

1. The use of experts;
2. The supply of machinery and equipment directly or under a contract which also provides for the TT;
3. The acquisition of patented technology through a licence agreement;
4. The use of turnkey contractors to set up and commission the plant;
5. The direct import of technologies by foreign companies

The following basic principles apply:

1. The technology-exporting country should furnish the recipient country with all standards, national regulations and other relevant information about the operation and development of the technology and why it is used.
2. The recipient country should compile from other sources all available HS information on the proposed technology.
3. The competent authorities in the recipient country should use the information thus collected to judge the safety and suitability of the proposed technology.
4. HS information compiled by the recipient country should be made public so that all concerned can deal expeditiously with initial proposals for the TT.
5. The technology-exporting country should not export technology involving processes, equipment or substances which are prohibited in its own territory because of their potential to cause serious risks to HS.
6. Imported technology should be subject to HS standards, regulations, practices or guidelines which are no less stringent than those applied to the same technology in the exporting country.

23.2.2 General factors to consider in TT

1. Any necessary modifications should be made to the original technology to take adequate account of the differences between the receiving and the supplying country.

2. Technology should not be selected for transfer on purely economic or technical data, but only after careful study of all factors affecting HS.
3. Operatives and maintenance staff in the recipient country who will have to operate and maintain the processes, plant and equipment must be properly trained so that they can work safely.
4. Proper maintenance and repair facilities should be available to or within the DC.
5. Factors to be considered in TT to tropical and sub-tropical regions are given in Appendix O.
6. Ergonomic considerations in TT are given in Appendix P.

23.2.3 Decisions to be made before TT

1. Technology-receiving countries should draw up lists of technologies whose import should be (a) prohibited and (b) subject to restrictions. In doing this, note should be taken of substances which are (a) prohibited or restricted in industrialised countries and (b) subject to stringent HS precautions.
2. Before importing any technology, the recipient country should ensure that its HS and social insurance infrastructure are sufficiently developed to provide the necessary medical surveillance, treatment and compensation for any resulting occupational injuries and diseases.
3. HS implications should be considered in the following particulars of choosing and implementing a foreign technology:
    o Alternative technologies which serve the same purpose (with a view to selecting the safest),
    o Pre-investment studies including feasibility and environmental impact,
    o Process and manufacturing studies,
    o Layout, design and engineering studies and machinery specifications,
    o Equipment selection, plant construction and start-up,
    o Personnel selection needed for the imported technology,
    o Technical assistance needed for training, commissioning and various management aspects.

23.2.4 HS-related standards, risk appraisal, consultants, Tchecklists and regulations

1. The suppliers of technology, plant and equipment should clearly inform the purchaser which HS-related technical standards have been used in the design. This information should also be given to the relevant workers' organisation (where there is one). Internationally recognised standards should be used wherever possible.
2. Qualified HS experts should be associated with the design work.
3. The location and design of the plant should be subjected to independent risk appraisal.
4. HS checklists should be used as an aid to risk appraisal for all relevant aspects of the project, including plant location, design and the materials and chemicals used in the manufacture.

5. Regulations adopted should refer to and comply with ILO Conventions and Regulations and codes of practice and/or national statutes and regional directives issued by inter-governmental agencies.

23.2.5 Major hazard installations requiring special HS attention

1. Major hazard installations are defined as those which by the nature or quantity of dangerous substances present could cause a major accident in one of the following categories:
   ○ Release in tonnage quantities of toxic gases which are lethal or harmful at considerable distances from the point of release,
   ○ Release in kilogram quantities of extremely toxic substances which are lethal or harmful at considerable distances from the point of release,
   ○ Release in tonnage quantities of flammable liquids or gases which form a large cloud which in turn burns or explodes,
   ○ The presence of unstable or highly reactive materials which may explode.
2. The supplier of a technology which requires the storage, processing or production of dangerous substances should inform the technology-receiving country whether the technology involves activities which are classified as a major hazard in the supplier's or any other country.

The code gives design principles for major hazard installations and lists the information which the technology supplier should provide to the recipient country and action required by the recipient country. These matters are dealt with in more detail in ILO's manual on major hazard control[5], and throughout this book.

23.2.6 Administrative and institutional arrangements

1. Legal standards governing HS and working conditions in TT should be linked with existing HS legislation and be enforced by a competent authority.
2. Licence agreements for TT should state whether the legal standards and regulations of the licensee's or licensor's country should apply, but in general the more stringent of the two should apply.
3. The validity of such licence agreements should be subject to approval by the HS authorities of the technology-receiving country.
4. Such licence agreements should cover appropriate HS aspects including training of national personnel.
5. New techniques incorporated into renewed agreements should comply with all relevant HS rules and regulations.
6. The granting of patents should stipulate that the technology-receiving country must be kept fully informed on all relevant HS provisions and means of hazard assessment and control used in the production of the patented item. The same should apply to the granting of trademarks.
7. Each country should establish the necessary institutional arrangements to ensure that HS aspects are considered in TT.

These institutional arrangements should include the preparation and harmonisation of national standards and regulations.

23.2.7 Training and education

1. Training programmes should be specifically adapted to the needs of technology-receiving countries.
2. Cultural aspects have a strong influence on attitudes towards risk. They should therefore be recognised in their entirety and taken into account by training organisers and trainers prior to training.
3. Trainers must be trained to the required level of expertise.
4. Workers should not pay for their training within industry.
5. The training of designers should include consideration of factors in technology-receiving countries such as climate which may affect design.
6. 'The training of engineering students from developing countries who study at universities and colleges in industrialised countries should emphasise the adaption of technology to local conditions. To promote this training DCs should be given the chance to contribute to the curricula of universities and colleges in industrialised countries.'
7. 'The understanding of the problems related to TT should be promoted by means of training material and special publications, and other measures such as courses, discussions and seminars. These promotion efforts should be directed at policy-makers, industrial planners, management in private and public enterprises, supervisors and foremen, workers and trade union officials, the staff of labour, medical and factory inspectorates, occupational hygienists, economists, engineers, chemists, safety officers, vocational and safety trainers, and agricultural and other workers.'

23.2.8 Collection and use of information

1. Technology suppliers should provide all HS information relevant to the technology to the authorities in the recipient countries and to the users. This should be updated periodically and when changes are made to the technology.
2. Such information should be drafted in an agreed language which is understood by the users of the technology. It should take into account all factors which influence the use of the technology in the recipient country and be supported by case studies and experience gained in the application of the technology.
3. The technology supplier should be consulted before the user makes any modification or adaption of the technology. This should be specified in the technology supply contract.
4. All available information and expertise should be shared with national tripartite safety councils and similar non-governmental organisations in the user country.
5. Multinational enterprises should make available information on relevant HS standards which they observe in other countries, particularly special hazards and related protective measures associated with new products and processes. They should cooperate with international organisations in preparing international HS standards.
6. Technology-receiving countries should be encouraged to exchange HS information including their field experience, the successful adaptation,

modification of imported technologies, and also to exchange technical personnel.

23.2.9 Action at company level

1. The technology selected should take HS aspects fully into account and the influences which local climate and cultural factors have on them.
2. During planning studies the technology supplier should consult the recipient country to obtain all necessary information required for design and should provide to that country all information needed for proper planning.
3. Planning should include studies of similar existing technologies to note (among other things) the effect of the technology on the environment and social system of the country of origin.
4. A competent representative of the technology-receiving country who will be involved in the operation of the process and plant using the technology should be present during the design of the process and plant.
5. The technology supplier should develop as part of the project documentation a safety specifications book containing specific information about the safe operation of the process and plant. This should include details of all hazard analyses provided by the licensor and the technical codes and standards used during design and construction. (The ILO manual includes a checklist to be used in preparing the safety specifications book.)
6. Personnel selected to operate the new technology should understand it well, and be professionally and technically qualified and motivated to work in the DC.
7. Technical advisers (generally from the technology supplier) should be employed for long enough for full TT, including responsibility for management, HS and working conditions, to local personnel.
8. Job descriptions of the technical advisers should detail their duties and responsibilities in HS matters.
9. Top management of the enterprise should formulate a written HS policy and ensure that a safety manual highlighting operation and maintenance is written and made available with the policy to all within the enterprise.
10. Where technology is received in 'package' form, personnel of the recipient country should be trained to fully understand all HS aspects rather than merely following instructions mechanically.
11. Occupational hygiene standards should be adapted to local conditions.
12. In packaging chemicals and other materials used in the process, conditions of transit and the handling and storage conditions in the recipient country should be considered.
13. The directors of the enterprise should maintain an adequate HS programme (similar to that discussed in Chapter 19 of this book).
14. Safety instructions and other notices should be in the languages of the workers employed, and easily understandable symbols should be used. Texts should be displayed in a durable form and protected against damage.

15. Employers should equip their workers with personal protective clothing and equipment which fits their physique and suits the prevailing climatic conditions.

23.2.10 Action at the national level

1. A technology-receiving country should develop the necessary occupational HS infrastructure to deal adequately with all the problems related to HS involved in TT. (This may include the setting up of a special national standards body where none existed before.)
2. National negotiators for TT should have been adequately trained in HS requirements in order to ensure the inclusion of these matters in the TT process.
3. 'Where policies for the progressive take-over of foreign enterprises by national interests are adopted by the national authorities, care should be taken that the resulting mixed or national enterprises have the full background knowledge, information, experience and competence, including staff skills, to deal with safety and health and working conditions aspects, as well as the ability to handle all emergencies.'

Other measures include the promotion of consultancy services, the sponsorship and publication of technical journals and the formation and development of professional organisations within DCs.

23.2.11 Action at the regional level

This includes cooperation between national organisations in different countries, the establishment of regional technology centres and the pooling and interchange of technical expertise at the regional level to assist in the diagnosis, identification and solution of relevant HS problems.

23.2.12 Role of international organisations

1. The ILO should continue its efforts in the dissemination of relevant technical information, in the promotion of its exchange, in the development of training and training materials, in strengthening existing national HS institutions and in assisting DCs through its technical cooperation programme.
2. The work of other international organisations in the HS field should include the provision of technical information, the maintenance of lists of suitable consultants, the provision of advice and assistance to DCs, especially technical assistance in the development of hazard-control systems, and international standard-setting activities such as the preparation of relevant conventions and codes of practice.
3. Projects financed by international agencies should include HS requirements in their guidelines. The various international organisations concerned with TT should cooperate more in HS aspects of their work.

[Such organisations include the United Nations Industrial Development Organisation (UNIDO), the United Nations Conference on Trade and Development (UNCTAD), the World Bank and various regional development banks. The World Health Organisation (WHO) and the United Nations Environment Programme (UNEP) already have a joint International Programme with the ILO on Chemical Safety (IPCS).

23.3 Examples of the spread of hazardous technologies

Many products and intermediates have been belatedly found to cause disease at low levels of exposure. Several years have sometimes elapsed before the hazards were recognised and appropriate measures introduced to protect workers and the public. Examples are asbestos, benzidine, benzene, vinyl chloride, cadmium and beryllium. Increased awareness of other health risks such as arsenic, chromium, mercury and lead has also resulted in tighter controls in industrialised countries. It is thus not surprising that some companies have closed down some of their more hazardous manufacturing operations in their home countries and set them up in poorer ones where controls are less stringent (see para. 5 of 23.2.1). This has even been encouraged by the governments of some developing countries, which, to attract foreign investment, set up 'export processing zones' with special concessions for foreign firms, including exemption from certain industrial safety legislation.

The following examples of such spread of hazardous technologies are due to Castleman[6].

23.3.1 Asbestos textiles

The USA was long a world leader in the manufacture of asbestos textiles, although most of the asbestos used was mined in Quebec. Although general recognition of the health hazards of asbestos is comparatively recent, some American insurance companies selling workers' compensation insurance appear to have recognised the risk as far back as 1918. By 1965, the high incidence of cancer among asbestos workers was well established. In 1972 the Occupational Safety and Health Administration (OSHA) set a temporary standard of 5 million fibres/m^3 in workroom air. This was reduced in 1976 to 500 000 fibres/m^3 and further later.

The industry faced considerable expense and difficulty in meeting even the 1976 standard of 500 000 fibres/m^3. The US industry declined while imports from Mexico, Taiwan and Brazil rose from almost nothing to 4.5 million pounds in 1976.

23.3.2 Arsenic

Arsenic is used to make pesticides, herbicides, wood preservatives (still in some countries) and glass. It is present in the ores of many metals, particularly copper, and unless special extraction equipment is installed, much of the arsenic in the ore is discharged into the atmosphere during smelting as arsenious oxide. Research in the USA from 1950 showed that deaths from lung cancer in counties where copper, lead and zinc ores were smelted were significantly higher than elsewhere and that airborne arsenic was the principal cause. The situation became so serious that in 1975

OSHA proposed to reduce the limit for airborne arsenic in the working atmosphere from $500 \, mg/m^3$ to $4 \, mg/m^3$.

The high cost of complying with the new limits caused US smelters to restrict their use of ores of high arsenic content and switch where possible to domestic ores of lower content. Copper ores and concentrates from several countries (Peru, Mexico, the Philippines and Namibia) which were once imported in large quantities into the USA have high arsenic contents. Partly because of this, new export-oriented smelters were constructed in these countries, where there was little regulation of the arsenic and other pollutants discharged to the atmosphere. A plant producing arsenic was shut down in the USA and new ones were built in Mexico and Peru.

23.3.3 Benzidine dyes

A useful range of cheap dyes based on benzidine or 4:4'-diaminodiphenyl was invented in the last century and widely used in textiles. Later an unusually high incidence of bladder cancer was found among benzidine workers. A retrospective study of benzidine workers in one US company showed that 17 out of 76 had developed bladder cancer. The manufacture of benzidine dyes was subsequently banned in a number of industrialised countries, many of which, however, still import and use them. The principal sources of benzidine dyes in the mid-1970s (when their manufacture in the USA virtually ceased) were Romania, Poland, India and France. A large new export-based plant came into operation in South Korea. The dyes themselves are apparently safe provided their free benzidine content is controlled at less than 20 ppm.

23.4 Problems of culture, communication and language[7]

Established cultural norms and attitudes vary widely throughout the world. Unless properly understood, they can aggravate the HS problems of importing new technologies. Thus the colour green, which is associated with 'safe' in the West, can have different associations in the Orient. A shake of the head which means '*No*' in the West means '*Yes*' in parts of Asia. In some parts of the world the new day starts at sunset. While these cultural differences can be bridged by care and local knowledge, attitudes fostered by religion can present greater difficulties.

In societies where events are thought to be controlled by divine fate rather than human action, attempts to implement safety practices may be met with complete incomprehension or (worse) by suspicion that the implementer is attempting to interfere with the Divine Will. The wearing of safety hats contravenes the tenets of one religion. Islam has strictly enforced periods of fasting, mourning and celebration during which its followers tend to become more accident prone. Road accidents are more numerous during the month of Ramadan in Moslem countries, when no food or drink is taken between sunrise and sunset.

Excessive politeness and fear of 'losing face' are widespread attitudes in several societies and tend to inhibit the free flow of information. In some friendly South-east Asian countries, many people, often to their own cost, are too polite to give '*No*' for an answer, while fear of losing face may inhibit the speaker from admitting that anything is amiss, and hence enabling corrective action to be taken.

The concept of maintenance is lacking in some pre-industrial societies. The writer has seen several factories under local management in two African countries which ground to a halt due to lack of maintenance and failure to purchase spares.

23.4.1 Language

Language can be a major barrier to communications. We in the English-speaking world, which has developed a wide technical vocabulary and has a relatively simple grammar, are fortunate in the widespread use of English as a common language. Yet within departments of many English-speaking enterprises special meanings have developed for familiar words and entirely new ones have been coined which are not to be found in any dictionary. The meaning of words often depends on their context, their intonation, and the gestures that accompany them, and can only be grasped after lengthy initiation.

Technical comprehension and communication are thus harder for those brought up in many other languages, particularly those of small countries which are little spoken outside their borders. Even Spanish, the principal language of most of South and Central America, has a limited technical vocabulary. Several Peruvian engineers have told the writer that they find it more ambiguous than English, even for non-technical discussions. Attempts to translate technical terms can be quite hilarious. Thus the meaning of a Peruvian term used for 'heat exchanger' was given in the writer's Spanish dictionary as 'brothel keeper'.

To overcome problems in communicating safety messages in Third World Countries, some experts recommend the use of pictures and visual aids, without written words. The pictures of workers should be ones with which the locals can identify, and not copies of posters showing workers in Western clothing. 'Match-stick' figures seem to be universally acceptable[7] (Figure 22.8).

23.5 Problems of standards in developing countries[8]

The role of standards in enabling industries to develop efficiently and safely has been discussed earlier [2.7]. Standards which we take for granted are most appreciated when they are lacking. They play important roles in all technologies and in the cultural infrastructures to which they relate. Until the new technology arrives, there is often little need for particular standards, but once it comes, it may do so in a 'standards vacuum' [23.2.1, para. 6].

Sometimes the problem is partly solved by creating a special 'multinational enclave' in the developing country around the industrial enterprise where the new technology is being introduced. In this enclave, a set of standards borrowed from the country of origin of the new technology is enforced, while the rest of the country carries on as before. Problems then arise when the enterprise is taken over by the developing country. Only a handful of people who held responsible positions when the enterprise was under foreign control may be aware of the standards in use and appreciate their importance. This is not a new twentieth century problem but is as old as history and was encountered in Britain when the Romans left.

A frequent question for countries receiving new technologies is whether they should wait for international standards to be agreed and published before creating their own, or whether to adopt or adapt standards from the country from which they received the technology. In the first case the country may have no effective standard for many years. In the second it risks dependence on the technology-exporting country for the supply of plant, equipment and/or materials. In this situation it is probably best to solve the problem at company level first, by adopting the most technically satisfactory, if restrictive, foreign standard. The creation of a national standard can then be postponed, perhaps until there is an adequate international one on which it can be based.

The following examples of standard-related problems in developing countries are taken from first-hand experience.

23.5.1 An LPG storage and distribution depot

One of several LPG (propane) depots built in the 1980s in a DC included pressure storage spheres, a road tank-car filling station and a cylinder-filling shed with equipment for filling several sizes of portable cylinder. In Europe it would have rated as a Major Hazard Installation. The depot included a 'stenching unit' which was designed to inject a chemical with a powerful and offensive odour into the incoming propane to enable customers who purchased it in rented cylinders to detect leaks. The contract called for the pressure relief system to be designed to API standards (RP 520), while the LPG was to be stenched to meet the UK specification for 'commercial propane' (BS 4250). These standards had no status in the country where the depots were built.

Fluids passing the relief valves should have been led through pressure 'knockout' vessels to separate liquid propane before the gas passed to a high-level vent pointing upwards, well away from possible sources of ignition. No knockout vessels were provided and any fluid discharged would have passed through short vertical lines terminating with inverted U bends, whose open ends pointed downwards over a road used by LPG road tank-cars.

An even more serious mistake was that other pressure relief lines discharged directly into a large shed for cylinder filling, where up to 100 people would normally be working. This was discovered when relief valves discharged propane into the shed while the plant was being commissioned.

The plant was far from leak-tight when liquid propane was first fed into the storage spheres and the stenching unit was started. So much stenched LPG entered the atmosphere that the plant manager was inundated with complaints from neighbouring residents. The stenching unit was therefore shut down and commissioning was continued!

23.5.2 A national institute for occupational safety and health

In the outskirts of a large Middle East city, a modern four-storey purpose-built building had been donated to the nation by an international agency to house a new national institute for occupational safety and health, whose functions included advising industry on fire protection. It had laboratories on two floors in which flammable organic liquids and LPG were used, and about 100 offices, with a staff of about 500. The building had two stairwells at opposite ends, one landing in the main foyer at the front of the building, the other leading to double doors which opened onto waste ground at the rear. These doors had only conventional locks. 'Panic' latches (Figure 23.1) which would have allowed the doors to be locked to outsiders but easily opened from the inside had not been fitted. The two stairwells were required as alternate fire exits, which under most countries' fire regulations should have been enclosed.

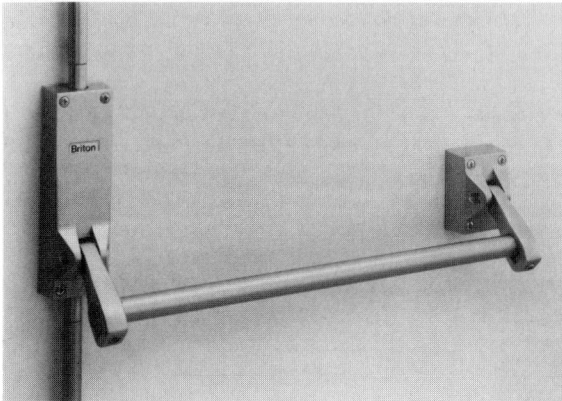

Figure 23.1 Panic bolt (courtesy Newman Tonks Engineering Ltd)

As a result of break-ins, the security chief of the enterprise had ordered these doors to be kept locked. This made the lowest flight of stairs unusable. The cleaning contractors, who previously had to take their daily harvest of waste paper several kilometres to the city refuse dump, now found a handy dumping space within the building itself. When this writer inspected the stairwell the bottom flight was filled with waste paper and laboratory rubbish.

Situations such as these which cause nightmares to safety advisers from industrialised countries are all too common in DCs where shortage of foreign currency inhibits the import of appropriate safety equipment.

23.6 Uganda 1976

This writer had the unusual experience of spending several months in Uganda in 1976, the fifth year of Amin's rule, as a consultant employed through the United Nations Industrial Development Organisation

(UNIDO) to study and advise on the rehabilitation of the country's non-agricultural industries. These had been shattered not by war but through government policies – part of the growing pains of national independence. Although HS did not feature in our terms of reference, the gradual breakdown of imported technology caused serious accidents and illness. The fragility of technological implants in an isolated and primarily agricultural society became only too apparent.

The causes of the industrial collapse were twofold:

1. The recent exodus of most technical and managerial staff of Indian and British origin, coupled with an acute shortage of trained and experienced Africans;
2. Restrictions in the supply of imported machinery, spare parts, fuel oil and raw materials, most of which came overland through Kenya. The reasons for this were partly political and partly economic.

By 1970 Uganda had several industrial projects based on imported Western technologies, mostly owned by subsidiaries of British and Kenyan-Indian companies. Much of the trade and many of the technical jobs in the country (such as electrical and instrument maintenance) were done by descendants of Indian workers brought in sixty years earlier to build a railway from the port of Mombasa to the capital, Kampala, later extended west to the copper mines at Kilembe. The success of this Indian minority and their discouragement of would-be African sons-in-law caused pent-up frustrations among Africans which encouraged Amin to expel them.

By the time our team arrived, several of the largest and most technically advanced factories had been forced to close down, while others were still running precariously at well below their rated capacities. Other less mechanised factories which had been specially designed for ease of maintenance were, however, still running smoothly. Of the many industries we studied, only a few which had special HS problems are discussed here.

23.6.1 Cement and brick production

Uganda had two cement plants, one at Tororo in the east of the country making a slow-setting cement used to construct the Nile dam and hydro-electric station at Jinga, and a larger one built later at Hima near the Kilembe copper mines in the west. These had been effectively nationalised by Amin. Their management in 1971 was mainly British, with Indians in many skilled jobs. In the next two years most British and Indian staff had left, leaving only a handful of trained Africans.

Early casualties of both works were the electrostatic precipitators which removed fine dust from the flue gases. At Tororo a layer of fine cement dust quickly settled on the flat roof of the raw materials yard. Not appreciating the hazard, nobody bothered to remove it. When rain came, the dust set hard and gradually the roof became increasingly thicker until the weight was too much for the supports, and it collapsed early in 1973, killing several workers.

The lesson failed to reach or was not learnt at Hima, where the main factory roof collapsed from the same cause in August 1973, killing more workers, damaging the gantry crane and halting production for several months.

Both works were shut down during our visit in 1976. They were in desperate need of maintenance and spares of all kinds, short of fuel (oil and charcoal from Kenya) and subject to long and frequent power cuts. There was little likelihood of restarting the Tororo plant. We visited Hima during an unsuccessful attempt to start the plant, which lay at the centre of a cloud of choking dust. Workers with inflamed eyes wearing strips of cloth over their mouths and noses were manhandling hot, dusty process materials because of the breakdown of mechanical handling equipment. None of the temperature-measuring instruments were working. After spending several hours at the works the writer became quite ill and had to return quickly to Kampala for medical treatment (thus missing the opportunity of visiting Kilembe Mines which had similar problems on a larger scale, and the Murchison Falls).

In contrast to the continuous and mechanised cement works, the brick and tile industry which used local clays, and coffee husks as fuel, was in a relatively healthy state. A semi-continuous Hoffmann kiln designed by Swiss engineers had been built almost entirely from local materials on a deposit of ball clay near Kampala in 1958. All equipment had been selected for ease of maintenance. The only imported machine was an Morondo extruder which formed the clay into the required shapes before these were air-dried and loaded daily into a fresh section of the kiln. It produced a range of hollow bricks, blocks and floor sections, roofing tiles, ridges, channels and grilles, and was working almost at full capacity.

23.6.2 Glass bottles

A small glass factory making bottles (for beer and soft drinks) and tumblers had been set up by an Indian group in 1968. This had an oil-heated tank furnace and used local sand, the fuel and other raw materials being imported. In 1972 the company was nationalised, when the Indian owners and staff left. Little maintenance was done during the next three years. The annealing section first broke down, as a result of which some 25% of the beer bottles made burst after filling, causing injuries. Shortly before our arrival the brickwork of the furnace collapsed, spilling its entire contents of molten glass onto the floor of the building, causing casualties and secondary damage.

23.6.3 Bottled beverages

During the 1950s and 1960s many middle-class Ugandans developed a taste for bottled beer and aerated beverages, in preference to the traditional and more nutritious plantain wine known as pombé, which is fermented in earthenware jars and drunk while still fermenting through hollow bamboo canes. Two foreign-owned breweries (one of which was nationalised in 1972) and a bottling plant for non-alcoholic beverages were hence set up in the 1950s and 1960s. Most of their raw materials, malt, hops, sugar and flavouring materials were imported. Production rose steadily until 1973 after which it fell rapidly for several reasons. These included shortage of foreign exchange to buy raw materials, lack of maintenance and shortage of spare parts, loss of skilled expatriate personnel, and poor quality bottles whose breakage was accentuated by a shortage of bottle openers (in place of which teeth were often used and broken). The run-down of the breweries and bottling plant was accompanied by many accidents and injuries, including the death of at least one man who entered an apparently empty vat containing carbon dioxide gas.

23.6.4 Industrial gases

A factory set up as a subsidiary of a British company to produce industrial oxygen, nitrogen, acetylene, argon, nitrous oxide and hydrogen was still operating precariously in 1976 but suffering severely from lack of maintenance which affected its safety. The following examples were noted:

1. Holes in the floor which (it was said) could not be repaired due to lack of cement;
2. Gas cylinders not painted with their distinguishing colours due to lack of paint, leading to mistakes in identifying their contents;
3. Missing nuts and bolts (in short supply) on flanged joints;
4. Lack of rubber cushions for unloading gas cylinders from trucks;
5. Shortage of protective caps on cylinders;
6. Use of an improvised intercooler on a gas compressor to replace one which had failed. This was thought to be about to burst at any time.

Throughout the country there was an acute shortage of welding goggles. At one factory making hand hoes from scrap-iron, gas welders were working with no form of eye protection. Several appeared to be suffering from cataract and one was nearly blind.

23.7 Important lessons for technology importers

The interests of sellers and buyers of technology are usually different. Only those of the importing country which will have to live with the imported technology are considered here. Consultants engaged to carry out objective feasibility studies on projects in DCs, more often than not in the writer's experience, find themselves under pressure from one of the possible parties to the TT. The following important lessons are among those pointed out briefly in the ILO code of practice [23.2].

1. A common mistake is to opt for the latest and most sophisticated technology, regardless of the level of technical development and education in the country and of possible difficulties in being able to get spare parts and special process materials. In Burma, for example, a country which has been very isolated for the last 40 years, there was general praise for a large Chinese-built textile mill which was designed for ease of maintenance (like the brick factory in Uganda [23.6.1]).
2. The siting of major hazard installations well away from populated centres needs special vigilance, and strict measures are required to

Figure 23.2 A typical Brazilian favela: '. . . a jumble of rude shacks made of wood, cardboard, scraps of metal . . .' (Popperfoto)

ensure that unofficial settlements like Brazil's *favelas* (Figure 23.2) do not spring up on their doorsteps. The scale of the disasters at Bhopal and Mexico City in 1984 was caused by the close proximity of major hazards to population centres.

3. Where hazardous materials are used, whether highly flammable like LPG or highly toxic like chlorine, the quantities stored and their inventories in process should be restricted to the bare minimum.
4. Nationals of the technology-importing country should be trained so that they can take charge of, operate and maintain the installation from the earliest possible moment after commissioning.
5. Good maintenance, whilst important for all plant and storage installations, is even more so for its protective systems.
6. The dangers of a hazardous plant are particularly high when it is being started up, and when the manufacturing operation is being run down or in economic difficulties.

7. Pay special attention to the effects of climate when that in the technology-importing country differs significantly from that in the country where the technology was developed and is now used. These effects are far-reaching and easily overlooked. Remember, for example, that in a hot climate human work may be impossible near furnaces, etc. although this may present few problems in a cold one; materials thought of as liquids in a temperate climate may freeze or deposit solids in cold ones; bacteria and fungi which are dormant in a cold climate multiply fast and cause serious problems in hot and humid ones.

References

1. International Labour Office, *Safety, health and working conditions in the transfer of technology to developing countries – An ILO code of practice*, Geneva (1988)
2. *United Nations Demographic Yearbook 1986*, New York (1986)
3. International Labour Office, *ILO Statistical Yearbook, 1988*, Geneva
4. International Labour Office, *Inter-regional tripartite symposium on occupational safety, health and working conditions specifications in relation to transfer of technology to the developing countries*, Geneva (1981)
5. International Labour Office, *Major Hazard Control – A practical manual*, Geneva (1988)
6. Castleman, B. I., 'The export of hazardous factories to developing nations' (paper reported in reference 4)
7. Brown, D. H., 'Safety problems faced by UK firms working abroad', *Industrial Safety Data File*, United Trade Press, London (1984)
8. King, R. W., 'The role of standards in the safe transfer of technology to developing countries' (paper reported in reference 4)

Process industries in the UK and numbers employed

The following features are characteristic of the process industries:

- Large scale of operations;
- Continuous operation, with long runs and shift working;
- Each plant designed for a single product;
- Mechanical handling of process materials;
- Use of process temperatures and pressures above or below ambient;
- Use of chemicals as process materials;
- Total enclosure of process materials within pipes and equipment;
- Use of liquid and gaseous process materials;
- Solid process materials handled in suspension in liquids or gases;
- High degree of instrument control;
- Operators need intellectual rather than manual skills;

The process industries account for over 10% of all employment in the UK. Table A.1 gives a list of UK process industries[1] with the numbers employed in September 1981[2]. (More recent figures give far less detail.). The list is divided into four categories:

Category and description	Employees (thousands)
A Mainstream process industry (excluding nuclear fuel production)	1225.5
B Fringe process industry (e.g. cold drawing metal tubes)	1130.4
C Contracting or equipment supply	268.9
D Indirect involvement such as public administration or technical training	Number unknown

Of those employed in the 'mainstream' process industries, 55 900 or 4.6% were employed in the oil and gas industries and 300 000 or 24.5% in what this writer regards as chemical industries proper (including manufacture of fertilisers, plastics, synthetic rubber and soap and detergents). This compares with a figure of 400 000 employees in the chemical industry reported in the 1985 report of the Chief Inspector of Factories[3].

Table A.1 The process industries and their workers

SIC No. (1980)	Description	Employees (thousands)
Type A	**Mainstream process industry**	
1115	Manufacture of solid fuels	1.7
1200	Coke ovens	4.3
1300	Extraction of mineral oil and natural gas	26.1
140	Mineral oil processing	29.8
1520	Nuclear fuel production	15.8
2100	Extraction and preparation of metalliferous ores	1.9
2210	Iron and steel industry	124.7
2245	Aluminium and aluminium alloys	31.7
2246	Copper, brass and other copper alloys	27.6
2247	Other non-ferrous metals and their alloys	19.8
2330	Salt extraction and refining	0.4
2410	Structural clay products	18.3
2420	Cement, lime and plaster	15.7
2470	Glass and glassware	58.2
2511	Inorganic chemicals except industrial gases	66.8
2512	Basic organic chemicals except specialised pharmaceutical chemicals	12.9
2513	Fertilisers	7.7
2514	Synthetic resins and plastic materials	40.4
2515	Synthetic rubber	1.1
2516	Dyestuffs and pigments	10.4
2563	Chemical treatment of oils and tars	0.2
2564	Essential oils and flavouring materials	3.3
2565	Explosives	5.4
2567	Miscellaneous chemical products for industrial use	25.9
2568	Formulated insecticides	1.7
2570	Pharmaceutical products	81.9
2581	Soap and synthetic detergents	16.6
2591	Photographic materials and chemicals	8.3
2600	Production of man-made fibres	17.5
411	Organic oils and fats (other than crude animal fat production	3.3
413	Preparation of milk and milk products	44.6
4147	Processing of fruit and vegetables	38.1
4180	Starch	1.6
4196	Bread and flour confectionery	146.6
4197	Biscuits and crispbread	36.1
4200	Sugar and sugar by-products	9.3
421	Ice cream, cocoa, chocolate and sugar confectionery	67.8
422	Animal feeding stuffs	124.0
4240	Spirit distilling and compounding	26.2
4261	Wines, cider and perry	4.9
4270	Brewing and malting	62.1
4283	Soft drinks	26.1
4290	Tobacco industry	29.9
4710	Pulp, paper and board	44.5

Type B   Fringe process industry

1620	Public gas supply	104.7
1700	Water supply industry	66.1
2220	Steel tubes	32.8
223	Drawing, cold rolling and cold forming of steel	30.8
2310	Extraction of stone, clay, sand and gravel	36.7
2440	Asbestos goods	12.3
2460	Abrasive products	6.9
248	Refractory and ceramic goods	60.0
255	Paints, varnishes and printing ink	33.1
2562	Formulated adhesives and sealants	11.5
2569	Adhesive film, cloth and foil	3.1
223	Drawing, cold rolling and cold forming of steel	30.8
2582	Perfumes, cosmetics and toilet preparations	23.4
259	Specialised chemical products mainly for household and office use	14.1
2599	Chemical products n.e.s.	5.8
311	Foundries (ferrous and non-ferrous)	79.6
312	Forging, pressing and stamping	34.0
412	Slaughtering of animals and production of meat and by-products	98.1
4150	Fish processing	14.2
4160	Grain milling	11.0
4310	Woollen and worsted industry	53.1
432	Cotton and silk industries	46.7
433	Throwing, texturing, etc. of continuous filament yarn	0.4
436	Ready-mixed concrete	9.5
4410	Leather (tanning and dressing) and fellmongery	14.1
4620	Manufacture of semi-finished wood products and further processing and treatment of wood	6.2
472	Conversion of paper and board	114.9
481	Rubber products	74.7
483	Processing of plastics	120.1
9210	Sewage disposal	12.5

Type C   Contracting and equipment supply

3205	Boilers and process plant fabrications	27.0
3230	Textile machinery	14.6
324	Machinery for the food, chemical and related industries; process engineering contractors	47.1
3255	Mechanical lifting and handling equipment	55.5
327	Machinery for the printing, paper, wood, leather, rubber, glass and related industries; laundry and dry cleaning machinery	13.7
3283	Compressors and fluid power equipment	55.3
3284	Refrigerating machinery, space heating, ventilating and air conditioning equipment	42.9
3287	Pumps	7.7
3287	Industrial valves	5.1

Type D   Indirect involvement

820	Insurance
8370	Professional and technical services
91	Public administration

References

1. *What's New in Processing*, a 'media pack' published by Morgan-Grampian (Process Press) Ltd, London (*ca* 1985)
2. 'Census of employment final results for September 1981', *Employment Gazette Occasional Supplement No. 2* (December 1983)
3. *Report by HM Chief Inspector of Factories 1985*, HMSO, London

NFPA classification of hazardous materials[1]

This provides a numerical rating of 0 to 4 for three regular hazards of many thousands of materials (mainly chemicals) – health, flammability and reactivity – plus an indication whether the material is radioactive or not, and whether water may be applied to spillages or fires. The scheme is intended for quick recognition in case of a fire emergency and is widely used for labelling materials in transport and storage both inside and outside the USA. For this the ratings are shown in coloured squares on an (up-ended) square label reproduced here as Figure B.1. Table B.1 gives a key to the ratings while Table B.2 gives full details of the health hazard ratings.

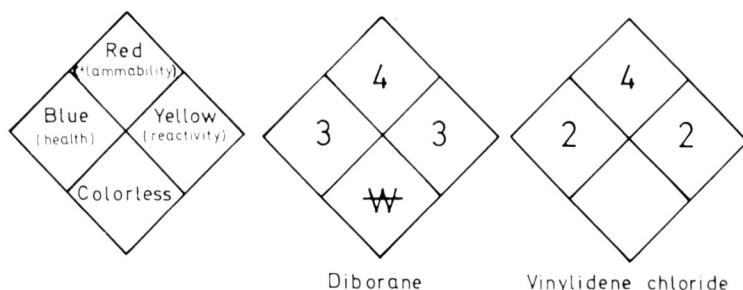

Figure B.1 Fire hazards of materials: NFPA label 704 (courtesy National Fire Protection Association, Boston)

Reference

1. National Fire Protection Association, *Standard 704 M, Identification systems for fire hazards of materials*, NFPA, Boston, Mass. (1975)

Table B.1 Key to NFPA hazard rating of materials[1]

Health hazard – Blue square – Type of injury

4 Materials which on very short exposure could cause death or major residual injury even though prompt medical treatment were given.
3 Materials which on short exposure could cause serious temporary or residual injury even though prompt medical treatment were given.
2 Materials which on intense or continued exposure could cause temporary incapacitation unless prompt medical treatment is given.
1 Materials which on exposure would cause irritation but only minor residual injury even through no treatment is given.
0 Materials which on exposure under fire conditions would offer no hazard beyond that of ordinary combustible material.

Flammability – Red square – Susceptibility of materials to burning

4 Materials which will rapidly or completely vaporise at atmospheric pressure and normal ambient temperature, or which are readily dispersed in air and which will burn readily.
3 Liquids and solids that can be ignited under almost all ambient temperature conditions.
2 Materials which must be moderately heated or exposed to relatively high ambient temperatures before ignition can occur.
1 Materials that must be preheated before ignition can occur.
0 Materials which will not burn.

Reactivity – Yellow square – Susceptibility to release of energy

4 Materials which in themselves are readily capable of detonation or of explosive decomposition or reaction at normal temperatures and pressures.
3 Materials which in themselves are capable of detonation or explosive reaction but require a strong initiating source or which must be heated under confinement before initiation or which react explosively with water.
2 Materials which in themselves are normally unstable and readily undergo violent chemical change but do not detonate. Also materials which may react violently with water or which may form potentially explosive mixtures with water.
1 Materials which in themselves are normally stable, but which can become unstable at elevated temperatures and pressures or which may react with water with some release of energy, but not violently.
0 Materials which in themselves are normally stable, even under fire exposure conditions, and which are not reactive with water.

Table B.2 NFPA Health hazard rating of materials

NFPA health hazard degree 4

- Materials which on very short exposure could cause death or major residual injury even though prompt medical treatment were given, including those which are too dangerous to be approached without specialised protective equipment. This degree should include:
- Materials which can penetrate ordinary rubber or synthetic protective clothing;
- Materials which under normal conditions or under fire conditions give off gases which are extremely hazardous (i.e. toxic or corrosive) through inhalation or through contact with or absorption through the skin.

NFPA health hazard degree 3

- Materials which on short exposure could cause serious temporary or residual injury even though prompt medical treatment were given, including those requiring protection from all bodily contact;
- Materials which give off highly toxic combustion products;
- Materials corrosive to living tissues or toxic by skin absorption.

NFPA health hazard degree 2

- Materials which on intense or continued exposure could cause temporary incapacitation or possible residual injury unless prompt medical treatment is given, including those requiring use of respiratory protective with independent air supply. This degree should include:
- Materials giving off toxic combustion products;
- Materials giving off highly irritating combustion products;
- Materials which either under normal conditions or under fire conditions give off toxic vapours lacking warning properties.

NFPA health hazard degree 1

- Materials which on exposure would cause irritation but only minor residual injury even if no treatment is given, including those which require the use of an approved canister gas mask. This degree should include:
- Materials which under fire conditions would give off irritating combustion products;
- Materials which on the skin could cause irritation without destruction of tissue.

NFPA health hazard degree 0

- Materials which on exposure under fire conditions would offer no hazard beyond that of ordinary combustible material.

Appendix C

Material safety data sheets

Form Approved
OMB No. 44-R1387

U.S. DEPARTMENT OF LABOR
Occupational Safety and Health Administration

MATERIAL SAFETY DATA SHEET

Required under USDL Safety and Health Regulations for Ship Repairing,
Shipbuilding, and Shipbreaking (29 CFR 1915, 1916, 1917)

SECTION I

MANUFACTURER'S NAME	EMERGENCY TELEPHONE NO.
ADDRESS *(Number, Street, City, State, and ZIP Code)*	
CHEMICAL NAME AND SYNONYMS	TRADE NAME AND SYNONYMS
CHEMICAL FAMILY	FORMULA

SECTION II - HAZARDOUS INGREDIENTS

PAINTS, PRESERVATIVES, & SOLVENTS	%	TLV (Units)	ALLOYS AND METALLIC COATINGS	%	TLV (Units)
PIGMENTS			BASE METAL		
CATALYST			ALLOYS		
VEHICLE			METALLIC COATINGS		
SOLVENTS			FILLER METAL PLUS COATING OR CORE FLUX		
ADDITIVES			OTHERS		
OTHERS					

HAZARDOUS MIXTURES OF OTHER LIQUIDS, SOLIDS, OR GASES	%	TLV (Units)

SECTION III - PHYSICAL DATA

BOILING POINT (°F.)		SPECIFIC GRAVITY ($H_2O=1$)	
VAPOR PRESSURE (mm Hg.)		PERCENT, VOLATILE BY VOLUME (%)	
VAPOR DENSITY (AIR=1)		EVAPORATION RATE (_____ =1)	
SOLUBILITY IN WATER			
APPEARANCE AND ODOR			

SECTION IV - FIRE AND EXPLOSION HAZARD DATA

FLASH POINT (Method used)	FLAMMABLE LIMITS	Lel	Uel
EXTINGUISHING MEDIA			
SPECIAL FIRE FIGHTING PROCEDURES			
UNUSUAL FIRE AND EXPLOSION HAZARDS			

SECTION V · HEALTH HAZARD DATA

THRESHOLD LIMIT VALUE

EFFECTS OF OVEREXPOSURE

EMERGENCY AND FIRST AID PROCEDURES

SECTION VI · REACTIVITY DATA

| STABILITY | UNSTABLE | | CONDITIONS TO AVOID | |
| | STABLE | | | |

INCOMPATABILITY *(Materials to avoid)*

HAZARDOUS DECOMPOSITION PRODUCTS

| HAZARDOUS POLYMERIZATION | MAY OCCUR | | CONDITIONS TO AVOID | |
| | WILL NOT OCCUR | | | |

SECTION VII · SPILL OR LEAK PROCEDURES

STEPS TO BE TAKEN IN CASE MATERIAL IS RELEASED OR SPILLED

WASTE DISPOSAL METHOD

SECTION VIII · SPECIAL PROTECTION INFORMATION

RESPIRATORY PROTECTION *(Specify type)*

| VENTILATION | LOCAL EXHAUST | | SPECIAL | |
| | MECHANICAL *(General)* | | OTHER | |

| PROTECTIVE GLOVES | EYE PROTECTION |

OTHER PROTECTIVE EQUIPMENT

SECTION IX · SPECIAL PRECAUTIONS

PRECAUTIONS TO BE TAKEN IN HANDLING AND STORING

OTHER PRECAUTIONS

PAGE (2)   Form OSHA-20

Draft headings (June 1990) for proposed EC material safety data sheets

1. Chemical product and company identification.
2. Composition/ingredient information.
3. Hazards identification (Hazards summary).
4. First aid measures.
5. Fire fighting measures.
6. Accidental release measures.
7. Storage and handling.
8. Exposure controls/personal protection.
9. Physico-chemical properties.
10. Stability and reactivity.
11. Toxicological information.
12. Ecological information.
13. Disposal considerations.
14. Transport information.
15. Regulatory information.
16. Other information.

Appendix D

Vapour cloud explosions up to 1983

This review is based Davenport's 1983 update[2] of his 1977 survey[1] of vapour cloud explosions [10.5.4] up to those years. In nearly all incidents some degree of confinement was involved, e.g. from building walls, pipe racks, process structures, rail cars or earth contours. BLEVEs are excluded from this review.

Of the 71 incidents surveyed in the 1983 update, abstracts of 29 of the most severe ones each causing known property damage in excess of £10 million are given in Table D.1 where they are listed in five categories: petrochemical plant, other chemical plant, oil refineries, transport and natural gas plant.

Hydrocarbons were involved in all but two of the incidents, the exceptions being ethylene oxide and dimethyl ether. The hydrocarbons and the number of times they were involved were ethylene (6), propane/propylene (6), methane (3), butadiene (2), isobutene (1), isobutane (1), C_5 hydrocarbons (1), cyclohexane (1), hexane (1), unspecified light hydrocarbons (2) and C_{10} and heavier hydrocarbons (1).

While at least 5 t of material were generally released, ethylene escapes from HP polyethylene plants were a notable exception since as little as 200 kg produced a serious explosion. Two escapes of cold methane (one at $-164°C$) produced VCEs, although escapes of methane at ambient temperature usually disperse upwards before they can form explosive clouds.

Excluding the two accidents at Ludwigshafen in the 1940s and two others for which fatalities are not known, we are left with 25 incidents each with property losses over £10 million. These caused 101 fatalities and a total property loss of £1047 million at 1983 values. This gives an average of four fatalities and £25 million property loss per incident and an average of £10 million of property loss per fatality.

References

1. Davenport, J. A., 'A study of vapour cloud incidents – an 'update', paper in *4th International Symposium on Loss Prevention and Safety Promotion in the Process industries, Harrogate*, I. Chem. E., Rugby (1983)
2. Davenport, J. A., 'A study of vapour cloud incidents', paper in *AIChE Loss Prevention Symposium*, Houston, Texas (March 1977)

Table D.1 Severe vapour cloud explosions up to 1983

(1) Place	(2) Date (d/m/y)	(3) Type of process	(4) Material released	(5) Amount released (t)	(6) Fatalities	(7) Damage (1983) ($10 million)
1. Petrochemical plant						
Bradenburg, Kentucky	17/4/62	Ethylamine	Ethylene oxide	16	1	27
Ammonia entered a tank containing ethylene oxide causing it to rupture releasing several tonnes which formed an explosive cloud with air and ignited immediately. The resulting explosion had a TNT equivalence of 16 t.						
Orange, Texas	25/10/64	HP Polyethylene	Ethylene	0.2	2	11
Failure of a 10 mm fitting on an HP ethylene line led to a partially confined explosion with a TNT equivalence of 180–270 kg.						
Antwerp, Belgium	10/2/65	HP Polyethylene	Ethylene	2.5	6	89
Fatigue failure of a liquid ethylene header caused discharge of HP ethylene through a 25 mm opening for 20–70 seconds. The explosion was mainly in the open but partly under a concrete structure.						
Lake Charles, Louisiana	13/7/65	Olefins	Methane or ethylene	Unknown	0	11
Cold fluid from an overpressurised demethaniser column discharged into an MS flare header which ruptured leading to a VCE.						
Raunheim, West Germany	19/1/66	Olefins	Methane	1 to 2	3	16
Failure of a blowdown line from the top of a demethaniser column led to the escape of methane at −164°C which ignited about 60 m from its source. TNT equivalence 1–1.5 t.						
Texas City	3/10/69	Butadiene	Butadiene	Unknown	3	24
Vinyl acetylene accumulated in a distillation column and detonated rupturing the column and releasing a large quantity of butadiene which mixed with air causing a VCE [9.8.2].						

Location	Date	Plant/Product	Material	Quantity	Deaths	No.
Longview, Texas	25/2/71	HP Polyethylene	Ethylene	0.5	3	16

Ethylene released from a leak in a high-pressure pipe caused a VCE.

| Tokuyama City, Japan | 7/7/73 | Ethylene | Ethylene | Unknown | 1 | 39 |

An acetylene hydrogenation reactor overheated causing a flange failure, fire, and a pipe failure. The escaping gas mixed with air and exploded.

| Flixborough, England | 1/6/74 | Cyclohexane oxidation | Cyclohexane | 36 | 28 | 140 |

See Chapter 4.

| Beaumont, Texas | 29/11/74 | Isoprene | C_5 + HCs | 8 | 2 | 27 |

A joint failed on a pump suction line. Hydrocarbons released exploded with air knocking out the deluge system. Serious fires followed.

| Beek, Holland | 7/11/75 | Olefins | Propylene | 5 | 14 | 71 |

The failure of a level controller caused cold liquid to enter an MS flare header which cracked. The cloud ignited after about 2 minutes and exploded with a TNT equivalence of about 2 t.

| Longview, Texas | 15/10/76 | Ethanol | Ethylene | Unknown | 1 | 22 |

An equipment failure led to the release of a jet of ethylene into a process area and its explosion with air.

| Brindisi, Italy | 8/12/77 | Olefins | Light HCs | Unknown | 0 | 44 |

A massive leak in a cold plant led to an explosion with air in less than one minute.

| Newcastle, Delaware | 21/10/80 | Polypropylene | Hexane | 13 | 5 | 54 |

Faulty maintenance caused the blow-out of a 4-inch plug from a valve and the release of hot hexane at high pressure. The cloud exploded within 4 minutes of the release.

Table D.1 (*Continued*)

(1) Place	(2) Date (d/m/y)	(3) Type of process	(4) Material released	(5) Amount released (t)	(6) Fatalities	(7) Damage (1983) ($10 million)
2. Other chemical plant						
Ludwigshafen, West Germany	29/7/43	Tank car in plant	Butadiene	16	60–80	53
An overfilled tank car without a relief valve was overheated by the sun and split.						
Ludwigshafen, West Germany	28/7/48	Tank car in plant	Dimethyl ether	30	209	25
The cause was similar to the last case. The TNT equivalence exceeded 20 t.						
3. Oil refineries						
Lake Charles, Louisiana	9/8/67	Alkylation	Isobutene	9	7	112
The failure of a 250 mm gate valve led to a VCE with a TNT equivalence of 11 t.						
Linden, New Jersey	5/12/70	H-Oil Unit	$>C_{10}$ HCs $+ H_2$	114	0	83
The failure of a reactor due to local overheating released a large HC cloud which exploded with a TNT equivalence of 45 t.						
Denver, Colorado	3/10/78	Polymer gasoline	Propane	Unknown	3	32
A pipe failure released an HC cloud which was ignited by a heater 35 m away.						
Pitesti, Romania	30/10/78	Depropaniser	Propane-propylene	Unknown	?	19
The failure of a column overhead line led to a large VCE.						
Texas City	21/7/79	Alkylation	Propane	3	0	29
A leak in a 300 mm elbow below a vessel led to a cloud which drifted into an adjacent catalytic cracking unit and exploded within about 3 minutes with a TNT equivalence of 0.9 t.						

Location	Date	Process/Type	Material	Quantity	?	Deaths
Borger, Texas	20/1/80	HF alkylation	Light HCs	Unknown	?	60

Few details known.

4. Transport

Location	Date	Process/Type	Material	Quantity	?	Deaths
Laurel, Massachussets	25/1/69	Rail accident	Mostly propane	63	2	18

An immediate explosion followed the accident.

Location	Date	Process/Type	Material	Quantity	?	Deaths
E. St Louis, Illinois	22/1/72	Railroad switchyard	Propylene	54	1	19

Two rail cars collided in a 'humping' operation. The resulting VCE may have been a detonation and had a TNT equivalence of 2.5 t.

Location	Date	Process/Type	Material	Quantity	?	Deaths
Decatur, Illinois	19/7/74	Rail yard	Isobutane	69	7	37

The accident caused an HC release which ignited to form a VCE within 8 to 10 minutes, with a TNT equivalence of 20–125 t.

Location	Date	Process/Type	Material	Quantity	?	Deaths
Houston, Texas	21/9/74	Rail yard	Butadiene	>80	1	27

A 'humping' accident in a railyard punctured a large railtank. The escaping butadiene was ignited by a locomotive 180 m away after 2 to 3 minutes, exploding with a TNT equivalence of 20–57 t.

5. Natural gas plant

Location	Date	Process/Type	Material	Quantity	?	Deaths
UMM Said, Qatar	3/4/77	Low temperature storage	Propane	23 000	7	124

The massive failure of an atmospheric pressure refrigerated propane tank caused propane to flow over bunds into a process area with a fire and probable explosion.

Location	Date	Process/Type	Material	Quantity	?	Deaths
Dhahran, Saudi Arabia	15/4/77	Gas/oil separation	Methane and LPG	Unknown	4	77

An unsupported 560 mm pipeline failed releasing methane with a fire causing the rupture of a 1600 m^3 spheroid containing LPG. A large VCE followed with a TNT equivalence of 17–24 t.

Largest losses in the hydrocarbon/ chemical industries 1958–1987

The data given here have been abstracted by kind permission of Marsh and McLennan Protection Consultants from the eleventh and twelfth annual editions of their 30-year world survey (excluding communist countries) of the largest property damage losses in the hydrocarbon/chemical industries[1]. Only individual losses in excess of US$10 million based on 1988 values are included. These are discussed and analysed in section 1.6 of this book.

The installations where the losses occurred fall under eight categories – oil refineries, petrochemical plants, plastic and rubber plants, chemical plants, natural gas plants, terminals and bulk plants, pipelines and miscellaneous. For each category, the date, location, type of installation, material involved and dollar loss (1988 basis), types of incident and initiating causes (where known) are given in Table E.1. The abbreviations LPG, LNG and NGL are used for liquefied petroleum gas, liquefied natural gas and natural gas liquids. The following abbreviations are used for types of incident:

BLEVE	Boiling liquid expanding vapour explosion
BO	Boil-over
E	Explosion (type unspecified)
EnCtm	Environmental contamination
F	Fire (type unspecified)
FB	Fireball
FF	Flash fire
IE	Internal explosion (e.g. in vessel or furnace)
Im	Implosion
ME	Mechanical explosion
PoF	Pool fire
TF	Tank fire
VCE	Vapour cloud explosion

The following abbreviations are used for causes, the initiating ones being given first:

BFa	Brittle failure
CM	Control malfunction
CoErr	Contractor error
Crn	Corrosion
EF	Electrical failure
EI	Electrical ignition
Eq	Earthquake
Er	Erosion
ES	Electrical storm
ExR	Exothermic reaction
Fat	Fatigue
GFa	Gasket failure
HEm	Hydrogen embrittlement
HeEL	Heat exchanger leak
IHPF	Ignition by hot particles from flare
L	Leak (from unknown cause)
MaErr	Management error
MFa	Mechanical failure
MI	Mechanical ignition
NF	Navigational failure
OpErr	Operational error
OP	Over-pressure
PFa	Pipe failure
PU	Process upset
SE	Static electricity
(SU)	(During start-up)
TeA	Terrorist attack
TO	Tank overfilling
U	Unknown cause
UM	Unsafe maintenance
Van	Vandalism
VFa	Valve failure
VeFa	Vessel failure
WFa	Weld failure
WFr	Water freezing

Losses of £50 million and more at 1988 prices are set in **bold** type in the table.

Reference

1. Garrison, W. G., *One hundred largest losses – A thirty year review of property damage losses in the hydrocarbon-chemical industries*, 11th and 12th edns, Marsh & McLennan Protection Consultants, 222 South Riverside Plaza, Chicago, Illinois (1988 and 1989)

Table E.1 100 largest property damage losses in the hydrocarbon/chemical industries, 1958–1987 (From M&M Protection Consultants[a]. Losses trended to 1988 values)

Date (d/m/y)	Location	Plant etc.	Material involved	Loss ($ million)	Incident type	Initiating causes

A. *Oil refineries*

Date (d/m/y)	Location	Plant etc.	Material involved	Loss ($ million)	Incident type	Initiating causes
25/10/88	Pulau Merlimau, Singapore	Tank farm	Naphtha	12	PoF	MFa
8/6/88	Port Arthur, Texas	Tank farm	Propane	16	VCE, F	PFa
5/5/88	Norco, Louisiana	Fluid cat cracker	C_3 hydrocarbons	300	VCE, Fs	Crn, P.Fa
24/11/87	Torrance, California	HF Alkylation	Propane	15	BLEVE, Fs	MFa
23/6/87	Mississauga, Ontario	Hydrotreater	H_2/lube oil	20	F	VFa
22/03/87	Grangemouth, Scotland	Hydrocracker	H_2/distillate	80	IE, Fs	CM
23/1/85	Wood River, Illinois	Deasphalter	Propane	30	PoF, VCE	WFr, PFa
13/12/84	Las Piedras, Venezuela	Hydrodesulphuriser	Hot oil	65	F, MEs	PFa (Fat, HEm?)
15/8/84	Ft. McMurray, Alberta	Fluid cat cracker	HCs	79	VCE, PoF	Er, PFa
23/7/84	Romeoville, Illinois	Oil refinery	HCs	132	BLEVEs, VCE, F	WFa
8/3/84	Kerala, India	Oil refinery	Naphtha etc.	12	VCE, PoF	HeEL?
30/8/83	Milford Haven, Wales	Crude oil tank	Crude oil	16	TF, BO	IHPF?
7/4/83	Avon, California	Fluid cat cracker	HCs	52	F	PFa
31/3/82	Kashima, Japan	Desulphuriser	Fuel oil	15	F	HEm, PFa
20/1/82	Ft. McMurray, Alberta	Cat refiner	Hydrogen, HCs	22	F	ES
20/8/81	Shuaiba, Kuwait	Oil tank farm	Naphtha	48	TF	U
31/12/80	Corpus Christi, Texas	Hydrocracker	HCs	22	F	MFa
26/6/80	Sydney, Australia	Deasphalter	Butane	23	IE, F	U (SU)
20/1/80	Borger, Texas	HF alkylation	C_3s, etc.	44	VCE, F	WFr?
8/1/80	Avon, California	Fluid coker	Heavy oils	19	?	Van, CM
11/12/79	Geelong, Australia	Oil distillation	Crude oil	16	F	MFa
1/9/79	Deer Park, Texas	Oil tanker	Ethanol & HCs	97	IE, TF, PoF	ES
30/8/79	Good Hope, Louisiana	LPG barge	Butane	15	MEs, Fs	NF
21/7/79	Texas City, Texas	Alkylation	C_3s, etc.	34	VCE, Fs	PFa
30/3/79	Linden, New Jersey	Fluid cat cracker	C_3s, C_4s	25	FF, IE	PFa
3/10/78	Denver, Colorado	Polymerisation	C_3s	34	VCE, F	PFa
30/5/78	Texas Ci, Texas	Tanks & spheres	C_4s, etc.	85	ME, FB, BLEVEs	CM

Date	Location	Unit	Material	No.		
17/7/77	Baton Rouge, Louisiana	Fluid cat cracker	Fuel gas	16	IEs, F	EI
24/9/77	Romeoville, Illinois	Storage tanks	Oil products	13	IE, TFs	ES
17/8/75	Philadelphia, Pa	Oil tanks	Various oils	24	IEs, TF, FF	TO
16/3/75	Avon, California	Fluid coker	Heavy oil	19	Im, Fs	OpErr
31/1/75	Marcus Hook, Pa	Oil tanker	Crude oil	15	IEs, F	NF
24/8/73	St Croix, Virgin Isles	Hydrodesulphuriser	H2/HCs	26	Fs,	WFa
30/3/72	Rio de Jnro, Brazil	LPG storage	LPG	12	FF, BLEVE, TFs	OpErr
29/2/72	Delaware Cty, Del	Fluid coker	Heavy oil	15	F	PFa
5/12/70	**Linden, New Jersey**	**Hydrocracker**	**Various oils**	77	IE	MFa
17/9/70	Beaumont, Texas	Oil tank	Various oils	19	TFs	ES
6/3/69	Prto La Cruz, Venezuela	Crude still	Crude oil	14	FF, IEs, TFs	PFa or VFa
20/1/68	**Pernis, Netherlands**	**Oil tank**	**Crude oil**	89	VCE, F	OpErr
19/12/67	El Segundo, California	Oil storage pond	Fuel oil	18	PoF	ES
8/8/67	**Lake Charles, Louisiana**	**Alkylation**	**Isobutane**	58	VCE, Fs	VFa
4/1/66	Fézin, France	**LPG sphere**	**LPG**	63	FF, BLEVEs	OpErr
16/6/64	**Niigata, Japan**	**Whole refinery**	**Various oils**	80	TFs, IE	Eq

B. Petrochemical plants

Date	Location	Unit	Material	No.		
8/9/88	Rafnes, Norway	Vinyl chloride	Vinyl chloride	11	VCE, F	GFa
14/11/87	**Pampa, Texas**	**Acetic acid**	**Butane**	184	IE, VCE, Fs	U
6/7/85	Clinton, Iowa	Ammonia	Synthesis gas	13	E	WFa
19/5/85	**Priola, Sicily**	**Ethylene**	**Propylene etc.**	65	F, BLEVE	CM
4/10/82	Freeport, Texas	Transformer	Mineral oil	15	F, E	EF
23/7/80	Seadrift, Texas	Ethylene oxide	Ethylene oxide	15	IE	ES, CM
11/12/79	Ponce, Puerto Rico	Dimerisation	HCs	21	F, E	VeFa
8/12/77	Brindisi, Italy	Ethylene	HC gases	46	VCE, F	U
4/5/76	Plaquemine, Louisiana	Ethylene oxide	EO/PO	21	F	HeEL
24/5/76	Geismar, Louisiana	Polyglycol ethers	EO/glycols	16	IE, F	CM?
7/11/75	Beek, Netherlands	Olefins	Propylene	42	VCE, F	BFa, PFa
1/6/74	**Flixborough, UK**	**Cyclohexanol/one**	**Cyclohexane**	378	VCE, Fs	PFa
8/7/73	Tokuyama, Japan	Ethylene	Hydrogen, ethylene	37	F, E	CM, ExR
7/11/71	Morris, Illinois	Ethylene oxide	Oxygen, ethylene	15	IE, F	CM
23/10/69	Texas City, Texas	Butadiene	Vinyl acetylene	24	IE	OpErr, ExR
19/1/66	Raunheim, W Germany	Ethylene	Methane	24	E	BFa
3/4/63	Plaquemine, Louisiana	Ethylene	Cracked gases	15	E, F	U

Table E.1 (*Continued*)

Date (d/m/y)	Location	Plant etc.	Material involved	Loss ($ million)	Incident type	Initiating causes
C. Plastics and rubbers						
11/7/83	Port Arthur, Texas	Polyethylene	Polyethylene	16	F	U
18/4/82	Edmonton, Alberta	LD polyethylene	Ethylene	22	E	L
21/10/80	**New Castle, Delaware**	**Polypropylene**	**Propylene**	58	VCE, F	UM
10/2/75	**Antwerp, Belgium**	**LD polyethylene**	**Ethylene**	53	E	Fat
8/10/73	Goi, Japan	Polyethylene	Propylene, hexane	17	VCE?, F	EF
26/2/71	Longview, Texas	LD polyethylene	Ethylene	15	VCE?, F	PFa
12/8/69	Flemington, New Jersey	PVC	VCM	13	VCE, F	OpErr, SE
13/10/66	La Salle, Quebec	Polystyrene	Styrene	14	E, F	ExR
25/8/65	Louisville, Kentucky	Neoprene	Vinyl acetylene	36	E, F	MFa
12/1/65	Attleboro, Mass	PVC	VCM	18	E, F	UM
D. Chemical plants						
9/3/82	Philadelphia, Pa	Phenol	Cumene hydroperoxide	27	IE, F	ExR
19/7/81	Greens Bayou, Texas	Herbicide	Terephthaloyldichloride	12	IE	ExR
11/2/81	Chicago Heights, Illinois	Resins	Catalyst?	16	E, F	ExR
17/5/80	Deer Park, Texas	Phenol/acetone	Cumene	19	F	MFa
28/7/79	Sauget, Illinois	Fine chemicals	Nitrodiphenylamine	11	IE, F	F, EF, ExR
27/4/62	Marietta, Ohio	Phenol	Benzene	18	E, F	OP
17/4/62	Brandenburg, Kentucky	Ethanolamine	Ethylene oxide	18	IE	OP, ExR
4/10/60	Kingsport, Tennessee	Aniline	Nitrobenzene/nitric acid	36	IE	PU
E. Natural gas plants						
15/8/87	**Ju'aymah, Saudi Arabia**	**Gas fractionation**	**Propane**	61	VCE	EF
21/11/85	Tioga, North Dakota	Gas processing	Light hydrocarbons	10	E, F	OP, MFa
30/9/84	Basile, Louisiana	Gas producing	Oil + gas	31	VCE, F	PFa
14/4/83	Bontang, Indonesia	LNG	Light hydrocarbons	34	IE, F	OpErr, MaErr
15/4/78	**Abqaiq, Saudi Arabia**	**Gas processing**	**Light hydrocarbons**	83	VCEs, F	Crn, PFa
3/4/77	**Umm Said, Qatar**	**Refrig. LPG storage**	**Propane, butane**	128	FF, PoF	WFa

F. Terminals

2/1/88	Floreffe, California	Petroleum products	Diesel oil	13	EnCtm	WFa?, BFa?
21/12/85	Naples, Italy	Petroleum products	Gasoline, fuel oil	43	VCE, PoF	TO
5/1/85	Mont Belvieu, Texas	NGL	Light hydrocarbons	42	VCE, F	CoErr
19/11/84	Mexico City, Mexico	LPG	LPG	21	BLEVEs, FB, F	PFa
7/1/83	Newark, New Jersey	Petroleum products	Gasoline	37	VCE, PoF	TO
19/4/79	Port Neches, Texas	Crude oil port	Crude oil	45	Es, F	ES
8/1/79	Bantry Bay, Ireland	Crude oil port	Crude oil	29	IEs, F	U
17/12/76	Los Angeles, California	Crude oil port	Crude oil	21	IE, F	EI?
4/8/72	Trieste, Italy	Crude oil port	Crude oil	27	Es, F	TeA

G. Miscellaneous

11/10/87	Ft. McMurray, Alberta	Tar sand plant	Diesel oil/tar sand	36	F	MI
26/5/83	Prudhoe Bay, Alaska	Oil flow station	NGL, crude oil	37	IE, F	OP, VeFa
26/2/80	**Near Brooks, Alberta**	**Compressor station**	Natural gas	51	E, F	PFa
19/12/82	**Tacoa, Venezuela**	**Oil tank farm**	**Fuel oil?**	53	IE, BO, F	MaErr

H. Pipelines

8/7/77	**Near Fairbanks, Alaska**	**Pumping station**	**Crude oil**	66	F	OpErr, UM
4/6/77	Abqaiq, Saudi Arabia	Pipeline	Light hydrocarbons	18	VCE	PFa

Some details given in the CPL regulations[1] and Approved List[2]

The written information given on the supply label must include the following details:

- The name of the substance;
- Its hazard classification(s), each of which consists of a listed adjective with or without a listed adverb;
- A code number to indicate particular risks (Table F.3);
- A second code number to indicate safety precautions (Table F.4).

Substances to which the supply regulations apply are classified as explosive, oxidising, extremely flammable, highly flammable, flammable, very toxic, toxic, harmful, corrosive and irritant. Their characteristic properties are shown in Table F.1. The last five classifications apply to 'substances hazardous to health' whose use is dealt with in the COSHH regulations. Corrosive here means corrosive to the skin rather than to metals. Criteria for very toxic, toxic and harmful substances are given in Table F.2. Since these classifications are not mutually exclusive, the label may have to show more than one hazard classification, although only the highest degree has to be shown for hazards of the same type.

Table F.1 Classification and characteristic properties of dangerous substances

Classification	Characteristic properties
Explosive	A substance which may explode under the effect of flame or which is more sensitive to shocks or friction than dinitrotoluene.
Oxidising	A substance which gives rise to highly exothermic reaction when in contact with other substances, particularly flammable substances.
Extremely flammable	A liquid having a flash-point of less than 0°C and a boiling point of less than or equal to 35°C.
Highly flammable	A substance which: (a) May become hot and finally catch fire in contact with air at ambient temperature without any application of energy; (b) Is a solid and may readily catch fire after brief contact with a source of ignition and which continues to burn or to be consumed after removal of the source of ignition; (c) Is gaseous and flammable in air at normal pressure; (d) In contact with water or damp air, evolves highly flammable gases in dangerous quantities; or (e) Is a liquid having a flash-point below 21°C.
Very toxic	A substance which if it is inhaled or ingested or if it penetrates the skin may involve extremely serious acute or chronic health risks and even death.
Toxic	A substance which if it is inhaled or ingested or if it penetrates the skin may involve serious acute or chronic health risks and even death.
Harmful	A substance which if it is inhaled or ingested or if it penetrates the skin may involve limited health risks.
Corrosive	A substance which may on contact with living tissues destroy them.
Irritant	A non-corrosive substance which, through immediate, prolonged or repeated contact with the skin or mucous membrane, can cause inflammation.

Table F.2 Criteria for the classification of substances as very toxic, toxic or harmful

Category	Median lethal dose (LD_{50})		Median lethal concentration (LC_{50}) absorbed by inhalation in rat (mg/litre) (4 hours)
	Absorbed orally in rat (mg/kg)	Absorbed percutaneously in rat or rabbit (mg/kg)	
Very toxic	≤25	≤50	≤0.5
Toxic	>25 to 200	>50 to 400	>0.5 to 2
Harmful	>200 to 2000	>400 to 2000	>2 to 20

Table F.3 Risk phrases for dangerous substances with examples from Approved List[2]

Number	Risk phrase	Examples
R1	Explosive when dry	Wet guncotton
R2	Risk of explosion by shock, friction, fire or other sources of ignition	Some peroxides
R3	Extreme risk of explosion by shock, friction, fire or other sources of ignition	Dibenzoyl peroxide
R4	Forms very sensitive explosive metallic compounds	Picric acid
R5	Heating may cause an explosion	Perchloric acid >50%
R6	Explosive with or without contact with air	Acetylene
R7	May cause fire	Fluorine
R8	Contact with combustible material may cause fire	Sodium nitrate
R9	Explosive when mixed with combustible material	Potassium chlorate
R10	Flammable	Kerosene
R11	Highly flammable	Xylene, methanol
R12	Extremely flammable	Isopentane
R13	Extremely flammable liquefied gas	LPG
R14	Reacts violently with water	Sodium, sulphur trioxide
R15	Contact with water liberates highly flammable gases	Calcium carbide, lithium metal
R16	Explosive when mixed with oxidising substances	Red phosphorus
R17	Spontaneously flammable in air	Yellow phosphorus
R18	In use may form flammable/explosive vapour–air mixture	Some paints
R19	May form explosive peroxides	1,4-dioxan
R20	Harmful by inhalation	Methyl chloride
R21	Harmful in contact with skin	Iodine
R22	Harmful if swallowed	Ethylene glycol
R23	Toxic by inhalation	Aniline
R24	Toxic in contact with skin	Phenol
R25	Toxic if swallowed	Arsenic
R26	Very toxic by inhalation	Carbon tetrachloride
R27	Very toxic in contact with skin	Nicotine
R28	Very toxic if swallowed	Mercuric chloride
R29	Contact with water liberates toxic gas	Aluminium phosphide
R30	Can become highly flammable in use	
R31	Contact with acids liberates toxic gas	Sodium hypochlorite
R32	Contact with acids liberates very toxic gas	Sodium cyanide
R33	Danger of cumulative effects	(being replaced by 7LR48 where appropriate)
R34	Causes burns	Sodium
R35	Causes severe burns	Sodium hydroxide
R36	Irritating to eyes	Sulphur dioxide
R37	Irritating to respiratory system	Hydrogen chloride
R38	Irritating to skin	Methyl methacrylate
R39	Danger of very serious irreversible effects	Benzene
R40	Possible risk of irreversible effects	Hydrazine
R41	Risk of serious damage to eyes	
R42	May cause sensitization by inhalation	Maleic anhydride
R43	May cause sensitization by skin contact	Formalin
R44	Risk of explosion if heated under confinement	
R45	May cause cancer	
R46	May cause heritable genetic damage	
R47	May cause birth defects	
R48	Danger of serious damage to health by prolonged exposure	

Table F.4 Safety phrases for dangerous substances with examples from Approved List[2]

Number	Safety phrase	Example
S1	Keep locked up	Sodium cyanide
S2	Keep out of reach of children	Hydroquinone
S3	Keep in a cool place	
S4	Keep away from living quarters	
S5	Keep contents under . . . § (liquid)	Sodium
S6	Keep under . . . § (inert gas)	
S7	Keep container tightly closed	Methanol
S8	Keep container dry	Calcium carbide
S9	Keep container in a well-ventilated place	Propylene
S12	Do not keep the container sealed	
S13	Keep away from food, drink and animal feeding stuffs	Pyrethrins
S14	Keep away from . . . § (incompatible material)	4-Chlorobenzoyl peroxide
S15	Keep away from heat	Chloramine T, sodium salt
S16	Keep away from sources of ignition – no smoking	Heptane
S17	Keep away from combustible material	
S18	Handle and open container with care	Methacrylonitrile
S20	When using do not eat or drink	
S21	When using do not smoke	
S22	Do not breathe dust	Aluminium phosphide
S23	Do not breathe gas/fumes/vapour/spray §	
S24	Avoid contact with skin	Chlorobenzene
S25	Avoid contact with eyes	Acetyl acetone
S26	In cases of contact with eyes, rinse immediately with plenty of water and seek medical advice	Acetic acide >25%
S27	Take off immediately all contaminated clothing	Acetonitrile
S28	After contact with skin, wash immediately with plenty of . . . § (liquid)	Sodium hypochlorite solution (> 10% active chlorine)
S29	Do not empty into drains	Acrolein
S30	Never add water to this product	
S33	Take precautionary measures against static discharges	Propane
S34	Avoid shock and friction	
S35	This material and its container must be disposed of in a safe way	
S36	Wear suitable protective clothing	Toluidine
S37	Wear suitable gloves	Sodium hydroxide
S38	In case of insufficient ventilation wear suitable respiratory equipment	
S39	Wear eye/face protection	Sodium peroxide
S40	To clean the floor and all objects contaminated by this material, use . . . §	
S41	In case of fire or explosion do not breathe fumes	Dichloroisocyanuric acid
S42	During fumigation/spraying wear suitable respiratory equipment §	
S43	In case of fire use . . . § (particular fire-fighting equipment)	Potassium
S44	If you feel unwell seek medical advice (show the label where possible)	Sodium nitrite
S45	In case of accident or if you feel unwell seek medical advice (show the label where possible)	Anisidines
S46	If swallowed, seek medical advice immediately and show this container or label	

Table F.4 (*Continued*)

Number	Safety phrase	Example
S47	Keep at temperature not exceeding . . . §°C	
S48	Keep wetted with . . . §	
S49	Keep only in the original container	
S50	Do not mix with . . . §	
S51	Use only in well-ventilated areas	
S52	Not recommended for interior use on large surface areas	

§ = word(s) or figure specified by manufacturer

References

1. S.I. 1984, No. 1244, *The Classification, Packaging and Labelling of Dangerous Substances Regulations 1984*, HMSO, London
2. HSC, *Information Approved for the Classification, Packaging and Labelling of Dangerous Substances*, 2nd edn, HMSO, London (1987)

Some details of the COSHH regulations and Approved Codes of Practice[1]

The COSHH regulations aim to:

1. Deal with the whole subject in one set of regulations;
2. Set out the principles to be followed;
3. Allow for future changes in standards and new technology;
4. Enable the UK to fulfil the terms of Directive/80/1107/EEC and ratify the ILO convention No. 139 on carcinogenic substances;
5. Simplify the law by revoking former legislation.

Two supporting ACOPs are issued with the regulations, a general one and one for the Control of Carcinogenic Substances. Two further ACOPs are issued separately, one for the Control of Vinyl Chloride Monomer (VCM) and one on Fumigation. The regulations apply to all substances which are hazardous to health, and to dust clouds of all kinds. The former are classified into the same five categories which apply in CPLR [2.5.4]. The import of several substances into the UK is prohibited [G.1]. Employers are required to make an assessment (see HSE Guidance Note EH 42) before work which may involve exposure to a substance hazardous to health is started. Appropriate selection, use and maintenance of means of controlling exposure (HSE Guidance Notes EH 40 and 42) which do not depend on the use of personal protective equipment are required. The latter will only be considered appropriate when other means are not reasonably practicable. There are requirements for the health surveillance of employees who may be exposed to certain substances hazardous to health [G.2], and for the collection, storage and proper use of these data. Special attention is required to prevent or control occupational exposure to carcinogenic substances.

G.1 Substances whose import into the UK is prohibited for all purposes (Regulation 4(1), Schedule 2, Item No. 1)

2-naphthylamine, benzidene, 4-amino biphenyl, 4-nitrobiphenyl, their salts and any substance containing any of these compounds, except, as the by-product of a chemical reaction, in any other substance in a total concentration exceeding 0.1%.

695

G.2 Substances for which medical surveillance is appropriate (Regulation 11(2)(a) and (5), Schedule 5)

Substances	*Processes*
Vinyl chloride monomer (VCM)	In manufacture, production, reclamation, storage, discharge, transport, use or polymerisation.
Nitro or amino derivatives of phenol and of benzene or its homologues	In the manufacture of nitro or amino derivatives of phenol and of benzene or its homologues and the making of explosives with the use of any of these substances.
Potassium or sodium chromate or dichromate	In manufacture.
1-Naphthylamine and its salts O-Toluidine and its salts Dianisidine and its salts Dichlorobenzidine and its salts	In manufacture, formation or use of these substances.
Auramine, Magenta	In manufacture.
Carbon disulphide, disulphur dichloride, benzene including benzol, carbon tetrachloride, trichloroethylene	Processes in which these substances are used, or given off as vapour, in the manufacture of indiarubber or of articles or goods made wholly or partially of indiarubber.
Pitch	In manufacture of blocks of fuel consisting of coal, coal dust, coke or slurry with pitch as a binding substance.

Reference

1. S.I. 1988, No. 1657, *The Control of Substances Hazardous to Health Regulations 1988* and Approved Codes of Practice, *Control of Substances Hazardous to Health* and *Control of Carcinogenic Substances*, HMSO, London

Important Codes of Practice and British Standards[1,2]

H.1 List of Approved Codes of Practice issued under section 16 of HSWA 1974

No. and title	*Regulation supported (year and S.I. No.)*
1. Safety representatives and safety committees: brown booklet.	1977/500
2. Time off for the training of safety representatives: leaflet HSC 9.	1977/500
3. Control of lead at work: unnumbered booklet.	1980/1248
4. Work with asbestos insulation and asbestos coating: booklet COP 3.	1969/690
5. Health and Safety (First-Aid) Regulations 1981: booklet COP 4.	
6. Classification of dangerous substances for conveyance in road tankers and tank containers: booklet COP 5.	1981/1059
7. Petroleum-Spirit (Plastic Containers): Regulations 1982. Requirements for testing and marking or labelling: booklet COP 6.	1982/630
8. Principles of good laboratory practice. Notification of New Substances Regulations 1982: booklet COP 7.	1982/1496
9. Methods for the determination of ecotoxicity. Notification of New Substances Regulations 1982: booklet COP 8.	1982/1496
10. Methods for the determination of physico-chemical properties. Notification of New Substances Regulations 1982: booklet COP 9.	1982/1496
11. Methods for the determination of toxicity. Notification of New Substances Regulations 1982: booklet COP 10.	1982/1496
12. BS 697: 1977 'Specification for Rubber Gloves for Electrical purposes'.	
13. BS 1870 Part 1: 1979 'Specification for Safety Footwear other than all rubber and all-plastic moulded types'.	

No. and title	*Regulation supported (year and S.I. No.)*
14. BS 5426: 1976 'Specification for Footwear'.	
15. BS 5169: 1975 'Specification for Fusion welded steel air receivers'.	
16. Operational provisions of the Dangerous Substances (Conveyance by road in Road Tankers and Tank Containers Regulations) Regulations 1981. COP 11.	
17. BS 1870 Part 2: 1976 Specification for lined Rubber Safety Boots.	
18. BS 1870 Part 3: 1981 Specification for Polyvinyl Chloride Moulded Safety Footwear.	
19. Packaging of dangerous substances for conveyance by road. Classification, Packaging and Labelling of Dangerous Substances regulations 1984: unnumbered booklet.	1982/1496 and 1984/1244
20. Classification and labelling of substances dangerous for supply and/or conveyance by road. Notification of New Substances Regulations 1982. Classification, Packaging and Labelling of Dangerous Substances regulations 1984: unnumbered booklet.	1982/1496 and 1984/1244
21. Road Tanker Testing. Examination, testing and certification of the carrying tankers used for the conveyance of dangerous substances by road: unnumbered booklet.	1981/1059
22. Zoos – Safety, Health and Welfare standards for Employers and Persons at Work.	
23. The protection of persons against ionising radiation arising from any work activity: unnumbered booklet.	1985/1333

H.2 Selection of British standards considered significant to HS

BS No.	*Title*
470	Access and inspection openings for pressure vessels
767	Hydro-extractors and centrifugal machines
1651	Industrial gloves
1710[a]	Identification of pipelines
2915	Bursting discs and bursting disc assemblies
3351[a]	Piping systems for petroleum refineries and petrochemical plants
4343[a, b]	Industrial plugs, socket-outlets and couplers for a.c. and d.c. supply
4752[a, b]	Specification for switchgear and control gear for voltages up to and including 1000 a.c. and 1200 d.c.
4994	Vessels and tanks in reinforced plastics
5240[a, b]	General purpose industrial safety helmets

BS No.	Title
5343[a]	Gas detector tubes
5500[a]	Unfired fusion welded pressure vessels
6129	Code of Practice for the selection and application of bellows expansion joints for use in pressurised systems
381C[a]	Specifications for colours for identification, coding and special purposes
4275[a]	Recommendations or the selection, use and maintenance of respiratory protective equipment
5304[a]	Code of Practice for the safeguarding of machinery
5502[a]	Code of Practice for protective coating of iron and steel structures against corrosion
5908[a]	Code of Practice for fire precautions in chemical plant
5958[a]	Code of Practice for the control of undesirable static electricity
3923	Methods for ultrasonic examination of welds
4871[a]	Approval testing of welders working to approved welding procedures
5760	Reliability of systems, equipment and components

[a] = referred to in HSE guidance publications.
[b] = being considered for approval under section 16 of HSWA.

References

1. HSC Consultative Document: *Reference to standards in safety at work*, HMSO, London (1982)
2. HSE, *Standards significant to health and safety at work*, HSE, Bootle (1985)

Questionnaire for designers to ensure safe maintainability

It is assumed that the plant under design consists of several process units, each within its own battery limits. As a minimum safeguard that any unit can be maintained safely when it is shut down, the design should allow each unit to be isolated completely from process streams and materials, electricity, other services, effluents and vapour-relief systems. All fluid-handling lines should be isolated by a combination of valves with locks at the battery limits, with flanged joints between the valve and the unit where approved spectacle plates or line blinds are fitted. Further precautions must be taken where it may be necessary to open up and maintain some parts of a process unit while other parts are still operating and/or contain process fluids.

To facilitate safe maintenance and vessel entry, designers should ask themselves the following questions.

Q1A Which plant items may have to be entered during the working life of the plant, and for what purpose?

Q1A Will hot work be required inside the item to be entered?

Q1B Will it be necessary to enter these items while the rest of the unit still contains process materials, or would this only be necessary when the entire unit has been shut down, isolated, emptied, vented, cleaned, purged and the atmosphere in it certified safe to enter?

Q2   (Mainly if answer to Q1B is 'Yes'.) List and consider each branch or connection into the item through which material might enter; state the nature and hazards of the material and the method of isolation provided for in the design (Figure J.1[1]), e.g.

    (a) Valve and lock only (only permissible for cold and clean air and water at low pressure);

    (b) Lockable double block and bleed valves (only permissible for clean, non-corrosive fluids of low hazard at moderate temperature and pressure, such as compressed air, hot water, low-pressure steam);

    (c) Valve and lock with spectacle plate or line blind between valve and item (generally used for non-corrosive fluids of low to medium hazard at ambient temperature and low pressure);

    (d) Valve and lock with removable spool piece and blank flange between valve and item (alternative to (c));

(e) Line blind valve (alternative to (c) when frequent isolation is required);

(f) Combination of (b) with (c) or (d) (used for all fluids at high temperatures and/or pressures, and for all high-hazard materials).

(Note: Special problems arise when pressure relief devices discharge directly into a common unvalved relief line [15.5 and 16.6.7].)

Q3 What materials will be left in the item to be entered when it has been shut down, emptied as far as possible without disconnecting any joints, and vented, and roughly how much of each. What are their expected hazards? Are they:

(a) Gas, liquid, solid?

(b) Flammable, combustible, unstable, toxic, corrosive?

(c) Soluble in water, insoluble in but lighter/heavier than water?

(d) Volatile in air, volatile in steam, non-volatile?

(e) Will non-volatile materials be left after steaming out?

(f) If so, will they adhere tenaciously to the internals of the equipment?

(g) Will they decompose when heated, giving off flammable/toxic vapours?

Q4 What methods will be required to remove residual materials from the item after shutdown, emptying to process, and venting?

(a) Dissolve in hot/cold water,

(b) Displace upwards with water,

(c) Disperse in hot/cold water with detergent and wash out with water,

(d) Break up with high-pressure water jet and wash out,

(e) Steam out,

(f) Use (specified) solvent,

(g) Other (specified) means,

(h) Will a temporary stirrer be required to assist removal of residual materials? If so, what type would be used, and what provision is there for installing and driving it?

Q5 Will it be safe to release gases vented, liquids drained, and solids removed from shutdown equipment and pipes connected to it, to the surroundings, or will special vent and/or drain lines be needed for safe disposal? If so, does the design include them?

Q6 Can the item be made safe to enter without breathing apparatus or protective clothing, or will these usually have to be used by maintenance workers? If so, explain why, and specify what protective clothing and devices should be worn.

Q7 What internal devices (stirrers, heating/cooling coils, baffles, distillation trays, etc.) are there in the item which will cause problems in emptying and cleaning, or hazards to a person working inside the item? If so, can these problems be overcome by modifying the design?

Q8 Will the equipment be entered through a manhole or other opening? Will the openings be large enough and suitably placed for a person inside who is wearing the protective equipment specified in Q6 to be rescued (e.g. by hauling with a life-line attached to a safety harness)?

Criterion	Line blind valve	Spectacle plate, or line blind	Double block and bleed	Removable spool
Relative overall cost	Least expensive	Medium expense depending on frequency of changeover		Most expensive
Manhours for double changeover	Negligible	1 to 3	Negligible	2 to 6
Initial cost	Fairly high	Low	Very high	High
Certainty of shut-off	Complete	Complete	Doubtful	Complete
Visual indication?	Yes	Yes	Yes, but suspect	Yes
Who operates?	Plant operator	Pipefitter	Plant operator	Pipefitter

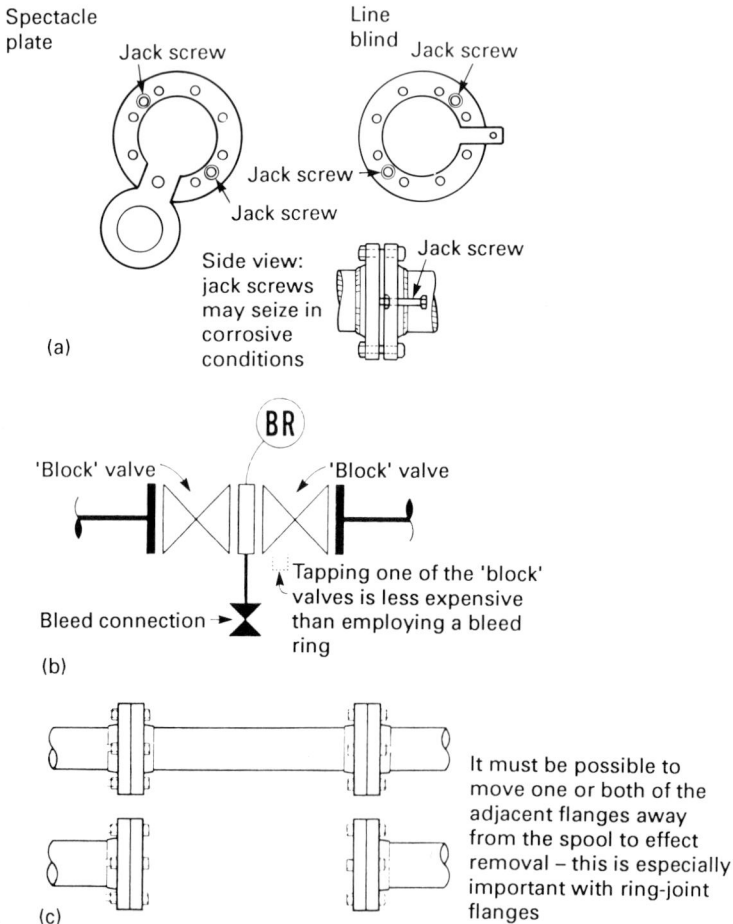

Figure J.1 Types of in-line closure: (a) spectacle plate and line blind; (b) double block and bleed; (c) removable spool. (From Sherwood and Whistance[1], courtesy Chapman and Hall)

Q9   What access is provided/needed both at the entrance of the item and inside it to its floor and any parts of it where work may have to be done? (If temporary access platforms, ladders, etc. have to be used, care must be taken to ensure that suitable ones are provided on site, that there is space for them and those using them inside the item, and that they can be fitted securely and used safely.)

Q10   What davits, winches and other gear are provided/needed to rescue a person who collapses inside the item? Can these be readily and securely fixed before anyone enters? (Note: One person outside a vessel cannot be expected to haul an unconscious person inside it to safety by a life-line only. A suitably placed winch or block and tackle is needed.)

Q11   Might the atmosphere surrounding the item present hazards to maintenance workers outside and inside it? For work inside the equipment, especially hot work, what means of ventilating it with uncontaminated air are provided/needed?

Q12   Can the item be isolated from electrical supply and all other services (including refrigeration, heating oil, hydraulic and pneumatic power) without affecting their supply to other items of the unit where they may still be needed while the item is being maintained?

Despite the number of these questions, they cannot claim to be comprehensive, and others will occur to readers. It is suggested to those readers who are concerned with the problem in a particular process industry that they consider and list questions which they think ought to be raised in plant design to ensure that it can be maintained safely. They should then take them up with those responsible for design in their own organisations.

The questions raised here apply not only to the initial design of the plant but also to the design of later modifications and revamps. If field changes have to be made to a plant during construction and commissioning, the same questions should be raised again then.

Reference

1. Sherwood, D. S. and Whistance, D. J., *The Piping Guide* (figures 2.59 to 2.61), Chapman and Hall, London (1979)

Checklist to test the safety policy statement[1]

The following checklist which appears as Appendix 2 in HSE's publication *Effective policies for health and safety*[1] was designed to probe the applicability, strengths and weaknesses of organisations' safety policies. The extent to which each section applies depends on the type of organisation and the hazards with which it has to cope.

The policy statement

1.1 Does it give a clear unequivocal commitment to safety?

1.2 Is it authoritative? Is it signed and dated by a director? Has it been agreed by the board?

1.3 Is the policy to be regularly reviewed? If so, by whom and how often?

1.4 Has it been agreed with the trade union representatives?

1.5 Are there effective arrangements to draw it to the attention of employees?

1.6 Does it state that its operation will be monitored at workplace, divisional and group level?

The organisation for health and safety

2.1 Is the delegation of duties logical and successive throughout the organisation?

2.2 Is final responsibility placed on the relevant director?

2.3 Are the responsibilities of senior managers written into the policy or specified in job descriptions?

2.4 Is the safety performance of managers an ingredient of their annual review?

2.5 Are the qualifications of managers where relevant to health and safety considered when making appointments?

2.6 Are the following key functional managers identified?
(a) Safety Manager
(b) Hygiene Manager
(c) Radiation Officer
(d) Engineering Manager
(e) Electrical Manager
(f) Training manager
Are their duties clearly understood?

2.7   Do managers understand the extent of their discretion to vary from systems and procedures?

2.8   Do they understand the consequences of failure to implement the policy in their area of responsibility?

2.9   Are there adequate arrangements for liaison with contractors' managers and others who come onto the site?

2.10 Are there adequate arrangements for consultation with the workforce?

Arrangements for health and safety

Training
3.1 Is there a system for the identification of training needs?
3.2 Is the responsibility for training properly allocated?
3.3 Does training cover all levels from senior managers to new entrants?
3.4 Are special risk situations analysed for training?
3.5 Are refresher courses arranged?

Safe systems of work
4.1 Are those tasks for which a safe system of work is required properly identified?
4.2 Are identified systems properly catalogued?
4.3 Are the systems monitored?
4.4 Are there systems to deal with temporary changes in the work?
4.5 Are there proper systems of work for maintenance staff?

Environmental control
5.1 Is the working environment made as comfortable as is reasonably practicable? Does it meet statutory requirements?
5.2 Is sufficient expertise available to identify the problems and reach solutions?
5.3 Is sufficient instrumentation available?
5.4 Are there arrangements to monitor the ventilation systems?
5.5 Are temperature/humidity levels controlled?
5.6 Is there adequate lighting provided? Are there satisfactory arrangements for replacement and maintenance?

Safe place of work
6.1 Are there arrangements to keep workplaces in a clean, orderly and safe condition?
6.2 Are walkways, gangways, paths and roadways clearly marked?
6.3 Are there arrangements for clearing hazards, e.g. substances likely to cause slipping, from the floors?
6.4 Is safe means of access provided to all working areas?
6.5 Are staircases, landings, teagles and openings in the floor protected?
6.6 Is storage orderly, safe and provided with easy access?
6.7 Are flammable, toxic and corrosive substances used safely and without hazard to health?
6.8 Are permit-to-work systems operated and monitored?

Machinery and plant
7.1 Is new machinery and plant vetted for health and safety prior to being brought onto site?
7.2 Is there a system of inspection to identify and safeguard dangerous machinery?
7.3 Is there a system for vetting plant and machinery after modifications?
7.4 Is there a routine check on interlocking devices?
7.5 Is pressurised plant subject to inspection and test?
7.6 Are monitoring systems and alarms tested at regular intervals?
7.7 Are lifting machines and tackle subject to regular inspection?

Noise
8.1 Are noise risks assessed and danger areas notified?
8.2 Is there a programme of noise reduction/control?
8.3 Is personal protection provided and worn?
8.4 Are the requirements of the Code of Practice for Reducing the Exposure of Employed Persons to Noise being met? Is there a risk from vibration?

Radiation
9.1 Is a competent person nominated to oversee the use of equipment and materials which may pose a radiation hazard?
9.2 Is adequate monitoring equipment available?
9.3 Are records kept in accordance with statutory regulations?

Dust
10.1 Do the arrangements for the control of dust meet statutory regulations?

Toxic materials
11.1 Are there adequate arrangements in the purchasing, stores, safety, medical and production departments for the identification of toxic chemicals and specifying necessary precautions?
11.2 Are storage areas adequately protected?
11.3 Are emergency procedures for handling spillage/escape laid down, known and tested?
11.4 Are there proper precautions for labelling?
11.5 Are there adequate arrangements for the issue, maintenance and use of respiratory protection where it is found to be necessary?

Internal communications
12.1 Is the role of safety representatives agreed?
12.2 Is there a properly constituted safety committee?
12.3 Is the level of management participation appropriate?
12.4 Is there a system for stimulating and maintaining interest in health and safety?
12.5 What arrangements are there to advise workers about the standard of the organisation's performance in health and safety?
12.6 Are there adequate means of communication from shop floor to management on health and safety matters?

12.7 Is there scope for joint management/shop floor inspections?

12.8 Are there efficient arrangements to process action on communication from the enforcing authorities?

Fire

13.1 Who is nominated to coordinate fire-prevention activities? Does he have sufficient authority?

13.2 What arrangements are there for fire fighting?

13.3 Is there an adequate fire-warning system? Is it regularly checked?

13.4 Are fire drills held and checked for effectiveness?

13.5 What arrangements are made to check compliance with the statutory fire certificate?

13.6 Are means of escape regularly checked and properly maintained?

13.7 Is there a proper system to account for staff and visitors in the event of an evacuation of the building being required?

13.8 Are flammable and explosive materials stored and used in compliance with statutory requirements?

Medical facilities and welfare

14.1 Are there adequate facilities for first-aid treatment?

14.2 Are sufficient persons trained in first-aid?

14.3 What arrangements are there for medical advice?

14 4 Are there adequate facilities to admit proper medical supervision particularly where this is a statutory requirement?

14.5 What medical records are needed and are they adequately kept?

14.6 Are the washing and sanitary facilities adequate?

14.7 Are cloakrooms and messrooms adequate?

Records

15.1 Are there adequate arrangements for the keeping of statutory records?

15.2 Are the records vetted for efficiency and accuracy?

15.3 Is sufficient use made of the information in the records to identify areas of strength and weakness (e.g. accident and ill-health experience or training needs)?

15.4 Is there sufficient access to records of performance by those with a legitimate interest?

15.5 Are copies of all the relevant statutory requirements and codes of practice available on site?

Emergency procedures

16.1 Are the areas of major hazard identified and assessed by qualified staff?

16.2 Are there procedures for dealing with the worst foreseeable emergency?

16.3 Have these procedures been promulgated and tested?

16.4 Are there adequate arrangements for liaison with other parties who may be affected or whose help may be required?

16.5 Are there arrangements to protect sensitive installations from malicious damage or hoax threats?

16.6 Do the above arrangements cover weekend/holiday periods?

Monitoring at the workplace

17.1 Is it understood that monitoring will be carried out?

17.2 Are there sufficient staff with adequate facilities to carry out the monitoring?

17.3 Are the standards expected known and understood?

17.4 Is there a system for remedying identified deficiencies within a given timescale?

17.5 Is the monitoring scheme sufficiently flexible to meet changes in conditions?

17.6 Are all serious mishaps investigated?

17.7 In the event of a mishap is the performance of individuals or groups measured against the extent of their compliance with the safety policy objectives?

17.8 Is monitoring carried out within the spirit as well as the letter of the written policy document?

Reference

1. HSE, *Effective policies for health and safety*, HMSO, London (1980)

Appendix L

Summary of incidents which have to be reported under RIDDOR[1]

(a) The death of any person as a result of an accident arising out of or in connection with work;
(b) Any person suffering any of the following injuries or conditions as a result of an accident arising out of or in connection with work;
   ○ Fracture of any bone except in the hand or foot,
   ○ Amputation of a hand or foot, or a finger, thumb or toe if completely severed,
   ○ The loss of sight of an eye, a penetrating eye injury or a chemical or hot metal burn to the eye,
   ○ Injury requiring prompt medical treatment or loss of consciousness resulting from an electric shock,
   ○ Loss of consciousness resulting from lack of oxygen,
   ○ Decompression sickness,
   ○ Acute illness or loss of consciousness caused by absorption of any substance by inhalation, ingestion or through the skin,
   ○ Acute illness thought to have resulted from exposure to a pathogen or infected material,
   ○ Any other injury causing the inured person to remain in hospital for more than 24 hours,
(c) Any of the dangerous occurrences listed in Schedule 1, part 1 of RIDDOR, a summary of which follows later,
(d) An employee, a self-employed person or a trainee who is incapacitated and off work for more than three days as a result of an accident at work,
(e) The death of an employee following not more than one year after an accident at work.

Any of these must be reported in writing to the enforcing authority, and those of type (a), (b) or (c) must first be reported immediately by telephone.

The dangerous occurrences which must be reported under RIDDOR include:

1. Collapses or overturning of failure of any load-bearing parts of lifts, hoists, cranes, derricks or mobile powered access platforms, excavators, and pile-driving frames or rigs more than seven metres high when working;

2. Various incidents at a fun fair;
3. Explosion, collapse or bursting of any closed vessel which resulted in stopping the plant involved for more than 24 hours and which could have caused death, or an injury or condition reportable under RIDDOR;
4. Electrical short circuit or overload accompanied by fire or explosion with the same possible consequences as under (3);
5. An explosion or fire due to the ignition of process materials, their by-products or finished products with the same possible consequences as under (3);
6. The sudden, uncontrolled release of one tonne or more of a highly flammable liquid or gas or a flammable liquid above its boiling point;
7. A collapse or partial collapse of certain scaffolds;
8. A collapse or partial collapse of certain buildings or parts of buildings;
9. The uncontrolled or accidental release of any substance or pathogen with the same possible consequences as under (3);
10. Any unintentional ignition or explosion of explosives;
11. Failure of any freight container while suspended;
12. Certain pipeline accidents;
13. Certain accidents involving road tankers or tank containers;
14. Certain accidents involving vehicles carrying dangerous substances by road;
15. Certain malfunctions of breathing apparatus;
16. Incidents involving unintentional contact with or discharge from high-voltage overhead cables;
17. Accidental collisions between locomotives or trains and other vehicles at factories or docks with the same possible consequences as under (3).

Certain diseases contracted by employees, trainees and self-employed persons through exposure to particular substances at work must be reported to the appropriate authority, but only if a written diagnosis has been received from a doctor. The reports should be made by the employer, the person responsible for providing the training and the self-employed person respectively.

1. Substances causing poisoning through any work activity are:
   (a) Acrylamide monomer,
   (b) Arsenic or an arsenic compound,
   (c) Benzene or a benzene homologue,
   (d) Beryllium or a beryllium compound,
   (e) Cadmium or a cadmium compound,
   (f) Carbon disulphide,
   (g) Dioxane,
   (h) Ethylene oxide,
   (i) Lead or a lead compound,
   (j) Manganese or a manganese compound,
   (k) Mercury or a mercury compound,
   (l) Methyl bromide,
   (m) Nitrobenzene or a nitro- or amino- or chloro- derivative of benzene or a benzene homologue,
   (n) Oxides of nitrogen,
   (o) Phosphorus or a phosphorus compound.

Certain skin diseases caused by work involving the following exposures are reportable:

2.   Chromic acid or other chromium compounds;
3,4,5. Mineral oil, tar, pitch or arsenic;
6.   Ionising radiation

The following lung diseases are reportable when associated with particular work or work involving particular exposures:

7. Occupational asthma;	14. Cancer of a bronchus or lung;
8. Extrinsic alveolitis;	15. Leptospirosis;
9. Pneumoconiosis;	16. Hepatitis;
10. Byssinosis;	17. Tuberculosis;
11. Mesothelioma;	18. Any pathogenic illness;
12. Lung cancer;	19. Anthrax.
13. Asbestosis;	

The following other conditions are reportable when associated with particular work exposures:

20,21.	Bone cancer	Ionising radiation
22.	Cataract	Electromagnetic radiation
23,24.	Decompression sickness	Breathing at increased pressure
25.	Nasal and sinus cancer	Wood and leather dust, volatile nickel compounds
26.	Angiosarcoma of the liver	Vinyl chloride polymerisation
27.	Urinary tract cancer	Naphthylamines and certain other substituted aromatic compounds
28.	Vibration white finger	Hand-held chain saws and certain other vibratory tools.

Reference

1. S.I. 1985, No. 2023, *The Reporting of Injuries, Diseases and Dangerous Occurrences Regulations*, HMSO, London

Sources of HS training and information (mainly UK)

This appendix is a selection of mainly British sources of HS training, training materials and information relevant to the process industries, but not including academic courses, books and journals. The role of the City and Guilds of London Institute (referred to in 21.4), 76 Portland Place, London WIN 4AA (tel 071-278 2468) is, however, mentioned for the benefit of non-UK readers. It provides syllabuses for training schemes to meet various industrial needs and also assessment and certification of trainees. The training is carried out at numerous centres (e.g. technical colleges) in English-speaking countries. I apologise to other sources which I have omitted to mention. This is not deliberate and merely reflects the limits of my knowledge and memory.

M.1 Non-statutory training organisations (NSTOs) in the UK

The role of NSTOs, in particular the Chemical and Allied Industries Training Review Council (CAITREC) and the Petroleum Training Federation, in training within industry in the UK was referred to in 21.4.2. CAITREC was set up at the joint initiative of the Chemical Industries Association and the Association of the British Pharmaceutical Industry, both of whom have their own training departments, and has since been joined by the Proprietary Association of Great Britain which has a training representative.

The names and addresses of NSTOs[1] for industries with a process content follow:

- The **Biscuit, Cake, Chocolate and Confectionery** Alliance, 1 Green Street, London W1Y 3RF
  Tel 071-629 8971
- The British **Cement** Association, Wexham Springs, Slough, Berks SL3 6PL
  Tel 0753 662727
- The **Chemical Industries** Association Ltd, and The **Chemical and Allied Industries** Training Review Council, King's Buildings, Smith Square, London SW1P 3JJ
  Tel 071-834 3399

- English **Clay** Industries Training Board, John Keay House, St Austell, Cornwall PL25 4DJ
  Tel 0726 74482
- **Dairy Trade** Federation, 19 Cornwall Terrace, London NW1 4QP
  Tel 071-486 7244
- **Drinks** Industries Training Association, Saxon House, Heritage Gate, Derby DE1 12NL
  Tel 0332 371980
- British **Fibreboard Packaging** Employers' Association, Sutherland House, 5/6 Argyle Street, London W1V 1AD
  Tel 071-434 3851
- **Fibre Cement** Manufacturers' Association, The Gersham Centre, Great Ashfield, Walsham-le-Willows, Bury St Edmunds, Suffolk IP30 9HS
  Tel 0359 259379
- **Food Manufacturers'** Industrial Group, 6 Catherine Street, London WC2B 5JJ
  Tel 071-836 2460
- UK Association of **Frozen Food** Producers, 1 Green Street, London W1Y 3RG
  Tel 071-629 0655
- British **Gas** plc, Rivermill House, 152 Grosvenor Road, London SW1B 3JL
  Tel 071-821 1444
- **Glass Training** Ltd, BGIRA Building, Northumberland Road, Sheffield S10 2DA
  Tel 0742 661494
- British **Leather** Federation, Leather Trade House, King's Park Road, Moulton Park, Northampton NN3 1JD
  Tel 0604 494131
- **Man-made Fibres** Producing Industry Training Board, 40 High Street, Rickmansworth, Herts WD3 1ER
  Tel 0932 778371
- National Association of British and Irish **Millers** Ltd, 21 Arlington Street, London SW1A 1RN
  Tel 071-493 2521
- **Offshore** Petroleum ITB, Fortes Road, Montrose, Angus, Scotland
  Tel 0674 72230
- The Flexible **Packaging Association**, 4 The Street, Shipton Moyne, Tetbury, Gloucestershire GL11 8PN
  Tel 0666 88406
- **Paintmakers'** Association of Great Britain, Alembic House, 93 Albert Embankment, London SE1 7TY
  Tel 071-582 1185
- The **Petroleum Training Federation**, 168 Regent Street, London W1
  Tel 071-439 2632
- The Association of the British **Pharmaceutical** Industry, 12 Whitehall, London SW1A 2DY
  Tel 071-930 3477
- **Plastics Processing** ITB, Training Centre, Halesfield 7, Telford, Shropshire TF7 4NA
  Tel 0952 684466

- The **Proprietary Association of Great Britain**, Vernon House, Sicilian Avenue, London WC1
  Tel 071-242 8331
- **Quarry Products** Training Council, 27 Crendon Street, High Wycombe, Bucks HP13 6LJ
  Tel 0494 34124
- **Refractories** Clay Pipes and Allied Industries Training Council, c/o The University of Sheffield, School of Materials, Elmfield, Northumberland Road, Sheffield S10 2TZ
  Tel 0742 78555
- British **Rubber** Industry Training Organisation (BRITO), Scala House, Holloway Circus, Birmingham B1 1EQ
  Tel 021-643 9599
- **Silica and Moulding Sands** Association, 19 Warwick Street, Rugby, Warwickshire CV21 3DH
  Tel 0788 73041
- **Soap and Detergent** Industry Association, Hayes Gate House, 27 Uxbridge Road, Hayes, Middlesex UB4 0JD
  Tel 081-573 7992
- The National Association of **Soft Drinks** Manufacturers, 6 Catherine Street, London WC2B 5UA
  Tel 071-379 5737
- British **Steel** Corporation Central Training Unit, Ashborne Hill College, Ashborne Hill, Leamington Spa, Warwickshire CV33 9GW
  Tel 0926 651321
- The **Sugar** Bureau, 120 Duncan House, Dolphin Square, London SW1V 3PW
  Tel 071-828 9465
- British **Textile** Employers' Association, Reedham House, 31 King Street West, Manchester M3 2PF
  Tel 061-834 7871
- **Water** Authorities' Association, 1 Queen Anne's Gate, London SW1H 9BT
  Tel 071-222 8111
- Confederation of British **Wool** Textiles, 60 Toller Lane, Bradford BD8 9BZ
  Tel 0274 491241

M.2 UK sources of visual aids for safety training

Safety training films (16 mm, colour, with optical soundtrack) which are also generally available as videocassettes (with choice of U-matic, VHS and usually Betamax format) as well as other visual aids can be purchased or hired from several other UK organisations. Some of relevance to the process industries are listed below with titles and brief details.

M.2.1 Longman Training

Cullum House, North Orbital Road, Denham, Middlesex UB9 5HL, Tel 0895 834142

Films and videocassettes available for purchase or hire from this company, launched in September 1990, include those made by the former Millbank Films Ltd, and several international films.

- *If you only knew* (24 min) – safe handling of chemicals.
- *Rescue team alert* (20 min) – works rescue teams.
- *What happens next* (18 min) – actions after the accident, with a Trainer's Guide.
- *One million hours* (22 min) – accident investigation.
- *Blind man's bluff* (19 min) – hazard spotting, with illustrated Trainer's Guide and score cards for trainees.
- *Flashpoint* (21 min) – safety in the laboratory, with a Leader's Guide.
- *Nobody's fault* (20 min) – a powerful lesson in the dangers of carelessness, with discussion notes and training recommendations. This film which won several awards in 1974 has proved very successful with industrial audiences.
- *Out of order* (23 min) – hazards arising from maintenance with Discussion Leader's Guide.
- *Is there anything I've forgotten?* (21 min) – permit-to-work systems, with handbook for discussion and training (an excellent film).
- *Permit to work* (23 min) – problems of contractors working in process areas, with a Trainer's Guide.
- *I can't see* (22 min) – a dramatic film on eye protection.
- *It need not happen* (21 min) – protective clothing and equipment.
- *Air to breathe* (20 min) – breathing apparatus, with a Trainer's Guide.
- *First thoughts* (20 min) – for new trainees.
- *Something to do with Safety Reps* (20 min), safety and industrial relations, with a Leader's Guide.
- *Talking of Safety* (23 min) – safety committees, with a Discussion Guide.
- *The Contract* (22 min) – contractors working safely on site, with a Leader's Guide.
- *Toxic hazards in industry* (22 min) – how to control them, with booklet of discussion notes.
- *Better than cure* (19 min) – occupational health, with a Leader's Guide. (Commissioned by the Chemical and Allied Products Industry Training Board.)
- *Emergency eye wash* (13 min).
- *Understanding your lungs* (20 min).
- *COSHH in practice* (21 min).
- *Unreasonably dead* (25 min video plus 8 min tape) – electrical safety and legislation.
- *Heard all about it?* (25 min main video, 7 min video on legislation) – noise and the Noise at Work Regulations 1989.

M.2.2 HSE/CFL Vision

Chalfont Grove, Gerrards Cross, Bucks SL9 8TN, Tel 0240 74433

The HSE film catalogue lists upwards of 40 safety films/videocassettes for purchase or hire. These include:

- *Health at work* (30 min) – shows the Employment Medical Advisory Service in action.
- *Don't tell the lads* (26 min) – about lead poisoning in a factory.
- *Watch that space: confined space hazards in factories* (19 min) – shows how three men were nearly killed when entering a reactor vessel through lack of a proper safety procedure.

M.2.3 RoSPA (Live Action Communications)

113 Humber Road, London SE3 7LW, Tel 081-853 4488
Also from RoSPA Film Library, Cannon House, The Priory, Queensway, Birmingham B4 6BS, Tel 021-233 2461

RoSPA/LAC have now 16 safety films/videocassettes, for purchase or hire, and five slide-tape programmes for purchase only. They also offer a full range of programme-making facilities. Their films include:

- *Nowhere Man* (25 min) – about a fire and explosion in which a man was killed through the fault of his friend.
- *Fire Drill* (20 min) – demonstrates how proper training can prevent a minor fire from becoming a major catastrophe.

M.2.4 The Fire Protection Association

140 Aldersgate Street, London EC1A 4HX, Tel 071-606 3757

The current FPA list nine films/videocassettes for purchase or hire and eleven slide sets for purchase only. All deal with fire safety. Their films/videos include:

- *In the event of fire* (13 min) – about the need for a fire plan. It shows how to raise the alarm, call the fire brigade, tackle the fire and evacuate the premises.
- *LPG Safety* (20 min) – brings out the hazards of LPG and describes the main safety precautions needed.

M.2.5 Video Arts Ltd

Dumbarton House, 68 Oxford Street, London W1N 9LA, Tel 071-637 7288

The current catalogue offers 89 training films on a variety of (mainly business) subjects, for sale or hire, together with supporting material (e.g. a discussion leader's guide).
Two films deal with safety topics and another is about stress in management:

- *Stress* (25 min) – key points: take exercise, learn to relax, establish priorities, manage your time, delegate, communicate and avoid isolation.
- *Oh what the hell* in two parts (18 and 15 min) – safety attitudes.
- *Think yourself lucky* (24 min) – an introduction to safety training programmes.

M.2.6 BBC Enterprises Ltd

Education and Training Sales, Woodlands, 80 Wood Lane, London W12 0TT, Tel 081-743 5588

A single safety videocassette is offered for purchase only for showing to non-paying audiences for training purposes.

- *Save a life* (in 6 parts each of 10 min) gives instruction on what to do if someone stops breathing, with a book and poster on resuscitation.

Videos on alcoholism, drug addiction, smoking and healthy eating habits are also available.

M.2.7 The Institution of Chemical Engineers

Bernard Hancock, Head of Safety and Loss Prevention, 165–171 Railway Terrace, Rugby, Warwickshire CV21 3HQ, Tel 0788 78214

Besides offering several training modules with visual aids and a computer program, the Institution is active in publishing and organising courses and symposia on safety topics in the process industries, further details of which are given in M.6.1.

The following video training modules are available for purchase or hire:

- *Inherent safety* by Trevor Kletz consists of a video (35 min) in four sections, 70 colour slides, a study guide for the presenter and a book.
- *Safe handling of LPG* consists of a video (45 min) in two main sections, with guidance notes and eight detailed case studies with 90 colour slides.
- *Preventing emergencies in the process industries* consists of a four-part video structured to include three discussion intervals, with guidance notes for the discussion leader.
- *Safer piping – Awareness training for the process industries* consists of a video (30 min), 90 slides and guidance notes for the discussion leader.

The following computer training package is now available for purchase:

- *Emergency simulation program* for use with an IBM PC/XT/AT computer with a minimum memory of 640K and an Epson-compatible printer. Trainer's Guides on emergency planning, pre-plan responses and an operations manual form part of the package.

Seven hazard training slide modules are also available. Each consists of a comprehensive guide for the discussion leader, case study notes, sets of colour slides relevant to the case studies and (with some modules) 25 copies of a safety guide to be distributed to participants. The modules are:

1. Hazards of over- and under-pressurising of vessels 34 slides
2. Hazards of plant modifications   38 slides
3. Fires and explosions   45 slides
4. Preparation for maintenance   29 slides
5. Furnace fires and explosions   23 slides
7. Work permit systems   45 slides
8. Human error   54 slides

M.3 The UK Health and Safety Commission and Executive

The HS Commission is a coordinating enforcement authority which was established by the HSW Act of 1974. It is made up of eight representatives of trade unions, employers and local authorities and a full-time chairman appointed by the Secretary of State for Employment. Its functions and policies are carried out by the HS Executive.

The HSE is a statutory body consisting of a Director General and two other people appointed by the Commission. It has a special responsibility to ensure that the HSW Act and other HS law are observed[2]. It liaises with other bodies which have HS responsibilities including local authorities, the Railway Inspectorate, and the Petroleum Engineering Division of the Department of Energy in respect to HS on off-shore oil and gas platforms. In 1987 the Industrial Air Pollution Inspectorate was transferred from HSE to the Department of Environment.

HSE inspectors operate through a network of 20 UK area offices which are listed with their addresses and phone numbers in Table M.1. They visit and review a wide range of work activities, give advice and guidance, and, where necessary, issue enforcement notices and institute prosecutions. HSE contains the following inspectorates:

- **HM Factory Inspectorate** (HMFI)
- **HM Agricultural Inspectorate**
- **HM Explosives Inspectorate**
- **HM Mines and Quarries Inspectorate**
- **HM Nuclear Installations Inspectorate**.

HS matters in the process industries are mainly dealt with by HMFI. Figure M.1 is an organisation chart for HSC and HSE.

The Technical, Scientific and Medical Group of HSE provides technical and medical back-up to its field work and policy making.

The Technology Division of this group consists of specialists with headquarters in Bootle[3]. Here there are several units which are classified under **Engineering** and **Chemicals**. It also has seven field consultant groups which cover the UK. Its organisation is shown in Figure M.2.

The Employment Medical Advisory Service (EMAS) was discussed in 7.1.1.

Most HSE and HSC publications are sold by HMSO at its bookshop at 49 High Holborn, London WC1, Tel 071-211 5656 and 071-873 9090 (postal sales). These include research papers, codes of practice, guidance notes and annual HSC and HMFI reports. The HSE maintains the following public inquiry points for its library and information services:

- Baynards House, 1 Chepstow Place, Westbourne Grove, London W2 4TF, Tel 071-221 0870 ext. 6721/2
- Broad Lane, Sheffield S3 7HQ, Tel 0742 768141
- St Hughes House, Stanley Precinct, Bootle, Merseyside, L20 3QY, Tel 051-951 4381.

Table M.1 The UK area offices of the HSE

Area	Address	Telephone no.
South-west	Inter-City House, Mitchell Lane, Victoria Street, Bristol BS1 6AN	0272 290681
South	Priestley House, Priestley Road, Basingstoke RG24 9NW	0256 473181
South-east	3 Grinstead House, London, Road, East Grinstead, West Sussex RH19 1RR	0342 26922
London north	Maritime House, 1 Linton Road, Barking, Essex IG11 8HF	081-594 5522
London south	1 Long Lane, London SE1 4PG	071-407 8911
East Anglia	90 Baddow Road, Chelmsford, Essex CM2 0HL	0245 284661
Northern Home Counties	14 Cardiff Road, Luton, Beds LU1 1PP	0582 34121
East Midlands	Belgrave House, 1 Greyfriars, Northampton NN1 2BS	0604 21233
West Midlands	McLaren Bldg, 2 Masshouse Circus, Queensway, Birmingham B4 7NP	021-200 2299
Wales	Brunel House, 2 Fitzalan Road, Cardiff CF2 1SH	0222 473777
Marches	The Marches House, Midway, Newcastle-under-Lyme, Staffs ST5 1DT	0782 717181
North Midlands	Birbeck House, Trinity Square, Nottingham, NG1 4AU	0602 470712
South Yorkshire	Sovereign House, 40 Silver Street, Sheffield S1 2ES	0742 739081
West and North Yorkshire	8 St Paul's Street, Leeds LS1 2LE	0532 446191
Greater Manchester	Quay House, Quay Street, Manchester M3 3JB	061-831 7111
Merseyside	The Triad, Stanley Road, Bootle L20 3PG	051-922 7211
North-west	Victoria House, Ormskirk Road, Preston PR1 1HH	0772 59321
North-east	Arden House, Regent Centre, Regent Farm Road, Gosforth, Newcastle-upon-Tyne, NE3 3JN	091-284 8448
Scotland east	Belford House, 59 Belford Road, Edinburgh EH4 3UE	031-225 1313
Scotland west	314 St Vincent Street, Glasgow G3 8XG	041-204 2646

M.4 The International Labour Office (ILO)

The ILO, which was founded in 1919 and is now a specialised agency of the United Nations, is based in Geneva. It is the only UN agency with a tripartite basis bringing together representatives of governments, workers and employers. An essential task under its constitution is the protection of the worker against 'sickness, disease and injury arising out of his

```
┌─────────────────┐
│ Health and      │
│ Safety          │
│ Commission      │
└────────┬────────┘
         │
┌────────┴────────┐
│ Health and      │
│ Safety          │
│ Executive       │
└────────┬────────┘
         │
┌────────┴────────┐
│ Director General│
└─────────────────┘
```

Health Policy Division

Special Hazards Division

HM Nuclear Installations Inspectorate

Safety and General Policy Division

Resources and Planning Division

Deputy Director General

Technical, Scientific and Medical Group, comprising:

Director of Field Operations and HM Chief Inspector of Factories

Technology Division (see Figure M.2) including HM Explosives Inspectorate and Major Hazards Assessment Unit (Bootle and outstations)

Research and Laboratory Services Division (Sheffield, Buxton, Cricklewood and outstations)

HM Mines Inspectorate

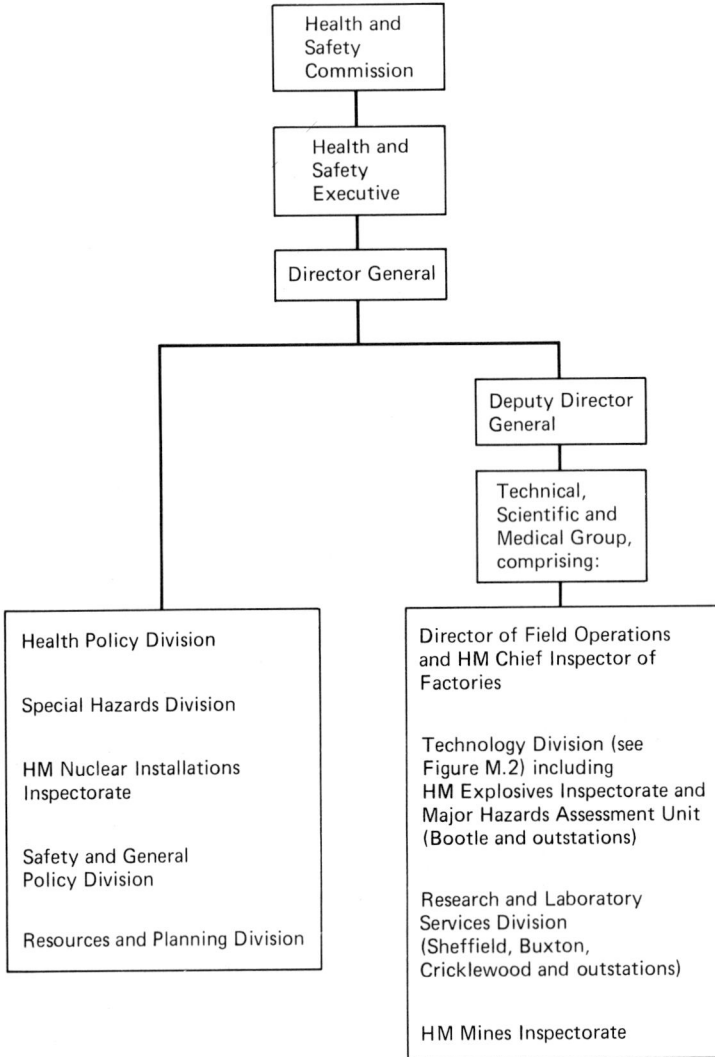

Figure M.1 The Health and Safety Commission and Health and Safety Executive

employment'. In this it works closely with the World Health Organisation (WHO) in preparing international standards and codes of practice, assisting member states in the drafting of legislation, in technical and medical inspection, in the establishment and strengthening of national HS institutes and infrastructure, and in training and dissemination of information. Although its role is mainly advisory, it has been responsible for a number of conventions and recommendations (subsequently ratified by most of its member states) on matters such as guarding of machinery, air pollution, noise and vibration, benzene and occupational cancer. It

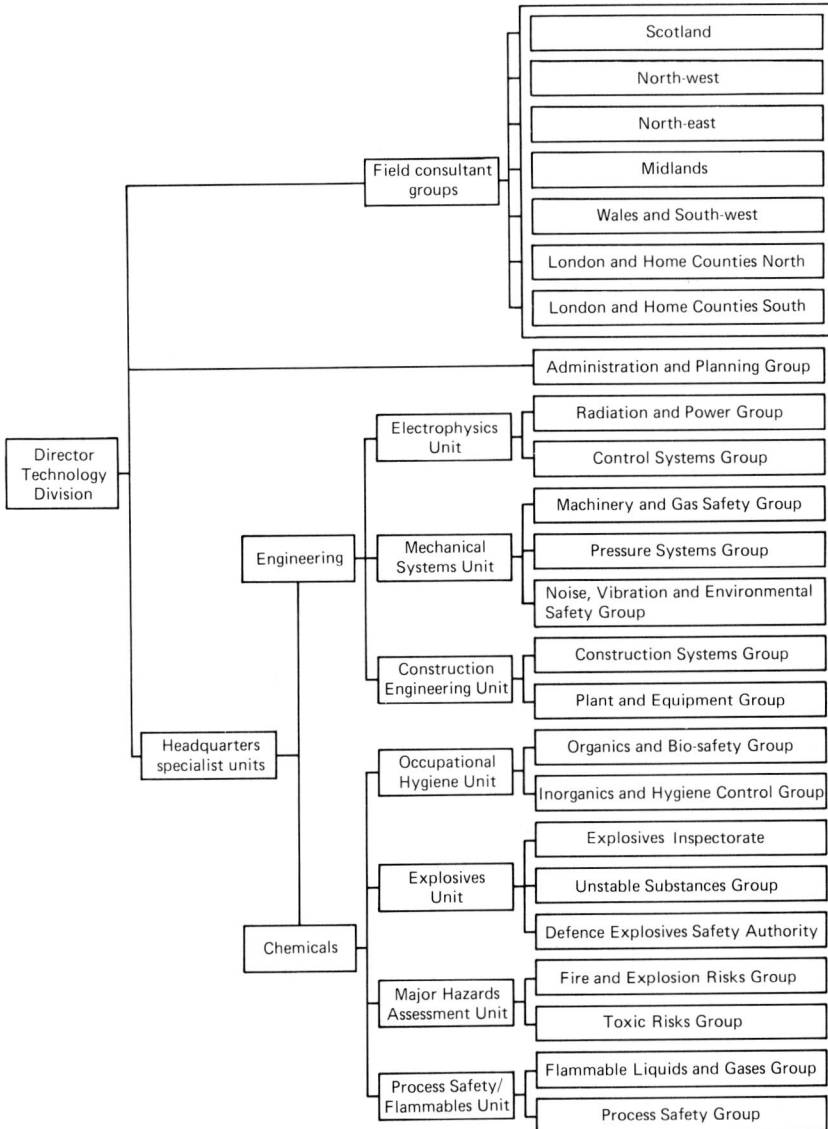

Figure M.2 HSE Technology Division

employs a number of staff for field and central office work, supplemented by the short-term employment of outside experts and consulting organisations for specific tasks. Its London office at Vincent House, Vincent Square, London SW1, Tel 071-828 6401, has a bookshop and a library of ILO publications. Some of the ILO's work on hazard control in Third World countries is discussed in 23.2. Two ILO activities are mentioned here.

M.4.1 The International Occupational Safety and Health Information Centre (CIS)

This keeps an up-to-date computerised data bank of health and safety-related publications world-wide, from which it publishes regular abstracts in several languages. This is one of several computerised HS data banks listed in M.12.

M.4.2 The International Occupational Safety and Health Hazard Alert System

This is a network system started in 1981 in which governments participate by designating a competent authority to issue and receive information about newly recognised occupational health and safety hazards. Many of these alerts relate to particular chemicals.

M.5 Professional HS organisations

M.5.1 The Institution of Occupational Safety and Health (IOSH)

222 Uppingham Road, Leicester LE5 0QG, Tel 0533 768424

Most UK safety professionals belong to IOSH. It has five grades of membership of which the first two, Fellow (FIOSH) and Member (MIOSH), are corporate grades. The other three grades are Graduate, Associate and Affiliate. Certificates and diplomas in occupational safety and health are awarded by the National Examination Board in Occupational Safety and Health (NEBOSH). The certificate leads to associateship and the diploma to full membership. Examinations for both grades comprise papers in the following subjects:

- Law
- Risk management
- Occupational health and hygiene
- Safety technology.

Courses leading to NEBOSH examinations are offered at a number of centres at certificate level and at a few centres at diploma level. The latter are:

- Barking College of Technology, Dagenham Road, Romford RM7 0XU Tel 0708 766841, Contact Mr Harrison, ext. 250
- Southwark College, Waterloo Branch, The Cut, London SE1 8LE Tel 071-928 9561, Contact Mr W. Aslett
- Coventry Management Training Centre, Woodland Grange, Old Milverton Lane, Leamington Spa CV32 6RN, Tel 0926 36621, Contact Ms Sandra Lynch
- North Notts College of Further Education, Carlton Road, Worksop S81 7HP, Tel 0909 473561, Contact Mr B. Stones
- Stoke-on-Trent College, Management Department, Couldon Campus, Stoke Road, Shelton, Stoke-on-Trent ST4 2DG. Tel 0782 208208, Contact Mr P. Bridden

- Central Manchester College (Openshaw), Whitworth Street, Manchester MI1 2WH, Tel 061-223 8282, Contact Mrs S. Hulby
- Wigan College of Technology (Management), Ashfield House, Pagefield Building, PO Box 53, Parson's Walk, Wigan WN1 1RR, Tel 0942 494911, Contact Mr E. Mooney
- Wirral Metropolitan College, Carlett Park, Easham, Wirral L62 0AY, Tel 051-327 4331, Contact Mr O. Williams

There is also an Offshore Specialist Examination for which training is offered at Aberdeen and Teesside.

IOSH has three specialist divisions, for Construction, Municipal and Public Services, and Offshore. It has a number of branches throughout the UK. Its journal *The Safety Practitioner* is published monthly.

M.5.2 The British Health and Safety Society

c/o Health and Safety Unit, Aston University, Aston Triangle, Birmingham B4 7ET, Tel 021-359 3611 ext. 4653

This society aims to promote research into HS, to influence HS policy, to increase awareness of HS issues and to act as a focus for information exchange between HS professionals in the UK and overseas. It has four grades of membership, full members being mostly graduates or diploma holders in HS or closely related subjects. It has about 350 members, holds conferences and seminars and publishes a newsletter and book reviews for its members.

M.5.3 The British Occupational Hygiene Society

1 St Andrews Place, Regent's Park, London NW1 4LB, Hon. Sec. Dr Derek Turner, Tel 071-486 4860

M.5.4 The Institute of Occupational Hygienists

Hon. Sec. Dr Derek Burns (home) 132 Oxgangs Road, Edinburgh EH10 7AZ, Tel 031-445 1032

Both M.5.3 and M.5.4 were discussed briefly in 7.1.2. Both hold conferences and meetings and can provide the names of consultants for atmospheric monitoring and other hygiene work.

M.5.5 The Fire Protection Association (FPA)

140 Aldersgate Street, London EC1, Tel 071-606 3757

Besides providing visual training aids [M.2.4], the FPA is a focal point for information on fires and fire protection. It keeps a list of consultants with special expertise whose services can be called upon. Other fire safety associations include:

M.5.6 The Institution of Fire Engineers

148 New Walk, Leicester LE1 7QB, Tel 0533 553654

M.6 UK Professional and industrial associations

Many industrial associations involved in the process industries were listed in M.1 under NSTOs.

M.6.1 The Institution of Chemical Engineers

As well as providing the audio-visual training aids listed in M.2.7, the I. Chem. E. organises frequent training courses, symposia and conferences on process hazards and safety. It publishes a bi-monthly *Loss Prevention Bulletin* which provides a forum for exchange of safety information within the process industries, as well as many safety booklets written by working parties of its members. Several I. Chem. E. publications are referred to in this book. The I. Chem. E. is a member of the European Federation of Chemical Engineering, which organises European symposia on loss prevention in member countries. It also has close connections with the American Institute of Chemical Engineers whose publications it can provide to its own members.

M.6.2. The Chemical Industries Association

King's Buildings, Smith Square, London SW1P 3JJ, Tel 071-834 3399

As well as its training activities [21.4.2 and M.1], the CIA has a special HS council, the *Chemical Industry Safety and Health Council (CISHEC)* which shares the same address.

M.6.3 The Society of Chemical Industry

14–15 Belgrave Square, London SW1 8PS, Tel 071-235 3681

This society has its own *Health and Safety Group* at the same address.

M.6.4 The Royal Society of Chemistry

Burlington House, Piccadilly, London W1V 0BN, Tel 071-437 8656

HS matters are dealt with by the *Health, Safety and Environment Committee* of the society's Professional Affairs Board.

M.6.5 The Confederation of British Industry

Centre Point, New Oxford Street, London WC1, Tel 071-379 7400

The CBI is active in HS matters which affect its members and has an efficient information service on a wide range of subjects.

M.6.6 The Institute of Petroleum

61 New Cavendish Street, London W1M 8AR, Tel 081-636 1004

Working groups have been responsible for several (IP) codes of safe practice including electrical safety [12.3] and pressure vessel inspection

[17.3.2] as well as collaborating with HSE and other bodies in drafting HS Guidance Notes, e.g. on the hazards of H_2S in residual fuel oils.

M.6.7 The Welding Institute

Abington Hall, Abington, Cambridge WC1E 7HT, Tel 0223 891162

The role of the Welding Institute in the training and examination of welders and welding inspectors is referred to in 17.2.

M.6.8 Others[4]

- British Chemical Engineering Contractors Association, 1-3 Regent Street, London SW1Y 4NR, Tel 071-839 6514
- Construction Industry Computing Association (CICA), Guildhall Place, Cambridge CB2 3QQ, Tel 0223 311246
- Computer Aided Design (CAD) Centre Ltd, Madingley Road, Cambridge CB3 0HB, Tel 0223 314848
- Institution of Mechanical Engineers [3.3.1], 1 Birdcage Walk, London SW1H 9JJ, Tel 071-222 7899
- Oil and Chemical Plant Constructors Association, Kent House, 87 Regent Street, London W1R 7HF, Tel 071-734 5246
- Process Plant Association, 8 Leicester Street, London WC2H 7BN, Tel 071-437 0678
- Industrial Safety (Protective Equipment) Manufacturers' Association (ISPEMA), 69 Cannon Street, London EC1N 5AB, Tel 071-248 4444
- Safety Equipment Distributors' Association, 50 High Street, Birmingham B4 7SY, Tel 021-643 6271
- Chemical Recovery Association, Kendal, Barnhill Road, Ridge, Wareham, Dorset BH20 5BG, Tel 09295 51295
- Plastics and Rubber Institute, 11 Hobart Place, London SW1W 0HL, Tel 071-245 9555
- Institute of Metals, 1 Carlton House Terrace, London SW1Y 5DB, Tel 071-839 4071
- Institution of Mining and Metallurgy, 44 Portland Place, London W1N 4BR, Tel 071-580 3802
- Institute of Measurement and Control, 87 Gower Street, London WC1E 6AA, Tel 071 387 4949
- SIRA Ltd, South Hill, Chislehurst, Kent, BR7 5EX, Tel 01-467 2636 (SIRA is an instrument technology centre with a laboratory and engineering facilities for R&D and a technical services division which can make safety assessments.)

M.7 UK sources of national and international standards

Standards and the organisations which produce them have been discussed in 2.7/8.

  The British Standards Institution, 61 Green Street, London W1, is the

London sales office for British and International (ISO, CEN) Standards and their annual catalogues. Inquiries should be addressed to BSI, Linford Wood, Milton Keynes MK14 6LE, Tel 0908 221166.

London Information (ILI Ltd), Index House, St George's Lane, Ascot, Berks SL5 7EU, Tel 0990 23377, is the UK selling agent for most relevant American standards including ANSI, API, ASME, IEEE, ISA and NFPA.

M.8 Consulting engineers and scientists

A number of consultants and consultancy firms offer services in the HS field.. Those needing such services are advised to enquire from the professional organisation which covers the field in question. They should also try:

- The British Consultants Bureau, 1 Westminster Palace Gardens, Artillery Row, London SW1, Tel 071-222 3561
- The Association of Consulting Engineers, Alliance House, Caxton Street, London SW1, Tel 071-222 6557
- The Register of Consulting Scientists, available from Adam Hilger Ltd, Techno House, Redcliffe Way, Bristol BS1 6NX, Tel 0272 297481

M.9 Insurance and inspection organisations

The insurance world includes brokers, underwriters and loss adjusters. Some specialise in process plant risks and have their own technical departments. Some offer consultancy services. A few insurance organisations, particularly American ones, publish technical guides on particular hazards, some of which are referred to elsewhere in this book. A short and arbitrary list of insurance and inspection organisations follows, with notes on the kind of help available from each:

- British Engine Insurance Ltd, 34 Lime Street, London EC3, Tel 071-623 5341, for inspection of pressure vessels and complete plant
- Factory Mutual International, Southside, 105 Victoria Street, London SW1, Tel 071-828 7799, for access to publications on oil and chemical hazards by FM's US associates
- Industrial Risk Insurers, 85 Woodland Street, Hartford, Connecticut, USA (incorporating Oil Insurance Association of Chicago), for publications on oil and petrochemical fire and explosion risks
- International Oil Insurers, 84 Fenchurch Street, London EC3, Tel 071-488 2703, for advice on fire and explosion risks
- National Vulcan Engineering Insurance Group Ltd, 18 Mansell Street, London E1, Tel 071-481 3155, for inspection of pressure vessels and complete plant
- Plant Safety Ltd, Parklands, Wilmslow Road, Didsbury, Manchester M20 8RE, Tel 061-434 9771, for plant inspection, materials testing, HS consultancy and training

M.10 Trade unions and allied organisations

Trade unions which represent employees in the process industries have a natural interest in their safety and health. They try to inform their members about many workplace hazards to which they may be exposed and to build up a climate of public opinion in favour of protective HS legislation (such as COSHH [2.5.6]). The need to consult trade unionists on health and safety matters which was recognised in HSWA 1974 was taken a step further in 1977 with the Safety Representatives and Safety Committee Regulations [2.5.8].

The TUC Centenary Institute of Occupational Health (London School of Hygiene and Tropical Medicine, Keppel Street, Gower Street, London WC1E 7HT) carries out research and education on HS. British trade unions organise training for safety representatives both as individual unions and through the Trade Union Congress, Congress House, Great Russell Street, London WC1B 3LS. The 1987 report of the General, Municipal, Boilermakers and Allied Trades Union[5] (Thorne House, Ruxley Ridge, Claygate, Esher, Surrey KT10 0TL, Tel 0372 62081) carried lively articles with such titles as:

- How safe is the chemical industry?
- What kinds of accidents happen to our members in the chemical industry?
- What information do we need?
- What particular hazards should we start with?

M.11 Equipment suppliers

Technical Indexes Ltd, Willoughby Road, Bracknell, Berkshire, RG12 1NS, Tel 0344 426311, publish seven useful indexes of engineering and laboratory equipment and computer hardware and their suppliers. These are frequently updated and cover:

- Construction and civil engineering
- Electronic engineering
- Engineering component materials
- Information technology computer and communications hardware
- Laboratory equipment
- Manufacturing and materials handling
- Process engineering.

Industrial Safety (Protective Equipment) Manufacturers Association (ISPEMA), 69 Cannon Street, London EC4N 5AB, Tel 071-248 4444, publishes a useful reference book[6] [22] with a list of British protective equipment suppliers and their products.

M.12 On-line computer data bases

There are now a number of computerised data bases on occupational health and safety subjects. These can be searched rapidly and the information extracted on a subscription or fee-paying basis via suitable computer terminals and telephone links[7]. Most of these use the English language and are broadly of three types:

1. Comprehensive abstracts of published HS information on a world-wide basis
2. Information on particular subjects or issues which generally apply particularly to one country (generally the USA)
3. The entire contents of a particular journal or newsletter on some HS topic. These data bases are all centred in the USA.

Only the first two are listed here.

M.12.1 Comprehensive abstracts of HS information

Data base: CISDOC
Producer and on-line service: The International Occupational Safety and Health Information Centre (CIS) of the International Labour Office
On-line service: CIS
Content: Abstracts of world-wide literature on occupational safety and health.

Data base: HSELINE
Producer: HSE library and information service
On-line services: DATA-STAR, Pergamon InfoLine
Content: Similar to CISDOC but with emphasis on the UK.

Data base: NIOSHTIC
Producer: US Department of Health and Human Resources Public Health Services (NIOSH)
On-line services: several including Pergamon InfoLine
Content: Similar to CISDOC but with emphasis on US sources.

M.12.2 Information on particular subjects or issues

Data base: AMILIT
Producer: The Swedish National Board of Occupational Safety and Health
On-line service: ARAMIS
Content: Swedish research in occupational safety and health.

Data base: Chemical Hazards in Industry
Producer: Royal Society of Chemistry
On-line services: DATA-STAR, Pergamon InfoLine
Content: Literature abstracts on data base title.

Data base: Chemical Hazards Response Information
Producer: US Coast Guard
On-line services: Chemical Information Systems Inc.
Content: Information on chemicals for use in spill situations.

Data base: Chemical Right to Know Requirements
Producer: Bureau of National Affairs Inc.
On-line service: Executive Telecommunication System Inc.
Content: Legal and practical aspects of disclosure of information on hazardous chemicals required of (US) government and business agencies.

Data base: Chemical Safety Data Guide
Producer: The Bureau of National Affairs
On-line service: Executive Telecommunication System Inc.
Content: Full text of Chemical Safety Data Guide – identification, handling and regulations of substances covered by OSHA.

Data base: ECDIN (Environmental chemicals data and information network)
Producer: Commission of the European Communities, Joint Research Centre
On-line service: DataCentralen
Content: All about chemical substances in the environment.

Data base: Industry File Index System
Producer: US Environmental Protection Agency (EPA)
On-line service: Chemical Information Systems Inc.
Content: Summaries of EPA regulations on specific chemicals and on the chemical industry.

Data base: International Health Physics Data Base
Producer: Creative Information Systems Inc.
On-line service: General Electric Information Services Company (GEISCO)
Content: Information on employee exposure to environmental health hazards.

Data base: OHS MSDS (Material Safety Data Sheets)
Producer and on-line service: Occupational Health Services Inc.
Content: Details about over 75 000 chemical substances which require documentation by chemical manufacturers under OSHA's standard on hazard communication and labelling.

M.12.3 Addresses of on-line services

- ARAMIS, Library information service, 17184 Solna, Sweden, Tel 730-90 00
- Chemical Information Systems Inc., 7215 York Road, Baltimore, MD 21212, USA, Tel 301/321-8440
- CISILO, 1211 Geneva 22, Switzerland, Tel 41(22)99 6740
  DataCentralen, Langdlystvej 40, 2650 Hvidorre, Copenhagen, Denmark, Tel 45(1) 75 81 22
- DATA-STAR, D-S Marketing Ltd, Plaza Suite, 114 Jermyn Street, London SW1Y 6HJ, Tel 071-930 5503

- Executive Telecommunication System Inc., Human Resource Information Network, 9585 Valparaiso Court, Indianapolis, IN 46268, USA, Tel 317/872-2045
- General Electric Information Services Company (GEISCO), 401 North Washington Street, Rockville, MD 20850, USA, Tel 301/294-5405
- Occupational Health Services Inc., 400 Plaza Drive, PO Box 1505, Secaucus, NJ 07094, USA, Tel 201/865-7500
- Pergamon InfoLine, Pergamon ORBIT InfoLine Ltd, Achilles House, Western Avenue, London W3 0UA, Tel 081-992 3456

References

1. *Industry Training Organisations*, Manpower Services Commission, IB5, Moorfoot, Sheffield S1 4PQ
2. *Working to make work safer*, Leaflet HSE 20, HMSO, London (1987)
3. HSE Technology Division *The specialists*, Leaflet HSE 22, HMSO, London (1988)
4. *The Directory of British Associations*, CBD Research Ltd, 154 High Street, Beckenham, Kent BR3 1EA
5. *Health and safety in the chemical and allied industries 1987*, Report by the General, Municipal, Boilermakers and Allied Trades Union, Thorne House, Ruxley Ridge, Claygate, Surrey KT10 0TL
6. Industrial Safety (Protective Equipment) Manufacturers Association (ISPEMA), *Reference Book of Protective Equipment*, 7th edn, 69 Cannon Street, London EC4 (1987)
7. *Directory of on-line data bases*, Cuadra/Elsevier, New York

Standards (mainly British) relating to personal protective equipment and clothing[1]

Type of protection	BS No.	Title
Eyes	2092: 1987	Specification for industrial eye-protectors[a]
	679: 1959 (1977)	Specification for filters for use during welding and similar industrial operations[a]
Eyes, face and neck	1542: 1982	Specification for equipment for eye, face and neck protection against non-ionising radiation arising during welding and similar operations[a]
Respiratory	4275: 1974	Recommendations for the selection, use and maintenance of respiratory protective equipment
	2091: 1969	Specification for respirators for protection against harmful dusts, gases and scheduled agricultural chemicals
	4555: 1970	Specification for high efficiency dust respirators
	4558: 1970	Specification for positive pressure, powered dust respirators
	6016: 1980	Specification for filtering facepiece dust respirators
Respiratory and skin	4771: 1971	Specification for positive pressure, powered dust hoods and blouses
Respiratory (breathing apparatus)	4667	Breathing apparatus
	Part 1: 1974	Closed circuit breathing apparatus
	Part 2: 1974	Open-circuit breathing apparatus
	Part 3: 1974	Fresh air hose and compressed air line breathing apparatus

Type of protection	BS No.	Title
	Part 4: 1982	Specification for escape breathing apparatus
Head	5240	Industrial safety helmets
	Part 1: 1987 ≠ ISO 3873	Specification for construction and performance
	4033: 1966 (1978) ≡ ISO 4869	Specification for industrial scalp protectors (light duty)
Hearing	5108: 1983	Method of measurement of sound attenuation of hearing protectors
	6344	Industrial hearing protectors
	Part 1: 1984	Specification for ear muffs
	Part 2: 1988	Specification for ear plugs
Hand	1651: 1986	Specification for industrial gloves
Foot	953: 1979	Method of test for safety and protective footwear
	1870	Safety footwear
	Part 1: 1979	Specification for safety footwear other than all-rubber and all-plastics moulded types
	Part 2: 1976 (1986)	Specification for lined rubber safety boots
	Part 3: 1981	Specification for polyvinyl chloride moulded safety footwear
	4972: 1973	Specification for women's protective footwear
	5145: 1984 ≠ ISO 2023	Specification for lined industrial vulcanised rubber boots
	5451: 1977	Specification for electrically conducting and antistatic rubber footwear
	6159	Polyvinyl chloride boots
	Part 1: 1981 ≠ ISO 4643	Specification for general industrial lined or unlined boots
Fall arrest and rescue	1397: 1979	Specification for industrial safety belts, harnesses and safety lanyards
	3367: 1980	Specification for fire brigade and industrial ropes and rescue lines
	5062	Self locking anchorages for industrial use
	Part 1: 1985	Specification for self locking safety anchorages and associated anchorage lines
	Part 2: 1985	Recommendations for selection, care and use
	5845: 1980(1986)	Specification for permanent anchors for industrial safety belts and harnesses

Type of protection	BS No.	Title
Clothing	4724	Resistance of clothing materials to permeation by liquids
	Part 1: 1986	Method for the assessment of breakthrough time
	6408: 1983	Specification for clothing made from coated fabrics for protection against wet weather

[a]   Under consideration (1989) for approval under the Approval of Safety Standards Regulations 1987.
≡   Identical to.
≠   Similar but not identical to.

Reference

1. Industrial Safety (Protective Equipment) Manufacturers Association (ISPEMA), *Reference Book of Protective Equipment*, 7th edn, London (1987)

Factors to be considered in setting up industries and transferring technologies to tropical and sub-tropical regions[1]

1. The effect of heat on the skin;
2. The acceptability of PPE/C and the protection provided by it;
3. The effect of high temperatures on the rate of absorption of toxic substances through the intact skin;
4. The effect of high levels of sunlight;
5. Heat stress problems in non-acclimatised persons, particularly when they have to wear PPE/C;
6. The effect of climate on the stability of chemical substances;
7. The effect of climate on equipment operation and maintenance;
8. The effect of climate on sampling and monitoring equipment and results;
9. The combined effect of the increased respiratory rate, the absorption of chemicals and the altered level of normal bodily functions resulting from work at high temperatures;
10. Parasitic, bacterial, viral and other biological conditions;
11. The physiological characteristics of workers in tropical regions;
12. The effect of climate on occupational exposure limits developed and established in temperate climates;
13. Special precautions to protect HS monitoring and analytical instruments and to ensure their proper operation and accuracy.

Reference

1. International Labour Office, *Safety, health and working conditions in the transfer of technology to developing countries – An ILO code of practice*, Geneva, (1988)

Ergonomic and anthrometric factors to be considered in setting up industries and transferring technologies[1]

1. The energy requirements for heavier work and the need for machines to prevent undue fatigue;
2. The efficienct and economy of physical work, especially lifting;
3. The appropriate design for seated and standing work taking posture and body movements into account;
4. Instrument dials and displays to suit the worker, taking cultural factors into account;
5. Face and head shapes and dimensions to ensure proper fit of PPE;
6. Aspects of body size, reach, grasp and muscular strength of machine operators to ensure that machines and plant, dials, control levers and panels suit the workers who will use them;
7. Environmental conditions such as temperature, air movement, humidity, noise, vibration, lighting, air contaminants and radiation to ensure that these do not stress workers unduly or damage their health;
8. Reduction in the length of the working day when the technology transfer results in environmental conditions which have an adverse cumulative effect;
9. The provision of adequate relief personnel to allow rest periods in cases where continuous work is required;
10. The provision of rest booths or rooms protected from adverse conditions of the working environment, when warranted;
11. The provision of emergency showers, special washing facilities and other facilities as required;
12. The prohibition of any payment scheme providing incentives for unsafe operation of a transferred technology.

Reference

1. International Labour Office, *Safety, health and working conditions in the transfer of technology to developing countries – An ILO code of practice*, Geneva, (1988)

List of abbreviations

Abbreviations given here do not include:

- Letters used only for quantities in equations,
- Those in common English usage,
- Symbols for SI and derived units,
- Symbols for items shown in figures accompanying the text.

ABPI	Association of British Pharmaceutical Industry
ABS	acrylonitrile-butadiene-styrene (copolymer)
abs.	absolute (pressure)
a.c.	alternating current
ACGIH	American Conference of Governmental Industrial Hygienists.
ACOP	Approved Code of Practice
ACTS	Advisory Committee on Toxic Substances (HSCs)
AD	Appointed Doctor (in UK)
AMOCO	American Oil Company
ANSI	American National Standards Institute
APAU	Accident Prevention Advisory Unit (of HMFI)
APC	accident prevention committee
API	American Petroleum Institute
ARC	accelerating rate calorimetry
ASME	American Society of Mechanical Engineers
ASR	accident severity rate
ASTM	American Society for Testing and Materials
BA	breathing apparatus
BASEEFA	British Approvals Service for Electrical Equipment in Flammable Atmospheres
BASF	Badishe Anilin und Soda Fabrik
BLEVE	boiling liquid expanding vapour explosion
BP	boiling point
BS	British Standard
CAD	computer-aided design

736

CAITREC	Chemical and Allied Industries Training Review Council (UK)
CAL	computer-assisted learning
CAM	computer-aided manufacture
CBI	Confederation of British Industry
CEN	European Committee for Standardisation
CENELEC	European Committee for Electrotechnical Standardisation
CFC	chlorofluorocarbon
CHETAH	chemical thermodynamic and energy hazard evaluation
CHP	cumene hydroperoxide
CIA	Chemical Industries Association (UK)
CIMAH	Control of Industrial Major Accident Hazards (UK Reg.)
CISHEC	Chemical Industry Safety and Health Council (UK)
CL	confidence level
CM	condition monitoring
CML	computer-managed learning
COP	code of practice
COSHH	Control of Substances Hazardous to Health (UK Reg.)
CPLR	Classification, Packaging and Labelling (of Dangerous Substances) Regulations (UK)
CRUNCH	dispersion model for continuous releases of a denser-than-air vapour into the atmosphere
CSWIP	Certification Scheme for Welding Inspection Personnel (UK)
dB'A'	decibel, 'A' scale
d.c.	direct current
DC	developing country
DENZ	computer program for the calculation of the dispersion of dense toxic or explosive gases in the atmosphere
DDT	2,2-bis [p-chlorophenyl] 1,1,1-trichloroethane
DHSS	Department of Health and Social Security (pre-1989) (UK)
d.p.	differential pressure (cell)
DSC	differential scanning calorimetry
DSM	Dutch State Mines
DSS	Department of Social Security (UK)
DTA	differential thermal analysis
DTI	Department of Trade and Industry (UK)
EDNA	ethylenediamine dinitramine
EEC	European Economic Community (also EC)
EH	Environmental Hygiene (series of HSE Guidance Notes)
EMA	Employment Medical Advisor (UK)
EMAS	Employment Medical Advisory Service (UK)
e.m.f.	electro-motive force
EML	estimated maximum loss
EO	ethylene oxide
F&E	fire and explosion (in Chapter 12)
F&EI	fire and explosion index (in Chapter 12)

FAR	fatal accident rate
FIFR	fatal injury frequency rate
FMEA	failure modes and effect analysis
FOC	Fire Offices' Committee (UK)
FPA	Fire Protection Association (UK)
FR	frequency rate
FRP	fibre-reinforced plastic
g.	gauge (pressure)
GLC	gas–liquid chromatography
GRP	general-purpose synthetic rubber (styrene–butadiene)
GS	general series (of HSE Guidance/Technical Notes)
HAZAN	hazard analysis (fault tree type)
HAZOP	hazard and operability (study)
HC	hydrocarbon
HD	high density (polyethylene)
HMFI	Her Majesty's Factory Inspectorate
HS	health and safety (in a general sense)
HSC	Health and Safety Commission (UK)
HSE	Health and Safety Executive (UK)
HSWA	Health and Safety at Work etc. Act
Hz	Hertz (cycles per second)
I. Chem. E.	Institution of Chemical Engineers (UK)
ICI	Imperial Chemical Industries
IEC	International Electrotechnical Committee
ILO	International Labour Office (or Organisation)
IOI	International Oil Insurers (London)
IOSH	Institution of Occupational Safety and Health
IPCS	International Programme for Chemical Safety
IR	infra-red
IR	incidence rate
IRI	Industrial Risk Insurers (Chicago)
ISO	International Standards Organisation
ISPEMA	Industrial Safety (Protective Equipment) Manufacturers Association
JIT	job instruction training
JSA	job safety analysis
LC_{50}	lethal concentration (for 50% of rat population)
LD_{50}	lethal dose (for 50% of rat population)
LNG	liquefied natural gas
LPG	liquefied petroleum gas
MAWP	maximum allowable working pressure
MDHS	methods for the determination of hazardous substances
MEL	maximum exposure limit
MF	material factor (in Chapter 12)

MHAU	Major Hazards Assessment Unit (of HSE)
MIC	methyl isocyanate
MP	member of parliament
MP	melting point
MPPD	maximum probable property damage
MTBF	mean time between failures
MTTF	mean time to failure
NADOR	Notification of Accidents and Dangerous Occurrences Regulations (UK)
NCB	National Coal Board (now British Coal)
NDT	non-destructive testing
NFPA	National Fire Protection Association
NIG	National industry group
NIHHS	Notification of Installations Handling Hazardous Substances (UK Reg.)
NIOSH	National Institute for Occupational Safety and Health (USA)
NJAC	National Joint Advisory Council (UK)
NPF	nominal protection factor
NSTO	non-statutory training organisation
OEL	occupational exposure limit
OES	occupational exposure standard
OM	organisational misconception
OSHA	Occupational Safety and Health Administration (USA)
P	proportional control (single mode)
P&ID	piping and instrumentation diagram
P&V	pressure and vent/vacuum (valve)
PCN	personnel certification in NDT
PE	programmable electronics
PES	programmable electronic system
PETN	pentaerythritol tetranitrate
PFD	process flow diagram
PI	proportional plus integral control (two mode)
PID	proportional plus integral plus derivative control (three mode)
PPC/E	personal protective clothing/equipment
ppm	parts per million
psi	pounds per square inch
PVC	polyvinylchloride
QC	Queen's Counsel (leading UK barrister)
RAM	random-access memory
RDX	cyclotrimethylamine trinitramine
ROM	read-only memory
RoSPA	Royal Society for the Prevention of Accidents
RPE	respiratory protective equipment

S.G.	specific gravity
S.I.	Statutory Instrument
SI	Système International d'Unités
SIC	standard industrial classification (UK)
SOAP	spectrometric oil analysis procedures
SRD	Safety and Reliability Directorate (UK)
SWT	School of Welding Technology (Welding Institute's) (UK)
TCDD	2,3,7,8, tetrachlorodibenzoparadioxin ('dioxin')
TCP	2,3,5 trichlorophenol
tetryl	tetranitroaniline
TIG	tungsten inert gas (welding)
TLV	threshold limit value (USA)
TNT	trinitro toluene
TT	technology transfer
TWA	time-weighted average
UCC	Union Carbide Corporation
UCIL	Union Carbide India Ltd
UFD	utility flow diagram
UNCTAD	United Nations Conference on Trade and Development
UNEP	United Nations Environment Programme
UNIDO	United Nations Industrial Development Organisation
UP&ID	utility piping and instrument diagam
UPS	uninterruptible power supply
UV	ultra-violet
VCE	vapour cloud explosion
VCM	vinyl chloride monomer
VDU	visual display unit
VGS	vent gas scrubber
WHO	World Health Organisation

Index

The following notes are intended to help the user of this index:

1. As themes have been used for the wording of entries, you will not always find their precise wording on the pages referred to in the text.
2. To save space, abbreviations only, as given in the preceding list, have been used exclusively, where appropriate, for entries and parts of entries.
3. Page references are shown in **bold** type for entries where about 30 or more lines of text are devoted to the subject of the entry. Subjects referred to only in passing have not been indexed.
4. All titles and honorifics have been omitted but entries for authors and other persons are followed by a word or two to indicate the subject of the reference.
5. Cross-references (*see also* . . .) are usually only made to the first word of the relevant entry. Cross-references are not given to entries whose first word is the same as that of the current one.